Tiny Machine Learning

IEEE Press
445 Hoes Lane
Piscataway, NJ 08854

Tiny Machine Learning

Design Principles and Applications

Edited by

Agbotiname Lucky Imoize
University of Lagos
Akoka, Lagos
Nigeria

Dinh-Thuan Do
University of Mount Union
Alliance
OH, USA

Houbing Herbert Song
University of Maryland
Baltimore County, Baltimore
USA

IEEE*Press*
WILEY

Published by John Wiley & Sons, Inc., Hoboken, New Jersey.

Library of Congress Cataloging-in-Publication Data Applied for:

Hardback ISBN: 9781394294541

Cover Design: Wiley
Cover Image: © Bokehstore/stock.adobe.com

Set in 9.5/12.5pt STIXTwoText by Straive, Chennai, India

Contents

About the Editors

Agbotiname Lucky Imoize received the BEng degree (Hons.) in Electrical and Electronics Engineering from Ambrose Alli University, Nigeria, in 2008 and the MSc degree in Electrical and Electronics Engineering from the University of Lagos, Nigeria, in 2012. He is a Lecturer in the Department of Electrical and Electronics Engineering at the University of Lagos, Nigeria. Before joining the University of Lagos, he was a Lecturer at Bells University of Technology, Nigeria. He was, until recently, a research scholar at the Ruhr University Bochum, Germany, under the sponsorship of the Nigerian Petroleum Technology Development Fund (PTDF) and the German Academic Exchange Service (DAAD) through the Nigerian-German Postgraduate Program. He was awarded the Fulbright Fellowship as a visiting researcher at the Wireless@VT Laboratory, Bradley Department of Electrical and Computer Engineering, Virginia Tech, USA, where he worked under the supervision of Prof. R. Michael Buehrer from 2017 to 2018. He worked as a Core Network Products Manager at ZTE, Nigeria, and as a Network Switching Subsystem Engineer at Globacom, Nigeria. His research interests cover the fields of 6G wireless communication, wireless security systems, and artificial intelligence. He is the vice chair of the IEEE Communication Society Nigeria chapter, a registered engineer with the Council for the Regulation of Engineering in Nigeria, and a member of the Nigerian Society of Engineers. He is a senior member of IEEE.

Dinh-Thuan Do is an assistant professor at the School of Engineering, University of Mount Union, Alliance, Ohio, USA. He was a research scientist at the Electrical Engineering Department of the University of Colorado Denver, Denver, USA. He was formerly a Research Scientist in the Department of Electrical and Computer Engineering at the University of Texas at Austin, USA. Prior to joining The University of Texas at Austin, he was an Assistant Professor at Asia University in Taiwan and a Research Assistant Professor at Ton Duc Thang University in Vietnam. His research interests include signal processing in wireless communications networks, non-orthogonal multiple access, full-duplex transmission, and reconfigurable intelligent surfaces (RIS). He received the Golden Globe Award from the Vietnam Ministry of Science and Technology in 2015 (top ten excellent scientists nationwide). He is currently serving as Editor of Computer Communications, Associate Editor of EURASIP Journal on Wireless Communications and Networking, Associate Editor of Electronics, Associate Editor of ICT Express, and Editor of KSII Transactions on Internet and Information Systems. His publications include over 100 SCIE/SCI-indexed journal articles and over 40 international conference papers. In addition, he is the author of two textbooks and six book chapters. He holds a PhD degree in communications engineering from the Vietnam National University (VNU-HCMC), Vietnam. He is a senior member of IEEE.

Houbing Herbert Song (M'12–SM'14-F'23) received his PhD degree in electrical engineering from the University of Virginia, Charlottesville, VA, in August 2012.

He is currently a Professor and the Founding Director of the NSF Center for Aviation Big Data Analytics (Planning), and the Director of the Security and Optimization for Networked Globe Laboratory (SONG Lab, www.SONGLab.us), University of Maryland, Baltimore County (UMBC), Baltimore, MD. He is currently the Co-Editor-in-Chief (Co-EiC) of IEEE Transactions on Industrial Informatics (2025–present). He has been the Founding Chair of Trustworthy Internet of Things (TRUST-IoT) Working Group within IEEE IoT Technical Community since 2024. He was a Distinguished Visiting Fellow of the Scottish Informatics and Computer Science Alliance (SICSA) in 2024. He serves as an Associate Editor for *IEEE Transactions on Artificial Intelligence (TAI)* (2023–present), *IEEE Transactions on Intelligent Transportation Systems* (2021–present), and *IEEE Journal on Miniaturization for Air and Space Systems (J-MASS)* (2020–present). He was an Associate Technical Editor for *IEEE Communications Magazine* (2017–2020) and an Associate Editor for *IEEE Internet of Things Journal* (2020–2024). He has served as the General Chair of IEEE UIC 2025 and IEEE SmartData 2022 and as the Technical Program Committee (TPC) Chair of IEEE AIoT 2025, NSS 2024, and IEEE GEM 2023. He is the editor of more than 10 books, the author of more than 100 articles, and the inventor of 2 patents. His research interests include neuro-symbolic AI, anomaly detection, Artificial Intelligence of Things, autonomous systems, and Cyber Physical Systems (CPS). His research has been sponsored by federal agencies (including National Science Foundation, National Aeronautics and Space Administration, US Department of Transportation, and Federal Aviation Administration, among others) and industry. His research has been featured in popular news media outlets such as IEEE Spectrum, IEEE GlobalSpec's Engineering360, IEEE Transmitter, insideBIGDATA, StateTech Magazine, Association for Uncrewed Vehicle Systems International (AUVSI), Security Magazine, CXOTech Magazine, Solutions Review, EdTech Magazine, PYMNTS, Fox News, U.S. News & World Report, The Washington Times, and New Atlas.

Dr. Song is an IEEE Fellow, an ACM Distinguished Member, and a Full Member of Sigma Xi. Dr. Song has been a Highly Cited Researcher identified by Web of Science since 2021. He is an ACM Distinguished Speaker (2020–present), an IEEE Computer Society Distinguished Visitor (2024–present), an IEEE Communications Society (ComSoc) Distinguished Lecturer (2024–present), an IEEE Intelligent Transportation Systems Society (ITSS) Distinguished Lecturer (2024–present), an IEEE Vehicular Technology Society (VTS) Distinguished Lecturer (2023–present), and an IEEE Systems Council Distinguished Lecturer (2023–present). Dr. Song received the Research.com Rising Star of Science Award in 2022, the IEEE 2021 Harry Rowe Mimno Award, and the 2024 IEEE Computational Intelligence Society (CIS) Publication Spotlight, and also received more than Best Paper Awards from major international conferences. He has been an IEEE Impact Creator since 2023.

List of Contributors

Mumin Adam
Department of Computer Engineering
King Fahd University of Petroleum
and Minerals (KFUPM)
Dhahran
Saudi Arabia

Jelil O. Agbo-Ajala
Centre for Augmented Intelligence
and Data Science
School of Computing
University of South Africa
Johannesburg
South Africa

Hasnain Ahmad
Department of Computer Science
COMSATS University
Islamabad-Lahore Campus
Lahore
Pakistan

Lateef A. Akinyemi
Centre for Augmented Intelligence
and Data Science
School of Computing
University of South Africa
Johannesburg
South Africa

and

Department of Electrical and Smart
Systems Engineering
School of Computing
University of South Africa
Johannesburg
South Africa

Emmanuel Alozie
Department of Information
Technology
Sule Lamido University
Kafin Hausa
Jigawa
Nigeria

Samson Y. Amare
Leiden University Medical Center
Leiden University
Leiden
Netherlands

Marco Arnesano
Department of Intelligent Software
Systems and Technologies
eCampus University
Novedrate
Italy

and

Department of Theoretical and
Applied Sciences
eCampus University
Novedrate
Italy

Ikechukwu Ignatius Ayogu
Department of Computer Science
Federal University of
Technology
Owerri
Nigeria

Gulvira Bekeshova
Department of Information Security
L.N. Gumilyov Eurasian National
University
Astana
Republic of Kazakhstan

Abeeb Akorede Bello
Department of Autonomous and
Electric Vehicle Engineering
Faculty of Electrical and Computer
Engineering
Ecole Centrale of Engineering
Mahindra University
Hyderabad
India

Chidi Ukamaka Betrand
Department of Computer Science
Federal University of
Technology
Owerri
Nigeria

Tomoki Chiba
Graduate School of Informatics and
Engineering
The University of
Electro-Communications
Chōfu
Tokyo
Japan

Anh Khoa Dang
Perkiomen School
Pennsburg
PA
USA

Olufisayo S. Ekundayo
Centre for Augmented Intelligence
and Data Science
School of Computing
University of South Africa
Johannesburg
South Africa

Nasir Faruk
Department of Information Technology
Sule Lamido University
Kafin Hausa
Jigawa
Nigeria

and

Directorate of Communication and Information Technology
Sule Lamido University
Kafin Hausa
Jigawa
Nigeria

Emanuele Frontoni
Department of Political Sciences
Communication and International Relations
University of Macerata
Macerata
Italy

Salisu Garba
Department of Software Engineering
Sule Lamido University
Kafin Hausa
Jigawa
Nigeria

Peter Anuoluwapo Gbadega
Department of Electrical and Electronic Engineering Science
University of Johannesburg
Johannesburg
South Africa

Agbotiname Lucky Imoize
Department of Electrical and Electronics Engineering
Faculty of Engineering
University of Lagos
Lagos
Nigeria

and

Department of Electrical Engineering and Information Technology
Institute of Digital Communication
Ruhr University
Bochum
Germany

Putu H. P. Jati
Leiden University Medical Center
Leiden University
Leiden
Netherlands

Richard Govada Joshua
Comprobase Inc.
Sterling
VA
USA

Ilker Kara
Department of Computer Engineering
Faculty of Engineering
Çankırı Karatekin University
Çankırı
Turkey

and

Department of Medical Services and Techniques
Çankırı Karatekin University
Çankırı
Turkey

Sergii Kavun
Computer Information Systems and Technologies Department
Interregional Academy of Personnel Management
Kyiv
Ukraine

Boris Khruskov
Department of Cyber Security and Information Technology
University of Customs and Finance
Dnipro
Ukraine

Evangelia Konstantopoulou
SCYTALE Group
Computer Engineering and Informatics Department
University of Patras
Patras
Greece

Oleksandr Kuznetsov
Department of Intelligent Software Systems and Technologies
eCampus University
Novedrate
Italy

and

Department of Information and Communication Systems Security
V. N. Karazin Kharkiv National University
Kharkiv
Ukraine

and

Department of Theoretical and Applied Sciences
eCampus University
Novedrate
Italy

Kateryna Kuznetsova
VRAI – Vision
Robotics and Artificial Intelligence Lab
Ancona
Italy

Yelyzaveta Kuznetsova
Department of Information and Communication Systems Security
V. N. Karazin Kharkiv National University
Kharkiv
Ukraine

Andronikos Kyriakou
SCYTALE Group
Computer Engineering and Informatics Department
University of Patras
Patras
Greece

Roman Minailenko
Department of Cyber Security and Software
Central Ukrainian National Technical University
Kropyvnytskyi
Ukraine

Eftychia Mistillioglou
SCYTALE Group
Computer Engineering and Informatics Department
University of Patras
Patras
Greece

Ernest Mnkandla
Centre for Augmented Intelligence and Data Science
School of Computing
University of South Africa
Johannesburg
South Africa

Kinzah Noor
Office of Research Innovation and Commercialization
University of Management and Technology
Lahore
Pakistan

Francisca Onyiyechi Nwokoma
Department of Computer Science
Federal University of Technology
Owerri
Nigeria

Euphemia Chioma Nwokorie
Department of Computer Science
Federal University of Technology
Owerri
Nigeria

Abiodun Temitope Odewale
Cybersecurity Consultant
TCPCyber Consulting
Calgary
Alberta
Canada

Juliet Nnenna Odii
Department of Computer Science
Federal University of Technology
Owerri
Nigeria

Akihiko Ohsuga
Graduate School of Informatics and Engineering
The University of Electro-Communications
Chōfu
Tokyo
Japan

Hawau I. Olagunju
Department of Information Technology
Sule Lamido University
Kafin Hausa
Jigawa
Nigeria

Ayobami P. Olatunji
Department of Computer Science
Western Illinois University
Macomb
IL
USA

Ruduan B. F. Plug
Leiden University Medical Center
Leiden University
Leiden
Netherlands

Yuichi Sei
Graduate School of Informatics and Engineering
The University of Electro-Communications
Chōfu
Tokyo
Japan

Wasswa Shafik
Dig Connectivity Research Laboratory (DCRLab)
Kampala
Uganda

Aigul Shaikhanova
Department of Information Security
L.N. Gumilyov Eurasian National University
Astana
Republic of Kazakhstan

Nicolas Sklavos
SCYTALE Group
Computer Engineering and Informatics Department
University of Patras
Patras
Greece

Oleksii Smirnov
Department of Cyber Security and Software
Central Ukrainian National Technical University
Kropyvnytskyi
Ukraine

Yasuyuki Tahara
Graduate School of Informatics and Engineering
The University of Electro-Communications
Chōfu
Tokyo
Japan

Samuel Oluwatobi Tofade
CleanTech Renewable Energy
Lagos
Nigeria

Pavlo Usik
Department of Cyber Security and Software
Central Ukrainian National Technical University
Kropyvnytskyi
Ukraine

Mirjam van Reisen
Leiden University Medical Center
Leiden University
Leiden
Netherlands

and

Tilburg School of Humanities and Digital Sciences
Tilburg University
Tilburg
Netherlands

Duy Nam Khanh Vu
Lê Quý Đôn High School
Ho Chi Minh City
Vietnam

Preface

Tiny Machine Learning (TinyML) enables the execution of machine learning (ML) models on low-power embedded devices. With the aid of these TinyML devices, deep learning models could be squeezed into billions of IoT devices and microcontrollers units (MCUs), expanding the scope of applications and enabling ubiquitous intelligence for a better society. On a global scale, shipments of TinyML devices are expected to reach 2.5 billion by 2030, with an economic value of over US$70 billion. However, the limited memory resources of TinyML cannot efficiently support deep learning models designed for cloud and mobile platforms, which are critical for edge applications. Edge devices must collect data using external sensors with limited power and energy resources. Additionally, TinyML has limited compiler and inference engine support for bare-metal devices, which limits their applications in critical systems that require extensive computations. To address these issues, it is imperative to codesign the algorithm and system stack to enhance the capabilities of TinyML models.

This book explores critical concepts, design principles, applications, and key issues related to TinyML. The book sheds light on the recent progress in TinyML and deep learning on MCUs. Additionally, the book introduces new TinyML concepts, including TinyML on IoT devices with system-algorithm codesign, emphasizing their applications in various domains. Specifically, the book explores TinyML paradigms, TinyML enablers, TinyML for anomaly detection, learning panorama under TinyML, TinyML devices and tools, codesign for TinyML, power consumption and memory in IoT MCUs, lightweight frameworks for TinyML, TinyML for real-time and environmental applications, TinyML inference, training, and validation, security and privacy of TinyML devices, interrelationship between TinyML and LargeML models, power consumption and memory in IoT MCUs, security and privacy of TinyML models, TinyML for federated and transfer learning, TinyML research targeting microcontrollers of the future, and more.

Additionally, the book discusses tiny on-device training techniques, regulatory frameworks, and social and ethical considerations, and presents future directions in the interesting domain of TinyML. The book offers industry personnel, researchers, and academics new insights into the real-world scenarios of the technological trend, deployment, application, and associated benefits of TinyML. Also, the book discusses critical security and privacy issues in practical TinyML-based designs and applications. Lastly, the book would be helpful to industry professionals, academics, researchers, scientists, engineers, lecturers, and advanced students in ICTs (including networking, computing, data science, Artificial Intelligence (AI), ML, communication), TinyML, federated learning, modeling, and emerging technologies. The book is structured into nineteen chapters, outlined as follows:

Chapter 1 introduces TinyML, starting with an annotated overview of the technology and its central philosophy. It then delves into a detailed discussion of its evolution, focusing on key timelines and milestones, current trends, and state-of-the-art in TinyML. Next, Chapter 2 dwells on learning panorama under TinyML. The chapter contributes to understanding the main issues relating to the design and optimization of TinyML models, and explores wide-ranging learning algorithms. Chapter 3 focuses on TinyML for anomaly detection, particularly within time-series data, and explores its potential in real-time monitoring of industrial machinery, infrastructure systems, and IoT networks. Chapter 4 examines TinyML power consumption and memory usage in IoT MCUs. The chapter discusses the scalability of TinyML for ultralow-power AI deployments in large-scale IoT networks, emphasizing its potential to support billions of connected devices while maintaining low latency and energy efficiency. Chapter 5 investigates efficient data cleaning and anomaly detection in IoT devices using TinyCleanEDF. The study proposes a compatible and scalable solution for various IoT applications ranging from smart home systems to industrial automation, including environmental monitoring.

Chapter 6 presents TinyML devices and tools. The chapter introduces advanced techniques in model quantization, pruning, and neural architecture search. It shows their effectiveness in obtaining highly optimized model sizes with energy efficiency, without losing as much accuracy as other lightweight models. Chapter 7 investigates privacy-preserving techniques in TinyML for IoT. The chapter provides a comprehensive analysis of frameworks such as secure machine learning (SecureML), secure neural network (SecureNN), and Falcon, highlighting their contributions to efficient and secure computation in limited-resource settings, and discusses the trade-offs between security guarantees, computational efficiency, and communication overhead, which are crucial for TinyML applications. Chapter 8 focuses on enhancing cybersecurity in TinyML with lightweight cryptographic algorithms. The chapter further addresses the integration of these cryptographic tools into existing TinyML frameworks, exemplifying their real-world applicability and effectiveness in securing TinyML applications against increasingly sophisticated cyber threats.

Chapter 9 explores TinyML for enhanced edge intelligence. It introduces TinyML-as-a-Service (TMLaaS), a hybrid architecture that extends the traditional TinyML paradigm by distributing ML workloads across edge devices and gateway systems. The chapter focuses on design trade-offs, integration with ML operations (MLOps), and validation through a case study involving datasets obtained under different scenarios. Chapter 10 sheds light on advanced security schemes for TinyML devices, emphasizing the need for chip-level or language-level privacy. The study dramatically expands the accepted attack landscape on TinyML devices and discusses extant work related to the threat model and system under discussion. Chapter 11 critically examines robust ground truth data mining for enhanced privacy and accuracy in noisy TinyML environments. The proposed model applies traditional differential privacy (DP) principles to the "ground truth," which is unknown to the data owner or anonymizer, rather than the erroneous "sensor data." The method is beneficial for TinyML, where optimizing data accuracy and privacy while minimizing resource usage is critical.

Chapter 12 presents the security and privacy of TinyML devices, addressing the unique challenges of securing TinyML models on microcontrollers and other low-power devices vulnerable to cyber-physical threats. Chapter 13 elaborates on the semantic management of TinyML for industrial applications. The study provides a framework that leverages semantic web technologies to enable large-scale collaborative management of TinyML models and IoT devices. Chapter 14 demonstrates fighting poison with poison, exploiting the resilience of TinyML against poisoning attacks. The study proposes a defense method to mitigate any degradation to the accuracy of a ML model caused by poisoning attacks. Chapter 15 delves into TinyML for real-time medical image classification and diagnosis. The study showed that TinyML models, such as MobileNet, ShuffleNet, and SqueezeNet, achieve high accuracy and efficiency on resource-constrained devices. This indicates that TinyML can significantly reduce healthcare costs and improve patient outcomes.

Chapter 16 thoroughly examines biometric authentication in TinyML, exploring its vast opportunities and challenges. The chapter discusses the potential of multimodal biometric authentication and the integration of biometric techniques with other security mechanisms to enhance the overall robustness and reliability of TinyML systems. Chapter 17 examines the secure deployment of TinyML applications, exploring new strategies and best practices. The chapter comprehensively analyzes current methodologies for secure bootstrapping, secure key storage, and firmware updates, illustrating their significance in maintaining the integrity and confidentiality of TinyML applications. Chapter 18 highlights recent advances in TinyML for environmental applications. The chapter examines TinyML's potential as a contributor to various environmental applications and represents as a technological advancement in the widely recognized field of ML. Finally, Chapter 19 concludes the book, benchmarking TinyML encrypted federated learning with secret sharing in medical computer vision. Specifically, the chapter benchmarks the performance impact of gradient encryption and secret sharing in federated learning through ablation studies. The results highlight an explicit trade-off between security guarantees and computational efficiency, guiding practical optimizations for deploying secure federated learning in privacy-sensitive settings.

The book serves as an ideal reference for practitioners and researchers in the design, inference, training, testing, validation, implementation, and applications of TinyML models, including deep reinforcement learning, deep transfer learning, federated learning, and transfer learning. It is also a suitable textbook for graduate and senior undergraduate courses in these subjects.

The editors would like to thank the reviewers of the original book proposal for their constructive suggestions, all the authors of the chapters for their valuable contributions, and the reviewers and editorial assistants at Wiley and IEEE Press for their cooperation and support.

November 2025

Agbotiname Lucky Imoize
Gelsenkirchen
North Rhine-Westphalia
Germany

Dinh-Thuan Do
Alliance, Ohio, USA

Houbing Herbert Song
Baltimore, Maryland, USA

1

Introduction to TinyML

Francisca Onyiyechi Nwokoma, Chidi Ukamaka Betrand, Juliet Nnenna Odii, Euphemia Chioma Nwokorie, and Ikechukwu Ignatius Ayogu

Department of Computer Science, Federal University of Technology, Owerri, Nigeria

1.1 Introduction

Machine learning (ML) has emerged as a key component of contemporary technology, propelling advancements in a wide range of industries, including healthcare [1–3], climate action [4], entertainment [5], and transportation [6]. ML models have historically required large amounts of computational power, usually supplied by high-performance computer clusters or cloud-based servers [7, 8]. However, there is a growing need to move ML closer to the network edge due to the proliferation of Internet of Things (IoT) devices and the rising demand for on-site data processing [9]. TinyML, focused on deploying ML algorithms on devices with low processing power and memory (edge computing devices), emerged to satisfy this requirement.

TinyML, which prioritizes efficiency, low latency, and privacy, is a paradigm shift in the development and application of ML models [10]. By enabling ML on edge devices, TinyML facilitates quicker decision-making, less dependence on cloud connectivity, and enhanced data privacy [11, 12]. Because of these benefits, TinyML is especially well-suited for applications in fields where low power consumption and real-time processing are essential, such as healthcare, agriculture, smart cities, and industrial automation.

TinyML has witnessed accelerated development, over the past 10 years, due to unprecedented advancements in software frameworks, hardware, and algorithms [13, 14]. Since early designs of microcontrollers and other resource-constrained devices were historically known to be incapable of executing complicated computations due to their limited processing power and memory, the first focus was on optimizing existing ML models to run on these devices [15]. However,

Tiny Machine Learning: Design Principles and Applications, First Edition.
Edited by Agbotiname Lucky Imoize, Dinh-Thuan Do, and Houbing Herbert Song.

the possibility of incorporating ML into these devices became more evident as the Internet of Things (IoT) gained popularity and the need for intelligent edge devices grew [10, 16, 17]. The increasing demand for real-time data processing, lower latency, enhanced privacy, and lower energy consumption – all of which are not always possible with cloud-based solutions – has spurred this change.

Reducing the size and complexity of conventional ML models to fit within the limited resources of microcontrollers was the focus of early TinyML research [7]. Among the earliest methods to be investigated were knowledge distillation, pruning, and model quantization. By decreasing the accuracy of the numbers used in the models, quantization lowers the computing demands and model size. In the case of microcontrollers, this approach worked very well since it made it possible for smaller models to function inside the constrained memory and processing power of these devices without noticeably sacrificing accuracy [17, 18]. Similarly, pruning reduces a neural network's overall size and computing cost while preserving performance by removing less significant weights or neurons [19, 20]. On the other hand, knowledge distillation efficiently transfers knowledge in a manner appropriate for deployment on devices with limited resources by enabling smaller models and students to learn from the outputs of bigger, pretrained models and teachers [21, 22].

In order to overcome the limitations of traditional microcontrollers and enable more complex and computationally demanding applications to be executed at the edge, specialized hardware accelerators were developed specifically for TinyML applications [23]. These accelerators, like Google's Edge TPU, Nvidia's Jetson Nano, and Intel's Movidius Myriad X, provide significant computational power at a fraction of the energy cost, making them ideal for running ML models on low-power devices [24]. The introduction of these accelerators marked a pivotal point in the evolution of TinyML, as they established a new platform for researchers and developers to explore the capabilities of ML on edge devices.

Software frameworks have greatly aided the development and uptake of TinyML in addition to hardware developments. ML model deployment on smartphones with limited resources has been made easier by frameworks such as TensorFlow Lite, uTensor, and Edge Impulse. For example, TensorFlow Lite provides several tools and modules for ML model optimization, such as hardware acceleration, pruning, and quantization support [18, 25]. TensorFlow models can be deployed on microcontrollers with no change thanks to uTensor, a microcontroller-focused ML framework [26]. Conversely, Edge Impulse provides an end-to-end development platform that makes it easier for developers with little ML experience to create and implement edge-device models [16, 27, 28]. These software frameworks have democratized access to TinyML, enabling a wider range of developers to build and deploy ML models on edge devices.

Another significant area of growth of research within TinyML is the development of new algorithms tailored explicitly for resource-constrained environments. Traditional ML algorithms are often too computationally expensive to run on edge devices, necessitating the creation of more efficient alternatives. Researchers have focused on developing lightweight versions of popular algorithms, such as convolutional neural networks (CNNs) and recurrent neural networks (RNNs), which are optimized for speed and energy efficiency [17, 29, 30]. These lightweight algorithms are designed to perform well under limited memory and processing power constraints, making them suitable for deployment on a wide range of edge devices.

Haoyu et al. [31] proposed a framework using Semantic Web technologies to enable the joint management of TinyML models and IoT devices at scale to facilitate TinyML component exchange and reuse. In recent years, there has also been a growing interest in the use of TinyML for applications beyond simple classification or regression tasks. For example, TinyML is increasingly used in applications requiring more complex processing, such as object detection, natural language processing, and anomaly detection [32]. These applications often require real-time processing and low-latency responses, which are difficult to achieve with traditional ML models running on cloud servers. By moving the computation to the edge, TinyML enables these applications to operate more efficiently and with lower latency, providing a better user experience.

In situations where data privacy is an issue, TinyML has also demonstrated a lot of promise, such as improving privacy and security by processing data locally on the device and eliminating the need to send sensitive data to the cloud. This is especially crucial in industries like healthcare, where the highest level of care must be taken while handling patient data [7, 10, 33]. With TinyML, privacy-preserving ML applications can be created that can function within the bounds of data protection laws, including the General Data Protection Regulation (GDPR) in Europe [34].

Even though TinyML has advanced significantly, there are still a lot of issues that need to be resolved. One of the primary obstacles is the absence of consistent benchmarks for assessing TinyML models' performance across different hardware platforms and use scenarios [10]. Also, more thorough and representative benchmarks are required to test the performance of TinyML models in practical situations. This is without disregarding the fact that several benchmark suites (e.g., MLPerf Tiny) have been offered. Furthermore, research efforts are still geared toward developing new algorithms and optimization strategies that could further lower ML models' memory and processing demands. As the field of TinyML continues to improve, we will likely see breakthroughs in both hardware and software that will enable even more sophisticated applications to be deployed at the edge.

1.1.1 Key Contributions of the Chapter

This chapter comprehensively explores TinyML, detailing its evolution, current advancements, and future potential. By examining key concepts, technological milestones, challenges, and cost–benefit considerations, it serves as a foundational resource for researchers, developers, and industry practitioners. The following key contributions highlight the chapter's significance in advancing the understanding and practical implementation of TinyML.

1) **Provides an In-depth Overview of the Fundamental Concepts and Principles of TinyML:** By dissecting its basic ideas, such as adapting ML models to edge devices with constrained memory, compute capability, and energy resources, this chapter provides a foundation for comprehending TinyML. It examines why TinyML is essential to contemporary computing, highlighting how it improves data privacy, permits real-time data processing, and lessens dependency on cloud computing. The main architectural elements of TinyML systems, including specialized hardware accelerators, lightweight software frameworks, and efficient neural network models, will become clearer to readers. The chapter guarantees that novices and seasoned experts can develop a solid fundamental understanding of TinyML by providing an organized introduction.
2) **Discusses the Historical Evolution and Key Milestones in the Development of TinyML:** The chapter follows TinyML's development from its beginnings in low-power computers and embedded systems to its present position in edge artificial intelligence (AI). It looks at how early ML models were modified for situations with limited resources and how innovations like knowledge distillation, quantized neural networks, and model pruning have influenced the discipline. Important turning points are also emphasized, such as the creation of TensorFlow Lite, the launch of hardware accelerators like Google Edge TPU, and the expansion of TinyML in IoT applications. Readers can better appreciate the technological developments that have made TinyML a workable and popular option by comprehending this historical viewpoint.
3) **Highlights the Current Trends and State-of-the-Art Technologies in TinyML:** This chapter explores the most recent developments influencing TinyML in light of the growing demand for edge-based intelligence. In-depth discussions are held on subjects like federated learning at the edge, ultralow-power AI models, and real-time data processing. This chapter also covers software and hardware developments, such as new frameworks like uTensor and Edge Impulse that make creating and implementing TinyML applications easier. It also emphasizes the quest for further interoperability across edge devices and integrating TinyML with next-generation AI models,

such as tiny transformers. Thanks to these insights, readers may stay current with the quickly changing TinyML landscape.

4) **Explores the Challenges and Bottlenecks Associated with TinyML Deployment:** TinyML has many deployment issues, such as memory limitations, processing power limitations, and energy efficiency issues, despite its potential. This chapter thoroughly analyzes these constraints and explains how they affect TinyML applications' viability and efficiency. Potential mitigation solutions include lightweight neural architectures, model compression approaches, and dedicated low-power hardware. The chapter discusses practical issues like maintenance needs, security threats, and the trade-offs between computational efficiency and model correctness. When creating TinyML solutions, developers and researchers can make wise choices if they are aware of these difficulties.
5) **Examines the Cost–Benefit Analysis of Various TinyML Approaches and Their Applications:** TinyML adoption necessitates thoroughly assessing the trade-offs between energy usage, computational efficiency, and model complexity. A systematic cost–benefit analysis is presented in this chapter, contrasting several TinyML techniques according to criteria including power efficiency, precision, latency, and hardware requirements. It talks about how these aspects are prioritized according to the needs of various application domains, including healthcare, smart cities, and industrial automation. The chapter also examines how companies and developers can use scalable deployment techniques, effective software frameworks, and hardware accelerators to maximize TinyML implementations. When selecting the appropriate TinyML method for their projects, readers may make educated decisions by carefully weighing these trade-offs.

1.1.2 Chapter Organization

Section 1.2 presents the foundational concepts of TinyML, focusing on its evolution from embedded systems to modern applications, emphasizing low-power and real-time processing capabilities. Section 1.3 explores the historical milestones of TinyML, such as the introduction of quantized neural networks and specialized hardware accelerators like Google Edge TPU and Intel Movidius, which have driven advancements in edge computing. It also highlights the current trends in TinyML, including real-time data processing, model pruning, and developing frameworks like TensorFlow Lite and Edge Impulse, which enhance accessibility and efficiency.

Section 1.4 provides a detailed analysis of the materials and methods for TinyML system development, outlining the integration of hardware platforms, such as Arduino Nano 33 BLE Sense and Raspberry Pi Pico, with optimized software frameworks designed for edge deployment. Section 1.5 discusses the challenges and bottlenecks associated with deploying TinyML, including processing power, memory limitations, and the trade-offs between model complexity and efficiency. Section 1.6 examines the cost–benefit analysis of TinyML, focusing on balancing performance, power consumption, and application-specific needs. The key findings are presented in Section 1.7. Section 1.8 highlights the limitations of TinyML; Section 1.9 concludes the chapter by summarizing the significance of TinyML in real-world applications, discussing its transformative potential in various industries, and emphasizing the need for continued research to address unresolved challenges and expand the boundaries of TinyML.

1.2 Evolution of TinyML

Early TinyML solutions consisted of simple ML models like decision trees and support vector machines (SVMs) [14]. But since then, the field has progressed to deep learning-based models, typically small, efficient CNNs. The evolution of TinyML can be traced back to the core ideas of embedded systems, a domain where the focus has traditionally been on executing specialized, frequently repeated tasks with minimal computing and memory resources. During the early days, the emphasis was on designing efficient, low-power devices capable of running straightforward algorithms and handling basic data processing. Incorporating ML into more complex embedded systems became a key area of innovation [35]. This is depicted in Figure 1.1.

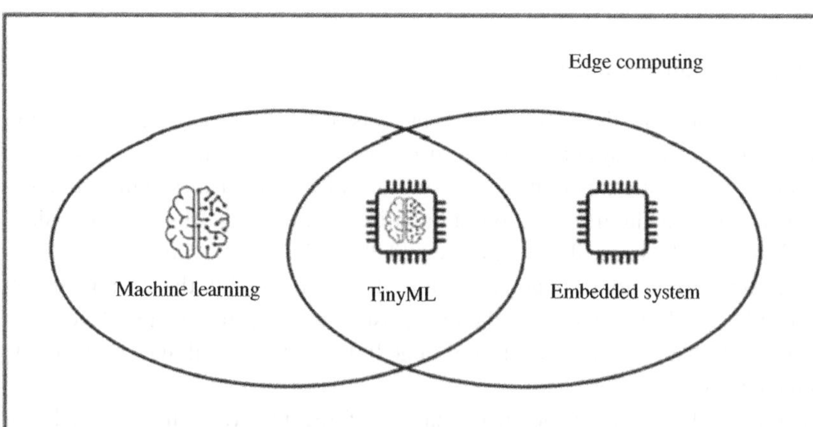

Figure 1.1 TinyML as a combination of machine learning and an embedded system.

The creation of quantized neural networks, a revolutionary turning point in the TinyML space, sparked this move toward ML [30, 36]. Quantized neural networks significantly reduced the memory footprint and computational requirements of ML models based on their adoption of low-precision arithmetic operations, making them suitable for deployment on microcontrollers. Since microcontrollers are usually limited by memory and power, the efficiency that quantization brought about greatly benefited these devices, creating new opportunities for directly deploying more intelligent functionalities on edge devices.

Another factor fuelling TinyML's evolution was the development of hardware accelerators specifically made to improve the performance of ML algorithms on edge devices, as opposed to general-purpose processors. For instance, Google's Edge tensor processing unit (TPU) and Intel's Movidius accelerators were revolutionary because they offered dedicated support for running neural networks at the edge, significantly increasing the speed and efficiency of these computations while maintaining low power consumption [36–38].

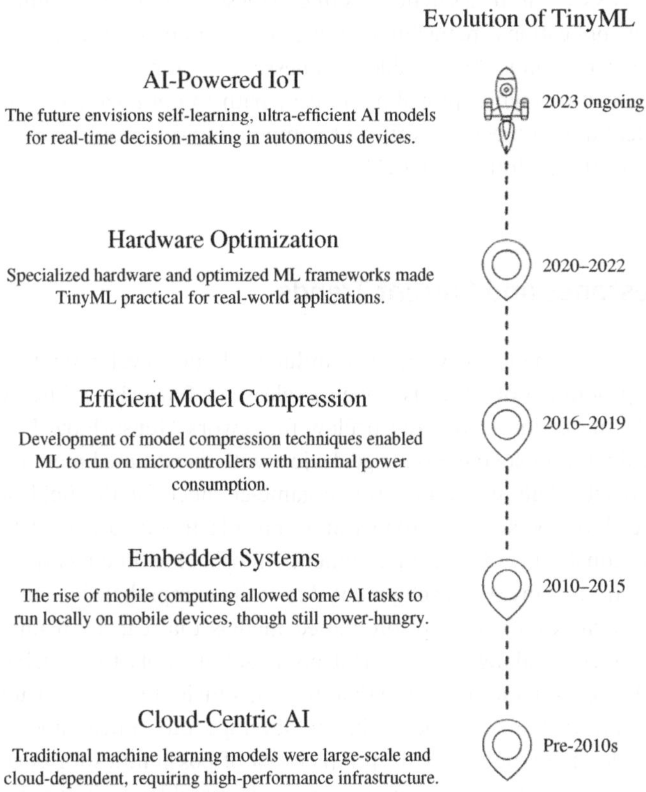

Figure 1.2 The evolution of TinyML.

These hardware advancements allowed for a significant leap in the complexity of the ML models that could be run on small, power-constrained devices without significantly depleting their batteries or requiring supplementary heat dissipation. This was crucial for applications like real-time image recognition, voice processing, and other sensor-based data analysis tasks, which are now commonly performed on edge devices in various IoT scenarios [39–41]. The combination of quantized neural networks and specialized hardware accelerators has thus been instrumental in the rapid advancement of TinyML, allowing for the development of more sophisticated, intelligent applications that operate efficiently at the edge [42]. This evolution reflects a broader trend toward decentralizing computation and embedding intelligence directly into everyday devices, paving the way for a new era of intelligent, interconnected systems. The key events in the evolution of TinyML are outlined in Figure 1.2.

There are still challenges to overcome. Small devices' limited memory and processing power constrain the complexity of models that can be used. Tiny transformers and RNNs are used less frequently due to these memory constraints. However, emerging applications are pushing this area of research. Regarding software, TinyML frameworks must be portable, seamlessly compiling and running across various architectures while providing an easy path to optimize code for platform-specific features to meet the stringent compute latency requirements. Table 1.1 summarizes the evolution of TinyML.

1.3 Key Milestones and Current Trends

A few significant developments have greatly influenced the development of TinyML, the most prominent of which is Google's release of TensorFlow Lite. A simplified, lightweight version of the TensorFlow framework, TensorFlow Lite was created especially for deployment on edge devices with constrained memory and processing capacity. This innovation was a game-changer for the field of TinyML, providing developers with a robust and versatile toolkit designed to cater to the unique constraints of edge environments. TensorFlow Lite has since become a cornerstone in TinyML development due to its comprehensive suite of features, which includes a variety of pretrained models that can be readily deployed on edge devices, alongside powerful optimization tools that enable fine-tuning of model performance to ensure that they run efficiently on hardware with restricted resources [36–38]. The ability to leverage pretrained models and perform on-device inference opened up a plethora of possibilities for edge computing applications, from smart home devices and wearables to industrial IoT systems.

Table 1.1 Evolution of TinyML.

Phase	Key features	Technological milestones	Applications	Challenges	References
Early solutions	Simple ML models like decision trees and SVMs. Focus on low-power, efficient systems.	Embedded systems performing specialized, repetitive tasks. Basic algorithms for data processing.	Basic data filtering. Signal processing.	Limited computational and memory resources.	[7, 8, 14]
Quantized neural networks	Introduction of low-precision arithmetic. Reduced memory and computation requirements. Compact models.	Development of quantized neural networks. Optimized ML models for microcontroller use. Efficient model deployment on edge devices.	Sensor data analysis. Basic IoT applications.	Reduced accuracy due to quantization. Balancing size and performance.	[18, 30, 38]
Hardware accelerators	Specialized hardware for ML (e.g., Google Edge TPU, Intel Movidius). High efficiency and low power.	Hardware designed to accelerate neural networks. Significant improvement in computation speed and energy efficiency. Support for complex ML models.	Real-time image recognition. Voice processing. IoT edge computing.	High initial cost. Compatibility with diverse ML frameworks.	[24, 33, 43]
Current trends	Incorporation of deep learning-based models. Focus on edge intelligence. Applications in IoT.	Support for convolutional neural networks. Emerging interest in Tiny transformers and RNNs. Framework advancements (e.g., portability and platform optimization).	Industrial automation. Smart cities. Healthcare monitoring.	Memory and processing power constraints. Optimization challenges for platform-specific features.	[10, 12, 15]
Future directions	Decentralized computation. Intelligent, interconnected systems.	Broader deployment of sophisticated TinyML applications. Development of highly portable frameworks. Integration of advanced models under constrained resources.	Autonomous systems. Privacy-sensitive applications.	Push for more efficient hardware. Increasing complexity of application scenarios.	[10, 25, 44]

Table 1.2 Summary of key milestones and current trends in TinyML.

Category	Key milestones/trends	Description	Applications	References
Key milestone: TensorFlow lite	Release of TensorFlow Lite by Google.	A lightweight, optimized version of the TensorFlow framework for edge devices. Provides a versatile toolkit, pretrained models, and optimization tools to fine-tune performance for hardware with limited memory and processing capacity. Facilitates on-device inference for a wide range of applications.	Smart home devices. Wearable technologies. Industrial IoT systems.	[7, 18]
Trend: Real-time processing	Increased focus on real-time data analysis and decision-making on edge devices.	Addresses latency issues by enabling immediate data processing and action at the source. Reduces dependence on cloud servers. Particularly critical in time-sensitive applications where instant response is essential.	Autonomous vehicles. Medical monitoring devices. Industrial automation systems.	[8, 10, 44]
Technique: Model pruning	Optimization technique to reduce neural network complexity.	Systematically removes redundant weights and neurons from neural networks. Decreases computational load and memory usage. Enhances speed and reduces power consumption, making models ideal for battery-powered and resource-constrained devices.	Battery-operated IoT devices. Real-time sensor data processing. Lightweight ML applications.	[19, 20, 38]
Trend: Expanding ecosystem	Development of tools and frameworks for broader TinyML accessibility.	Increasing number of frameworks to simplify model deployment and optimize performance across a variety of edge devices. Emphasis on models that work efficiently in diverse environmental conditions and resource-constrained scenarios.	Edge-based anomaly detection. Energy-efficient smart cities solutions. Personalized wearables.	[3, 12, 15]
Trend: Efficiency and accessibility	Enhancing ML models' accessibility and efficiency on low-power devices.	Focus on making TinyML models usable across a wide array of devices. Emphasizes energy efficiency, compactness, and adaptability to diverse use cases.	Consumer electronics. Predictive maintenance. Environment monitoring systems.	[10, 13, 25]

As tools and frameworks advance, TinyML is emphasizing improving ML models for real-time processing and decision-making. The need to conduct real-time data analysis and actions on the edge without the latency in sending data to a centralized server for processing has prompted this change [44]. The demand for such capabilities is particularly crucial in cases where time-sensitive judgments must be made, such as in autonomous vehicles, medical monitoring devices, and responsive industrial automation systems [10, 45]. To satisfy these requirements, new methods and algorithms have been created to reduce the latency and power usage of ML models, allowing them to operate efficiently within the strict energy limits of edge devices.

One prominent technique that has gained traction is model pruning, which reduces the complexity of neural networks by systematically removing redundant or less significant weights and neurons. By pruning away these unnecessary components, the neural network's overall size is decreased, reducing its computational load and memory usage. This enhances the speed at which the model can process data and lowers its power requirements, making it more suitable for deployment on battery-powered devices where conserving energy is a primary concern [43].

Furthermore, these developments in TinyML are a part of a larger movement to increase ML's effectiveness and accessibility across a larger range of devices and applications. An expanding ecosystem of tools and methods has been developed as the technology advances making it easier to deploy ML models on edge devices while guaranteeing that these models can function in a variety of resource-constrained and environmental scenarios (Table 1.2).

1.4 TinyML System Development

TinyML is a multidisciplinary field that combines various aspects of hardware design, software engineering, and algorithm development to facilitate the implementation of ML models on edge devices with limited computational resources. This integration is crucial for creating systems that can process data locally, directly on the device, rather than relying on cloud-based servers. By doing so, TinyML improves data processing speed, reduces latency, minimizes the consumption of energy, and protects data privacy by keeping information on the device [46]. Since these devices are associated with very little RAM and storage, selecting parts that can work efficiently without consuming too much power is essential. Hence, to optimize the execution of ML algorithms, hardware accelerators like digital signal processors (DSPs) and TPUs are commonly used. These devices offer notable speedups and lower energy consumption for operations like matrix multiplications and convolutions, which are common in neural network models.

Utilizing specialized frameworks and tools created to enhance ML models for edge deployment is an aspect of the software side of TinyML development. Developers can compress and transform models trained in a conventional ML framework into formats that can be used on edge devices with the help of these frameworks (such as TensorFlow Lite, PyTorch Mobile, and Edge Impulse). Techniques like pruning, which eliminate superfluous parameters from the model, and quantization, which lowers the precision of the numbers used from floating-point to integers, are frequently used to minimize model size and computing requirements [10, 45, 47]. Without appreciably compromising accuracy or performance, these optimizations are essential for fitting models into edge hardware's constrained memory and processing power.

Algorithmically, TinyML requires the development of novel models and methods that are specifically tailored to the constraints of edge devices. This includes designing lightweight neural networks and ML models that can operate effectively with fewer parameters and lower precision [21]. Researchers and engineers in the field of TinyML are constantly exploring new ways to improve the efficiency and robustness of these models, ensuring they can handle a variety of tasks, from image and speech recognition to predictive maintenance and anomaly detection, all while running on hardware with stringent resource limitations.

A battery-operated IoT sensor for environmental monitoring might prioritize ultralow-power consumption and use highly compressed models to extend battery life. In contrast, an application that requires real-time image processing might prioritize a hardware setup with a dedicated graphics processing unit (GPU) or TPU paired with software optimized for high-speed data throughput. TinyML's versatility and adaptability allow it to serve a wide range of applications, making it a versatile solution for bringing ML capabilities to the edge (whether hardware components, software frameworks, or algorithms). As technology continues to evolve, we can expect further hardware, software, and algorithm innovations, driving even more efficient and capable TinyML systems that can seamlessly integrate into our everyday lives and industries.

1.4.1 Hardware Platforms

A key element in the field's advancement has been the creation of hardware platforms specifically designed for TinyML applications. These platforms offer a variety of choices that satisfy different requirements regarding power consumption, computing capability, and memory limitations. These platforms enable developers to apply ML algorithms to frequently resource-constrained devices, including wearables, embedded systems, and IoT devices [7].

Among the most prominent platforms is the Arduino Nano 33 BLE Sense (Figure 1.3), which has become well-known due to its small size and low power

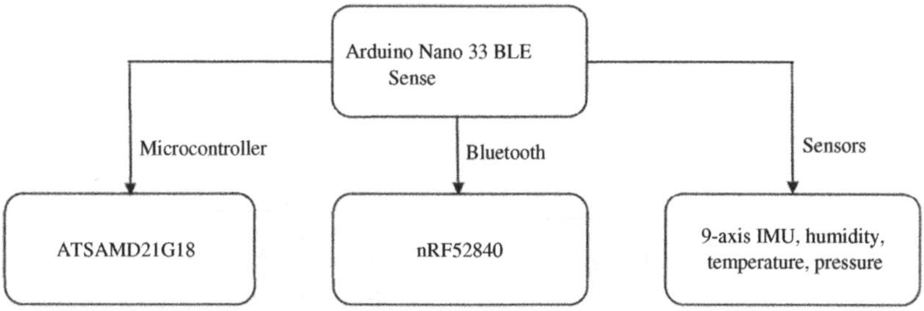

Figure 1.3 Architecture of the Arduino Nano 33 BLE sense.

consumption. Equipped with a 32-bit ARM Cortex-M4 CPU, this board is designed to handle IoT applications that balance energy efficiency and computational power. The ARM Cortex-M4 processor is optimized for signal processing tasks, crucial in many TinyML applications, like audio processing and sensor data analysis. The Arduino Nano 33 BLE Sense also has several onboard sensors, including temperature, humidity, barometric pressure, and motion sensors, making it an excellent option for developers who wish to prototype and implement ML models in settings where these kinds of data are vital.

Another widely used platform is the Raspberry Pi Pico (Figure 1.4), which features a dual-core ARM Cortex-M0+ processor. This microcontroller is designed for simplicity and affordability, making it an excellent option for basic ML tasks that do not require extensive computational power. The Raspberry Pi Pico's dual-core architecture allows it to handle multiple tasks simultaneously, which can be beneficial in scenarios where data acquisition and processing must occur in parallel. Although the Raspberry Pi Pico lacks the advanced features and higher processing power of other platforms, its cost-effectiveness and ease of use make it a popular choice for educational purposes, hobbyist projects, and applications where only modest ML capabilities are needed.

For more demanding ML tasks, specialized hardware accelerators like the Google Edge TPU offer significant advantages [48]. The Edge TPU is designed to run ML models at high speeds while maintaining low power consumption. A graphical overview of the Google Edge TPU is presented in Figure 1.5. This specialized accelerator can execute billions of operations per second, making it ideal for real-time data processing and decision-making applications. The Edge TPU's architecture is optimized for running deep learning models, particularly those heavily reliant on matrix multiplications, common in CNNs used for image and video analysis. By offloading these intensive computations to the TPU, developers can achieve faster inference times and lower power usage than running the same models on general-purpose central processing units (CPUs) or GPUs.

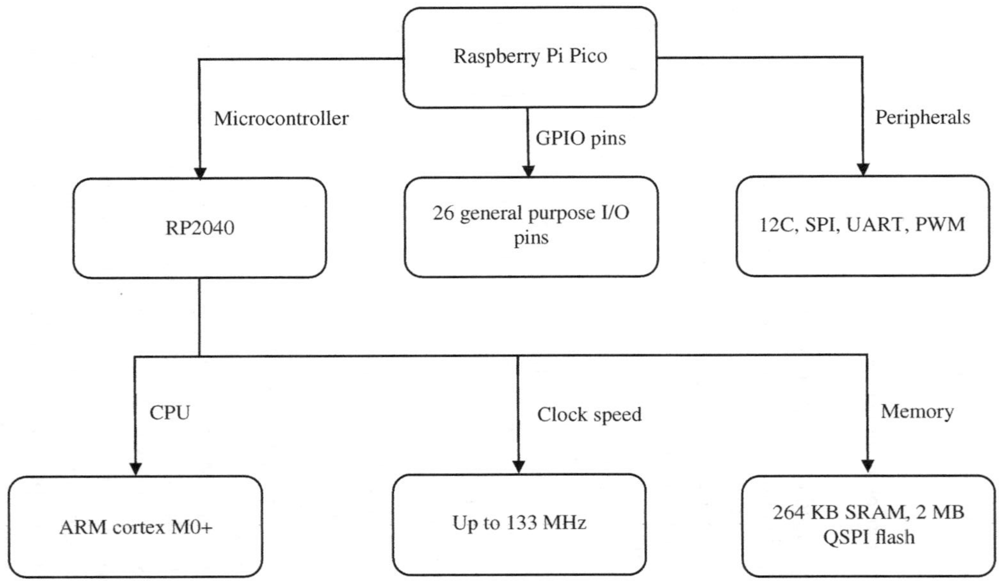

Figure 1.4 Structure of the Raspberry Pi Pico board.

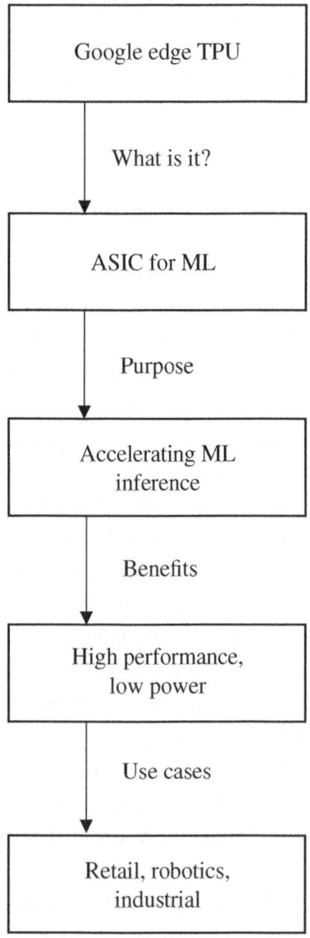

Figure 1.5 Overview of the Google Edge TPU.

Each of these hardware platforms has its own strengths and weaknesses, which makes the choice of platform heavily dependent on the application's specific requirements. For instance, the Arduino Nano 33 BLE Sense may be the best choice for a battery-powered environmental sensor that can operate for extended periods without recharging. On the other hand, the Raspberry Pi Pico might be more suitable for a classroom setting where cost and ease of use are more important than processing power. For applications that require fast, real-time processing of complex models, such as autonomous vehicles or advanced robotics, the Google Edge TPU would be the preferred choice due to its superior speed and efficiency.

Ultimately, choosing a TinyML platform requires carefully weighing the trade-offs between cost, memory, processing speed, and power usage. Developers must consider these aspects when selecting the best hardware for their TinyML applications, considering their unique application requirements. We may anticipate significant advancements in hardware platforms as the TinyML space develops, giving us even more choices for implementing ML at the edge in various sectors and use cases.

1.4.2 Software Frameworks

The development of several specialized software frameworks that facilitate the deployment of ML models on edge devices has been a major driving force behind the advancement of TinyML [18]. These frameworks are designed to function efficiently within low power availability limitations, limited computational resources, and minimal memory, which are edge device characteristics. They have enabled developers to deploy sophisticated ML applications in environments where traditional, resource-heavy models would be difficult, if not impossible, to implement. The typical workflow for TinyML development using TensorFlow and TensorFlow Lite is depicted in Figure 1.6.

TensorFlow Lite, an edge-friendly version of TensorFlow, is a widely used framework for TinyML tool development. One of the primary characteristics of TensorFlow Lite is its capacity to optimize models using methods like quantization, pruning, and compression. These strategies reduce the size and enhance model performance, making them more suited for deployment on devices with restricted computational capabilities. TensorFlow Lite also offers a runtime interpreter that facilitates fast inference and minimal latency, which are crucial for real-time applications. Its wide range of tools and pretrained models provides developers with a robust foundation for building efficient and effective TinyML applications.

Another popular framework is uTensor, which is optimized explicitly for microcontrollers. uTensor is a microcontroller-optimized variant of TensorFlow, designed to function on devices that have minimal memory and processing power. Unlike TensorFlow Lite, which is aimed at slightly more capable edge devices like

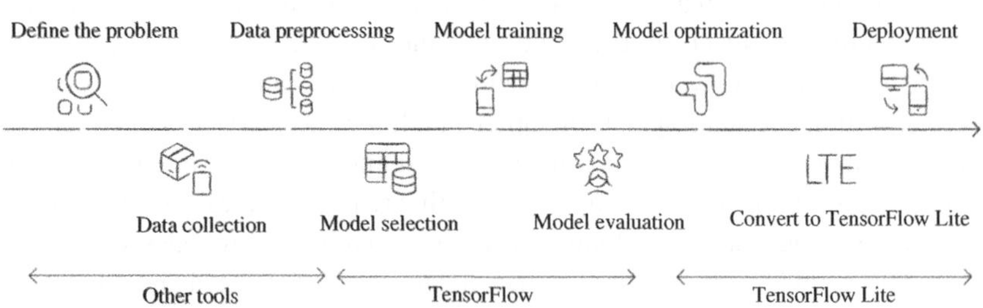

Figure 1.6 Workflow for TinyML development via TensorFlow and TensorFlow Lite.

smartphones and embedded systems, uTensor targets the most constrained environments, such as those in small, battery-powered IoT devices. The framework employs static memory planning and operator fusion techniques to minimize memory usage and enhance performance. uTensor's focus on efficiency and its minimalistic footprint make it ideal for scenarios where energy efficiency is paramount, such as remote monitoring systems, environmental sensors, and wearable devices. By bringing ML capabilities to microcontrollers, uTensor has expanded the potential use cases for TinyML, enabling intelligent processing and decision-making at the very edge of the network.

Another cutting-edge platform that makes it easier to develop and implement ML models on edge devices is Edge Impulse [49]. Edge Impulse provides a more intuitive graphical user interface that enables developers to create models with less code knowledge, unlike typical frameworks that demand a high level of coding skill. This platform is adaptable for computer vision, sensor fusion, and audio processing applications since it supports a wide range of sensor inputs, such as accelerometers, microphones, and cameras. In an integrated environment, Edge Impulse offers training, labeling, data gathering, and model optimization tools. Even those with little experience with ML may design complex TinyML apps because of this end-to-end support, which streamlines the development process. Additionally, Edge Impulse's cloud-based platform allows for rapid iteration and testing, making it an attractive option for developers looking to deploy models quickly and efficiently.

These frameworks – TensorFlow Lite, uTensor, and Edge Impulse – provide the essential tools and libraries required to deploy ML models on resource-constrained devices. They cater to different levels of device capability and developer expertise, democratizing the TinyML field and making it accessible to a broader audience. By lowering the barriers to entry and providing efficient, optimized solutions, these frameworks have played a pivotal role in the rapid growth of TinyML, enabling developers to bring intelligence to the edge across a wide range of applications and industries.

1.5 Challenges and Bottlenecks

TinyML has numerous advantages, but to reach its full potential, a number of issues must be resolved. Deploying TinyML on IoT and edge devices in resource-constrained environments introduces several complicated challenges that affect the performance and feasibility of ML models. These challenges include, but are not limited to, limited processing power, memory limitations, and power consumption, which are some of the most significant issues with TinyML; each presents different difficulties that developers must get beyond [10, 50].

1) **Limited Processing Power:**
   Edge devices, the primary platforms for TinyML applications, typically have limited processing power [51]. Unlike traditional computing devices such as desktop computers and servers, equipped with powerful CPUs and GPUs capable of performing billions of calculations per second, edge devices are designed to be small, efficient, and low-cost. This design philosophy means they are often equipped with simple, low-power microcontrollers without the computational heft to execute complex ML models. As a result, developers must carefully optimize their models to ensure they can run efficiently on these devices. Techniques such as model quantization, pruning, and knowledge distillation are commonly employed to reduce the computational burden [52]. Even with these optimizations, there is often a trade-off between the complexity of the model and its ability to run in real-time, meaning developers must make careful choices about which features to include and which to omit.
2) **Memory Constraints:**
   Besides limited processing power, many edge devices also have very limited memory. The size and complexity of the ML models that can be deployed on these devices may be restricted by memory constraints [53]. This memory restriction is not associated with traditional computing environments, where gigabytes of RAM are normal, and edge devices might only have a few kilobytes or megabytes of RAM available. To overcome this, developers are advised to use highly efficient data structures and algorithms to minimize memory usage. Techniques such as model compression, where redundant information in the model is eliminated, and weight sharing, where multiple connections in the model share the exact weights, can help reduce memory requirements [54]. Furthermore, developers often use specialized memory management techniques to ensure that the limited memory available is used as efficiently as possible. For instance, they may load and unload parts of the model as needed instead of loading the entire model into memory at once. Notwithstanding, memory constraints remain a major challenge in the development of complex models on edge devices.
3) **Power Consumption:**
   This is another critical challenge for TinyML applications. Many edge devices are battery-powered or rely on energy harvesting, so they must operate within strict power constraints. ML models, particularly deep neural networks, can be computationally intensive and require significant power to run, which can quickly drain a portable device's battery. This necessitates the development of models that are not only computationally efficient but also energy-efficient [52, 54]. Techniques such as dynamic voltage and frequency scaling, where the processor's power consumption is reduced by lowering its operating frequency, and event-driven computing, where the device only processes data when an

Table 1.3 A breakdown of challenges and bottlenecks of TinyML.

Challenge	Description	Implications	Mitigation strategies	References
Limited processing power	Edge devices lack powerful CPUs and GPUs, using low-power microcontrollers. Unable to execute complex machine learning models efficiently.	Difficulty running real-time models. Trade-offs between model complexity and runtime efficiency.	Model quantization, pruning, and knowledge distillation. Optimizing algorithms to reduce computational demands. Selecting essential model features.	[7, 8, 10]
Memory constraints	Edge devices have minimal memory, often only a few kilobytes or megabytes of RAM. Limits the size and complexity of deployable models.	Restricts the deployment of large, complex models. Hinders handling of memory-intensive tasks.	Model compression and weight sharing. Efficient memory management (e.g., dynamic loading/unloading). Specialized data structures and algorithms.	[14, 18, 52]
Power consumption	Edge devices are often battery-powered or rely on energy harvesting. Deep neural networks require significant power to operate.	Drains battery life quickly. Limits the operational lifespan of devices. Trade-offs between performance and power efficiency.	Dynamic voltage and frequency scaling. Event-driven computing. Low-power hardware accelerators like Google Edge TPU and Intel Movidius. Energy-efficient models.	[33, 38, 47]

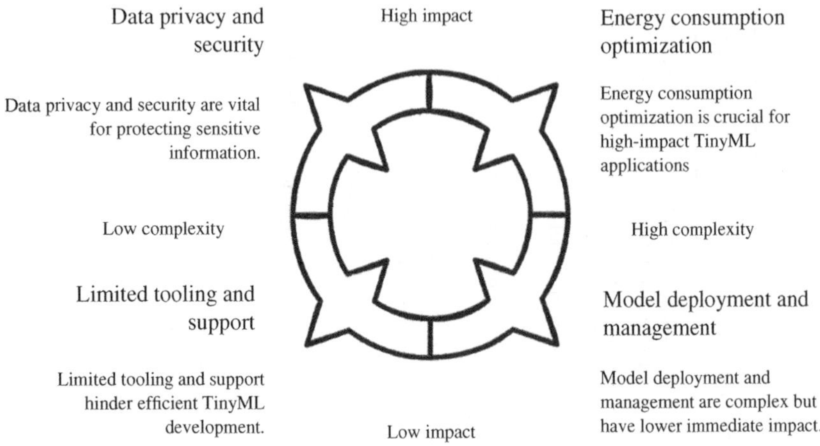

Figure 1.7 Challenges and bottlenecks in TinyML development.

event of interest is detected, are often used to reduce power consumption. Additionally, developers might use low-power hardware accelerators, such as Google's Edge TPU or Intel's Movidius, specifically designed to run low-power ML models. Even with these optimizations, there is often a trade-off between the model's performance and power consumption, and developers must carefully balance these factors to achieve the desired outcome. The challenges and bottlenecks of TinyML and probable mitigation strategies are summarized in Table 1.3, while Figure 1.7 highlights the challenges and bottlenecks.

1.6 Cost–Benefit Analysis

The cost–benefit analysis of TinyML entails evaluating the trade-offs between model complexity, performance, and power consumption, which are essential factors influencing the deployment and efficacy of ML models on edge devices. Consideration for the choice of TinyML in a given scenario is typically hinged on several issues, specifically, as depicted in Figure 1.8.

Generally, more intricate models – like deep neural networks with multiple layers and parameters – perform better in handling challenging tasks and predicting outcomes. Since these models can learn complex patterns and relationships within data, they are suitable for applications requiring high precision, such as sophisticated anomaly detection, image recognition, and natural language processing. However, for these improved capabilities, increasing the computational resources needed to run these models becomes a challenge. For edge devices like microcontrollers or low-power CPUs that lack the computing capabilities of

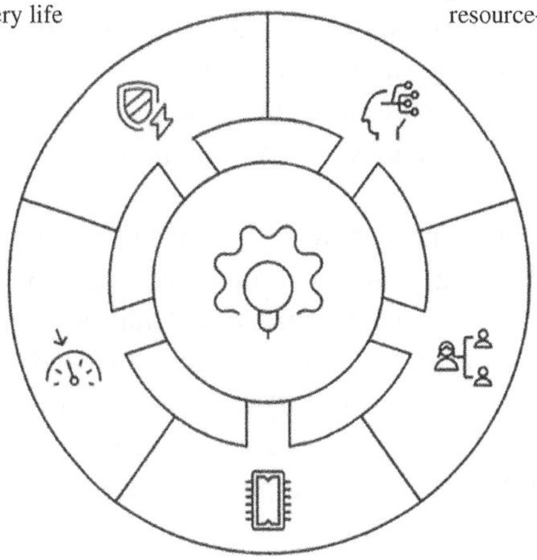

Figure 1.8 Issues in TinyML cost–benefit analysis.

conventional servers or cloud-based systems, creating complex models becomes a challenge because they require more processing resources to carry out the required computations.

These complex models also use adequate memory to store their parameters and intermediate computations during inference. This requirement can exceed the memory capacity for many edge devices, which typically have limited RAM and storage space. As a result, deploying complex models on edge devices can lead to quicker battery depletion and potential overheating, reducing the device's operational life and reliability.

In terms of memory and computation, simpler models seem to be more efficient. Examples of models with fewer parameters and lower processing overhead include decision trees, linear regression, and shallow neural networks. These models are perfect for applications where energy efficiency is a top priority, like wearable technology, remote monitoring systems, or other IoT applications where long battery life is essential.

However, the drawback of this trade-off for this efficiency is that simpler models will likely not give the same level of accuracy or capability as their more complex counterparts [54]. In healthcare diagnostics or autonomous driving, where high accuracy is paramount, the reduced performance of simpler models could lead to suboptimal decisions or outcomes. In some applications, the rewards of implementing a more complex model may surpass the expenses linked to elevated power consumption and computational demands. For instance, the model's ability to identify threats accurately is crucial in a security camera system that detects intruders. This could support using a more complex model, even though it requires more resources. To lessen the effect on power and performance in these situations, developers may choose to employ specialized hardware accelerators or other optimizations.

Conversely, a simpler model may be a better option for applications where power efficiency and long battery life are critical, and with a lower accuracy threshold. For instance, a smart thermostat that adjusts room temperature based on basic patterns and user preferences might not need a complex model to perform effectively. In such cases, a lightweight model that runs continuously without significantly draining the battery or increasing power consumption would be ideal. TinyML, therefore, emphasizes that efficiency does not depend on the size of the technology is [51].

Therefore, developers carefully consider the potential costs of updating and maintaining models deployed on edge devices. Complex models may need frequent updates for accuracy and performance maintenance, particularly in dynamic environments where data distribution can change over time. This regular update maintenance requirement can increase the cost and complexity of maintaining a TinyML deployment, especially when devices are deployed in hard-to-reach or remote locations.

In conclusion, the cost–benefit analysis of TinyML is a complex process that involves weighing the needs and limitations of the application against the conflicting demands of model complexity, performance, and power consumption. Developers may optimize TinyML's advantages and minimize its disadvantages by carefully weighing these trade-offs and selecting the best model and deployment approach. This equilibrium guarantees that the deployed models function effectively within the hardware and power limitations of the target edge devices and satisfy the application's performance requirements. New methods and tools to better handle these trade-offs will probably surface as TinyML develops, allowing for the deployment of increasingly complex and powerful models on a wider variety of edge devices.

TinyML opens new possibilities for ML model deployment on devices with limited resources. In contrast to traditional ML, TinyML is centered on incorporating ML on edge devices, such as microcontrollers and other embedded systems that have limited computing power and memory capacity. This is a ground-breaking move in edge computing since it reduces latency and the need for data transmission to centralized servers, allowing ML models to operate in real-time, closer to the location where the data is generated. This is an essential feature for applications that need to make decisions instantly, such as driverless cars, smart cameras, wearable health monitors, and IoT devices. Hence, the possible strengths of TinyML are demonstrated in power efficiency, real-time processing, and data privacy. Therefore, by carrying out inference immediately on the device where data is collected, TinyML reduces the latency in transmitting data in and out of the cloud.

For example, the safe functioning of autonomous driving relies solely on a swift reaction to sensor inputs. This implies that preserving control and accuracy requires quick processing for a good outcome in industrial automation application scenarios. Although there are still some unresolved challenges of TinyML, such as limited and dynamic resources, security, and privacy, TinyML remains one of the cutting-edge technologies. Hence, to cope with the precincts of edge devices, traditional ML methods need to be redesigned completely or modified to a reasonable extent.

This implies that forthcoming TinyML research may focus on developing new methods and algorithms that maximize model performance while lowering power usage (Table 1.4).

This includes improvements in model compression strategies, making available more efficient quantization approaches that reduce model weight precision without losing accuracy, or pruning techniques that find and eliminate unnecessary model components. Researchers are also advised to direct their studies on investigating novel training techniques that generate models that are optimized for edge deployment from the start, or new kinds of neural network topologies that are intrinsically more efficient.

To assess the performance of TinyML models across different hardware platforms, it is also essential to set up comprehensive benchmarks and evaluation methods. Recent benchmarks focus on definite aspects of performance, such as inference time or accuracy, but do not always provide a holistic view of a model's suitability for edge deployment. Based on this, novel benchmarks, which are expected to provide developers with a clearer understanding of the trade-offs involved in deploying different models on different devices, helping them to make more informed decisions about which models to use for specific applications, should consider a broader range of factors, including power consumption, memory usage, and robustness to variations in input data or operating conditions.

In summary, TinyML is a foremost advancement in ML that will enable the application of AI to gadgets at the periphery of our technological environment. By enabling intelligent processing on the edge, TinyML has created fresh possibilities for innovation in a variety of sectors, including, but not limited to, consumer electronics, manufacturing, and healthcare.

Table 1.4 Cost–benefit analysis of TinyML.

Aspect	Details	Examples/applications	Challenges	References
Model complexity	Complex models (e.g., deep neural networks) handle challenging tasks and require significant computational resources. Provide high accuracy for intricate applications like anomaly detection, image recognition, and NLP.	Security camera systems detecting intruders. Sophisticated healthcare diagnostics. Autonomous driving.	High power consumption. Greater memory requirements. Regular updates increase maintenance costs.	[8, 10, 38]
Simpler models	Efficient with fewer parameters, lower computation, and memory demands. Ideal for applications prioritizing energy efficiency over precision.	Wearable health monitors. Smart thermostats adjusting temperatures. IoT devices in remote locations.	Lower accuracy compared to complex models. Unsuitable for high-stakes applications requiring precision.	[7, 14, 52]
Power consumption	Complex models consume more power and may lead to faster battery depletion and overheating. Simpler models are energy-efficient, enabling longer device operation.	Real-time industrial automation. Long-term monitoring systems.	Balancing power consumption with application demands.	[33, 38, 47]
Hardware innovations	Specialized hardware accelerators (e.g., Google Edge TPU, Intel Movidius) improve model performance and energy efficiency. Enable the use of complex models on constrained devices.	Smart cameras processing image data in real-time. IoT-enabled predictive maintenance systems.	Higher costs for hardware integration. Limited compatibility across different frameworks.	[18, 36, 37]

Real-time processing	Critical for latency-sensitive applications. Ensures faster decision-making closer to the data source. Reduces reliance on cloud transmission.	Driverless cars reacting to sensor inputs. Wearables providing immediate health feedback.	Balancing latency reduction with computational efficiency.	[12, 44, 45]
Key optimization techniques	Model pruning and quantization reduce the size and computational load without significant accuracy loss. Novel training techniques for edge-ready models.	Lightweight neural networks for real-time monitoring systems. Energy-efficient smart home devices.	Striking the balance between performance and energy savings. Refining techniques for diverse application needs.	[19–21]
Emerging needs	Comprehensive benchmarks to evaluate models holistically. Focus on factors like power consumption, robustness, and memory usage in addition to accuracy.	New benchmarks for evaluating TinyML's edge deployment feasibility.	Lack of universal standards across devices and applications.	[18, 23, 49]
Future directions	Advanced model compression techniques. Development of neural architectures inherently efficient for edge devices. Enhanced algorithms for resource optimization.	Broader adoption in consumer electronics, healthcare, and manufacturing.	Dynamic resource constraints and evolving application demands.	[28, 42, 48]

1.7 Key Findings

The significant findings presented in this chapter are organized as follows:

1) **Role of Model Optimization Techniques:** Techniques like quantization, pruning, and knowledge distillation are essential for adapting ML models to edge devices. These methods reduce computational, and memory demands and ensure energy efficiency, making TinyML viable for a broad range of applications.
2) **Role of Specialized Hardware in Advancing the Frontiers of TinyML:** Hardware accelerators such as Google Edge TPU and Intel Movidius have revolutionized the field by deploying more complex models on resource-constrained devices. Besides highlighting the interplay between hardware innovation and the current growth trend of TinyML, these accelerators are expected to continue to play a significant role in the accelerated development of TinyML applications.
3) **Trade-offs Between Model Complexity and Efficiency:** The literature underscores the need to balance model complexity and efficiency. While simpler models excel in energy conservation, they lack the precision for high-stakes applications. In contrast, complex models demand more resources but offer greater accuracy. Hence, determining the appropriate midpoint between the two ends of these competing requirements is a subsisting research problem.
4) **Edge Intelligence as a Game-Changer:** By enabling real-time data processing and decision-making directly at the edge, TinyML significantly reduces latency and dependency on cloud-based systems. This is vital for time-sensitive applications like autonomous vehicles and medical monitoring; it removes the potential for failures associated with data transmission, security vulnerabilities, and costs.
5) **Growing Ecosystem of Frameworks:** The development of tools like TensorFlow Lite and Edge Impulse has democratized TinyML, allowing developers with varying levels of expertise to build efficient edge solutions. This expansion reflects TinyML's increasing accessibility and adaptability, and it is expected to continue liberalizing the TinyML economy.
6) **Need for Standardized Benchmarks:** The absence of universal benchmarks for evaluating TinyML models is a significant bottleneck. Establishing standardized metrics that account for power consumption, latency, and memory usage alongside accuracy will be critical for future advancements.
7) **Potential for Novel Research Areas:** Emerging areas such as decentralized learning, federated learning, and swarm intelligence integrated with TinyML

offer exciting research opportunities. These approaches could address scalability and privacy concerns in IoT applications.

8) **Sustainability and Scalability:** TinyML aligns well with global sustainability goals by enabling low-power solutions. Its scalability across diverse industries, from smart cities to personal healthcare, demonstrates its transformative potential.

1.8 Limitations of TinyML

Despite the numerous advantages and innovations that TinyML brings to edge computing, its deployment and widespread adoption are hindered by several limitations. These constraints affect model performance, scalability, and practical usability in real-world applications. The primary limitations include restricted computational resources, memory constraints, power consumption challenges, model accuracy trade-offs, limited on-device learning, security vulnerabilities, and interoperability issues.

1) **Limited Computational Resources:** One of the most significant limitations of TinyML is the constrained computational capability of edge devices. Unlike cloud-based systems that leverage high-performance CPUs and GPUs, TinyML relies on microcontrollers and low-power processors without the hardware acceleration for complex ML tasks [10]. These resource limitations restrict the complexity and depth of neural networks that can be deployed, forcing developers to use lightweight models such as quantized deep learning models, pruned neural networks, and shallow architectures [19, 38]. While these optimizations improve efficiency, they often compromise accuracy and performance, especially for applications requiring high precision, such as medical diagnostics and autonomous navigation [14, 33].
2) **Memory Constraints:** TinyML operates on embedded devices with extremely limited memory, typically ranging from a few kilobytes to a few megabytes of RAM. This is significantly lower than conventional ML frameworks, which require gigabytes of memory to store model parameters, intermediate computations, and feature representations [52]. Memory constraints hinder the deployment of larger models and limit real-time processing capabilities. Techniques such as weight sharing, model quantization, and efficient memory allocation strategies are used to address this issue. Still, they often have trade-offs regarding processing speed and prediction accuracy [18, 52]. Additionally, real-time inference on limited-memory devices can lead to bottlenecks when handling large data streams, making TinyML less suitable for applications requiring complex feature extraction and analysis [23].

3) **High Power Consumption:** Many TinyML applications are deployed on battery-powered devices or rely on energy-harvesting techniques. Running deep learning models on such devices can quickly drain battery life due to continuous computations and sensor activity [10]. While optimizations such as event-driven computing, low-power hardware accelerators (e.g., Google Edge TPU, Intel Movidius), and dynamic voltage scaling help mitigate energy consumption, they do not eliminate power constraints, especially for applications requiring continuous monitoring and real-time inference [38, 47]. The trade-off between performance and energy efficiency remains a critical challenge, making it difficult to deploy complex TinyML models in resource-constrained environments such as wearable health monitors and industrial automation systems [28, 45].
4) **Model Accuracy and Trade-Offs:** TinyML models often experience accuracy degradation due to extreme compression techniques such as quantization and pruning. While these methods reduce model size and computational load, they can also introduce errors and reduce generalization capabilities, particularly in applications that require high precision [19, 20]. For example, in health-care applications, where small changes in input data can significantly impact diagnosis, a loss in accuracy due to compression techniques can lead to mis-diagnoses or false alerts [5]. This limitation necessitates balancing optimizing model size and maintaining reliable performance, an ongoing research challenge [23, 43].
5) **Limited On-Device Learning and Adaptability:** Most TinyML models are designed for inference-only operations, meaning they cannot learn from new data dynamically. Traditional ML models can be updated and retrained with new data, but TinyML models often require manual updates and retraining on external systems before redeployment [18]. This limitation is particularly problematic for applications that require adaptive learning, such as predictive maintenance in industrial settings, where environmental conditions change over time [44, 54]. Federated learning and continual learning strategies are emerging as potential solutions, but these methods still require substantial optimization to function effectively on ultralow-power devices [12].
6) **Security and Privacy Concerns:** Deploying TinyML models on edge devices introduces significant security and privacy risks. Since these models operate on resource-constrained devices with limited security mechanisms, they are vulnerable to adversarial attacks, model inversion, and data breaches [34, 46]. For example, an attacker could manipulate sensor data to mislead the model's predictions, leading to incorrect decisions in autonomous systems or fraudulent activities in financial applications [26]. Data privacy concerns arise when sensitive information, such as biometric data or medical records, is processed

on unsecured edge devices [10, 41]. Implementing robust encryption, secure boot mechanisms, and lightweight privacy-preserving ML techniques is necessary to mitigate these risks, but doing so often increases computational overhead and power consumption [27].

7) **Interoperability and Deployment Challenges:** Deploying TinyML models across diverse hardware architectures presents interoperability challenges. TinyML frameworks such as TensorFlow Lite, uTensor, and Edge Impulse must be optimized for different microcontrollers and embedded platforms, each with unique constraints [18, 49]. This lack of standardization complicates model portability and hinders the seamless deployment of ML applications across different devices [30]. Moreover, updating remote or embedded systems models can be complex and costly, particularly in industrial or agricultural applications where manual maintenance is impractical [9, 15].

1.9 Conclusion

The evolution and application of TinyML represent a paradigm shift in how ML is integrated into resource-constrained environments and systems. This chapter has provided a comprehensive overview of TinyML, discussing its foundational principles, historical milestones, core state-of-the-art technologies, and the challenges that influence its deployment. TinyML's ability to operate efficiently on low-power and low-memory devices positions it as a highly transformative technology with far-reaching implications across healthcare, agriculture, smart cities, and industrial automation. The insights presented in this chapter highlight the significance of TinyML in addressing real-world challenges through edge computing. Its potential to enable real-time decision-making, enhance data privacy, and reduce latency makes it indispensable in privacy-critical and time-sensitive applications. The cost–benefit analysis of TinyML emphasizes the need to balance model complexity with device capabilities to ensure that solutions are both practical and energy-efficient. TinyML's continued growth will hinge on advancements in hardware accelerators, optimization algorithms, and software frameworks. These innovations will broaden TinyML's applicability and enable its adoption in diverse environments. Researchers, practitioners, and policymakers can leverage the insights provided in this chapter to drive innovation and expand the practical use of TinyML. By building on these foundations, future developments in TinyML will contribute to advancing ML, enabling more intelligent and more interconnected systems worldwide. This chapter serves as both a reference and a call to action for those interested in this emerging field, encouraging further exploration, development, and adoption of TinyML to unlock its full potential.

References

1 Johnvictor, A.C., Poonkodi, M., Prem Sankar, N., and Vs, T. (2024). TinyML-based lightweight AI healthcare mobile Chatbot deployment. *Journal of Multidisciplinary Healthcare* 12 (4): 5091–5104.

2 Bhamare, M., Kulkarni, P.V., Rane, R. et al. (2024). TinyML applications and use cases for healthcare. *TinyML for Edge Intelligence in IoT and LPWAN Networks* 8 (2): 331–353.

3 Tsoukas, V., Boumpa, E., Giannakas, G., and Kakarountas, A. (2021). A review of machine learning and TinyML in healthcare. In Proceedings of the 25th Pan-Hellenic Conference on Informatics (November), 6 (11), 69–73.

4 Andrade, P., Silva, I., Silva, M. et al. (2022). A TinyML soft-sensor approach for low-cost detection and monitoring of vehicular emissions. *Sensors* 22 (10): 3838.

5 Hou, K.M., Diao, X., Shi, H. et al. (2023). Trends and challenges in AIoT/IIoT/IoT implementation. *Sensors* 23 (11): 5074.

6 Zhou, S., Du, Y., Chen, B. et al. (2022). An intelligent IoT sensing system for rail vehicle running states based on TinyML. *IEEE Access* 10 (5): 98860–98871.

7 Warden, P. and Situnayake, D. (2019). *TinyML: Machine Learning with Tensor Flow Lite on Arduino and Ultra-Low-Power Microcontrollers*, 1e, 26. O'Reilly Media.

8 Banbury, C., Reddi, V.J., Lam, M., et al. (2020). Benchmarking TinyML systems: challenges and directions. Proceedings of the International Conference on Machine Learning, 7 (3).

9 Murshed, M.S., Murphy, C., Hou, D. et al. (2021). Machine learning at the network edge: a survey. *ACM Computing Surveys (CSUR)* 54 (8): 1–37.

10 Abdussalam, E., Dini, P., Saponara, S., and Zheng, Q. (2024). Advancements in TinyML: applications, limitations, and impact on IoT devices. *Electronics* 13 (17): 3562. https://doi.org/10.3390/electronics13173562.

11 Xiaying W., Magno, M., Cavigelli, L., and Benini, L. (2022). FANN-on-MCU: an open-source toolkit for energy-efficient neural network inference at the edge of the Internet of Things. *arXiv:1911.03314v3* [cs.LG] 17 Feb 2022.

12 Evangelia, F. and Dimitrios, K. (2024). A joint survey in decentralized federated learning and TinyML: a brief introduction to swarm learning. *Future Internet* 2024 (16): 413. https://doi.org/10.3390/fi16110413.

13 Iqbal, Z. (2024). A TinyML-enabled approach to embed machine learning in avionics control systems. PhD thesis, University of Padova (p. 44). https://www.research.unipd.it/retrieve/b86378c6-5012-4c60-9f19-910467fee9fa/PhD_Thesis_Iqbal.pdf.

14 Dutta, L. and Bharali, S. (2021). TinyML meets IoT: a comprehensive survey. *Internet of Things* 16: 100461.

15 Rajapakse, V., Karunanayake, I., and Ahmed, N. (2023). Intelligence at the extreme edge: a survey on reformable TinyML. *ACM Computing Surveys* 55 (13s): 1–30.

16 Alexander W., Famouri, M., Pavlova, M., and Surana, S. (2020). TinySpeech: attention condensers for deep speech recognition neural networks on edge devices. *arXiv:2008.04245v6* [eess.AS] 12 Oct 2020.

17 Thakker, U., Beu, J. G., Gope, D., et al. (2019). Compressing RNNs for IoT devices by 15–38x using Kronecker products. CoRR abs/1906.02876 (2019). *arXiv preprint arXiv:1906.02876.*

18 David, R., Duke, J., Jain, A. et al. (2021). Tensorflow lite micro: embedded machine learning for TinyML systems. *Proceedings of Machine Learning and Systems* 3: 800–811.

19 Vadera, S. and Ameen, S. (2022). Methods for pruning deep neural networks. *IEEE Access* 10 (8): 63280–63300.

20 Liu, Z., Sun, M., Zhou, T., et al. (2019). Rethinking the value of network pruning. *ICLR*, 1.

21 Gou, J., Yu, B., Maybank, S.J., and Tao, D. (2021). Knowledge distillation: a survey. *International Journal of Computer Vision* 129 (6): 1789–1819.

22 Cho, J. H. and Hariharan, B. (2019). On the efficacy of knowledge distillation. Proceedings of the IEEE/CVF International Conference on Computer Vision, 14 (5), 4794–4802.

23 Ray, P.P. (2022). A review on TinyML: state-of-the-art and prospects. *Journal of King Saud University, Computer and Information Sciences* 34 (4): 1595–1623.

24 Samanta, A., Hatai, I., and Mal, A.K. (2024). A survey on hardware accelerator design of deep learning for edge devices. *Wireless Personal Communications* 137 (3): 1715–1760.

25 Giovanni, D., Mirri, S., Prandi, C., and Manzoni, P. (2023). An evaluation methodology to determine the actual limitations of a TinyML-based solution. *Internet of Things Journal* 19 (2): 100729. https://doi.org/10.1016/j.iot.2023.100729.

26 Profentzas, C., Almgren, M., and Landsiedel, O. (2021). Performance of deep neural networks on low-power IoT devices. Proceedings of the Workshop on Benchmarking Cyber-Physical Systems and Internet of Things (May) 5 (7), 32–37.

27 Zhanghao W., Liu, Z., Lin, J., et al. (2020). Lite transformer with long-short range attention. *arXiv, 2004.11886v1* [cs.CL]. ICLR, 2020, 1.

28 Subhrangshu, A., Subhayu, D., and Ashutosh, D. (2024). TinyWolf—efficient on-device TinyML training for IoT using enhanced grey wolf optimization. *Internet of Things* 28 (3): 101365.

29 Konkala, V.R., Reddy, B.K., Goutham, V. et al. (2024). Edge AI in sustainable farming: deep learning-driven IoT framework to safeguard crops from wildlife threats. *IEEE Access* 12 (6): 77707–77723. https://creativecommons.org/licenses/by-nc-nd/4.0/.

30 Lin, J., Chen, W.M., Lin, Y. et al. (2020). Mcunet: tiny deep learning on IoT devices. *Advances in Neural Information Processing Systems* 33: 11711–11722.

31 Haoyu R., Darko A., and Thomas A. R. (2022). How to manage tiny machine learning at scale – an industrial perspective. TinyML Research Symposium' 22, 3 (9).

32 Abadade, Y., Temouden, A., Bamoumen, H. et al. (2023). A comprehensive survey on TinyML. *IEEE Access* 15 (4): 1–31. https://doi.org/10.1109/ACCESS.2023.3294111.

33 Alajlan, N.N. and Ibrahim, D.M. (2022). TinyML: enabling of inference deep learning models on ultra-low-power IoT edge devices for AI applications. *Micromachines* 13 (6): 851.

34 Truong, N., Sun, K., Wang, S. et al. (2021). Privacy preservation in federated learning: An insightful survey from the GDPR perspective. *Computers & Security* 110: 102402.

35 Biglari, A. and Tang, W. (2023). A review of embedded machine learning based on hardware, application, and sensing scheme. *Sensors* 23 (4): 2131.

36 He, K., Zhang, X., Ren, S., and Sun, J. (2016). Deep residual learning for image recognition. Proceedings of the IEEE Conference on Computer Vision and Pattern Recognition, 9 (4), 770–778.

37 Yihui H., Lin, J., Liu, Z., et al. (2018). AMC: AutoML for model compression and acceleration on mobile devices. Springer Conference on Computer Vision and Pattern Recognition 5 (2). http://springer.com/conference/eccv/.

38 Han, S., Mao, H., and Dally, W. J. (2015). Deep compression: compressing deep neural networks with pruning, trained quantization and Huffman coding. *arXi.1510.00149*. Online. https://arxiv.org/abs/1510.00149.

39 Abdel Magid, S., Petrini, F., and Dezfouli, B. (2020). Image classification on IoT edge devices: profiling and modeling. *Cluster Computing* 23 (2): 1025–1043.

40 Matthieu C. (2016). Binarized neural networks: training neural networks with weights and activations constrained to +1 or −1. *arXiv:1602.02830v3* [cs.LG] 17 Mar 2016.

41 Sujin K. (2022). Joint data deepening-and-prefetching for energy-efficient edge learning. *arXiv:2211.07146v1* [cs.NI] 14 Nov 2022.

42 Terlizzi, M. A. (2023). Mixed-precision quantization and inference of MLPerf tiny DNNs on precision-scalable hardware accelerators. Doctoral dissertation, Politecnico di Torino. Online. http://webthesis.biblio.polito.it/id/eprint/26664.

43 Wang, Y., Qin, Y., Deng, D. et al. (2022). Trainer: an energy-efficient edge-device training processor supporting dynamic weight pruning. *IEEE Journal of Solid-State Circuits* 57 (10): 3164–3178.

44 Mishra, S. and Manda, J. (2024). Improving real-time analytics through the Internet of Things and data processing at the network edge. *Journal of AI-Assisted Scientific Discovery* 4 (1): 184–206.

45 Rakhee K., Pai, K., Raghuwanshi, P., et al. (2023). TinyML: tools, applications, challenges, and future research directions. *arXiv:2303.13569v1* [cs.LG] 23 Mar 2023.

46 Karras, A., Giannaros, A., Karras, C. et al. (2024). TinyML algorithms for big data management in large-scale IoT systems. *Future Internet* 16 (2): 42.

47 Cyient (2024). TinyML optimization approaches for deployment on edge devices. Healthcare and Life Sciences Whitepaper, 1–17. https://www.cyient.com/hubfs/2024/Whitepaper/WP%20-%20TinyML%20Optimization%20Approaches%20for%20Deployment%20on%20Edge%20Devices%20-%200424.pdf.

48 Boroumand, A., Ghose, S., Akin, B., et al. (2021). Google neural network models for edge devices: analyzing and mitigating machine learning inference bottlenecks. 2021 30th International Conference on Parallel Architectures and Compilation Techniques (PACT) (September), 159–172. IEEE.

49 Reddi, J., Elium, A., Hymel, S., et al. (2023). Edge impulse: an MLOps platform for tiny machine learning. Proceedings of Machine Learning and Systems, 5.

50 Adlakha, R. and Kabbar, E. (2023). The challenges of TinyML implementation: a literature review. Proceedings CITRENZ 2023 Conference, Auckland (27–29 September). https://doi.org/10.34074/proc.240120.

51 Immonen, R. and Hämäläinen, T. (2022). Tiny machine learning for resource-constrained microcontrollers. *Journal of Sensors* 2022 (1): 7437023.

52 Berthelier, A., Chateau, T., Duffner, S. et al. (2021). Deep model compression and architecture optimization for embedded systems: a survey. *Journal of Signal Processing Systems* 93 (8): 863–878.

53 Mohammad, G., Marimuthu, P., and Rajkumar, B. (2021). A distributed deep reinforcement learning technique for application placement in edge and fog computing environments. *IEEE Transactions on Mobile Computing* 2021 (22): 2491–2505.

54 Liebenwein, L., Baykal, C., Carter, B. et al. (2021). Lost in pruning: the effects of pruning neural networks beyond test accuracy. *Proceedings of Machine Learning and Systems* 3: 93–138.

2

Learning Panorama Under TinyML

Ikechukwu Ignatius Ayogu, Euphemia Chioma Nwokorie, Juliet Nnenna Odii, Francisca Onyiyechi Nwokoma, and Chidi Ukamaka Betrand

Department of Computer Science, Federal University of Technology, Owerri, Nigeria

2.1 Introduction

Tiny machine learning (TinyML) pushes the frontiers of machine learning (ML) applications to resource-constrained, battery-operated devices, with the potential to enable the ubiquity of artificial intelligence and increase the scale of innovation in ML applications. TinyML combines the gains of decades of intensive research and development in the domain of embedded systems, ML, Internet of Things (IoT), and software development [1, 2] to provide effective solutions that facilitate effective data processing at the edge, leveraging the IoT framework [3–6]. It has made significant positive impacts in many areas of application within the IoT and edge computing domains, including smart homes [7–9], smart agriculture [10–15], automobiles [16], wearable devices [17–19], and smart cities [20].

The key drivers for the accelerated development of TinyML include, in part, the unprecedented improvement in microcontroller units (MCUs), which continue to grow in computational power, memory capacity, and energy efficiency, as well as accelerated developments within the two fundamental building blocks: the ML algorithm and the hardware architecture. The rapid improvements in both algorithmic approaches and hardware architectures have led to improved performance of various ML models in the context of edge computing, especially in the TinyML domain. Algorithmic approaches, such as growing attention-based models for sequence data processing, do not rely on backpropagation for training and thus work on simpler algorithms and simpler hardware architectures, which are special considerations for edge ML.

Tiny Machine Learning: Design Principles and Applications, First Edition.
Edited by Agbotiname Lucky Imoize, Dinh-Thuan Do, and Houbing Herbert Song.

However, developing and deploying TinyML on microcontrollers presents significant challenges relating to design, optimization, and learning. Other challenges arise from the dearth of development frameworks – currently, TensorFlow Lite for MCUs and uTensor, and Edge Impulse are among the popular options available – as well as programming language restrictions, the lack of good support for various development boards, the choice of suitable sensors, the absence of standardized benchmarks, sensor selection issues, and the scarcity of development datasets [21–23]. These challenges need to be resolved in order for TinyML to realize its full potential. In Africa, where the adoption of artificial intelligence (AI) and embedded systems is still in its infancy, TinyML could address issues related to connectivity, energy, and costs [24]. Moreover, its potential to facilitate computation at the edge, reducing dependence on cloud-based operations, opens up new areas for energy-efficient, secure, and private data processing. It also ushers in a new era of novel localized applications in all sectors.

2.1.1 Chapter Contribution

This chapter presents a comprehensive examination of the processes by which ML models for tiny devices, a class of computational devices with limited computing, memory, and power resources, are created and deployed. A sequential knowledge presentation approach was adopted, highlighting challenges in TinyML design, opportunities for model design improvements, model compression and optimization techniques, development frameworks, tools, TinyML hardware, and opportunities for algorithmic innovations for emerging TinyML applications. The major contributions of this chapter are as follows:

1) It presents a clear outline of the major issues that are of importance to the design, training, and deployment of ML models on resource-constrained edge devices, as well as the opportunities for an improved model design using already established techniques. Thus, it serves as a resourceful and handy information source for experienced and upcoming TinyML practitioners. More importantly, it reduces the time required for a novice reader to get acquainted and also develop interest in the subject matter.
2) The chapter identifies the frontiers for TinyML development through model optimization approaches and also provides a detailed cost–benefit analysis of the major model compression techniques that are currently under active research. This elucidation further provides new entrants with balanced information and reasoning for selecting a given technique in a given application scenario.
3) This chapter presents an in-depth, informative discussion of the learning frameworks and tools for TinyML development. It devotes significant attention to the discussion of the key features, strengths, weaknesses, examples, and

use cases of the more popular tool framework currently used for developing models for tiny devices and key hardware platforms for deploying tiny models.

4) The chapter highlights avenues for algorithmic innovations for improved learning for TinyML. The chapter forecasts the importance of meta-learning and multimodal learning in developing TinyML applications that are both adaptive and extensible in the near future.

2.1.2 Chapter Organization

Section 2.2 presents an in-depth discussion of the major challenges facing the TinyML model design as well as the available avenues for overcoming them. In particular, the section analyzed the prominent challenges of TinyML design before delving into the opportunities for overcoming the identified bottlenecks using state-of-the-art Tiny model development techniques, including neural architecture search (NAS), neural attention mechanisms, depth-separable convolutions, and rational activation functions. Section 2.3 presents a detailed account of the frontiers in model optimization for TinyML development and deployment, viewing model compression as the principal paradigm for model development for TinyML. It examined quantization, knowledge distillation (KD), pruning, and low-rank decomposition as principal examples of learning a tiny model from large parameter models. It also presents a balanced cost–benefit analysis of these major model compression techniques. Section 2.4 provides a detailed discussion and analysis of learning frameworks and tools for TinyML development, focusing on the various development frameworks, tools, and major edge hardware. The section presents a detailed analysis of the strengths and weaknesses of representative TinyML development frameworks as well as key features, weaknesses, and use cases for some popular TinyML hardware systems. Section 2.5 discusses the frontiers of algorithmic innovations for TinyML, identifying multimodal learning and meta-learning as key contenders. Section 2.6 presents a high-level summary of the TinyML development process, highlighting model training, conversion, and inference as the major phases in the TinyML development process. Section 2.7 concludes the chapter by summarizing the main issues facing TinyML model development in the context of learning, compression, and inference.

2.2 Challenges and Opportunities for Improved TinyML Model Design

TinyML is a maturing field of study within the domain of AI and ML, with significant capacity for commercial impact. However, the design and deployment of effective ML models in the TinyML space present many challenges, mainly limited computational capacity, hardware resources, and power deficiencies [5, 25]. This peculiarity compromises lightweight models, where trade-offs between model accuracy and computational complexity/efficiency need to be made. Efficient inference in resource-constrained environments depends on a number of factors, including model complexity. Within the context of important applications such as healthcare and agriculture, there are further challenges arising from a dearth of high-quality datasets [26–28].

Designing ML models for edge devices is a challenging problem. TinyML model design involves working with devices with limited energy resources; to be of practical use, the new designs must be small and energy-efficient. Restricted energy resources usually motivate the use of less complicated algorithms, leading to considerable loss and poor application-level performance. Some of the main challenges affecting the applications of TinyML are captured in Figure 2.1, including lack of standards to promote interoperability, limited tooling and community support, computing, energy, data protection, scalability, and post-deployment management challenges.

Although TinyML enables the deployment of ML models on resource-constrained edge computing devices, its development continues to face several challenges. A brief outline of the key obstacles and examples is presented in Table 2.1.

In addition to devising innovative solutions to mitigate the shortage of computational resources, memory constraints must also be considered when designing

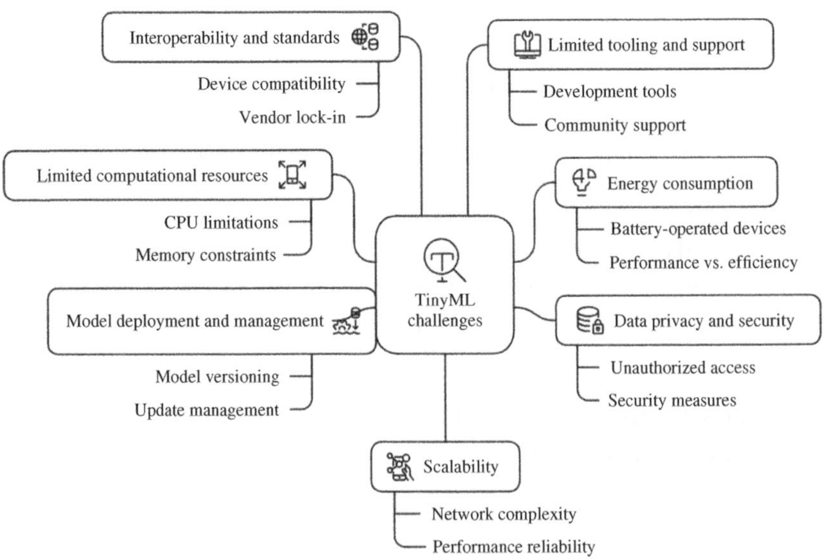

Figure 2.1 An overview of the major challenges of TinyML.

Table 2.1 The scope of some of the key challenges of TinyML.

Key challenges	Explanation	Example	References
Compute and memory constraint	Current TinyML models are constrained to run on microcontrollers with kilobytes to a few megabytes of RAM and low processing power. This renders the execution of complex models infeasible.	Running MobileNet on an Arduino Nano 33 BLE Sense requires extreme optimizations, such as quantization and pruning, to reduce the model's computing and memory requirements.	[29, 30]
Acute power constraints	TinyML applications typically run on battery-powered or energy-harvesting devices. This necessitates that the operations performed be minimally power-consuming.	A typical sensor must consume microwatts of energy while processing computations for extended battery life.	[31, 32]
Model size-performance trade-off	Striking a good balance between model miniaturization and performance (accuracy) is a challenge since excessive simplification leads to a remarkable loss in accuracy, a principal performance measure.	Quantizing FP32 to INT8 on a TinyML gesture recognition model reduces the memory requirement, but it has been noted to result in a significant number of misclassifications.	[33, 34]
Scarcity of dataset	The scarcity of data at the edge is a significant issue for TinyML applications. Collecting and using data on a device is challenging because it seems impossible to provide labels in real-time applications.	Critical but resource-challenged applications, such as speech recognition for low-resource languages, face performance issues due to the unavailability of labeled data.	[35, 36]
Complexity-induced challenges for on-device learning	The lack of cloud connectivity makes adapting models to new data difficult. This is due to memory-related limitations and the lack of complicated training algorithms on microcontrollers.	A personalized speech application that runs on an embedded device must learn from a user's voice samples without sending data to the cloud.	[37, 38]

(Continued)

Table 2.1 (Continued)

Key challenges	**Explanation**	**Example**	**References**
Lack of standardized frameworks	TinyML deployment requires custom optimizations for different hardware, making portability and scalability difficult.	A model optimized for TensorFlow Lite Micro may not run efficiently on Arm CMSIS-NN without further tuning.	[31, 39]
Security and privacy issues	On-device ML models are known to be vulnerable to adversarial attacks. Thus, when sensitive information is involved, data privacy concerns naturally arise.	A TinyML-powered baby monitor could be attacked with adversarial noise, triggering false alarms.	[40, 41]
Hardware compatibility issues	The diversity of TinyML model architectures, such as ARM Cortex-M, RISC-V, and ESP32, makes universal ML model deployment complex.	A LoRa-based TinyML sensor for air quality monitoring may require different optimizations compared to ESP32-based AI cameras.	[32, 42]

TinyML models. The RAM is required to store the model and input data processed by the model as it is run. ML models that run on devices with tight memory constraints run the risk of memory thrashing, which occurs when limited memory is used inefficiently, resulting in model slowdown [43]. Addressing these challenges is vital for successfully deploying deep learning models on edge devices. A successful TinyML model design must deliver the processing capabilities required while minimizing resource requirements, complexity, and computational effort. Some of the most popular techniques for improved TinyML model design are discussed in this section.

2.2.1 Neural Architecture Search

TinyML has drawn markedly high research interest in the past few years. However, searching for very accurate and resource-efficient models is still a manual investigation involving domain knowledge and assumptions [44]. This renders training and deployment of TinyML models difficult, resource-inefficient, and significantly more expensive in terms of training and development time. NAS is an effective method for automatically engineering various high-performance deep neural network architectures [45].

NAS aims to automate the design of neural network models by directly searching for optimal neural architectures. NAS typically has three components – the search space, the search strategy, and the evaluation strategy. A search space, normally very large, describes the set of all architectures available for the NAS algorithm to choose from. A search strategy, which may be black-box, is an optimization method for identifying a high-performing neural architecture within the defined search space. A performance estimation strategy, such as learning curve extrapolation, is used to speed up the process of evaluating the performance of neural architectures without completing the training of the model. NAS searches for the network architecture from scratch with parameters, layers, modules, and more fine-grained operations or connections [46]. TinyML, as an increasingly important paradigm, represents a potential, albeit critical, application area for NAS. As such, NAS is a promising method to remove the performance-resource bottleneck in TinyML by designing TinyML models that autonomously optimize model architectures through extensive searching. NAS has been applied to thoroughly search for a TinyML neural network for object detection [47].

The recent advancements in NAS could come in handy in offloading the challenges associated with the traditional heuristic approach to TinyML design, thus revolutionizing the TinyML landscape with the possibilities of more innovative solutions – the YOLO-NAS example is classic. The transformations built for optimizing tiny designs using NAS are agnostic and could be used for many TinyML applications. Tools and frameworks are increasing support and gaining popularity, attracting academic and industry professionals to engineer and design miniature ML models.

However, there are some drawbacks to using NAS to optimize TinyML. First, it requires a large amount of computational resources [48]. The implementation of NAS makes it necessary to train and evaluate millions of different designs sequentially to determine the highest-performing model. Secondly, scalability issues exist when applying the NAS methods [49]. The search space explodes quickly, such that the possibilities of selecting a suboptimal model increase significantly. Despite these challenges, the successful applications of NAS to TinyML design have been validated by a broad range of case studies, and the feasibility of improving the design of TinyML models has been demonstrated [50–53]. These applications entail developing models that can perform well in real-world circumstances when the models have been designed and pretrained. The application of NAS to TinyML design is operationalized in Figure 2.2. The first step is to define a search space, setting conditions and constraints for possible network architectures. In the second step, the search strategy is selected – popular choices are reinforcement learning, evolutionary or gradient-based techniques. Next is the actual architectural search, during which several candidate models are trained and evaluated. Upon the search's conclusion, the model that gives

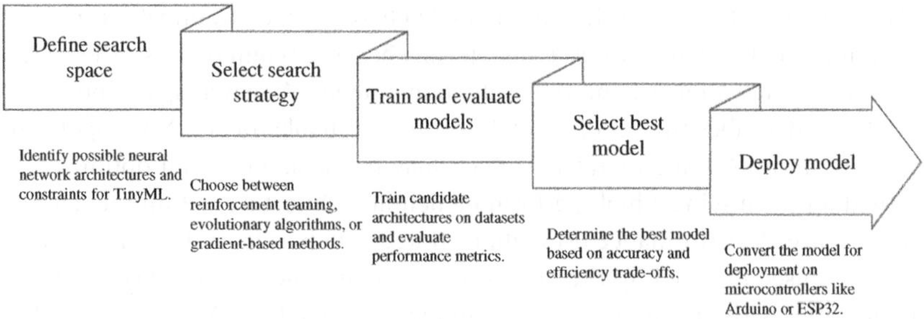

Figure 2.2 Typical workflow for NAS in TinyML development.

the best trade-off between efficiency and accuracy is selected and deployed on suitable hardware.

2.2.2 Neural Attention Mechanisms

Attention mechanisms are a key part of most modern deep learning research. The basic principle of attention is written into its name; these mechanisms allow the network to *pay attention* to specific parts of its input data points. This has applications in numerous domains and purposes, from localizing objects or parts of objects in visual inputs, weighting different parts of text inputs based on perceived importance, or altering the network structure used to process different sections of longer sequences [54]. In all of these cases, the underlying concept is to allow the network to focus on the most important parts of its input data for whatever task it is carrying out in order to optimize information processing, reduce errors, and improve accuracy. However, understanding and properly utilizing these concepts can be complex.

The TinyML community has demonstrated that attention mechanisms can be integrated within TinyML frameworks to obtain value not only in terms of model accuracy but also in terms of interpretability, as they reveal which input features the model actually *focuses on* to make its prediction. However, integrating attention in the TinyML domain also presents unique challenges: top-attention TinyML models with increased performance often come attached with proportionally higher computational and memory requirements [55]. This tightly constrains the feasibility of its real-world applications on devices with strict resource limitations and severe hardware constraints, especially compared to their non-attentional TinyML counterparts. It is, therefore, crucial to understand these trade-offs and weigh the costs versus the benefits with every decision around the model design when it comes to integrating attention.

When implementing attention, the TinyML community predominantly adopts the transformer and gated attention models for TinyML [56]. In the future, the TinyML community will need to research new adaptations to attention mechanisms for resource-constrained conditions, focusing on achieving more resource-efficient attention-driven TinyML models for cutting-edge applications in which there is the predominance of large-scale deep learning architectures. For instance, in the application of TinyML to the burgeoning object detection domain, consideration of the attention mechanism design and makeup of the visual architecture is of high importance; research needs to elucidate how adaptations change the visual representation of the system, trade off the level of accuracy improvement for a decrease in efficiency, or find an ideal compromise between both.

Adding attention mechanisms to TinyML models has benefits and challenges [57]. The integration of the attention mechanism benefits TinyML by making tiny models more adaptable and enabling them to focus on data features relevant to the application for better performance. Correctly infusing attention into tiny models makes the overall model size smaller because it does not require all data features at each layer level. However, it is essential to balance the benefits of adding attention with the computational demand and improvements to the accuracy of the model. As Figure 2.3 illustrates, the challenges of low model adaptability in basic TinyML models can be significantly overcome by incorporating attention technology – gated or transformer models. However, achieving the trade-off between high model adaptability and high computational demand is a delicate balance, as the performance of edge devices is severely degraded with increasing computational demand. Hence, it is imperative to carefully weigh the objectives and motivation for adopting TinyML prior to adoption for a given application scenario or use case.

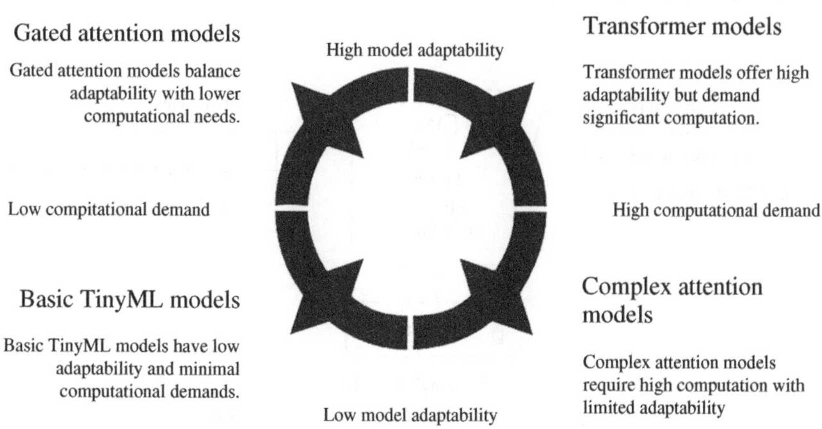

Figure 2.3 Balancing benefits and challenges of attention mechanisms in TinyML.

2.2.3 Depth-Separable Convolution

The convolution operation is a fundamental component of modern deep learning architectures. Depth-separable convolution was first introduced to boost the efficiency of convolution by dividing it into two steps called depthwise and pointwise convolutions. This is differentiated from the commonly used (two-dimensional) 2D convolutional filters that process all the input, which they share across the output channels simultaneously. As a result, 2D convolutional filters tend to have a large number of learned parameters. In contrast, pointwise convolution uses 1×1 filters across the channels that lie in the output space of the depthwise convolution operations and consequently reduces the computational burden in training and in the subsequent forward and backward passes drastically.

The advantage of using depthwise separable convolution is quite manifold. Firstly, it requires relatively fewer parameters [58, 59]. Hence, it reduces step computation and offers a quicker design implementation. Furthermore, depthwise separable convolution reduces the computational cost of initialization and increases the speed of convergence and the validation accuracy of the model. It splits the convolution into two steps, reducing the number of multiply-accumulate (MACs) operations to compute. These two approaches use significantly fewer parameters than traditional convolution to process images. Furthermore, by conveniently separating each output index for each input at the convolution output map, the depth-separable convolution strategies can have progressive advantages for TinyML-related applications [59]. This method of breaking down the standard convolutions, step by step, makes training the larger model and its inference more efficient in terms of the amount of memory and computational power used [60]. Figure 2.4 outlines the most essential points about applying depth-separable convolution for TinyML.

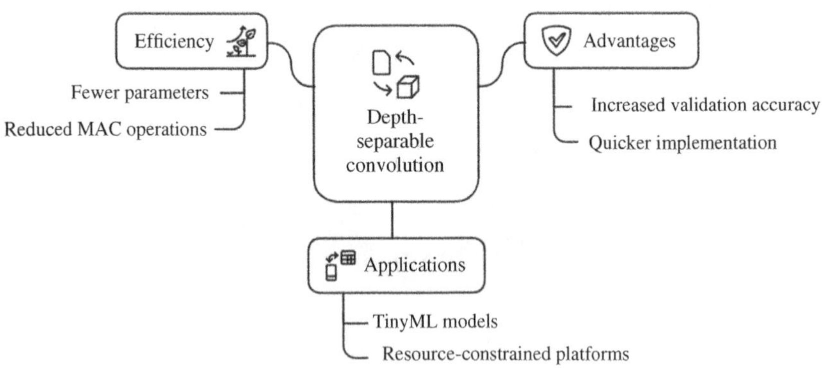

Figure 2.4 Depth-separable convolution for TinyML.

The interest, in most instances, is in reducing the number of MAC operations and the amount of memory required to keep each layer of the network while maintaining accuracy. Because depthwise convolutions and especially pointwise convolutions require fewer operational resources, spending more convolutions of the model's input data on these iterations leads to a proportionally lower amount of MAC convolutions, which account for a disproportionate amount of the computational power required, for instance [58]. Therefore, one benefit of depthwise and pointwise separable convolutions is enabling efficient training of larger models in high-capacity hardware and platforms. Moreover, since the use of depth-separable convolution allows designing smaller models with the same feature maps and operations, the approach is beneficial for reducing memory footprint and decreasing the amount of convolution operations on resource-constrained platforms for TinyML as well. The most significant advantages of using depth-separable convolutions in creating TinyML models are the efficient reduction of the number of parameters in the model, the number of floating-point operations necessary for inference, and finally, the amount of memory needed to store the weights of the neural network.

2.2.4 Rational Activation Functions

Activation functions are mathematical functions that can be applied to weighted sums of inputs and biases of neurons before they are used as inputs to other neurons or the output of a neural network [61]. The use of activation functions is a key part of deep learning systems. Rational activation functions (RAFs), expressed as ratios of polynomials, are innovative alternatives to the traditional activation functions such as the Sigmoid, ReLU, GeLU, and tanh. The nonlinearity offered by RAFs has the promise to be of significant benefit to TinyML – it represents researchers' attempts to optimize the activation mechanisms [62].

Traditional activation functions are known to suffer from several important issues, including high computational cost and the very popular vanishing gradients problem. RAFs are an emerging, promising alternative – they are flexible and have the potential for efficient approximation.

Given some numbers $a_0, \ldots, a_m, b_0, \ldots, b_m \in \mathbb{R}$, RAFs are mathematically defined as in equation (2.1) [63]:

$$f(x) = \frac{P(x)}{Q(x)} = \frac{a_0 + a_1 x + \cdots + a_m x^m}{b_0 + b_1 x + \cdots + b_n x^n} \tag{2.1}$$

where $P(x)$ and $Q(x)$ are polynomials of varying degrees; $b_j \neq 0$ for at least one j.

RAFs offer the flexibility of modeling complex nonlinearity without losing computational efficiency. Other motivations for RAFs in TinyML applications include compactness, efficient hardware implementation, and flexibility in designing TinyML models. RAFs are compact because they use fewer coefficients than piece-wise linear or transcendental functions; RAFs are compatible with fixed-point arithmetic and are, thus, well-suited for hardware-based acceleration, leading to possibilities of efficient hardware implementation. The RAF approach has the potential to significantly reduce overreliance on traditional activation functions and thus avoid computationally expensive operations that are associated with their use. Unlike the traditional activation functions, RAFs do not use extensive lookup tables – a memory-hungry structure – thus lowering memory requirements, which is critical for TinyML. RAFs facilitate faster inference since they predominantly perform addition and multiplication. This property enables RAFs to better adapt to scenarios where model size and parameter counts are limited. Computationally, the simplicity of RAFs also translates to lower energy requirements.

Although RAFs have significant chances to contribute to the design of an efficient TinyML system, there are a number of considerations that must be given careful attention while implementing RAFs [63]. First, the division operations inherent in RAFs are a potential source of numerical instability, which risk increases as the denominator approaches zero. To mitigate this, researchers have sought to constrain the polynomial coefficients. Secondly, parameter optimization is somewhat challenging with RAFs; there is no straightforward way to determine the optimal degrees of $P(x)$ and $Q(x)$ and their coefficients. A third concern is dependency on target hardware. The efficiency of RAFs depends on the capabilities of the target hardware; hardware that does not natively support division is bound to suffer performance bottlenecks with the use of RAF. The challenges and prospects of RAFs for TinyML are further expressed in Figure 2.5.

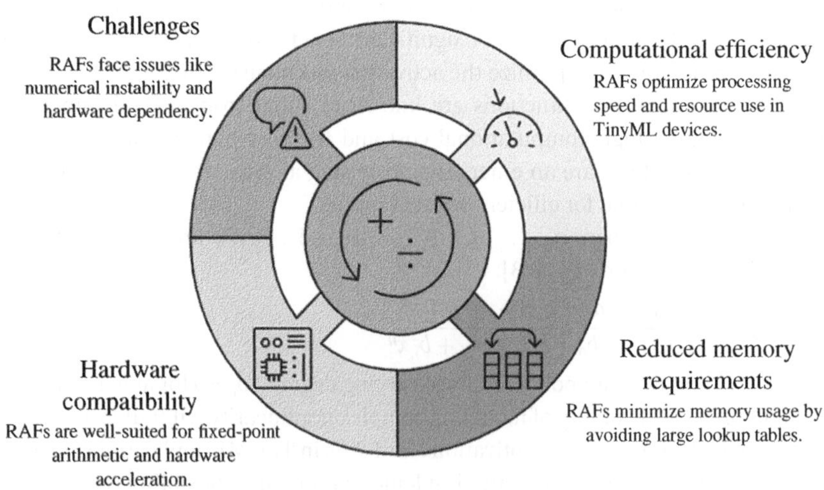

Figure 2.5 Benefits and challenges of RAFs for TinyML.

Although research on RAFs for TinyML is still in its infancy, there is evidence to suggest speedy development and adoption of RAF-based approaches for TinyML in the near future. The focus will be on developing automated design tools capable of generating and optimizing RAFs for specific hardware constraints, increasing framework integration by including RAFs as a standard option in deep learning libraries, exploring hybrid approaches, and developing methods to dynamically adjust the degrees of coefficients of RAFs during training or inference in changing workload scenarios.

RAFs promise compelling solutions to the challenges currently undermining the wide adoption and deployment of TinyML systems. Their promise lies in their ability to combine computational efficiency, representational power, and adaptability. While challenges like numerical stability and ecosystem support remain, ongoing research and development could make rational activation functions a cornerstone of future TinyML applications.

2.3 Frontiers in Model Optimization for TinyML

Model optimization is at the forefront of learning for TinyML at the moment. Emerging approaches in the development of the TinyML model significantly leverage the optimization and miniaturization of large parameter learning models that are primarily developed on regular, high-capacity computing hardware systems. In the last five years, model miniaturization techniques such as pruning, quantization, low-rank decomposition, and (KD) have greatly advanced, following major breakthroughs in optimization research. These techniques have been used to facilitate the development of accurate, resource-efficient, lightweight ML models that have shown good performance on resource-constrained devices. The literature has a rich account of research that illuminates the subject of ML model optimization for TinyML development and its bourgeoning applications in resource-limited environments. Model compression is a principal technique in the context of TinyML; many researchers have emphasized and demonstrated its growing importance for edge computing applications. This section examines the techniques of model pruning, quantization, low-rank decomposition, and KD as a high-level summary of the major neural network model compression for the development of TinyML systems. The major strengths of these four techniques are highlighted in Figure 2.6.

The efficiency and suitability of each of these popular model compression techniques vary according to application requirements. The choice of model compression approach impacts the cost, performance, and complexity of the final model. Before delving into the main discussion of the four major compression techniques discussed in this section, a brief cost–benefit analysis is presented in Table 2.2 to serve as a background to the reader.

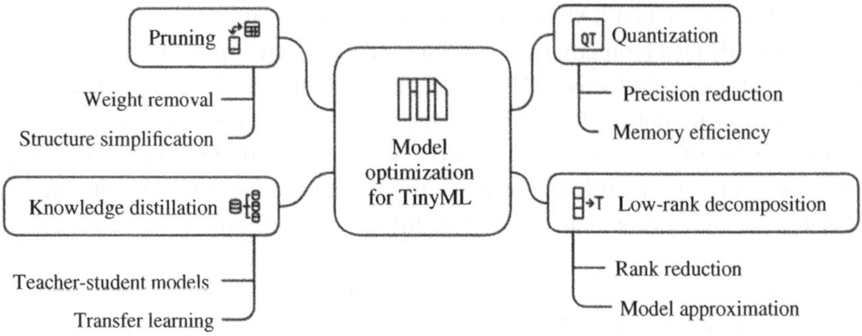

Figure 2.6 Model optimization techniques for TinyML.

2.3.1 Pruning

Liu et al. [74] extensively discussed model compression techniques, with a focus on using pruning techniques to facilitate the deployment of convolutional neural networks (CNNs) for edge applications. Pruning is a key approach to reducing the size of neural networks for TinyML applications; it creates a toned-down version of larger models by systematically dropping off nonessential or redundant neurons, thereby reducing model complexity at a small cost to model performance. One major issue common to all model compression techniques is the trade-off between accuracy and model complexity; therefore, it is important to carefully consider alternative options and employ the most promising approach, given a set of predefined objectives [75, 76]. A helpful pruning algorithm is expected to strike a good balance between efficiency in model size reduction, improved inference speed, and model performance. Focusing on these three issues is crucial in edge applications, which often necessitate real-time processing capabilities [76].

The process of pruning a large neural network involves, as shown in Figure 2.7, identifying redundant neurons, applying the chosen pruning technique, evaluating trade-offs, and optimizing the model to generate a simpler model.

Typically, a neural network model has a significant number of non-influential or redundant parameters that do not contribute significantly to the generalization capacity of the model. Hence, the main objective of model pruning is to exploit this opportunity to reduce the space and time complexity of the model while preserving model performance. This is achieved by removing unimportant parameter weights from a deep learning model. Neural network parameter counts can be reduced by over 90% through pruning without significant performance losses [69]. This makes pruning an essential approach.

In general, pruning focuses on the weights of the parameters; the biases are not dropped. Research has shown that pruning biases negatively affect the resulting lightweight model [76]. Frankle and Carbin [69], in the *lottery ticket hypothesis*,

Table 2.2 A cost–benefit summary of the major model compression techniques.

Compression technique	Benefits	Costs/challenges	References
Quantization – lowering the precision of weights and activations.	• Can reduce model size and memory requirement by up to a factor of 4. • Significant increase in inference speed on hardware that supports low-bit operations. • Lowers energy requirement, thus improving battery life for edge devices.	• Degrades model accuracy, especially significant in the aggressive case (FP32 to INT 4). • Often needs hardware-aware optimization since not all devices can perform efficient low-bit operations. • Sensitive to model architecture. Quantizing transformers is more challenging than CNNs.	[33, 34, 64]
Knowledge distillation (KD) (Based on teacher-student model. Transfer of knowledge from a large, complicated to a small, simple model).	• Derives a smaller, efficient model that retains the accuracy of the larger model. • More robust and generalized since the student model learns from soft labels. • It can be combined with other compression techniques (e.g., KD-quantization setup) to produce better models.	• KD incurs additional training costs as both the teacher and the student model must be trained. • Optimizing KD models is challenging. Performance is affected by the choice of teacher-student pair and the loss function. • Performance gain is task-dependent.	[65–67]
Pruning (Reduces complexity by removing redundant connections and weights).	• Achieves a significant reduction in model size. A 90% weight reduction can be achieved with structured pruning. • Improves inference speed on sparsity-optimized hardware. • It is highly accuracy-preserving when done properly.	• Incurs increased training time as fine-tuning is necessary to recover lost accuracy. • Sparse models can be pretty slow on some hardware due to workload requirements for sparse matrix computations. • High potential to degrade model interpretability as meaningful neurons could be pruned.	[68–70]
Low-rank decomposition (Factorizes weight matrices to lower computational requirement).	• Creates a more efficient model by reducing parameter count. • Favors lower latency in resource-limited environments. • It is well-suited for fully connected layers and transformers.	• It requires careful selection of decomposition rank. If the decomposition rank is too low, an approximation error is introduced, resulting in a significant loss in accuracy. • Pruning is less effective for already compact models.	[71–73]

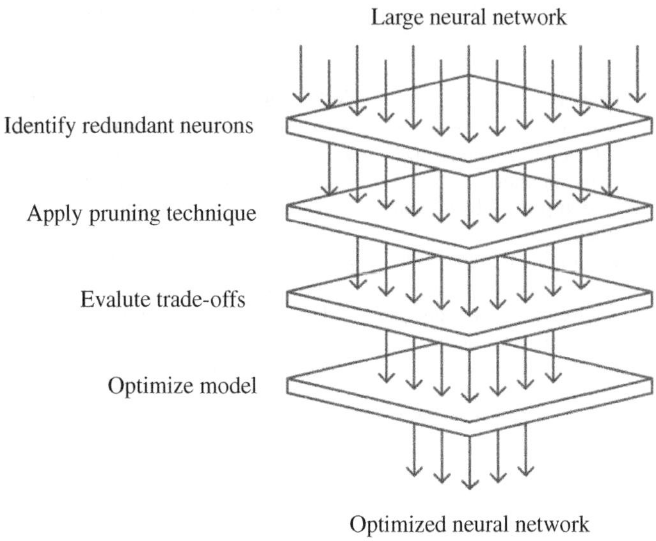

Figure 2.7 Pruning approach for TinyML.

have shown that pruning can be conceptualized as a search for subnetworks that have optimal initial weights that facilitate effective training in a similar number of iterations and with better performance than the original large parameter networks but with only about 10–20% of the size of the fully connected CNN. Pruning is preferred in situations where a balance between model size and accuracy is desired.

Pruning techniques are generally categorized into two, namely, train-time pruning and post-training pruning. Train-time pruning (TTP) removes less important connections in the network during training using L1 and L2 regularization, while post-training pruning (PTP) removes less important connections after training has been completed. While PTP is simple to implement, it is susceptible to suboptimal decisions. TTP, on the other hand, is more efficient because it often incorporates sparsity by default. Regarding implementation, PTP can be achieved via a structured or unstructured approach.

Structured pruning drops specific weight structures. In contrast, unstructured PTP, a simpler approach, drops individual model weight parameters by selecting individual parameters to prune using minimum threshold values of weights or their activations. This approach works by setting parameter weight values below the set threshold to zero, thus removing them from the active connections. The weight-zeroing operation necessitates recomputation of the weights each time, and thus, unstructured pruning does not have a significant positive impact on latency. It, however, has a positive impact on the reduction of noise from the model weights, thus facilitating more efficient inference performance.

Unstructured pruning is reported to achieve up to 50% model compression with less than 1% performance degradation [77].

Structured pruning offers greater improvements in model inference speed and model size compared to unstructured pruning. Structured pruning removes an entire group of weights, simplifying the architecture in fewer steps. The approach usually reduces the computations typically made in the forward pass through the neural network's weight matrices. Structured pruning has a strong capability for increased inference speed while maintaining accuracy in an architecture-agnostic fashion [78]. In spite of this, observed performance gains, experience, and good planning are key requirements for success when using structured pruning.

Pruning techniques are further categorized into local and global pruning depending on the focal point of the pruning operation on the network. Local pruning is targeted at individual neurons, weights, or connections within a layer of the neural network architecture. Low weight magnitude, low importance in the context of the given layer, or low contribution to the network's inference performance are the main yardsticks for local pruning. Local pruning methods such as unit magnitude-based pruning, weight magnitude pruning, and connection sensitivity-based pruning follow an iterative process that prunes weights or connections, one at a time, based on the set criteria. Global pruning removes entire neurons, layers, or sections of the neural architecture simultaneously. Thus, the global pruning approach considers the overall importance of neural nodes or layers over the entire network simultaneously. Furthermore, unlike local pruning, global pruning considers interactions and interdependencies between model segments. It considers the overall importance of neurons or layers across the entire network rather than focusing on specific parts within individual layers. Global pruning often involves more sophisticated techniques that consider the interactions and dependencies between different parts of the network. A typical example is the iterative magnitude pruning approach [79]. Although global pruning makes more impactful decisions than local pruning, given that it has more context, it can lead to suffering specific layers. Table 2.3 summarizes the main points about global, local, structured, and unstructured pruning approaches.

2.3.2 Quantization

Quantization, like other model compression techniques, aims to shrink the model size and enhance efficiency without significantly sacrificing performance. Quantization is a numerical approach for converting high-precision, compute- and memory-intensive data to low-precision, compute- and memory-light versions [86, 87]. Quantization can reduce power consumption by over 94% and increase the speed with which the quantized model transfers data by up to four times

Table 2.3 Comparative analysis of pruning strategies.

Pruning strategy	Method	Advantages	Disadvantages	References
Global pruning	It uses a global threshold to delete the lowest-magnitude weights across the entire network (model).	• There are no strict layer constraints; hence, it maximizes compression. • Produces very sparse models.	• The risk of removing essential weights from critical layers is a significant concern. • Fine-tuning is required for optimal performance.	[70, 80]
Local pruning	Removes a specific percentage of weights per layer, thereby maintaining relative sparsity across layers.	• Prevents over-pruning of critical layers; it maintains structured sparsity. • Easier to implement than global pruning.	• Achieves lower compression than global pruning. • Requires manual layer-wise tuning.	[81, 82]
Structured pruning	Drops entire neurons, channels, or filters rather than individual weights.	• Highly efficient for hardware acceleration such as CPUs, GPUs, and MCUs. • Ease of implementation.	• Offers lower flexibility than unstructured pruning. • Has the potential to degrade accuracy more than unstructured pruning.	[83, 84]
Unstructured pruning	Drops individual weights without considering structure. It uses magnitude-based thresholds.	• Achieves extreme model compression by maximizing sparsity. • Well-suited for applications deployed on memory-constrained devices.	• Difficult to accelerate on standard hardware. • Inefficient inference due to the influence of irregular weight patterns.	[69, 85]

the full-parameter model [88]. Neural quantization is, therefore, an important technique for accelerating the deployment of high-performing AI innovations on embedded devices (Figure 2.8).

The primary problem of quantization lies in minimizing the loss of precision resulting from the lowering of numerical bitwidth [87]. A substantial body of research has been undertaken to identify quantization strategies appropriate for various network architectures and application contexts, along with the best

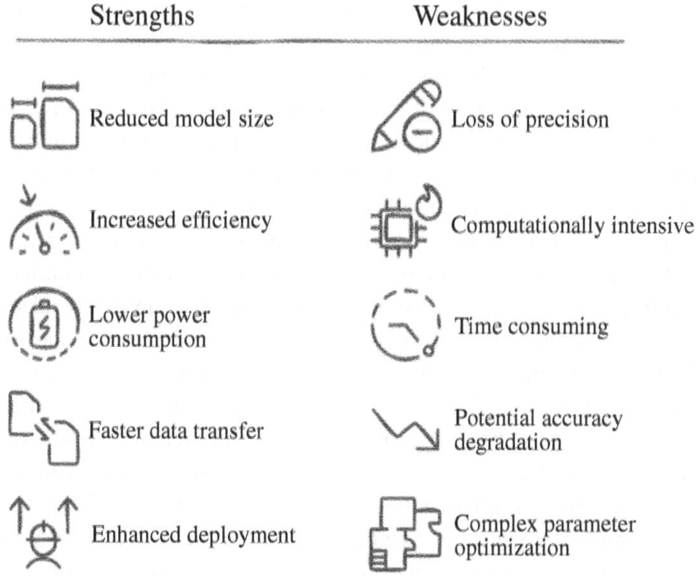

Figure 2.8 Strengths and weaknesses of neural model quantization.

quantization parameters for distinct quantization methods. However, quantization approaches still face significant challenges in a number of important ways, including how to optimally allocate quantization bitwidth, balance computational speed with accuracy, and optimize quantization parameters based on disparities in model and data distribution [88].

Researchers have developed various techniques to quantize neural network models, each geared toward addressing the identified challenges. The post-training quantization (PTQ) method quantizes a model post-training without necessitating retraining, thus facilitating swift deployment. Nevertheless, PTQ may result in significant accuracy degradation, particularly in models sensitive to precision [89]. In contrast, quantization-aware training (QAT) incorporates quantization into the training phase, thus creating better optimized models with minimal accuracy loss. However, QAT is computationally intensive and more time-consuming; it necessitates retraining. PTQ is applied to neural network models after training is completed. PTQ works best for Gaussian-like distributions, especially for complicated applications with higher accuracy requirements. The key issues for PTQ lie in determining the appropriate clip range for the optimal quantization parameters. Typically, the range of activations is computed using a series of input calibrations [33, 90]. A very common choice of the clip range is the use of the min/max value of the input data [33]. This has been shown to often increase the range to the detriment of quantization resolution. Other more intelligent approaches include the use of the *i*th largest or smallest value in the data and the use of information loss between original and quantized values [91], such as information derived using mean squared error (MSE), entropy, or Kullback–Leibler (KL) divergence. The challenges associated with computing optimal model quantization parameters and reducing the loss of accuracy after quantization remain key problems to be addressed by research.

In QAT, the quantization steps are integrated into the model training stage. In contrast to PTQ, QAT allows the model to learn and adapt the effects of quantization during training. The QAT procedure produces models that optimally adapt and preserve accuracy, notwithstanding reduced precision. However, the majority of QAT methods are susceptible to the effect of gradient error, which results from the use of a straight-through gradient estimation technique. Research has proposed a number of approaches to address this problem, including activation value trimming [92], the use of a differentiable tanh function to approximate the quantization function, pre-quantization tailoring of weight range [93], and in-training use of gradient update and weight adjustment process. QAT methods deliver superior quantitative performance. However, they require huge training datasets alongside high demand for compute resources, with excessively long training times, exceeding 100 GPU hours [94].

The quantization process applied to a neural network model involves constraining the activation and weight values to a discrete set of numbers, which exhibit various distributions, including uniform and nonuniform types [95]. The inference of neural network models on hardware platforms necessitates processing quantized computational results through specific inverse quantization steps to achieve an accurate floating-point output. Nonuniform quantization methods exhibit intricate inverse quantization computation processes, which may pose challenges for compatibility with hardware platforms. Current hardware predominantly supports uniform quantization methods. The predominant quantization scheme employed is the uniform distribution method, characterized by a consistent step size.

The primary concern in quantization involves developing an appropriate quantization mapping function and a practical approach for calculating quantization parameters. In the context of uniform quantization, most current methodologies employ either asymmetric or symmetric quantization mapping functions [96–98]. The literature appears to have been predominantly focused on symmetric quantization, while the investigation of asymmetric quantization has only recently gained research interest [99–101].

The distinction between the two strategies is based on treating positive and negative values. Symmetric quantization guarantees that both positive and negative values utilize an identical quantization resolution, whereas asymmetric quantization permits distinct scales for positive and negative values. As a result, symmetric quantization can result in increased quantization error, whereas asymmetric quantization may require additional resources to maintain two sets of parameters [97, 102]. Asymmetric quantization demonstrates superior performance in classification tasks, whereas symmetric quantization is favored for object detection and image-to-image regression applications [103–105].

The configuration of the quantization strategy is essential for optimizing the effectiveness of both quality preservation and resource efficiency. In the context of various natural language processing and computer vision models, the appropriateness of the scenario is significantly influenced by the model architecture and the dataset being utilized [97, 106]. Additionally, the implementation of various quantization strategies can result in a cascading impact on performance concerning precision-aware layers, necessitating a trade-off based on the permissible decline in these layers.

To understand the asymmetry characteristics in neural network quantization, it is important first to outline the distinctions between symmetric and asymmetric quantization methods. In symmetric quantization, both weights and activations are represented uniformly. This can be accomplished in a feedforward network by normalizing weights or activations. In contrast, asymmetric quantization permits the quantization of weights and activations through distinct methods [97, 102], potentially enhancing information encoding within the network. Furthermore, it facilitates enhanced reconstructions of prior activations based on the forward activations and weights, which can be utilized since the network weights can be identified.

The distinction between symmetric and asymmetric quantization can be analyzed through the lens of information-preserving quantization error, highlighting the limitations inherent in various representations. This error, associated with an approximation under constant top-norm, is commonly employed to measure the spectrum of feasible degrees of approximation attainable through quantization. In practical applications, this extreme approximation is rarely utilized, and the selection of quantization resolution is determined by the acceptable level of this approximation. The asymmetry in quantization can be addressed by adjusting the bias direction to minimize error, shifting focus from activations to weights, and ensuring fairness through the top-norm constraint on network depth [107, 108]. By permitting one of these dimensions to be significantly deeper, the theoretical difference can be minimized if the quantization of the other dimension is increased. Consequently, quantizing the larger dimension of weights may lead to more significant relative differences when the associated activations are quantized to the same level.

Quantization of neural networks represents a thoroughly researched area, with numerous optimization algorithms developed to enhance this process. These methods can primarily be categorized into gradient-based and heuristic-based [109–111]. The choice of optimization algorithm is critical during the quantization process, as both accuracy and speed directly impact the resulting quantization outcome. Several widely utilized optimization techniques for quantization include SGD, LBFGS, Adam, and Proximal Mapping [112–114]. However, the predominant method employed is quantization-aware training. In this approach, the loss function deviates from the standard categorical cross-entropy by incorporating additional quantization functions linked to the model weights. This modification compels the network to enhance its learning efficiency [106, 115, 116]. The parametrized quantization approach has recently gained significant popularity, attributed to its efficacy in enhancing quantization processes. The application of several standard optimization techniques, such as simulated annealing and k-means, has been studied for the quantization process [97, 102, 103]. Novel approaches that integrate optimization techniques with various other concepts are emerging. For example, the combination of the gradient-based method with the fixed-point number quantization error principle lowers bounding to streamline arithmetic operations.

Selecting the optimal bitwidth for quantization is a significant challenge. The use of high bitwidth leads to high accuracy with elevated memory and compute requirements. On the other hand, the low bitwidth approach favors memory and compute efficiency at the expense of model performance. Hence, researchers have investigated diverse approaches for selecting quantization bitwidth. Measuring quantization-induced loss is a significant issue in the quantization literature. There is a need to assess the performance loss due to quantization over a wide range of possibilities. Currently, effectively assessing the degree of accuracy loss of different neural network models in different computational task scenarios after quantization is a significant challenge that begs for a solution.

The above notwithstanding, several researchers have proposed wide-ranging criteria to assess accuracy losses due to quantization, primarily based on statistical principles. The principles of signal-to-noise ratio were extended to the context of quantization by Qualcomm [117]. In a similar vein, Feng et al. [118] proposed a method based on the intuition of mean squared error, called quantization mean square error, to determine the quantization coefficients. In [119], an overly computationally expensive predictor for quantization accuracy that is based on the flexible once-for-all network described in [120] for a direct prediction of model quantization accuracy is described. Beyond these, the assessment of the nature and scale of losses due to the quantization of deep learning models has not been adequately studied [88], and it requires research attention since developing a clear understanding of the origin and mechanism of losses could aid the development of better quantization approaches.

2.3.3 Low-Rank Decomposition

Low-rank decomposition (low-rank factorization) is a matrix decomposition technique for reducing the size of neural network models. Similar to other model compression techniques, it strives to achieve optimal model light-weighting while preserving model performance. This approach decomposes large, dense weight matrices characteristic of deep neural networks into smaller (usually two) low-rank matrices [121, 122]. The strength of this approach lies in its ability to lower model complexity by approximating the original large, dense matrix through the combination of two low-rank matrices. Furthermore, by capturing the most relevant information in the model's weights, the low-rank factorization approach creates a compact representation of the large model without significant loss in performance.

The two most prominent low-rank decomposition approaches are matrix and tensor decomposition techniques. Within this application, singular value decomposition (SVD) is a very common method for decomposing a dense matrix into three low-rank matrices and capturing the essential features of the original matrix while reducing storage and compute requirements of the resulting approximate model [123]. Tensor decomposition, on the other hand, focuses on tensors (multidimensional arrays) in deep neural networks. A popular and efficient tensor decomposition technique is the canonical polyadic low-rank decomposition. It optimizes the model by removing redundant weights and activations in the fully connected layers [121].

The potential of low-rank tensor decomposition to compress weight matrices and reduce the dimensionality of deep neural network model parameters is demonstrated in [124] using the Kronecker tensor decomposition technique. Despite the high potential of low-rank tensor decomposition, there are known issues with low-rank decomposition approaches; their applicability is limited to scenarios where there is significant information redundancy in the fully connected layers [121, 122]. Hence, the applicability of low-rank decomposition is somewhat constrained, showing limitations in neural network models that have varying degrees of architectural complexities or in those architectures that do not have appreciable redundancy in their fully connected layers [125]. These are the main issues in low-rank tensor decomposition that invite interest for ongoing research – there is a need to widen the scope and increase the effectiveness of low-rank decomposition techniques by exploring schemes that are applicable to a broader spectrum of deep neural network architectures without significantly trading off model performance.

2.3.4 Knowledge Distillation

KD is a predominant approach for deep neural network model compression and optimization [126, 127]. It is based on teacher-student knowledge transfer principles where the experienced large-scale model teaches the inexperienced small-scale student model. The purpose is to improve the student model's performance without inheriting the computational overheads of the teacher model [128]. KD fundamentally involves extracting the informative characteristics of a large model's behavior and transferring this knowledge to a smaller model to facilitate the preservation of excellent accuracy in the student model while markedly decreasing its size and complexity. Figure 2.9 presents a high-level overview of KD, highlighting its basic principles, techniques, challenges, and limitations.

In the teacher-student model framework underlying distillation, the large-scale, well-trained model is utilized to guide knowledge development in a smaller-scale model; the large-scale network offers direction to the small-scale network [129]. The objective is to replicate the performance of the large-scale model in the small-scale model while using fewer parameters and minimizing computational complexity. The small-scale model may be trained to align with the softmax output of the large-scale model or to correspond with its feature representations. Distillation loss is a quantity used to measure the small-scale model's understanding of the large-scale model's behavior, extending beyond simple categorical inference. Common loss functions exist to evaluate the alignment between small-scale and large-scale outputs [126].

The process of transferring knowledge from a teacher model to a smaller version, the student model, aims to optimally preserve vital information as decision-making abilities required for applications without jeopardizing the performance levels of the student model. Although small-scale models may not achieve the same performance as large-scale models in all tasks, the emphasis lies in attaining comparable performance on specific targeted tasks while leveraging

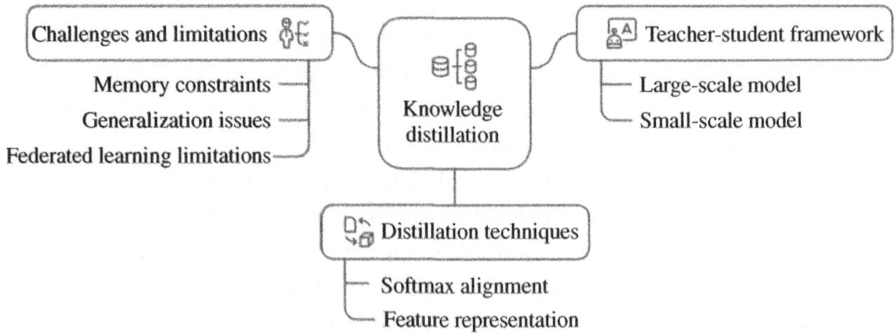

Figure 2.9 Overview of knowledge distillation.

smaller models' efficiency and speed benefits. The objective is to balance model size, computational efficiency, and task-specific performance rather than replicating the large model's exact performance. This makes KD an important technique for model compression and optimization in practical applications.

Distillation techniques have been shown to markedly improve the performance of smaller models, frequently surpassing those trained using conventional methods [130]. KD has been effectively utilized in computational vision, natural language processing, and speech recognition, showcasing its adaptability and efficacy. Intelligent selection of large and small-scale models is essential; an overly complex teacher can diminish the effectiveness of the distillation process. The architecture of both large and small-scale models significantly influences the success of the distillation process. Adjusting hyperparameters, including temperature in softmax and the weight of distillation loss, is essential for optimizing performance in the student model [131].

KD, recognized for its ability to condense the capabilities of bigger models into formats suitable for deployment in resource-constrained environments, encounters a range of complex obstacles and limitations. A significant challenge exists, for instance, in implementing large pretrained language models (PLMs) on devices with constrained memory, requiring a careful equilibrium to enhance performance without depleting system resources [132]. The generalization capacity of distillation models may be undermined when employing public datasets that diverge from the training datasets, diminishing the model's relevance and accuracy [133]. The limitations of current KD methods in federated learning highlight the necessity for novel strategies to tackle the lack of cross-lingual alignments for knowledge transfer and the risk of unreliable temporal knowledge discrepancies [134, 135]. These limitations underscore the imperative for research to improve KD methodologies to guarantee their efficacy in minimizing model size and complexity while preserving or potentially improving performance across various applications.

KD is an effective method in CNNs, where it has been used to facilitate model compression and improve the performance of smaller networks. The increasing demand for sophisticated models in resource-limited environments underscores the significance of KD as a critical area of research, capable of facilitating the advancement of more efficient and accessible AI applications [136].

2.4 Learning Frameworks and Tools for TinyML Development

The quest to attain the transformational effect of TinyML in edge computing and IoT application domains has led to increased efforts in developing

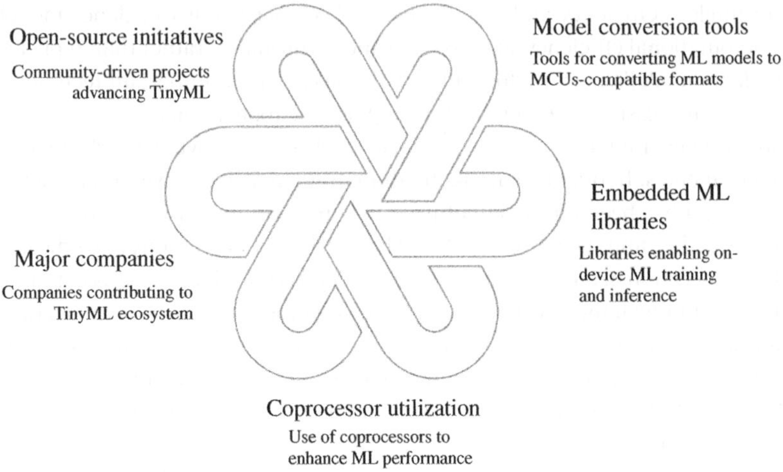

Figure 2.10 TinyML development ecosystem.

frameworks, libraries, and tools to facilitate the implementation of ML algorithms on constrained platforms.

TinyML development approaches may be grouped into three distinct categories (Figure 2.10). The predominant approach involves converting pretrained models to accommodate MCUs limitations. Consequently, these tools typically utilize inference engines produced by established ML libraries, including TensorFlow, Scikit-learn, and PyTorch, among others, and adapt their code for execution on resource-constrained, mostly battery-powered edge and IoT devices [22]. The second approach focuses on embedding ML libraries within MCUs, thereby enabling local model training and analysis. This allows for the immediate development of learning models from data collected by the device itself, thereby potentially enhancing the chances for improved model accuracy and facilitating the application of unsupervised techniques [22]. The final approach depends on a dedicated coprocessor to assist the primary computing unit with ML-specific tasks. This is a strategic way to enhance computational performance, though it is not quite familiar due to the significant increase in cost and complexity of the development platforms.

Research has explored the space of shallow ML algorithms and deep learning algorithms for developing tools and frameworks for TinyML development, with remarkable achievements. Deep learning has significantly influenced the landscape because of its unprecedented growth and adoption across the ML ecosystem. Tools for converting shallow ML techniques, such as decision trees, Naive Bayes classifier, and k-nearest neighbors (k-NNs), to edge or embedded device executable formats have been developed. MicroML [137] enables the conversion of support vector machine (SVM) and relevance vector machine (RVM) algorithms into C code, facilitating execution on various C-compatible MCUs. It interfaces with the commonly utilized Scikit-learn toolkit to modify the models produced by this library for execution on 8-bit microcontrollers with 2 KB of RAM.

Sklearn-porter [138] is a tool for the translation of trained estimators into Java, GO, C, Ruby, and JavaScript. Sklearn-porter is comprehensive due to its support for a wide array of languages and its compatibility with a variety of ML algorithms. Sklearn-porter can port Scikit-learn models, including decision trees, SVM, random forest (RF), and AdaBoost classifier, specifically for the C language. m2cgen [139] can also translate models trained with Scikit-learn into various native code formats, including Java, Python, C, Go, and JavaScript. The number of compatible algorithms and the target programming languages exceed those offered by sklearn-porter.

Weka-porter [140] is a more algorithm-specific tool that converts decision tree classifiers produced by the WEKA tool into C, Java, and JavaScript code. EmbML [141] is another library that enables the translation of models trained by WEKA or Scikit-learn into C++ source code files, which can then be compiled and executed on resource-constrained platforms. This library supports various ML algorithms, including decision trees, SVMs, and neural networks. Learn [142] is another shallow ML-targeted framework that generates portable C code from Python libraries, including Scikit-learn and Keras. This library supports models generated from various algorithms, including RF, decision trees, Naive Bayes, and linear models.

Major technology companies are facilitating the growth of the TinyML ecosystem by investing in and developing open-source development libraries. Google, for instance, has introduced TensorFlow Lite, which comprises a collection of tools designed to modify TensorFlow models for the execution of deep learning models on embedded or IoT devices. TensorFlow Lite consists of two primary components: the Converter and the Interpreter. The Converter translates models created with TensorFlow into optimized code suitable for execution on resource-constrained platforms. The Interpreter executes the code produced by the Converter. The resulting model is optimized and could have been developed using any or a combination of various algorithms from the neural network family and is compatible with diverse platforms, including smartphones, embedded Linux, and MCUs. Added to TensorFlow Lite, Google has also adapted the GO language for microcontrollers, creating TinyGo [143].

Microsoft has contributed the open-source embedded learning library (ELL) to the development of TinyML [144]. ELL enables the design and deployment of pretrained ML models on constrained platforms, such as ARM Cortex-A and Cortex-M architectures, including Arduino and Raspberry Pi. ELL functions as an optimizing cross-compiler operating on a standard desktop computer, producing C++ code prepared for execution on the designated single-board computer.

The ELL API is accessible through C++ or Python. It utilizes pretrained neural network models generated by any of the following: Microsoft Cognitive Toolkit, Darknet, or models in the Open Neural Network Exchange (ONNX) format [145].

At ARM, the efforts to incorporate ML in its devices have increased in recent times. The ARM-NN toolkit is a component of Linaro's ML initiative. It facilitates the incorporation of neural network workloads within ARM Mali GPUs, ARM Cortex-A CPUs, and Ethos neural processing units. The ARM-NN initiative is driven by an open-source philosophy and is compatible with popular ML frameworks, including TensorFlow, Caffe, and ONNX. To support its suite of devices, ARM has released the Cortex microcontroller software interface standard-NN (CMSIS-NN) [146]. CMSIS-NN facilitates the conversion of floating-point neural networks produced by widely used libraries such as TensorFlow, PyTorch, and Caffe into a fixed-point format suitable for execution on Cortex-M devices.

STMicroelectronics is another electronics manufacturer that has created specialized libraries for its devices. The STM32 Cube.AI toolkit [147] facilitates the integration of pretrained neural networks into STM32 ARM Cortex-M-based microcontrollers. The tool produces STM32-compatible C code from neural network models trained with TensorFlow and Keras. STM32Cube.AI enables the execution of large neural networks by utilizing external flash memory for weight storage and RAM for activation buffers.

The Artificial Intelligence for Embedded Systems (AIfES) library [148] developed by the Fraunhofer Institute for Microelectronic Circuits and Systems, is a C-based, platform-independent tool designed for generating neural networks compatible with various MCUs, such as Arduino UNO, STM32 F4 Series, and ATMega32U4. AIfES distinguishes itself from other frameworks by enabling the implementation of the neural network training process on embedded devices. It can also generate efficient codes for Windows, Linux platforms, and Raspberry Pi. Cartesiam's NanoEdge AI Studio is a commercial product that enables the creation of ML static libraries for embedding in Cortex-M-based microcontrollers. Like AIfES, it facilitates the execution of the training process, including unsupervised model training, on the MCU. This toolkit includes a microcontroller emulator for testing and debugging the resulting model before deployment. uTensor [149] and TinyMLgen [150] are libraries designed to enhance the compatibility of TensorFlow models by generating C code suitable for integration with various families of MCUs. The former emphasizes Mbed boards, whereas the latter exports the trained model to a standard C array, making it applicable to a broader spectrum of microcontrollers.

The implementation of diverse forms of deep neural networks on ARM Cortex-M processors is currently widespread. CMix-NN [151] is another open-source library designed for the deployment of mixed-precision quantized deep neural networks on MCUs. This tool provides for convolutional kernels with precisions of 2, 4, or 8 bits for any operand. FANN-on-MCU [152] is another open-source framework developed from the fast artificial neural network (FANN) library [153]; it is designed to execute lightweight neural network models on microcontrollers. This toolkit processes multilayer perceptrons trained with FANN. It also generates code operating on the ARM Cortex-M and parallel ultralow-power platforms [154].

To further increase understanding of the frameworks and hardware tools for TinyML development, Table 2.4 compares some of these TinyML development frameworks, indicating their key features and weaknesses. Meanwhile, Table 2.5 presents a comparative review of TinyML hardware platforms in terms of features, weaknesses, and use cases.

2.5 Frontiers for Algorithmic Innovations for TinyML

Studies for the development of learning algorithms for TinyML have gained significant traction in recent years as researchers strive to address the unique challenges and take advantage of the opportunities in deploying ML on resource-constrained devices. To overcome these challenges, Rashid et al. [159] introduced TinyM^{2}Net, a novel multimodal learning framework codesigned with system algorithms. This framework incorporates various model compression techniques, effectively addressing the issue of large model sizes and making multimodal learning feasible on tiny devices. The innovative approach presented in TinyM^{2}Net represents a significant advancement in the field, providing a pathway for future research to optimize multimodal learning for resource-constrained environments.

Ray [160] points out the issues related to limited-size memory, which complicates the implementation of ML algorithms. This limitation is particularly pertinent in the context of the growing demand for more sophisticated models that require substantial computational resources. The discussion on performance variability among TinyML technologies is also noteworthy, as it highlights the infancy of this field and the ongoing need for advancements in ML algorithms, development environments, and hardware compatibility.

Ren et al. [60] present a significant advancement in the realm of TinyML, particularly through its introduction of a lightweight meta-learning algorithm designed for resource-constrained IoT devices. The authors highlight the challenges posed by the heterogeneous nature of deployment environments in TinyML applications, emphasizing the necessity for tailored ML solutions for each device. This need is accentuated by the limitations of typical IoT devices, which often operate without an operating system and primarily support inference rather than on-device learning. In the following subsections, multimodal ML and meta-learning approaches for TinyML are highlighted as principal frontiers for learning in TinyML.

Table 2.4 Key features and weaknesses of some TinyML development frameworks.

Framework	Key features	Weaknesses	References
TensorFlow Lite Micro (TFLM)	• Highly optimized for MCUs and power-constrained devices. • Provides support for 8-bit quantization. • Runs on bare metal; no operating system is required. • Has good support for multiple hardware: ARM Cortex-M, ESP32.	• It does not provide GPU or TPU acceleration. • Does not support a wide range of models. • It requires manual memory management.	[39]
PyTorch mobile and edge	• Support for dynamic quantization and weight pruning. • Compatible with Android/iOS/edge devices. • Integrates TorchScript for deployment.	• It is not optimized for MCUs. • Has higher memory requirement than TFLM, • Provides limited support for bare metal MCUs.	[155]
MCU AI framework (CMSIS-NN)	• Optimized for ARM Cortex-M processors. • Has highly efficient low-level kernels for neural networks. • It is lightweight; it does not need an operating system.	• Does not have any other hardware than ARM Cortex-M devices. • Requires manual optimization for new models. • Does not provide support for GPUs or accelerators.	[156]
Tensor Virtual Machine (TVM)/μTVM	• Provides auto-tuning for optimizing ML models on embedded devices. • Support for LLVM-based optimizations. • Compatible with RISC-V, ARM, and x86 architectures.	• Requires deep ML compiler knowledge. • No direct integration with Arduino/ESP32. • Long compilation times.	[157]
Edge impulse	• Supports end-to-end TinyML development (data collection, training, deployment). • Provides no-code GUI for easy model training • Has support for Arduino, STM32, and ESP32.	• Cloud-dependent for model training. • Limited support for custom models. • Not as flexible as raw TensorFlow/PyTorch.	[158]
Sony neural network libraries	• It is optimized for low-power embedded AI applications. • It provides support for QAT. • Well-suited for ARM Cortex-M and DSPs.	• No elaborate community support. • Limited to Sony-supported hardware.	[159]

Table 2.5 Overview of selected TinyML hardware platforms.

Hardware platform	Key features	Weaknesses	Use case	References
Arduino Nano 33 BLE Sense	• ARM Cortex-M4. • Built-in 9-axis Inertial measurement unit (IMU), microphone, and temperature sensor. • Provides support for TensorFlow Lite Micro. • Low power consumption.	• Limited RAM. • No hardware acceleration. • Not suitable for deep learning models.	Activity recognition using IMU sensors.	[160]
Raspberry Pi 4	• Quad-core ARM Cortex-A72 (1.5 GHz, up to 8GB RAM). • Supports TensorFlow and PyTorch. • GPU acceleration via OpenCL is available. • Suitable for edge AI applications.	• High power consumption. • Requires active cooling for sustained workloads. • Larger than MCU boards.	Real-time image classification for security cameras.	[161]
Google Coral Dev Board	• Edge TPU accelerator. • Supports TensorFlow Lite EdgeTPU models. • Optimized for real-time inference. • Low power consumption.	• Limited to EdgeTPU-compatible models. • Higher cost than MCU-based boards. • Limited flexibility.	Object detection in smart surveillance systems.	[162]
STM32 Nucleo Boards	• ARM Cortex-M series (M4, M7, etc.). • Supports CMSIS-NN and TensorFlow Lite Micro. • Energy-efficient.	• Limited RAM. • No native GPU or TPU acceleration. • Requires manual optimization.	Keyword spotting for low-power IoT devices.	[156]
ESP32	• Dual-core Xtensa LX6 (240 MHz, 520 KB RAM). • Built-in WiFi and Bluetooth. • Supports TinyML models on Edge Impulse. • It is energy-efficient.	• Limited RAM. • No native hardware acceleration. • Not ideal for large ML models.	Voice-controlled smart home automation.	[157]
Kendryte K210	• RISC-V Dual-core (400 MHz, 8 MB SRAM). • Built-in KPU AI accelerator (0.8 TOPS). • Supports CNN-based models. • Low power consumption.	• Limited support for development frameworks – no PyTorch. • Smaller developer community. • Optimization is challenging.	Face detection for access control systems.	[158]

2.5.1 Multimodal Machine Learning for TinyML

Multimodal ML has emerged as a helpful approach in AI. It encompasses the procedures for training robust ML models by fusing multiple types of data – text, images, audio, and even raw sensor inputs. While this paradigm has shown immense promise across domains like healthcare, autonomous systems, and human-computer interaction, its application in the TinyML domain has remained a challenge and an opportunity at the same time. Since TinyML primarily aims at enabling intelligent processing on edge devices, typically characterized by limited computing, storage, and energy resources, integrating multimodal learning into mainstream TinyML modeling has the potential to significantly enhance context awareness and robustness in edge applications [18].

Although integrating information from multiple modalities has the potential to increase computational workload if not carefully designed, multimodal learning holds good promise for TinyML. The promise of multimodal ML for TinyML stems directly from its potential to significantly increase the quantity and quality of data available to train the models at the edge. By fusing multiple data streams, multimodal algorithms can compensate for noise or missing information in individual modalities, ensuring robust performance in dynamic environments. Thus, multimodal TinyML systems are advantageously positioned to achieve higher accuracy and adaptability by leveraging complementary information from diverse data sources. A wearable device, for example, can combine readings from an accelerometer and heart rate data to generate more accurate insights about a patient's fitness. In a similar fashion, a smart home security device may combine audio, visual, and environmental sensor data to improve efficiency. More importantly, processing multimodal data locally on edge devices enhances data privacy and enables real-time decision-making, which is known to be particularly valuable in safety-critical applications such as medical monitoring, where low-latency and secure processing is a critical requirement [162].

There are clear obstacles to multimodal learning in TinyML despite its potential. The high computational overhead of the multimodal learning approach is a problem; only lightweight and energy-efficient models can truly be used on resource-constrained edge devices. It is well known that real-time alignment and synchronization of data from various modalities can further complicate data handling procedures, especially when working with disparate data rates and formats. Therefore, creative approaches like dynamic modality prioritization and modular architectures are needed to efficiently integrate multimodal data without drastically raising compute and storage demands to the point where implementation on edge devices becomes impractical.

Recent advancements in TinyML and multimodal ML are progressively addressing these challenges. Model compression, quantization, and pruning have increasingly become preferred methods for minimizing multimodal models' size and power demands. KD and NAS also positively contribute to developing lightweight yet highly efficient multimodal architectures. Edge-specific algorithms such as dynamic modality selection, resource-adaptive multimodal inference, hierarchical modality fusion, edge-aware multimodal optimization, NAS with modality adaptation, and attention-based multimodal pruning, which dynamically select the most informative modalities based on available resources, have demonstrated potential in enhancing efficiency.

The advancement of multimodal TinyML has significant potential to transform the implementation of TinyML-driven applications across various sectors, including healthcare, where wearable devices with multimodal functionalities can provide real-time diagnostics, and in industrial IoT, where edge device applications must amalgamate data from multiple sensors to improve predictive maintenance, for instance. Multimodal ML for TinyML represents an emerging frontier necessitating enhanced research and development initiatives.

Although enormous technical challenges subsist, ongoing innovations in model optimization through compression, pruning and adaptation, hardware development, and algorithm design are paving the way for robust, real-time, and resource-efficient multimodal systems, through which multimodal TinyML is shaping the future of intelligent, context-aware devices.

2.5.2 Meta-Learning Algorithms for TinyML

Meta-learning is an ML modeling approach that creates models capable of rapidly adapting to new tasks without needing a substantial amount of data on the new task [47, 162]. Since data is often a challenge for many critical edge applications where insights are required at intervals, meta-learning is particularly relevant for TinyML. By leveraging meta-learning principles, TinyML applications can significantly improve performance and flexibility without the need for extensive computational resources or retraining.

Meta-learning algorithms are designed to solve new tasks using knowledge from previous experiences. The key principles for meta-learning for TinyML include Task generalization through the design of algorithms that can adapt to a range of tasks, such as extrapolating gesture recognition from motion detection [162], efficient learning, such as in few-shot learning, where the objective is to lower computational overhead by focusing on lightweight operation, and optimization across constraints – where an objective to balance accuracy, memory usage, and energy efficiency is considered key to achieving practical deployment on resource-constrained devices.

Meta-leaning techniques for TinyML have been developed and are in use. These include model-agnostic meta-learning (MAML), few-shot learning, reptile, and NAS. MAML is widely applied to develop models that are easily adapted to new tasks. The adaptation of MAML for TinyML focuses mainly on reducing the complexity of inner-loop optimization to increase the efficiency of gradient computation. MAML is useful for task-specific fine-tuning, such as adapting automatic speech recognition models to new accents, without model retraining.

Recent research has advanced the course of low-power NAS techniques, and this technique has, in turn, offered improvement for the design of edge-computing tailored lightweight meta-models by optimizing for latency and energy efficiency. Although applications of meta-learning in TinyML can range widely, including in personalized healthcare, smart home systems, industrial IoT, and autonomous systems, its successful application in the TinyML domain faces several challenges occasioned by the resource-limited nature of the target devices, limited availability of datasets, and uncertainties around hardware optimization for multiple platforms.

A meta-learning approach represents a promising route to achieving efficient TinyML systems that are adaptable, efficient, intelligent, and suitable for resource-constrained environments. Through dedicated research and attention to lightweight techniques and innovative optimization strategies, the barriers to TinyML can be overcome.

2.6 TinyML Development Process

The process of developing a practical TinyML application can be divided into three phases: training, conversion, and inference [163]. The principal activities involved in the training phase are data collection by the sensors, data preprocessing, feature extraction, and model training, as elaborated in Figure 2.11.

In the data collection step, data is collected from the application scenario using sensors. The data collected, which can be in the form of text, audio, or image, is then stored locally or sent to cloud storage. Data cleaning is carried out in the preprocessing step. Data cleaning is essential for removing of noise, handling of missing values, and removing unwanted data elements. The clean data is then transformed into other needed formats. Common data preprocessing operations for image data include resizing, conversion to grayscale, and normalization. The feature extraction step is crucial; it identifies the most relevant features in the data.

Model training refers to a range of activities carried out to fit the learning algorithm to the dataset. Supervised, unsupervised, and reinforcement learning algorithms are among the prominent paradigms used to train learning models.

The conversion phase is critical for TinyML applications. The model resulting from the training phase would always require some downsizing to make it deployable on tiny devices. To achieve this, any or a combination of the optimization approaches described in the preceding Sections 2.5 of this chapter may be

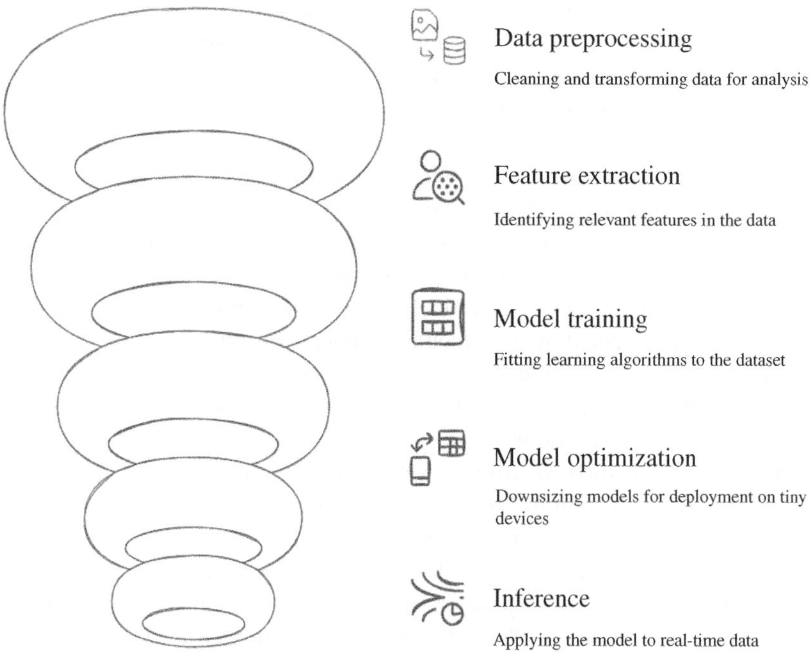

Figure 2.11 TinyML development process.

employed, depending on the objectives. The optimization techniques applicable to deep learning-based models include quantization, compression, pruning, low-rank decomposition, and KD, which have been discussed extensively in this chapter. Finally, the optimized model is converted to an appropriate form for the target device or application.

The final phase concerns the application's use in a real application. Here, the real-time data from the sensors are channeled through the preprocessing and feature expression pipeline, and the model on the device examines it and produces an output, which may be as expected or otherwise.

2.7 Key Findings

The major findings in this chapter are organized as follows:

1) **The Key Drivers of Increasing TinyML Adoption:** The chapter notes that TinyML has enabled widespread intelligence and innovation by bringing ML to resource-constrained, battery-powered devices and, in so doing, combines the benefits of advancements in embedded systems, ML, IoT, and software for

edge computing applications. The key drivers include improved microcontroller performance and advancements in algorithms (e.g., incorporation of attention-based models into tiny models for edge devices) and hardware.

2) **The Major Barriers to Success:** Significant challenges remain in terms of development, optimization, and learning. TinyML, while promising, requires innovative approaches to surmount significant hurdles that currently face the design and deployment of effective ML models on tiny devices. This often forces compromises on model accuracy for efficiency. Other important challenges include thc inadequacy of standards, limited tooling, community support, compute and energy limitations, data protection concerns, scalability, and post-deployment management. Memory constraints are also a major factor, with the risk of memory thrashing impacting model speed. Successful TinyML design requires balancing processing power with minimal resource use, complexity, and computation.
3) **Model Design Optimization:** TinyML specialized literature reveals that efficient model design optimization techniques are emerging strongly due to active TinyML dedicated research activities and advances in hardware design capabilities in the past five years. Techniques such as NAS, depth-separable convolution, low-rank decomposition, rational activation functions, and neural attention mechanisms for TinyML applications became feasible and burgeoned as a result. Hence, one major expectation is a consistent advancement in the development and application of these techniques, irrespective of the challenges that currently limit their use.
4) **Datasets, Development Frameworks, Tools, and Hardware:** The chapter revealed the existence of significant challenges related to these key aspects of the TinyML tooling system. Nonetheless, there has been considerable improvement in the last two to three years, following which there has been remarkable speed in the emergence of development tools such as TensorFlow Lite, Edge Impulse, and others. TinyML hardware design has also benefited from growing hardware design technologies, with each of the major players launching highly competitive products within the last two years. The current accelerated growth is expected to continue into the near future.

2.8 Conclusion

Current events in the AI/ML-enabled innovations and their applications are clearly indicative of a future that is predominantly run on tiny devices. This study has revealed that developing directly on tiny devices is still significantly challenging due to the unique resource limitations embodied by edge computing hardware. Learning for TinyML majorly consists of developing regular models using regular capable hardware and miniaturizing the model for deployment on tiny devices. To realize this objective, researchers have been able to develop a wide array of efficient approaches for converting large models to tiny models that can run efficiently on resource-constrained devices.

This chapter approached the subject of learning for TinyML by examining the main challenges faced in developing and deploying ML models on edge devices. The major challenges relate to design, optimization, and learning. Helpful insights for understanding the main issues relating to the design and optimization of TinyML models have been presented with a focus on opportunities that abound in learning algorithms designed to facilitate performance-preserving model compression and optimization. The challenges and opportunities for improving TinyML model design using NAS, rational activation function, depth-separable convolution, and the attention mechanism were discussed before examining important frontiers in TinyML model optimization via pruning, quantization, KD, and related techniques. A systematic study of learning algorithms and the process for TinyML development was also presented. Finally, the chapter highlights the key drivers for the increasing adoption of TinyML in current and emerging applications, major barriers to success, model design optimizations and advances, and drawbacks associated with datasets, standardization, development frameworks, tools, and hardware. These issues are expected to shape the foreseeable future of the development and deployment of TinyML applications in current and emerging application scenarios.

References

1 Giordano, M., Baumann, N., Crabolu, M. et al. (2022). Design and performance evaluation of an ultralow-power smart IoT device with embedded TinyML for asset activity monitoring. *IEEE Transactions on Instrumentation and Measurement* 71: 1–11.

2 Oliveira, F., Costa, D.G., Assis, F., and Silva, I. (2024). Internet of intelligent things: A convergence of embedded systems, edge computing and machine learning. *Internet of Things* 25: 101153.

3 Karras, A., Giannaros, A., Karras, C. et al. (2024). TinyML algorithms for Big Data Management in large-scale IoT systems. *Future Internet* 16 (2): 42.

4 Zaidi, S.A.R., Hayajneh, A.M., Hafeez, M., and Ahmed, Q.Z. (2022). Unlocking edge intelligence through tiny machine learning (TinyML). *IEEE Access* 10: 100867–100877.

5 Schizas, N., Karras, A., Karras, C., and Sioutas, S. (2022). TinyML for ultra-low power AI and large scale IoT deployments: A systematic review. *Future Internet* 14 (12): 363.

6 Han, H. and Siebert, J. (2022). TinyML: A systematic review and synthesis of existing research. *Proceedings of the 2022 International Conference on Artificial Intelligence in Information and Communication (ICAIIC)*, Jeju Island, Republic of Korea (21–24 February 2022), pp. 269–274.

7 Sheikhi, N. (2023). Evaluating incremental machine learning for smart home adaptation with embedded systems. Student thesis, Malmö University, Faculty of Technology and Society (TS), Department of Computer Science and Media Technology.

8 Domb, M. (2019). Smart home systems based on internet of things. *Internet of Things (IoT) for Automated and Smart Applications*. London, UK: IntechOpen.

9 Warriach, E.U. (2013). State of the art: embedded middleware platform for a smart home. *International Journal of Smart Home* 7 (6): 275–294.

10 Hayajneh, A.M., Aldalahmeh, S.A., Alasali, F. et al. (2024). Tiny machine learning on the edge: A framework for transfer learning empowered unmanned aerial vehicle assisted smart farming. *IET Smart Cities* 6 (1): 10–26.

11 Adeola, J. O., Degila, J., and Zennaro, M. (2022). Recent advances in plant diseases detection with machine learning: Solution for developing countries. *2022 IEEE International Conference on Smart Computing*, IEEE, 374–380.

12 Tsoukas, V., Gkogkidis, A., Kakarountas, A. (2023). TinyML-based system for smart agriculture. *Proceedings of the 26th Pan-Hellenic Conference on Informatics*, New York, NY (25–27 November), 207–212.

13 Nicolas, C., Naila, B., Amar, R.C. (2022). TinyML smart sensor for energy saving in Internet of Things precision agriculture platform. *Proceedings of the 2022 Thirteenth International Conference on Ubiquitous and Future Networks (ICUFN)*, Barcelona, Spain (5–8 July), 256–259.

14 Nicolas, C., Naila, B., Amar, R.C. (2022). Energy efficient firmware over the Air Update for TinyML models in LoRaWAN agricultural networks. *Proceedings of the 2022 32nd International Telecommunication Networks and Applications Conference (ITNAC)*, Wellington, New Zealand (30 November–2 December), 21–27.

15 Viswanatha, V., Ramachandra, A.C., Hegde, P.T., et al. (2023). Implementation of smart security system in agriculture fields using embedded machine learning. *Proceedings of the 2023 International Conference on Applied Intelligence and Sustainable Computing (ICAISC)*, Zakopane, Poland (18–22 June), 1–6.

16 Flores, T., Silva, M., Azevedo, M., et al. (2023). TinyML for safe driving: The use of embedded machine learning for detecting driver distraction. *Proceedings of the 2023 IEEE International Workshop on Metrology for Automotive (MetroAutomotive)*, Modena, Italy (28–30 June), 62–66.

17 Savanna, R.L., Hanyurwimfura, D., Nsenga, J.; Rwigema, J. A. (2023). Wearable device for respiratory diseases monitoring in crowded spaces. Case Study of COVID-19. *Proceedings of the International Congress on Information and Communication Technology*, London, UK (20–23 February 2023), Springer: Berlin/Heidelberg, Germany, 515–528.

18 Gibbs, M., Woodward, K., and Kanjo, E. (2023). Combining multiple TinyML models for multimodal context-aware stress recognition on constrained microcontrollers. *IEEE Micro* 44 (3): 67–75.

19 Nguyen, H.T., Mai, N.D., Lee, B.G., and Chung, W.Y. (2023). Behind-the-ear EEG-based wearable driver drowsiness detection system using embedded tiny neural networks. *IEEE Sensors Journal* 23: 23875–23892.

20 Hammad, S.S., Iskandaryan, D., and Trilles, S. (2023). An unsupervised TinyML approach applied to the detection of urban noise anomalies under the smart cities environment. *Internet of Things* 23: 100848.

21 Elhanashi, A., Dini, P., Saponara, S., and Zheng, Q. (2024). Advancements in TinyML: Applications, limitations, and impact on IoT devices. *Electronics* 13 (17): 3562.

22 Sanchez-Iborra, R. and Skarmeta, A.F. (2020). Tinyml-enabled frugal smart objects: Challenges and opportunities. *IEEE Circuits and Systems Magazine* 20 (3): 4–18.

23 Oufettoul, H., Chaibi, R., and Motahhir, S. (2024). TinyML applications, research challenges, and future research directions. *2024 21st Learning and Technology Conference (L&T)*. IEEE, 86–91.

24 Ooko, S.O., Muyonga Ogore, M., Nsenga, J., and Zennaro, M. (2021). TinyML in Africa: Opportunities and challenges. *Proceedings of the 2021 IEEE Globecom Workshops (GC Wkshps)*, Madrid, Spain (7–11 December 2021), pp. 1–6.

25 Prakash, S., Stewart, M., Banbury, C. et al. (2023). Is TinyML sustainable? *Communications of the ACM* 66 (11): 68–77.

26 Delnevo, G., Mirri, S., Prandi, C., and Manzoni, P. (2023). An evaluation methodology to determine the actual limitations of a tinyml-based solution. *Internet of Things* 22: 100729.

27 Hayajneh, A.M., Hafeez, M., Zaidi, S.A., and McLernon, D. (2024). TinyML empowered transfer learning on the edge. *IEEE Open Journal of the Communications Society* 5: 1656–1672.

28 Kang, P., Moosmann, J., Bian, S., and Magno, M. (2024). On-device training empowered transfer learning for human activity recognition. *arXiv preprint arXiv:2407.03644*.

29 Warden, P. and Situnayake, D. (2019). *TinyML: Machine Learning with TensorFlow Lite on Arduino and Ultra-Low-Power Microcontrollers*. O'Reilly Media.

30 Banbury, C., Reddi, V. J., Torelli, P., et al. (2021). Mlperf tiny benchmark. *arXiv preprint arXiv:2106.07597*.

31 Banbury, C.R., Reddi, V.J., Lam, M., et al. (2020). Benchmarking tinyml systems: Challenges and direction. *arXiv preprint arXiv:2003.04821.*

32 Sze, V., Chen, Y.H., Yang, T.J., and Emer, J.S. (2017). Efficient processing of deep neural networks: A tutorial and survey. *Proceedings of the IEEE* 105 (12): 2295–2329.

33 Jacob, B., Kligys, S., Chen, B., et al. (2018). Quantization and training of neural networks for efficient integerarithmetic-only inference. Proceedings of the IEEE Conference on Computer Vision and Pattern Recognition, Salt Lake City, UT (18–22 June), 2704–2713.

34 Lin, S., Ji, R., Yan, C., Zhang, B., et al. (2019). Towards optimal structured cnn pruning via generative adversarial learning. *Proceedings of the IEEE/CVF Conference on Computer Vision and Pattern Recognition*, 2790–2799.

35 Reitmaier, T., Wallington, E., Kalarikalayil Raju, D., et al. (2022). Opportunities and challenges of automatic speech recognition systems for low-resource language speakers. *Proceedings of the 2022 CHI Conference on Human Factors in Computing Systems*, 1–17.

36 Ren, H., Anicic, D., Li, X., and Runkler, T. (2024). On-device online learning and semantic management of TinyML systems. *ACM Transactions on Embedded Computing Systems* 23 (4): 1–32.

37 Cai, H., Gan, C., Zhu, L., and Han, S. (2020). TinyTL: Reduce memory, not parameters for efficient on-device learning. *Advances in Neural Information Processing Systems* 33: 11285–11297.

38 Hu, E. J., Shen, Y., Wallis, P., et al. (2021). Lora: Low-rank adaptation of large language models. *arXiv preprint arXiv:2106.09685.*

39 David, R., Duke, J., Jain, A. et al. (2021). Tensorflow lite micro: Embedded machine learning for tinyml systems. *Proceedings of Machine Learning and Systems* 3: 800–811.

40 Preuveneers, D., Verheyen, W., Joos, S. and Joosen, W. (2023). On the adversarial robustness of full integer quantized TinyML models at the edge. *Proceedings of the 2nd International Workshop on Middleware for the Edge* (December), 7–12.

41 Zhao, J., Mortier, R., Crowcroft, J., and Wang, L. (2018). Privacy-preserving machine learning based data analytics on edge devices. *Proceedings of the 2018 AAAI/ACM Conference on AI, Ethics, and Society*, 341–346.

42 Han, S., Mao, H., and Dally, W. J. (2015). Deep compression: Compressing deep neural networks with pruning, trained quantization and Huffman coding. *arXiv preprint arXiv:1510.00149.*

43 Klein, J., Boybat, I., Qureshi, Y.M. et al. (2022). Alpine: Analog in-memory acceleration with tight processor integration for deep learning. *IEEE Transactions on Computers* 72 (7): 1985–1998.

44 Wu, G. (2024). Streamlining TinyML lifecycle with large language models: A framework for automation. PhD thesis, University of Helsinki.

45 Zhou, K., Huang, X., Song, Q. et al. (2022). Auto-GNN: Neural architecture search of graph neural networks. *Frontiers in Big Data* 5: 1029307.

46 White, C., Safari, M., Sukthanker, R., et al. (2023). Neural architecture search: Insights from 1000 papers. *arXiv preprint arXiv:2301.08727*.

47 Chitty-Venkata, K.T. and Somani, A.K. (2022). Neural architecture search survey: A hardware perspective. *ACM Computing Surveys* 55 (4): 1–36.

48 Jin, H., Song, Q., and Hu, X. (2019). Auto-keras: An efficient neural architecture search system. *Proceedings of the 25th ACM SIGKDD International Conference on Knowledge Discovery & Data Mining*, 1946–1956.

49 Zhang, W., Shen, Y., Lin, Z., et al. (2022). Pasca: A graph neural architecture search system under the scalable paradigm. *Proceedings of the ACM Web Conference 2022*, 1817–1828.

50 Bharat S. C., Sheetal N. G., Marco, Z. et al. (2024). TinyML applications and use cases for healthcare. In: Bhamare, M., Kulkarni, P. V., Rane. R., Bobde, S. and Patankar, R. *TinyML for Edge Intelligence in IoT and LPWAN Networks*, 331–353. Academic Press.

51 Ahmed, H. H., Ahmed, Z., Choden, T., and Chaudhary, N. (2024). TinyML for emotion detection in voice signals: evaluating and proposing algorithms for IoT wearable devices. Doctoral dissertation, Brac University.

52 Asritha, A., Dusyanth, C., Sianyengele, O., et al. (2023). Efficient speech emotion recognition for resource-constrained devices. *2023 Second International Conference on Augmented Intelligence and Sustainable Systems*, IEEE, 765–772.

53 Tharian, J., Nandakrishnan, R., Sajesh, S., Arun, A. V., and Jayadas, C. K. (2022). Automatic emotion recognition system using tinyML. *2022 International Conference on Futuristic Technologies*, IEEE 1–4.

54 Vaswani, A., Shazeer, N., Parmar, N., et al. (2017). Attention is all you need. *Advances in neural information processing systems, 30.*

55 Brauwers, G. and Frasincar, F. (2021). A general survey on attention mechanisms in deep learning. *IEEE Transactions on Knowledge and Data Engineering* 35 (4): 3279–3298.

56 Wong, A., Famouri, M. and Shafiee, M.J. (2020). AttendNets: Tiny deep image recognition neural networks for the edge via visual attention condensers. *arXiv:2009.14385.*

57 Tan, M., Chen, B., Pang, R., et al. (2018). Mnasnet: Platform-aware neural architecture search for mobile. *arXiv preprint arXiv:1807.11626, 2018.*

58 Xu, K., Li, Y., Zhang, H., et al. (2022). EtinyNet: Extremely tiny network for tinyML. *Proceedings of the AAAI Conference on Artificial Intelligence*, 36 (4), 4628–4636.

59 Garai, S., and Samui, S. (2024). Exploring TinyML frameworks for small-footprint keyword spotting: A concise overview. *2024 IEEE International Conference on Signal Processing and Communications*, 1–5.

60 Rashid, H.A., Kallakuri, U., and Mohsenin, T. (2024). TinyM2Net-V2: A compact low-power software hardware architecture for multi-modal deep neural networks. *ACM Transactions on Embedded Computing Systems* 23 (3): 1–23.

61 Dubey, S.R., Singh, S.K., and Chaudhuri, B.B. (2022). Activation functions in deep learning: A comprehensive survey and benchmark. *Neurocomputing* 503: 92–108.

62 Peiris, V.I. (2022). Rational activation functions in neural network with uniform norm based loss function and its application in classification. *Communications in Optimization Theory* 1–25.

63 Trimmel, M., Zanfir, M., Hartley, R., and Sminchisescu, C. (2022). ERA: Enhanced rational activations. In: *European Conference on Computer Vision*, 722–738. Springer Nature Switzerland: Cham.

64 Banner, R., Nahshan, Y., and Soudry, D. (2019). Post training 4-bit quantization of convolutional networks for rapid-deployment. *Advances in Neural Information Processing Systems* 32: 1–9.

65 Hinton, G., Vinyals, O., and Dean, J. (2015). Distilling the knowledge in a neural network. NeurIPS Workshop.

66 Gou, J., Yu, B., Maybank, S.J., and Tao, D. (2021). Knowledge distillation: A survey. *International Journal of Computer Vision* 129 (6): 1789–1819.

67 Polino, A., Pascanu, R., and Alistarh, D. (2018). Model compression via distillation and quantization. *arXiv preprint arXiv:1802.05668*.

68 Han, S., Liu, X., Mao, H. et al. (2016). EIE: Efficient inference engine on compressed deep neural network. *ACM SIGARCH Computer Architecture News* 44 (3): 243–254.

69 Frankle, J., and Carbin, M. (2018). The lottery ticket hypothesis: Finding sparse, trainable neural networks. *arXiv preprint arXiv:1803.03635*.

70 Blalock, D., Gonzalez Ortiz, J.J., Frankle, J., and Guttag, J. (2020). What is the state of neural network pruning? *Proceedings of Machine Learning and Systems* 2: 129–146.

71 Denton, E. L., Zaremba, W., Bruna, J., et al. (2014). Exploiting linear structure within convolutional networks for efficient evaluation. *Advances in Neural Information Processing Systems*, 27.

72 Jaderberg, M., Vedaldi, A., and Zisserman, A. (2014). Speeding up convolutional neural networks with low rank expansions. *arXiv preprint arXiv:1405.3866*.

73 Tai, C., Xiao, T., Zhang, Y., and Wang, X. (2015). Convolutional neural networks with low-rank regularization. *arXiv preprint arXiv:1511.06067*.

74 Liu, J., Tripathi, S., Kurup, U., and Shah, M. (2020). Pruning algorithms to accelerate convolutional neural networks for edge applications: A survey. *arXiv preprint arXiv:2005.04275.*

75 Lagunas, F., Charlaix, E., Sanh, V., and Rush, A. M. (2021). Block pruning for faster transformers. *arXiv preprint arXiv:2109.04838.*

76 Cheng, H., Zhang, M., and Shi, J.Q. (2024). A survey on deep neural network pruning: Taxonomy, comparison, analysis, and recommendations. *IEEE Transactions on Pattern Analysis and Machine Intelligence* 46 (12): 10558–10578.

77 Lazarevich, I., Kozlov, A., and Malinin, N. (2021). Post-training deep neural network pruning via layer-wise calibration. *Proceedings of the IEEE/CVF International Conference on Computer Vision*, 798–805.

78 Ma, X., Fang, G., and Wang, X. (2024). Deepcache: Accelerating diffusion models for free. *Proceedings of the IEEE/CVF Conference on Computer Vision and Pattern Recognition*, 15762–15772.

79 Chen, T., Frankle, J., Chang, S., et al. (2021). The lottery tickets hypothesis for supervised and self-supervised pre-training in computer vision models. *Proceedings of the IEEE/CVF Conference on Computer Vision and Pattern Recognition*, 16306–16316.

80 Han, S., Pool, J., Tran, J., and Dally, W. (2015). Learning both weights and connections for efficient neural network. *Advances in Neural Information Processing Systems*, 28.

81 Liu, Z., Sun, M., Zhou, T., et al. (2018). Rethinking the value of network pruning. *arXiv preprint arXiv:1810.05270.*

82 Molchanov, D., Ashukha, A., and Vetrov, D. (2017). Variational dropout sparsifies deep neural networks. *International Conference on Machine Learning*, PMLR, 2498–2507.

83 Luo, J. H., Wu, J., and Lin, W. (2017). Thinet: A filter level pruning method for deep neural network compression. *Proceedings of the IEEE International Conference on Computer Vision*, 5058–5066.

84 He, Y., Kang, G., Dong, X., et al. (2018). Soft filter pruning for accelerating deep convolutional neural networks. *arXiv preprint arXiv:1808.06866.*

85 Gale, T., Elsen, E., and Hooker, S. (2019). The state of sparsity in deep neural networks. *arXiv preprint arXiv:1902.09574.*

86 Hong, J., Duan, J., Zhang, C., et al. (2024). Decoding compressed trust: Scrutinizing the trustworthiness of efficient LLMs under compression. Computing Research Repository. *arXiv 2024, arXiv:2403.15447.*

87 Agustsson, E. and Theis, L. (2020). Universally quantized neural compression. *Advances in Neural Information Processing Systems* 2020 (33): 12367–12376.14.

88 Wei, L., Ma, Z., Yang, C., and Yao, Q. (2024). Advances in the neural network quantization: A comprehensive review. *Applied Sciences* 14: 7445. https://doi.org/10.3390/app14177445.

89 Yuan, Z., Liu, J., Wu, J., et al. (2023). Benchmarking the reliability of post-training quantization: a particular focus on worst-case performance. *arXiv preprint arXiv:2303.13003.*

90 Yao, Z., Dong, Z., Zheng, Z., et al. (2020). Hawqv3: Dyadic neural network quantization. *arXiv:2011.10680.*

91 McKinstry, J.L., Esser, S.K., Appuswamy, R., et al. (2018). Discovering low-precision networks close to full-precision networks for efficient embedded inference. *arXiv:1809.04191.*

92 Choi, J., Wang, Z., Venkataramani, S., et al. (2018). Pact: Parameterized clipping activation for quantized neural networks. *arXiv:1805.06085.*

93 Zhou, S., Wu, Y., Ni, Z., et al. (2016) Dorefa-net: Training low bitwidth convolutional neural networks with low bitwidth gradients. *arXiv:1606.06160.*

94 Diao, H., Li, G., Xu, S. et al. (2024). Attention round for post-training quantization. *Neurocomputing* 565: 127012.

95 Frantar, E., Ashkboos, S., Hoefler, T. and Alistarh, D. (2023). GPTQ: Accurate post-training quantization for generative pre-trained transformers. *arXiv:2210.17323.*

96 Gholami, A., Kim, S., Dong, Z., et al. (2021). A survey of quantization methods for efficient neural network inference. *arXiv 2021, arXiv:2103.13630.*

97 Liang, T., Glossner, J., Wang, L. et al. (2021). Pruning and quantization for deep neural network acceleration: A survey. *Neurocomputing* 461: 370–403.

98 Vindas, Y., Roux, E., Guépié, B.K. et al. (2024). An asymmetric heuristic for trained ternary quantization based on the statistics of the weights: an application to medical signal classification. *Pattern Recognition Letters* 188: 37–45.

99 Li, Y., Shen, M., Ma, J., et al. (2021). Mqbench: Towards reproducible and deployable model quantization benchmark. *arXiv preprint arXiv:2111.03759.*

100 Dettmers, T. (2024). Accessible foundation models: Systems, algorithms, and science. Doctoral dissertation, University of Washington.

101 Benedetti, V., Casini, H., and Magan, J. M. (2023). ABJ anomaly as a U (1) symmetry and Noether's theorem. *arXiv preprint arXiv:2309.03264.*

102 Gholami, A., Kim, S., Dong, Z., et al. (2022). A survey of quantization methods for efficient neural network inference. *Low-Power Computer Vision,* 291–326.

103 Ma, L., Luo, X., Hong, H. et al. (2024). Logit variated product quantization based on parts interaction and metric learning with knowledge distillation for fine-grained image retrieval. *IEEE Transactions on Multimedia* 26: 10406–10419.

104 Qi, Q., Lu, Y., Li, J. et al. (2021). Learning low resource consumption CNN through pruning and quantization. *IEEE Transactions on Emerging Topics in Computing* 10 (2): 886–903.

105 Li, Z., Wang, Y., Zhang, N. et al. (2022). Deep learning-based object detection techniques for remote sensing images: A survey. *Remote Sensing* 14 (10): 2385.

106 Chen, M., Shao, W., Xu, P., et al. (2024). Efficientqat: Efficient quantization-aware training for large language models. *arXiv preprint arXiv:2407.11062.*

107 Huh, M., Cheung, B., Agrawal, P., and Isola, P. (2023). Straightening out the straight-through estimator: Overcoming optimization challenges in vector quantized networks. *International Conference on Machine Learning*, 14096–14113.

108 Wu, H., He, R., Tan, H. et al. (2023). Vertical layering of quantized neural networks for heterogeneous inference. *IEEE Transactions on Pattern Analysis and Machine Intelligence* 45 (12): 15964–15978.

109 Huang, X., Fan, J., Deng, Z. et al. (2021). Efficient IoT data management for geological disasters based on big data-turbocharged data lake architecture. *ISPRS International Journal of Geo-Information* 10: 743.

110 Alnemari, M. and Bagherzadeh, N. (2024). Ultimate compression: Joint method of quantization and tensor decomposition for compact models on the edge. *Applied Sciences* 14 (20): 9354.

111 Liu, D., Kong, H., Luo, X. et al. (2022). Bringing AI to edge: From deep learning's perspective. *Neurocomputing* 485: 297–320.

112 Bai, Y., Chen, W., Chen, J., and Guo, W. (2020). Deep learning methods for solving linear inverse problems: Research directions and paradigms. *Signal Processing* 177: 107729.

113 Wang, M., Fu, W., He, X. et al. (2020). A survey on large-scale machine learning. *IEEE Transactions on Knowledge and Data Engineering* 34 (6): 2574–2594.

114 Nguyen, T. T., Huynh, T. T., Nguyen, P. L., et al. (2022). A survey of machine unlearning. *arXiv preprint arXiv:2209.02299.*

115 Liu, J., Cai, J., and Zhuang, B. (2021). Sharpness-aware quantization for deep neural networks. *arXiv preprint arXiv:2111.12273.*

116 Kulkarni, U., Meena, S. M., Joshua, P., et al. (2021). Performance improvements in quantization aware training and appreciation of low precision computation in deep learning. *Advances in Signal Processing and Intelligent Recognition Systems: 6th International Symposium, SIRS 2020*, Chennai, India (14–17 October 2020), Revised Selected Papers 6, 90–107.

117 Sheng, T., Feng, C., Zhuo, S., et al. (2018). A quantization-friendly separable convolution for mobilenets. *Proceedings of the 2018 1st Workshop on*

Energy Efficient Machine Learning and Cognitive Computing for Embedded Applications (EMC2), Williamsburg, VA, USA (25 March 2018).

118 Feng, P., Yu, L., Tian, S.W. et al. (2022). Quantization of 8-bit deep neural networks based on mean square error. *Journal of Computational Design and Engineering* 43: 1258–1264.

119 Wang, T., Wang, K., Cai, H., et al. (2020). APQ: Joint search for network architecture, pruning and quantization policy. *Proceedings of the IEEE Computer Society Conference on Computer Vision and Pattern Recognition*, Seattle, WA, USA (13–19 June 2020), Volume 2006.08509, 2075–2084.

120 Wang, P., Yang, A., Men, R., et al. (2022). Unifying architectures, tasks, and modalities through a simple sequence-to-sequence learning framework. *Proceedings of the International Conference on Machine Learning*, Baltimore, MD, USA (17–23 July 2022), 23318–23340.

121 Hu, Z., Nie, F., Wang, R., and Li, X. (2021). Low rank regularization: A review. *Neural Networks* 136: 218–232. https://doi.org/10.1016/j.neunet.2020.09.021.

122 He, S., Li, Z., Tang, Y. et al. (2020). Parameters compressing in deep learning. *Computers Materials and Continua* 62 (1): 321–336. https://doi.org/10.32604/cmc.2020.06130.

123 Goyal, S., Choudhury, A.R. and Sharma, V. (2019). Compression of deep neural networks by combining pruning and low rank decomposition. *2019 IEEE International Parallel and Distributed Processing Symposium Workshops*, 952–958. https://doi.org/10.1109/ipdpsw.2019.00162.

124 Lin, S., Ji, R., Chen, C. et al. (2019). Holistic CNN compression via low-rank decomposition with knowledge transfer. *IEEE Transactions on Pattern Analysis and Machine Intelligence* 1 (12): 2889–2905. https://doi.org/10.1109/tpami.2018.2873305.

125 Yu, X., Liu, T., Wang, X. and Tao, D. (2017). On compressing deep models by low rank and sparse decomposition. *Proceedings of the IEEE Conference on Computer Vision and Pattern Recognition*, 7370–7379. https://doi.org/10.1109/cvpr.2017.15.

126 Ji, M., Heo, B., and Park, S. (2021). Show, attend and distill: Knowledge distillation via attention-based feature matching. *Proceedings of the AAAI Conference on Artificial Intelligence* 35 (9): 7945–7952. https://doi.org/10.1609/aaai.v35i9.16969.

127 Suau, X., and Apostoloff, N. (2020). Filter distillation for network compression. *2020 IEEE Winter Conference on Applications of Computer Vision*, 3129–3138. https://doi.org/10.1109/wacv45572.2020.9093546.

128 Yang, Z., Zhang, Y., Sui, D. et al. (2023). Explanation guided knowledge distillation for pre-trained language model compression. *ACM Transactions* *on Asian and Low-Resource Language Information Processing* 23 (2): 1–19. https://doi.org/10.1145/3639364.

129 Kim J, Chang S, and Kwak N (2021). PQK: Model compression via pruning, quantization, and knowledge distillation. In: ISCA (ed) Interspeech 2021. https://doi.org/10.21437/interspeech.2021-248.

130 Li, Y., Hu, F., Liu, Y. et al. (2023). A hybrid model compression approach via knowledge distillation for predicting energy consumption in additive manufacturing. *International Journal of Production Research* 61 (13): 4525–4547. https://doi.org/10.1080/00207543.2022.2160501.

131 Xu, Q., Wu, M., Li, X. et al. (2023). Contrastive distillation with regularized knowledge for deep model compression on sensor-based human activity recognition. *IEEE Transactions on Industrial Cyber-Physical Systems* 1: 217–226. https://doi.org/10.1109/ticps.2023.3320630.

132 Tan S., Tan, S., Tam, W.L., et al. (2023). Gkd: A general knowledge distillation framework for large-scale pre-trained language model. *arXiv preprint arXiv:2306.06629*. https://doi.org/10.48550/arXiv.2306.06629.

133 Ravikumar D., Saha G., Aketi S.A., and Roy K. (2023). Homogenizing non-IID datasets via in-distribution knowledge distillation for decentralized learning. *arXiv. 2304.04326*. https://doi.org/10.48550/arXiv.

134 Wu Z, Sun S, Wang Y, et al. (2023). Survey of knowledge distillation in federated edge learning. *arXiv*. https://doi.org/10.48550/arXiv.2301.05849.

135 Hou, Y., Zhu, X., Ma, Y., Loy, C.C. and Li, Y., (2022). Point-to-voxel knowledge distillation for lidar semantic segmentation. *Proceedings of the IEEE/CVF Conference on Computer Vision and Pattern Recognition*, 8479–8488. https://doi.org/10.1109/cvpr52688.2022.00829.

136 Li, Z., Xu, P., Chang, X. et al. (2023). When object detection meets knowledge distillation: A survey. *IEEE Transactions on Pattern Analysis and Machine Intelligence* 45 (8): 10555–10579. https://doi.org/10.1109/tpami.2023.3257546.

137 MicroML (2020). [Online]. https://github.com/eloquentarduino/micromlgen.

138 Sklearn-porter (2020). [Online]. https://github.com/nok/sklearn-porter.

139 m2cgen (2020). [Online]. https://github.com/BayesWitnesses/m2cgen.

140 Weka-porter (2020). [Online]. https://github.com/nok/weka-porter.

141 da Silva, L. T., Souza, V. M., and Batista, G. E. (2019). EmbML Tool: Supporting the use of supervised learning algorithms in low-cost embedded systems. *2019 IEEE 31st International Conference on Tools with Artificial Intelligence*, 1633–1637. https://ieeexplore.ieee.org/document/8995408/.

142 Nordby, J. (2019) emlearn: Machine Learning inference engine for microcontrollers and embedded devices, [Online]. https://github.com/emlearn/emlearn.

143 TinyGo. (2020). [Online]. https://tinygo.org/.

144 Embedded Learning Library (2020). [Online]. https://microsoft.github.io/ELL/.

145 Open Neural Network Exchange (2020). [Online]. https://onnx.ai/.

146 CMSIS-NN. (2020). [Online]. https://arm-software.github.io/CMSIS_5/NN/html/.

147 STM32Cube.AI (2020). [Online]. https://www.st.com/en/embedded-software/x-cube-ai.html.

148 Wulfert, L., Kühnel, J., Krupp, L. et al. (2024). AIfES: A next-generation edge AI framework. *IEEE Transactions on Pattern Analysis and Machine Intelligence* 46 (6): 4519–4533. https://www.ims.fraunhofer.de/en/Business_Units_and_Core_Competencies/Electronic-Assistance-Systems/Technologies/Artificial-Intelligence-for-Embedded-Systems-AIfES.html.

149 uTensor. (2020). [Online]. https://github.com/uTensor/uTensor.

150 TinyMLgen. (2020). [Online]. https://github.com/eloquentarduino/tinymlgen.

151 Capotondi, A., Rusci, M., Fariselli, M., and Benini, L. (2020). CMix-NN: Mixed low-precision CNN library for memory-constrained edge devices. *IEEE Transactions on Circuits and Systems II: Express Briefs* 67 (5): 871–875. https://ieeexplore.ieee.org/document/9049084/.

152 Wang, X., Magno, M., Cavigelli, L., and Benini, L. (2020). FANN-on-MCU: An open-source toolkit for energy-efficient neural network inference at the edge of the Internet of Things. *IEEE Internet of Things Journal* 7 (5): 4403–4417. https://ieeexplore.ieee.org/document/9016202/.

153 Fast Artificial Neural Network Library (FANN). (2020). [Online]. http://leenissen.dk/fann/wp/.

154 Rossi, D., Conti, F., Marongiu, A., et al. (2015). PULP: A parallel ultra low power platform for next generation IoT applications. *2015 IEEE Hot Chips 27 Symposium*, 1–39. http://ieeexplore.ieee.org/document/7477325/.

155 Paszke, A., Gross, S., Massa, F., et al. (2019). Pytorch: An imperative style, high-performance deep learning library. *Advances in Neural Information Processing Systems*, 32.

156 Lai, L., Suda, N., and Chandra, V. (2018). CMSIS-NN: Efficient neural network kernels for arm cortex-m cpus. *arXiv preprint arXiv:1801.06601.*

157 Chen, T., Moreau, T., Jiang, Z., et al. (2018). TVM: An automated end-to-end optimizing compiler for deep learning. *13th USENIX Symposium on Operating Systems Design and Implementation*, 578–594.

158 Janapa Reddi, V., Elium, A., Hymel, S. et al. (2023). Edge impulse: An mlops platform for tiny machine learning. *Proceedings of Machine Learning and Systems* 5: 254–268.

159 Narihira, T., Alonsogarcia, J., Cardinaux, F., et al. (2021). Neural network libraries: A deep learning framework designed from engineers' perspectives. *arXiv preprint arXiv:2102.06725.*

160 Banzi, M. and Shiloh, M. (2020). *Arduino: The Official Guide to Arduino Nano 33 BLE Sense*. Arduino Press.

161 Upton, E. and Halfacree, G. (2021). *Raspberry Pi User Guide*, 4e. Wiley Publishing.

162 Jouppi, N.P., Young, C., Patil, N., et al. (2017). In-datacenter performance analysis of a tensor processing unit. *Proceedings of the 44th Annual International Symposium on Computer Architecture* (June), 1–12.

163 Yelchuri, H. (2023). Workflow of TinyML. Medium. [Online]. https://medium.com/@harshayelchuri2000/workflow-of-tinyml-60bf181257e5.

3

TinyML for Anomaly Detection

Richard Govada Joshua[1], *Peter Anuoluwapo Gbadega*[2], *Agbotiname Lucky Imoize*[3], *and Samuel Oluwatobi Tofade*[4]

[1] *Comprobase Inc., Sterling, VA, USA*
[2] *Department of Electrical and Electronic Engineering Science, University of Johannesburg, Johannesburg, South Africa*
[3] *Department of Electrical and Electronics Engineering, Faculty of Engineering, University of Lagos, Lagos, Nigeria*
[4] *CleanTech Renewable Energy, Lagos, Nigeria*

3.1 Introduction

When class labels are not available for data instances, a model can be trained using existing datasets to classify most cases as standard while identifying potential anomalies. This approach allows the system to distinguish between normal and abnormal cases, even in the absence of predefined labels. Anomaly detection queries are crucial in monitoring applications, enabling the identification of irregular patterns that may indicate system failures, fraudulent activities, or other unexpected behaviors [1]. In many cases, domain knowledge and predefined thresholds can effectively detect most abnormalities. However, some anomalies remain undetected by conventional methods, making anomaly detection techniques particularly valuable. These techniques go beyond traditional monitoring approaches by identifying subtle, complex, or previously unknown patterns that could signal emerging threats or inefficiencies. Additionally, anomaly detection plays a crucial role in data cleansing and noise reduction, ensuring that datasets remain accurate, reliable, and free from misleading outliers.

The primary goal of anomaly detection is to determine whether a given data instance belongs to the standard (normal) class or represents an outlier (anomaly). This capability is invaluable across various industries. In finance, anomaly detection helps uncover fraudulent transactions, preventing financial losses and

Tiny Machine Learning: Design Principles and Applications, First Edition.
Edited by Agbotiname Lucky Imoize, Dinh-Thuan Do, and Houbing Herbert Song.

enhancing security. In cybersecurity, it detects potential breaches, unauthorized access, and malware activities, protecting sensitive information. In manufacturing, predictive maintenance powered by anomaly detection minimizes downtime by identifying equipment failures before they occur. In healthcare, the early detection of anomalies in medical data facilitates accurate disease diagnosis, ultimately leading to improved patient outcomes. Integrating anomaly detection into business operations enables organizations to proactively mitigate risks, prevent failures, and maintain system integrity. This results in enhanced security, improved operational efficiency, and significant cost savings while ensuring customer satisfaction. Additionally, anomaly detection empowers organizations to make data-driven decisions, optimize performance, and adapt to evolving challenges in an increasingly data-centric world [2].

Tiny machine learning (TinyML) represents a groundbreaking advancement in artificial intelligence, enabling machine learning (ML) algorithms to operate efficiently on low-power, resource-constrained devices such as microcontrollers and embedded systems. Unlike traditional ML models, which require significant computational power and cloud-based processing, TinyML brings intelligence directly to the edge of the network. This shift reduces dependence on centralized data centers, lowers latency, conserves bandwidth, and enhances real-time decision-making. As a result, TinyML has become a crucial component in modern intelligent systems, particularly for applications that require on-device anomaly detection. With the rapid growth of the Internet of Things (IoT) and edge computing (EC), there is an increasing demand for smart devices that can perform advanced analytics without relying on continuous cloud connectivity. TinyML meets this demand by deploying optimized ML models that can process and interpret data in real-time. This capability is crucial for various industries where detecting irregularities early can prevent failures, enhance safety, and improve overall operational efficiency. From industrial monitoring to healthcare, cybersecurity, and environmental management, TinyML-driven anomaly detection is transforming how organizations manage risk and maintain system integrity [3].

One of the most promising applications of TinyML is in predictive maintenance, where embedded sensors continuously monitor equipment performance. Traditional maintenance strategies often rely on scheduled service or reactive repairs following a failure. However, TinyML-enabled devices can analyze sensor data in real-time, identifying subtle patterns and deviations that signal potential equipment failure. This proactive approach enables organizations to anticipate issues before they become critical, resulting in reduced downtime, optimized maintenance schedules, and extended asset lifespans. For industries that depend on complex machinery, such as manufacturing, transportation, and energy, this translates to significant cost savings and enhanced reliability [4]. Beyond predictive maintenance, TinyML-based anomaly detection plays a crucial role in healthcare applications. Wearable devices equipped with TinyML models can continuously analyze physiological signals, detecting irregular heart rates, respiratory issues, or signs of neurological disorders in real-time. This enables early diagnosis and intervention, potentially saving lives while reducing the burden on healthcare systems. Similarly, in cybersecurity, TinyML-powered intrusion detection systems can monitor network traffic patterns, identifying unusual behaviors that may indicate security breaches or malware attacks, all while operating efficiently on edge devices without requiring constant cloud communication.

The integration of TinyML into environmental monitoring systems further showcases its versatility. Smart sensors deployed in remote areas can analyze air quality, detect gas leaks, monitor soil conditions, or predict wildfires, all without requiring continuous internet connectivity. These real-time insights support sustainable development, disaster prevention, and climate resilience initiatives. As the field of TinyML continues to evolve, advancements in model optimization techniques, such as quantization, pruning, and knowledge distillation, are further enhancing its feasibility for real-world applications. By embedding intelligent learning capabilities into IoT devices, organizations can harness the power of data-driven decision-making, improving efficiency, security, and overall system performance. The future of TinyML holds endless possibilities, making it a crucial technology for the next generation of smart, autonomous, and energy-efficient systems [5].

TinyML not only enables real-time decision-making at the edge but also significantly enhances operational efficiency while reducing potential safety risks. By embedding intelligence into low-power devices, TinyML allows for immediate responses to environmental changes, system malfunctions, and security threats without requiring constant communication with cloud-based systems. This localized intelligence ensures faster detection of anomalies, quicker corrective actions, and improved overall system performance, making it a game changer across multiple industries. A key advantage of TinyML is its ability to support continuous learning, allowing devices to evolve and adapt based on real-world data. Unlike traditional rule-based systems that rely on predefined logic, TinyML-powered devices operate dynamically, refining their models over time as they process new inputs. This adaptability is crucial in unpredictable environments, such as autonomous vehicles navigating busy streets, industrial sensors detecting early signs of equipment failure, and wearable health monitors identifying subtle physiological changes. The ability to learn and improve makes these systems smarter, more responsive, and increasingly efficient in handling complex scenarios beyond conventional approaches.

TinyML transforms edge devices from passive data collectors into intelligent decision-makers, enabling them to anticipate user needs and optimize functionality in real-time. By processing data locally, these devices extract valuable insights and take immediate action without relying on cloud-based systems [6]. In smart homes, TinyML-powered thermostats, lighting systems, and security cameras learn user behavior, detect anomalies, and autonomously adjust settings for enhanced convenience and security. Wearable health monitors utilize TinyML to track sleep patterns, detect abnormal heart rates, and provide real-time health alerts, thereby improving early disease detection and personalized healthcare. Industrial IoT sensors use predictive maintenance techniques to analyze equipment performance, detecting potential failures before they occur, thus reducing downtime and maintenance costs. Meanwhile, autonomous vehicles integrate TinyML for real-time object recognition, route optimization, and enhanced safety features, minimizing dependence on remote data processing. By embedding ML into ultralow-power devices, TinyML is revolutionizing human-technology interactions, making systems smarter, more efficient, and seamlessly interconnected.

Anomaly detection is essential across various industries, helping identify irregular data patterns that may indicate system malfunctions, security threats, or emerging phenomena. Traditionally, this process depends on centralized data processing, which demands high computational power and energy consumption. However, with the rapid growth of IoT networks and the increasing demand for decentralized, scalable solutions, TinyML has emerged as a game-changing alternative. By embedding lightweight ML models directly into IoT devices, TinyML enables real-time anomaly detection with minimal power consumption and reduced reliance on cloud computing. This advancement enhances cybersecurity, where TinyML-powered security systems detect unauthorized access, malware activity, and network anomalies independently. In healthcare, continuous monitoring devices can flag abnormal physiological changes, facilitating early medical intervention. Industrial safety benefits from smart sensors that analyze temperature, pressure, and vibrations to predict equipment failures and prevent accidents. Additionally, environmental monitoring systems leverage TinyML to detect wildfire risks, gas leaks, and pollution spikes before they escalate. This chapter explores the integration of TinyML in anomaly detection, discussing its advantages, limitations, and future potential. As TinyML continues to evolve, it is redefining edge intelligence by delivering unparalleled efficiency, adaptability, and security, making it a critical tool for future data-driven applications [7].

3.1.1 Research Concepts and Motivation

ML has revolutionized numerous fields, enabling automation, intelligent decision-making, and advanced data analytics. However, its traditional reliance on high-performance computing resources has limited its widespread adoption in low-power, resource-constrained environments. As the IoT ecosystem rapidly expands, the need for intelligent devices that can process data locally has become increasingly critical. TinyML addresses this challenge by allowing ML models to be deployed on microcontrollers and ultralow-power devices, often operating within a power budget of just a few milliwatts. Unlike conventional ML approaches that rely on cloud-based computation, TinyML enables these devices to operate autonomously, offering significant advantages in latency reduction, enhanced privacy, and improved energy efficiency. Among the various applications of TinyML, anomaly detection stands out as a crucial area where real-time, on-device intelligence can drive transformative change. Anomaly detection is widely used to identify irregularities or faults in various fields, including predictive maintenance, fraud detection, network security, and healthcare diagnostics. Traditional anomaly detection systems rely heavily on cloud-based infrastructures, requiring IoT devices to transmit data to centralized servers for analysis. While this method offers high computational power, it presents several challenges, including high latency, increased energy consumption, potential security vulnerabilities, and a reliance on stable network connectivity. These limitations have spurred growing interest in decentralized, on-device anomaly detection solutions, making TinyML an ideal alternative [8].

The increasing reliance on IoT devices across industries has created a pressing need for efficient, scalable, and real-time anomaly detection solutions. Traditional methods depend on centralized cloud-based systems, where data is collected from edge devices and processed remotely. While these systems offer high computational power, they introduce significant challenges, including high latency, energy consumption, and security risks. This has led to the exploration of TinyML as a viable alternative, enabling lightweight ML models to be deployed directly on microcontrollers and ultralow-power devices [9]. One of the primary motivations for integrating TinyML into anomaly detection is scalability. As the number of IoT-connected devices continues to grow into billions, relying on centralized data processing architectures becomes impractical. Transmitting massive volumes of data to the cloud for anomaly detection can overburden networks and increase operational costs. TinyML enables each device to handle its computations, reducing dependency on centralized infrastructure and allowing for a more distributed and efficient approach to anomaly detection. Another critical factor is latency and real-time processing.

In fields such as healthcare, industrial automation, and cybersecurity, timely anomaly detection can mean the difference between preventing system failures and catastrophic losses. Cloud-based anomaly detection introduces delays due to the time required for data transmission and processing. TinyML eliminates these delays by performing anomaly detection locally, ensuring that abnormal patterns are identified instantly, which leads to faster decision-making and response times. Energy efficiency is another driving force behind the adoption of TinyML in anomaly detection. Many IoT devices operate in remote, battery-powered environments where minimizing energy consumption is crucial. Traditional ML models, designed for robust computing systems, are resource-intensive and quickly drain battery life. TinyML models are optimized for ultralow-power consumption, enabling devices to operate efficiently for extended periods without requiring frequent recharging or battery replacements. This makes it an ideal solution for applications such as predictive maintenance in industrial settings, where sensors must operate continuously. Additionally, privacy and security concerns motivate the shift toward TinyML.

Cloud-based anomaly detection requires continuous data transmission, increasing the risk of data breaches and unauthorized access. By processing data locally on the device, TinyML significantly reduces the exposure of sensitive information, enhancing data privacy and making it a more secure solution for healthcare, finance, and cybersecurity applications. Lastly, cost-effectiveness plays a crucial role in motivating the adoption of TinyML. Cloud computing services and high-end ML infrastructure require substantial investment in both hardware and operational costs. TinyML leverages existing low-power microcontrollers, eliminating the need for expensive infrastructure upgrades and reducing dependency on cloud-based services. This makes it a cost-effective solution, especially for large-scale IoT deployments. By addressing these challenges, TinyML is revolutionizing anomaly detection, offering a scalable, real-time, energy-efficient, privacy-conscious, and cost-effective alternative to traditional methods. As research and innovation in this field continue to progress, TinyML-based anomaly detection systems will play a crucial role in making IoT ecosystems more intelligent, autonomous, and resilient [10, 11].

3.1.2 Importance of TinyML in Anomaly Detection

The integration of TinyML into anomaly detection offers transformative benefits, reshaping how irregularities are identified and addressed in various domains. The following sections outline its importance [10, 12].

In numerous mission-critical applications, such as predictive maintenance in industrial settings and patient monitoring in healthcare, timely anomaly detection is crucial to avoid catastrophic failures and provide immediate interventions. TinyML facilitates real-time anomaly detection by executing ML models directly on resource-constrained edge devices, circumventing the latency introduced by network communications. This direct processing capability ensures that any irregularities can be detected and acted upon instantaneously, which is particularly vital in environments where delays can lead to severe operational or safety risks.

Furthermore, by processing data locally, TinyML enables continuous surveillance without relying on a constant data stream to central servers, optimizing both responsiveness and resource utilization.

Traditional ML models, due to their computational intensity, are often ill-suited for deployment on edge devices with limited power budgets. TinyML, on the other hand, is designed specifically for low-power hardware, ensuring that anomaly detection algorithms can run continuously without rapidly draining battery life. Through techniques such as model pruning, quantization, and efficient algorithm design, TinyML models are optimized to operate within the strict power limitations of devices like IoT sensors, wearables, and remote monitoring systems. This energy efficiency is crucial for applications in remote or difficult-to-access locations, where frequent battery replacements or network connectivity are impractical, such as environmental monitoring stations, agricultural sensors, or space exploration equipment.

As the need for sensitive data processing grows, especially in sectors such as healthcare and finance, ensuring privacy and data security becomes increasingly critical. TinyML addresses these concerns by enabling local data processing directly on edge devices, which eliminates the need to transmit raw data to centralized servers or cloud platforms. This local processing architecture mitigates potential exposure of sensitive information to security vulnerabilities during transmission, offering a robust defense against data breaches and unauthorized access. Additionally, TinyML aligns with strict regulatory frameworks such as HIPAA and GDPR by ensuring that sensitive data remains on-device, thereby maintaining compliance with privacy standards and reducing the risk of costly penalties associated with data misuse.

3.1.3 Key Contributions

With the recent availability of numerous robust metrics, learning, and inference methods, as well as intensive learning and inference for various learning models, anomalies in some symbolic or face-value representations can be identified. However, library data processing often compromises the interpretability and practical benefits of these methods. Since the amount of data transferred between the sensing site and an analytics center could be a problem, even with a good communication link, these data systems can only be fully practically feasible. In addition, for online inference, the model running on the computing infrastructure will require real-time performance magnitudes less than the tolerable time delays for corrective actions [4].

We identify ways that ML, particularly anomaly detection, can be more effectively applied to the explosive growth of sensors by creating computationally efficient models designed for specific applications. Anomaly detection in time-series data is crucial in monitoring and diagnosing electronic, mechanical, and civil infrastructure systems. Typically, the detection function is performed using statistical-based methods, knowledge-based decision protocols, expert systems, and other similar techniques [13]. The data are usually collected by multiple embedded sensors, providing real-time system status on an incredibly fast nanosecond to millisecond timescale. These high-frequency measurements enable the capture of intricate system dynamics and the early detection of abnormal behavior. By leveraging the power of ML algorithms, we can develop advanced models that can learn from these complex patterns and accurately detect anomalies, even in vast and diverse sensor networks. These models can effectively handle the massive influx of sensor data while maintaining computational efficiency, enabling real-time and scalable anomaly detection solutions for various domains [14]. By tailoring the models to specific applications, we can achieve higher accuracy and reliability in identifying irregularities, allowing for timely interventions and proactive maintenance. This approach opens new possibilities for leveraging ML in the era of sensor-driven systems, ensuring their seamless integration into our everyday lives [14, 15].

- **Availability of Advanced Methods:** Recent advancements in powerful metrics learning and inference methods, as well as intense learning, have enabled the discovery of anomalies in symbolic or face-value representations.
- **Library Data Processing Issues:** While these methods are powerful, library data processing often compromises their interpretability and practical benefits.
- **Data Transfer Challenges:** Even with a good communication link, the data transfer between the sensing site and an analytics center can be problematic.
- **Practical Application:** These data systems can only be efficient if they manage the data transfer effectively.
- **Real-Time Performance Requirement:** For online inference, the model running on the computing infrastructure must perform in real-time, with performance requirements significantly lower than the minimum tolerable time delays for corrective actions.
- **Enhancing Machine Learning Applicability:** To make ML, particularly anomaly detection, more applicable to the rapid growth of sensors, we need to develop computationally efficient models tailored to specific applications.
- **Importance of Anomaly Detection:** Anomaly detection in time-series data is crucial for monitoring and diagnosing electronic, mechanical, and civil infrastructure systems.
- **Traditional Detection Methods:** Anomaly detection is typically performed using statistical-based methods, knowledge-based decision protocols, or expert systems.
- **Sensor Data Collection:** Numerous embedded sensors collect data, providing system status at nanosecond to millisecond time scales.

3.1.4 Chapter Organization

This chapter is organized into several key sections that explore various facets of anomaly detection in constrained devices, with a focus on IoT and edge devices. "Related Work" provides an overview of existing research in anomaly detection, explicitly focusing on resource-efficient algorithms designed for constrained devices. It discusses notable works in the field, including convolutional neural network (CNN)-based approaches for industrial machinery monitoring, PCA and MEM-based online anomaly detection systems for IoT devices, and data-driven solutions for cyberattack detection in IoT systems. Additionally, the section introduces the potential application of lightweight deep learning (DL) models for anomaly detection on embedded devices. Anomaly Detection in IoT and Edge Devices examines the challenges and solutions involved in detecting anomalies in IoT and edge devices. It covers the limitations of traditional cloud computing methods in handling the high latency, computational demands, and power consumption of large-scale IoT systems. The section also introduces fog computing as a promising architecture to address these issues by enabling partial data processing at intermediate IoT devices. ML models for anomaly detection explores the role of ML in anomaly detection, emphasizing the importance of rapid convergence for models deployed on IoT endpoints. The section emphasizes the need for accurate prediction models that strike a balance between false positives and false negatives when interpreting real-time sensor data effectively.

TinyML applications in anomaly detection presents an innovative approach to anomaly detection using TinyML, a cutting-edge technology that enables the execution of ML models on resource-constrained devices. It introduces a one-class neural network model trained on publicly available image data from social networks and X-ray scans, demonstrating the potential of TinyML for enhancing anomaly detection in dynamic IoT environments. "Lessons Learned" reflects the critical insights gained from developing and deploying anomaly detection algorithms in the context of IoT and EC. This section discusses the technical, operational, and strategic lessons from working with constrained devices and provides valuable guidance for future research. "Future Scope and Conclusion" concludes the chapter by discussing the future potential of anomaly detection in IoT ecosystems, with a focus on integrating TinyML into edge devices. The section highlights the opportunities for improving scalability, adaptability, and efficiency in anomaly detection models and proposes avenues for further research to address the dynamic needs of IoT systems.

3.2 Context and Literature Review

The integration of TinyML into modern technological ecosystems is gaining momentum as the demand for edge intelligence increases. TinyML's ability to deliver ML capabilities in resource-constrained environments aligns with the growing proliferation of IoT devices and the need for real-time data processing. This section provides a detailed contextual analysis and a review of relevant literature to establish the foundation and significance of TinyML in anomaly detection, highlighting key developments, challenges, and research gaps. EC has emerged as a critical paradigm in the era of IoT, where data is generated at an unprecedented scale. Traditional cloud-centric approaches to data processing are often inadequate due to latency, bandwidth limitations, and concerns about privacy. EC addresses these issues by processing data closer to its source, enabling faster decision-making and reducing dependency on centralized infrastructure [16].

TinyML extends this concept by embedding ML capabilities into edge devices, making them intelligent and self-sufficient. Moreover, anomaly detection plays a crucial role in various domains, including industrial automation, healthcare, and cybersecurity. Detecting irregularities or deviations from standard patterns helps prevent potential failures, identify fraudulent activities, and ensure the reliability of systems. Traditional anomaly detection techniques rely on cloud-based systems that may not meet the real-time requirements of critical applications. TinyML provides a transformative solution by enabling anomaly detection directly on edge devices, ensuring timely interventions while preserving energy efficiency and data privacy. Recent advancements in hardware, such as the development of low-power microcontrollers and software, including lightweight ML frameworks like TensorFlow Lite for Microcontrollers, have accelerated the adoption of TinyML. These technologies enable the deployment of complex ML models on devices with limited computational and memory resources, making them suitable for applications that require real-time anomaly detection [17].

3.2.1 TinyML: An Overview

ML techniques can identify irregularities in datasets, including accelerometer readings. Furthermore, advancements in hardware technology have enabled embedded microcontroller units (MCUs) to process sophisticated ML algorithms on energy-efficient embedded systems. TinyML represents one such innovative framework, facilitating the deployment of advanced neural network models while utilizing only milliwatts of power. As illustrated in Figure 3.1, TinyML lies at the crossroads of embedded system engineering and ML methodologies. This framework enables the adaptation of ML models to low-power embedded devices, including consumer products like Alexa, as well as more compact, sensor-based systems such as controllers for autonomous vehicles. At the most energy-efficient end of this spectrum are applications operating below 1 mW, which can function for extended periods using a tiny coin-cell battery.

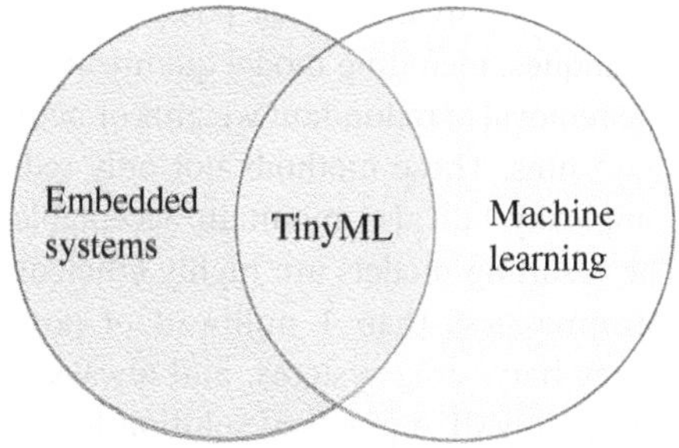

Figure 3.1 Frameworks for machine learning models.

TinyML encompasses ML structures, methodologies, tools, and strategies designed for conducting on-device analytics across diverse sensing modalities (e.g., vision, audio, motion) on battery-powered, low-energy devices. Transmitting continuous microcontroller data to centralized data centers for processing and feedback is often impractical due to network bandwidth limitations and the significant battery consumption involved in sending and receiving data packets over network interfaces [18, 19]. By enabling AI functionality directly at the data source, TinyML mitigates these challenges, significantly reducing communication overhead while addressing concerns related to bandwidth and energy consumption. This paradigm emphasizes integrating ML-driven mechanisms within compact, MCU-powered devices. TinyML is a subfield of ML that focuses on deploying ML models on ultralow-power hardware, such as microcontrollers and embedded systems. These devices typically operate with limited computational resources and memory, making traditional ML techniques impractical. TinyML bridges the gap by optimizing models to run efficiently on these constrained devices, enabling intelligent decision-making directly at the edge. This overview delves into the definition and key features of TinyML, its advantages for edge devices, and the limitations and challenges associated with its implementation [20].

3.2.2 Definition and Key Features

TinyML can be defined as the deployment of ML algorithms on resource-constrained devices that consume minimal power, often measured in milliwatts. These devices are typically equipped with microcontrollers that operate in environments where network connectivity may be intermittent or unavailable. TinyML enables local data processing, making it ideal for applications requiring low latency, real-time inference, and energy efficiency [21, 22].

TinyML brings significant advantages in resource efficiency and energy-aware computing by enabling ML inference directly on ultralow-power microcontrollers and edge devices. These models are specifically engineered to operate within severe constraints of memory (often <256 KB) and processing power (typically

<100 MHz). To make this possible, developers utilize a suite of optimization techniques, including model quantization (e.g., 8-bit or binary weights), pruning (the removal of redundant weights or neurons), and advanced model compression algorithms. These methods not only reduce the model size and computational complexity but also maintain acceptable levels of accuracy for targeted tasks. The resulting models are highly efficient, allowing execution on hardware that consumes less than 1 milliwatt of power, ideal for battery-operated devices, energy-harvesting systems, and always-on sensing applications. This efficiency makes TinyML a practical solution for deploying intelligent systems in environments where conventional DL models would be computationally or economically infeasible.

A key benefit of TinyML is its ability to perform on-device inference, eliminating the need to continuously transmit raw data to cloud-based servers for processing. This localized computation significantly reduces communication overhead, bandwidth usage, and latency factors that are critical for time-sensitive applications such as anomaly detection, predictive maintenance, and real-time health monitoring. Furthermore, real-time decision-making becomes feasible because data is processed as it is generated, enabling immediate responses to anomalous conditions or critical events. The decentralized nature of TinyML also enhances system scalability, allowing intelligent models to be deployed across millions of edge devices without overloading centralized infrastructure. In addition, by keeping sensitive data on the device, TinyML supports robust privacy and security postures, mitigating the risk of data interception or breaches during transmission. This is particularly important in domains like healthcare and industrial monitoring, where data confidentiality and integrity are paramount.

3.2.3 Advantages of TinyML for Edge Devices

The integration of TinyML into edge devices offers several advantages, which make it a transformative technology for various applications. Some of these advantages are detailed below [20, 23].

TinyML enables real-time inference by allowing ML models to process data directly on edge devices, eliminating the latency associated with cloud-based analytics. This capability is vital in latency-sensitive applications, such as industrial automation, where the immediate detection of anomalies, like unusual vibrations or temperature spikes, can prevent equipment failure and unplanned downtime. By executing inference on-device with minimal delay, TinyML ensures timely decision-making, thereby improving operational efficiency and safety. In applications like autonomous drones, medical monitoring, and smart surveillance, this real-time responsiveness is not just beneficial but often mission-critical, enabling systems to adapt instantly to dynamic environments.

Another key advantage of TinyML is its ultralow power consumption, which enables the prolonged operation of battery-powered devices in the field. TinyML models are specifically optimized to run on microcontrollers and specialized AI accelerators that consume minimal energy, often in the milliwatt range. This energy efficiency is particularly valuable in IoT deployments across remote or hard-to-reach environments, such as agricultural fields, wildlife reserves, and environmental monitoring stations, where frequently changing or recharging batteries is impractical. Additionally, by processing data locally, TinyML reduces the need for continuous wireless communication, further conserving power. This autonomy not only extends device lifespan but also promotes sustainability by lowering the energy demands of large-scale distributed sensor networks.

Beyond performance and efficiency, TinyML delivers strategic benefits in terms of infrastructure, privacy, and scalability. By offloading computation from centralized servers to edge nodes, TinyML significantly reduces dependence on cloud infrastructure, minimizing bandwidth usage, cloud storage costs, and vulnerability to connectivity disruptions. This decentralization enhances system resilience, especially in settings with unreliable or intermittent internet access. Local data processing also offers enhanced privacy and data security, which are critical in sensitive domains such as healthcare, smart homes, and financial services, by avoiding the transmission of personally identifiable information over networks. Furthermore, TinyML's lightweight footprint makes it highly scalable, enabling seamless deployment across millions of heterogeneous devices without centralized orchestration. Its flexibility allows developers to customize models for domain-specific use cases, ensuring performance optimization even within stringent hardware constraints. Collectively, these advantages make TinyML a cost-effective, environmentally responsible, and highly adaptable solution for deploying intelligent, real-time systems across diverse sectors.

3.2.4 Limitations and Challenges of TinyML

While TinyML offers substantial advantages for edge devices, its adoption and effectiveness come with a set of challenges that need to be addressed. These challenges primarily stem from the inherent limitations of the hardware, the complexities of model optimization, and the constraints on data handling, deployment, and security [24, 25].

3.2.4.1 Hardware Constraints

TinyML, the deployment of ML models on ultralow-power microcontrollers and edge devices, introduces a unique set of technical challenges stemming from the extreme constraints on computational resources, memory capacity, and power consumption. These devices often operate with limited CPU speeds – typically in the range of tens to hundreds of megahertz, and may provide only a few hundred kilobytes of RAM and flash memory. As a result, only highly efficient, lightweight models can be executed, severely limiting the feasibility of deploying large-scale architectures such as deep CNNs or recurrent neural networks (RNNs) commonly used in conventional anomaly detection.

To accommodate these constraints, developers must employ a range of model compression and optimization strategies, including quantization (e.g., 8-bit or lower precision weights and activations), pruning (removing redundant parameters), knowledge distillation (transferring knowledge from a larger teacher model to a smaller student model), and architecture design tailored for embedded inference (e.g., MobileNet, Tiny-YOLO, or custom compact neural networks). While these techniques significantly reduce memory footprint and computational demands, they can also lead to degradation in model fidelity and generalization performance, particularly when the data distribution is nonstationary or complex, as is often the case in real-world anomaly detection scenarios.

Moreover, the quality and consistency of input data, typically captured from onboard sensors, directly influence the efficacy of TinyML models. Sensor imperfections such as drift, limited resolution, or susceptibility to environmental interference (e.g., temperature, electromagnetic noise, or mechanical vibration) can introduce variability and noise into the data stream. Given that TinyML models are often trained on relatively small and static datasets due to connectivity and storage constraints, they are especially vulnerable to such variability. This increases the risk of false positives or negatives in anomaly detection, especially when models are deployed in dynamic environments without robust mechanisms for online learning or adaptation.

Addressing these challenges requires a holistic codesign approach that aligns model architecture, sensor selection, data preprocessing pipelines, and hardware capabilities. Techniques such as adaptive thresholding, on-device filtering, and hierarchical inference (offloading complex decisions to more capable nodes when necessary) are increasingly being explored to balance performance with resource efficiency. Ultimately, while TinyML holds significant promise for enabling pervasive, real-time anomaly detection at the edge, realizing its full potential demands careful engineering trade-offs and rigorous optimization across the entire stack, from data acquisition to inference execution.

3.2.4.2 Model Optimization Challenge

One of the most critical and technically demanding challenges in TinyML lies in achieving an optimal trade-off between model accuracy and computational efficiency. Deploying ML models on ultralow-power, resource-constrained devices, such as microcontrollers with limited memory (often less than 256 KB of RAM), processing capacity, and energy budgets, necessitates aggressive optimization strategies. These strategies, while essential for feasibility, often compromise the model's representational capacity and generalization performance.

Model compression techniques such as pruning, quantization, and knowledge distillation are frequently employed to reduce model size, latency, and power consumption. Pruning involves removing redundant or less significant weights or neurons, thereby reducing the model's complexity. Quantization converts model weights and activations from high-precision (e.g., 32-bit floating-point) to lower-precision formats (e.g., 8-bit integers), significantly decreasing memory footprint and computational demand. Knowledge distillation transfers knowledge from a larger, more expressive model (the teacher) to a smaller, more efficient model (the student). While these methods are effective in enabling model deployment on edge hardware, they often introduce trade-offs that can lead to degraded performance in terms of accuracy, robustness, or inference consistency, especially in tasks such as anomaly detection, where even minor deviations can be critical.

Moreover, the development of lightweight yet effective models for TinyML environments requires a nuanced understanding of both advanced ML principles and embedded systems architecture. Engineers must account not only for algorithmic considerations but also for hardware-specific constraints such as memory access patterns, processor instruction sets, and energy consumption per operation. Techniques like model architecture search, hardware-aware neural architecture optimization, and in-situ training are emerging areas of research aimed at bridging this gap.

The complexity of designing TinyML algorithms that maintain high inference fidelity under strict resource limitations is compounded by the lack of standardization across embedded platforms and the difficulty of real-time benchmarking. As such, ensuring that models retain sufficient predictive power without exceeding computational budgets remains a major research frontier. This challenge underscores the necessity for interdisciplinary collaboration and continued innovation in model compression, compiler optimization, hardware acceleration (e.g., via DSPs or NPUs), and codesign methodologies that jointly optimize hardware and software for TinyML use cases.

3.2.4.3 Data Availability and Preprocessing Constraints

Another significant challenge in deploying TinyML models on edge devices is the limited availability and quality of data, particularly for anomaly detection tasks in specialized or domain-specific applications. Data collection in such environments often requires extensive domain knowledge, prolonged monitoring periods, and customized sensor setups to capture relevant events, which may occur infrequently. Furthermore, labeled datasets are especially scarce for anomalies, as these events are by nature rare and diverse, making supervised learning approaches difficult to scale without synthetic data generation or expert annotation, both of which can be resource-intensive.

In addition to data scarcity, the constrained computational and memory capacities of edge devices severely limit the extent and sophistication of on-device preprocessing. Traditional preprocessing pipelines such as normalization, filtering, dimensionality reduction, and time-series transformations (e.g., FFT, wavelet decomposition), can be prohibitively expensive in terms of processing power and memory footprint. Consequently, many TinyML deployments are forced to rely on minimally processed or even raw sensor data, which can degrade model performance due to the presence of noise, outliers, or redundant features.

On-device feature extraction techniques, which are crucial for enhancing signal-to-noise ratio and reducing input dimensionality, must be carefully optimized to fit within stringent resource budgets. This includes employing lightweight statistical features (e.g., mean, variance, and peak-to-peak) or using low-complexity embedded algorithms such as moving averages or fixed-size sliding windows. However, the trade-off between preprocessing fidelity and computational feasibility poses a significant bottleneck. The inability to effectively process data before inference may lead to lower anomaly detection accuracy, higher false positives, and reduced model robustness.

To address these constraints, emerging strategies include hierarchical processing architectures, where lightweight feature extraction is performed at the edge, with more complex analytics off-loaded to gateway or cloud infrastructure when needed. Additionally, research is ongoing into ultra-efficient preprocessing algorithms tailored for MCUs and co-optimization approaches that jointly design preprocessing and model architectures to maximize performance under strict energy and memory constraints.

3.2.4.4 Deployment Challenges

The deployment of TinyML models in real-world environments introduces a range of complex technical challenges that must be carefully addressed to ensure reliability, scalability, and security. One of the primary concerns lies in ensuring cross-platform compatibility across diverse hardware architectures. Edge devices often utilize a heterogeneous mix of microcontrollers (e.g., ARM Cortex-M series, RISC-V cores) with significant variation in processing capabilities, available RAM/flash memory, and supported instruction sets. This diversity complicates the development and optimization of TinyML models, as models must be tailored to operate efficiently within the strict resource constraints of each target platform. Furthermore, hardware-specific dependencies may limit portability, necessitating the use of abstraction layers or hardware-agnostic frameworks such as TensorFlow Lite for Microcontrollers or CMSIS-NN to streamline deployment.

Another critical challenge involves the scalability of TinyML deployments in distributed IoT networks. While TinyML enables large-scale, decentralized analytics, managing these deployments across thousands of geographically dispersed devices introduces significant operational overhead. Over-the-air (OTA) model updates, firmware management, and configuration synchronization become increasingly complex when dealing with constrained devices that may have intermittent connectivity, limited bandwidth, or strict power budgets. Ensuring consistency in model performance and reliability across such a fragmented landscape requires robust orchestration mechanisms and lightweight update protocols tailored for embedded environments.

Additionally, security considerations are paramount in TinyML applications. Although local on-device inference enhances data privacy by minimizing the transmission of sensitive information to centralized servers, it introduces new attack vectors. Physical access to devices exposes them to potential threats such as side-channel attacks, firmware extraction, or model inversion. Adversaries may attempt to tamper with the model, inject adversarial inputs, or manipulate sensor data to produce incorrect inferences. To mitigate these risks, hardware-based security modules (e.g., secure enclaves, trusted execution environments), lightweight cryptographic algorithms, and secure boot mechanisms must be integrated into the device lifecycle. Furthermore, models themselves must be hardened against adversarial attacks, ensuring robustness even in the presence of noisy or manipulated inputs.

Collectively, these challenges underscore the importance of a holistic deployment strategy that incorporates hardware-aware optimization, scalable system design, and end-to-end security practices. Addressing these concerns is essential to unlocking the full potential of TinyML in mission-critical anomaly detection and edge intelligence applications.

3.2.4.5 Development and Testing Challenges

Developing and testing TinyML applications necessitate specialized tools and frameworks that are still in the process of maturation. Unlike traditional ML platforms, which benefit from well-established and feature-rich development environments, TinyML faces significant limitations due to the constraints of edge devices. These limitations, such as reduced computational power, memory, and storage capacity, exacerbate the complexities of debugging and testing ML models. Currently, TinyML development lacks the maturity of its mainstream counterparts, resulting in the absence of integrated, unified environments that can seamlessly support the end-to-end development lifecycle – from model training to deployment and testing.

In the context of TinyML, debugging and testing are further complicated by the real-time constraints inherent to EC. Gathering feedback from these resource-constrained devices is often challenging, as developers are unable to rely on the standard debugging tools commonly used in desktop or cloud-based ML environments. In addition, the devices themselves are typically deployed in distributed and physically remote locations, making direct access for real-time monitoring difficult.

The small form factor of TinyML devices means they often have limited input/output interfaces and reduced debugging capabilities. As a result, developers are unable to easily examine intermediate model states, which makes pinpointing issues like memory overflows, inaccurate predictions, or computational bottlenecks more time-consuming and error-prone. Furthermore, the absence of high-fidelity data logs, often a staple in traditional ML workflows, forces developers to rely on more manual, time-intensive methods for gathering insights into model performance and system behavior.

Additionally, the need to optimize for low power consumption and real-time processing further complicates the testing process. Since the models must be highly efficient, achieving a balance between model accuracy, resource usage, and inference speed introduces a set of trade-offs that are difficult to evaluate and iterate on rapidly. TinyML frameworks must also cater to hardware-specific constraints, meaning that performance optimization is often tied to specific device architectures, which introduces a degree of fragmentation in development.

To address these challenges, developers are increasingly adopting novel techniques such as in-situ validation, where small portions of the model or data are tested on actual devices in real-time environments. Some frameworks integrate hardware-in-the-loop (HIL) testing, which allows simulated inputs to be processed by real hardware to verify model accuracy in a controlled environment before full deployment. Additionally, lightweight, cloud-connected debugging tools are becoming increasingly important for monitoring deployed TinyML models remotely, enabling more effective debugging without the need for physical access.

Ultimately, the iterative development cycle in TinyML is slower and more complex than in traditional ML, requiring developers to adopt agile, resource-efficient testing strategies. As TinyML frameworks continue to evolve, we can expect the emergence of more sophisticated debugging tools and improved integration with edge device architectures, leading to faster development cycles and more robust deployment of ML models in resource-constrained environments.

3.2.5 Related Work

The concept of deploying ML on resource-constrained devices dates to efforts aimed at optimizing ML algorithms for embedded systems. Initial research focused on model compression techniques such as pruning and quantization to reduce model size and computational demands. Antonini et al. [26] pioneered this approach by introducing a pruning method that significantly reduced the size of neural networks without compromising accuracy. Several frameworks have been developed to facilitate the deployment of TinyML models. Among them, TensorFlow Lite for Microcontrollers, introduced by Google, is one of the most widely used frameworks. It supports the development of lightweight ML models tailored for microcontrollers. Other frameworks, such as Edge Impulse and MicroTVM, offer tools for data preprocessing, model training, and deployment, catering to the unique requirements of edge devices. These tools have been instrumental in advancing TinyML's capabilities for anomaly detection [27].

In industrial settings, TinyML has been applied to detect equipment malfunctions and optimize maintenance schedules. Genge et al. [28] demonstrated the use of TinyML for vibration analysis in rotating machinery. Their study revealed that locally processed data could identify anomalies in real-time, thereby reducing downtime and maintenance costs. TinyML has shown promise in healthcare applications, particularly in wearable devices. Šabić et al. [29] developed a TinyML-based system for detecting cardiac arrhythmias using electrocardiogram (ECG) signals. Their findings highlighted the potential of on-device intelligence to provide timely alerts and reduce dependency on centralized monitoring systems. In environmental systems, TinyML has been used to detect anomalies in parameters such as air quality and temperature. A study by Nguyen et al. [30] implemented TinyML on environmental sensors to identify pollution events. Localized processing reduced data transmission costs and enabled rapid responses to environmental changes. Hayajneh et al. [22] explored the application of TinyML in less-studied domains, such as smart agriculture and space exploration. They demonstrated how anomaly detection models could be tailored for domain-specific challenges, broadening the scope of TinyML research.

Guo et al. [31] focused on energy-efficient implementations of TinyML for detecting anomalies in IoT devices. Their study integrated energy-harvesting technologies with TinyML to enable sustainable deployments in remote areas. Alwaisi et al. [32] investigated the security challenges associated with deploying TinyML models for anomaly detection. They identified vulnerabilities to physical and adversarial attacks and proposed strategies for enhancing model robustness. Njor et al. [33] explored the use of TinyML for predictive maintenance in manufacturing. They developed lightweight models to monitor machine health and predict failures, demonstrating the scalability of TinyML for industrial applications. Andrade et al. [34] proposed an adaptive optimization technique for TinyML models used in anomaly detection. Their method dynamically adjusted model parameters based on available resources, improving efficiency without compromising accuracy. Baciu et al. [35] introduced a benchmarking framework for evaluating TinyML models across various tasks, including anomaly detection.

Their work provided a standardized approach for comparing model performance, addressing a critical gap in the field.

Acquaah and Kaushik [36] implemented TinyML on environmental sensors to detect anomalies in air quality and temperature. Their system showcased the advantages of localized processing in reducing data transmission costs and enabling rapid responses to environmental changes. Rostami et al. [37] applied TinyML in wearable healthcare devices to detect cardiac arrhythmias using ECG signals. Their approach demonstrated the feasibility of on-device intelligence for continuous monitoring, enabling timely alerts without reliance on centralized systems. Chen et al. [38] developed a TinyML-based solution for detecting anomalies in rotating machinery through vibration analysis. Their study highlighted the real-time capabilities of TinyML in identifying faults, reducing downtime, and minimizing maintenance costs in industrial environments. Liang et al. [39] introduced techniques for model compression, including pruning and quantization, to reduce the size of neural networks. While their work was not explicitly focused on anomaly detection, it laid the groundwork for deploying ML models on resource-constrained devices, forming the basis for TinyML applications.

Anomaly detection on constrained devices is also an active area of research. As shown in Table 3.1, several existing works have been utilized to reduce the resource requirements of traditional anomaly detection algorithms. For example, A. Rehman et al. proposed a CNN-based approach to monitor industrial machinery, E. Park et al. designed an online anomaly detection system based on PCA and MEM for IoT devices. The techniques are primarily implemented on application-specific hardware such as specialized sensing nodes, FPGAs, or GPUs. Therefore, the potential application of these methods in general IoT devices, such as microcontrollers, has not been extensively studied. With the development of lightweight DL models specifically designed for embedded devices, our work has shown the potential of adding advanced anomaly detection algorithms to the edge of IoT infrastructure [44].

Detecting anomalies in time-series data is crucial in various real-world applications, including equipment maintenance and fraud detection. Many time-series anomaly detection algorithms have been proposed in the literature. We can divide these algorithms into two categories based on whether they use a labeled training dataset (termed supervised), such as in the multi-scale convolutional-LSTM networks (MTAD) or the LSTM_Autoencoder, or an unlabeled training dataset (termed unsupervised), such as in the denoising-LSTM-autoencoder (DLSTMAE) or the convolutional autoencoder with novelty detection (AD-CAE) [45].

3.2.6 Anomaly Detection in IoT and Edge Devices

Numerous efforts have been made to define and characterize anomalous data. One widely accepted definition comes from Hawkins, who describes an outlier as an

Table 3.1 Comprehensive summary of key related works in anomaly detection for constrained and IoT devices.

Category	Description	Key related works	References
Anomaly detection on constrained devices	Explores reducing the resource requirements for anomaly detection algorithms, primarily focusing on hardware-specific implementations	A CNN-based approach for industrial machinery monitoring. Online anomaly detection using PCA and MEM for IoT devices. Specialized nodes, FPGAs, and GPUs	[40]
Time-series anomaly detection	It addresses real-world applications, such as equipment maintenance and fraud detection, using both labeled (supervised) and unlabeled (unsupervised) training datasets	Supervised: Multi-scale convolutional-LSTM networks (MTAD), LSTM_Autoencoder. Unsupervised: Denoising-LSTM-autoencoder (DLSTMAE), AD-CAE (convolutional autoencoder with novelty detection)	[41]
IoT and edge devices	It focuses on real-time anomaly detection for IoT devices, leveraging local computation through TinyML to reduce dependency on cloud resources while enhancing privacy and security	Lightweight models for real-time, on-device anomaly detection. Models for low power and memory (e.g., RNNs, CNNs for complex patterns). Reducing model size	[42]
TinyML applications	Innovative deployment of machine learning models for edge-based anomaly detection, particularly for low-power IoT devices, emphasizing methods like quantization for enhanced efficiency	One-class neural network models trained on public datasets (e.g., JSA database, social network image data). Industrial equipment monitoring and sensor data analysis	[20]
Zero-configuration computing	It highlights the role of zero-configuration computing in security in IoT systems, enabling self-adaptive and responsive network adjustments to safeguard data transfer	Adds security by dynamically identifying anomalies and optimizing resource allocation and network adjustments	[43]

observation that significantly deviates from the rest of the dataset, to the extent that it raises suspicion of having been generated by a different underlying mechanism. Similarly, Barnett and Lewis define an outlier as a data point or subset of data that appears inconsistent with the general pattern observed in the dataset. These definitions emphasize the idea that anomalies, or outliers, represent data points that differ considerably from expected trends, often indicating an irregularity in the data collection process or a significant change in the system being monitored. In the context of the IoT, anomalies are particularly important as they reflect measurable deviations from a system's expected state. These deviations may occur due to unexpected environmental changes, system faults, cyberattacks, or other disruptions. Unlike static datasets, IoT-generated data streams are dynamic, meaning that the definition of "normal" behavior is not fixed but evolves over time. This evolving nature of IoT data presents a unique challenge for anomaly detection, as methods must adapt to shifts in what is considered normal operation while still identifying truly abnormal events. Anomalies in time-series data can generally be classified into three main categories as depicted in Figure 3.2 [46].

1) **Point Anomalies:** These occur when a single observation significantly deviates from the normal pattern of the dataset. Such anomalies are isolated data points that stand out in contrast to surrounding values. For instance, in network traffic analysis, a sudden spike in data transfer rates could indicate unauthorized access or a cyberattack.
2) **Contextual Anomalies:** These anomalies depend on the context in which the data is observed. A data point may appear normal in one scenario but be

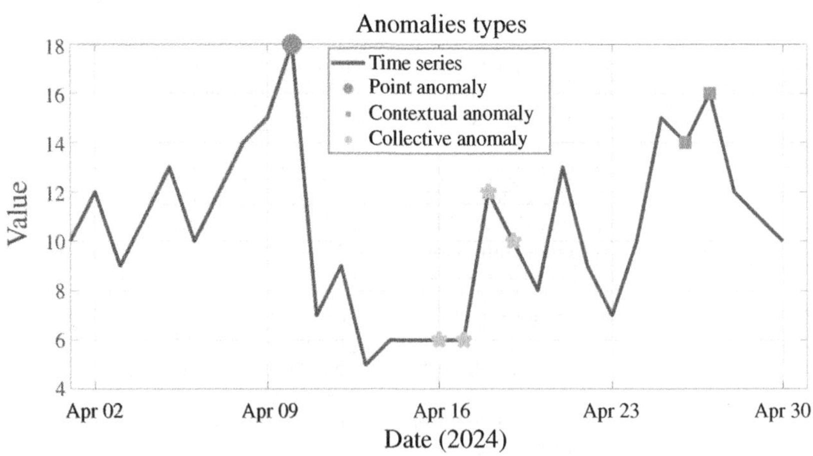

Figure 3.2 Detection of anomalies in time-series data: highlighting point, contextual, and collective anomalies over time.

considered anomalous in another. For example, a high temperature reading from a sensor might be normal during summer but anomalous during winter. Contextual anomalies require understanding the temporal or spatial context to make an accurate assessment.

3) **Collective Anomalies:** Unlike point anomalies, collective anomalies involve a group of data points that together form an unusual pattern, even though individual points within the group may not appear anomalous on their own. This type of anomaly is often seen in sequential data, such as a series of irregular fluctuations in a stock market trend or repeated failures in an industrial system.

To effectively identify these anomalies in time-series data, researchers have developed various detection techniques. While some approaches blend multiple strategies, most methods can be categorized into distinct groups.

- **Statistical and Probabilistic Methods:** These techniques rely on historical data to construct models of expected behavior. Any new observations that significantly deviate from this model are flagged as anomalies. These methods are particularly useful for detecting rare but extreme events.
- **Pattern Matching:** This approach involves creating a database of labeled anomalies and comparing new observations against these pre-identified patterns. Supervised learning techniques often leverage pattern matching for accurate anomaly detection in domains where labeled data is available.
- **Distance-Based Methods:** These methods use mathematical distance metrics to measure the similarity between new observations and previously recorded data points. If a new data point is significantly different from past observations, it is classified as an anomaly.
- **Clustering Techniques:** Clustering methods group similar data points together in a multidimensional space. Data points that do not belong to any dense cluster are considered anomalies. This technique is particularly useful for unsupervised anomaly detection, where labeled data is unavailable.
- **Predictive Models:** These models use regression techniques to forecast future values based on recent trends. When a new observation deviates significantly from the predicted value, it is flagged as an anomaly. Predictive modeling is effective in applications such as financial forecasting and predictive maintenance.
- **Ensemble Methods:** These approaches combine multiple anomaly detection techniques to improve accuracy. By using a voting mechanism, ensemble methods enhance robustness but may introduce increased computational complexity.

The choice of an anomaly detection method depends on the nature of the dataset, the environment in which the detector is deployed, and the specific characteristics of the anomalies being targeted. As IoT systems continue to generate vast amounts of real-time data, advanced anomaly detection techniques are becoming increasingly essential for ensuring system reliability, security, and efficiency.

Edge devices and gateways serve as critical components in generating substantial volumes of data within IoT ecosystems. These devices, often characterized by their resource constraints, are responsible for collecting data from sensors and actuators distributed across various environments, from industrial settings to smart cities. However, traditional cloud computing architectures are not well-equipped to handle the unique challenges posed by IoT systems, such as high latency, significant computational load, and the power consumption associated with processing vast amounts of data in real time. In many IoT applications, sending all data to the cloud for centralized processing introduces unacceptable delays, as well as heavy bandwidth usage, which could overwhelm both the network infrastructure and cloud resources. Additionally, the high energy consumption of cloud-based systems becomes impractical for large-scale IoT deployments that require continuous, low-power operation.

To mitigate these issues, fog computing architectures have been explored as a more effective solution. Fog computing extends the cloud computing paradigm by introducing a decentralized computing layer, where intermediate devices – often termed "fog nodes" perform partial or full data processing closer to the source of data generation. These edge or fog nodes are strategically positioned between IoT devices and the cloud, allowing for quicker localized data analysis, reducing latency, conserving bandwidth, and alleviating cloud processing bottlenecks. This intermediate layer enhances the ability to handle the real-time processing demands of IoT systems, as well as enabling the deployment of lightweight, context-aware applications on resource-constrained edge devices. Despite the adoption of fog computing for distributed data processing, a substantial volume of data still requires transfer to the cloud for deeper analytics, long-term storage, and further decision-making processes. This data transfer between the edge and cloud introduces significant challenges related to security and privacy. The sheer scale and heterogeneity of IoT devices increase the potential attack surface, making IoT systems more vulnerable to cyber threats such as data breaches, unauthorized access, and malicious manipulation. Moreover, privacy concerns arise due to the sensitive nature of the data collected from various sources, including health monitors, industrial sensors, and smart home devices.

To address these pressing concerns, zero-configuration computing has emerged as a promising solution. Zero-configuration computing focuses on simplifying device onboarding and management in dynamic IoT environments while ensuring that security is not compromised. This approach enables IoT devices to operate autonomously without the need for complex setup or manual configuration, which is especially critical in environments where scalability and rapid deployment are essential. The security aspect of zero-configuration computing involves anomaly detection mechanisms that identify and mitigate potential security risks at the edge, reducing the need for data to be transmitted to centralized systems for processing. These mechanisms may include ML models that detect abnormal patterns in data, monitor system behavior for unusual activities, and provide adaptive responses to emerging threats. By integrating zero-configuration computing within IoT and fog computing architectures, the security and privacy of data can be enhanced, ensuring that sensitive information remains protected as it travels between devices and the cloud. This approach not only strengthens the integrity of the system but also improves the overall resilience of IoT networks, enabling faster identification of malicious activity or system failures. Furthermore, zero-configuration computing facilitates more efficient use of network resources by preventing unnecessary data transfers, thereby contributing to improved system performance and reduced energy consumption.

The combination of edge and fog computing, alongside the integration of zero-configuration computing, provides a robust framework for addressing the unique challenges posed by IoT systems. It enables secure, low-latency, and energy-efficient processing at the edge while maintaining the scalability and analytical capabilities provided by the cloud, ensuring that IoT ecosystems can operate securely and effectively at scale. This, in turn, results in optimized resource allocation from the cloud and network adjustments that are more responsive and adaptive to varying circumstances [47].

One potentially promising approach to anomaly detection on the IoT and edge devices is TinyML. This revolutionary field encompasses deploying ML models directly onto the devices themselves. This groundbreaking methodology enables efficient and real-time data analysis of the devices' data. It drastically reduces the need for constant communication with cloud servers, providing a significant advantage in terms of privacy, security, and cost-effectiveness. By empowering IoT and edge devices with the ability to perform sophisticated ML tasks locally, the use of TinyML brings about a paradigm shift in anomaly detection, unlocking immense potential for a range of applications such as predictive maintenance, fault detection, and optimization of IoT networks [48].

With the integration of ML models at the edge, these devices can autonomously detect and flag anomalies or irregularities within their operational parameters, enabling prompt and targeted responses. Furthermore, deploying TinyML enables devices to learn and adapt continuously to evolving patterns, ensuring long-term accuracy and reliability in anomaly detection. As the power and capabilities of edge devices continue to expand, the field of TinyML presents an exciting frontier with immense possibilities for enhancing anomaly detection in the IoT ecosystem. One potential and highly promising approach to anomaly detection in the IoT and edge devices is the utilization of TinyML [27]. Figure 3.3 illustrates how to enhance IoT network efficiency through autonomous anomaly detection.

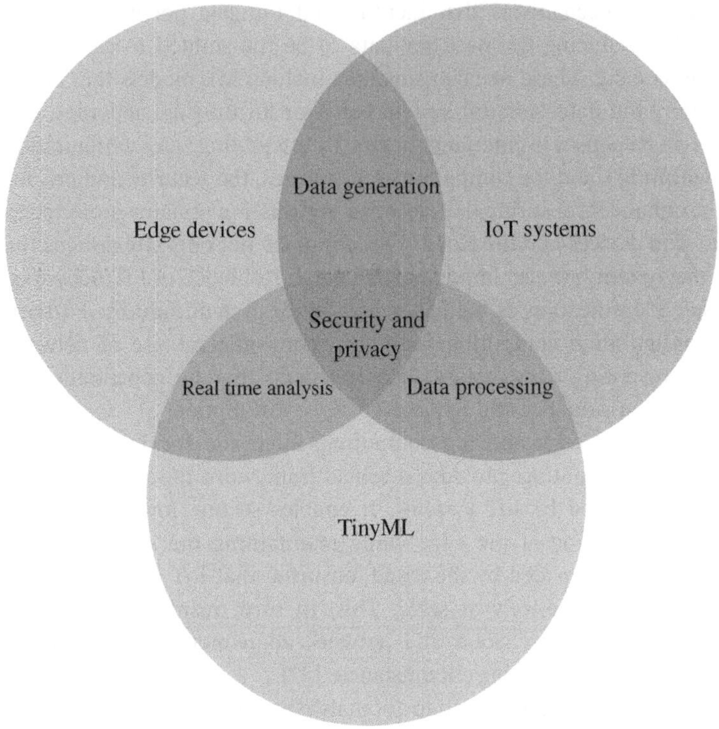

Figure 3.3 Enhancing IoT network efficiency with autonomous anomaly detection.

This revolutionary and groundbreaking field encompasses deploying ML models directly onto the devices themselves, enabling efficient and real-time data analysis of these devices' data. This methodology revolutionizes the approach to anomaly detection, ushering in a paradigm shift throughout the entire IoT ecosystem. One potentially promising approach to anomaly detection on the IoT and edge devices is TinyML. This revolutionary and groundbreaking field encompasses deploying ML models directly onto the devices themselves, enabling efficient and real-time data analysis of these devices' data. This methodology revolutionizes the approach to anomaly detection, ushering in a paradigm shift throughout the entire IoT ecosystem. By enabling real-time anomaly detection directly on edge devices, TinyML allows immediate responses to potential threats and reduces the need for constant data transmission to a central server. This can significantly improve response times and reduce the burden on network infrastructure. By empowering IoT and edge devices to perform sophisticated ML tasks locally, TinyML unlocks immense potential for various applications, such as predictive maintenance, fault detection, optimization of IoT networks, and proactive anomaly identification [49].

The integration of cutting-edge ML algorithms at the edge allows these devices to autonomously and intelligently detect and flag anomalies or irregularities within their operational parameters with unprecedented accuracy and timeliness. This level of autonomy and intelligence significantly reduces the dependence on constant communication with cloud servers, enhancing privacy, security, and cost-effectiveness. Moreover, deploying TinyML on IoT and edge devices enables them to learn and adapt to evolving patterns continuously in real-time, ensuring long-term accuracy, reliability, and robustness in anomaly detection [50]. With the exponential growth in the power and capabilities of edge devices, the field of TinyML presents an exciting frontier with limitless possibilities for enhancing anomaly detection in the IoT ecosystem. It enables organizations to harness and leverage the full potential of TinyML, thereby empowering them with remarkable improvements in efficiency, reliability, and intelligence within their IoT deployments. This translates into proactive identification and mitigation of anomalies, drastically reducing downtime, optimizing resource allocation, and improving overall operational efficiency. TinyML represents a groundbreaking approach that reshapes the landscape of anomaly detection in the IoT and edge device domains. Its deployment enables edge devices to autonomously and intelligently analyze, detect, and address anomalies with unparalleled accuracy and speed [51].

By eliminating the need for constant reliance on cloud servers, TinyML enhances privacy, security, and cost-effectiveness while enabling continuous learning and adaptation to evolve patterns. As the future unfolds and opportunities in the IoT ecosystem expand, TinyML stands at the forefront, providing organizations with a unique competitive edge in terms of efficiency, reliability, and intelligence. The potential of TinyML to revolutionize anomaly detection in the IoT ecosystem is undeniable, making it a vital and transformative field that holds immense promise for the future of IoT deployments. Integrating TinyML with edge devices enables real-time anomaly detection and response, thereby enhancing overall system reliability and security. This transformative power has the incredible potential to revolutionize numerous industries, elevating them to unprecedented heights and fundamentally improving our everyday lives in unimaginable ways.

3.2.7 Structured Literature Mapping Approach

A structured and rigorous methodology is essential when conducting a systematic literature mapping (SLM) to ensure that the gathered data is reliable and useful for addressing research questions. By following a strict approach, the review process becomes quantifiable, reproducible, systematic, clearly articulated, and comprehensive within the scope of a given research topic. This systematic approach enhances the credibility and transparency of the findings, allowing for a well-organized evaluation of existing literature. For this study, a systematic review was conducted using the Preferred Reporting Items for Systematic Reviews and Meta-Analyses (PRISMA) methodology. PRISMA provides a structured checklist that enables other researchers to replicate the review process, ensuring consistency and accuracy in data collection for future studies. While PRISMA has been widely adopted in the medical and physical sciences, its application in computer science research is still evolving. However, integrating a PRISMA-based approach into an SLM enhances the quality and credibility of the review by making the methodology transparent and replicable. This allows researchers to critically evaluate the strengths and limitations of the study while also providing a structured reference for other professionals working in the field. A key benefit of using the PRISMA methodology in SLM is its ability to standardize the review process. It ensures that the structure and format of the study are clearly defined, making it easier for other researchers to replicate and build upon the findings. The accuracy and reliability of an SLM are significantly improved when following the PRISMA technique, as it minimizes biases and ensures a structured assessment of existing research. In this study, PRISMA was employed to analyze the current evidence regarding anomaly detection techniques, particularly focusing on their feasibility for deployment on resource-constrained embedded devices (MCUs) at the edge of the network. The review process followed in this study is visually represented in Figure 3.4. A fundamental aspect of conducting an SLM is the formulation of well-defined research questions. Clearly articulating research questions is crucial for addressing specific concerns and eliminating potential biases by maintaining a

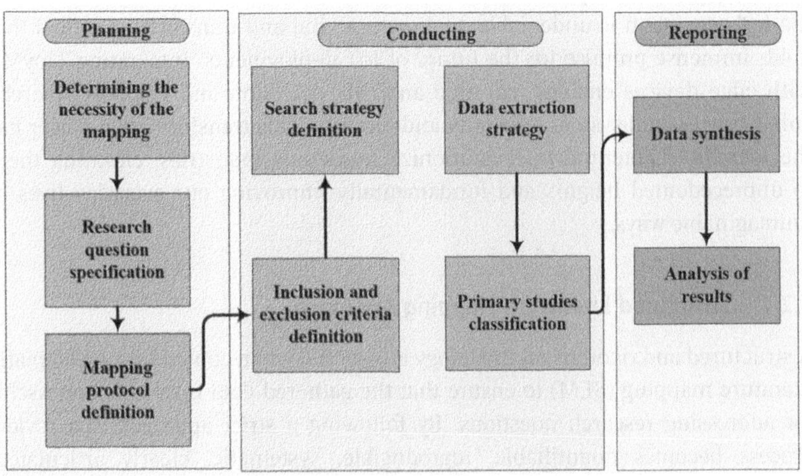

Figure 3.4 The process of the structured literature mapping research approach.

transparent and objective approach. As highlighted in previous research, defining the research problem is one of the most critical steps in a systematic mapping study. The formulation of research questions serves as a guiding framework to identify relevant primary studies, determine key data extraction criteria, and establish a structured methodology for synthesizing information. Finally, the mapping process involves implementing a predefined protocol, ensuring that the study is carried out systematically and aligns with the objectives set forth in the research framework. By adhering to a structured review methodology, such as PRISMA within an SLM, researchers can enhance the quality, reproducibility, and transparency of their findings. This methodological rigor is essential for making meaningful contributions to the field, particularly in emerging areas such as anomaly detection in embedded systems and EC environments.

3.2.7.1 Understanding PICOC in the Context of Systematic Literature Mapping

When formulating research questions in a systematic review or mapping study, a structured framework such as PICOC is often employed to guide the process. Originally developed for evidence-based practices in the medical field, PICOC stands for population, intervention, comparison, outcome, and context. According to the authors in [52], this framework helps ensure that the research questions are comprehensive and focused. Each component serves a specific purpose as follows:

- **Population** refers to the group or elements being studied.
- **Intervention** identifies the specific action, treatment, or technique being evaluated.
- **Comparison** represents the alternative to the intervention, which might be another technique or a control condition.
- **Outcome** describes the results or effects expected from the intervention.
- **Context** defines the environment or conditions under which the study is conducted.

These five components collectively help in crafting clear and well-structured research questions. Table 3.1 in the study outlines how each PICOC element is defined and applied to the current research focus. In this particular study, the population comprises IoT devices, with a special emphasis on MCUs and TinyML-enabled systems. The Intervention involves implementing ML and DL algorithms for anomaly detection on these constrained devices. Because the focus is on exploring possibilities rather than comparing multiple alternatives, the comparison element is not applicable in this context. The outcome of interest is the identification and analysis of various solutions, methods, and advancements in deploying ML/DL techniques on IoT edge devices. This includes a range of anomaly detection strategies that are specifically adapted to operate within the

Table 3.2 Interpretation of PICOC elements in the context of systematic literature mapping.

Elements	Description
Population (P)	Artificial intelligence of things or tiny machine learning
Intervention (I)	Anomaly detection
Comparison (C)	Not applicable
Outcomes (O)	Solution, method, approach, or development
Context (C)	Internet of Things

computational and memory limitations of MCUs. Finally, the Context of the study is centered on EC within IoT ecosystems, particularly involving fog and cloud computing architectures. The research is scoped to understand how ML/DL algorithms can be effectively integrated into these environments to enhance real-time anomaly detection without relying heavily on centralized systems. By applying the PICOC framework, this study ensures a clear and methodical approach to reviewing the literature, guiding the identification of relevant works, and framing the scope of the investigation. This structured methodology not only improves the transparency of the review process but also supports the replicability and relevance of the findings in the broader field of AI-driven IoT applications. Table 3.2 presents the definitions of the PICOC elements as applied to the current study.

3.2.7.2 Formulating Research Questions Based on the PICOC Framework

After defining the key elements of the PICOC structure, the next critical step is to develop the research questions that this study seeks to address. The structured nature of PICOC, comprising population, intervention, comparison, outcome, and context, provides a solid foundation for constructing clear and purposeful research enquiries. These questions are not only intended to guide the SLM process but also to ensure that the study remains aligned with its core objectives. As emphasized by the authors in reference [53], crafting precise and well-targeted research questions is essential in ensuring the success of a systematic review or mapping study. Well-defined questions help to narrow the focus, reduce bias, and provide a clear path for data extraction and analysis. Without this clarity, the risk of diverging from the original research goals increases, potentially compromising the reliability and relevance of the findings. In line with this, a series of mapping questions (MQs) has been developed to serve as the backbone of the current research. These questions are directly informed by the PICOC elements previously defined, ensuring consistency and relevance across all stages of the study. Each MQ is designed to address a specific aspect of the research,

Table 3.3 Catalog of mapping questions addressed in the research.

Question No.	Questions
MQ – 1	How many studies have been published over the years, and how was this distribution constructed?
MQ – 2	Which ML/DL techniques are applied to detect anomalies through TinyML?
MQ – 3	Which data types are used to detect anomalies?
MQ – 4	Which ML/DL evaluation metrics are used for investigation?
MQ – 5	Which IoT devices (MCUs) are used by the researchers?
MQ – 6	Which IoT conditions in terms of power and connectivity are used?
MQ – 7	In which use cases and domains are they applied?

from the types of AI algorithms used in IoT devices to their performance and limitations in EC environments. These mapping questions guide the systematic search for relevant literature, help determine the inclusion and exclusion criteria, and shape the synthesis of existing knowledge. Table 3.3 presents the complete list of MQs formulated for this study, outlining the specific areas of inquiry that will be explored throughout the literature mapping process. By grounding the research questions in a structured framework like PICOC, this study ensures a comprehensive, transparent, and replicable approach to investigating the application of AI techniques in resource-constrained IoT environments.

3.2.7.3 Inclusion/Exclusion Criteria

Inclusion and exclusion criteria are essential tools in the research process, serving to determine which studies are relevant and suitable for addressing specific research questions. These criteria help researchers filter through a large pool of potential studies, ensuring that only those that meet specific standards are selected for review or analysis. The inclusion criteria, outlined in Table 3.4, specify the characteristics that a research paper must possess to be considered in the study. By adhering to these standards, researchers ensure that the selected papers align with the goals and scope of the research. On the other hand, exclusion criteria serve as the opposite of inclusion criteria. They define the attributes or conditions that would disqualify a paper from being included in the review. However, in this instance, the exclusion criteria are not listed, as their purpose is inherently to exclude certain studies, and their detailed description is not necessary for the selection process outlined here. Essentially, by focusing on the inclusion criteria, the selection process is streamlined to gather the most relevant and valuable sources for the research.

Table 3.4 Criteria for inclusion in the research.

No.	Inclusion criteria
IC – 1	Articles that propose a research/empirical study and apply an ML/DL algorithm to detect anomalies
IC – 2	Articles that apply an ML/DL algorithm on the EC layer
IC – 3	Articles that necessarily have a title, abstract, and keywords
IC – 4	Articles published in English
IC – 5	Articles published at peer-reviewed events, such as journals, conferences, or workshops

3.2.7.4 Search Strategy

The search strategy is a critical component of the research methodology, outlining how relevant studies are identified and selected for inclusion. This section provides a detailed explanation of the approach used, which includes selecting appropriate academic libraries and utilizing an effective string search technique. The process aims to ensure that the studies retrieved align closely with the research objectives.

i) **Source Selection:** The research queries, along with the inclusion and exclusion criteria, guide the selection of source libraries. Four key academic databases were chosen for the search process: the ACM Digital Library, Web of Science, IEEExplore, and Scopus. These libraries were selected based on specific criteria: first, they are widely recognized for their comprehensive collections of research materials relevant to the subject of the review; second, they support advanced search capabilities, such as the use of logical operators like "AND" and "OR"; and third, they allow searches within both the document body and metadata fields, including the title, keywords, and abstract. These features make the selected databases ideal for conducting thorough and precise searches related to the research topic.
ii) **String Search:** Table 3.5 illustrates the key terms used in the string search, designed to identify papers that apply ML and DL models for detecting anomalies within the contexts of TinyML and AIoT (Artificial Intelligence of Things). The string search incorporates the two key components from the PICOC framework, which defines the population relevant to the study, using both acronyms and full terms to ensure a comprehensive search. The intervention discussed above is also referenced in the string search, specifically focusing on the aspect of the investigation. However, terms related to the outcomes of the studies are intentionally omitted from the search criteria in order to maximize the number of relevant results. Furthermore, the search includes terms that specify the context of the research, such as "Internet of Things" (IoT), as well as its

Table 3.5 Search string used in the research.

No.	Search terms
SC – 1	"tinyml" OR "tiny ml" OR "aiot" OR "tiny machine learning" OR "Artificial Intelligence of Things"
SC – 2	"event detection" OR "anomaly detection" OR "fault detection" OR "abnormal behavior" OR "abnormal behaviour" OR "anomaly prediction" OR "anomalous behavior recognition" OR "anomalous behaviour recognition" OR "outlier detection"

subsets "TinyML" and "AIoT." Including these terms ensures that the search retrieves the most pertinent results. This step is carefully executed to ensure that only studies presenting relevant results are selected, and the entire process, from search to paper review, is designed to ensure the inclusion of studies that meet the predefined criteria. By focusing on these aspects, the search strategy aims to yield a collection of high-quality sources that contribute to the overall objectives of the research.

3.2.7.5 Data Extraction Process

The data extraction process is crucial to ensure that only the most relevant and high-quality studies are selected for in-depth analysis. This process is organized into four main stages. These stages allow for a systematic and thorough approach to gathering and refining the data.

i) **Stage 1: Search and Retrieval:** In the initial stage, we conducted a search across four academic databases, restricting the search to metadata fields such as the title, keywords, and abstract. However, a distinctive feature of the ACM library allowed us to expand the search, as it included a field called "Anywhere," which permitted a search throughout the entire document. This adjustment allowed us to retrieve a broader range of studies, as we did not impose a restriction on the publication year due to the recent nature of the research in question. As a result, a total of 162 papers were initially retrieved from the four databases.

ii) **Stage 2: Initial Screening and Refinement:** Once the papers were retrieved, the next step was to screen them for relevance. Through this process, 18 studies were identified as meeting the criteria for a deeper analysis. These papers were selected based on their relevance to the subject of our research. The most recent studies from the past three years were prioritized, while those that did not align with the study's scope were excluded. Additionally, any duplicate entries were identified and removed, leading to the elimination of 55 duplicated files.

iii) **Stage 3: Application of Exclusion Criteria:** After eliminating duplicates, we applied the exclusion criteria to further narrow down the selection. This step was essential to ensure that the studies included in the final set truly aligned with the research focus. Following this process, we ended up with a total of 89 articles, which were then subjected to a more detailed examination. This involved reviewing the titles, abstracts, and keywords of each article to ensure they adhered to the predefined inclusion and exclusion criteria.

iv) **Stage 4: Final Selection and Data Extraction:** The final stage involved a thorough review of the remaining articles. After applying the inclusion/exclusion criteria, we selected 18 studies that addressed the key aspects of our research topic. The data extracted from each of these studies included the type of publication (whether it was a journal, conference paper, or book chapter), the authors and their institutional affiliations, the country of publication, the study abstract, keywords, and the publication's URL, ISBN/ISSN/DOI. This dataset was then published on Zenodo, making the information available for further research and analysis. The entire search and extraction process was carried out on March 31, 2025, ensuring that the most up-to-date studies were included in the review. Through this rigorous approach, we were able to collect a set of high-quality studies that provide valuable insights into the research topic.

3.2.7.6 Distribution of Research Studies

Figure 3.5 illustrates the initial outcomes of the literature search, presenting the distribution of articles across various academic libraries and publication years.

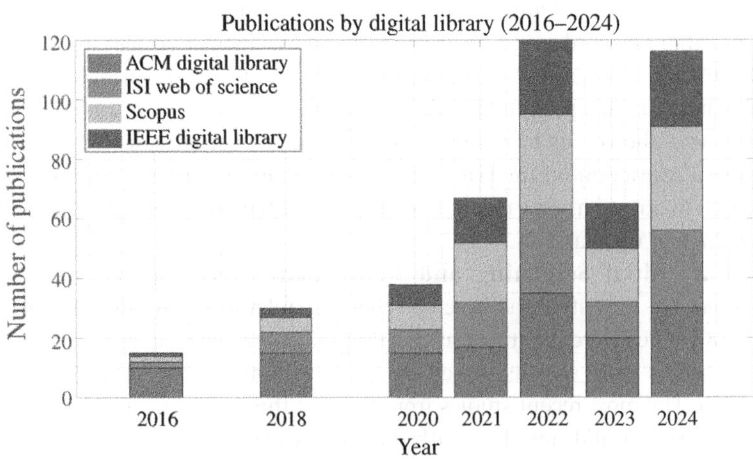

Figure 3.5 Publications by academic library source and publication year.

Through the methodological process, a total of 20 relevant research papers were identified. These studies specifically employed ML and DL techniques for anomaly detection in MCUs. All selected publications fall within the timeframe of 2020 to 2024, reflecting the recent and evolving nature of this research domain. Despite the novelty of the topic, there has been a noticeable and steady rise in the number of related publications each year. In fact, the volume of studies appears to have approximately doubled annually, with over 77% of the identified articles released within the last two years. This growth trend is particularly notable considering that the literature review was conducted in March 2025. Even within the early months of that year, a significant number of papers had already been published, emphasizing the accelerating interest in this field. Figure 3.6 provides a breakdown of publication counts by year, further supporting these observations. From the data visualized in Figure 3.6, the ISI Web of Science emerges as the academic database contributing the most articles. Seven out of the total 20 studies were sourced from this platform, accounting for nearly 39% of all selected works. Of the identified publications, 15 appeared in peer-reviewed journals, while the remaining five were presented at academic conferences. A substantial portion of the journal articles, six in total, representing 40%, were published in the Sensors journal (MDPI). Additionally, Future Generation Computer Systems published two of the identified papers, which constitute approximately 67% of the journals content beyond Sensors.

In the reviewed studies, researchers employed a total of 12 distinct ML and DL algorithms for anomaly detection. Figure 3.7 presents the proportional distribution of these algorithms based on their frequency of use. Among them, the CNN

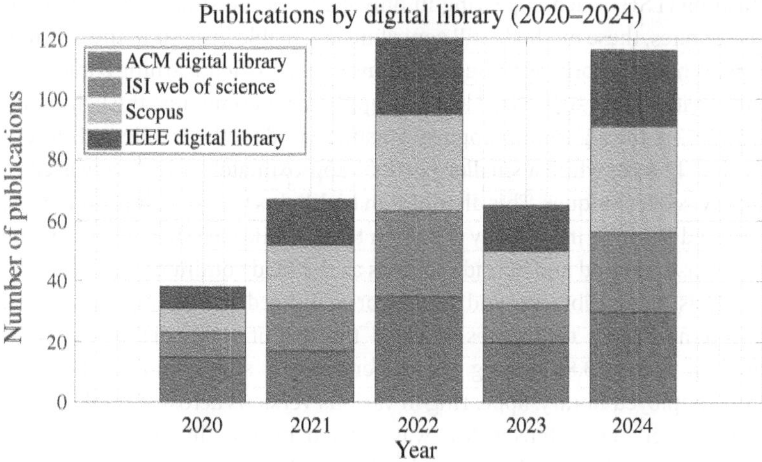

Figure 3.6 Selection of articles by academic libraries across publication years.

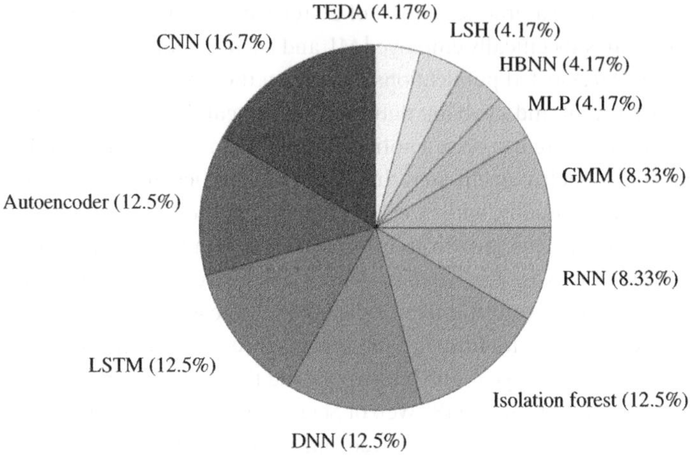

Figure 3.7 The proportional distribution of ML/DL algorithms based on their frequency of use.

emerged as the most frequently adopted method, representing 16.7% of all implementations across the selected studies. Several other algorithms were also widely utilized, including autoencoders, long short-term memory (LSTM) networks, deep neural networks (DNNs), isolation Forest, and Gaussian mixture models (GMM), each accounting for approximately 12.5% of the total. On the other hand, a number of algorithms were used less frequently. These include the multilayer perceptron (MLP), RNNs, deep convolutional neural networks (DCNNs), Type-2 ellipsoidal data analyzer (TEDA), hybrid binary neural networks (HyBNNs), and locality sensitive hashing (LSH), each appearing in only 4.17% of the studies. Despite their lower usage rates, these methods still contributed valuable insights to the research landscape. When categorizing these algorithms based on their learning paradigms, it becomes evident that supervised learning approaches dominated the field, comprising 52.36% of the algorithms applied. Unsupervised methods followed closely, representing 42.86%, while a smaller portion, approximately 4.76%, consisted of semi-supervised techniques. This distribution highlights the prevailing preference for supervised learning in anomaly detection tasks, while also showing a growing interest in unsupervised and hybrid methods as the field continues to evolve.

An analysis of the libraries and frameworks utilized in the implementation of the aforementioned techniques revealed the use of seven distinct tools, as illustrated in Figure 3.8. Among these, TensorFlow stands out as the most frequently employed library, appearing in various versions across several studies. The standard version of TensorFlow was applied in four different cases, each requiring relatively high-performance hardware platforms. Specifically, it was used on devices such as the Raspberry Pi in three instances and the Nvidia

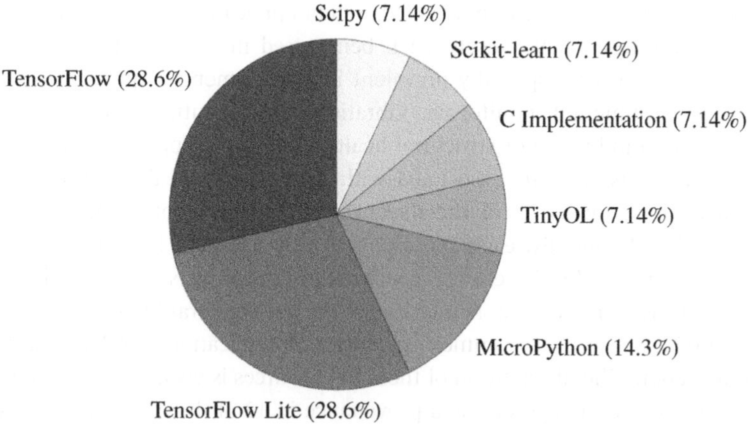

Figure 3.8 The proportional distribution of ML/DL libraries based on their frequency of use.

Jetson Nano in one. In addition to the traditional version, TensorFlow Lite also featured prominently, being implemented in four separate cases. This lightweight variant is particularly well-suited for resource-constrained environments and was deployed across a diverse range of hardware, including ESP32 boards, Arduino devices, Wio modules, and the Raspberry Pi. The versatility and broad hardware compatibility of TensorFlow Lite make it a popular choice for embedded ML applications. The second most frequently used framework was MicroPython, which was adopted in two studies. Both instances involved the ESP32 family of microcontrollers, reflecting the framework's efficiency and suitability for lightweight, embedded systems. TinyML development also included the use of the TinyOL framework, which appeared once in connection with the Arduino Nano 33 BLE Sense board, a microcontroller designed specifically for low-power AI applications. Other libraries, such as SciPy and Scikit-learn, were each used once, both in contexts involving more capable hardware like the Raspberry Pi. These tools are well-established in the data science community and are typically favored for their robust ML and statistical analysis capabilities. Interestingly, one study implemented the algorithm without relying on any external ML library. Instead, the solution was developed entirely in the C programming language and tailored specifically for the ESP32 microcontroller. This approach demonstrates a highly customized, lightweight implementation suitable for environments where external dependencies must be minimized.

An evaluation of the selected studies highlights that the inertial measurement unit (IMU) is the most commonly used data source for anomaly detection in TinyML-based applications, appearing in 20% of the reviewed cases. This is likely due to the IMU's ability to capture dynamic motion-related data, making it especially useful in a range of embedded system applications. Image data and temperature readings follow closely, each being used in 16.7% of the studies. These types of input are especially prevalent in environments requiring visual analysis or environmental monitoring. Vibration data was utilized in 13.3% of the cases, often in industrial or structural health monitoring applications, while pulse data, which is typically associated with biometric or health monitoring systems, accounted for 6.67% of the instances. A variety of other data types appeared far less frequently, each representing only 3.33% of the total. These include body glucose levels, energy consumption, bearing status, humidity, proximity sensing, air pressure, sound data, and network traffic, all of which suggest a broader, albeit less common, spectrum of applications for TinyML in anomaly detection. The distribution of these data sources is visually represented in Figure 3.9. Further analysis of data provenance in 12 of the examined works reveals a mix of custom and publicly available datasets. Some researchers opted to develop their own datasets tailored to their specific use cases, reflecting the highly customized nature of embedded applications. Others utilized well-known public datasets to train and evaluate their models. Notable examples include the CWRU dataset, which was used in two studies for bearing fault detection, as well as IMS, UCE, ShanghaiTech, and several image-based datasets like MNIST, FashionMNIST, CIFAR-10 (used in two works), and NEU64. The diversity of dataset sources underscores the adaptability of TinyML approaches to a wide range of domains and data types.

The evaluation metrics employed to assess the performance of anomaly detection methods in TinyML studies are summarized in Figure 3.10. This figure illustrates how frequently different metrics were used across the analyzed

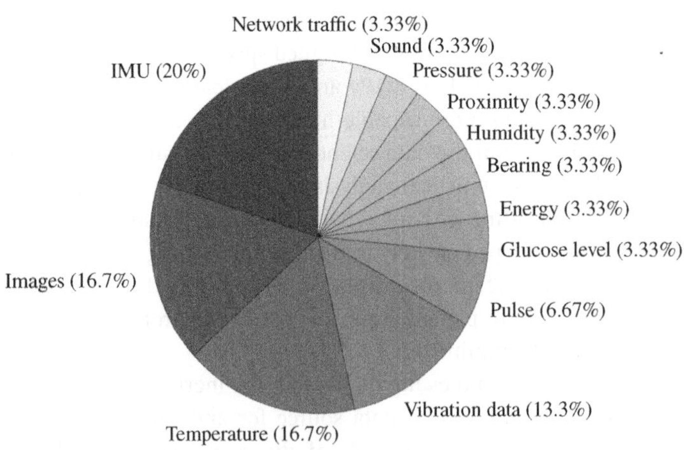

Figure 3.9 The proportional distribution of articles by data types.

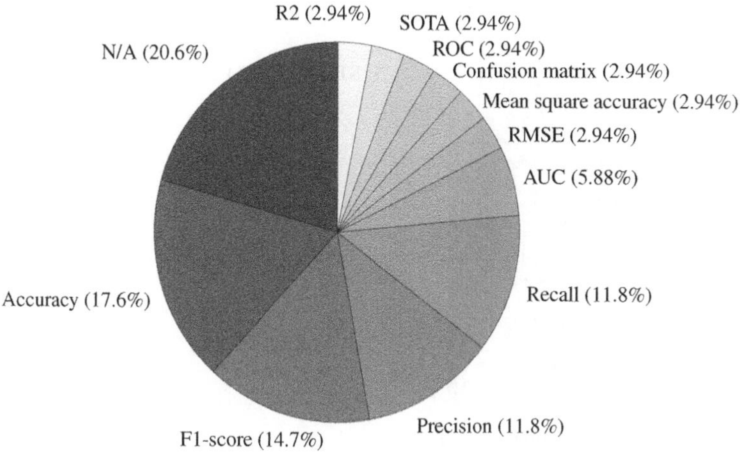

Figure 3.10 The proportional distribution of articles by the evaluation metrics.

publications. Accuracy (ACC) emerged as the most commonly reported performance indicator, appearing in eight of the reviewed studies. Its popularity is likely due to its straightforward interpretation and general applicability for classification tasks. To further evaluate the reliability and robustness of the models, the F1-score was applied in five studies. This metric is particularly useful in cases where the data is imbalanced, as it considers both precision and recall in its calculation. Precision and recall were individually reported in three publications, offering a more granular view of model performance in terms of false positives and false negatives. Other evaluation metrics were less frequently mentioned. Root mean squared error (RMSE) and area under the curve (AUC) were each reported in two studies, providing insight into regression performance and classification trade-offs, respectively. Meanwhile, mean squared error (MSE), coefficient of determination (R^2), specificity, and sensitivity were each used in just one study. These metrics, though valuable, may have been less applicable depending on the problem domain or the type of data used. Interestingly, seven studies did not specify the evaluation metrics used to validate their models. The absence of this information suggests a lack of standardization in reporting and highlights the need for clearer methodological transparency in future TinyML research.

Figure 3.11 presents the breakdown of IoT devices used across the 18 reviewed studies. Among the various MCUs and edge devices employed, the Raspberry Pi platform was the most widely adopted, appearing in eight studies, which accounts for approximately 42.1% of the total. Its popularity can be attributed to its versatility, ease of use, and relatively powerful processing capabilities for embedded applications. Notably, a wide range of Raspberry Pi models were utilized,

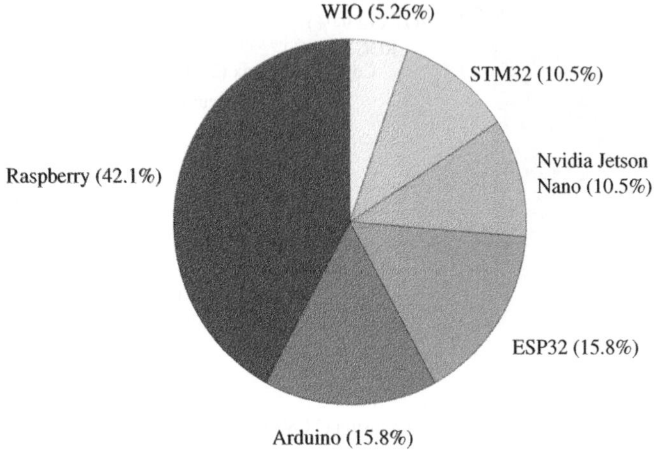

Figure 3.11 The proportional distribution of articles by hardware.

including the Raspberry Pi 1, 2B, 3B, and 3 B+, as well as the more advanced Raspberry Pi 4 and 4 B. The 3 B+ and 4B variants were each featured in two separate studies, indicating their particular suitability for TinyML deployments. In addition to the Raspberry Pi, the Arduino Nano 33 BLE was used in three studies, representing 15.8% of the total. This specific Arduino model is optimized for low-power ML tasks, making it a fitting choice for embedded AI applications. Similarly, the ESP32 microcontroller was implemented in three different research projects. Of these, two studies utilized the ESP32 DevKit V4, while the third made use of the V1 version. The ESP32's built-in Wi-Fi and Bluetooth capabilities, along with its dual-core processor, make it a strong candidate for real-time IoT solutions. Other devices were used less frequently. The Nvidia Jetson Nano, a more powerful edge AI device, appeared in two studies (10.5%), showcasing its relevance for computationally intensive models. Likewise, STM32 microcontrollers were also used in two instances, accounting for another 10.5%. The specific versions cited were the STM32F401RE and STM32H743, both of which were used within a single study. Finally, the Wio platform was mentioned in one study (5.26%), reflecting a smaller but still relevant presence in the research landscape. This distribution highlights the growing diversity of hardware platforms used in TinyML applications and suggests that device selection often depends on the computational needs, energy constraints, and intended application context of the model being deployed.

The connectivity used to transmit the results of ML and DL models in the reviewed studies varies depending on the microcontroller in use, as each device typically supports different types of connectivity options. Figure 3.12 illustrates

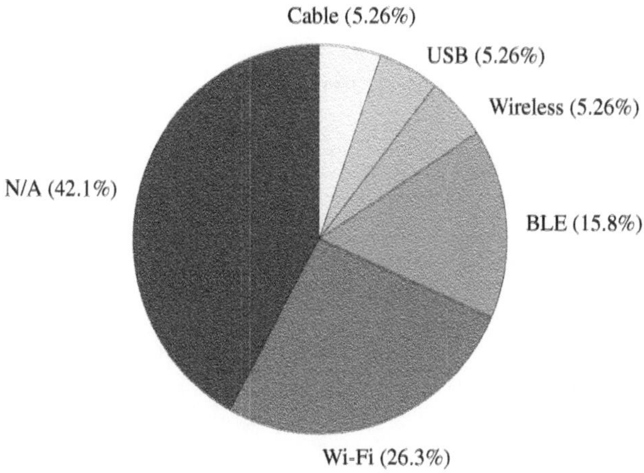

Figure 3.12 The proportional distribution of articles by power and connectivity.

the distribution of connectivity types employed across the studies. Wireless connectivity is the most common method, utilized in nine studies, representing 47.36% of the total. Among these, Wi-Fi is the dominant wireless technology, used in five studies. Bluetooth is also frequently employed, appearing in three papers, while one study did not specify the exact type of wireless connection used. On the other hand, wired connectivity is less common but still present. One study used a USB connection for data transfer, while another employed Ethernet connectivity. This suggests that while wireless options are generally preferred for their flexibility and ease of integration, some applications still rely on wired methods for stability and higher data throughput. Notably, eight studies did not disclose the type of connectivity used, indicating a gap in methodological transparency. In terms of power supply, only five studies explicitly mentioned that their IoT devices are powered through a cable connection to an external energy source. Interestingly, none of the studies referenced the use of batteries as a power source for the devices, which may suggest a preference for continuous, reliable power in TinyML applications. This could be due to the power demands of ML/DL models, which often require a steady power supply, especially during extended processing tasks.

The applications and use cases of TinyML in the reviewed studies are categorized into four main areas, as shown in Figure 3.13. The predominant application of TinyML is in the industry and business sector, which accounts for 61.1% of the studies, or eleven out of the total number. This highlights the growing interest and potential of TinyML in optimizing industrial processes, improving business

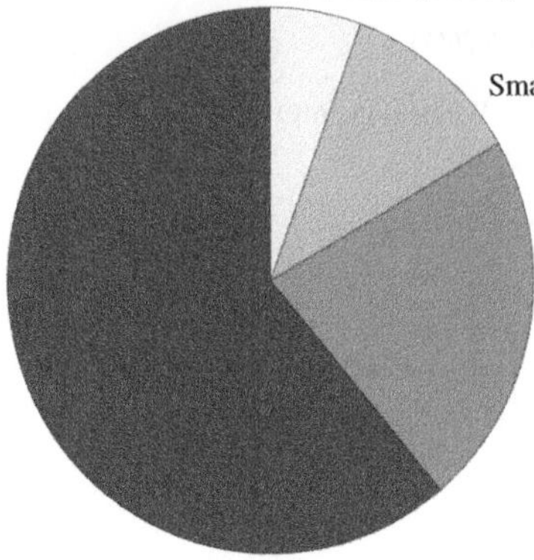

Figure 3.13 The proportional distribution of articles by use case and domain.

operations, and enabling automation and predictive maintenance in various industries. The healthcare sector follows as the second most common application, representing 22.2% of the studies. TinyML in healthcare can be used for real-time health monitoring, diagnostics, and even personalized treatment, showing its value in improving patient care and medical outcomes. The smart city domain, with 11.1% of the studies, indicates the use of TinyML for enhancing urban infrastructure, optimizing traffic flow, improving security, and enabling various IoT-based services. Lastly, the environmental sector is the least represented, accounting for only 5.56% of the studies. In this field, TinyML can be applied to monitor environmental conditions, predict natural disasters, and enhance sustainability efforts. These findings suggest that while TinyML holds promise across multiple sectors, its most significant impact has been in industrial applications, followed by healthcare and urban development.

Incorporating TinyML technologies opens a world of possibilities, enabling organizations to optimize and streamline their operations, improve automation processes, drive exceptional levels of precision, and push the boundaries of what is currently considered feasible [54]. The remarkable impact of this groundbreaking innovation extends far beyond traditional boundaries, ushering in a transformative new era of unparalleled technological advancement poised to shape the future profoundly. With the continued advancement of TinyML and its integration with IoT and edge devices, we are poised to witness unparalleled levels of efficiency, reliability, and intelligence that will redefine how we perceive and interact with the world around us [19]. Let's understand the following case study that brings out the solutions and recommendations that can be made for similar industrial settings seeking to implement edge-based anomaly detection

3.2.7.7 Case Study 1: Industrial Anomaly Detection Using Edge Computing and TinyML

Introduction and Background Industrial environments are highly dynamic, with equipment such as motors, turbines, and pumps constantly generating real-time data through sensors embedded in various parts of the machinery. The volume and complexity of this data necessitate the use of advanced technologies to detect anomalies and prevent unplanned downtime, which can lead to costly repairs and loss of productivity. Traditional anomaly detection systems often rely on cloud-based solutions, which can suffer from latency issues and high bandwidth requirements, making them less suitable for real-time applications. To address these challenges, an innovative solution leveraging EC and TinyML has been implemented for industrial anomaly detection. This approach allows ML models to run directly on resource-constrained edge devices, significantly reducing latency, optimizing bandwidth, and enhancing overall system responsiveness. This case study focuses on the application of such a solution in an industrial setting, examining how it has transformed anomaly detection and predictive maintenance practices.

Technical Implementation The system architecture is designed to enable real-time anomaly detection in industrial equipment, utilizing a combination of edge devices, gateways, and cloud infrastructure. Edge devices, equipped with various sensors, collect operational data such as temperature, vibration, pressure, and sound from industrial machines. This data is often large and complex, making it challenging to process in real-time without efficient handling strategies. To address this, the system relies on edge gateways positioned between the IoT devices and the cloud, responsible for preprocessing the sensor data before it is processed further. Edge gateways run TinyML models, which are optimized for low-power devices and capable of performing anomaly detection directly at the edge. These models analyze the collected sensor data and identify patterns or deviations, such as abnormal temperature or vibration spikes, which may indicate equipment failure. By processing this data locally, the system minimizes computational overhead and ensures faster detection, reducing the need for cloud connectivity and ensuring low-latency decision-making.

While edge devices handle most of the data processing, cloud infrastructure plays a key role in model refinement, long-term data storage, and trend analysis. AWS SageMaker is used for training and fine-tuning ML models based on historical data, while Amazon Timestream stores time-series data for predictive maintenance and trend monitoring. When an anomaly is detected at the edge, relevant data is sent to the cloud for further analysis, allowing the system to continually improve and adapt while maintaining efficient, real-time anomaly detection at the edge.

Case Study Analysis This case study examines the deployment of the edge-based anomaly detection system in a large-scale industrial environment. The system was deployed across multiple machines in a manufacturing plant that previously struggled with frequent, unanticipated downtimes due to equipment failure. The main goals of implementing this solution were as follows:

The primary objective of the system is real-time anomaly detection, enabling the identification of issues as soon as they arise. This approach minimizes the time between the onset of a problem and the corrective action, allowing for faster responses and reducing the likelihood of equipment failures or extended downtime. In addition to real-time detection, the system focuses on optimizing bandwidth usage. By processing most of the data at the edge, the amount of information sent to the cloud is significantly reduced. This not only alleviates network congestion but also ensures quicker processing times, as critical data can be analyzed locally without delays caused by cloud communication.

Cost efficiency is another key consideration in this solution. The use of low-cost, energy-efficient edge devices allows for anomaly detection without the need for expensive centralized cloud infrastructure. This approach helps minimize operating costs while still delivering robust performance, making it a scalable and cost-effective solution for industrial environments.

The results from the deployment of this system were measured against these goals, and the analysis yielded the following outcomes:

- **Reduced Downtime:** The system successfully detected early signs of equipment failure, such as abnormal vibration patterns in motors, before they escalated into costly breakdowns. This proactive approach allowed the plant to schedule maintenance activities during nonoperational hours, thereby minimizing production disruption.
- **Improved Operational Efficiency:** By continuously monitoring equipment performance and detecting issues in real time, the system helped the plant operate more efficiently. Anomalies were addressed promptly, and the equipment was maintained in optimal conditions, leading to improved performance and reduced energy consumption.
- **Optimized Bandwidth Usage:** By processing data at the edge, the system significantly reduced the amount of data transmitted to the cloud. Only critical anomaly alerts and relevant sensor data were sent, which lowered network congestion and optimized bandwidth resources.
- **Cost Savings:** The use of resource-efficient edge devices with TinyML models significantly reduced the need for expensive cloud infrastructure. Additionally, the ability to perform real-time anomaly detection without needing to send large volumes of data to the cloud helped reduce operating costs.

Solutions and Recommendations Based on the results of this case study, several solutions and recommendations can be made for similar industrial settings seeking to implement edge-based anomaly detection as follows:

1) **Adopt Edge Computing for Low-Latency Applications:** For real-time decision-making, EC is essential. Anomaly detection performed at the edge significantly reduces the delay caused by cloud communication, enabling faster identification of issues. This is particularly important in critical applications where delays could lead to system failures or safety risks.
2) **Optimize Machine Learning Models for Edge Devices:** Utilizing TinyML for resource-constrained devices ensures that ML models can run efficiently without requiring excessive computational power. As sensor data in industrial environments can be vast and high-dimensional, TinyML models should be specifically optimized to handle large datasets with minimal resource usage, enabling scalable anomaly detection across a wide range of industrial applications.
3) **Leverage Cloud for Continuous Model Improvement:** While most anomaly detection can occur at the edge, the cloud still plays a vital role in refining models and performing long-term data analysis. Periodically training models with data from various machines and equipment can enhance their ability to detect more complex patterns and improve accuracy over time. Cloud-based platforms like AWS SageMaker offer the tools necessary to perform such updates and optimizations.

This case study demonstrates the transformative power of combining EC with TinyML for anomaly detection in industrial environments. By deploying ML models on resource-constrained edge devices, the system can perform real-time data analysis with low latency and minimal bandwidth usage. The approach has proven to be highly effective in reducing unplanned downtime, improving operational efficiency, and optimizing costs in industrial settings. As IoT devices continue to proliferate in industries ranging from manufacturing to energy, the integration of TinyML and EC will become increasingly critical for realizing autonomous, intelligent systems that can detect and address anomalies before they escalate into major failures. With the ongoing advancement of ML models and EC technologies, the potential for scaling these solutions across a wide range of industrial applications is immense, providing both operational and economic benefits.

3.2.7.8 Case Study 2: TinyML for Anomaly Detection in Industrial Equipment Monitoring

TinyML has emerged as a transformative approach in embedding intelligence into edge devices with minimal computational resources. One of its most promising applications is anomaly detection, particularly in environments where latency, connectivity, and power constraints limit the use of traditional cloud-based AI solutions. Anomaly detection involves identifying patterns in data that do not conform to expected behavior, which is critical in fields such as industrial monitoring, healthcare, and cybersecurity. By deploying lightweight models directly on microcontrollers, TinyML enables real-time, on-device anomaly detection, reducing the need for constant connectivity and enhancing response times. A compelling case study highlighting the efficacy of TinyML for anomaly detection is the deployment of a vibration-based predictive maintenance system in a manufacturing plant. In this scenario, traditional predictive maintenance systems required large volumes of sensor data to be transmitted to cloud servers for analysis, incurring high bandwidth costs and latency. To overcome these challenges, engineers integrated a TinyML model onto an ARM Cortex-M4-based microcontroller attached to the motors and pumps within the factory floor. The model was trained to recognize normal vibration patterns during different operational states and identify deviations that could signal mechanical issues, such as imbalance, misalignment, or bearing failure. The training process involved collecting labeled vibration data under normal and faulty conditions, followed by preprocessing and feature extraction (e.g., Fast Fourier Transform for frequency domain features). A lightweight neural network classifier was then trained using TensorFlow Lite for Microcontrollers. The final model, compressed to fewer than 20 KB, was deployed on the microcontroller along with real-time signal processing logic. This allowed the system to continuously monitor equipment, detect anomalies within milliseconds, and alert technicians without needing to send data to a central server. Over a six-month trial period, the TinyML-based system successfully identified early signs of wear and tear in multiple machines, reducing unplanned downtimes by 30% and maintenance costs by 20%. Importantly, it also operated on battery-powered sensors, proving both the energy efficiency and feasibility of deploying ML models at the edge. This case demonstrates how TinyML for anomaly detection can not only improve operational efficiency and safety but also enable scalable, low-cost monitoring solutions in environments where traditional approaches are impractical [26, 55].

3.2.7.9 Case Study 3: TinyML for Anomaly Detection in Smart Agriculture

In smart agriculture, maintaining optimal environmental conditions is vital for crop health and yield. Anomaly detection plays a key role in identifying irregularities in temperature, humidity, or soil moisture that could lead to plant stress or disease outbreaks. In a pilot project conducted in a vineyard in Spain, a team of researchers deployed a network of ultralow-power microcontroller-based sensor nodes throughout the fields. These nodes were equipped with temperature, humidity, and soil moisture sensors, as well as a TinyML model trained to detect abnormal environmental conditions. The model was trained on historical environmental data collected during healthy crop cycles and during known periods of disease or irrigation failure. Using unsupervised learning techniques like autoencoders, the system learned to recognize normal patterns and flag outliers in real time. Once deployed, the model detected anomalies such as sudden drops in soil moisture and spikes in humidity, which correlated with irrigation system malfunctions and early signs of fungal infections. The early alerts allowed farmers to take preventive action, ultimately reducing crop loss and improving yield consistency. With solar-powered sensors and long-range wireless communication (LoRa), the system operated autonomously and efficiently, showcasing TinyML's potential in remote and data-scarce agricultural environments [56, 57].

3.2.7.10 Case Study 4: TinyML for Anomaly Detection in Smart Homes

Smart home security systems are increasingly incorporating ML to improve intrusion detection and fault identification. In a real-world deployment in a residential complex in South Korea, a research team implemented a TinyML-powered solution for real-time anomaly detection on door and window sensors. These sensors use microcontrollers to process motion and vibration data locally, recognizing normal behaviors such as a door opening during specific hours and detecting unusual activity like forced entry or glass breakage. The TinyML model used on-device time-series analysis to distinguish between benign and suspicious events. It was trained on various patterns of user behavior, noise profiles, and environmental disturbances, making it robust against false positives. The system worked without relying on Wi-Fi or cloud connectivity, preserving privacy and ensuring consistent functionality during power or internet outages. During the first three months of operation, the system detected and prevented two attempted intrusions and also identified faulty windows with abnormal vibrations caused by mechanical issues. The case highlighted how TinyML enhances home security systems by making them smarter, faster, and more self-reliant [2].

3.2.7.11 Case Study 4: TinyML for Anomaly Detection in Wearable Health Devices

Healthcare wearables are becoming more prevalent, offering continuous monitoring of physiological parameters like heart rate, body temperature, and motion. A notable case study involved the use of a TinyML model embedded in a wrist-worn device for detecting cardiac arrhythmias in elderly patients. Traditional wearable solutions often transmit raw ECG data to the cloud for analysis, but this raises concerns over data privacy and power consumption. To address this, a lightweight TinyML model based on decision trees and temporal pattern recognition was developed and integrated into a low-power microcontroller (such as an nRF52 series chip). The model was trained on thousands of annotated ECG signals to classify normal rhythms and detect anomalies such as atrial fibrillation and premature ventricular contractions. Once deployed, the wearable device was able to perform real-time analysis locally and issue alerts to users and caregivers via Bluetooth if an anomaly was detected. The system successfully identified irregular heart activity in multiple patients during a six-month clinical study and operated with a battery life exceeding 10 days, a significant improvement over traditional solutions. This case demonstrated how TinyML can provide accessible, reliable, and privacy-conscious health monitoring at scale, especially in underserved or rural areas. These case studies illustrate how TinyML for Anomaly Detection is revolutionizing various industries by enabling intelligent, low-power, and real-time solutions at the edge. Whether in agriculture, home security, healthcare, or manufacturing, TinyML offers a scalable and cost-effective way to monitor systems, protect assets, and improve decision-making [3, 58].

3.2.8 Machine Learning Models for Anomaly Detection

With the help of the initial model iteration, it is crucial to ensure fast convergence when deploying these models in IoT endpoint devices. Only then, at this precise juncture, will training the TinyML model on edge offer the utmost precision in interpreting the real-time distribution of input data at the model input. Notably, the prediction model for the sensor's readings necessitates an unparalleled level of accuracy, alongside a meticulously balanced ratio of false negatives and false positives for both the case detection and prediction of the noise models. Achieving this delicate equilibrium is of paramount importance. Inference-free probability prediction for the samples in the datasets is the most optimal and reliable metric to utilize [59]. For instance, let's consider a scenario where we develop models to accurately identify anomalies associated with various sensors and their respective readings from IoT devices. In such cases, the model output must correspond to the noise probabilities, specifically the likelihood of a faulty sensor.

When dealing with an anomaly detection model, each time a new sample is obtained from an IoT sensor, the model can effectively leverage the sensor's readings within one minute to forecast the likelihood of a sensor noise or anomaly occurrence. Given that the model produces a probability threshold for identifying anomalies, it exclusively subjects the new sample to its classification API for further verification [60]. This signifies that if the model's output probability falls below the established threshold value, the anomaly detection model and API will conclude "NO NOISE" as the final output. Such an approach represents an inference-free anomaly detection model that ensures impartiality in its results. With the aid of these models, it is possible to efficiently monitor the noise level of any IoT sensor, even when the available data is highly sporadic and widely dispersed. This can be accomplished by training the ML model on the sensor data and using it to detect real-time anomalies. Additionally, these models can be optimized for low-power and low-memory devices, making them suitable for TinyML applications [61].

ML models, with their ability to effectively reduce the occurrence of false positives and enhance prediction accuracy, have experienced a profound surge in popularity across various industries and applications. Their remarkable capabilities have positioned them as a compelling and practical alternative to traditional rules-based approaches, igniting a paradigm shift in anomaly detection techniques. On IoT devices, where TinyML revolutionizes how we perceive and harness data, experts and I firmly advocate for adopting a thorough and intricate threefold process that transcends conventional practices. This comprehensive approach encompasses numerous meticulously designed and strategically implemented steps to ensure the utmost success and effectiveness of anomaly detection models specifically tailored for IoT devices [62].

The ML models used for anomaly detection in TinyML are specifically designed to address the unique constraints and requirements of IoT devices, including limited power and memory resources, as well as the need for real-time processing. These models are optimized for low-power consumption and can run on microcontrollers with limited computational capabilities. The first step involves acquiring and preprocessing data from diverse IoT sensors to commence this meticulous process. This includes gathering information from various sources, ranging from temperature and humidity sensors to motion detectors and accelerometer data. Collecting a comprehensive dataset encompassing multiple dimensions and attributes can establish a robust foundation for anomaly detection [12]. Once the data acquisition phase is completed, the second step entails the meticulous design and training of ML models.

Leveraging cutting-edge algorithms and techniques, experts strive to develop models that can discern standard patterns while accurately identifying and flagging anomalous events. The models can effectively capture intricate temporal dependencies and complex patterns that may signify anomalies by employing advanced architectures such as RNNs or CNNs. This allows for more accurate and reliable detection of abnormalities in the data. However, the process does not halt solely at model training. The third step encompasses meticulous model evaluation and optimization to refine the anomaly detection capabilities continuously [63]. Figure 3.14 presents the key elements of ML models for efficient and optimized IoT applications.

Through rigorous testing and validation against both known anomalies and unseen scenarios, the models can be fine-tuned and calibrated to achieve optimal performance. Expert data scientists conduct meticulous analyses of various performance metrics, including precision, recall, and F1-score, to assess the robustness and accuracy of the models. This process typically involves tuning hyperparameters and selecting the most suitable algorithm for the specific anomaly detection

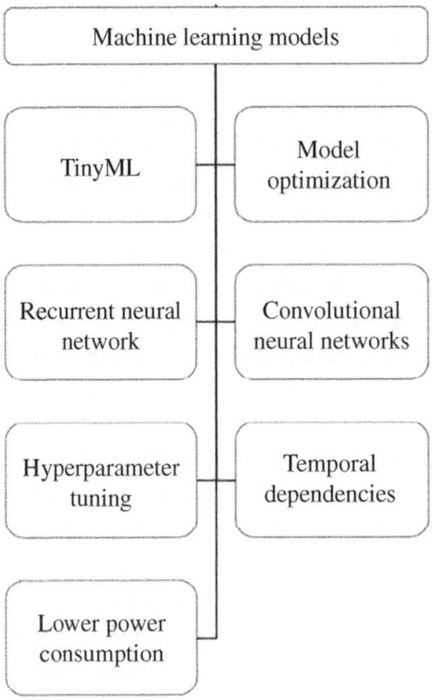

Figure 3.14 Key elements of machine learning models for efficient and optimized IoT applications.

task. In the rapidly evolving field of IoT devices, TinyML has revolutionized how we perceive and harness data [64]. As experts in the field, we firmly advocate adopting a thorough and intricate threefold process that transcends conventional practices, ensuring the utmost success and effectiveness of anomaly detection models tailored explicitly for IoT devices. By following this comprehensive approach, we can effectively address the unique constraints and requirements of IoT devices, including limited power and memory resources, as well as the need for real-time processing in the context of anomaly detection in TinyML. The first step in this meticulous process is acquiring and preprocessing data from diverse IoT sensors. This includes gathering information from various sources, ranging from temperature and humidity sensors to motion detectors and accelerometer data. By collecting a comprehensive dataset that encompasses multiple dimensions and attributes, we establish a robust foundation for anomaly detection. This gives us a holistic data view, enabling us to detect potential anomalies with greater accuracy and reliability [65]. This provides us with a holistic view of the data, allowing us to detect potential anomalies with greater accuracy and reliability.

Additionally, by utilizing TinyML for anomaly detection, we can ensure real-time monitoring and quick response to deviations from standard patterns. Once the data acquisition phase is complete, we proceed to the second step, which involves the meticulous design and training of ML models. Leveraging cutting-edge algorithms and techniques, our experts strive to construct models that possess the aptitude to discern standard patterns while accurately identifying and flagging anomalous events. To achieve this, we employ advanced architectures, such as RNNs or CNNs, capable of capturing intricate temporal dependencies and complex patterns that may signify anomalies [66]. By employing these advanced models, we enhance the accuracy and reliability of anomaly detection in TinyML. However, the process does not stop at model training. The third step involves meticulous model evaluation and optimization to continually refine anomaly detection capabilities. Rigorous testing and validation against known anomalies and unseen scenarios allow us to fine-tune and calibrate the models for optimal performance. Our expert data scientists conduct meticulous analyses of various performance metrics, including precision, recall, and F1-score, to assess the robustness and accuracy of the models. This iterative process ensures that our anomaly detection models are highly effective and reliable. It can also accurately identify abnormal patterns in the data [67]. Figure 3.15 highlights the key elements of anomaly detection in IoT systems.

Moreover, we employ continuous monitoring and update mechanisms to ensure the longevity and adaptability of the anomaly detection models in the rapidly evolving IoT landscape. Regular retraining of the models, incorporating

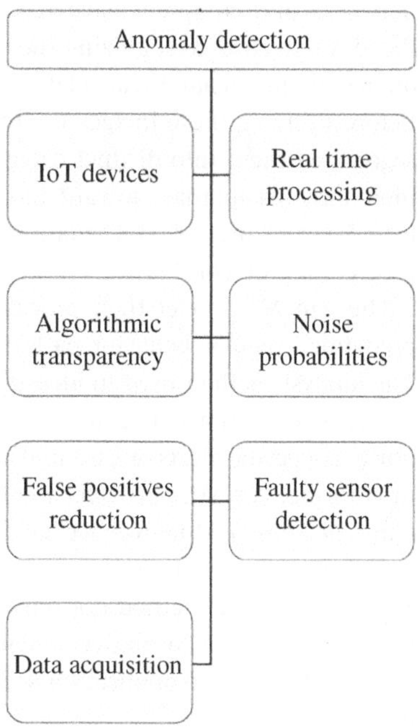

Figure 3.15 Key elements of anomaly detection in IoT systems.

newly acquired data, and accounting for shifting patterns and emerging threats guarantees their ability to stay ahead of the curve and effectively detect anomalies in real-time. We enable efficient anomaly detection in resource-constrained environments by integrating advanced algorithms such as LSTM networks and CNNs into TinyML devices. We usher in a paradigm shift by adopting this meticulous and intricate threefold process for developing IoT anomaly detection models using TinyML. We pave the way for highly effective and robust anomaly detection solutions through the meticulous acquisition and preprocessing of data, the design and training of advanced ML models, and rigorous performance evaluation and optimization. With these solutions, IoT devices can safeguard against potential threats, enabling a future where seamless connectivity and secure operations thrive together [52].

3.2.9 TinyML Applications in Anomaly Detection

In this groundbreaking and innovative work, we propose a cutting-edge and revolutionary method for abnormality detection by training a state-of-the-art one-class neural network model. Our model leverages the power of public image data collected from various social network services and taps into the vast potential of the JSA database. To ensure accuracy and reliability, we have meticulously integrated image data from X-ray scans of passenger baggage shared on social networks and comprehensive records of illegal transportation incidents from the JSA database. Recognizing the sensitive nature of X-ray scan images, which often contain crucial threat-related information, we have taken extra precautions. Before utilizing these images to create a dataset, our dedicated team manually segregated them into distinct categories of threat and non-threat, facilitating a more targeted approach toward anomaly detection. These carefully sorted images have been securely stored, guaranteeing utmost privacy and adherence to data protection protocols [68].

The TinyML model then processes these images and compares them to a predefined regular behavior pattern, allowing for real-time anomaly detection. The analysis is then used to identify deviations from the norm, triggering alerts or necessary actions. The analysis is then used to identify deviations from the norm, triggering alerts or essential actions. For example, in industrial equipment monitoring, TinyML can detect abnormal patterns in sensor data and alert maintenance personnel. Moreover, we have implemented a groundbreaking quantization technique to fully harness the immense capabilities of advanced neural networks. This cutting-edge method not only enhances our model's performance but also leverages the transformative potential of TinyML technology [69]. Table 3.6 presents the diverse application scenarios of TinyML in anomaly detection and beyond.

Table 3.6 Diverse application scenarios of TinyML in anomaly detection and beyond.

Application scenario	Description	Related work	References
Industrial equipment monitoring	Detects abnormal patterns in sensor data to alert maintenance personnel for proactive intervention	The TinyML model processes sensor data and identifies deviations from predefined norms	[70]
Healthcare monitoring	It enables continuous monitoring of vital signs through wearable devices, facilitating the early detection of abnormalities and enhancing patient care	Real-time data analysis for personalized treatment plans and early intervention	[71]
Environmental preservation	Analyzes data from sensors, satellite imagery, and audio recordings to detect ecological imbalances, aiding biodiversity conservation and promoting sustainability	Deployment of TinyML in ecological monitoring systems for enhanced environmental analysis	[10]
Robotics and automation	Integrating TinyML into robotic systems improves autonomous decision-making, dynamic adaptation, and precise task execution	Enhances manufacturing with real-time anomaly detection and logistics with supply chain security	[4]
Manufacturing and predictive maintenance	Assists autonomous navigation and equipment anomaly detection, enhancing the reliability of space missions	TinyML for equipment monitoring and navigation in space exploration contexts	[72]

By utilizing quantization, we have effectively optimized the performance and efficiency of our neural network, pushing the boundaries of what is possible in anomaly detection. In our rigorous and comprehensive experimentation, we conducted a detailed comparison between the average inference performance of our model, both with and without quantization. Our meticulous analysis revealed that MobileNet, renowned for its exceptional inference speed and compact size, is an ideal benchmark for our evaluation. Without quantization, our model showcases unparalleled accuracy, achieving an impressive 83.8%

accuracy. However, to overcome the limitations of storage space, we exploited the power of quantization, ultimately reducing the model's size by a staggering 4.5 times. This reduction in storage space results in a slight decrease in accuracy, equivalent to 18.8% lower than the original inference model. Nevertheless, even with this reduction, our model still delivers reliable results, ensuring the utmost precision in abnormality detection.

Our groundbreaking work presents a landmark contribution to the field of abnormality detection [73]. We have developed a highly accurate and efficient one-class neural network model by leveraging the immense potential of social network image data and the extensive JSA database while incorporating meticulous measures to address sensitivity. The implementation of quantization and the utilization of TinyML technology further enhance the capabilities of our model, propelling it to the forefront of anomaly detection research. Through our feasibility study and comparative analysis, we have demonstrated the outstanding performance of our model, even with a slight trade-off in accuracy. This work lays the groundwork for future advancements in anomaly detection, revolutionizing the fields of security and public safety. This work lays the groundwork for future advancements in anomaly detection, revolutionizing the fields of security and public safety. It also opens new possibilities for real-time monitoring and proactive intervention. It also opens new possibilities for real-time monitoring and proactive intervention in various industries such as healthcare, manufacturing, and transportation [5].

TinyML has recently been applied in multiple fields, resulting in numerous fascinating works. The detection of abnormalities and anomalies has driven these impressive achievements. Among the intriguing works, we can find the application of TinyML in rust detection utilizing on-site YOLO combined with FFT spectroscopy. Another compelling example is the utilization of TinyML in gas leakage detection, where a temperature sensor's temperature pattern is combined with a CNN to achieve remarkable results.

Additionally, TinyML has been employed in animal movement analysis, leveraging accelerometer data to gain valuable insights. It is worth noting that, in addition to developing anomaly detection applications, substantial efforts are being devoted to constructing high-quality datasets. In this regard, companies specializing in anomaly detection have taken the initiative to utilize public data to develop innovative and advantageous services. These companies recognize the potential of public datasets and advocate for their use in building sophisticated models. To improve anomaly detection in various industries and applications. However, it's essential to acknowledge that certain limitations may arise when working with analysis algorithms, such as constraints in computing power and the duration required for data analysis [74]. Figure 3.16 depicts the applications of TinyML in real-time anomaly detection and predictive monitoring across diverse domains.

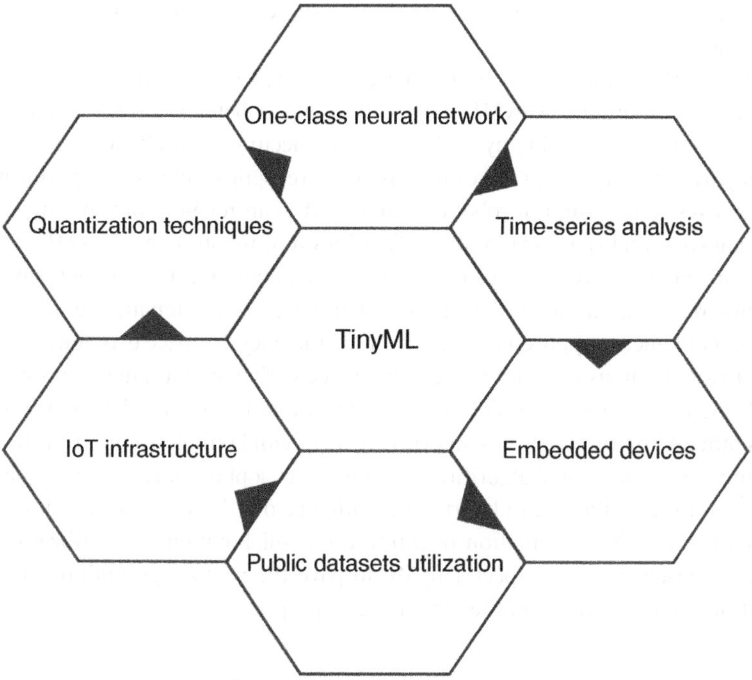

Figure 3.16 Applications of TinyML in real-time anomaly detection and predictive monitoring across diverse domains.

Considering these constraints, a viable solution is to leverage the vast resources available in public datasets and utilize them to build robust models. One can overcome these limitations by using public datasets, unlocking the potential for more advanced analysis, and developing innovative solutions. Furthermore, one can envision even more possibilities for TinyML soon. For example, researchers are exploring the potential of TinyML in weather forecasting, where real-time data from various sensors combined with ML algorithms can provide accurate and localized weather predictions. Another exciting avenue is the application of TinyML in healthcare, which enables the continuous monitoring of vital signs through wearable devices, leading to the early detection of abnormalities and ultimately improving the overall quality of patient care. This can be achieved using advanced ML algorithms and real-time data analysis, enabling proactive interventions and personalized patient treatment plans. In addition to these advancements, the scope of TinyML is expanding into environmental preservation. By deploying TinyML in ecological monitoring systems, it becomes possible to analyze and interpret data from various sources, such as audio recordings, satellite imagery, and environmental sensors. This can help detect and address ecological imbalances, support biodiversity conservation, and promote sustainable practices [75].

Moreover, integrating TinyML with robotics opens new prospects in the automation field. Incorporating TinyML capabilities into robotic systems enhances autonomous decision-making by making it more accurate and efficient. Robots with TinyML can adapt to dynamic environments, recognize objects, and perform complex tasks. This fusion of robotics and TinyML can revolutionize manufacturing, logistics, and space exploration. This fusion of robotics and TinyML can revolutionize manufacturing, logistics, and space exploration. In manufacturing, it enables real-time anomaly detection and predictive maintenance, while in logistics, it enhances supply chain security and efficiency. In space exploration, it can aid in autonomous navigation and anomaly detection for equipment and systems. While TinyML has made significant contributions to anomaly detection and various other domains, it remains an evolving field with boundless opportunities. As technology advances and algorithms become more sophisticated, the potential for TinyML to revolutionize industries and enhance our daily lives continues to grow. With the right combination of innovation, collaboration, and utilization of public datasets, we can expect TinyML to pave the way for groundbreaking applications and discoveries in the coming years [76].

3.2.10 Practical Applications of TinyML in Anomaly Detection

Anomaly detection is an essential component in maintaining the reliability, security, and operational efficiency of modern systems across diverse sectors, including industrial operations, healthcare, cybersecurity, and financial services. It involves identifying irregular patterns or deviations from expected behavior, which can indicate potential failures, security breaches, or fraud. As the complexity and scale of data continue to grow, particularly with the increasing proliferation of IoT devices and the widespread adoption of EC, traditional anomaly detection methods that rely on centralized processing or heuristic rule-based systems are encountering significant limitations. These approaches often struggle to process the massive volumes of data generated by millions of connected devices in real-time, leading to delays in detecting anomalies and an increase in system vulnerability.

TinyML offers a transformative approach to anomaly detection by enabling the deployment of efficient, lightweight ML models directly onto resource-constrained edge devices. These models are designed to operate on devices with limited processing power, memory, and storage, addressing the challenges posed by traditional cloud-based solutions. TinyML leverages a variety of techniques, such as model quantization, pruning, and hardware acceleration, to optimize the performance of ML models while keeping their resource consumption minimal. This allows for real-time data processing and anomaly detection at the point of data generation, reducing the need for continuous data transmission to centralized servers and improving response times.

The integration of TinyML in edge devices significantly enhances the capability of anomaly detection systems by enabling them to operate autonomously in dynamic environments. These edge-based models are capable of monitoring and analyzing streams of sensor data – such as temperature, pressure, or network traffic – on devices like microcontrollers, wearables, or industrial sensors. The ability to perform anomaly detection locally, without relying on cloud infrastructure, not only reduces latency and bandwidth requirements but also improves privacy and security by ensuring that sensitive data does not need to be transmitted over networks.

Moreover, TinyML facilitates the scalability of anomaly detection systems by supporting the deployment of models across many devices, each performing real-time analysis and flagging anomalous patterns as they occur. This distributed processing approach ensures that systems can scale efficiently to handle large-scale IoT deployments, where each edge device functions as an intelligent node capable of detecting anomalies in isolation or in collaboration with other devices. TinyML's application in anomaly detection is particularly advantageous in scenarios where resources such as computational power, memory, and energy are constrained. It enables real-time monitoring and decision-making, which is critical for systems that require immediate responses to detect and mitigate potential threats or failures. Additionally, the ability to deploy these models on edge devices makes it possible to run anomaly detection algorithms in environments where cloud connectivity is intermittent or unreliable, further enhancing the robustness and resilience of the system.

As TinyML continues to evolve, its potential for enhancing anomaly detection in edge devices grows. With ongoing advancements in hardware design and ML techniques, TinyML offers an increasingly viable solution for deploying intelligent, scalable, and energy-efficient anomaly detection systems capable of operating autonomously at the edge of the network. This subsection explores the practical applications of TinyML in anomaly detection across various domains, as defined in Figure 3.5, emphasizing its transformative potential in addressing challenges related to time-sensitive and resource-efficient anomaly detection [12].

3.2.10.1 Industrial Machinery Monitoring

One of the most transformative applications of TinyML in anomaly detection is in the real-time monitoring of industrial equipment. Industrial systems, which are often integral to manufacturing, energy production, and other critical sectors, demand continuous and uninterrupted operation. Consequently, the early detection of equipment malfunctions or operational anomalies becomes paramount to prevent costly downtime, improve safety, and avert potentially catastrophic failures. TinyML offers a compelling solution by enabling the deployment of ML models directly on edge devices, such as low-power microcontrollers or specialized embedded systems, which are seamlessly integrated with industrial machinery. These models are trained to process sensor data – including vibrations, temperature, pressure, humidity, and acoustic signals – locally, often in real-time, without the need for constant data transmission to centralized cloud or server systems. This distributed approach minimizes the dependency on stable network connectivity, which can be prone to interruptions in remote or harsh industrial environments.

For example, vibration analysis of rotating machinery, such as motors, turbines, or pumps, can reveal microscale deviations in frequency or amplitude that indicate incipient mechanical failure, such as bearing wear, misalignment, or imbalance. Traditional anomaly detection systems rely on centralized architectures, where raw sensor data is transmitted to a cloud-based platform for processing. This process introduces inherent latency, which can be detrimental in high-stakes environments, where delays in detecting faults may result in expensive repairs, production halts, or even system-wide shutdowns. In contrast, TinyML allows for the preprocessing and analysis of sensor data at the source, enabling edge devices to identify abnormalities, diagnose issues, and trigger proactive maintenance alerts immediately – often before a failure is perceptible to human operators.

Additionally, TinyML models can be optimized for resource-constrained environments through techniques such as model quantization, pruning, and knowledge distillation, ensuring that they maintain high levels of accuracy without overwhelming the limited processing power and memory of edge devices. Furthermore, by deploying anomaly detection models directly on the edge, manufacturers gain enhanced privacy, as sensitive operational data never needs to leave the device, improving data security and reducing the risks associated with data transmission and storage.

Incorporating TinyML into industrial anomaly detection systems therefore addresses critical challenges: it reduces latency, enhances operational efficiency, and enables real-time, actionable insights that empower maintenance teams to perform predictive and preventive actions. This approach not only improves equipment uptime but also extends the lifespan of industrial assets, ultimately contributing to the optimization of overall system performance, cost savings, and safety. The low power consumption of TinyML-enabled devices ensures continuous operation without frequent battery replacements, making it particularly suitable for remote or hard-to-reach industrial settings [77].

3.2.10.2 Infrastructure Health Monitoring

The monitoring of structural health in critical infrastructure, such as bridges, tunnels, and buildings, is a vital application of anomaly detection, particularly in the context of maintaining public safety and optimizing maintenance costs. Traditional methods of structural health monitoring (SHM) often rely on centralized systems that transmit large volumes of data to remote servers for processing. This approach, while effective, can introduce significant latency, communication overhead, and energy consumption, especially when dealing with real-time monitoring of large-scale infrastructures. TinyML, by contrast, offers a powerful alternative by enabling the deployment of lightweight, energy-efficient ML models directly on low-power edge devices, such as microcontrollers and sensors embedded within the infrastructure itself. These sensors can continuously capture critical data points, such as strain, displacement, vibration, temperature, or acoustic emissions, and process this information locally to detect anomalous patterns that may indicate potential structural issues.

For example, microcontrollers equipped with TinyML models can analyze strain gauges and accelerometer data in real-time to detect subtle changes in stress and vibration patterns, which may signal early signs of structural fatigue, material degradation, or the formation of cracks in a bridge or tunnel. By leveraging advanced anomaly detection techniques, such as unsupervised learning algorithms or one-class classifiers, these models can differentiate between normal operating conditions and potentially hazardous deviations. This enables the identification of structural anomalies long before they escalate into significant failures, allowing for early detection of issues such as crack propagation, corrosion, or excessive load-bearing stress.

In addition to anomaly detection, TinyML can also contribute to predictive maintenance by forecasting potential future failures based on historical data and trends. For example, by continuously monitoring the acoustic signals emitted by a bridge under stress, TinyML-based models can learn to recognize the early sounds of material degradation, such as the formation of microcracks, and predict when maintenance or repairs may be necessary. This predictive capability further enhances the cost-effectiveness of infrastructure management by allowing for scheduled, proactive interventions rather than reactive, emergency repairs. The key advantage of TinyML in this context lies in its ability to perform data processing and analysis locally, thus significantly reducing the need for continuous data transmission to central servers, which can be bandwidth-intensive and time-consuming. This localized data processing reduces latency, ensures real-time analytics, and minimizes the energy consumption of the system, making it ideal for monitoring remote or difficult-to-reach infrastructure in environments where power and connectivity are limited. Moreover, by utilizing techniques such as model quantization and pruning, TinyML models can operate efficiently on resource-constrained devices while maintaining high levels of accuracy in anomaly detection.

Overall, the integration of TinyML into structural health monitoring systems offers a transformative solution for ensuring the longevity and safety of critical infrastructure. By enabling continuous, real-time detection of potential structural issues, TinyML empowers engineers and maintenance teams to take proactive steps in preserving infrastructure integrity, thereby enhancing public safety, reducing downtime, and lowering the overall cost of maintenance and repairs. The localized processing of data by TinyML minimizes the bandwidth required for transmitting sensor readings and enhances privacy by avoiding the need to send raw data to the cloud [7].

3.2.10.3 Cybersecurity in IoT Networks

IoT networks are intrinsically susceptible to cybersecurity threats, primarily due to their decentralized architecture, heterogeneous devices, and resource-constrained nature. These networks typically operate across large, distributed environments, often with limited computational power, storage, and energy resources, which makes them particularly vulnerable to malicious activities. TinyML has emerged as a promising solution to address these vulnerabilities by enabling local anomaly detection on edge devices within the IoT ecosystem. TinyML leverages lightweight ML models that can run on resource-constrained devices, offering a powerful approach for real-time anomaly detection without the need for constant communication with centralized servers. By detecting abnormal patterns in data locally at the edge, TinyML can identify potential security breaches as they occur, minimizing latency and reducing the risk of widespread attacks.

For example, in the context of smart home devices, TinyML models can be deployed to monitor user behavior patterns and network traffic to identify deviations from baseline behaviors. This includes detecting unusual login times, access attempts to restricted functionalities, or anomalous communication patterns indicative of potential cyberattacks such as distributed denial of service (DDoS), botnet activities, or unauthorized access. Additionally, TinyML models can monitor communication anomalies, such as unexpected device interactions or traffic spikes, which may suggest a security threat like a compromised device or network intrusion. Implementing anomaly detection directly at the edge offers several advantages in terms of security. Edge-based anomaly detection allows for immediate action when suspicious activities are detected. For instance, if an unusual access pattern or deviation from expected behavior is identified, the edge device can immediately block the suspicious activity or alert the user, preventing the propagation of threats across the IoT network. This localized, real-time decision-making significantly reduces the reliance on cloud-based processing, thereby mitigating latency, conserving bandwidth, and enhancing privacy by avoiding the transmission of sensitive data to centralized servers.

Moreover, TinyML's ability to run on low-power, low-memory devices ensures that even IoT devices with minimal resources can participate in advanced anomaly detection processes without significantly impacting their operational performance. This is particularly crucial for IoT networks that operate in remote or embedded environments where computational resources are scarce, and the costs of over-the-air data transfer and processing in centralized cloud infrastructures can be prohibitive. The use of TinyML for cybersecurity in IoT networks also aligns with the principle of distributed trust, as it allows each device to independently assess its security posture, reducing single points of failure. As IoT ecosystems continue to expand, leveraging TinyML for anomaly detection will be essential in creating more resilient, self-aware, and secure networks that can proactively identify and mitigate cybersecurity threats at the earliest possible stage. Moreover, this approach alleviates the need for constant connectivity to centralized servers, improving security and reducing latency [78]. Figure 3.17 presents the applications of TinyML in real-time anomaly detection across diverse domains.

3.2.10.4 Environmental Monitoring

In environmental monitoring, anomaly detection in sensor data is crucial for identifying critical events such as natural disasters, pollution, or irregular weather patterns. These events often require timely detection to enable rapid response and mitigation efforts. TinyML, a specialized subset of ML designed for deployment on resource-constrained devices, offers significant advantages in this context. By leveraging TinyML, lightweight, efficient models can be deployed on distributed sensors placed in remote, hard-to-reach locations, enabling real-time data analysis at the point of collection. This reduces the need for constant data transmission to centralized servers, thus conserving both energy and bandwidth while enabling low-latency decision-making.

For example, in wildfire detection, a network of TinyML-powered sensors distributed across forested areas can continuously monitor environmental variables such as temperature, humidity, and gas levels. The models embedded in these sensors can detect subtle changes in data patterns that may indicate the onset of a wildfire, such as rapid increases in temperature combined with a drop in humidity, or the detection of specific gases associated with combustion. By processing this data locally, these sensors can trigger immediate alerts, enabling early intervention to prevent the spread of wildfires and protect lives and property.

Similarly, in urban environments, TinyML-enabled sensors can be deployed across a city's infrastructure to monitor air quality and detect anomalous

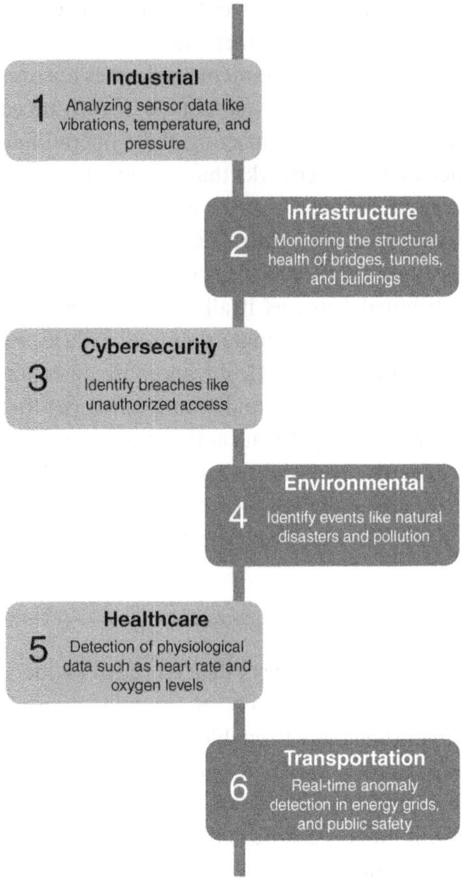

Figure 3.17 Applications of TinyML in real-time anomaly detection across diverse domains.

fluctuations in pollutant concentrations, such as sudden spikes in nitrogen dioxide (NO_2), sulfur dioxide (SO_2), or particulate matter (PM2.5). These anomalies could signal a potential environmental or public health hazard, such as a chemical spill, industrial accident, or heavy traffic congestion. Through real-time data analysis on edge devices, these systems can provide immediate notifications to local authorities or the public, prompting swift action to mitigate exposure to harmful pollutants, enact traffic management measures, or activate emergency response protocols.

In both cases, TinyML allows for the integration of sophisticated ML models with minimal computational resources, ensuring that environmental monitoring systems can scale efficiently while maintaining high levels of accuracy in anomaly detection. Additionally, TinyML models can be optimized using techniques like model pruning, quantization, and edge-specific optimization, ensuring that even in low-power, memory-constrained devices, performance remains robust. This approach empowers a wide array of environmental monitoring systems to become more autonomous, responsive, and scalable, contributing significantly to both disaster prevention and environmental health protection. The compact and energy-efficient nature of TinyML models makes them ideal for long-term deployment in battery-powered environmental monitoring systems [17].

3.2.10.5 Healthcare and Wearable Devices

In healthcare, the integration of TinyML with wearable devices is revolutionizing real-time health monitoring and early anomaly detection. TinyML models, optimized for low-power, resource-constrained edge devices, can analyze continuous physiological data directly from wearable sensors, enabling the detection of subtle deviations from normal patterns that could indicate underlying health issues. These models, running on devices like smartwatches or fitness trackers, provide immediate, on-device analysis, reducing the need for constant cloud connectivity and minimizing latency in decision-making. For example, a TinyML model implemented on a smartwatch can continuously monitor heart rate variability and detect arrhythmic patterns indicative of potential cardiac issues such as atrial fibrillation or ventricular tachycardia. By recognizing these irregularities in real-time, the system can alert the wearer to the abnormal rhythm, prompting timely medical intervention, and potentially preventing life-threatening complications. The ability to monitor these conditions in real-time offers a significant advantage over traditional healthcare systems that rely on sporadic, in-clinic assessments.

In addition to cardiovascular monitoring, TinyML can be employed to analyze motion sensor data from wearable devices to detect abnormal gait patterns. Such abnormalities may signal the early onset of neurological conditions like Parkinson's disease or indicate an increased risk of falls in elderly individuals. By continuously tracking movement patterns, these devices can detect subtle changes in walking speed, stride length, or symmetry that may not be immediately noticeable to the wearer or healthcare providers. For instance, a slight alteration in gait, such as a shuffling step or uneven weight distribution, could indicate an elevated risk of falls or the progression of a neurological disorder. In these cases, the TinyML model could send an alert to caregivers or family members, enabling proactive measures to prevent injury. Furthermore, TinyML enhances the personalization of healthcare by adapting its anomaly detection models to individual baseline health metrics. For example, a wearable device could learn a user's normal heart rate variability over time, and then detect any significant deviations based on personalized thresholds. This individualized approach ensures that the system remains sensitive to the specific needs of each wearer while minimizing false positives that could arise from generalized models.

The advantages of TinyML in healthcare extend beyond real-time detection. They also contribute to reducing the overall healthcare burden by enabling continuous monitoring outside clinical settings, reducing hospital readmissions, and providing healthcare professionals with more accurate, real-time data to inform clinical decisions. Additionally, by operating at the edge, TinyML models preserve user privacy, as personal health data can be processed locally, minimizing the need for sensitive data transmission.

However, several challenges remain, including the need for advanced data preprocessing and model training to ensure high accuracy with limited computational resources. Ensuring the robustness and reliability of TinyML models across diverse environmental conditions, such as varying sensor quality or user activity levels, also remains an ongoing area of research. Despite these challenges, the ongoing advancements in TinyML offer significant potential to enhance the accessibility, efficiency, and personalization of healthcare services, paving the way for smarter, more proactive health management systems. By processing data locally, TinyML ensures privacy while delivering timely insights that can enhance patient outcomes [79].

3.2.10.6 Smart Cities and Transportation Systems

TinyML has the potential to revolutionize real-time anomaly detection in smart cities, particularly within critical infrastructures such as transportation systems, energy grids, and public safety frameworks. By integrating ML models directly on resource-constrained edge devices, TinyML enables the deployment of intelligent, low-latency solutions that can process data locally, reducing the need for constant communication with centralized servers and thus alleviating bandwidth limitations and improving responsiveness. In transportation systems, for instance, TinyML-powered traffic sensors can continuously monitor vehicle flow, road conditions, and environmental factors in real-time. These sensors, equipped with on-device anomaly detection models, can identify irregular patterns such as sudden stops, unusual congestion, or even traffic accidents. By recognizing these anomalies as they occur, TinyML models can trigger immediate interventions such as dynamic traffic signal adjustments or real-time rerouting instructions for drivers, thus enhancing the overall flow of traffic and minimizing delays. This immediate, localized response is crucial in urban environments, where time-sensitive decisions can significantly impact traffic management and public safety.

Additionally, in energy grids, TinyML enables real-time monitoring of power distribution systems, detecting anomalies such as voltage irregularities, sudden load spikes, or equipment malfunctions. By deploying anomaly detection models at the edge, near power meters or distribution units, potential failures can be identified and addressed before they escalate, preventing widespread outages and improving the resilience of smart grids. This decentralized approach reduces the reliance on cloud-based systems for real-time decision-making, which is often hindered by latency issues, and ensures a more robust, scalable infrastructure for critical energy systems. In the context of public safety infrastructures, TinyML can support surveillance and environmental monitoring systems by detecting unusual behaviors, such as loitering in restricted areas or sudden environmental changes like chemical leaks or fires. Real-time anomaly detection can trigger immediate alerts to emergency response teams, enabling faster deployment of resources and reducing response times during critical situations. Furthermore, TinyML models deployed on surveillance cameras or environmental sensors can continuously learn and adapt to evolving patterns, improving their accuracy and efficiency over time.

By leveraging TinyML for anomaly detection, smart cities can enhance operational efficiency, reduce risks, and improve the quality of life for citizens. TinyML's ability to process data locally on edge devices not only ensures faster, more responsive actions but also promotes the scalability and sustainability of smart city solutions, which are increasingly essential as urban populations grow and infrastructure demands intensify. However, the adoption of TinyML also requires overcoming challenges such as optimizing model accuracy within constrained hardware resources, ensuring privacy and security in data processing, and managing the complexity of continuously evolving anomaly patterns in dynamic environments. Despite these challenges, the integration of TinyML presents an exciting opportunity for intelligent, autonomous urban management. In energy grids, TinyML models can identify unusual power consumption patterns or detect faults in distribution lines, ensuring a stable energy supply and reducing the risk of blackouts [80].

3.3 Lessons Learned

Developing and deploying anomaly detection algorithms on constrained devices, particularly in IoT and EC environments, has unveiled several critical insights. These lessons span technical, operational, and strategic domains, reflecting the complexities and opportunities inherent in this evolving field.

3.3.1 Adaptability of Models to Constrained Environments

One of the most significant realizations is the need for algorithms to be inherently adaptable to constrained environments. While effective in high-resource settings, traditional ML models often fail to meet the stringent requirements of IoT devices, such as limited memory, processing power, and energy efficiency.

Techniques such as model quantization, pruning, and knowledge distillation have emerged as indispensable tools. Simplifying models without compromising their predictive performance ensures that anomaly detection systems remain feasible for deployment on microcontrollers and other resource-constrained devices. This underscores the importance of prioritizing lightweight, energy-efficient models in IoT applications.

3.3.2 Importance of Real-Time Processing

IoT systems operate in real-time environments where latency can significantly impact system reliability and user experience. Lessons from deploying models like TinyML underscore the importance of reducing dependency on cloud-based computations for enabling real-time anomaly detection. This shift not only enhances responsiveness but also mitigates challenges related to bandwidth constraints and intermittent connectivity. Implementing on-device inference proved pivotal in achieving real-time detection, allowing systems to react swiftly to potential threats or irregularities.

3.3.3 Balancing Precision and Efficiency

Striking the right balance between precision and efficiency in anomaly detection models was a recurring challenge. High precision is necessary to minimize false alarms, while efficiency ensures the feasibility of deploying models on constrained devices. Achieving this balance involved carefully tuning hyperparameters, using advanced optimization techniques, and adopting hybrid approaches that combine traditional and modern ML methods. This balance is crucial for ensuring the accuracy of anomaly detection and maintaining trust and reliability in IoT ecosystems.

3.3.4 Data Quality and Preprocessing

The diversity and noise in IoT sensor data often pose challenges to effective anomaly detection. Preprocessing steps, such as filtering outliers, normalizing data, and handling missing values, proved essential for enhancing model accuracy. Moreover, integrating domain knowledge into the preprocessing pipeline significantly enhanced the relevance of the extracted features. This lesson highlights the vital role of high-quality data preparation in the success of ML applications within IoT systems.

3.3.5 Privacy and Security Concerns

Privacy and security emerged as primary considerations, particularly given the sensitive nature of data processed by IoT devices. On-device anomaly detection using TinyML offered a compelling solution by reducing data transmission to external servers, thereby minimizing exposure to potential breaches. However, the need for secure model updates and robust encryption mechanisms during deployment remains an ongoing concern. Addressing these issues is crucial for fostering user confidence and ensuring the adoption of anomaly detection technologies in IoT environments.

3.3.6 Scalability and Deployment Challenges

Scaling anomaly detection solutions to accommodate diverse IoT ecosystems presents unique challenges. Each deployment scenario required tailored configurations to account for variations in sensor types, environmental conditions, and usage patterns. Lessons learned from these experiences emphasized the importance of modular and flexible architectures that can be easily customized for different applications. Additionally, robust deployment strategies, including over-the-air updates and remote diagnostics, were critical for maintaining system reliability and performance.

3.3.7 Continuous Learning and Adaptation

IoT environments are inherently dynamic, with data distributions and operational conditions evolving. Models deployed in these settings must support continuous learning and adaptation to maintain relevance and accuracy. Techniques such as incremental learning and unsupervised anomaly detection effectively addressed this challenge. These approaches allowed models to update themselves based on new data without requiring complete retraining, enhancing their longevity and utility.

3.3.8 Collaborative and Interdisciplinary Approaches

Implementing anomaly detection in IoT systems often requires collaboration across multiple disciplines, including ML, embedded systems, and domain-specific expertise. Engaging stakeholders from diverse backgrounds facilitated a deeper understanding of the problem space and enabled the design of more effective solutions. This lesson highlights the importance of interdisciplinary collaboration in addressing complex technological challenges.

3.3.9 Transformative Potential of TinyML

The integration of TinyML into anomaly detection workflows highlighted its transformative potential. By enabling advanced analytics directly on edge devices,

TinyML improved operational efficiency and opened new avenues for innovation. Applications such as predictive maintenance, real-time fault detection, and autonomous system optimization have demonstrated the profound impact of this technology. This lesson reinforces the importance of investing in and advancing TinyML technologies to unlock their full potential.

3.3.10 Open Challenges

Despite significant progress, several challenges remain open. Developing more robust models that can generalize across a wide range of scenarios without extensive customization is a key area for future research. Additionally, enhancing the interpretability of anomaly detection models to provide actionable insights is critical for end-user adoption. Addressing these challenges will require sustained efforts and collaboration between researchers, developers, and industry stakeholders.

3.4 Future Scope

The continued evolution of IoT ecosystems and the increasing capabilities of edge devices present exciting opportunities for advancing anomaly detection methodologies. Integrating TinyML with IoT and edge devices has unlocked immense potential, but there is still significant room for improvement and exploration. Future research should focus on enhancing the scalability, adaptability, and efficiency of TinyML models to cater to IoT systems' diverse and dynamic requirements [81].

3.4.1 Model Optimization and Transfer Learning

As IoT devices become increasingly heterogeneous, developing generalized TinyML models that are capable of transfer learning can address the need to deploy anomaly detection across various hardware platforms. Future work can emphasize creating modular and lightweight models that can be easily retrained or fine-tuned on new datasets without requiring extensive computational resources. Model optimization involves designing lightweight, efficient models that can operate effectively within the constraints of resource-constrained devices, such as those with limited memory and processing power. Optimizing models for TinyML involves techniques such as pruning, quantization, and distillation to reduce the size and complexity of DL models while maintaining their performance. On the other hand, transfer learning enables a model trained on one dataset or device to be adapted for use on another dataset or device with minimal retraining. In the context of anomaly detection, transfer learning can significantly reduce the time and resources required to develop models for new or unseen environments. By leveraging pretrained models on similar tasks or datasets, adapting them to new IoT devices or new anomaly detection scenarios becomes more efficient.

3.4.2 Energy-Efficient Algorithms

Energy consumption remains a critical concern for IoT edge devices. Advancements in energy-efficient algorithms, including pruning, quantization, and neural architecture search tailored for constrained devices, are essential. Techniques that reduce computational complexity without compromising model accuracy could revolutionize anomaly detection in battery-operated devices. Key techniques driving energy efficiency in IoT devices include pruning, quantization, and neural architecture search (NAS), all of which contribute to reducing the computational load of ML models. Pruning involves removing redundant or less significant weights from neural networks, thereby reducing the number of computations required during inference. By eliminating unnecessary parameters, pruning reduces power consumption and accelerates processing, making the model more suitable for real-time applications.

3.4.3 Federated and Collaborative Learning

Federated learning represents a transformative approach for privacy-preserving anomaly detection in IoT and edge devices, particularly in scenarios where data sensitivity and user privacy are paramount. In traditional ML paradigms, raw data from devices or users is sent to a centralized server, where models are trained and refined. However, this centralized approach raises significant concerns about data security and privacy, as sensitive user information may be exposed during the data transfer. Federated learning addresses these concerns by allowing models to be trained locally on edge devices, where data remains on the device. Instead of transmitting raw data, only the model updates or gradients are sent to the central server, ensuring that sensitive information is never shared or exposed. This privacy-preserving feature of federated learning makes it an ideal solution for anomaly detection in IoT systems, where data may include private or confidential information, such as health metrics in wearable devices or security data from smart homes. By utilizing federated learning, IoT devices can collaboratively enhance their anomaly detection models by learning from the data they collect while maintaining the data's privacy and security on the devices themselves. This collaborative model enables the continuous improvement of detection algorithms, allowing them to adapt to new types of anomalies and enhance overall accuracy.

3.4.4 Real-Time and Adaptive Detection

IoT environments are inherently dynamic, characterized by constantly changing operational patterns and unexpected conditions. Devices in these environments continuously collect data, and the nature of the data and its underlying distribution can shift over time due to various factors, such as changes in user behavior, environmental influences, or updates to the system's configuration. Traditional anomaly detection models often rely on static assumptions about data distributions, making them ill-suited to adapt to these evolving conditions without significant retraining or human intervention. Therefore, there is a pressing need to develop adaptive anomaly detection models that can self-calibrate to new patterns and data distributions in real time, thereby maintaining ongoing accuracy in anomaly identification. These adaptive systems should not only detect outliers or irregularities but also understand the context of the data and the system's changing state, making them more effective at distinguishing between normal fluctuations and true anomalies.

3.4.5 Multimodal Data Fusion

In IoT systems, data is often generated from various sensors, devices, and other sources, each capturing different aspects of the system's environment. For example, sensors may collect data from temperature, humidity, motion, pressure, and vibration, among others, depending on the application. However, relying on a single data source to detect anomalies may not always provide a comprehensive understanding of the underlying system behavior. This is where multimodal data fusion becomes particularly valuable. By integrating data from multiple sensor types or sources, IoT systems can create a more holistic and accurate view of the system's status, significantly improving the robustness of anomaly detection.

3.4.6 Security and Robustness

As IoT continues to proliferate, it becomes an increasingly attractive target for cyberattacks, ranging from data breaches to denial-of-service attacks. The rise of sophisticated attack vectors poses a significant challenge to the integrity and trustworthiness of IoT systems, making security a top priority. Traditionally seen as a safeguard against system faults, anomaly detection must also evolve to detect and mitigate malicious activities that exploit vulnerabilities in these interconnected devices. As cyberattacks become more advanced and targeted, ensuring that anomaly detection models can withstand adversarial manipulation and false positives is critical. Adversarial attacks, which involve intentionally modifying inputs to confuse or mislead ML models, present a substantial risk to the reliability of IoT systems. To address this, future research should focus on developing robust anomaly detection models that are resilient to such attacks.

3.4.7 Integration with Emerging Technologies

Integrating anomaly detection with emerging technologies such as 5G, blockchain, and quantum computing presents exciting opportunities to address key challenges that have traditionally hindered real-time anomaly detection in large-scale IoT networks. These technologies offer unique capabilities that can enhance the performance, security, and efficiency of anomaly detection systems, facilitating more scalable and adaptive solutions for IoT ecosystems. One of the primary benefits of 5G technology is its ultralow latency and high data transfer rates, which are essential for enabling real-time anomaly detection in IoT networks. With IoT devices generating vast amounts of data, the ability to transmit this data quickly and efficiently is crucial for promptly identifying anomalies. Security and data integrity are significant concerns in IoT systems, where devices often operate autonomously and in decentralized environments.

Blockchain technology can provide a robust solution by offering a transparent and immutable ledger for recording data transactions. Quantum computing holds the potential to revolutionize anomaly detection in IoT networks by providing massive computational power that can process complex data sets at speeds far exceeding those of classical computers. IoT systems often generate highly complex and voluminous data, and quantum algorithms could enhance the ability to identify subtle, low-probability anomalies that traditional methods may overlook. Quantum computing could accelerate ML models for anomaly detection, enabling more sophisticated and accurate models that can adapt in real-time to changing conditions in IoT environments.

3.4.8 Applications in New Domains

As anomaly detection continues to evolve, its potential extends beyond its current applications in predictive maintenance and cybersecurity. Future research can explore its transformative impact in emerging domains such as healthcare IoT, smart agriculture, and autonomous systems, each presenting its unique challenges and opportunities. In healthcare IoT, integrating anomaly detection can lead to significant advancements in patient care. IoT devices, such as wearable sensors and remote monitoring systems, continuously collect patient data, including vital signs, glucose levels, and heart rate. By leveraging anomaly detection algorithms, healthcare providers can identify early signs of potential health issues, such as sudden changes in a patient's condition or abnormal trends in their vital metrics. This enables proactive intervention, reducing emergency incidents, hospital readmissions, and improving patient outcomes. However, healthcare IoT presents challenges related to data privacy, regulatory requirements, and the need for highly accurate and reliable anomaly detection models that operate in real-time and with limited computational resources.

3.5 Conclusion

This chapter underscores the transformative potential of TinyML in anomaly detection for IoT and edge devices. By enabling local, real-time, and autonomous data analysis, TinyML provides an efficient solution for resource-constrained devices, overcoming the limitations of traditional anomaly detection methods. These conventional approaches often struggle with IoT's unique challenges, such as limited computational power, high data transmission costs, and privacy concerns. TinyML addresses these issues by allowing IoT devices to perform sophisticated data analysis locally, reducing reliance on centralized cloud systems and ensuring quicker responses to detected anomalies. The chapter also highlights significant advancements in energy-efficient algorithms, adaptive learning techniques, federated learning, and security measures, all of which contribute to enhancing the scalability and accuracy of anomaly detection. These innovations are expected to drive the future of anomaly detection in IoT, with emerging technologies such as 5G, blockchain, and quantum computing opening up new opportunities. TinyML is poised to foster innovation in fields like predictive maintenance, cybersecurity, and healthcare. Its integration with IoT systems will continue to redefine how anomalies are detected and managed in real-time, ultimately improving operational efficiency and system resilience. Furthermore, the systematic literature mapping of 20 research papers from 2021 to 2024 provides a comprehensive overview of anomaly detection in IoT devices using TinyML and MCUs. This research sheds light on key aspects of anomaly detection, including ML and DL techniques, data sources, hardware, power supply methods, and communication technologies. The study also develops a taxonomy of algorithms used for anomaly detection and presents a balanced view of both the benefits and challenges associated with implementing TinyML in constrained environments. As the IoT ecosystem continues to evolve, continued collaboration among researchers, developers, and industry stakeholders will be essential to fully realize the potential of TinyML, paving the way for smarter, more resilient infrastructures.

References

1 You, Z., Cui, L., Shen, Y. et al. (2022). A unified model for multi-class anomaly detection. *Advances in Neural Information Processing Systems* 35: 4571–4584.

2 Hammad, S.S., Iskandaryan, D., and Trilles, S. (2023). An unsupervised TinyML approach applied to the detection of urban noise anomalies under the smart cities environment. *Internet of Things* 23: 100848.

3 Wang, Y. and Sun, L. (2024). Energy-efficient dynamic sensor time series classification for edge health devices. *Computer Methods and Programs in Biomedicine* 254: 108268.

4 Dutta, L. and Bharali, S. (2021). Tinyml meets iot: a comprehensive survey. *Internet of Things* 16: 100461.

5 Jhaveri, R.H., Chi, H.R., and Wu, H. (2025). TinyML for empowering low-power IoT edge consumer devices. *IEEE Transactions on Consumer Electronics* 70 (4): 7318–7321.

6 Raza, W., Osman, A., Ferrini, F., and Natale, F.D. (2021). Energy-efficient inference on the edge exploiting TinyML capabilities for UAVs. *Drones* 5 (4): 127.

7 Khatoon, A., Wang, W., Ullah, A. et al. (2024). Optimized binary neural networks for road anomaly detection: a TinyML approach on edge devices. *Computers, Materials & Continua* 80 (1): 526–546.

8 Siang, Y.Y. and Ahmad, M.R. (2024). Modified over-complete autoencoder for anomaly detection based on TinyML. *IEEE Sensors Letters* 8 (10): 1–4.

9 Siddique, T. and Mahmud, M.S. (2024). Physics-enhanced TinyML for real-time detection of ground magnetic anomalies. *IEEE Access* 12: 25372–25384.

10 Karras, A., Giannaros, A., Karras, C. et al. (2024). TinyML algorithms for big data management in large-scale IoT systems. *Future Internet* 16 (2): 42.

11 Akram, J., Anaissi, A., Othman, W. et al. (2024). Dronessl: self-supervised multimodal anomaly detection in internet of drone things. *IEEE Transactions on Consumer Electronics* 70 (1): 4287–4298.

12 Trilles, S., Hammad, S.S., and Iskandaryan, D. (2024). Anomaly detection based on artificial intelligence of things: a systematic literature mapping. *Internet of Things* 25: 101063.

13 Blázquez-García, A., Conde, A., Mori, U., and Lozano, J.A. (2021). A review on outlier/anomaly detection in time series data. *ACM computing surveys (CSUR)* 54 (3): 1–33.

14 Baraneetharan, E. (2020). Role of machine learning algorithms intrusion detection in WSNs: a survey. *Journal of Information Technology* 2 (03): 161–173.

15 Alajlan, N.N. and Ibrahim, D.M. (2022). TinyML: enabling of inference deep learning models on ultra-low-power IoT edge devices for AI applications. *Micromachines* 13 (6): 851.

16 Slimani, C., Morge-Rollet, L., Lemarchand, L. et al. (2025). A study on characterizing energy, latency and security for Intrusion Detection Systems on heterogeneous embedded platforms. *Future Generation Computer Systems* 162: 107473.

17 Andrade, P., Silva, I., Silva, M. et al. (2022). A tinyml soft-sensor approach for low-cost detection and monitoring of vehicular emissions. *Sensors* 22 (10): 3838.

18 Tsoukas, V., Gkogkidis, A., Boumpa, E., and Kakarountas, A. (2024). A review on the emerging technology of TinyML. *ACM Computing Surveys* 56 (10): 1–37.

19 Kallimani, R., Pai, K., Raghuwanshi, P. et al. (2024). TinyML: tools, applications, challenges, and future research directions. *Multimedia Tools and Applications* 83 (10): 29015–29045.

20 Elhanashi, A., Dini, P., Saponara, S., and Zheng, Q. (2024). Advancements in TinyML: applications, limitations, and impact on IoT devices. *Electronics* 13 (17): 3562.

21 Thuppari, C., Jannu, S., Edla, D.R. et al. (2024). Energy-aware compression and consumption algorithms for efficient TinyML model using aquila optimization in industrial IoT. *IEEE Transactions on Consumer Electronics* 71 (1): 334–342.

22 Hayajneh, A.M., Hafeez, M., Zaidi, S.A., and McLernon, D. (2024). TinyML empowered transfer learning on the edge. *IEEE Open Journal of the Communications Society* 5: 1656–1672.

23 Oliveira, F., Costa, D.G., Assis, F., and Silva, I. (2024). Internet of Intelligent Things: a convergence of embedded systems, edge computing and machine learning. *Internet of Things* 26: 101153.

24 Müller, K., Weidner, J., Franchi, N., and Wägemann, P. (2024). TinyEP: TinyML-enhanced energy profiling for extreme edge devices. *IEEE Access* 12: 193747–193762.

25 Ficco, M., Guerriero, A., Milite, E. et al. (2024). Federated learning for IoT devices: enhancing TinyML with on-board training. *Information Fusion* 104: 102189.

26 Antonini, M., Pincheira, M., Vecchio, M., and Antonelli, F. (2023). An adaptable and unsupervised TinyML anomaly detection system for extreme industrial environments. *Sensors* 23 (4): 2344.

27 Ray, P.P. (2022). A review on TinyML: state-of-the-art and prospects. *Journal of King Saud University-Computer and Information Sciences* 34 (4): 1595–1623.

28 Genge, B., Haller, P., and Enăchescu, C. (2019). Anomaly detection in aging industrial internet of things. *IEEE Access* 7: 74217–74230.

29 Šabić, E., Keeley, D., Henderson, B., and Nannemann, S. (2021). Healthcare and anomaly detection: using machine learning to predict anomalies in heart rate data. *Ai & Society* 36 (1): 149–158.

30 Nguyen, V.K., Renault, É., and Milocco, R. (2019). Environment monitoring for anomaly detection system using smartphones. *Sensors* 19 (18): 3834.

31 Guo, H., Zhou, Z., Zhao, D., and Gaaloul, W. (2024). EGNN: energy-efficient anomaly detection for IoT multivariate time series data using graph neural network. *Future Generation Computer Systems* 151: 45–56.

32 Alwaisi, Z., Kumar, T., Harjula, E., and Soderi, S. (2024). Securing constrained IoT systems: a lightweight machine learning approach for anomaly detection and prevention. *Internet of Things* 28: 101398.

33 Njor, E., Hasanpour, M.A., Madsen, J., and Fafoutis, X. (2024). A holistic review of the TinyML stack for predictive maintenance. *IEEE Access* 12: 184861–184882.

34 Andrade, P., Silva, M., Medeiros, M. et al. (2024). TEDA-RLS: a TinyML incremental learning approach for outlier detection and correction. *IEEE Sensors Journal* 24 (22): 38165–38173.

35 Baciu, V.-E., Stiens, J., and da Silva, B. (2024). MLino bench: a comprehensive benchmarking tool for evaluating ML models on edge devices. *Journal of Systems Architecture* 155: 103262.

36 Acquaah, Y.T. and Kaushik, R. (2024). Normal-only anomaly detection in environmental sensors in CPS: a comprehensive review. *IEEE Access* 12: 191086–191107.

37 Rostami, A., Tarvirdizadeh, B., Alipour, K., and Ghamari, M. (2024). Real-time stress detection from raw noisy PPG signals using LSTM model leveraging TinyML. *Arabian Journal for Science and Engineering* 50: 1–23.

38 Chen, Z., Gao, Y., and Liang, J. (2023). LOPdM: a low-power on-device predictive maintenance system based on self-powered sensing and TinyML. *IEEE Transactions on Instrumentation and Measurement* 72: 1–13.

39 Liang, T., Glossner, J., Wang, L. et al. (2021). Pruning and quantization for deep neural network acceleration: a survey. *Neurocomputing* 461: 370–403.

40 Fährmann, D., Martín, L., Sánchez, L., and Damer, N. (2024). Anomaly detection in smart environments: a comprehensive survey. *IEEE Access* 12: 64006–64049.

41 Ge, D., Dong, Z., Cheng, Y., and Wu, Y. (2024). An enhanced spatio-temporal constraints network for anomaly detection in multivariate time series. *Knowledge-Based Systems* 283: 111169.

42 Sharma, P., Sharma, S.K., and Dani, D. (2024). Edge-assisted federated learning for anomaly detection in diverse IoT network. *International Journal of Information Technology* 17: 1–11.

43 Sun, P., Wan, Y., Wu, Z., and Fang, Z. (2024). A survey on security issues in IoT operating systems. *Journal of Network and Computer Applications* 231: 103976.

44 Villa-Pérez, M.E., Alvarez-Carmona, M.A., Loyola-González, O. et al. (2021). Semi-supervised anomaly detection algorithms: a comparative summary and future research directions. *Knowledge-Based Systems* 218: 106878.

45 Choi, K., Yi, J., Park, C., and Yoon, S. (2021). Deep learning for anomaly detection in time-series data: review, analysis, and guidelines. *IEEE access* 9: 120043–120065.

46 Li, C., Li, Y., Cao, Y. et al. (2025). Anomaly detection of cross-operational state streaming data in an edge computing platform. *IEEE Transactions on Industry Applications* 1–16.

47 Tawalbeh, L.A., Muheidat, F., Tawalbeh, M., and Quwaider, M. (2020). IoT privacy and security: challenges and solutions. *Applied Sciences* 10 (12): 4102.

48 Babun, L., Denney, K., Celik, Z.B. et al. (2021). A survey on IoT platforms: communication, security, and privacy perspectives. *Computer Networks* 192: 108040.

49 Raschka, S., Patterson, J., and Nolet, C. (2020). Machine learning in python: main developments and technology trends in data science, machine learning, and artificial intelligence. *Information* 11 (4): 193.

50 Bibi, R., Saeed, Y., Zeb, A. et al. (2021). Edge AI-based automated detection and classification of road anomalies in VANET using deep learning. *Computational Intelligence and Neuroscience* 2021 (1): 6262194.

51 Hoang, T.M., Nam, S.H., and Park, K.R. (2019). Enhanced detection and recognition of road markings based on adaptive region of interest and deep learning. *IEEE Access* 7: 109817–109832.

52 Ariyaluran Habeeb, R.A., Nasaruddin, F., Gani, A. et al. (2022). Clustering-based real-time anomaly detection—a breakthrough in big data technologies. *Transactions on Emerging Telecommunications Technologies* 33 (8): e3647.

53 Babu, A. and Bagubali, A. (2025). Federated learning with sailfish-optimized ensemble models for anomaly detection in IoT edge computing environment. *IEEE Access* 13: 53171–53187.

54 Sanchez-Iborra, R. and Skarmeta, A.F. (2020). Tinyml-enabled frugal smart objects: challenges and opportunities. *IEEE Circuits and Systems Magazine* 20 (3): 4–18.

55 Katib, I., Albassam, E., Sharaf, S.A., and Ragab, M. (2025). Safeguarding IoT consumer devices: deep learning with TinyML driven real-time anomaly detection for predictive maintenance. *Ain Shams Engineering Journal* 16 (2): 103281.

56 Karras, A., Karras, C., Giannaros, A. et al. (2023). TinyML-based event detection: an edge-cloud approach for smart agriculture over LoRa WSNs. *2023 8th South-East Europe Design Automation, Computer Engineering, Computer Networks and Social Media Conference (SEEDA-CECNSM)*, 1–10. IEEE.

57 Kulkarni, S. and Bhudhwale, S. (2024). Real-time environmental monitoring in smart agriculture using TinyML and machine learning. *2024 International Conference on Intelligent Systems and Advanced Applications (ICISAA)*, 1–6. IEEE.

58 Elallam, O., Jami, O., and Zaki, M. (2024). TinyML and IoT based early Covid19 detection wearable system. In: *International Conference on Smart Applications and Data Analysis*, 268–280. Springer.

59 Rousopoulou, V., Vafeiadis, T., Nizamis, A. et al. (2022). Cognitive analytics platform with AI solutions for anomaly detection. *Computers in Industry* 134: 103555.

60 Caiazzo, B., Murino, T., Petrillo, A. et al. (2023). An IoT-based and cloud-assisted AI-driven monitoring platform for smart manufacturing: design architecture and experimental validation. *Journal of Manufacturing Technology Management* 34 (4): 507–534.

61 Siniosoglou, I., Radoglou-Grammatikis, P., Efstathopoulos, G. et al. (2021). A unified deep learning anomaly detection and classification approach for smart grid environments. *IEEE Transactions on Network and Service Management* 18 (2): 1137–1151.

62 Pang, G., Shen, C., Cao, L., and Hengel, A.V.D. (2021). Deep learning for anomaly detection: a review. *ACM Computing Surveys (CSUR)* 54 (2): 1–38.

63 Onan, A. (2022). Bidirectional convolutional recurrent neural network architecture with group-wise enhancement mechanism for text sentiment classification. *Journal of King Saud University-Computer and Information Sciences* 34 (5): 2098–2117.

64 Chandrasekhar, N. and Peddakrishna, S. (2023). Enhancing heart disease prediction accuracy through machine learning techniques and optimization. *Processes* 11 (4): 1210.

65 Almazroi, A.A., Aldhahri, E.A., Bashir, S., and Ashfaq, S. (2023). A clinical decision support system for heart disease prediction using deep learning. *IEEE Access* 11: 61646–61659.

66 Yaseen, A. (2023). The role of machine learning in network anomaly detection for cybersecurity. *Sage Science Review of Applied Machine Learning* 6 (8): 16–34.

67 Lian, L., Zhu, T., Zhu, F., and Zhu, H. (2021). Deep learning for caries detection and classification. *Diagnostics* 11 (9): 1672.

68 Akcay, S. and Breckon, T. (2022). Towards automatic threat detection: a survey of advances of deep learning within X-ray security imaging. *Pattern Recognition* 122: 108245.

69 Lv, Z., Wang, Y., Chen, J. et al. (2020). Semiconductor quantum dots for memories and neuromorphic computing systems. *Chemical Reviews* 120 (9): 3941–4006.

70 Barbieri, L., Brambilla, M., Stefanutti, M. et al. (2023). A tiny transformer-based anomaly detection framework for IoT solutions. *IEEE Open Journal of Signal Processing* 4: 462–478.

71 Rajapakse, V., Karunanayake, I., and Ahmed, N. (2023). Intelligence at the extreme edge: a survey on reformable tinyml. *ACM Computing Surveys* 55 (13s): 1–30.

72 Zeynali, M., Alipour, K., Tarvirdizadeh, B., and Ghamari, M. (2025). Non-invasive blood glucose monitoring using PPG signals with various deep learning models and implementation using TinyML. *Scientific Reports* 15 (1): 581.

73 Guo, L., Choe, W., and Lin, F.X. (2023). Sti: Turbocharge NLP inference at the edge via elastic pipelining. In: *Proceedings of the 28th ACM International Conference on Architectural Support for Programming Languages and Operating Systems*, vol. 2, 791–803. ACM.

74 Villamizar-Medina, B.A., Vergel, A.J.S., Medina-Delgado, B. et al. (2025). Neural network quantification for solar radiation prediction: an approach for low power devices. *AiBi Revista de Investigación, Administración e Ingeniería* 13 (1): 11–19.

75 Bi, Q., Zhou, B., Ji, W., and Xia, G.-S. (2025). Universal fine-grained visual categorization by concept guided learning. *IEEE Transactions on Image Processing* 34: 394–409.

76 Allioui, H. and Mourdi, Y. (2023). Unleashing the potential of AI: investigating cutting-edge technologies that are transforming businesses. *International Journal of Computer Engineering and Data Science (IJCEDS)* 3 (2): 1–12.

77 Pereira, E.S., Marcondes, L.S., and Silva, J.M. (2023). On-device tiny machine learning for anomaly detection based on the extreme values theory. *IEEE Micro* 43 (6): 58–65.

78 Tekin, N., Acar, A., Aris, A. et al. (2023). Energy consumption of on-device machine learning models for IoT intrusion detection. *Internet of Things* 21: 100670.

79 Abadade, Y., Benamar, N., Bagaa, M., and Chaoui, H. (2024). Empowering healthcare: TinyML for precise lung disease classification. *Future Internet* 16 (11): 391.

80 Zaidi, S.A.R., Hayajneh, A.M., Hafeez, M., and Ahmed, Q.Z. (2022). Unlocking edge intelligence through tiny machine learning (TinyML). *IEEE Access* 10: 100867–100877.

81 Hayajneh, A.M., Aldalahmeh, S.A., Alasali, F. et al. (2024). Tiny machine learning on the edge: a framework for transfer learning empowered unmanned aerial vehicle assisted smart farming. *IET Smart Cities* 6 (1): 10–26.

4

TinyML Power Consumption and Memory in IoT MCUs

Peter Anuoluwapo Gbadega[1], Agbotiname Lucky Imoize[2], Richard Govada Joshua[3], and Samuel Oluwatobi Tofade[4]

[1] *Department of Electrical and Electronic Engineering Science, University of Johannesburg, Johannesburg, South Africa*
[2] *Department of Electrical and Electronics Engineering, Faculty of Engineering, University of Lagos, Lagos, Nigeria*
[3] *Comprobase Inc., Sterling, VA, USA*
[4] *CleanTech Renewable Energy, Lagos, Nigeria*

4.1 Introduction

The proliferation of the Internet of Things (IoT) has led to the development of countless smart devices, each tasked with processing and transmitting data to enhance human life. Whether through wearable health monitors, smart home automation, or industrial sensors, IoT has become a critical component of personal and professional environments. The effectiveness of IoT systems largely depends on the performance of their foundational hardware, Microcontroller Units (MCUs) [1]. However, with the increasing demand for smaller, more energy-efficient devices, optimizing power consumption and memory in these MCUs has become a significant research focus. One of the most promising advancements in addressing these challenges is the integration of Tiny Machine Learning (TinyML) into IoT devices. TinyML is a branch of machine learning (ML) focused on deploying ML models on low-power, resource-constrained devices. It enables IoT devices to process data locally, reducing the reliance on cloud computing and thereby optimizing both power and memory usage. This research investigates the role of TinyML in addressing the power and memory challenges of IoT MCUs, with a specific focus on ultralow-power smart devices [2, 3].

Tiny Machine Learning: Design Principles and Applications, First Edition.
Edited by Agbotiname Lucky Imoize, Dinh-Thuan Do, and Houbing Herbert Song.

4.1.1 Background and Context of IoT MCUs

The concept of the IoT revolves around the interconnection of physical devices that can collect, transmit, and analyze data over networks without human intervention. From environmental monitoring to industrial automation, IoT systems play a critical role in streamlining operations and improving efficiencies across numerous sectors. IoT devices generally rely on MCUs for their operation. MCUs are compact, integrated circuits designed to execute real-time tasks, such as monitoring sensors, controlling actuators, and managing communication with other devices or cloud servers. As the complexity and number of IoT devices increase, so do the demands placed on MCUs [4]. In a typical IoT system, MCUs are responsible for handling a wide range of tasks, from processing sensor data to controlling device behavior and managing communication protocols. These tasks require considerable computational power and memory, which are often limited to the small, energy-constrained devices that form the backbone of IoT ecosystems. Furthermore, many IoT devices are battery-operated, imposing stringent power consumption limitations. As a result, reducing the energy consumption and optimizing memory usage of IoT MCUs has become a pressing concern [5].

4.1.2 Emergence of TinyML in IoT Systems

TinyML is an emerging field of ML that focuses on running ML models directly on low-power devices, such as microcontrollers and edge sensors. It represents a significant breakthrough for IoT systems, enabling devices to perform complex computations locally without relying on cloud-based services. This local processing reduces latency, enhances privacy, and, crucially, lowers power consumption by minimizing the need for data transmission over networks. The application of TinyML in IoT devices addresses several key challenges. First, by performing ML inference at the edge, TinyML reduces the need to transmit large amounts of data to cloud servers for processing. This reduction in data transmission directly correlates to reduced power consumption, which is essential for battery-powered IoT devices. Second, TinyML models are optimized to require minimal memory, making them suitable for the resource-constrained environments in which IoT devices operate [6]. Finally, TinyML enables IoT devices to make real-time decisions based on the data they collect, allowing for more responsive and adaptive systems. TinyML has already demonstrated significant potential in a wide range of applications, from predictive maintenance in industrial settings to real-time health monitoring in wearable devices. However, integrating TinyML into ultralow-power IoT devices presents several challenges, particularly in terms of power consumption and memory management. This research explores these challenges and investigates the potential solutions offered by TinyML [7, 8].

4.1.3 Power Consumption Challenges in IoT MCUs

Power consumption is one of the most critical concerns in the design and operation of IoT devices. Many IoT devices are deployed in remote or hard-to-reach locations, where access to power is limited and battery replacement is costly or impractical. As a result, IoT devices must operate for extended periods on minimal power, often relying on energy-harvesting techniques to sustain their operation. The power consumption of an IoT device is influenced by several factors, including the complexity of the tasks it performs, the frequency with which it transmits data, and the efficiency of its hardware components [9]. MCUs, as the central processing units of IoT devices, play a significant role in determining overall power consumption. However, MCUs are typically designed to prioritize computational performance over energy efficiency, which can result in excessive power usage, particularly in devices that are required to process large amounts of data. One of the primary ways to reduce power consumption in IoT MCUs is to optimize how data is processed and transmitted. TinyML offers a promising solution by enabling IoT devices to perform local data processing, thereby reducing the need for energy-intensive data transmissions. In addition, TinyML models can be designed to operate efficiently within the limited computational and memory resources of MCUs, further reducing power consumption [10, 11].

4.1.4 Memory Constraint in IoT MCUs

Memory is another critical constraint in IoT devices. Unlike traditional computing systems, which often have access to abundant memory resources, IoT devices are typically equipped with limited memory due to size, cost, and power constraints. This limited memory capacity can pose significant challenges, particularly as IoT devices are required to process and store increasing amounts of data [12]. Efficient memory management is essential for ensuring the optimal performance of IoT MCUs. In many cases, IoT devices must balance competing demands for memory, including the storage of sensor data, the execution of algorithms, and the management of communication protocols. TinyML can help address these challenges by providing memory-efficient ML models specifically designed to operate within the constraints of IoT MCUs. One of the key advantages of TinyML is its ability to perform inference on small, lightweight models that require minimal memory. This is achieved through techniques such as model pruning, quantization, and compression, which reduce the model size without sacrificing accuracy. By optimizing memory usage this way, TinyML enables IoT devices to perform complex computations while operating within their limited memory resources [13, 14].

4.1.5 Ultralow-Power Smart IoT Devices with Embedded TinyML

Ultralow-power IoT devices are designed to operate on minimal energy, often relying on energy-harvesting techniques or long-lasting batteries to sustain their operation. These devices are typically used in applications where power availability is limited, such as environmental monitoring, agriculture, and remote sensing. The integration of TinyML into ultralow-power IoT devices offers several advantages, including improved decision-making capabilities, reduced power consumption, and enhanced memory efficiency. In ultralow-power IoT devices, the power consumption of the MCU is a critical factor in determining the overall energy efficiency of the system [15]. By embedding TinyML into the MCU, IoT devices can perform real-time data processing and decision-making locally without relying on cloud-based services. This reduces the need for data transmission, which is one of the most power-intensive activities in IoT systems.

Additionally, TinyML models can be optimized to run efficiently on low-power MCUs, reducing energy consumption. Examples of ultralow-power IoT devices with embedded TinyML include wearable health monitors, environmental sensors, and smart agriculture devices. These devices use TinyML to analyze data in real time, enabling them to respond to changing conditions and optimize their operation. For example, a smart agriculture device equipped with TinyML can analyze soil moisture levels and adjust irrigation schedules in real time, thereby reducing water usage and conserving energy [16, 17].

4.1.6 Data Management Techniques Utilizing TinyML in IoT Systems

Data management is one of the core challenges in IoT systems due to the massive influx of data generated by devices across various application domains. Whether monitoring environmental conditions, managing smart cities, or tracking health data from wearable devices, the amount of data collected by IoT devices can quickly become overwhelming. In traditional IoT architectures, much of this data is transmitted to centralized cloud servers for processing, increasing network traffic and consuming significant amounts of power in both transmission and reception [18]. TinyML offers a paradigm shift in how data is managed within IoT systems by allowing for local data processing on the devices themselves. This local processing reduces the amount of data that needs to be sent to the cloud, thereby decreasing energy consumption and improving system efficiency. Furthermore, TinyML models can be specifically designed to prioritize critical data, process it in real-time, and discard irrelevant or redundant information. This selective data management not only optimizes memory usage but also enhances overall system performance [8, 15]. Key techniques for data management using TinyML include the following:

1) **Edge Processing:** One of the primary benefits of TinyML is its ability to bring ML to the edge of IoT networks. By performing data analysis and decision-making directly on the device, edge processing reduces the need for continuous data transmission to centralized servers. This is particularly beneficial in applications where network connectivity may be unreliable or costly, such as in remote or rural areas. TinyML models are typically lightweight and resource-efficient, making them ideal for edge processing tasks like anomaly detection, predictive maintenance, and real-time monitoring [19].
2) **Data Prioritization and Filtering:** Not all data collected by IoT devices is equally important. TinyML models can be designed to filter and prioritize data based on specific criteria, ensuring that only the most relevant information is processed and stored. For example, in a health-monitoring device, TinyML can be used to prioritize abnormal readings (such as irregular heartbeats) over normal, routine data. By focusing on the most critical data, TinyML helps to optimize both power and memory usage in IoT MCUs [20].
3) **Data Compression and Optimization:** In addition to filtering data, TinyML techniques can be employed to compress data before it is stored or transmitted. Data compression algorithms can reduce the size of datasets without sacrificing essential information, thereby conserving memory space and minimizing the energy required for data transmission. Techniques such as quantization, where the precision of the data is reduced, are commonly used in TinyML applications to achieve this goal [21].
4) **Efficient Data Storage:** Memory is a limited resource in IoT MCUs, and managing it effectively is crucial to ensuring the longevity and performance of IoT devices. TinyML models are typically designed to operate within limited memory constraints, often using techniques such as model pruning (removing unnecessary parameters from the model) and weight sharing to minimize their memory footprint. This allows IoT devices to store and process more data without exhausting their memory resources [22].

Data management in IoT systems utilizing TinyML is critical for optimizing the performance of resource-constrained devices. By enabling local processing, prioritization, compression, and efficient data storage, TinyML significantly reduces power consumption and memory usage, making it a valuable tool for the next generation of IoT devices.

4.1.7 TinyML for Real-Time Low-Power IoT Circuit Modeling

One of the most critical aspects of IoT devices is their ability to operate in real-time environments where decisions must be made instantaneously based on incoming data. In many cases, these devices are responsible for controlling physical systems, such as power grids, industrial equipment, or healthcare devices, where delays in decision-making can have serious consequences [23]. Achieving real-time performance while maintaining low power consumption and minimal memory usage is a major challenge in IoT system design. TinyML can be leveraged to meet this challenge by enabling real-time modeling of IoT circuits and systems. By embedding ML models directly into IoT devices, TinyML enables these devices to monitor their own operations and adjust their behavior in real-time. This not only improves the performance of the IoT system but also helps to reduce power consumption by enabling more efficient use of system resources [8, 24]. Key applications of TinyML in real-time IoT circuit modeling are as follows:

1) **Predictive Maintenance:** One of the most promising applications of TinyML in IoT systems is predictive maintenance, where ML models are used to predict when a piece of equipment is likely to fail and to schedule maintenance before the failure occurs. By analyzing sensor data in real time, TinyML models can detect patterns and anomalies that indicate potential issues with the equipment. This allows maintenance to be performed only when necessary, reducing downtime and minimizing the energy consumed by unnecessary maintenance operations [25].
2) **Power Management:** IoT devices often operate in energy-constrained environments, where minimizing power consumption is essential. TinyML models can be used to monitor the power usage of different components of an IoT system in real-time and adjust the system's behavior to optimize energy efficiency. For example, a TinyML model could be used to dynamically adjust the operating frequency of an MCU based on the workload, reducing power consumption during periods of low activity [26].
3) **Fault Detection and Correction:** In many IoT applications, it is critical to detect faults in the system as soon as they occur to prevent damage or failures. TinyML models can be used to monitor the operation of an IoT device in real time and detect faults or anomalies in the system's behavior. Once a fault is detected, the TinyML model can either trigger an alert or automatically adjust the system's operation to correct the fault, minimizing the impact of the failure [27].
4) **Sensor Calibration:** Many IoT devices rely on sensors to collect data from the environment. However, these sensors can drift over time, leading to inaccurate data. TinyML models can be used to monitor sensor data in real time and detect when a sensor is no longer providing accurate readings. The model can then

trigger a recalibration of the sensor, ensuring that the data collected by the IoT device remains accurate and reliable [23, 27].

Hence, by integrating TinyML into IoT systems for real-time circuit modeling, designers can create devices that are more responsive, energy-efficient, and capable of operating autonomously in dynamic environments. This represents a significant advancement in the design and operation of low-power IoT systems, particularly in applications where real-time decision-making is critical.

4.1.8 TinyML for Ultralow-Power AI in Large-Scale IoT Deployments

As IoT systems continue to scale, with deployments in smart cities, agriculture, transportation, and more, the need for ultralow-power artificial intelligence (AI) becomes more apparent. Large-scale IoT networks often consist of hundreds or thousands of devices, each collecting data and performing tasks that contribute to the overall operation of the system [28]. Managing such a large network of devices requires efficient use of both power and memory resources, especially as AI-based applications become more widespread in IoT environments. TinyML, focusing on deploying lightweight ML models on resource-constrained devices, is ideally suited for large-scale IoT deployments. By enabling AI capabilities on low-power IoT devices, TinyML allows these devices to perform intelligent tasks, such as data analysis, anomaly detection, and predictive maintenance, without constant communication with cloud servers. This reduces the power consumption of the devices themselves and alleviates the strain on network infrastructure, making large-scale IoT deployments more feasible [7, 25]. The key considerations for scaling TinyML in large IoT deployments are as follows:

1) **Scalability of AI Models:** One of the key challenges in scaling AI in IoT systems is ensuring that the ML models themselves are scalable. Traditional AI models are often too large and computationally intensive for resource-constrained IoT devices. However, TinyML models are specifically designed to be lightweight and efficient, making them suitable for deployment across large networks of IoT devices [29].
2) **Power Efficiency in Distributed Networks:** In large-scale IoT deployments, the power consumption of each device contributes to the overall energy usage of the system. By using TinyML to perform local data processing, IoT devices can significantly reduce their power consumption, as they no longer need to transmit data to the cloud constantly. This is particularly important in applications such as smart agriculture, where devices are often deployed in remote locations and must operate on minimal power [30].
3) **Data Processing at the Edge:** In large IoT networks, the volume of data generated by devices can be overwhelming. TinyML enables data to be processed

locally at the edge, reducing the need for data transmission and making the system more efficient. This is especially beneficial in applications where real-time decision-making is required, such as in traffic management systems, where data from sensors must be processed and acted upon immediately [31].

4) **Security and Privacy Considerations:** As IoT networks scale, security and privacy become increasingly important concerns. By processing data locally using TinyML, IoT devices can reduce the amount of sensitive data that needs to be transmitted over the network, thereby enhancing security and privacy. Additionally, TinyML models can be designed to detect security threats, such as unauthorized access or anomalous behavior, in real time, further enhancing the security of the IoT system [32].

In large-scale IoT deployments, the use of ultralow-power AI powered by TinyML enables devices to operate autonomously and intelligently, reducing the need for human intervention and improving the overall efficiency of the system. As the number of IoT devices continues to grow, TinyML will play a crucial role in ensuring that these devices can operate efficiently and sustainably, even in resource-constrained environments. TinyML represents a promising solution to the challenges of power consumption, memory efficiency, and real-time processing in IoT systems. By enabling local AI capabilities on resource-constrained devices, TinyML can significantly reduce the energy required for data processing and transmission, making IoT systems more sustainable and scalable.

As IoT continues to expand into new domains and applications, the integration of TinyML will be critical to ensuring the success and efficiency of these systems. Despite significant advances in IoT technology, several challenges remain in the areas of power consumption, memory efficiency, and real-time processing, particularly in resource-constrained environments [25]. The integration of AI into IoT systems introduces additional complexity, as traditional AI models are often too computationally intensive to run on low-power devices. This research aims to address these challenges by exploring the use of TinyML to enable ultralow-power AI capabilities in IoT devices. The primary objectives of this research are as follows:

1) To investigate the power consumption and memory efficiency of IoT MCUs with embedded TinyML models.
2) To explore data management techniques that optimize both power and memory usage in IoT systems.
3) To analyze the use of TinyML for real-time circuit modeling and decision-making in IoT devices.
4) To assess the scalability of TinyML in large-scale IoT deployments and its impact on power consumption, network efficiency, and security.
5) To develop a framework for the integration of TinyML into IoT systems that optimizes performance, power, and memory across different applications.

4.1.9 Research Contributions

The research offers significant contributions to academia and industry by addressing the critical challenges of deploying TinyML in IoT systems. The study provides a comprehensive framework for optimizing power consumption, memory usage, real-time processing, and security in resource-constrained environments within academic research. This contributes to the broader field of IoT and AI, offering new insights into how ML can be deployed efficiently on low-power devices.

For the industry, the findings of this research have practical applications in sectors such as smart cities, agriculture, healthcare, and industrial automation. The development of energy-efficient, memory-optimized TinyML models opens up new possibilities for deploying AI at the edge, reducing operational costs, and improving the performance and reliability of IoT systems [33].

The remaining part of this chapter is structured as follows: Section 4.2 explores the latest research and methodologies in optimizing power consumption and memory management in IoT MCUs. Section 4.3 outlines a detailed methodology employed to systematically explore the integration of TinyML with ultralow-power IoT devices. Section 4.4 discusses the results and findings of the bibliometric analysis conducted in this study, providing insights into the research landscape of TinyML and its integration with ultralow-power IoT systems. Finally, Section 4.5 concludes with recommendations for future research.

4.2 Context and Literature Review

IoT has witnessed exponential growth, with applications ranging from industrial automation to smart homes and wearable health monitoring. This proliferation of IoT devices has placed immense pressure on hardware requirements, particularly MCUs that serve as the computing backbone for many IoT applications. IoT MCUs must balance power consumption, memory allocation, processing speed, and environmental durability while meeting the constraints of battery-powered or energy-harvesting devices [34]. Since IoT devices are often deployed in energy-constrained environments, minimizing power consumption is crucial for extended battery life. Simultaneously, MCUs with limited memory must efficiently manage data processing and storage without compromising computational performance. This section introduces these dual challenges of power consumption and memory optimization, which are interlinked and form the basis for the research. The literature review will discuss the latest research and methodologies in optimizing power consumption and memory management in IoT MCUs. Techniques such as low-power design, TinyML, edge computing, and memory-efficient data management have effectively addressed the unique challenges of IoT MCUs. As IoT continues to expand, so will the need for innovative solutions in MCU design and operation to ensure sustainable and efficient performance [16, 35].

4.2.1 Power Consumption in IoT MCUs

Various studies have discussed energy-saving techniques for MCUs, emphasizing low-power operation modes, dynamic voltage scaling (DVS), and adaptive duty cycling. Al-Kofahi et al. [36] explored DVS as a key strategy for power reduction in MCUs, where the voltage and frequency of the MCU are adjusted according to the processing requirements to conserve power. Protocols like Bluetooth Low Energy (BLE), Zigbee, and LoRa have been optimized for low power, enabling IoT devices to transmit data with minimal energy expenditure. According to [37], implementing BLE on low-power MCUs achieved up to 80% reduction in power consumption during data transmission, which is critical for battery-dependent IoT sensors. To predict and manage power usage effectively, researchers have developed power consumption models that assess the impact of various operations on MCU power draw.

Tekin et al. [10] proposed a comprehensive model that accounts for processor activity, peripheral usage, and sleep cycles to predict energy consumption more accurately. These models allow for fine-tuning of software algorithms to reduce power consumption. Techniques such as edge computing, TinyML (ML at the edge), and event-driven processing are frequently employed to reduce data transmission frequency and thereby conserve power. Yugank et al. [38] illustrated how edge processing reduces energy consumption by locally processing data and sending only necessary information to cloud servers.

4.2.2 Memory Constraints in IoT MCUs

Due to cost and space constraints, IoT MCUs are designed with limited RAM and flash memory. As IoT applications grow increasingly data-intensive, efficient memory utilization becomes essential. Adegbija et al. [39] reviewed memory optimization techniques like data compression and dynamic memory allocation to improve memory efficiency without additional hardware. With limited storage, managing data becomes a challenge in IoT MCUs. Studies such as [40] have suggested memory-efficient data management techniques, including in-memory databases and data compression algorithms, and prioritized caching to handle data more effectively on constrained MCUs.

IoT developers have employed lightweight databases and file systems such as SQLite and TinyDB to support local data storage. These databases are optimized for small memory footprints and minimal power usage, allowing IoT devices to store and retrieve data efficiently. Sheng et al. [41] demonstrated that integrating a lightweight database can extend the battery life of IoT MCUs by reducing data transmission frequency and energy consumption.

4.2.3 Integration of Power and Memory Optimization Techniques

Power consumption and memory usage are interdependent; optimizing one can affect the other. Leghari et al. [42] focused on integrated approaches where software algorithms are optimized to reduce memory footprints and power draw. For instance, utilizing a compressed data structure minimizes memory usage and reduces the processing load, leading to power savings. TinyML has emerged as a viable approach for real-time data analysis on IoT devices. TinyML algorithms are specially optimized to run on memory- and power-constrained MCUs, performing tasks like anomaly detection and image recognition without needing a cloud connection.

According to [43], implementing TinyML models on MCUs significantly reduced energy consumption, as tasks that would traditionally require constant data transmission to a central server were processed locally. Event-driven programming, where the MCU wakes up only upon receiving specific signals, has been shown by Kong et al. [44] to minimize energy and memory use. This strategy and edge computing capabilities reduce data processing demands on the MCU, as data is analyzed locally, minimizing both memory requirements and energy consumption for data transmission.

4.2.4 Comparative Analysis of MCU Architectures for IoT

Different MCU architectures offer varied capabilities in terms of power and memory management. The study by Silva et al. [45] provides a benchmarking analysis of popular MCUs (e.g., ARM Cortex-M series, ESP32, and RISC-V), comparing them based on their energy efficiency, memory capacity, and computational speed in IoT applications. These benchmarks assist in selecting suitable MCUs based on the specific requirements of an IoT application. Recent studies suggest incorporating energy harvesting technologies like solar and kinetic energy to reduce battery dependency. These techniques complement low-power MCUs, enabling sustainable energy use in remote IoT applications.

According to Ku et al. [46], integrating energy harvesting systems with IoT MCUs reduced battery reliance by 50% in off-grid applications. Zhang [47] explored quantum-inspired algorithms and neuromorphic computing as ways to further reduce power consumption while enhancing processing capabilities in IoT devices. Although still in the nascent stage, these approaches have shown promise in reducing the memory and energy footprints of complex tasks, particularly in AI-driven IoT applications. Power and memory constraints often limit the ability to implement strong security protocols on IoT MCUs. Lightweight encryption and secure memory management techniques are being researched to ensure that low-power MCUs can still maintain robust security, as seen in studies by Ray et al. [48], which proposes a cryptography method with minimal computational overhead suitable for constrained IoT devices.

To critically analyze past literature effectively, a comparative analysis of different studies, including evaluation metrics, reduction achieved, savings achieved, use case scenarios, and limitations, is conducted, as shown in Table 4.1. Table 4.1 depicts the comparative analysis of past literature on power consumption and memory optimization in IoT MCUs. This analysis offers a nuanced view of past studies by examining the power and memory efficiency gains through specific techniques and contextualizing their benefits and limitations within real-world IoT scenarios. It also highlights the diversity in approaches depending on specific IoT applications and hardware configurations. Table 4.2 shows selective studies of recent literature on power consumption and memory optimization in IoT MCUs. This table synthesizes and categorizes past research, facilitating the understanding of key themes and advancements in power and memory management within IoT MCUs.

4.3 Methodology

This section outlines a detailed methodology employed to systematically explore the integration of TinyML with ultralow-power IoT devices, focusing on four core stages: conducting a comprehensive literature review, establishing selection criteria for relevant studies, developing a comparative analysis framework, and synthesizing findings to derive actionable insights. The methodology is tailored to address the research subtopics: ultralow-power smart IoT devices with embedded TinyML, data management techniques utilizing TinyML in IoT systems, TinyML for real-time low-power IoT circuit modeling, and TinyML for ultralow-power AI and large-scale IoT deployments. The initial step, a thorough literature review, involved gathering a foundational understanding of the intersection between TinyML and IoT systems. Priority was given to peer-reviewed journal articles, conference proceedings, white papers, and technical reports published within the last decade to ensure the inclusion of cutting-edge research. Search queries were formulated strategically using keywords like "TinyML," "ultralow-power IoT," "real-time IoT modeling," and "large-scale IoT AI integration." Boolean operators such as AND, OR, and NOT were used to refine searches, ensuring comprehensive coverage while excluding irrelevant content. Following the search, selection criteria were applied to filter studies based on relevance, scientific rigor, and

Table 4.1 Comparative analysis of past literature on power consumption and memory optimization in IoT MCUs.

Ref./Year	Evaluation metrics	Reduction achieved	Savings achieved	Use case scenarios	Limitations
Al Mamun and Yuce [49]/2019	Battery longevity, power in sleep mode	Up to 60% with ultralow-power sleep modes	Minimal (optimized for power)	Wearable health devices and environmental sensors	Limited application due to wake-up latency; effectiveness depends on sleep mode duration
Khalifeh et al. [50]/2022	Energy per instruction, idle power	ARM Cortex-M had up to 50% lower idle power	10% more efficient in memory use with RISC-V	Environmental monitoring, portable IoT devices	Findings vary significantly by workload; optimal MCU depends on the specific use case
Adegbija et al. [39]/2017	Combined memory and power use	30% overall reduction	20%	General IoT scenarios need balanced performance	Complex to implement; increased development time due to dual optimization
Chen et al. [51]/2018	Data transmission reduction	35% with local processing	N/A	Industrial IoT with edge computing	Limited benefit if data processing exceeds edge capabilities
Balasubramanian and Elangeswaran [52]/2024	Database query time, memory use	10–15%	25% using in-memory DB	Smart agriculture with real-time data	May face challenges with larger data sets; not suited for high-throughput data

(Continued)

Table 4.1 (Continued)

Ref./Year	Evaluation metrics	Reduction achieved	Savings achieved	Use case scenarios	Limitations
Kane et al. [53]/2020	Memory utilization efficiency	N/A	30% improvement with data compression	IoT nodes with limited RAM, like ESP8266	Overhead from compression algorithms; limited to low-data applications
Rosero-Montalvo et al. [54]/2023	Energy consumption in data processing	40% reduction with TinyML	15–20%	Anomaly detection in IoT devices	Requires specific MCU models with ML support; limited by processing power
Diab and Rodriguez-Villegas [55]/2022	Predictive power draw accuracy	25% reduction through state-based optimization	N/A	Predictive modeling in wearables	Model accuracy dependent on specific workload conditions
Chien et al. [2]/2016	Transmission energy consumption, latency	Up to 80% with BLE	N/A	Remote monitoring and low-power sensors	BLE range limitations; higher latency in Zigbee
Ali et al. [56]/2021	Energy consumption per operation cycle	20–30% reduction with adaptive duty cycling and DVS	N/A	IoT sensors in smart home settings	Limited to specific voltage levels; may not adapt well to all IoT workloads

Table 4.2 Selective studies of recent literature on power consumption and memory optimization in IoT MCUs.

Ref./Year	Focus of study	Optimization techniques	Methodology	Key findings	Contributions
Radfar et al. [57]/2019	Power management in IoT MCUs for prolonged battery life	Dynamic Voltage Scaling (DVS) and adaptive duty cycling	Experimental testing on IoT sensors with varying workloads	Achieved significant energy savings with adaptive duty cycling and DVS	Demonstrated how DVS effectively lowers power consumption without sacrificing performance
Ayılmış et al. [58]/2022	Evaluating the impact of communication protocols on power in IoT MCUs	Bluetooth Low Energy (BLE), Zigbee, LoRa	Comparative analysis of power usage across protocols in IoT	BLE outperformed other protocols by reducing power consumption by up to 80% during data transmission	Provided a guideline on protocol selection for low-power IoT applications
Schaarschmidt et al. [59]/2021	Power consumption modeling in IoT MCUs	Power consumption models based on activity states	Developed a predictive model considering processor activity, peripherals, and sleep cycles	Enhanced accuracy in predicting power draw, assisting in software adjustments for power savings	Introduced a practical model for energy management in IoT development
Schizas et al. [7]/2022	TinyML implementation for energy efficiency in IoT MCUs	TinyML algorithms for local data processing	Simulation of TinyML for anomaly detection on MCU-based IoT devices	Reduced data transmission needs, lowering energy consumption substantially	Validated TinyML as a tool for extending battery life in IoT applications through local processing
Yun et al. [60]/2022	Memory management in memory-constrained IoT MCUs	Data compression, dynamic memory allocation	Evaluation of memory-saving methods in RAM-limited MCUs	Demonstrated up to 30% improvement in memory utilization through data compression	Provided solutions for efficient memory management in low-memory MCUs

(Continued)

Table 4.2 (Continued)

Ref./Year	Focus of study	Optimization techniques	Methodology	Key findings	Contributions
Zhuang et al. [61]/2018	Lightweight database usage for memory and power savings	In-memory databases, TinyDB, SQLite	Benchmarking different lightweight databases on MCUs	Showed efficient data storage with minimal memory usage and power draw	Recommended lightweight databases for IoT MCUs in storage-limited environments
Capra et al. [62]/2019	Edge computing for reducing data transmission in IoT MCUs	Event-driven processing, edge computing	Implementation of edge computing for IoT tasks	Significantly reduced energy consumption by processing data locally	Established the benefits of edge computing in minimizing data transfer power needs
Chéour et al. [63]/2020	Integrated power and memory optimization in IoT MCUs	Combined software optimizations for power and memory	Analysis of compressed data structures and reduced processing load	Achieved power savings by minimizing memory usage alongside computational load	Highlighted the interplay between memory and power optimizations for IoT MCUs
Götz et al. [64]/2020	Benchmarking IoT MCUs for power and memory efficiency	Comparative analysis of ARM, ESP32, RISC-V MCUs	Power and memory performance tests across MCU types	Identified ARM Cortex-M as superior in low-power operations; RISC-V was more memory-efficient	Provided benchmarking data to guide MCU selection based on IoT requirements
Lueders et al. [65]/2013	Ultralow-power sleep modes and memory controllers in MCU design	Sleep modes, nonvolatile memory options	Tested ultralow-power features on ARM Cortex and ESP series	Achieved extended battery life by leveraging ultralow-power modes	Validated advanced MCU design features for improved IoT power and memory management

recency. Studies directly addressing TinyML's application to IoT, emphasizing energy efficiency, real-time performance, or scalability, were prioritized [17].

A comparative analysis framework was then developed to systematically evaluate findings. This framework grouped studies by subtopic, defined metrics such as energy consumption, latency, and scalability for evaluation, and performed cross-study comparisons to identify trends, best practices, and gaps. For instance, under ultralow-power IoT devices, techniques like model pruning, quantization, and compression were analyzed for their effectiveness in reducing power consumption. For data management, strategies like federated learning and decentralized processing were assessed for their potential to optimize IoT networks while preserving security and efficiency. Studies evaluating TinyML's capacity to deliver low-latency responses in constrained hardware were critically reviewed regarding real-time low-power IoT circuit modeling.

Hierarchical and distributed AI architectures were examined for large-scale deployments to understand trade-offs between computational complexity and energy efficiency. The final stage involved synthesizing the gathered insights into a cohesive understanding of TinyML's role in advancing IoT technologies. This synthesis emphasized actionable trends, such as the growing adoption of on-device learning for resource-constrained IoT systems and the critical need for scalable AI frameworks to manage large networks. Validation measures, including quality assessment checklists and expert consultations, were incorporated to ensure the reliability of interpretations. By employing this structured methodology, the research effectively bridges knowledge gaps. It lays the groundwork for future innovations in integrating TinyML with ultralow-power IoT devices, facilitating advancements in energy-efficient, real-time, and scalable IoT systems [15, 17].

4.3.1 Research Design: Literature Review Framework

The literature review framework serves as the cornerstone of this study, ensuring a systematic approach to identifying, analyzing, and synthesizing scholarly works that align with the research objectives. This framework outlines the key steps for conducting a comprehensive literature review, including defining the scope, developing selection criteria, implementing search strategies, and synthesizing findings. The ultimate goal is to ensure that the review provides a robust foundation for understanding the integration of TinyML with ultralow-power IoT systems, real-time modeling, and large-scale AI deployments.

4.3.2 Defining the Scope of the Literature Review

Defining the scope of the literature review was a critical step in framing the boundaries and focus areas of the research, thereby ensuring a targeted and meaningful investigation of TinyML applications in ultralow-power IoT systems. The scope was shaped by the research objectives and aligned with four core subtopics: (1) ultralow-power smart IoT devices with embedded TinyML, (2) data management techniques utilizing TinyML in IoT systems, (3) TinyML for real-time low-power IoT circuit modeling, and (4) TinyML for ultralow-power AI and large-scale IoT deployments. This delineation allowed for a concentrated exploration of the intersection between TinyML and IoT technologies, emphasizing energy efficiency, real-time functionality, and scalability. The scope was further refined to include studies published within the last decade, reflecting contemporary advancements while incorporating seminal works to establish foundational knowledge.

Peer-reviewed journal articles, conference papers, technical reports, and white papers were prioritized to ensure scientific rigor. Geographically, the review focused on global contributions to capture diverse perspectives and innovations. By defining these boundaries, the review excluded general IoT studies without a TinyML focus and nontechnical articles, emphasizing TinyML-driven transformations in ultralow-power IoT ecosystems. This precise scoping facilitated a systematic and impactful exploration of the research topic, ensuring relevance and depth in addressing key challenges and opportunities in the field.

The first step was to establish the boundaries of the literature review, guided by the specific subtopics under investigation as follows:

1) **Ultralow-Power Smart IoT Devices with Embedded TinyML:** Focused on exploring how TinyML enables IoT devices to achieve significant energy savings while maintaining functionality.
2) **Data Management Techniques Utilizing TinyML in IoT Systems:** Investigated methods for secure, efficient, and low-energy data processing, storage, and transfer in TinyML-enabled IoT networks.
3) **TinyML for Real-Time Low-Power IoT Circuit Modeling:** Examined applications of TinyML in modeling IoT circuits for real-time analysis and control with minimal power consumption.
4) **TinyML for Ultralow-Power AI and Large-Scale IoT Deployments:** Addressed scalability challenges and strategies for integrating TinyML in vast IoT ecosystems.

4.3.3 Developing Selection Criteria

Selection criteria were established to ensure the relevance, quality, and scientific rigor of the reviewed studies.

4.3.3.1 Inclusion and Exclusion Criteria

The inclusion and exclusion criteria are pivotal in ensuring that the literature review systematically identifies and evaluates the most relevant and high-quality studies aligned with the research objectives. **The inclusion criteria** were designed to prioritize studies that specifically addressed the integration of TinyML in ultralow-power IoT systems. Studies on subtopics such as TinyML-enabled ultralow-power smart IoT devices, data management techniques utilizing TinyML, real-time low-power IoT circuit modeling, and scalable TinyML-based AI for large IoT deployments were included. Furthermore, the research was restricted to peer-reviewed journal articles, conference proceedings, white papers, and technical reports published within the last ten years to ensure that the findings were current and reflective of ongoing advancements. High scientific rigor was a nonnegotiable criterion, favoring studies with clear methodologies, robust experimental setups, and well-documented results. Additionally, publications written in English were selected to ensure consistency in analysis.

Conversely, the **exclusion criteria** eliminated studies that lacked direct relevance to the research focus, such as those addressing general IoT technologies without specific mention of TinyML applications or those that discussed TinyML but not in the context of energy efficiency, real-time functionality, or scalability. Non-peer-reviewed articles, blogs, and opinion pieces were excluded to maintain scientific integrity. Studies with ambiguous methodologies, insufficient data, or lack of replicability were also disregarded. Additionally, unless considered seminal works offering foundational insights, outdated studies were excluded to ensure the research remained relevant to contemporary technological landscapes. By systematically applying these criteria, the review focused on credible and impactful contributions, enabling a comprehensive yet targeted understanding of TinyML's transformative role in ultralow-power IoT systems.

4.3.4 Search Strategy Implementation

The search strategy implementation involved a systematic and comprehensive approach to identifying relevant literature on the integration of TinyML with ultralow-power IoT systems. The process began with the formulation of targeted search queries combining key terms such as "TinyML," "ultralow-power IoT," "real-time IoT modeling," "TinyML scalability," and "energy-efficient IoT systems." Boolean operators like AND, OR, and NOT were utilized to refine the search scope, ensuring the inclusion of relevant studies while filtering out unrelated content. For example, queries like "TinyML AND ultralow-power IoT" or "Real-time TinyML OR low-power IoT circuits" captured studies relevant to energy-efficient and real-time applications. The search extended across reputable academic databases, including IEEE Xplore, ScienceDirect, SpringerLink, and Google Scholar, to ensure comprehensive coverage of peer-reviewed journal articles, conference proceedings, and technical reports.

Backward and forward citation tracking was employed to discover influential papers referenced by or citing the initially identified studies, thus broadening the review's scope. Filters were applied to limit results to publications in English from the past decade, with exceptions for seminal works that significantly contributed to foundational knowledge. Metadata such as authorship, publication year, journal source, and study focus were documented during the search to aid organization and analysis. Advanced features like database alerts were set to capture emerging publications during the review period. By adopting this rigorous and iterative search strategy, the study ensured the inclusion of high-quality, diverse, and up-to-date literature, forming a robust foundation for the analysis and synthesis of TinyML's applications in ultralow-power IoT systems.

4.3.5 Data Extraction and Organization

The data extraction and organization process was meticulously designed to ensure systematic collection, categorization, and analysis of information from selected studies on the integration of TinyML with ultralow-power IoT systems. Once the relevant literature was identified, key data points were extracted, focusing on metadata (e.g., authorship, publication year, journal, and study type) and technical content such as objectives, methodologies, results, and conclusions. Specific metrics relevant to the research subtopics were also captured, including energy consumption, latency, scalability, model accuracy, and computational efficiency.

A coding framework was established to facilitate thematic analysis, categorizing studies under predefined subtopics: ultralow-power IoT devices, TinyML-based data management, real-time IoT circuit modeling, and scalable TinyML applications for large IoT networks. A spreadsheet was employed to organize this data, enabling quick retrieval and cross-referencing. Thematic grouping allowed for the identification of patterns, trends, and gaps within the literature while tagging specific studies with unique identifiers to ensure traceability during analysis. Standardized templates were used during the extraction process to maintain consistency and avoid biases, and quality checks were performed periodically. This structured approach not only streamlined the synthesis of findings but also provided a robust foundation for deriving actionable insights into TinyML's transformative applications in ultralow-power IoT systems.

4.3.6 Synthesizing Findings

The synthesis process involved integrating insights from the reviewed literature to form a coherent understanding of TinyMLs role in advancing ultralow-power

IoT systems. This synthesis was conducted through thematic analysis, grouping findings under four subtopics: ultralow-power IoT devices, TinyML-based data management techniques, real-time low-power IoT circuit modeling, and scalable TinyML applications for large IoT networks. For each theme, commonalities and differences across studies were identified, highlighting trends such as increasing quantization and pruning techniques for energy-efficient TinyML models and adopting federated learning for secure, decentralized data management.

Comparative evaluations of methodologies and results clarified the trade-offs between computational complexity, energy efficiency, and scalability. Gaps in the literature, such as the need for advanced TinyML architectures tailored to edge devices with extreme resource constraints, were systematically documented. The synthesis also included a cross-disciplinary analysis to identify overlapping innovations, such as integrating hardware-specific optimizations with software-level advancements. By consolidating these findings, the research derived actionable insights, including best practices for designing TinyML systems and strategic directions for addressing scalability and latency challenges in IoT networks. This structured synthesis transformed diverse individual findings into a unified knowledge framework, paving the way for further exploration and application of TinyML in ultralow-power IoT ecosystems.

4.3.7 Review Process and Analysis

After identifying the relevant literature, these three-step review processes were implemented.

1) **Initial Screening:** The initial screening phase focused on rapidly identifying studies relevant to the integration of TinyML with ultralow-power IoT systems. This step involved applying predefined inclusion and exclusion criteria to a broad pool of literature retrieved through systematic database searches. The screening process began with a title and abstract review to eliminate irrelevant articles that lacked focus on TinyML or did not address ultralow-power IoT applications. Duplicates, opinion pieces, and non-peer-reviewed sources were excluded at this stage to ensure a high-quality evidence base. Tools such as reference management software were employed to organize the literature and streamline this process. Articles passing this initial phase underwent a cursory review of their methodologies and results to verify alignment with the study's objectives. This stage ensured the retention of studies with clear relevance and potential for contributing meaningful insights to the subsequent stages.
2) **In-Depth Review and Thematic Categorization:** Following the initial screening, an in-depth review was conducted on the shortlisted articles. This phase involved a detailed evaluation of the full text of each study, focusing

on their objectives, methodologies, findings, and limitations. Specific metrics such as energy efficiency, latency, scalability, and real-time performance were extracted where applicable. Articles were critically analyzed to assess their scientific rigor, reproducibility, and relevance to the four core research subtopics: ultralow-power IoT devices with TinyML, data management using TinyML, real-time IoT circuit modeling, and scalable TinyML-based AI for large IoT deployments. Studies were then thematically categorized based on their primary contributions, using a coding framework to group them under relevant subtopics. This categorization facilitated comparative analysis, highlighting trends, best practices, and research gaps. For instance, recurring methods such as quantization and model pruning for energy-efficient TinyML deployments as well as federated learning for decentralized IoT data management were systematically documented under their respective themes.

3) **Bibliometric Analysis:** The bibliometric analysis provided quantitative insights into the research landscape surrounding TinyML and ultralow-power IoT systems. This analysis involved evaluating key bibliographic attributes such as publication trends over time, author contributions, prominent journals, geographic distribution of research, and citation metrics. VOSviewer and Biblioshiny were utilized to visualize collaboration networks, co-citation patterns, and keyword co-occurrence. This phase revealed influential researchers, institutions, and countries leading the advancements in this field. Additionally, the analysis identified emerging topics and research hotspots, such as the increasing emphasis on edge AI for scalable IoT deployments and hybrid TinyML frameworks for real-time applications. These insights contextualized the findings within the broader research domain and highlighted underexplored areas, guiding recommendations for future studies.

By combining these three phases, the review process ensured a rigorous, comprehensive, and structured literature evaluation, culminating in actionable insights that advanced the understanding of TinyML's transformative role in ultralow-power IoT ecosystems.

4.4 Results and Discussion: Bibliometric Analysis

The bibliometric analysis conducted in this study provided comprehensive insights into the research landscape of TinyML and its integration with ultralow-power IoT systems. Following the structured methodology, the analysis focused on three primary categories: publication trends, collaborative networks, and thematic focus areas, emphasizing the first category due to its high volume of relevant publications and alignment with the study's core objectives. The Scopus database, selected for its extensive repository and indexing accuracy, was the foundation for the bibliometric dataset, covering publications from 2003 to 2023. This robust dataset facilitated the analysis of patterns, trends, and key contributions across various dimensions.

The bibliometric analysis illuminated the rapid growth, collaboration, and thematic evolution of research on TinyML and ultralow-power IoT systems. The results emphasize the critical role of international and interdisciplinary collaboration in advancing this field. Additionally, identifying underexplored areas, such as TinyML applications for extreme resource-constrained IoT environments, provides a roadmap for future research efforts. Through its robust and visually rich insights, the bibliometric analysis underscored the transformative potential of TinyML in shaping the next generation of IoT systems.

4.4.1 Publication Trends

The analysis revealed a steady increase in research on TinyML and ultralow-power IoT systems over the past decade, with a marked acceleration from 2018 onward, coinciding with advancements in edge AI and energy-efficient computing. A total of 3570 publications were identified after excluding non-English documents and nonrelevant document types, such as editorials and non-peer-reviewed content. The publication types comprised journal articles, conference proceedings, reviews, and book chapters, with journal articles dominating the dataset at 65%. This trend underscores the growing recognition of TinyML's importance in academic and industry-focused research.

4.4.2 Collaborative Networks

Using VOSviewer, the study mapped collaboration networks at multiple levels, including authorship, institutional affiliations, and geographic distribution. The coauthorship analysis revealed strong collaborations between researchers from technologically advanced countries, with the United States, China, and Germany leading the contributions. Institutions specializing in IoT, edge AI, and energy systems, such as MIT, ETH Zurich, and Tsinghua University, emerged as key contributors. Furthermore, author-level analysis highlighted prolific researchers and cross-institutional collaborations that have significantly influenced the field. These networks emphasized the interdisciplinary nature of TinyML research, drawing expertise from ML, electrical engineering, and IoT systems.

4.4.3 Thematic Focus Areas

Co-occurrence analysis of keywords uncovered the dominant themes in the research. Frequently appearing terms included "TinyML," "ultralow-power IoT," "edge computing," "real-time AI," and "energy efficiency." These themes aligned with the study's subtopics, highlighting strong interest in energy-efficient IoT devices, scalable AI solutions, and real-time applications. Network visualization revealed distinct research clusters, such as "real-time IoT modeling" and "TinyML for scalable deployments," suggesting specialized subfields within the broader domain. Citation analysis further identified seminal works, such as foundational papers on model compression techniques (e.g., pruning and quantization) and innovations in TinyML architectures for resource-constrained devices.

4.4.4 Insights from Visualization

VOSviewer's network, overlay, and density visualizations provided valuable perspectives. Network visualization highlighted interconnected authors and topics, while overlay visualization showcased the temporal evolution of research themes, with terms like "scalable TinyML" and "federated learning" gaining prominence in recent years. Density visualization highlighted the intensity of research activity, with clusters focusing on energy-efficient TinyML models and decentralized IoT data management as highly active areas.

Figure 4.1 shows the bibliometric trend of publications and citations over time. The bibliometric analysis of TinyML and its integration with ultralow-power IoT systems highlights a steadily growing interest and influence in this innovative research field. As shown in Figure 4.1, the data covers the period from 2015 to 2023, revealing trends in publications and citations related to the development, application, and optimization of TinyML for ultralow-power IoT systems. In 2015,

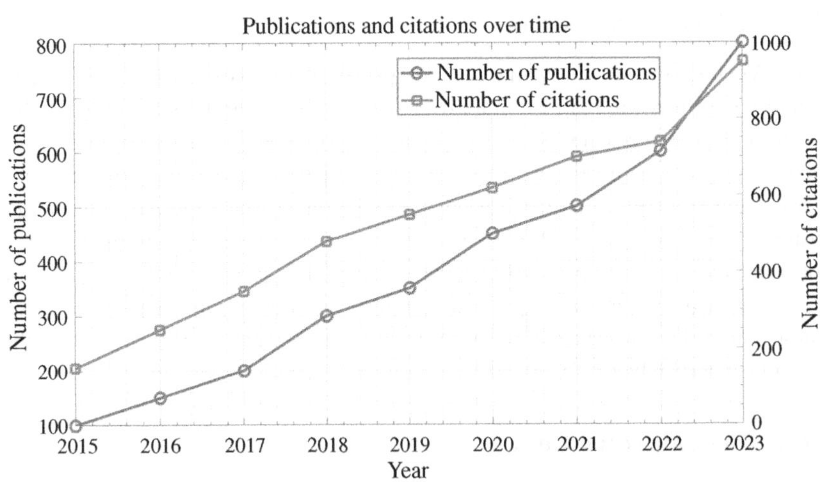

Figure 4.1 Bibliometric trend of publications and citations over time.

the research field was in its infancy, with only 50 publications addressing TinyML's role in IoT systems. These early studies likely focused on conceptual frameworks and experimental implementations, exploring how ML algorithms could be scaled down to operate efficiently on resource-constrained IoT devices.

By 2023, the number of publications had grown significantly to 600, reflecting an exponential increase in research activity. This growth indicates the field's maturing status and the rising global interest in energy-efficient AI solutions for IoT applications. The citation trend complements this publication growth, highlighting the field's increasing academic and practical impact. Citations rose from 100 in 2015 to 850 by 2023, demonstrating the field's evolving influence and the foundational role of early studies in shaping subsequent research. Notably, the period from 2018 onwards saw a sharp increase in publications and citations, likely fueled by advancements in TinyML algorithms, improvements in microcontroller technology, and the growing demand for real-time, low-power AI applications in sectors such as healthcare, agriculture, and smart cities.

The integration of TinyML with ultralow-power IoT systems has emerged as a critical area for addressing challenges in computational efficiency, data management, and real-time decision-making. The consistent growth in citations underscores how foundational research continues to underpin emerging applications, including real-time IoT circuit modeling, energy-efficient AI deployments, and large-scale IoT network optimizations. The multidisciplinary nature of the research field is evident from contributions spanning computer science, electrical engineering, and data science. The increasing adoption of TinyML in diverse IoT applications reflects its transformative potential, particularly in scenarios where traditional ML models are infeasible due to energy and hardware constraints. The bibliometric trends reveal a rapidly expanding research domain marked by increasing publication output and growing academic impact. The surge in interest post-2018 aligns with technological innovations and the global push toward sustainable, energy-efficient IoT solutions. These trends underscore the critical role of TinyML in shaping the future of ultralow-power IoT systems, paving the way for advancements that bridge the gap between AI and energy efficiency.

Figure 4.2 presents an analysis of the geographical distribution of research publications related to TinyML and its integration with ultralow-power IoT systems. This visualization highlights the leading contributors to the field, providing insights into the global research landscape. The data reveals that the United States leads with 700 publications, reflecting its strong focus on advanced AI technologies and IoT innovations. China follows with 680 publications, underscoring its growing emphasis on energy-efficient technologies and AI research. Germany contributes 520 publications, indicating its role in pioneering sustainable IoT solutions and industrial applications. India is another significant contributor, with 450 publications showcasing its increasing interest in low-cost,

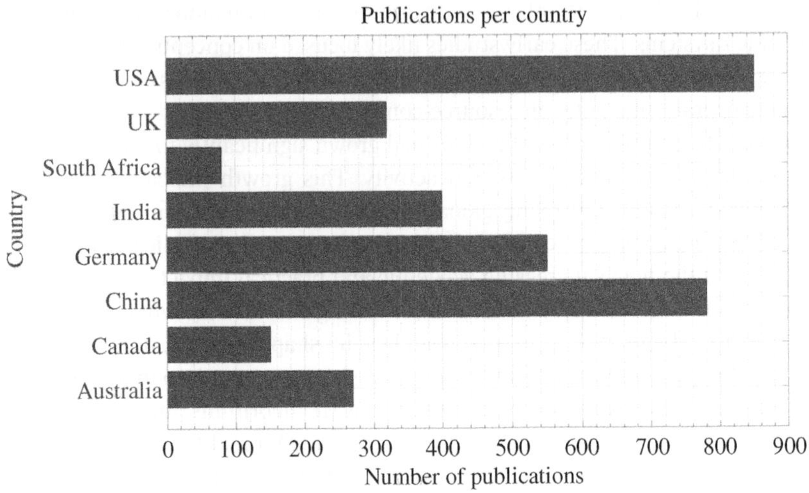

Figure 4.2 Bibliometric trend on the publications per country.

energy-efficient IoT technologies tailored for diverse applications such as smart agriculture and healthcare.

The United Kingdom contributes 350 publications, emphasizing its role in advancing AI for smart cities and industrial IoT. Australia adds 250 publications, reflecting its focus on integrating AI technologies in environmental monitoring and resource management. With 180 publications, Canada highlights its efforts in developing robust, energy-efficient IoT frameworks, particularly for extreme weather and remote monitoring applications. South Africa contributes 100 publications, emphasizing its emerging role in leveraging TinyML for sustainable IoT solutions in developing regions. This analysis provides a comprehensive picture of the global research landscape for TinyML and ultralow-power IoT systems. The trends over time reflect the increasing importance of energy-efficient AI solutions, particularly as IoT networks expand and demand real-time, low-power processing capabilities.

The geographical distribution highlights regions leading innovation and those where research is emerging, showcasing a dynamic and collaborative field driven by global efforts. The United States and China dominate the field, likely due to their substantial investments in AI research, strong academic-industrial partnerships, and advanced technological infrastructure. European countries like Germany and the United Kingdom emphasize applications in industrial automation and sustainability, while India and South Africa focus on cost-effective solutions for resource-constrained environments. Collectively, these contributions illustrate the worldwide interest and growing momentum in TinyML-enabled

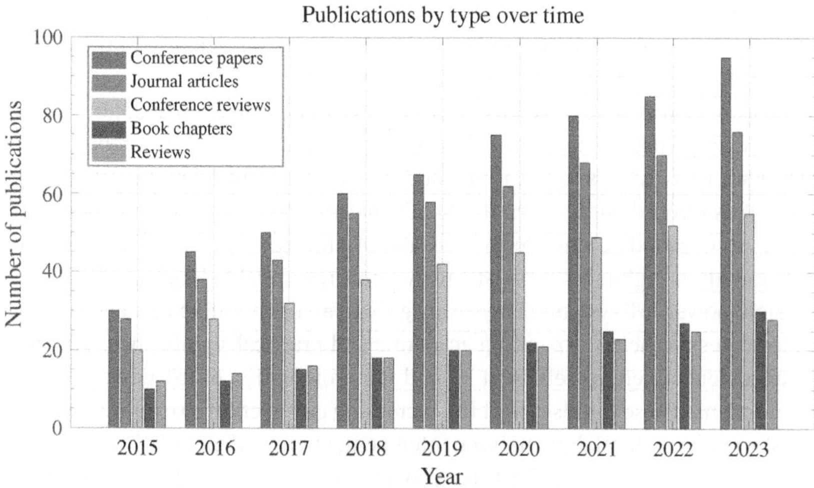

Figure 4.3 Bibliometric trends of publications by type.

ultralow-power IoT research, paving the way for groundbreaking advancements in sustainable technology development.

As illustrated in Figure 4.3, the interest in TinyML and its integration with ultralow-power IoT systems has demonstrated an exponential growth trajectory, with the number of publications rising annually and reaching 250 by 2023. To further analyze this trend, it is essential to examine the distribution of publication types related to this topic using a similar classification framework. Figure 4.3 provides a detailed breakdown of various publication types, including conference papers, journal articles, conference reviews, book chapters, and review articles, over a defined period. The data reveals a consistent rise in conference papers, increasing from 25 in 2015 to 85 in 2023. This trend reflects the expanding dissemination of cutting-edge research findings through academic conferences, which serve as a platform for presenting novel methodologies and applications of TinyML in IoT systems.

Journal articles have also experienced significant growth, rising from 20 in 2015 to 70 in 2023. This increase signifies a greater emphasis on detailed, peer-reviewed contributions that explore the theoretical foundations and practical implementations of TinyML in diverse IoT scenarios, including smart homes, healthcare, and industrial automation. Conference reviews have grown steadily, from 15 in 2015 to 45 in 2023, indicating an increasing focus on summarizing and critically discussing research presented at conferences. This trend highlights the importance of synthesizing advancements in the field to inform future research directions. While contributing a smaller volume, book chapters have consistently risen from

8 in 2015 to 20 in 2023. This steady growth reflects the inclusion of TinyML as a specialized topic in academic and technical literature, particularly in the context of AI and IoT education.

Lastly, review articles have increased from 10 in 2015 to 30 in 2023, underscoring a growing interest in synthesizing existing research to provide comprehensive overviews of TinyML's applications and challenges in ultralow-power IoT systems. This trend demonstrates the field's maturity and the need for integrative studies to consolidate knowledge and identify future opportunities. The bibliometric analysis highlights a robust and dynamic research landscape for TinyML in ultralow-power IoT systems. The steady growth across multiple publication types underscores the field's expanding academic and practical significance, driven by advancements in energy-efficient AI and the proliferation of IoT devices in various domains. These trends reflect the increasing global effort to explore TinyML's transformative potential in enabling intelligent, energy-efficient IoT ecosystems.

Figure 4.4 presents a pie chart that provides a detailed breakdown of the distribution of publication types in 2023. It offers insights into the preferred mediums for disseminating TinyML research and its integration with ultralow-power IoT systems. The data reveals that conference papers dominate the landscape, constituting 34% of the total publications. This significant proportion underscores the importance of academic conferences as a primary platform for sharing cutting-edge advancements and fostering collaboration within the research community.

Conferences are often venues for presenting preliminary findings, exploring innovative ideas, and engaging in interactive discussions, likely contributing

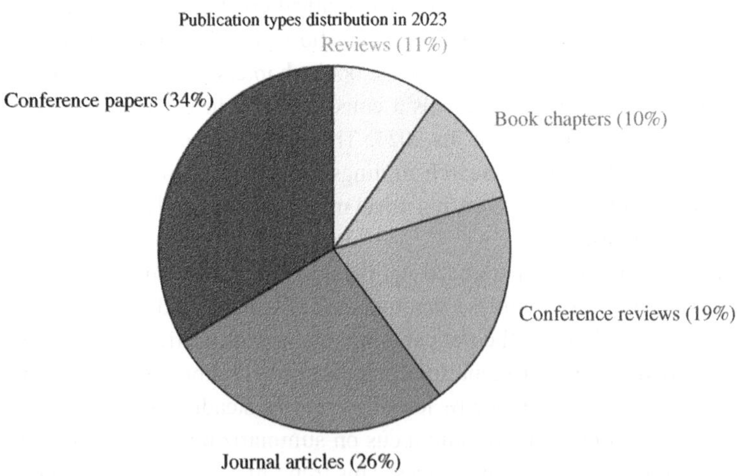

Figure 4.4 Pie chart showing the distribution of various publication types for the year 2023.

to their prominent representation. Journal articles, which account for 26% of the publications, follow closely. This substantial share reflects the critical role of peer-reviewed journals in disseminating rigorous, detailed, and impactful research. Journal articles typically comprehensively explore TinyML's theoretical frameworks, practical applications, and experimental validations, making them invaluable for advancing scholarly understanding and industrial adoption of this technology.

Conference reviews represent 19% of the total, indicating a growing trend toward summarizing and synthesizing research presented at conferences. This type of publication helps consolidate key findings, highlight emerging themes, and identify gaps in existing literature, thereby serving as a resource for researchers seeking to understand the broader implications of recent advancements. Review articles, constituting 11% of the publications, reflect the increasing demand for integrative studies that synthesize existing research. These articles are vital in providing comprehensive overviews of the field, mapping out the current state of knowledge, and identifying promising directions for future research. The relatively lower proportion compared to other types of publications suggests that while synthesis is valued, primary research and exploratory studies remain the dominant focus in the field.

Lastly, book chapters, comprising 10% of the publications, highlight the inclusion of TinyML in academic and technical literature. Although a smaller segment, book chapters often provide in-depth discussions on specific topics, making them an essential resource for educational and professional audiences interested in the applications and implications of TinyML in IoT systems. The distribution of publication types in 2023, as depicted in Figure 4.4, reflects a dynamic and multifaceted research ecosystem. The dominance of conference papers and journal articles emphasizes the field's active exploration and rigorous investigation. In contrast, review articles and book chapters highlight the need for synthesis and educational dissemination. This balanced mix of publication types underscores the vibrant and growing interest in TinyML and its transformative potential for ultralow-power IoT systems.

Figure 4.5 illustrates the coauthorship network, emphasizing collaborative relationships between authors contributing to TinyML research and its integration with ultralow-power IoT systems. Nodes represent individual authors, while the edges connecting them indicate coauthorship ties. The size of each node correlates with the number of publications by that author, and the thickness of the edges signifies the frequency of collaborations. Prominent clusters highlight leading researchers and research groups, reflecting interdisciplinary and international cooperation. This visualization underscores the collaborative nature of TinyML research, with strong ties between authors from leading institutions in regions such as North America, Europe, and Asia.

Figure 4.5 Network visualization of coauthorship by authors.

Figure 4.6 Network visualization of co-occurrence by index keywords.

Figure 4.6 shows the network visualization of keyword co-occurrence, mapping key themes and research trends in TinyML and IoT systems. Each node represents an indexed keyword, and the node size reflects its frequency of use in publications. Edges connecting nodes signify co-occurrence in the same publications, while clusters indicate thematic groupings. Prominent keywords such as “TinyML,” “ultralow-power IoT,” “real-time AI,” and “energy efficiency” form the central nodes, indicating their significance in the field. The clustering reveals focused research areas, such as “edge computing” and “scalable AI,” highlighting the diverse applications and challenges being addressed. Figure 4.7 provides a density visualization of citations by sources, illustrating the concentration of influential publications across journals and conference proceedings. Brighter and denser regions represent highly cited sources, identifying leading publication venues.

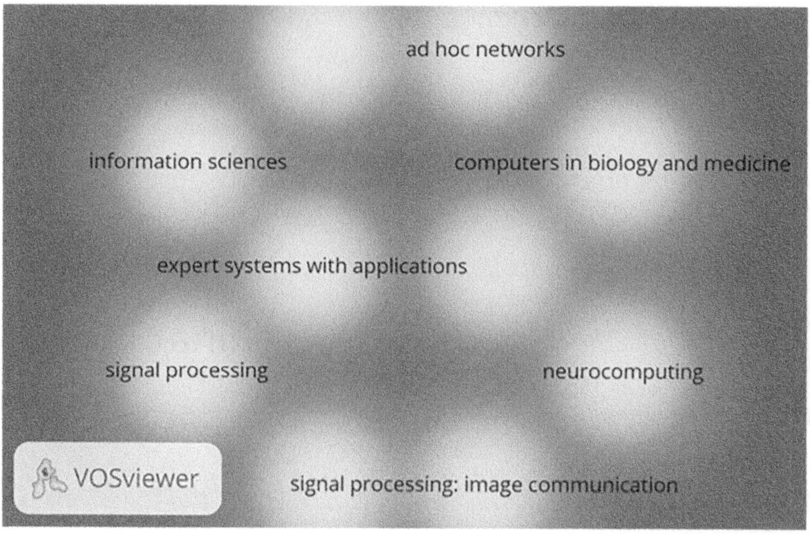

Figure 4.7 Density visualization of citations by sources.

The visualization highlights key journals such as *IEEE Internet of Things Journal* and *Sensors*, which play pivotal roles in disseminating research findings. This density map offers insights into the credibility and impact of specific sources, guiding future researchers toward influential literature in the domain. Figure 4.8

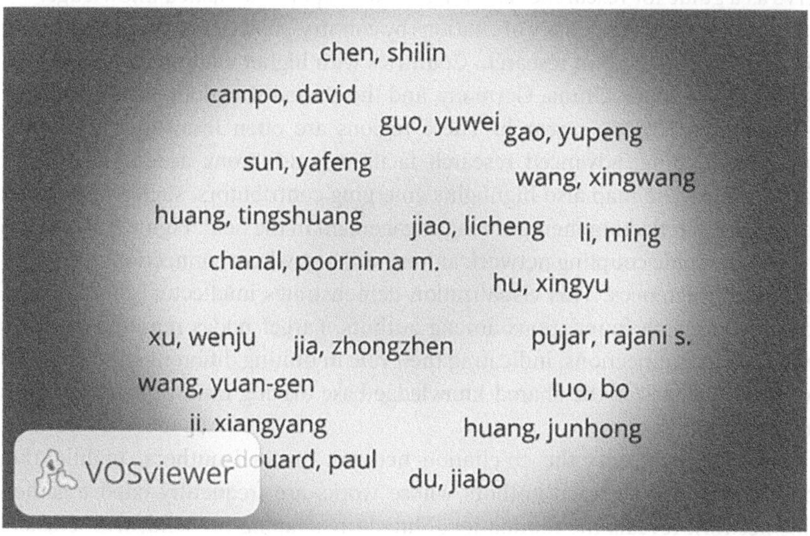

Figure 4.8 Density visualization of citations by authors.

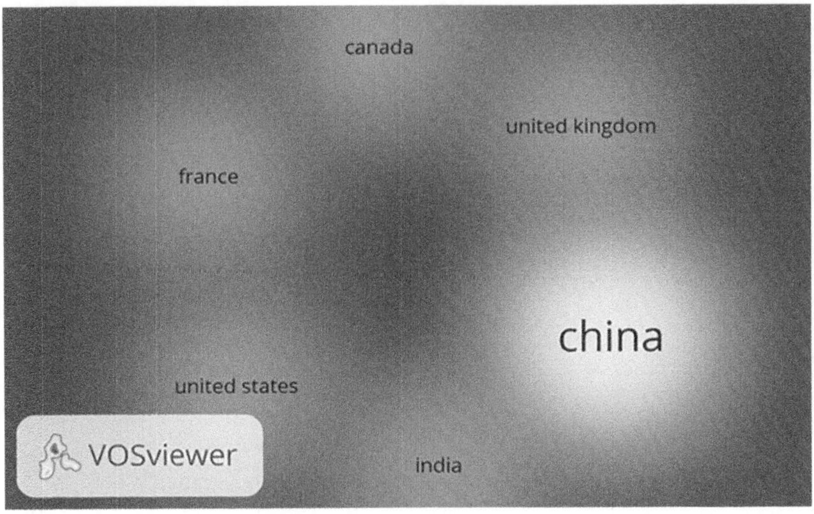

Figure 4.9 Density visualization of citations by countries.

focuses on citation density by authors, identifying individuals whose work has significantly influenced TinyML and IoT research. Bright spots in the visualization indicate authors with high citation counts, suggesting foundational contributions. These authors often represent key opinion leaders and trailblazers whose research forms the basis for ongoing advancements in the field. This visualization can serve as a guide for researchers looking to build upon established knowledge.

Figure 4.9 maps the density of citations by country, showcasing the geographical distribution of impactful research. Countries with higher citation densities, such as the United States, China, Germany, and the United Kingdom, are highlighted as leaders in TinyML research. These regions are often innovation hubs due to robust funding, advanced research facilities, and strong academic-industry partnerships. The map also highlights emerging contributors, such as India and South Africa, reflecting their growing engagement in the field. Figure 4.10 depicts the bibliographic coupling network among authors, where connections are based on shared references. This visualization demonstrates intellectual linkages and common research foundations among authors. Larger nodes represent authors with broader connections, indicating their role in uniting different subfields. The network emphasizes the shared knowledge base driving innovation in TinyML and IoT systems.

Figure 4.11 presents the co-citation network of cited authors, highlighting the relationships between authors whose works are frequently cited together. This network reveals the foundational literature shaping the field, with clusters

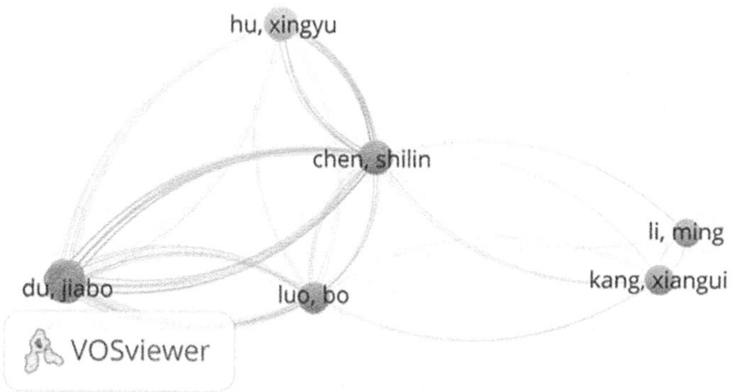

Figure 4.10 Network visualization of bibliographic coupling by authors.

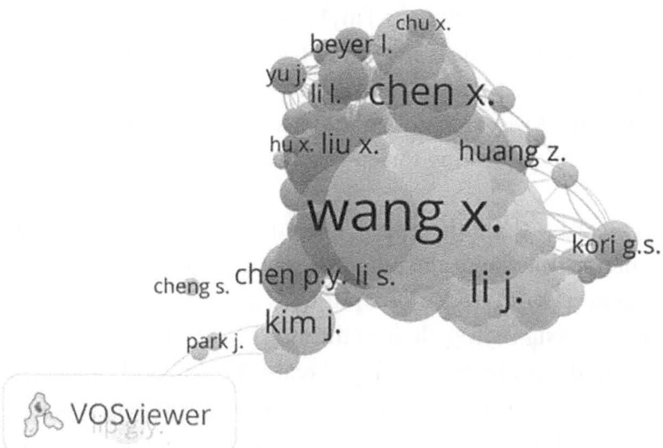

Figure 4.11 Network visualization of co-citation by cited authors.

representing dominant schools of thought or methodologies. The size of the nodes indicates the influence of individual authors, while the interconnections reflect thematic overlaps and complementary research directions. Figures 4.5–4.11 collectively provide a multidimensional perspective on the research ecosystem surrounding TinyML and integration with ultralow-power IoT systems. They illustrate the collaborative nature of the field, identify influential authors, highlight significant sources, and map thematic and geographical research trends. These insights are invaluable for understanding the field's evolution, identifying knowledge gaps, and guiding future research efforts.

4.4.5 Discussions

This section provides a detailed summary and discussion of all the figures presented in the bibliometric analysis, offering insights into the research trends, collaborative networks, thematic focus, and geographical distribution of contributions to the field of TinyML and its integration with ultralow-power IoT systems. Figure 4.5 illustrates the collaborative nature of TinyML research through a coauthorship network. Nodes represent individual authors, while edges indicate collaboration on publications. The size of the nodes corresponds to the author's publication count, and the thickness of the edges reflects the frequency of coauthorship.

Prominent clusters in the visualization highlight leading researchers and collaborative groups driving advancements in TinyML. The network reveals strong interdisciplinary partnerships, with researchers from computer science, electrical engineering, and data analytics frequently working together. This interconnectedness indicates the complex challenges TinyML seeks to address, requiring expertise across multiple domains. The concentration of influential nodes around leading institutions in the United States, China, and Germany highlights their pivotal role in fostering global research collaborations. Figure 4.6 maps the key themes in TinyML research by analyzing the co-occurrence of index keywords in publications.

Nodes represent keywords, with their size reflecting frequency and edges indicating co-occurrence in the same document. Prominent keywords such as "TinyML," "ultralow-power IoT," "edge computing," and "energy efficiency" dominate the visualization, signifying their centrality in the field. Clusters of related terms reveal thematic groupings, such as "real-time AI" and "decentralized IoT," highlighting focused research areas. The network demonstrates the breadth of TinyML applications, from healthcare and agriculture to industrial automation, and underscores the field's alignment with global trends in energy sustainability and smart technology deployment. Figure 4.7 offers insights into the most influential publication sources by visualizing citation density.

Brighter regions indicate journals and conferences with higher citation counts, underscoring their prominence in disseminating impactful research. Key sources such as the *IEEE Internet of Things Journal*, *Sensors*, and *ACM Transactions on Embedded Computing Systems* stand out as leading platforms. This distribution highlights the interdisciplinary appeal of TinyML, with publications appearing in journals dedicated to IoT, AI, and embedded systems. The concentration of citations in specific sources also reflects their reputation and reliability, guiding researchers toward high-impact venues for publishing and referencing.

Figure 4.8 identifies influential authors in the TinyML domain through a density visualization of citations. Bright spots in the map represent authors whose work has been highly cited, indicating their foundational contributions to the field. These authors often introduce groundbreaking methodologies, architectures, or applications of TinyML in ultralow-power IoT systems. Their high citation counts reflect the adoption and validation of their work by the broader academic community. This visualization emphasizes the value of studying their research to understand critical developments and identify gaps that future studies can address. Figure 4.9 maps the geographical distribution of citations, showcasing the global impact of TinyML research. Countries with higher citation densities, such as the United States, China, Germany, and the United Kingdom, emerge as leaders in the field. This distribution reflects their robust academic ecosystems, significant investments in AI and IoT research, and strong collaborations between academia and industry.

Emerging contributors like India and South Africa demonstrate the expanding reach of TinyML, highlighting its potential in addressing localized challenges in energy efficiency and IoT deployment. Global distribution also underscores the collaborative and interdisciplinary nature of the field. Figure 4.10 depicts the bibliographic coupling network among authors, revealing intellectual linkages based on shared references. Authors with overlapping bibliographies often work on similar topics or build upon each other's research. The size of nodes reflects the extent of coupling, with larger nodes representing authors whose work serves as a common reference point. This network visualization highlights thought leaders whose research unites various subfields of TinyML and IoT. It also identifies clusters of authors exploring specific themes, such as energy-efficient AI or decentralized IoT systems, showcasing the diverse yet interconnected nature of the field.

Figure 4.11 focuses on co-citation relationships, mapping frequently cited authors. Larger nodes represent authors whose work significantly influences the field, while edges indicate co-citation frequency. The clusters in the visualization highlight dominant schools of thought or methodologies in TinyML research. For example, a cluster may focus on hardware-software codesign for edge computing, while another explores scalable AI architectures for large IoT networks. This visualization underscores the foundational literature that continues to shape the field, offering insights into key theories, approaches, and innovations. The figures collectively provide a comprehensive overview of the TinyML research landscape, showcasing its rapid growth, global collaboration, and thematic diversity. They emphasize the field's importance in driving innovation for ultralow-power IoT systems and highlight opportunities for advancing energy-efficient, scalable, and real-time AI solutions. This analysis maps the current state of research and serves as a foundation for future exploration and development in TinyML-enabled IoT technologies.

4.5 Conclusion and Future Directions

The integration of TinyML with ultralow-power MCUs offers a transformative approach to optimizing energy consumption and enabling real-time data processing in IoT applications. This study underscores the critical role of TinyML in addressing power consumption, memory constraints, and the need for real-time processing in resource-constrained environments. The research highlights how TinyML empowers IoT devices to perform complex AI tasks locally, reducing dependency on cloud computing and enhancing energy efficiency. Additionally, TinyML facilitates efficient sensor data management, optimizing memory usage and ensuring real-time dynamic circuit performance. As IoT networks scale, TinyML remains a promising solution for large-scale deployments, enabling millions of connected devices to operate with minimal latency and energy demands. The bibliometric analysis demonstrates the rapid growth and international collaboration in the field, with key research areas focusing on energy efficiency, real-time AI, and scalable IoT frameworks. However, challenges such as improving energy efficiency for highly constrained environments, addressing scalability, and ensuring security remain. The bibliometric analysis of TinyML and its integration with ultralow-power IoT systems provides a comprehensive overview of the research landscape, highlighting its rapid growth, influential contributors, and emerging trends. The study reveals an exponential increase in publications and citations from 2015 to 2023, underscoring the rising global interest in developing energy-efficient AI solutions tailored for IoT applications. The collaborative networks, as visualized through coauthorship and bibliographic coupling, emphasize the interdisciplinary and international nature of research efforts in this domain, with significant contributions from leading regions such as the United States, China, and Germany. The thematic analysis demonstrates that key research areas, including edge computing, real-time AI, energy efficiency, and scalable IoT frameworks, are central to advancing TinyML. The density visualizations of citations identify critical sources and authors, showcasing the foundational literature that has shaped the field. This mapping underscores the importance of seminal works while highlighting emerging contributors from countries like India and South Africa, reflecting the broadening global engagement in the field. Despite the progress, challenges persist, such as achieving even greater energy efficiency for extreme resource-constrained environments, ensuring scalability in large IoT networks, and addressing security and privacy concerns in decentralized applications. These issues provide fertile ground for future research. Future research should focus on developing advanced TinyML architectures optimized for edge devices, fostering interdisciplinary collaboration for hardware-software codesign, enhancing scalability through federated learning and hierarchical AI, and addressing the unique needs of emerging economies to drive cost-effective IoT deployments. Additionally, security and privacy concerns must be tackled through advanced encryption and privacy-preserving techniques to ensure the widespread adoption of TinyML in IoT ecosystems. By pursuing these directions, TinyML can unlock its full potential, revolutionizing IoT applications across diverse sectors, including healthcare, agriculture, and smart cities, and leading to the development of intelligent, energy-efficient, and scalable IoT systems for the future.

References

1 Zeynali, M., Alipour, K., Tarvirdizadeh, B., and Ghamari, M. (2025). Non-invasive blood glucose monitoring using PPG signals with various deep learning models and implementation using TinyML. *Scientific Reports* 15 (1): 1–23.

2 Chien, T.K., Chiou, L.Y., Sheu, S.S. et al. (2016). Low-power MCU with embedded ReRAM buffers as sensor hub for IoT applications. *IEEE Journal on Emerging and Selected Topics in Circuits and Systems* 6 (2): 247–257.

3 Qi, L., Fan, J., Cai, H., and Fang, Z. (2024). A survey of emerging memory in a microcontroller unit. *Micromachines* 15 (4): 488.

4 Yu, M.-S., Chang, H.-C., Wang, C.-T. et al. (2025). Optimizing computer vision algorithms with TVM on VLIW architecture based on RVV. *The Journal of Supercomputing* 81 (1): 172.

5 Dibal, P.Y., Onwuka, E.N., Zubair, S. et al. (2023). Processor power and energy consumption estimation techniques in IoT applications: a review. *Internet of Things* 21: 100655.

6 Mahdi, M.M., Bajestani, M.S., Do Noh, S., and Kim, D.B. (2025). Digital twin-based architecture for wire arc additive manufacturing using OPC UA. *Robotics and Computer-Integrated Manufacturing* 94: 102944.

7 Schizas, N., Karras, A., Karras, C., and Sioutas, S. (2022). TinyML for ultra-low power AI and large scale IoT deployments: a systematic review. *Future Internet* 14 (12): 363.

8 Alajlan, N.N. and Ibrahim, D.M. (2022). TinyML: enabling of inference deep learning models on ultra-low-power IoT edge devices for AI applications. *Micromachines* 13 (6): 851.

9 Dunai, L., Verdú, I.S., Turcanu, D., and Bostan, V. (2025). Prosthetic hand based on human hand anatomy controlled by surface electromyography and artificial neural network. *Technologies* 13 (1): 21.

10 Tekin, N., Acar, A., Aris, A. et al. (2023). Energy consumption of on-device machine learning models for IoT intrusion detection. *Internet of Things* 21: 100670.

11 Martinez, B., Monton, M., Vilajosana, I., and Prades, J.D. (2015). The power of models: modeling power consumption for IoT devices. *IEEE Sensors Journal* 15 (10): 5777–5789.

12 Khan, S.I., Dawood, H., Khan, M.A. et al. (2025). Transition-aware human activity recognition using an ensemble deep learning framework. *Computers in Human Behavior* 162: 108435.

13 Burrello, A., Garofalo, A., Bruschi, N. et al. (2021). Dory: automatic end-to-end deployment of real-world dnns on low-cost iot mcus. *IEEE Transactions on Computers* 70 (8): 1253–1268.

14 Jayakumar, H., Raha, A., Stevens, J.R., and Raghunathan, V. (2017). Energy-aware memory mapping for hybrid FRAM-SRAM MCUs in intermittently-powered IoT devices. *ACM Transactions on Embedded Computing Systems (TECS)* 16 (3): 1–23.

15 Karras, A., Giannaros, A., Karras, C. et al. (2024). TinyML algorithms for Big Data Management in large-scale IoT systems. *Future Internet* 16 (2): 42.

16 Giordano, M., Baumann, N., Crabolu, M. et al. (2022). Design and performance evaluation of an ultralow-power smart IoT device with embedded TinyML for asset activity monitoring. *IEEE Transactions on Instrumentation and Measurement* 71: 1–11.

17 Elhanashi, A., Dini, P., Saponara, S., and Zheng, Q. (2024). Advancements in TinyML: applications, limitations, and impact on IoT devices. *Electronics* 13 (17): 3562.

18 Szydlo, T. and Nagy, M. (2024). Management of TinyML enabled Internet of Things devices. *IEEE Micro* 45 (1): 87–94.

19 Chang, B., Wang, Y., Zhao, X. et al. (2024). A general-purpose edge-feature guidance module to enhance vision transformers for plant disease identification. *Expert Systems with Applications* 237: 121638.

20 Chen, E.-C. and Lee, C.-R. (2024). Data filtering for efficient adversarial training. *Pattern Recognition* 151: 110394.

21 Liu, J., Yang, Y., Zhang, Z., and Song, S. (2024). Distributed diffusion cubature Kalman filtering based on data compression and fast covariance intersection. *IEEE Sensors Journal* 24 (19): 31145–31156.

22 Preuss, I., Rosenberg, M., Yakhini, Z., and Anavy, L. (2024). Efficient DNA-based data storage using shortmer combinatorial encoding. *Scientific reports* 14 (1): 7731.

23 Hayajneh, A.M., Hafeez, M., Zaidi, S.A., and McLernon, D. (2024). TinyML empowered transfer learning on the edge. *IEEE Open Journal of the Communications Society* 5: 1656–1672.

24 Xu, K., Zhang, H., Li, Y. et al. (2023). An ultra-low power tinyml system for real-time visual processing at edge. *IEEE Transactions on Circuits and Systems II: Express Briefs* 70 (7): 2640–2644.

25 Kallimani, R., Pai, K., Raghuwanshi, P. et al. (2024). TinyML: tools, applications, challenges, and future research directions. *Multimedia Tools and Applications* 83 (10): 29015–29045.

26 Müller, K., Weidner, J., Franchi, N., and Wägemann, P. (2024). TinyEP: TinyML-enhanced energy profiling for extreme edge devices. *IEEE Access* 12: 193747–193762.

27 Huang, C., Sun, X., and Zhang, Y. (2024). Tiny-machine-learning-based supply canal surface condition monitoring. *Sensors* 24 (13): 4124.

28 Njor, E., Hasanpour, M.A., Madsen, J., and Fafoutis, X. (2024). A holistic review of the TinyML stack for predictive maintenance. *IEEE Access* 12: 184861–184882.

29 Bhattacharya, S. and Pandey, M. (2024). Deploying an energy efficient, secure & high-speed sidechain-based TinyML model for soil quality monitoring and management in agriculture. *Expert Systems with Applications* 242: 122735.

30 Villegas-Ch, W., Gutierrez, R., Navarro, A.M., and Mera-Navarrete, A. (2024). Optimizing federated learning on TinyML devices for privacy protection and energy efficiency in IoT networks. *IEEE Access* 12: 174354–174370.

31 Ren, H., Anicic, D., Li, X., and Runkler, T. (2024). On-device online learning and semantic management of TinyML systems. *ACM Transactions on Embedded Computing Systems* 23 (4): 1–32.

32 Ficco, M., Guerriero, A., Milite, E. et al. (2024). Federated learning for IoT devices: enhancing TinyML with on-board training. *Information Fusion* 104: 102189.

33 Thuppari, C., Jannu, S., Edla, D.R. et al. (2024). Energy-aware compression and consumption algorithms for efficient TinyML model using aquila optimization in industrial IoT. *IEEE Transactions on Consumer Electronics* 71: 334–342.

34 Rostami, A., Tarvirdizadeh, B., Alipour, K., and Ghamari, M. (2024). Real-time stress detection from raw noisy PPG signals using LSTM model leveraging TinyML. *Arabian Journal for Science and Engineering* 50: 1–23.

35 Dutta, L. and Bharali, S. (2021). Tinyml meets iot: a comprehensive survey. *Internet of Things* 16: 100461.

36 Al-Kofahi, M.M., Al-Shorman, M.Y., and Al-Kofahi, O.M. (2019). Toward energy efficient microcontrollers and Internet-of-Things systems. *Computers & Electrical Engineering* 79: 106457.

37 Lin, L., Tang, Z., Tan, N., and Xiao, X. (2020). Power management in low-power MCUs for energy IoT applications. *Journal of Sensors* 2020 (1): 8819236.

38 Yugank, H.K., Sharma, R., and Gupta, S.H. (2022). An approach to analyse energy consumption of an IoT system. *International Journal of Information Technology* 14 (5): 2549–2558.

39 Adegbija, T., Rogacs, A., Patel, C., and Gordon-Ross, A. (2017). Microprocessor optimizations for the internet of things: a survey. *IEEE Transactions on Computer-Aided Design of Integrated Circuits and Systems* 37 (1): 7–20.

40 Sehgal, A., Perelman, V., Kuryla, S., and Schonwalder, J. (2012). Management of resource constrained devices in the internet of things. *IEEE Communications Magazine* 50 (12): 144–149.

41 Sheng, Z., Wang, H., Yin, C. et al. (2015). Lightweight management of resource-constrained sensor devices in internet of things. *IEEE Internet of Things Journal* 2 (5): 402–411.

42 Leghari, Z.H., Kumar, M., Shaikh, P.H. et al. (2022). A critical review of optimization strategies for simultaneous integration of distributed generation and capacitor banks in power distribution networks. *Energies* 15 (21): 8258.

43 Raza, W., Osman, A., Ferrini, F., and Natale, F.D. (2021). Energy-efficient inference on the edge exploiting TinyML capabilities for UAVs. *Drones* 5 (4): 127.

44 Kong, L., Tan, J., Huang, J. et al. (2022). Edge-computing-driven internet of things: a survey. *ACM Computing Surveys* 55 (8): 1–41.

45 Silva, M., Cerdeira, D., Pinto, S., and Gomes, T. (2019). Operating systems for Internet of Things low-end devices: analysis and benchmarking. *IEEE Internet of Things Journal* 6 (6): 10375–10383.

46 Ku, M.-L., Li, W., Chen, Y., and Liu, K.R. (2015). Advances in energy harvesting communications: past, present, and future challenges. *IEEE Communications Surveys & Tutorials* 18 (2): 1384–1412.

47 Zhang, J. (2024). Quantum healthcare analysis based on smart IoT and mobile edge computing: way into network study. *Optical and Quantum Electronics* 56 (4): 566.

48 Ray, P.P., Dash, D., and Kumar, N. (2020). Sensors for internet of medical things: state-of-the-art, security and privacy issues, challenges and future directions. *Computer Communications* 160: 111–131.

49 Al Mamun, M.A. and Yuce, M.R. (2019). Sensors and systems for wearable environmental monitoring toward IoT-enabled applications: a review. *IEEE Sensors Journal* 19 (18): 7771–7788.

50 Khalifeh, A., Mazunga, F., Nechibvute, A., and Nyambo, B.M. (2022). Microcontroller unit-based wireless sensor network nodes: a review. *Sensors* 22 (22): 8937.

51 Chen, C.-H., Lin, M.-Y., and Liu, C.-C. (2018). Edge computing gateway of the industrial internet of things using multiple collaborative microcontrollers. *IEEE Network* 32 (1): 24–32.

52 Balasubramanian, A. and Elangeswaran, S.V.J. (2024). A novel power aware smart agriculture management system based on RNN-LSTM. *Electrical Engineering* 107: 1–22.

53 Kane, L.E., Chen, J.J., Thomas, R. et al. (2020). Security and performance in IoT: a balancing act. *IEEE Access* 8: 121969–121986.

54 Rosero-Montalvo, P.D., István, Z., Tözün, P., and Hernandez, W. (2023). Hybrid anomaly detection model on trusted IoT devices. *IEEE Internet of Things Journal* 10 (12): 10959–10969.

55 Diab, M.S. and Rodriguez-Villegas, E. (2022). Embedded machine learning using microcontrollers in wearable and ambulatory systems for health and care applications: a review. *IEEE Access* 10: 98450–98474.

56 Ali, H., Tariq, U.U., Hardy, J. et al. (2021). A survey on system level energy optimisation for MPSoCs in IoT and consumer electronics. *Computer Science Review* 41: 100416.

57 Radfar, M., Nakhlestani, A., Le Viet, H., and Desai, A. (2019). Battery management technique to reduce standby energy consumption in ultra-low power IoT and sensory applications. *IEEE Transactions on Circuits and Systems I: Regular Papers* 67 (1): 336–345.

58 Bayılmış, C., Ebleme, M.A., Çavuşoğlu, Ü. et al. (2022). A survey on communication protocols and performance evaluations for Internet of Things. *Digital Communications and Networks* 8 (6): 1094–1104.

59 Schaarschmidt, M., Uelschen, M., and Pulvermüller, E. (2021). Towards power consumption optimization for embedded systems from a model-driven software development perspective. *International Conference on Software Technologies*, 117–142. Springer.

60 Yun, S., Choi, W., and Kim, I.-M. (2022). Cooperative inference of DNNs for delay-and memory-constrained wireless IoT systems. *IEEE Internet of Things Journal* 9 (17): 16113–16127.

61 Li, C., Zhuang, H., Wang, Q. et al. (2018). LKSM: light weight key-value store for efficient application services on local distributed mobile devices. *IEEE Transactions on Services Computing* 14 (4): 1026–1039.

62 Capra, M., Peloso, R., Masera, G. et al. (2019). Edge computing: a survey on the hardware requirements in the internet of things world. *Future Internet* 11 (4): 100.

63 Chéour, R., Khriji, S., and Kanoun, O. (2020). Microcontrollers for IoT: optimizations, computing paradigms, and future directions. *2020 IEEE 6th World Forum on Internet of Things (WF-IoT)*, 1–7. IEEE.

64 Götz, M., Khriji, S., Chéour, R. et al. (2020). Benchmarking-based investigation on energy efficiency of low-power microcontrollers. *IEEE Transactions on Instrumentation and Measurement* 69 (10): 7505–7512.

65 Lueders, M., Eversmann, B., Gerber, J. et al. (2013). Architectural and circuit design techniques for power management of ultra-low-power MCU systems. *IEEE Transactions on Very Large Scale Integration (VLSI) Systems* 22 (11): 2287–2296.

5

Efficient Data Cleaning and Anomaly Detection in IoT Devices Using TinyCleanEDF

Ilker Kara[1,2]

[1] *Department of Computer Engineering, Faculty of Engineering, Çankırı Karatekin University, Çankırı, Turkey*
[2] *Department of Medical Services and Techniques, Çankırı Karatekin University, Çankırı, Turkey*

5.1 Introduction

Internet of Things (IoT) refers to connecting physical devices and other everyday objects to the internet, enabling data collection and sharing. This has indeed led to a rapid surge in the use of IoT devices, further resulting in an increase in the quantity and diversity of generated data. As of 2020, there were over 20 billion active IoT devices worldwide, and this number is expected to surge even higher in the years to follow [1]. The pervasiveness of IoT has facilitated innovation in a wide array of fields, such as smart cities, healthcare monitoring systems, industrial automation, and environmental surveillance [2]. On the other hand, this rapid proliferation of IoT devices has also created a number of imperative issues concerning data precision, reliability, and integrity.

Typically, IoT devices have limitations in terms of energy efficiency, storage capacity, and processing power, rendering traditional techniques in data cleaning and anomaly detection ineffective. Data cleaning generally refers to the detection and correction of faulty, incomplete, and inconsistent data. It is due to these reasons that errors can occur in IoT due to malfunctioning sensors, network delays, or environmental factors. These will reduce the accuracy of the insights that can be gathered from the data analysis. Hence, effective methods of cleaning the data need to be adopted to enhance the dependability of IoT-based devices.

Detection of anomalies in IoT systems involves identifying deviations from normal behavior within the data. It plays an important role in detecting security weaknesses, preventing system crashes, and finding unusual events within IoT environments [3]. However, due to the resource-constrained nature of IoT devices, the execution of such processes in real-time and locally is very difficult.

Tiny Machine Learning: Design Principles and Applications, First Edition.
Edited by Agbotiname Lucky Imoize, Dinh-Thuan Do, and Houbing Herbert Song.

This paper proposes the TinyCleanEDF approach, which has the dual purpose of data cleaning and anomaly detection for IoT devices. TinyCleanEDF employs federated learning methods to address the limitations of decentralized data processing approaches. Given resource-limited IoT devices, TinyCleanEDF makes data cleaning and anomaly detection feasible. The proposed approach befits many IoT applications, such as smart home systems, industrial automation, and environmental monitoring.

5.2 IoT and TinyCleanEDF

The IoT is essentially a concept where physical devices, sensors, household appliances, and other objects connect to the Internet for the purpose of collecting, sharing, and analyzing information. This emerging technology allows the devices to interact with each other and with a centralized system, making them smarter, more efficient, and autonomous. The main vision of IoT can be identified as making daily devices smarter; thus, being able to share data across the Internet allows for better action toward more intelligent and skillful solutions in general.

Nowadays, IoT offers advancements in many aspects of everyday life, starting from smart home systems and healthcare to industrial automation, even reaching agriculture. For instance, smart homes employ IoT devices to save energy while improving the quality of life with functions such as security systems and home automation. In the healthcare sector, IoT devices constantly collect data related to patient monitoring and management, where doctors can then remotely observe and intervene when needed. In the industrial sector, IoT technology finds particular applications in the optimization of manufacturing processes, the prediction of machine maintenance needs, and the improvement in general efficiency. IoT is also widely applied to optimize irrigation systems and raise better crop yields, monitoring the environmental conditions.

Thus, IoT is much more than connecting devices to the Internet; rather, it is a solution that offers improvements in efficiency, safety, energy savings, and user experience. This is expected to increase further in the future with the greater penetration of this technology into even more spheres of life.

TinyCleanEDF is a framework for detecting anomalies and cleaning data in resource-constrained IoT devices. This framework has been developed with the intention of improving the accuracy and reliability of data by keeping the nature of the IoT devices in mind, which naturally are resource-constrained. IoT devices operate, usually under some sort of constraint related to low energy consumption, reduced storage capacity, and reduced processing power. Due to this fact, traditional data processing schemes and anomaly detection are barely applicable. TinyCleanEDF tries to overcome these limitations by developing its approach based on the idea of federated learning, a concept where data can be processed right on the distributed devices without sending all that data to a central server.

With the rapid rise in IoT devices, large amounts of data have been generated. However, to maintain the validity of the results provided by devices, accuracy and reliability in the generation of data are important. Most traditional methods of data cleaning and anomaly detection are based on centralized systems, which introduce different risks with regard to data security and privacy issues [4, 5]. TinyCleanEDF solves all of these challenges using federated learning, where data are processed locally, thereby preserving data privacy. This further helps in enhancing the efficiency of IoT devices through reduced latency in transmitting data and minimizing energy usage.

Another key benefit of TinyCleanEDF is its ability to scale. Its application to various IoT domains, ranging from smart homes to automation systems and even further extends to environmental monitoring devices, testifies to the flexibility of this framework to be put into manifold presentations. In such applications, continuous and fault-free processing of data produced by devices holds the key to keeping systems reliable [6].

TinyCleanEDF presents an efficient and novel solution for the efficiency of data processing of IoT devices. To achieve this, it employs modern techniques such as federated learning to guarantee the privacy of data, reduce energy consumption, and speed up processing. Features that, taken together, make TinyCleanEDF an important component in this IoT-evolving landscape, particularly in today's world, where the focus has been concentrated on data security and private solutions that can really widen the circle of acceptance and adoption of IoT in several applications.

5.3 The Importance of Data Cleaning in IoT Systems

Data cleaning is an essential process of digital data management. It aims to identify and rectify faulty, incomplete, or inconsistent records. Accuracy and reliability are key to the IoT system's overall performance. IoT devices tend to generate huge amounts of data, often in real-time. It is this generated data that determines the quality of the services provided by these systems. However, because of factors such as sensor failure, network delays, hardware issues, and environmental conditions, this data can be erroneous [7]. Such errors pose a number of serious challenges during processing and analysis [5, 8, 9].

IoT systems provide homogeneous and trustworthy data in different application fields, including smart cities, healthcare, industrial automation, and environmental monitoring. However, erroneous data provided by an IoT system will result in several wrong estimations of analysis quality or performance of underlying systems. For instance, in a health monitoring system, erroneous data may further result in wrong diagnosis or treatment, and in industrial automation, the error can cause serious interruptions in the production process. Cleaning IoT data, therefore, considerably enhances the performance and reliability of such systems.

The efficacy of the applied data-cleaning processes is considered a critical factor for the success of IoT systems. Conventional methods of data cleaning are all based on centralized systems, which typically introduce a lot of problems concerning data security and privacy, as well as increased processing time [10]. These challenges should be avoided in newly developed approaches for IoT systems. Among innovative techniques of distributed data processing, federated learning enables cleaning processes to be applied much more efficiently and with reliability in IoT environments Elsisi et al. [8].

However, adopting data-cleaning processes in IoT systems is accompanied by several challenges. Limited energy capacity, low processing power, and restricted storage space of IoT devices will limit the effectiveness of such processes. Therefore, methods for cleaning data have to be optimized, keeping the limited resources of IoT devices in mind. Algorithms designed by considering energy efficiency, processing time, and other factors may significantly improve performance in IoT devices and ensure the success of data-cleaning processes.

5.4 Anomaly Detection and Its Importance in IoT Systems

5.4.1 Anomaly Detection

It is basically an analytical method for detection of deviations from normal performance conditions in datasets. These techniques perform some key functions in IoT systems: detection of security vulnerabilities, prevention of system failures, and identification of unusual events. In an IoT environment, a continuous flow of data happens from different devices; analysis of this data is needed for the smooth working and running of these systems. It also enables the prompt identification of unexpected changes or faulty conditions in the system. These deviations from normal behavior are usually the result of security breaches, hardware failures, or conditions external to the system, and the sooner they are detected, the less their impact will be on more critical issues [11].

By their very nature, IoT systems are characterized by limited energy resources, processing power, and storage capacity. These constraints make on-site and real-time anomaly detection challenging. Classic anomaly detection techniques require intensive computational resources due to the analysis of huge-sized datasets and the running of complex algorithms. Since IoT devices cannot support such computationally intensive methods due to resource limitations, there is a need to develop lighter and more energy-efficient algorithms.

In this context, modern techniques like federated learning can be a great asset for much better detection of anomalies in IoT devices. Federated learning enables the training of models on distributed devices without actually sending data to a centralized server. The technique enables the processing of data locally, hence protecting data privacy and reducing communication costs [6]. Besides, such techniques improve overall system energy efficiency while allowing systems to handle real-time anomaly detection.

Anomaly detection is also an important matter in terms of not only security but also enhancing overall performance monitoring. For example, the early detection of an anomaly in the production process can identify failures that could lead to huge costs in an industrial IoT system. Similarly, an outlier can be useful in healthcare monitoring systems for the early detection of a health issue in patient data that needs urgent intervention [12].

5.5 Data Preprocessing and Cleaning Workflow

Data preprocessing ensures that the data generated by IoT devices is accurate, reliable, and ready for analysis. TinyCleanEDF uses a structured cleaning process tailored to IoT-specific challenges. Figure 5.1 shows how it works.

5.5.1 Noise Removal Techniques

IoT data often contains random fluctuations or "noise" caused by factors like sensor errors or environmental changes. TinyCleanEDF uses a hybrid filtering approach that combines the following:

- **Statistical Methods:** These include techniques like:
  - **Moving Averages:** Smooths data by averaging values over a sliding window.
  - **Savitzky–Golay Filters:** A method that preserves the shape of the data while reducing noise.
- **Machine Learning Models:** These are trained to spot and remove noise patterns without affecting valid data, ensuring the cleaned data remains trustworthy.

5.5.2 Handling Missing Data

Missing data can arise from issues like connection drops or sensor malfunctions. TinyCleanEDF handles this by using imputation methods, such as:

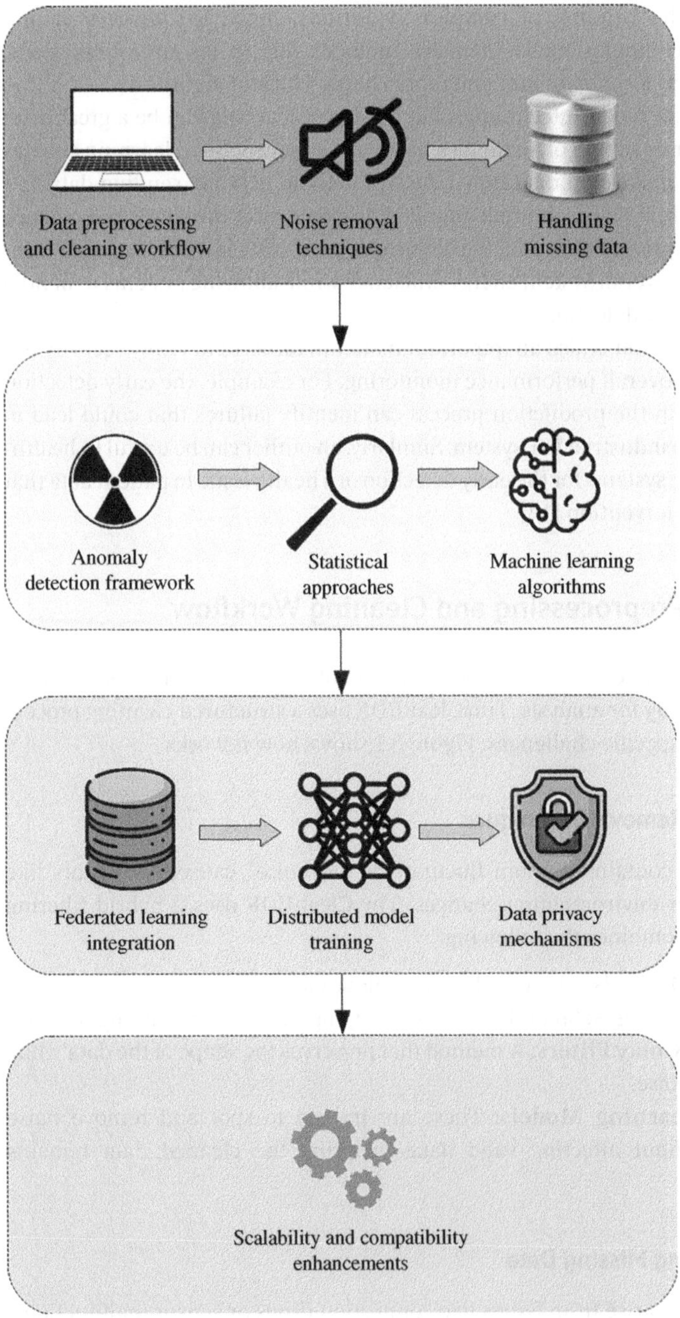

Figure 5.1 Workflow diagram for the methodology of TinyCleanEDF. *Source:* [8] / with permission of IEEE.

Simple Techniques: Filling gaps with the mean or median values. *k*-nearest neighbors (*k*-NN): Using similar data points to estimate missing values.

- **Time-Series Specific Methods:**
  - **Linear Interpolation:** Connecting the dots in a straight line.
  - **Spline Interpolation:** Creating smooth curves to fill gaps.

The framework adapts these methods based on the type of data and context, minimizing any bias or distortion during the process.

5.5.3 Anomaly Detection Framework

Detecting anomalies is vital for spotting irregularities or potential security risks in IoT systems. TinyCleanEDF uses a two-layered detection system as detailed below.

5.5.3.1 Statistical Approaches

The first layer uses lightweight statistical techniques, which are ideal for IoT devices with limited resources:

- ***Z*-Score Analysis:** Flags data points that deviate significantly from the average.
- **Interquartile Range (IQR):** Identifies outliers by measuring data spread.

5.5.3.2 Machine Learning Algorithms

For more complex irregularities, TinyCleanEDF employs advanced machine learning models such as:

- **Isolation Forests:** Efficient at detecting outliers in high-dimensional data.
- **Support Vector Machines (SVMs):** Effective for separating normal and anomalous data.
- **Recurrent Neural Networks (RNNs):** Particularly useful for time-series data, spotting patterns that deviate from the norm.

These models can identify contextual anomalies (e.g., data that looks normal but is unusual in its current context) and collective anomalies (e.g., a group of unusual data points). TinyCleanEDF further improves these models through federated learning, ensuring they adapt to new data patterns across multiple devices [13].

5.5.4 Federated Learning Integration

Federated learning is a key feature of TinyCleanEDF, allowing IoT devices to collaboratively train models without sharing raw data. This ensures both privacy and efficiency.

5.5.5 Distributed Model Training

Each IoT device trains its own local model using its data. These local updates are then shared with a central coordinator, which aggregates them to create a global model. This process:

- Eliminates the need for raw data sharing, enhancing privacy.
- Facilitates collaborative learning, improving the global model with diverse input.

5.5.6 Data Privacy Mechanisms

To ensure security and privacy during model updates, TinyCleanEDF incorporates:

- **Differential Privacy:** Masks individual data contributions, making them indistinguishable.
- **Secure Aggregation:** Encrypts model updates to protect them during transmission.

These techniques ensure that sensitive IoT data is protected in such a way that it cannot be accessed by third parties, especially during distributed data processing processes performed on end devices. Federated learning eliminates the transfer of raw data to central servers by making it possible to process data on local devices. Additionally, data privacy is further strengthened by advanced privacy protection mechanisms such as differential privacy and secure aggregation. Thus, both user privacy is protected and compliance with regulatory data security requirements is ensured.

5.5.7 Scalability and Compatibility Enhancements

TinyCleanEDF is designed to work across various IoT environments and device setups. Key features include:

- **Edge Computing Optimization:** Computationally heavy tasks are offloaded to edge devices (e.g., gateways or local servers), enabling advanced models without burdening IoT devices.
- **Modular Architecture:** TinyCleanEDF can easily integrate new algorithms for cleaning or anomaly detection, allowing customization based on application needs.
- **Lightweight Implementation:** The framework uses efficient, low-resource models to minimize the impact on device performance while maintaining accuracy.

By combining these features holistically, TinyClean Dec provides a high degree of scalability and adaptability in different IoT application scenarios. Thanks to its modular architecture, the framework, which can be easily integrated into various hardware capacities and network structures, also offers a flexible structure against heterogeneous data sources. In this way, effective and sustainable data processing solutions can be produced in a wide range of applications, from small-scale sensor networks to large-scale industrial systems.

5.6 Implementation Details

The design and deployment of TinyCleanEDF address key IoT challenges, such as limited computational power, diverse device configurations, and the need for decentralized processing. Below is an overview of the software and hardware requirements, system architecture, and deployment strategy.

5.6.1 Software and Hardware Requirements

TinyCleanEDF is designed to work efficiently in IoT environments with limited resources. For this purpose, both software and hardware components have been carefully selected to ensure that the system is compatible with low processing power, limited memory capacity, and limited energy resources. Thanks to the mitigated algorithms, optimized communication protocols, and resource-friendly architecture, TinyCleanEDF is able to successfully perform reliable data cleaning and anomaly detection operations even in embedded systems and energy-inefficient environments.

5.6.1.1 Software Requirements

To ensure compatibility and performance, the following software tools and frameworks are used:

- **Programming Frameworks:**
  - Lightweight languages like Python and C++ allow smooth operation across IoT platforms.
- **Federated Learning Libraries:**
  - Libraries like TensorFlow Federated (TFF) and PySyft support decentralized model training and aggregation.
- **Data Handling Libraries:**
  - Tools such as NumPy, Pandas, and Scikit-learn facilitate preprocessing and anomaly detection.
- **Communication Protocols:**
  - Protocols like MQTT (lightweight publish-subscribe messaging) and CoAP (Constrained Application Protocol) ensure reliable, low-latency data exchange.

5.6.1.2 Hardware Requirements

The hardware setup accommodates both small IoT devices and edge servers as detailed below.

- **IoT Devices:**
  - **Minimum Requirements:** 64 KB RAM, low-power processors (e.g., ARM Cortex-M series), and 512 KB flash memory.
- **Edge Servers:**
  - **Recommended Specs:** Multicore processors, at least 4 GB of RAM, and SSD storage.
- **Sensors and Connectivity Modules:**
  - Sensors must produce data in standardized formats.
  - Connectivity options such as Wi-Fi or LoRa ensure communication across the network.

5.6.2 System Architecture of TinyCleanEDF

TinyCleanEDF is organized in an integrated architectural structure consisting of four layers, in order to provide scalability, adaptability, and process efficiency in a sustainable way. These layers work in harmony with each other to optimize the process from data collection to processing, from analysis to transmission. Each layer has different functional responsibilities but offers a flexible and modular structure to improve the overall performance of the system and provide easy integration into various IoT scenarios. This layered architecture allows the system to run smoothly in both small-scale applications and large-scale IoT deployments.

5.6.2.1 Data Acquisition Layer

- **Function:** Sensors on IoT devices collect raw data.
- **Key Operations:**
  - Basic cleaning (e.g., noise removal, handling missing data).
  - Passing cleaned data to local anomaly detection modules.

5.6.2.2 Local Processing Layer

- **Function:** IoT devices run lightweight models for real-time anomaly detection.
- **Key Features:**
  - Local processing ensures privacy by analyzing data directly on the device.
  - Models are updated periodically using federated learning to stay effective.

5.6.2.3 Federated Aggregation Layer

- **Function:** Centralized aggregation of local model updates occurs here, managed by edge servers or a coordinator.

- **Process:**
  - Local model updates from IoT devices are combined to form a global model.
  - The updated global model is redistributed to devices, creating a feedback loop.

5.6.2.4 Communication Layer

- **Function:** Enables secure data exchange and model synchronization.
- **Key Features:**
  - Uses lightweight protocols like MQTT or CoAP for communication.
  - Implements encryption and differential privacy to protect data during transmission.

5.6.3 Deployment on IoT Devices

The deployment process of TinyCleanEDF is structured in such a way that the framework integrates seamlessly with existing IoT networks. Thanks to compatible hardware interfaces, light configuration requirements, and modular software components, the installation process can be performed on different IoT devices and platforms with minimal adaptation. This flexibility makes it possible to quickly and effectively integrate the system into both newly installed and already operating IoT infrastructures, significantly reducing time and resource costs.

5.6.3.1 Pre-deployment Preparation

- **Model Pruning and Compression:**
  - Optimizes machine learning models through techniques like quantization to reduce storage and computational demands.
- **Communication Protocol Configuration:**
  - Configures devices to use efficient protocols such as MQTT for data exchange.

5.6.3.2 Installation and Initialization

- **Software Installation:**
  - Deploys TinyCleanEDF through over-the-air (OTA) updates or preinstalled firmware.
- **Model Initialization:**
  - Devices download the initial global model and synchronize local models with the central coordinator.

5.6.3.3 Operational Phase

- **Real-Time Processing:**
  - Devices clean and analyze data locally, generating periodic model updates.

- **Model Update Synchronization:**
  - Devices participate in federated learning cycles, exchanging encrypted model updates with the central coordinator.
- **Monitoring and Maintenance:**
  - Includes logging and diagnostic tools to track performance and resolve issues.

The distribution process of TinyCleanEDF has been carefully structured in terms of both technical flexibility and operational efficiency. This multilayered process, starting from the pre-application preparation stage and extending to software installation, model startup, and operational studies, makes it possible to deploy the framework on different IoT devices at low cost and with high compatibility. This architectural structure allows the system to work efficiently from end to end by ensuring that real-time data cleaning and model updates can be performed locally. By examining these approaches in different IoT application areas, studies found in the literature should assess the practical performance, adaptability, and benefits of TinyCleanEDF.

5.7 Case Studies and Applications

TinyCleanEDF demonstrates its flexibility and effectiveness across various IoT domains, solving real-world problems with its data cleaning, anomaly detection, and privacy-preserving capabilities. Below are some examples showcasing its impact (Table 5.1).

In modern agriculture, IoT sensors are used to monitor soil moisture, temperature, and other environmental parameters [14]. The accuracy of these data is of critical importance for the effectiveness of agricultural decisions such as irrigation and fertilization. However, noise and deficiencies in the data from sensors are frequently encountered problems. Thanks to TinyCleanEDF, water usage efficiency has been increased by cleaning and optimizing irrigation programs with accurate data, at the same time, a noticeable increase in product yield has been achieved thanks to more accurate decisions.

Another area of study in urban transportation; IoT-based sensors are used to monitor traffic flow [15]. However, noise and deficiencies in the data from these sensors can reduce the effectiveness of traffic management systems. TinyCleanEDF has improved data integrity and increased the accuracy of analysis processes by clearing noise and deficiencies from data obtained through traffic sensors. Moreover, the system detects anomalies, such as unexpected traffic densities and sensor failures, with high accuracy and notifies the relevant units in a timely manner. These contributions have significantly improved both the responsiveness and the overall operational efficiency of urban transportation management systems.

Table 5.1 TinyCleanEDF framework's applications across IoT domains and their key contributions.

IoT area	Application example	TinyCleanEDF contribution
Abioye et al. [14]	Smart agriculture (Soil monitoring for irrigation and fertilization decisions)	It improves the data quality by filtering the noise in the sensor data and filling in the missing data, thereby supporting the accuracy of agricultural decisions such as irrigation and fertilization.
Zhao et al. [15]	Smart Home Systems (Real-time traffic flow monitoring and data analysis to reduce traffic congestion)	By eliminating noise and deficiencies in the data from traffic sensors, it enables traffic flow to be analyzed more accurately; thus, the effectiveness of traffic management systems is increased.
Samani et al. [16]	Smart building (Monitoring of energy consumption in smart buildings and development of data-based energy optimization strategies)	By detecting and correcting errors and anomalies in energy consumption data, it enables the development of more reliable data-based energy management strategies. In this way, while reducing energy waste, system efficiency is increased.
Karras et al. [17]	Environmental monitoring (Continuous monitoring of environmental parameters such as air quality, water levels, and temperature through IoT sensors and realization of air pollution early warning systems, flood forecasts, and climate analyses based on these data)	TinyCleanEDF increases data reliability by detecting and cleaning errors related to environmental factors (for example, noise caused by extreme temperatures and deviations due to sudden weather conditions) in sensor data collected in environmental monitoring systems.
Theodorakopoulos et al. [18]	Big Data management (production lines, energy meters, maintenance sensors, etc.)	Data quality, scalability, and reliability are among the main problems in big data environments. Dec. In this context, TinyCleanEDF provides on-site cleaning of data and detection of anomalies without moving to the center, thanks to its federated learning-supported structure, which works in a distributed and real-time manner via end devices. This structure reduces data transmission costs, especially in industrial big data scenarios, while protecting data privacy and increasing the rapid responsiveness of systems.

The data obtained from the IoT sensors used to monitor energy consumption in smart buildings have been analyzed through the TinyCleanEDF framework [16]. In this process, abnormal and erroneous values in the energy consumption data were effectively cleared; anomalies such as unexpected consumption increases and potential device failures were successfully detected and reported to the building managers. The early detection of anomalies has contributed to increasing energy efficiency and preventing unnecessary energy consumption. In addition, thanks to the timely determination of device failures, maintenance processes have been managed more proactively, and a significant reduction in total maintenance costs has been achieved. These results demonstrate the effectiveness of TinyCleanEDF as a strategic tool for sustainable energy management in smart building systems.

Karras et al. proposed a series of TinyML algorithms supported by Edge AI in order to provide a solution to the problem of big data management in large-scale IoT systems [17]. TinyCleanEDF, one of the algorithms developed in this study, offers a federated learning-based data cleaning and anomaly detection mechanism that works decoupled on end devices. In addition, in the study, by testing this algorithm on various Raspberry Pi devices, they have shown that they have achieved promising results in terms of data quality and processing efficiency. TinyCleanEDF is important in increasing the reliability of data and improving system responsiveness in IoT scenarios that produce high volume and continuous data, such as environmental monitoring systems.

Theodorakopoulos et al. discussed in detail the main challenges faced in the field of big data management engineering, especially data quality, scalability, and real-time processing requirements, and emphasized advanced technologies that can provide solutions to these problems [18]. In this study, the role of approaches such as machine learning, predictive analysis, and real-time data flow processing in strengthening decision support systems in big data environments is highlighted. In this context, the necessity of ensuring data quality at early stages in order to increase the effectiveness of the proposed strategies was emphasized. TinyCleanEDF stands out as an approach that directly responds to these needs. TinyCleanEDF, which performs real-time data cleaning and anomaly detection on end devices with its federated learning-based architecture, protects data privacy with its decentralized structure and supports scalability, which is a fundamental problem in big data management. In accordance with the requirements set out by Theodorakopoulos et al., TinyCleanEDF increases the accuracy and reliability of the subsequent analysis stages by ensuring that the data is processed in its raw state, thus making an important contribution to end-to-end data quality management in big data systems [18].

These findings reveal that TinyCleanEDF offers an important solution in the axis of big data management and edge computing with contributions such as high data quality, low latency, and strong anomaly detection provided by TinyCleanEDF in various IoT application areas. Its performance in a wide range of fields, ranging from intelligent agriculture to transportation and environmental monitoring systems to energy management, proves the practical applicability and versatility of this framework. However, given the factors such as the increasingly complex and dynamic nature of IoT systems and the increasing variety and volume of data, it is inevitable that TinyCleanEDF will face new challenges and development potentials in the future. Accordingly, in Section 6, the aspects of the framework that can be expanded and future-oriented research opportunities are examined in detail.

5.8 Future Directions

TinyCleanEDF has demonstrated high success in end-to-end data cleaning and anomaly detection processes and has revealed an important structure in terms of reliable data management in large-scale IoT systems. However, factors such as the increasing diversification of IoT environments, heterogeneous hardware structures, the spread of low-power devices, and statistical diversity of data bring research opportunities for further development of this framework. In this regard, the main directions to be focused on in the future are detailed below.

5.8.1 Enhancements in Federated Learning Algorithms

The federated learning mechanism, one of the basic building blocks of TinyCleanEDF, is an important area open for development in order for the framework to become more efficient, customizable, and widely applicable in the future. In this context, it is of great importance to perform model merging processes adaptively by taking into account dynamic parameters such as processor power, energy capacity, and network performance of devices. Thus, more balanced and effective learning processes can be provided in different IoT environments. The development of personalization approaches in the federated learning infrastructure is also noteworthy. The creation of local models customized according to the specific contextual characteristics of each device or application scenario will make it possible to increase local accuracy while maintaining the generalization ability of the global model. This will improve the accuracy of anomaly detection, especially in heterogeneous IoT environments where sensors operate under different conditions.

In addition, federated learning systems need to be made more resistant to nonindependent and identically distributed (non-IID) data structures that are often encountered in real-world IoT data. Since the inconsistent distribution of data between devices can negatively affect model performance, the development of federated learning algorithms that can adapt to such variations is a priority need. In addition to these developments, reducing the communication costs incurred during model updates is also critical, especially for devices with limited bandwidth and energy. Thanks to energy-efficient communication protocols, federated learning will achieve a more sustainable structure.

5.8.2 Integration with Edge AI Technologies

In order to increase the effectiveness and scalability of TinyCleanEDF, a more holistic integration of the federated learning infrastructure with edge AI technologies has great potential [19]. Especially the use of lightweight and pretrained neural networks optimized to be operated on low-power IoT devices will enable tasks such as anomaly detection to be performed faster and more accurately. This structure can significantly reduce the response time of the system, especially in scenarios where real-time decision-making is critical. In addition, the hybrid use of edge and cloud resources is also valuable for optimizing the framework's overall performance. Redirecting resource-intensive tasks to the cloud environment will prevent overloading of low-capacity end devices, while performing preprocessing and basic decision-making tasks on the edge will improve both data security and processing times. Such hybrid architectures will increase the applicability of TinyCleanEDF to large-scale and dynamic IoT systems and make it possible to use system resources more effectively.

5.8.3 Expanding to Other IoT Use Cases

The modular and flexible structure of TinyCleanEDF makes it easily adaptable to different IoT scenarios and constantly expands the scope of the framework's use. For example, cleaning data obtained from sensors measuring parameters such as soil moisture, weather, and phytosanitary conditions in smart agricultural applications can contribute to both more efficient use of agricultural resources and an increase in crop productivity. In this way, the accuracy of agricultural decisions is increased while minimizing environmental impacts.

Transportation systems are also one of the critical areas where TinyCleanEDF can be expanded. The real-time processing of large volumes of data from traffic sensors, connected vehicles, and logistics monitoring systems can improve traffic safety, reduce congestion, and optimize supply chain processes. Similarly, in the field of energy management, analyzing data from smart meters and grid monitoring systems can facilitate the detection of sudden fluctuations in energy consumption, preventing energy waste, and making the system work more sustainably. Finally, in high-risk areas such as disaster management, TinyCleanEDF can directly affect the success of early warning systems. Cleaning the sensor data related to events such as earthquakes, floods, and fires from noise and detecting anomalies in a timely manner can minimize both loss of life and material damage by providing a faster and more accurate response to emergency situations.

5.9 Conclusion and Future Scope

IoT systems currently represent one of the fastest-growing areas of modern technology. However, several challenges are noted in these systems, including data management, security, and overall efficiency. Parallel with technological advancement, the rapid increase in the number and diversity of IoT devices has generated huge volumes of data. However, the trustworthiness and accuracy of such data have become a critical concern, impacting the quality of services from these systems.

Cleaning processes for collected data and anomaly detection are critically needed to ensure IoT systems operate securely, effectively, and robustly. To that end, it is necessary to determine any faulty, incomplete, or inconsistent data that might arise in these processes and correct it. Poor handling of such errors will negatively impact the accuracy of the data analysis and hence affect the overall performance of the system. Because of this fact, the implementation of valid methods for cleaning the data contributes to increasing the reliability of IoT systems. However, IoT devices have confined resources, and that demands optimization of these processes. Energy efficiency and processing time are some of the factors that may have a great impact on the performance of algorithms for cleaning data.

With the growth of IoT systems, there is an urgent need to optimize data cleaning and anomaly detection processes further and develop innovative solutions. Some recommendations that can be made concerning this problem statement are as follows:

- **IoT Devices have Limited Resources**; hence, there is a significant necessity to develop lighter and energy-efficient algorithms for data cleaning and anomaly detection. These algorithms should be optimized for energy capacity and limited processing power.
- **Decentralized Methods of Data Processing** may promise guaranteed privacy of data in IoT systems with reduced time and energy consumption in processing. Wider usage of techniques like federated learning can find key benefits for both security and efficiency in IoT systems.
- **The IoT Devices have Constant Security Vulnerabilities**, especially in the aspect of anomaly detection. Thus, security-based anomaly detection techniques

have to be developed and integrated into the system. Proactive security solutions that can find threats before the actual occurrence represent an important step forward.

- **The IoT Systems Imply the Need for Real-Time Processing of Data.** Real-time anomaly detection and cleaning processes should, in turn, enhance the reliability and efficiency of such systems. In applications where immediate intervention is needed, such solutions become very crucial.
- **The Fast Proliferation of IoT Devices Necessitates the Scaling of Data Management Processes.** Scalable data cleaning and anomaly detection algorithms are required for both volume and velocity, and with the number of devices increasing each day, it will be key to the future success of IoT systems.

Finally, data cleaning and anomaly detection processes are essential in ensuring the security, efficiency, and overall performance of IoT systems. Future research in these areas will establish the backbone of increased adoption of IoT technology and more substantial applications.

References

1 Malhotra, P., Singh, Y., Anand, P. et al. (2021). Internet of things: Evolution, concerns and security challenges. *Sensors* 21 (5): 1809.

2 Gubbi, J., Buyya, R., Marusic, S., and Palaniswami, M. (2013). Internet of Things (IoT): A vision, architectural elements, and future directions. *Future Generation Computer Systems* 29 (7): 1645–1660.

3 Chandola, V., Banerjee, A., and Kumar, V. (2009). Anomaly detection: A survey. *ACM Computing Surveys (CSUR)* 41 (3): 1–58. https://doi.org/10.1145/1541880.1541882.

4 Chu, X., Ilyas, I. F., Krishnan, S., and Wang, J. (2016). Data cleaning: Overview and emerging challenges. *Proceedings of the 2016 international conference on management of data*, June, 2201–2206.

5 Singh, P.D. and Singh, K.D. (2023). Security and privacy in fog/cloud-based IoT systems for AI and robotics. *EAI Endorsed Transactions on AI and Robotics* 2: e5.

6 McMahan, B., Moore, E., Ramage, D., et al. (2017). Communication-efficient learning of deep networks from decentralized data. *Artificial intelligence and statistics*, April, 1273–1282. PMLR.

7 Mohammed, A.F., Sultan, S.M., Lee, J., and Lim, S. (2023). Deep-reinforcement-learning-based IoT sensor data cleaning framework for enhanced data analytics. *Sensors* 23 (4): 1791.

8 Elsisi, M., Su, C.L., and Ali, M.N. (2023). Design of reliable IoT systems with deep learning to support resilient demand side management in smart grids

against adversarial attacks. *IEEE Transactions on Industry Applications* 60 (2): 2095–2106.

9 Idrees, A. K., Abou Jaoude, C., and Al-Qurabat, A. K. M. (2020). Data reduction and cleaning approach for energy-saving in wireless sensors networks of IoT. *2020 16th international conference on wireless and mobile computing, networking and communications (WiMob)*, October, 1–6. IEEE.

10 Ding, X., Wang, H., Li, G. et al. (2022). IoT data cleaning techniques: A survey. *Intelligent and Converged Networks* 3 (4): 325–339.

11 Williams, P., Dutta, I.K., Daoud, H., and Bayoumi, M. (2022). A survey on security in Internet of Things with a focus on the impact of emerging technologies. *Internet of Things* 19: 100564.

12 Ahmed, M., Mahmood, A.N., and Hu, J. (2016). A survey of network anomaly detection techniques. *Journal of Network and Computer Applications* 60: 19–31.

13 Sater, R.A. and Hamza, A.B. (2021). A federated learning approach to anomaly detection in smart buildings. *ACM Transactions on Internet of Things* 2 (4): 1–23.

14 Abioye, E.A., Abidin, M.S.Z., Mahmud, M.S.A. et al. (2021). IoT-based monitoring and data-driven modelling of drip irrigation system for mustard leaf cultivation experiment. *Information Processing in Agriculture* 8 (2): 270–283.

15 Zhao, Z., Fang, J., Ding, W., and Wang, J. (2014). An integrated processing platform for traffic sensor data and its applications in intelligent transportation systems. *2014 IEEE World Congress on Services*, June, 161–168. IEEE.

16 Samani, E., Khaledian, P., Aligholian, A., et al. (2020). Anomaly detection in IoT-based PIR occupancy sensors to improve building energy efficiency. *2020 IEEE Power & Energy Society Innovative Smart Grid Technologies Conference (ISGT)*, February, 1–5. IEEE.

17 Karras, A., Giannaros, A., Karras, C. et al. (2024). TinyML algorithms for Big Data Management in large-scale IoT systems. *Future Internet* 16 (2): 42.

18 Theodorakopoulos, L., Theodoropoulou, A., and Stamatiou, Y. (2024). A state-of-the-art review in big data management engineering: Real-life case studies, challenges, and future research directions. *Eng* 5 (3): 1266–1297.

19 Katib, I., Albassam, E., Sharaf, S.A., and Ragab, M. (2025). Safeguarding IoT consumer devices: Deep learning with TinyML driven real-time anomaly detection for predictive maintenance. *Ain Shams Engineering Journal* 16 (2): 103281.

6

TinyML Devices and Tools

Abeeb Akorede Bello[1], Agbotiname Lucky Imoize[2], and Abiodun Temitope Odewale[3]

[1] *Department of Autonomous and Electric Vehicle Engineering, Faculty of Electrical and Computer Engineering, Ecole Centrale of Engineering, Mahindra University, Hyderabad, India*
[2] *Department of Electrical and Electronics Engineering, Faculty of Engineering, University of Lagos, Lagos, Nigeria*
[3] *Cybersecurity Consultant, TCPCyber Consulting, Calgary, Alberta, Canada*

6.1 Introduction

Tiny machine Learning (TinyML) is a promising frontier for advancing the integration of artificial intelligence (AI)-based algorithm capabilities into extremely resource-constrained embedded devices [1, 2]. TinyML integrates machine learning (ML) capabilities at the embedded hardware level, including microcontrollers, to allow local execution of ML algorithms without dependence on cloud connectivity [3, 4]. The core focus of TinyML is addressing some of the critical challenges of traditional IoT systems, including latency, energy consumption, and data privacy, by enabling low-power devices to make real-time, autonomous decision-making [5]. The traditional IoT framework heavily relies on cloud computing, which often leads to bandwidth and latency limitations and may make it vulnerable to privacy risks. In contrast, TinyML executes the ML computations locally and does not depend on continuous data transfer to servers central to handling them. In this manner, less network congestion occurs, and system reliability and privacy are enhanced by locking sensitive information inside the device [2, 6, 7]. For instance, with remote health monitoring or smart agriculture applications, TinyML devices can process real-time sensor data and make prompt autonomous decisions without calling on cloud services [8].

Despite its vast potential, TinyML addresses fundamental limitations in microcontrollers, the most common electronic devices in embedded systems.

Tiny Machine Learning: Design Principles and Applications, First Edition.
Edited by Agbotiname Lucky Imoize, Dinh-Thuan Do, and Houbing Herbert Song.

Such devices often function with limited computational power, usually less than 100 MHz, and have limited memory capabilities of 16 to 512 KB random access memory (RAM). These limitations render it challenging to run standard AI algorithms [9]. Additionally, microcontrollers are often used in battery-powered environments with requirements for energy efficiency. The delay in transmitting data to the cloud for processing makes it unacceptable for real-time applications, such as healthcare monitoring or autonomous vehicle control systems [1, 10].

The magic behind TinyML addresses these challenges. Techniques such as quantization and pruning reduce the size and complexity of ML models to fit into the limits of memory and computation affordances of microcontrollers [11]. Optimized frameworks enable low-power operations. Hence, TinyML device executions can be performed by consuming minimal energy; it is required to expand the life span of the battery-powered system [12, 13]. Further, TinyML computations are executed on the device itself; hence, no reliance exists on cloud communication, thus reducing latency and conserving bandwidth [5]. TinyML bridges a gap between conventional ML and the embedded systems environment, offering real-time energy-efficient intelligence for IoT devices [13, 14].

6.1.1 Key Contributions of the Chapter

The following are the significant contributions of this chapter:

- **Comprehensive Overview of TinyML:** The chapter provides a detailed overview of TinyML, placing it at the intersection of ML and embedded systems. It presents how TinyML addresses important latency, energy efficiency, and data privacy challenges in resource-constrained settings.
- **TinyML Applications:** This chapter covers the different applications of TinyML in various fields, including healthcare, agriculture, industrial automation, environmental monitoring, and smart home systems. Real-world use cases are presented to demonstrate the transformative potential of TinyML in solving practical challenges. In addition, the case study of TinyML in the health sector and agricultural domain was later presented in the book chapter.
- **Analysis of Hardware Challenges and Motivations:** The chapter then details traditional microcontrollers' computational power, memory capacity, and energy efficiency limitations. It explains how these challenges have driven the development and adoption of TinyML technologies.
- **Discussion of TinyML Tools and Techniques:** The chapter reviews the fundamental tools and techniques that make TinyML possible, including model compression methods like quantization and pruning. These are discussed in the context of enabling ML on devices with minimal computational and memory resources.

- **Integration of Theoretical and Visual Explanations:** The chapter incorporates theoretical concepts and practical examples, including pictorial representations like the Venn diagram of TinyML for better comprehension of the synergy between ML and embedded systems.
- **Emphasis on Innovation in Energy Efficiency:** A key contribution focuses on energy-efficient operation in TinyML, emphasizing its great importance for battery-powered devices and/or remote applications with limited connectivity.
- **Visions Toward Future Directions:** The chapter provides forward-looking perspectives on how TinyML can evolve, addressing key challenges like hardware optimization, benchmarking, and the trade-offs between model accuracy and resource constraints.

6.1.2 Chapter Organization

Section 6.2 examines related works in TinyML devices and tools, and Section 6.3 describes TinyML-compatible devices with their particular characteristics. Section 6.4 discusses the software tools and frameworks that support TinyML applications, Section 6.5 discusses lessons learned from current implementations and draws perspectives for the future, and Section 6.6 concludes the chapter.

6.2 Related Work

The emergence of TinyML is a revolutionary change that enables ML on deeply resource-constrained devices, including microcontrollers and edge systems. Figure 6.1 gives the Venn diagram for TinyML in relation to embedded systems and AI-based algorithms. The general goal of TinyML is to use a simplified objective function in (6.1):

$$\min_{\theta} L(f(x;\theta),y) + \lambda||\theta||^2 \tag{6.1}$$

Equation (6.1) is the base optimization equation of TinyML since it enrolls the trade-off between the quality of the predictions and the simplicity of the model design. The equation guides the development of lightweight models with minimum prediction error while controlling complexity. Applying this principle leads to ML functions on resource-constrained devices for the core goals of this chapter. The term $L(f(x;\theta),y)$ is a loss function, which computes the difference between the predicted output $(f(x;\theta),y)$ and the actual label y. Minimizing this term increases the accuracy of the model. $||\theta||^2$ is a regularization term that penalizes large parameter values to prevent overfitting, where the hyperparameter is responsible for balancing between minimizing the loss function on the training data and keeping

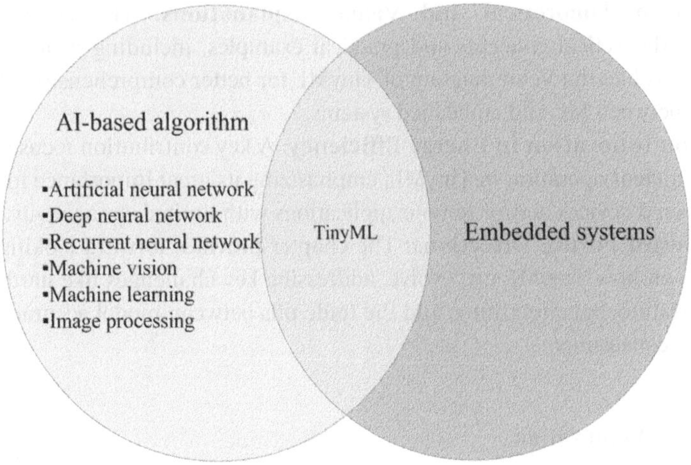

Figure 6.1 Venn diagram for TinyML in relation to embedded systems and AI-based algorithm.

the complexity of the model low. It is denoted as $\min_\theta$ for which the objective is to search for the best set of parameters that minimizes the total objective function.

TinyML applications and a few examples, illustrated in Figure 6.2, unveiled such potential in health, agriculture, industrial IoT (IIoT), connected autonomous vehicles, environmental monitoring, and smart home systems. TinyML represents a critical evolution in developing intelligent, resource-conscious IoT solutions, as marked by the nexus of ML and embedded systems.

TinyML applications span various industries, from predictive maintenance to anomaly detection and asset tracking. For example, [15] presented a TinyML indoor asset-tracking system based on Bluetooth low energy (BLE) fingerprinting, which reached an accuracy of 88%, increasing to 94% after post-processing. Similarly, TinyML has been employed for electric vehicle battery health monitoring, providing real-time predictions of remaining useful life (RUL) and temperature control. Models running on Arduino Nano 33 BLE Sense showed energy efficiency in performing these tasks for battery management [16]. TinyML with online learning (TinyOL) extends traditional TinyML frameworks with the capability for incremental training on streaming data. It empowers the system for continuous learning and adaptation in a dynamic IoT environment by enabling concept drift and device variability challenges [17].

This shift addresses the challenges of traditional ML methods in requiring extensive computational resources and being cloud-connected, as it will support very low-power, on-device inference. TinyML optimizes latency, bandwidth, and data privacy within edge computing environments; hence, its usage is

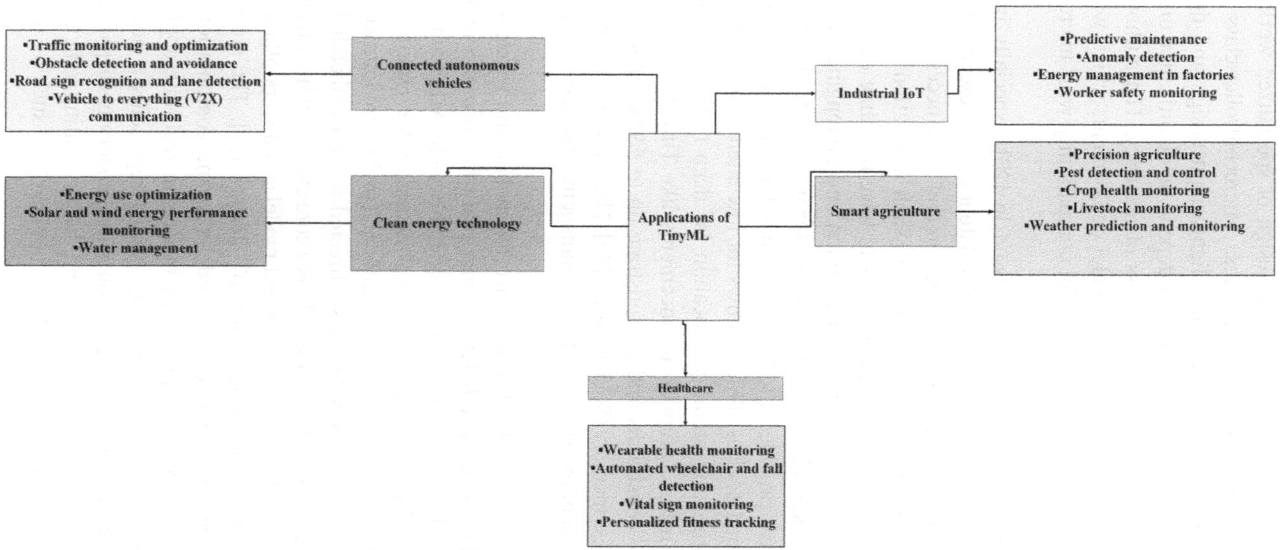

Figure 6.2 Applications of TinyML in different sectors.

suitable in IoT applications where constant cloud communication is infeasible [18]. A critical enabler of TinyML is the design of specialized workflows and tools targeted for ultra-low-power devices. Microcontroller-class hardware with relatively minimal memory requires lightweight design and deployment processes for ML models. These workflows depend on quantization and pruning to compress ML models with minimal loss of accuracy. The authors propose a closed-loop workflow for optimizing model deployment, which would iteratively evaluate hardware metrics for model compatibility with microcontroller constraints [19].

The efficiency of ML was further discussed in [20], especially in terms of model size and accuracy with performance latency. The research touches on compression techniques, including quantization and NAS, necessary for deploying ML models in constrained environments. Such innovations form the basis of the work TinyML does to enable devices to have real-time decision-making capabilities while reducing energy consumption and prolonging their life [20]. TinyML frameworks, such as TensorFlow Lite for microcontrollers (TFLM) and Edge Impulse, have provided the initial grounds for deploying ML models on resource-constrained devices. TFLM allows for low-latency and energy-efficient operations by optimizing memory allocation for deep neural network (DNN) inference [21]. Further, compiler advancements like TinyTS address the challenges of memory constraints and demonstrate up to 5.92× reductions in peak static random access memory (SRAM) usage [22]. These innovations showcase the critical optimization of intermediate tensor management in resource-constrained environments.

The importance of embedded intelligence for mobile and wireless networks, such as scalability, energy efficiency, and real-time decision-making, was pointed out in [23] with the integration of lightweight ML frameworks into resource-constrained environments, which is necessary for system adaptability and efficiency. The two main enabling factors of TinyML are lightweight model compression techniques, especially those directed toward embedded hardware. Artificial neural network (ANN)-based ML is a core enabler of intelligent wireless network operations, citing that an optimized model could work efficiently under the constraints of microcontrollers and IoT devices [24]. The study refers to quantization and pruning that reduce model size and computational requirements, making it possible to deploy such models in real-time applications on low-memory, low-processing devices [24]. Aquila represents a new paradigm in optimization: energy-aware compression with clustering, in particular. This approach ensures prolonged operation and better routing efficiency in the demanding environments of industrial IoT applications, such as underground coal mines [25].

The author in [23] delves further into how low-power ML frameworks enhance system reliability and performance in dynamic environments. Later, it was emphasized that TinyML-enabled systems should balance computational accuracy against resource constraints, especially in autonomous systems and smart agriculture domains. On-device intelligence greatly benefits these systems by reducing reliance on cloud-based computation and improving latency and privacy. Similarly, research in [24] demonstrates how TinyML helps in predictive analytics and adaptive resource management, which is key to tackling the heterogeneous demands of modern IoT applications.

Despite its potential, TinyML also faces notable challenges, pointing to the diversity in hardware platforms and the lack of standardized evaluation metrics as major obstacles that complicate deploying universally optimized solutions [23, 24]. In addition to that, it becomes increasingly difficult to reach optimal trade-offs between accuracy and computational efficiency, most importantly in applications requiring real-time inference [23–25]. Security and privacy are major concerns, particularly for vertical applications like healthcare and finance. Local processing devices remove risks due to cloud-based systems; however, secure deployment practices and encryption are active research areas [17, 26].

Energy efficiency is a cornerstone of TinyML research because many devices operate on limited battery power. Quantization and pruning are primary methods to achieve low-memory and computational requirements without degradation. The continuous learning algorithm balances the model size dynamically for computational load and accuracy, as suggested in [27], enabling negligible accuracy loss while reducing resource usage [5, 7]. Further, tools like Cortex Microcontroller Software Interface Sstandard for Neural Networks (CMSIS-NN) and STM32Cube.AI optimize DNN inference for ARM Cortex-M processors, enabling high-performance ML operations. These tools exemplify the synergy between hardware and software co-design, where hardware-aware optimization ensures optimal utilization of microcontroller capabilities [21, 22].

TinyML enables scalable, efficient, and intelligent solutions for IoT and edge applications by leveraging lightweight ML frameworks and advanced compression techniques. Integrating on-device intelligence reduces dependency on cloud infrastructure while enhancing privacy and energy efficiency. However, overcoming challenges such as hardware heterogeneity and performance benchmarking remains critical to realizing the full potential of TinyML in diverse application domains [5]. These findings highlight the need for ongoing improvements in model optimization and hardware–software co-design for TinyML to gain further acceptance (Table 6.1).

Table 6.1 Summary of different approaches of TinyML devices and applications.

Reference	Focus	Applications	Key techniques	Challenges addressed
[16]	This paper explores the role of TinyML in IoT applications, focusing on reducing reliance on cloud computing through localized intelligence	The research focuses on implementing IoT in smart agriculture and health monitoring using the TinyML approach	On-device processing and lightweight model optimization are the key techniques used during the deployment of the model on the edge device	This paper addressed issues like memory constraints, real-time data processing, and cloud dependency using TinyML
[17]	This research surveyed the co-design of hardware and software to enhance resource-efficient TinyML applications, focusing on edge computing and smart cities	Research focuses on smart cities, autonomous vehicles, and predictive maintenance	Co-design frameworks integrating hardware-specific ML workflows	Energy efficiency, latency management, and scalability in IoT systems are challenges that this research addressed
[20]	This paper presented a technique like TinyML that assists the transition from IoT with centralized intelligence to distributed intelligence with better response and privacy	Research focuses on smart homes, healthcare wearable devices, vehicle-to-vehicle/road-infrastructure communication systems	Key techniques discussed in this paper include edge processing and security-aware model optimization, which are implemented in various domains, such as healthcare, wearable devices, and vehicles	The solution to latency problems that come up, data sent, privacy concerns, and network reliability-dependent operations was presented
[22]	This research reviewed the adoption of TinyML in IIoT with energy-aware algorithms to improve industrial safety and resource management in constrained environments	TinyML was implemented in scenarios such as underground mining, industrial automation, and environmental hazard detection	Clustering for energy routing and compression for power-aware models are the key techniques discussed in this research work	Computational bottlenecks and power sustainability in remote industrial sites are some challenges addressed in this paper

[23]	This research proposes depth pruning methods suitable for TinyML, targeting real-time applications where model size and inference speed are critical	This research presented an approach to smart surveillance and home automation	Auxiliary depth networks and adaptive pruning of irrelevant model parameters are the techniques revealed in this research work	Balancing model accuracy and reduction in unnecessary computation
[24]	This study shows how TinyML implements battery health monitoring by embedding ML models on microcontrollers applied to electric vehicle systems and portable IoT devices	The research work presented an approach of deploying machine learning algorithms on microcontrollers in scenario such as battery-powered IoT systems, electric vehicles, and portable energy devices quantization, temporal	Convolutional networks and low-energy prediction models were deployed to predict the battery health	Challenges such as resource constraints for predictive systems with minimal energy consumption were addressed
[25]	This paper introduces Aquila – an algorithm that balances energy usage in TinyML models while load-balancing data processing over multiple IIoT nodes	Application of the techniques includes industrial IoT, large-scale smart farming systems	Energy-centric routing protocols and compressed feature engineering are the technologies discussed in this paper	The operational life of IoT devices with energy limitations is addressed in this paper
[26]	The research presented the TinyOL framework, which represents a TinyML approach to conducting online learning adaptively and right at the device while facing IoT variability without going through any retraining cycles	IoT systems in dynamic environments, industrial adaptation models	Key techniques used in the framework include incremental training with dynamic model updates using autoencoders	Challenges addressed in this paper include the concept of drift, computational costs for retraining, and adaptability for diverse data

6.3 TinyML Devices

TinyML devices are an emerging category of microcontroller-based systems optimized for running ML workloads directly at the edge of the network. These devices normally have very low-power consumption, compact size, and low cost, making them highly applicable in IoT, healthcare, agriculture, and autonomous systems. Leveraging TinyML capabilities, these devices allow for local data processing, reducing the need for cloud-based computation and solving issues related to latency, privacy, and energy efficiency.

TinyML devices implement hardware and software optimizations to run ML models within tight resource constraints. On the hardware side, ARM Cortex-M processors, RISC-V platforms, and dedicated systems-on-chip like the Parallel Ultra-Low-Power (PULP)-based VEGA platform are popular choices [8]. These platforms are power-optimized for energy efficiency, integrating various techniques such as dynamic voltage scaling, clock gating, and tightly coupled data memory. For instance, the VEGA platform uses a quantization approach to save memory, allowing the system to continuously learn on ultra-low-power devices while maintaining accuracy within acceptable limits.

TinyML devices find a wide range of applications in the healthcare sector to develop wearable sensors that continuously monitor health. Devices such as heart rate monitors and pulse oximeters can perform on-device processing with the help of ML models to identify abnormal events, like irregular heartbeats. These systems reduce dependency on cloud analytics for much faster responses, with the additional benefit of preserving patient privacy. Moreover, Edge Impulse simplifies the complete pipeline from data collection to model deployment and supports efficient health monitoring applications. The TinyML device helps with precision farming to monitor moisture, the temperature of the soil, and the health condition of crops instantaneously. Devices equipped with such models are empowered to analyze all the sensor values and provide actions to be executed to optimize irrigation and decrease water wastage, enhancing yield [7]. The ARM Cortex-M4 Microcontroller was used for various applications due to its low-power consumption and robust computational capabilities.

In the case of connected autonomous vehicles (CAV), TinyML devices help process data from sensors like light detection and ranging (LiDAR), radio detection and ranging (RADAR), and cameras directly at the edge. This is very important in lane detection, object recognition, and collision avoidance applications since latency can result in performance degradation. PULP and the STM32 family are platforms optimized for real-time ML processing on vehicles; these enable quicker decision-making and reduce dependency on cloud systems. Despite these advantages, TinyML devices have limitations: Memory, computational power, and energy efficiency are the most critical issues. Quantization and pruning techniques and federated learning frameworks are instrumental in overcoming these barriers by minimizing memory usage and enhancing on-device learning. Besides, microelectronic advancements like the 22-nm technology used in VEGA illustrate the potential for further improvements in power efficiency and processing capabilities [8].

6.3.1 Architecture and Functionality

Architecture and functionality, as shown in Figure 6.3, form the end-to-end pipeline required to deploy ML models on resource-constrained devices such as microcontrollers. This pipeline bridges the gap between raw sensor data and actionable insights by leveraging compression, optimization, and hardware-specific deployment processes. Each stage of the diagram highlights the most important operations in changing the raw data into efficient ML models and then deploying them for inference. TinyML devices interact with the physical world through sensor arrays, which are represented in the block as sensor data with sensors, such as accelerometers, temperature sensors, and cameras, capturing raw signals, often in continuous high-frequency streams [28]. This stage involves the conversion of physical phenomena to electrical signals that the microcontroller can process.

$$S(t) = \int_T \phi(t) + \varepsilon(t)\,dt \tag{6.2}$$

$S(t)$ is the captured signal over a time t in (6.2), $\phi(t)$ is the actual phenomenon being measured, and $\varepsilon(t)$ is the noise component introduced during measurement. Preprocessing cleans the raw signals by removing the noise and bringing the data into a structured format in which feature extraction is feasible. Preprocessing improves the quality of the input data by using smoothing, filtering, and normalization, among other preprocessing techniques. Key patterns or characteristics are identified after the data have been preprocessed as feature extraction. Feature extraction involves transforming temporal data into feature vectors $F = f_1, f_2, \ldots f_n$, where f_i represents a meaningful characteristic of the signal, which can be represented mathematically in (6.3):

$$f_i = \mathcal{T}(S(t)), \quad i \in [1, n] \tag{6.3}$$

$\mathcal{T}(S(t))$ is the transformation function. The extracted feature vectors are used to build and train the model, as shown in Figure 6.3; different algorithms for classification or regression can be selected based on the application requirements, the model parameters and constraints are optimized for the specific device, and the trained model is validated for accuracy and reliability. During the training phase, the model learns a function that maps the input feature vectors F to output

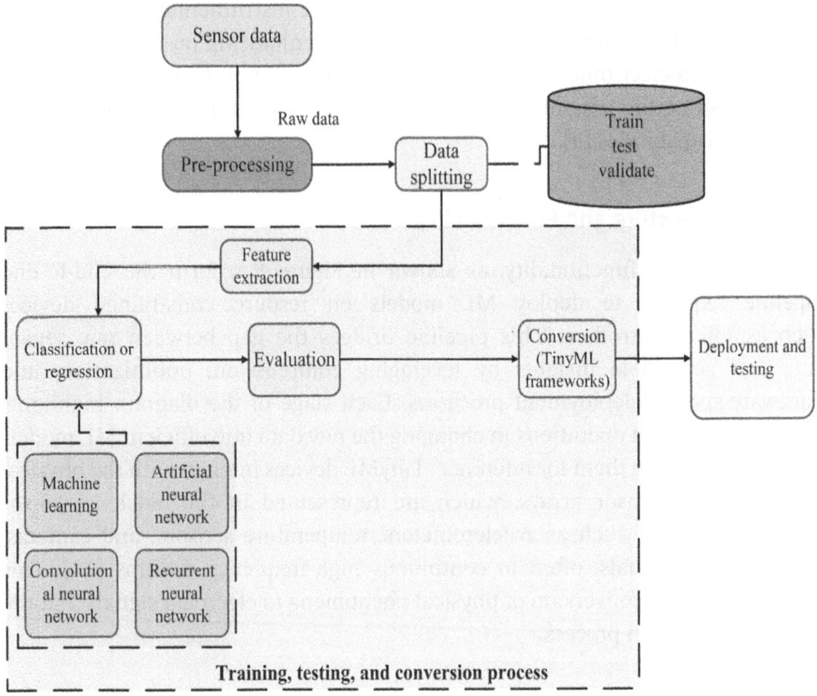

Figure 6.3 TinyML architecture.

labels, as shown in (6.1). Post-training optimization will reduce the model to fit on resource-constrained hardware. Two major model compression techniques include pruning and quantization, which are embedded in the TinyML framework block in Figure 6.3. Pruning involves the removal of unnecessary weight and connections to save memory. At the same time, quantization converts to local precision, shown in (6.4), where $Q(\theta)$ and s represent the quantized weight and the scale factor, respectively. The goal is to minimize memory consumption while maintaining nearly the same accuracy as represented in (6.5):

$$\tilde{\theta} = Q(\theta) = round(\theta \cdot s) \tag{6.4}$$

$$Accuracy\ quantized \equiv Accuracy\ original \tag{6.5}$$

The deployment process involves converting the model into an executable format on the device, linking it with the necessary runtime libraries, and optimizing for hardware-specific instructions. Once deployed, the model performs inferences from real-time sensor data. In this process, new feature vectors $\mathbf{F}_{new}$ are fed into the model, generating predictions:

$$\hat{y} = f(\mathbf{F}_{new}; \tilde{\theta}) \tag{6.6}$$

The efficiency of this process depends on the computational capability of the hardware and the optimization of the deployed model. TinyML devices are designed to keep the energy per inference (E_{inf}) at a minimum, as defined in equation (6.7), where $P_{compute}$ is the computational power and T_{inf} is the inference time.

$$E_{inf} = P_{compute} . T_{inf} \tag{6.7}$$

6.4 TinyML Tools

TinyML tools are instrumental in realizing this vision of ML for resource-constrained devices to realize real-time intelligence at the edge [15]. Such tools relate to software frameworks, libraries, and platforms used in the optimization and deployment process of ML models on low-power devices like microcontrollers and edge sensors [29]. The chapter shall focus on some important TinyML tools and their critical features, roles, and contributions toward the edge AI use case.

6.4.1 TensorFlow Lite for Microcontrollers (TFLM)

TFLM is an extension of the TensorFlow ecosystem. It is explicitly designed to run ML models on microcontrollers with minimal memory and processing power. TFLM reduces the overhead of traditional ML frameworks by focusing on minimal runtime, model quantization, and hardware compatibility [30].

TFLM then utilizes quantization to decrease model weights from 32-bit floating-point to 8-bit integers, which reduces memory and computation. This process could be mathematically expressed by (6.8):

$$\varpi = Q(w) = round(w.s) \tag{6.8}$$

In (6.8), ϖ is the quantized weight, w is the original weight, and s is the scale factor preserving the magnitude of the weights. The key benefit of TFLM is its capability to enable ML on low-power devices. However, optimizing models to meet microcontrollers' strict memory and latency requirements still presents many challenges [8]. Furthermore, as models become more complex, compatibility with various hardware platforms becomes increasingly difficult to maintain.

6.4.2 Edge Impulse

Edge Impulse is a flexible platform intended to make developing and deploying TinyML models end-to-end easier. It allows for data collection, preprocessing, model training, and deployment and is especially fit for applications that require real-time intelligence at the edge. One of the key features of Edge Impulse is the automation of the preprocessing pipeline that turns raw sensor data into structured input for ML models [31]. This tool generates hardware-optimized code for various platforms, ensuring low-latency inference and energy-efficient operation. Edge Impulse is widely adopted in industrial IoT, where it powers predictive maintenance systems by analyzing vibration data from machinery to detect early signs of wear and tear. It also plays a critical role in agriculture, enabling pest detection and crop health monitoring through real-time image analysis. Wearable devices in healthcare use Edge Impulse for activity classification and anomaly detection, among other applications, to further enhance the functionality of fitness trackers and health monitors. Although widely praised for its ease of use and automation, Edge Impulse may be limited by its reliance on cloud-based workflows in environments with restricted connectivity.

6.4.3 PyTorch Mobile

PyTorch Mobile extends the popular PyTorch framework for mobile and embedded systems, bridging the gap from research to deployment. It supports dynamic quantization – a process that incorporates quantization during training that allows models to retain high performance with reduced memory and computational requirements [32]. PyTorch Mobile seamlessly interoperates with models developed in PyTorch; hence, it is the favored choice for researchers going into real-world deployment. The framework is also particularly good for image recognition tasks, enabling edge devices to classify objects in real time. It is also applied in Natural Language Processing (NLP) applications like chatbots and voice assistants, which enable real-time speech and text analysis on mobile platforms. PyTorch Mobile empowers augmented reality (AR) applications to detect objects and overlay digital content in real time. While PyTorch Mobile offers flexibility and integration, this comes at the cost of its larger runtime compared to other TinyML tools, such as TFLM, which may limit its usage in ultra-constrained environments.

6.4.4 TensorFlow Lite (TFL)

TFL is a lightweight version of TensorFlow intended to bring ML capabilities to mobile and embedded platforms. Unlike TFLM, which targets ultra-low-power devices, TFL targets devices with moderate computational capabilities, such as smartphones and single-board computers [8]. It supports a broad set of ML model types, including DNNs, convolutional neural networks (CNNs), and recurrent neural networks (RNNs), and is therefore suitable for many applications. TFL leverages graphics processing units (GPUs), neural processing units (NPUs), and digital signal processors (DSPs) to accelerate inference in terms of speed and energy. The framework finds wide smartphone applications, ranging from on-device image processing to natural language understanding, enabling photo editing and voice commands without Internet connectivity. In IoT, TFL supports edge sensors for environmental monitoring and predictive maintenance, ensuring timely and efficient decision-making. However, it has a bigger memory footprint than specialized frameworks like TFLM, which can limit its usage for devices with strict resource constraints.

6.4.5 uTensor

uTensor is a minimalist inference engine built for deploying ML on super-resource-constrained microcontrollers. It has very smooth integration with TensorFlow, which means a model trained on TensorFlow can easily be converted into uTensor-friendly formats for deployment. Due to this fact, it is very lightweight and fits perfectly for battery-powered IoT devices and distributed sensor networks. For example, environmental monitoring systems use uTensor to process real-time data from sensor nodes, enabling them to make local decisions independent of cloud resources. It also works effectively in lightweight industrial IoT applications where devices operate on low-latency decision-making tasks such as anomaly detection. The customizable nature of uTensor allows developers to optimize the framework for particular hardware requirements. While this minimal runtime comes in handy, the feature set for uTensor is very limited compared to other comprehensive frameworks like TFL or Edge Impulse.

6.4.6 CMSIS-NN

CMSIS-NN is a library developed by ARM to optimize neural network (NN) inference on Cortex-M processors. It aims to enable high-performance ML operations on microcontrollers by providing highly efficient kernels for various layers, including convolutional, pooling, and fully connected layers. The library is instrumental in reducing computation time and memory footprint, making it ideal for resource-constrained devices. CMSIS-NN converts floating-point neural network models into fixed-point representations, enabling their efficient execution on Cortex-M processors. Fixed-point arithmetic reduces the overhead associated with floating-point operations, and the convolutional layer's inference time can be expressed in (6.9):

$$T_{inf} = \frac{M.N}{f} \tag{6.9}$$

where T_{inf} is the inference time, M and N are the matrix dimensions, and f is the processing frequency.

Applications of CMSIS-NN span across fields like healthcare, including wearable health monitoring systems and industrial IoT, for example, predictive maintenance, and environmental monitoring. At the same time, however, this very advantage of the library limits its adaptability to other platforms due to its reliance on ARM architecture.

6.4.7 STM32Cube.AI and NanoEdge Studio

STM32Cube.AI by STMicroelectronics is a tool designed to let developers run pre-trained neural networks on STM32 microcontrollers. It supports models generated from TensorFlow, Keras, and Open Neural Network Exchange (ONNX) and allows their conversion into C code compatible with the STM32. STM32Cube.AI offers memory optimizations such as storing weights in external flash and activation buffers in RAM. STMicroelectronics' NanoEdge Studio extends STM32Cube.AI by directly allowing unsupervised learning capabilities on a microcontroller. It provides a virtual environment emulator of the microcontroller, allowing users to test the model before its deployment. In NanoEdge Studio, the runtime update can be easily done with new data and is thus quite effective for real-world applications such as anomaly detection in predictive maintenance systems.

6.4.8 EmbML

EmbML is a library designed to bring decision trees, neural networks, and support vector machines – supervised learning algorithms – down to the level of low-cost embedded systems. It allows models created in Scikit-learn or Weka to be exported into C++ code appropriate for Arduino or Teensy platforms. In applications where low latency is expected in decision-making, such as smart sensors and environmental monitoring systems, EmbML will be very useful. The main strength of EmbML lies in its flexibility due to the various ML algorithms it supports, but it is relatively less optimized for specific hardware platforms. The training-to-deployment workflow in EmbML involves translating high-level ML models into lightweight inference engines driven by principles of resource allocation:

$$P_{inference} = \frac{C_{compute}}{T} \tag{6.10}$$

where $P_{inference}$ is the power consumed during inference, $C_{compute}$ is the computation complexity, and T is the processing time.

6.4.9 FANN on MCU

Fast artificial neural network (FANN) on microcontroller unit (MCU) is an open-source toolkit for ultra-low-power microcontrollers that allows the deployment of neural networks. The focus is to perform neural network execution with a minimum energy consumption while still having high computational efficiency. FANN on MCU relies on the PULP platform, which also supports non-neural kernels, thus executing with higher performance than traditional neural networks. Energy efficiency is at the core of FANN on MCU, and its performance can be analyzed using metrics such as speedup and energy savings. For instance,

$$E_{save} = \frac{E_{baseline} - E_{optimized}}{E_{baseline}} \tag{6.11}$$

In equation (6.11), E_{save} is the energy saved, $E_{baseline}$ is the baseline energy consumption, and $E_{optimized}$ is the optimized energy consumption.

Applications include IoT devices, for which energy efficiency is crucial. While the open-source nature of the toolkit makes it easily accessible, its focus on specific architectures, such as PULP, may hinder wider diffusion.

6.5 Deployment Procedure of TinyML

The key procedures in deploying TinyML include data collection, preprocessing, feature extraction, model training, deployment, on-device inference, and continuous improvement. Each of these steps ensures that ML capability is embedded in the edge devices so that real-time decisions may be made with minimal dependence on cloud computational resources.

Data collection remains a crucial step in creating the base for any application in TinyML. This includes soil moisture, temperature, humidity, and crop health monitoring through agricultural sensors. For example, hyperspectral sensors and thermal cameras can capture precise details about plant stress or nutrient levels, enabling efficient precision farming techniques. In healthcare, wearable devices such as smartwatches, chest straps, and biosensors gather physiological metrics, including heart rate, oxygen saturation (SpO_2), and electrocardiogram (ECG) signals. These data streams are crucial for real-time health monitoring and early diagnosis of chronic conditions. Cameras, LiDAR systems, and ultrasonic sensors in autonomous vehicles collect environmental data such as road conditions and distances of objects, which are crucial for navigation and obstacle detection [33, 34].

Collected data then go through preprocessing to enhance its quality and reliability. Raw data contain a lot of noise and inconsistencies that must be addressed to get an accurate analysis. Filtering, normalization, and data augmentation are some general techniques that could be used in this phase. In agriculture, filtering removes environmental noise from sensor readings, while normalization normalizes the output of various sensors. Wavelet transformation and low-pass filtering isolate meaningful features such as R–R intervals in ECG signals from healthcare data. Preprocessing in autonomous vehicles requires noise reduction and segmentation of frames for better visibility of objects detected via cameras or LiDAR. TinyML tools such as TFLM and CMSIS-NN perform these preprocessing steps directly on edge devices, providing an efficient implementation for these processes with minimum power consumption [35, 36].

Feature extraction is the subsequent crucial step, whereby the system identifies the most relevant attributes in the data for further analysis. It simplifies the data by focusing on specific elements crucial for the intended application. Features such as chlorophyll levels, temperature gradients, or pest infestation patterns are extracted in agriculture to monitor crop health and optimize irrigation. Examples of such extracted features in healthcare applications include heart rate variability (HRV), respiration rates, or SpO_2 levels used to characterize patients. Object boundaries, lane markers, and road curvature in autonomous vehicles are all feature types that provide the necessary functionality to drive safely. These would be provided as input into various ML models trained on particular tasks [37].

Model training comprises a big portion of the TinyML deployment pipeline. Training usually happens on high-performance computing systems with software such as TensorFlow, Edge Impulse, or PyTorch [38]. In agriculture, models get trained to classify plant diseases or predict optimal watering times. Healthcare models may involve the identification of irregular heart rhythms or the detection of respiratory anomalies. For autonomous vehicles, training involves object detection models that identify pedestrians, vehicles, and road signs. After training, models are optimized for edge deployment by quantization, which reduces the precision of model weights, and pruning, which removes redundant parameters. These optimizations significantly reduce the model size and computational requirements, making them suitable for deployment on microcontrollers or other resource-constrained devices [39].

Deployment involves integrating the optimized models into edge devices for real-time inference. TinyML tools such as TFLM, Edge Impulse, and NanoEdge AI Studio ease this with application programming interfaces (APIs) and frameworks tailored for low-power hardware. For instance, models deployed on agricultural microcontrollers independently run irrigation systems based on the soil moisture level. Models deployed in healthcare detect anomalies in heart rate or SpO_2 and alert caregivers in real time. These models enable real-time obstacle avoidance and lane navigation for autonomous vehicles, thus providing much safer driving experiences [40, 41].

On-device inference enables edge devices to independently process incoming data streams with minimum reliance on cloud resources. This is an essential ability for those applications that demand low latency and high reliability. In agriculture, a microcontroller with a deployed TinyML model can immediately

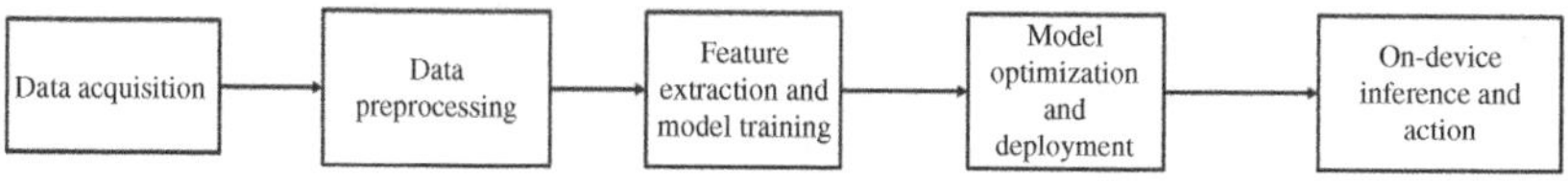

Figure 6.4 Steps involved in TinyML deployment.

adjust the irrigation or fertilization schedule without waiting for instructions from the cloud. Wearables in healthcare can preprocess or analyze data locally, for example, to identify irregular heartbeats, which are especially critical in emergencies. Autonomous vehicles make decisions such as applying brakes upon detection of obstacles with local inference [42].

Continuous monitoring and improvement are crucial to having an effective TinyML system. It involves assessing performance metrics such as accuracy, precision, recall, and F1 score. Any discrepancies or inefficiencies are updated in the models. Techniques like federated learning are employed to update models without compromising data privacy. Federated learning aggregates locally trained models from multiple devices, improving global model accuracy while keeping sensitive data on the device. This approach is of particular value in healthcare, where patient privacy is paramount. In agriculture, models are continuously updated to adapt to the new conditions of the environment, for example, outbreaks of pests or changes in weather conditions. Real-time updates enable autonomous vehicles to enhance their object detection and navigation in diverse driving environments [43, 44].

In sectors such as agriculture, healthcare, and autonomous vehicles, implementing TinyML will require a structured pipeline from data collection to data preprocessing, feature extraction, model training, deployment, on-device inference, and continuous improvement. By using tools such as TFL, Edge Impulse, and CMSIS-NN, the TinyML systems realize real-time, low-power intelligence on edge devices. Such a systematic approach reinforces better decision-making, efficiency in operations, and privacy across multiple domains – a testament to how this little ML can impact reality. Figure 6.4 gives the steps involved in TinyML deployment.

6.5.1 Case Studies of TinyML in the Health Sector Domain

Smart health monitoring systems (SHMSs) revolutionize healthcare by incorporating IoT with TinyML to process data on edge and improve health diagnostics further. Traditional healthcare relied on cloud solutions for data processing, presenting a couple of critical challenges: high latency, reliance on reliable network connections, and privacy exposure in sensitive health data transmission cases. With the advent of TinyML, healthcare applications have transitioned toward decentralized architectures that enable real-time processing and decision-making on resource-constrained devices.

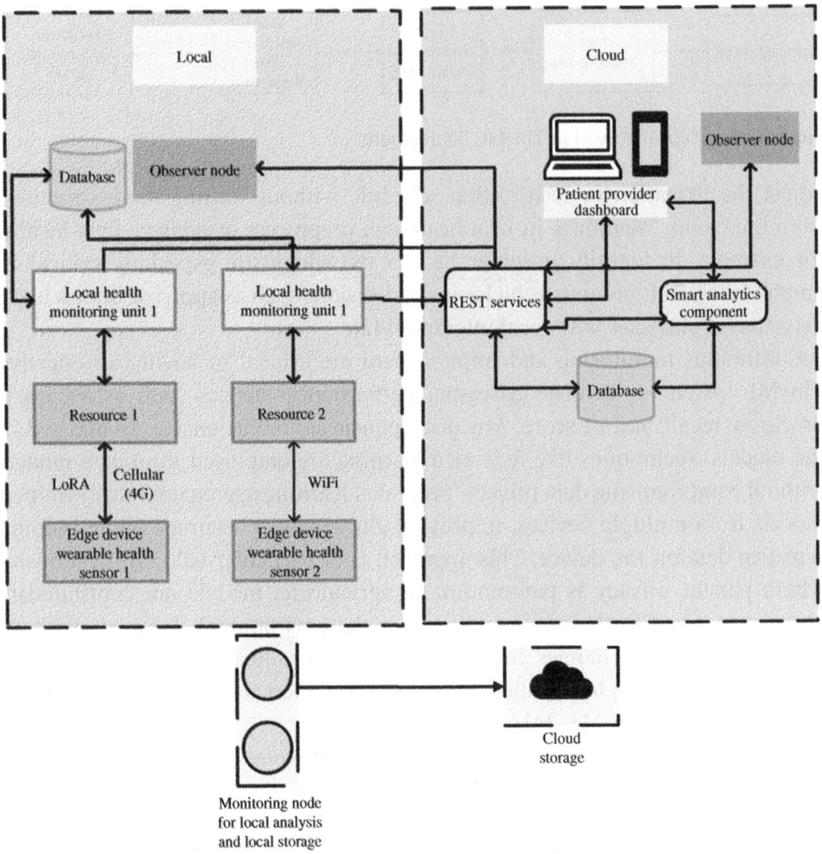

Figure 6.5 Proposed TinyML architecture for SHMS.

Figure 6.5 illustrates the proposed TinyML approach, which gives an overall view of the SHMS as an application in the health domain, combining local analytics and cloud-based intelligence. This, therefore, ensures that there is always some form of efficient and timely monitoring of patient health metrics with wearable sensors and TinyML tools and devices to provide an integrated system. TinyML tools such as TFLM, Edge Impulse, and CMSIS-NN will drive low-power, edge-level intelligence, latency minimization, and data privacy. These tools will make it possible to process health data locally while still harnessing the computational power of cloud analytics for advanced insights.

The architecture of the SHMS will be based on two main domains: the local domain, which is responsible for collecting data and preliminary analyses, and the cloud domain, which is responsible for advanced analytics, storage, and application-level integration. The local domain includes wearable health sensors such as heart rate monitors, pulse oximeters, and ECG devices that constantly collect physiological data. On-device processing units or microcontrollers become responsible for where the data execute the models of TinyML, taking real-time analysis. By utilizing ML lightweight models, deploy directly to microcontrollers (TFLM), this can implement real-time detection of heart-beating variability anomalies or extreme fluctuations in HRV. Edge Impulse further smooths this process by easing data collection, preprocessing, and model training; hence, it is effective in raw sensor data processing.

In the local domain, resources like microcontrollers or gateways that have TinyML capabilities provide a temporary node for preprocessing and real-time analysis. Such resources leverage the TinyML models trained with tools such as CMSIS-NN, an optimized runtime for neural network operations on ARM Cortex-M processors. For example, the model deployed on the Cortex-M4 microcontroller can carry out anomaly detection of heart rate and respiration waveforms by using compressed neural network inferences. These local insights are sent to controller services. Controller services handle data flow across the local to cloud domains. Controller services will also take responsibility for coordinating alerts and preliminary health assessments based on local inferences. For instance, if HRV falls below a critical threshold, the system can instantly alert caregivers via mobile notifications.

The cloud domain is important for the SHMS, as it involves complex analysis and long-term storage. Representational State Transfer (REST) ful APIs are adopted to transmit preprocessed data from the local domain to the cloud. TinyML tools like TFL enable seamless integration with cloud-based ML models, where advanced analytics algorithms analyze historical and real-time data to detect patterns indicative of chronic diseases or potential emergencies. For example, analytics in the cloud can identify relationships between HRV trends and stress levels, giving healthcare providers actionable insights. IoT intelligence-informed analytics unify these models into predictive and prescriptive analytics on patient care. Architecture also contains observer nodes that offer real-time visualization of the patients' metrics to healthcare providers and caregivers.

Applications powered by Edge Impulse enable these nodes to provide user-friendly dashboards in which vital health metrics and alerts are presented. Observer nodes can monitor vital signs like heart rate, SpO_2, and body temperature. Thus, caregivers will study health trends and quickly act as per alerts. These dashboards ensure healthcare providers have the correct and updated information to make proper decisions. The key differentiating factor of SHMS depends on edge analytics to a greater extent for minimum latency and dependency on cloud services. These TinyML models can be installed on monitoring nodes to make inferences with a local model, reducing continuous data transmission to the cloud. For example, a wearable heart rate monitor can do some preprocessing and directly calculate HRV from the raw data using a model compiled by TFL. For instance, if the computed HRV indicates a probable health issue, the system creates a local alert and enables immediate action. This is especially useful in time-critical applications, like detecting arrhythmias or sudden drops in oxygen levels.

Edge Impulse and TFL for TinyML enable the SHMS to perform dynamic model updates. In the presence of federated learning, the system aggregates models trained locally from various devices to update the global model with improved accuracy without exposing any raw patient data. This provides a method to enhance the models while keeping private health information confined to the local device.

Apart from accuracy, energy efficiency is another key success factor for SHMS. Given that wearable devices mostly depend on limited battery life, development tools like CMSIS-NN and TFL have used quantization and pruning methods to attain low computational loads and reduce model size. This will keep the device running for longer periods on a single battery charge without frequent recharges. Quantization, for example, reduces model weights from 32-bit floating-point to 8-bit integers, drastically reducing memory and power consumption. These optimizations are critical to extending the operating life of wearable sensors and microcontrollers in practical healthcare applications. Complementing real-time monitoring, cloud-based advanced analytics gives the SHMS retrospective and predictive insights into health data. Deep learning models analyze historical data stored in the cloud for long-term trends and correlations. For instance, models can be trained to detect arrhythmias or forecast the onset of cardiovascular diseases using a patient's ECG data. TinyML tools seamlessly integrate these cloud-based models with local devices to enable a cohesive system combining edge and cloud computing strengths.

The SHMS integrates a robust security architecture for sensitive health data protection. RESTful communications from/to the local to/from cloud domains are performed using encryption protocols, like transport layer security (TLS), which guarantees data privacy and is tamper-proof during transmission. Observer nodes and cloud dashboards use different authentication mechanisms to avoid unauthorized access, adding another layer of protection to sensitive patient information.

The application of TinyML tools in SHMS enhances technical capability and mitigates some of the challenges that most traditional health monitoring systems have been facing. Performing analytics locally minimizes latency and reduces reliance on cloud services, making it more reliable in scenarios with poor network connectivity. The energy-efficient design of TinyML models ensures that wearable devices can operate continuously without frequent interruptions for recharging.

The system will be able to support continuous model improvement through federated learning with no compromise on data privacy.

The integration of tools such as TFL, Edge Impulse, and CMSIS-NN in the architecture shows the transformational potential of TinyML in smart health applications. The application of these tools within SHMS enables real-time, scalable, and privacy-preserving health insights, enabling the healthcare provider to make proactive decisions that improve patient outcomes. The system relies on edge analytics, where critical health issues must be identified and fixed instantaneously, while cloud-based intelligence provides further detail and long-term monitoring capability.

Eventually, the system guarantees that healthcare solutions will be delivered with precision, speed, and security to the patient and caregivers via edge-cloud computing. Further evolution of TinyML technologies will unlock new possibilities in smart health applications and drive innovation in improving the quality of care worldwide.

6.5.2 Case Studies of TinyML in the Agriculture Domain

Smart and precision agriculture has emerged as one of the revolutionary ways to modern farming, adopting advanced technologies like the Internet of Things, ML, and cloud computing to maximize resources, enhance crop yield, and make it more sustainable. The architecture in Figure 6.6 provides an excellent framework for implementing this through the smooth integration of the local and cloud domains to allow for real-time data collection, analysis, and decision-making. This case study uses the IoT architecture in a precision agriculture system to demonstrate how sensor nodes, centralized controllers, and cloud-based analytics can be integrated to present actionable insights for efficient farming. This architecture enhances the efficiency of farming operations. It addresses key challenges such as water conservation, pest management, and predictive decision-making, making it an invaluable tool in the agricultural domain.

The architecture is tailored for smart and precision agriculture and provides a comprehensive framework for managing data flow, analytics, and decision-making in modern farming. The diagram represents a well-structured integration of local and cloud domains, enabling real-time monitoring and analytics to optimize agricultural practices. This case study explains the components of the architecture, their functions, and their importance in ensuring a scalable and efficient system for precision agriculture.

It originates in the local domain of the system, where data are collected directly from the agricultural environment. At this level, the main data acquisition devices are the sensor nodes, which constantly monitor key parameters regarding soil

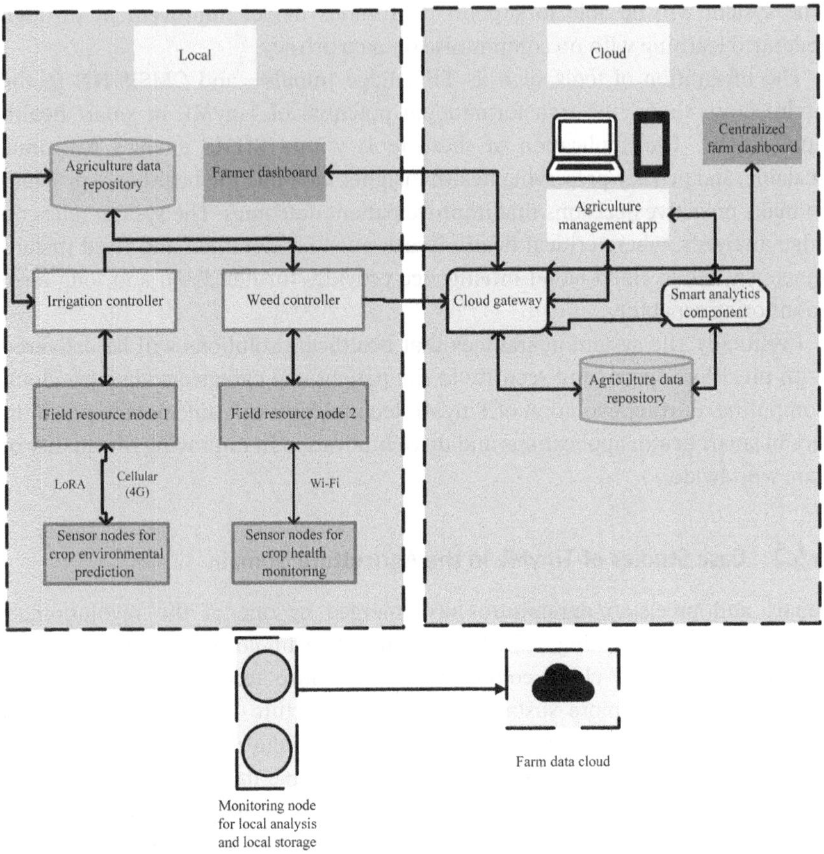

Figure 6.6 Proposed TinyML architecture for smart and precision agriculture.

moisture, temperature, humidity, and light. In general, sensor nodes form an integral part of precision agriculture since they provide, on the one hand, direct knowledge about the soil status in almost real time. For instance, soil moisture sensors placed across a farm in different plots collect granular data about water availability, thus enabling farmers to optimize irrigation schedules. Temperature and humidity sensors likewise provide vital information on threats like frost or heat stress, thus enabling farmers to take timely preventive measures.

The sensor nodes transmit the collected data to field resource nodes acting as local aggregators and preprocessors. These resource nodes act like brokers, combining data from several sensor nodes into something less redundant and more understandable, ready for analysis. For example, readings from moisture sensors spread out over different parts of the field can be combined to describe the overall soil water status. These resource nodes also conduct basic preprocessing tasks, such as filtering noise and normalizing sensor readings to ensure the data are accurate and usable.

6.5.2.1 Irrigation Controller

This module controls real-time irrigation decisions based on the data gathered and processed in the local domain. Analyzing moisture levels and weather data, the irrigation controller can switch the irrigation systems on or off accordingly. This system also offers more than the above; it ensures that should the soil moisture in any plot fall below a threshold, irrigation is automatically turned on to maintain optimum soil conditions. This reduces water loss and ensures crops have appropriate hydration levels, enhancing yield and sustainability.

The farmer dashboard acts as an interface between the system and the end user. Mobile or web applications allow the monitoring of all field data in real time, receiving alarms, and managing different systems like irrigation. The dashboard visualizes, in a user-friendly way, critical metrics such as soil moisture, temperature, and crop health to assist farmers in decision-making. The farmer can look into real-time data about the dry parts of the field and, from this dashboard, turn on irrigation in that part of the field.

The centralized controller bridges the local and cloud domains, serving as the core node to orchestrate data flow in the system. This layer handles the interaction between the field devices and the cloud infrastructure. For example, it gathers data from all resource nodes, conducts further preprocessing, and ensures that only relevant data are sent to the cloud. By reducing unnecessary data transmission, the centralized controller minimizes bandwidth use and ensures efficiency in the system's operation.

The cloud domain is where advanced analytics and long-term data storage occur. The cloud gateway facilitates secure communication between the centralized controller and the cloud infrastructure, using REST/WebSocket protocols to ensure reliable data transfer. This gateway is crucial for large-scale farming operations, where vast amounts of data are collected from multiple fields. For instance, data from various plots can be securely transmitted to the cloud for further analysis.

The database in the architecture represents the farm data cloud, which stores real-time and historical agricultural data down to the last detail of soil moisture, crop health metrics, and weather patterns. The availability of historical data provides a lot of value in identifying trends and making predictive decisions. For instance, historical weather data can be combined with current soil conditions to forecast irrigation needs for the days ahead. Furthermore, long-term information on crop health will enable farmers to detect recurring problems, such as pest infestation or nutrient deficiency, and thus take remedial measures in advance.

The crop health analytics component is powered by advanced ML algorithms combined with IoT intelligence. It analyzes data from the farm data cloud to present actionable insights. For example, predictive models analyze soil conditions and weather patterns to recommend optimal planting and harvesting times. Likewise, prescriptive analytics can help with specific recommendations regarding which fertilizers and when to apply them to maximize crop yield. This is critical in precision agriculture, where small changes in farming practices can substantially change productivity and sustainability.

The centralized farm dashboard, extended from the cloud domain, provides real-time visualization for farmers and agricultural managers based on data and analytics results. It summarizes data from various plots into a digestible form. Farmers, for example, can view line graphs of soil moisture over time, maps with field areas requiring attention, and warnings regarding drought or pests. The dashboard centralizes the information, allowing farmers to make confident decisions based on the data.

The integration of many monitoring nodes ensures the farm's comprehensiveness of the whole area. These nodes will be placed in different sections within the farm to monitor parameters such as soil fertility, water distribution, and pest activities. For instance, a node on one plot can report an outbreak of pests and send an alarm, which the farmer will work to address before the disease spreads to other areas.

The cloud storage and analytics module combines all elements for holistic insights. This module will integrate real-time sensor data with historical data to create comprehensive reports and forecasts. For example, integrating soil moisture data with weather forecasts allows farmers to plan irrigation events in consideration of forecasted rainfall. In the same way, data on crop health from previous seasons can be used to create better pest management strategies and efficient resource allocation.

6.6 Lesson Learned and Prospects

TinyML has changed how ML is deployed on resource-constrained devices, enabling a wide range of applications, from agriculture and healthcare to connected autonomous vehicles. The integration of lightweight models optimized for low-power operation granted TinyML real-time intelligence on edge devices, minimizing cloud computing while addressing key latency, energy efficiency, and data privacy challenges. This chapter provides insight into the lessons learned from the practical implementation of TinyML and speculates about its future across various domains. As the field matures, there is great potential to scale TinyML applications for global productivity and sustainability, addressing key challenges through emerging innovations.

6.6.1 Lesson Learned

Lessons learned underline the crucial optimization of models for resource-constrained devices, the viability of edge analytics for real-time decision-making, and the importance of energy efficiency toward long-term operations. Besides, privacy-preserving methods such as federated learning have been important in sustaining data security and have enabled continuous improvement. These lessons underline the revolutionary potential of TinyML while underlining further areas of innovation that will go a long way in maximizing impact for healthcare, agriculture, and autonomous systems. However, lessons learned from this book chapter are summarized as follows:

1) **Importance of Model Optimization:** One of the most important things learned in TinyML is the need for robust optimization techniques. Applications of TinyML operate on devices with stringent memory, computational, and energy constraints. Techniques such as quantization, pruning, and NAS have proven instrumental in reducing model size while maintaining accuracy. For example, quantization reduces model weights from 32-bit floating-point to 8-bit integers at a fraction of the memory and with limited loss in accuracy. The success of tools like TFLM and CMSIS-NN represents the value of those optimization strategies.
2) **Accuracy-Efficiency Trade-offs:** A common challenge when working with TinyML is balancing model accuracy and efficiency. Running high-accuracy models on resource-constrained hardware has been an inevitable trade-off during deployments, often impacting real-world performance. Industrial IoT applications have taught us that over-optimizing for efficiency can degrade a model's ability to handle edge cases, while prioritizing accuracy may lead to energy and latency issues. The importance of case-specific adaptive optimization techniques underlines this trade-off.
3) **Hardware–Software Co-design:** TinyML's success represents tight integration between hardware and software. Tools like STM32Cube.AI and NanoEdge Studio have shown that hardware-aware model design significantly improves performance. In alignment with the underlying hardware capabilities, software optimizations offer superior energy efficiency and processing speeds.
4) **Scaling Challenges:** Scaling TinyML solutions to various applications and environments remains very difficult. Experiences with recent deployments in smart agriculture, healthcare, and industrial IoT make one thing clear: A road to scalability will require adaptive frameworks that can deal with the challenge of diverse data sources and hardware platforms. The lack of common benchmarks further complicates model benchmarking across various domains.
5) **Real-Time Inference and Latency:** TinyML applications usually need real-time decision-making, especially in safety-critical domains such as

autonomous vehicles and medical monitoring. Lessons learned from these fields shed light on the necessity of optimizing model size and inference latency. Tools developed for this purpose, such as Edge Impulse, have been effective, but further developments are needed to accommodate ultra-low-latency applications.

6) **Data Privacy and Security:** TinyML has shown huge potential for enhancing data privacy by having devices process information locally instead of relying on cloud services. However, lessons learned from healthcare and finance reveal vulnerabilities in deploying the models securely. Protecting sensitive data while keeping model integrity intact is crucial and requires improvement on fronts like encryption and secure boot.
7) **Cost and Accessibility:** Cost is the main factor in democratizing AI. Tools and devices are becoming very affordable. However, various global deployments have shown that a prohibitively expensive pathway using highly specialized hardware, such as LiDAR or custom accelerators, would inhibit its broad-scale adoption, especially in developing parts of the world. Affordability and open-source approaches to TinyML are critical in its universal reach.

6.6.2 Prospects

The future of TinyML is promising, with many possibilities for technological advancement, application diversification, and societal influence. The following are some critical areas where TinyML will evolve in the coming years.

1) **Advanced Model Compression Techniques:** While quantization and pruning are some of the existing techniques that have already brought down model sizes by an order of magnitude, future innovations in model compression will be around adaptive and hardware-specific techniques. Sparsity-aware training and structured pruning will further improve compatibility with constrained devices. Federated learning combined with TinyML may even allow efficient training on edge devices without compromising privacy.
2) The integration of TinyML and IoT-Edge AI will continue to drive a fundamental role in the evolution of edge AI and IoT. Shortly, integrations of TinyML and IoT ecosystems will be faultless, and smart homes, cities, and industries will enable autonomous operations. Such integration could enable advanced sensor fusion that combines the outputs from LiDAR, cameras, and RADAR sensors to create an advanced understanding used by applications like autonomous cars.
3) **Energy Harvesting and Sustainable Systems:** Energy efficiency will be the core foundation upon which the development of TinyML will progress. Energy harvesting and integrating TinyML into such technologies as solar panels or piezoelectric sensors will drive research into developing self-sustaining systems

in the future. These advancements will be pivotal in the fields of environmental conservation and disaster management, where remote monitoring is essential.

4) **Improved Security and Privacy:** As TinyML applications expand into sensitive domains like healthcare and finance, ensuring data security and privacy will be paramount. Future developments will focus on robust encryption techniques, secure hardware modules, and privacy-preserving ML methods such as homomorphic encryption and differential privacy.
5) **Standardization and Benchmarks:** The absence of standardized benchmarks has been one of the longest-standing problems when benchmarking TinyML models. The prospects are establishing universal performance benchmarks targeted for edge devices, enabling fair comparisons and fostering innovation. Initiatives like MLPerf Tiny are a step in this direction.
6) **Democratization of TinyML:** Future efforts will be oriented toward making TinyML affordable and accessible to people worldwide. Open-source frameworks such as FANN on MCU and uTensor will play a vital role in enabling developers worldwide to build and deploy TinyML applications. The collaboration of academia, industry, and governments will bridge the gap in the digital divide, especially in underserved regions.
7) **Hybrid Architectures:** The future of TinyML will probably be hybrid architectures, where microcontrollers are combined with accelerators or even cloud-based processing. These architectures enable resource-intensive tasks to be offloaded in the cloud while retaining the benefits of local processing for critical decision-making.
8) **Collaboration Across Disciplines:** TinyML requires an evolution in ML, embedded systems, hardware design, and cybersecurity through interdisciplinary collaboration; this collaboration between fields brings forward innovative approaches in the use and enlargement of applications by addressing these limitations of TinyML.
9) **TinyML for Climate Action:** TinyML will play a major role in climate action by enabling low-power, distributed systems for environmental monitoring and resource management. Applications ranging from wildland fire detection to air quality and water conservation will be some segments that gain huge scalability and efficiency due to TinyML.

6.7 Conclusion

Advances within TinyML are rapid, illustrating very promising scenarios wherein AI at the edge and in resource-constrained environments could be game changers. By using optimized frameworks, tools, and techniques, TinyML allows real-time intelligence at edge microcontrollers and devices. It would address the important challenges for any of those applications' efficiency, latency, and data privacy. Domain integration, including healthcare, agriculture, IIoT, and smart cities, adds varied applicability of TinyML into practical solution spaces.

This paper has reviewed the fundamental tools and hardware that make TinyML possible, their limitations, lessons learned, and future directions. The key takeaways are that model optimization techniques, such as quantization and pruning, are critical to making TinyML possible. Despite the significant progress, challenges remain regarding the standardization, security, and adaptability of TinyML to diverse hardware platforms and various applications. Shortly, TinyML development will likely focus on model compression, energy harvesting, and security. Making the tools more affordable and using open-source resources are vital in the democratization of TinyML. Interdisciplinary collaboration, including hybrid architectures that merge the edge and cloud, will mark the new frontiers of TinyML. The conclusive development of TinyML presents a very important step toward allowing intelligent systems at the edge with extremely good energy efficiency. Therefore, continuous research and development promise great things for TinyML's prospects in shaping the future of embedded AI – scalable, available, and sustainable to handle most global challenges. It highlights how continued innovation and collaboration can lead to the full utilization of this technology.

References

1 Bharadwaj, H.K., Agarwal, A., Chamola, V. et al. (2021). A review on the role of machine learning in enabling IoT based healthcare applications. *IEEE Access* 9: 38859–38890.

2 Kallimani, R., Pai, K., Raghuwanshi, P. et al. (2024). TinyML: tools, applications, challenges, and future research directions. *Multimedia Tools and Applications* 83 (10): 29015–29045.

3 De Leon, J.D. and Atienza, R. (2022). Depth pruning with auxiliary networks for TinyML. In *ICASSP 2022–2022 IEEE International Conference on Acoustics, Speech and Signal Processing (ICASSP)*, 3963–3967. IEEE.

4 Zhao, T., Zhou, S., Guo, X., et al. (2015). A cooperative scheduling scheme of local cloud and internet cloud for delay-aware mobile cloud computing. In *2015 IEEE GlobeCom Workshops (GC Workshops)*, 1–6. IEEE.

5 Dutta, L. and Bharali, S. (2021). TinyML meets IoT: a comprehensive survey. *Internet of Things* 16: 100461.

6 Merenda, M., Porcaro, C., and Iero, D. (2020). Edge machine learning for AI-enabled IoT devices: a review. *Sensors* 20 (9): 2533.

7 Abadade, Y., Temouden, A., Bamoumen, H. et al. (2023). A comprehensive survey on TinyML. *IEEE Access* 11: 96892–96922. https://doi.org/10.1109/ACCESS.2023.3294111.

8 Giordano, M., Baumann, N., Crabolu, M. et al. (2022). Design and performance evaluation of an ultralow-power smart IoT device with embedded TinyML for asset activity monitoring. *IEEE Transactions on Instrumentation and Measurement* 71: 1–11.

9 Shamim, M.Z.M. (2022). Hardware deployable edge-AI solution for prescreening of oral tongue lesions using TinyML on embedded devices. *IEEE Embedded Systems Letters* 14 (4): 183–186.

10 Oufettoul, H., Chaibi, R., and Motahhir, S. (2024). TinyML applications, research challenges, and future research directions. In: *2024 21st Learning and Technology Conference (L&T)*, 86–91. IEEE.

11 Jacob, B., Kligys, S., Chen, B., et al. (2018). Quantization and training of neural networks for efficient integer arithmetic-only inference. In *Proceedings of the IEEE Conference on Computer Vision and Pattern Recognition*, 2704–2713.

12 Luigi Capogrosso, F., Cunico, D.S., Cheng, F.F., and Cristani, M. (2024). A machine learning-oriented survey on tiny machine learning. *IEEE Access* 12: 23406–23426. https://doi.org/10.1109/ACCESS.2024.3365349.

13 Liberis, E., Dudziak, L. and Lane, N.D. (2021). μNAS: constrained neural architecture search for microcontrollers. In *Proceedings of the 1st Workshop on Machine Learning and Systems*, 70–79.

14 Chen, S., Hui, X., Liu, D. et al. (2014). A vision of IoT: applications, challenges, and opportunities with China perspective. *IEEE Internet of Things Journal* 1 (4): 349–359.

15 Nasir Abbas, Yan Zhang, Amir Taherkordi, and Tor Skeie (2017). Mobile edge computing: a survey. *IEEE Internet of Things Journal* 5 (1): 450–465.

16 Weng, Y., Guan, W., and Ababei, C. (2024). Prediction of remaining useful life and cell temperature for li-ion batteries using TinyML. In: *2024 IEEE 67th International Midwest Symposium on Circuits and Systems (MWSCAS)*, 562–566. IEEE.

17 Ren, H., Anicic, D., and Runkler, T.A. (2021). TinyOL: TinyML with online-learning on microcontrollers. In *2021 International Joint Conference on Neural Networks (IJCNN)*, 1–8.

18 Schizas, N., Karras, A., Karras, C., and Sioutas, S. (2022). TinyML for ultra-low power AI and large scale IoT deployments: a systematic review. *Future Internet* 14 (12): 363.

19 Saha, S.S., Sandha, S.S., and Srivastava, M. (2022). Machine learning for microcontroller-class hardware: a review. *IEEE Sensors Journal* 22 (22): 21362–21390.

20 Menghani, G. (2023). Efficient deep learning: a survey on making deep learning models smaller, faster, and better. *ACM Computing Surveys* 55 (12): 1–37.

21 Alajlan, N.N. and Ibrahim, D.M. (2023). DDD TinyML: a TinyML-based driver drowsiness detection model using deep learning. *Sensors* 23 (12): 5696.

22 Liu, Y.-Y., Zheng, H.-S., Hu, Y.F., et al. (2024). TinyTS: memory-efficient TinyML model compiler framework on microcontrollers. In *2024 IEEE International Symposium on High-Performance Computer Architecture (HPCA)*, 848–860.

23 Zhang, C., Patras, P., and Haddadi, H. (2019). Deep learning in mobile and wireless networking: a survey. *IEEE Communications Surveys & Tutorials* 21 (3): 2224–2287.

24 Chen, M., Challita, U., Saad, W. et al. (2019). Artificial neural networks-based machine learning for wireless networks: a tutorial. *IEEE Communications Surveys & Tutorials* 21 (4): 3039–3071.

25 Thuppari, C.C., Jannu, S., Edla, D.R. et al. Energy-aware compression and consumption algorithms for efficient TinyML model using Aquila optimization in industrial IoT. *IEEE Transactions on Consumer Electronics* 71: 334–342. https://doi.org/10.1109/TCE.2024.3475393.

26 Ravaglia, L., Rusci, M., Nadalini, D. et al. (2021). A TinyML platform for ondevice continual learning with quantized latent replays. *IEEE Journal on Emerging and Selected Topics in Circuits and Systems* 11 (4): 789–802.

27 Rüb, M., Tuchel, P., Sikora, A., and Mueller-Gritschneder, D. (2024). A continual and incremental learning approach for TinyML on-device training using dataset distillation and model size adaption. In: *2024 IEEE 7th International Conference on Industrial Cyber-Physical Systems (ICPS)*, 1–8. IEEE.

28 Bechtel, M., Weng, Q.T., and Yun, H. (2022). Deeppicarmicro: applying TinyML to autonomous cyber physical systems. In 2022 *IEEE 28th International Conference on Embedded and Real-Time Computing Systems and Applications (RTCSA)*, 120–127.

29 Amor, H.B., Bernier, C., and Přikryl, Z. (2021). A risc-v isa extension for ultra-low power IoT wireless signal processing. *IEEE Transactions on Computers* 71 (4): 766–778.

30 Banbury, C. R., Reddi, V. J., Lam, M., et al. (2020). Benchmarking TinyML systems: challenges and direction. *arXiv preprint arXiv:2003.04821*.

31 Arthur, E.A.E., Wulnye, F.A., Gookyi, D.A.N., et al. (2024). Edge Impulse vs TensorFlow: a comparative analysis of TinyML platforms for maize leaf disease identification. In *2024 Conference on Information Communications Technology and Society (ICTAS)*, 1–6.

32 Hayajneh, A.M., Hafeez, M., Zaidi, S.A.R., and McLernon, D. (2024). TinyML empowered transfer learning on the edge. *IEEE Open Journal of the Communications Society* 5: 1656–1672.

33 Srinivasagan, R., Mohammed, M., and Alzahrani, A. (2023). TinyML-sensor for shelf life estimation of fresh date fruits. *Sensors* 23 (16): 7081.

34 Suresh, V.M., Sidhu, R., Karkare, P. et al. (2018). Powering the IoT through embedded machine learning and lora. In: *2018 IEEE 4th World Forum on Internet of Things (WF-IoT)*, 349–354. IEEE.

35 Lai, L., Suda, N., and Chandra, V. (2018). CMSIS-NN: efficient neural network kernels for arm Cortex-M CPUs. *arXiv preprint arXiv:1801.06601*.

36 Sarwar Murshed, M.G., Murphy, C., Hou, D. et al. (2021). Machine learning at the network edge: a survey. *ACM Computing Surveys (CSUR)* 54 (8): 1–37.

37 Saranya, T., Jeyamala, D., Sellamuthu, S. et al. (2024). A secure framework for MIoT: TinyML-powered emergency alerts and intrusion detection for secure real-time monitoring. In: *2024 8th International Conference on I-SMAC (IoT in Social, Mobile, Analytics and Cloud)(I-SMAC)*, 13–21. IEEE.

38 Akubathini, P., Chouksey, S., and Hariram Selvamurugan, S. (2021). Evaluation of machine learning approaches for resource constrained IIoT devices. In: *2021 13th International Conference on Information Technology and Electrical Engineering (ICITEE)*, 74–79. IEEE.

39 Pravallika, A., Hashmi, M.F., and Gupta, A. (2024). Deep learning frontiers in 3D object detection: a comprehensive review for autonomous driving. *IEEE Access* 12: 173936–173980. https://doi.org/10.1109/ACCESS.2024.3456893.

40 Tsoukas, V., Boumpa, E., Giannakas, G., and Kakarountas, A. (2021). A review of machine learning and TinyML in healthcare. In *Proceedings of the 25th Pan-Hellenic Conference on Informatics*, 69–73.

41 Akhtar, M.N., Shaikh, A.J., Khan, A. et al. (2021). Smart sensing with edge computing in precision agriculture for soil assessment and heavy metal monitoring: a review. *Agriculture* 11 (6): 475.

42 Liang, S., Hao, W., Zhen, L. et al. (2022). Edge yolo: real-time intelligent object detection system based on edge-cloud cooperation in autonomous vehicles. *IEEE Transactions on Intelligent Transportation Systems* 23 (12): 25345–25360.

43 Zhang, T., Gao, L., He, C. et al. (2022). Federated learning for the internet of things: applications, challenges, and opportunities. *IEEE Internet of Things Magazine* 5 (1): 24–29.

44 Benos, L., Tagarakis, A.C., Dolias, G. et al. (2021). Machine learning in agriculture: a comprehensive updated review. *Sensors* 21 (11): 3758.

7

Privacy-Preserving Techniques in TinyML for IoT

Oleksandr Kuznetsov[1,2], Emanuele Frontoni[3], Kateryna Kuznetsova[4], Marco Arnesano[1], and Pavlo Usik[5]

[1] *Department of Intelligent Software Systems and Technologies, eCampus University, Novedrate, Italy*
[2] *Department of Information and Communication Systems Security, V. N. Karazin Kharkiv National University, Kharkiv, Ukraine*
[3] *Department of Political Sciences, Communication and International Relations, University of Macerata, Macerata, Italy*
[4] *VRAI – Vision, Robotics and Artificial Intelligence Lab, Ancona, Italy*
[5] *Department of Cyber Security and Software, Central Ukrainian National Technical University, Kropyvnytskyi, Ukraine*

7.1 Introduction

The rapid proliferation of Internet of Things (IoT) devices has ushered in a new era of ubiquitous computing, where intelligent decision-making can occur at the edge of networks [1]. Tiny machine learning (TinyML), the field of deploying machine learning models on resource-constrained devices, has emerged as a powerful paradigm for enabling such edge intelligence [2, 3]. By bringing computation closer to the data source, TinyML offers benefits such as reduced latency, improved energy efficiency, and enhanced privacy [4]. However, the processing of potentially sensitive data on these devices raises significant privacy concerns that must be addressed to ensure widespread adoption and trust in TinyML systems.

The research landscape in privacy-preserving TinyML encompasses a wide range of challenges and opportunities. Key areas of investigation [2, 3] include

1) Adapting privacy-preserving techniques to resource-constrained environments
2) Balancing privacy guarantees with computational efficiency and model accuracy
3) Developing communication-efficient protocols for distributed TinyML systems

Tiny Machine Learning: Design Principles and Applications, First Edition.
Edited by Agbotiname Lucky Imoize, Dinh-Thuan Do, and Houbing Herbert Song.

4) Integrating privacy-preserving methods with model compression and quantization techniques
5) Exploring hardware–software co-design for privacy-preserving TinyML operations

This chapter contributes to the field by providing a comprehensive examination of state-of-the-art privacy-preserving techniques specifically tailored for TinyML environments. Our key contributions include

1) A systematic analysis of secure multi-party computation (MPC), homomorphic encryption (HE), and differential privacy (DP) approaches in the context of TinyML
2) An in-depth evaluation of recent frameworks such as SecureML, SecureNN, and Falcon, highlighting their advancements in efficient and secure computation for resource-constrained devices
3) A critical assessment of the trade-offs between privacy, performance, and model accuracy in TinyML applications
4) Identification of key challenges and promising future research directions in privacy-preserving TinyML

The chapter is structured as follows:

- Section 7.2 provides an overview of related works in the field of privacy-preserving machine learning, setting the context for our focused examination of TinyML techniques.
- Section 7.3 delves into HE in TinyML, covering its fundamentals, various encryption schemes, applications to TinyML tasks, and practical considerations.
- Section 7.4 explores DP for TinyML, discussing its principles, mechanisms, algorithms, and the challenge of balancing privacy and utility.
- Section 7.5 examines secure MPC (SMPC) in TinyML, covering its foundations, key techniques like secret sharing and garbled circuits, and efficiency optimizations.
- Section 7.6 presents case studies and applications of privacy-preserving TinyML, providing concrete examples of how these techniques are implemented in real-world scenarios.
- Section 7.7 offers a discussion of the current state of the field and outlines future research directions, addressing the specific challenges and opportunities in each privacy-preserving approach.
- Finally, Section 7.8 concludes the chapter, summarizing the key insights and emphasizing the importance of continued research in privacy-preserving TinyML.

By providing this comprehensive examination, we aim to equip researchers and practitioners with the knowledge necessary to advance the field and develop secure, efficient machine learning solutions for resource-constrained devices. As TinyML continues to expand its reach across various domains, from healthcare to industrial IoT, the development of robust privacy-preserving techniques will be crucial in realizing its full potential while safeguarding user privacy and data security.

7.2 Related Works

Privacy-preserving techniques for machine learning, particularly in resource-constrained environments, have garnered significant attention in recent years. This section provides an overview of the key related works that form the foundation for privacy-preserving TinyML.

HE has emerged as a powerful tool for privacy-preserving machine learning. Al Badawi et al. [5] made significant strides in this direction with their work on homomorphic convolutional neural networks (HCNNs). They demonstrated the feasibility of running CNNs on encrypted data using graphics processing units (GPUs), achieving classification latencies of 5.16 and 304.43 seconds for MNIST and CIFAR-10 datasets, respectively. This work represents a crucial step toward practical privacy-preserving deep learning on encrypted data.

In the realm of federated learning, Bonawitz et al. [6] presented a scalable production system for mobile devices. Their work addressed key challenges in federated learning, including communication efficiency, client availability, and privacy. By demonstrating the practicality of federated learning in large-scale mobile environments, they paved the way for privacy-preserving distributed learning in TinyML contexts.

DP has been widely explored as a means to provide strong privacy guarantees in machine learning. Ekenstedt et al. [7] investigated the relationship between DP and syntactic privacy, providing insights into how these models can be interpreted in terms of each other. This work is particularly relevant for TinyML applications, where multiple privacy models may need to be satisfied simultaneously.

The application of DP in social network analysis, as surveyed by Jiang et al. [8], offers valuable insights for TinyML in IoT environments. Their comprehensive review of DP applications in tasks such as degree distribution analysis and subgraph counting provides a foundation for privacy-preserving analysis of interconnected TinyML devices.

Kus and Levi [9] explored the application of DP in cryptocurrency systems, specifically Bitcoin. Their work on noise addition to transaction amounts and user graph perturbation demonstrates the potential of DP mechanisms in enhancing privacy in distributed, resource-constrained systems, which is directly applicable to TinyML scenarios.

Mahawaga Arachchige et al. [10] proposed LATENT, a local differentially private algorithm for deep learning in IoT environments. Their approach, which splits the architecture of a CNN into three modules, including a randomization layer, is particularly relevant for TinyML applications where data privacy must be preserved before leaving the device.

Pan et al. [11] provided a comprehensive survey on DP in deep learning, offering taxonomy of privacy attacks and preservation techniques. Their work highlights the importance of integrating DP with deep learning models, a crucial consideration for TinyML applications dealing with sensitive data.

In the context of SMPC, Wagh et al. [12] introduced Falcon, a framework for efficient private training and inference of large machine learning models. Falcon's support for high-capacity networks and batch normalization, coupled with its malicious security guarantees, makes it particularly relevant for TinyML applications requiring strong security assurances.

Zhao et al. [13] proposed a local DP-based federated learning approach for IoT, addressing both privacy concerns and communication efficiency. Their work on perturbing gradients and optimizing mechanisms for different privacy budgets offers valuable insights for implementing privacy-preserving federated learning in TinyML contexts.

Zheng et al. [14] presented Helen, a system for maliciously secure coopetitive learning of linear models. While not specifically designed for TinyML, Helen's approach to SMPC in the presence of malicious adversaries provides important considerations for TinyML systems operating in untrusted environments.

Zhou et al. [15] developed a differentially private federated learning model resistant to poisoning attacks in edge computing. Their weight-based anomaly detection algorithm and use of DP to protect both data and model privacy are particularly relevant for TinyML deployments in adversarial settings.

These works collectively form a rich foundation for privacy-preserving techniques in TinyML. They demonstrate the potential of HE, DP, and SMPC in addressing the unique challenges posed by resource-constrained environments. However, they also highlight the need for further research to adapt these techniques to the specific constraints and requirements of TinyML systems.

7.3 Homomorphic Encryption in TinyML

HE has emerged as a promising technique for enabling privacy-preserving computation on sensitive data, particularly in resource-constrained environments such as TinyML [5, 16]. By allowing computations to be performed directly on encrypted data without the need for decryption, HE enables TinyML devices to process and analyze data while maintaining the confidentiality of the underlying information. This section explores the fundamentals of HE, its variants, and its application to TinyML tasks, along with an evaluation of its performance and practical considerations.

7.3.1 Fundamentals of Homomorphic Encryption

HE is a cryptographic technique that enables computations to be performed on encrypted data without revealing the underlying plaintext. The core principle behind HE is the homomorphic property, which allows the encryption function to preserve certain mathematical operations performed on the plaintext. More formally, given an encryption function $E(\cdot)$, a decryption function $D(\cdot)$, and an operation $\odot$ on the plaintext space, an encryption scheme is said to be homomorphic with respect to $\odot$ if the following holds:

$$E(x \odot y) = E(x) \otimes E(y),$$

where $\otimes$ is a corresponding operation in the ciphertext space. This property enables computations to be performed directly on the encrypted data, and the result, when decrypted, matches the result of the same operations performed on the plaintext.

The concept of HE was first introduced by Rivest, Adleman, and Dertouzos in 1978 [17], but it was not until the groundbreaking work of Gentry in 2009 [18] that a fully HE (FHE) scheme was constructed. Gentry's scheme was based on ideal lattices and relied on a technique called bootstrapping to manage the noise growth in the ciphertext. The scheme consists of three main steps:

- **Key Generation:** Generate a secret key sk and a public key pk.
- **Encryption:** Given a plaintext message m, compute the ciphertext $c = E_{pk}(m)$.
- **Decryption:** Given a ciphertext c, compute the plaintext message $m = D_{sk}(c)$.

The homomorphic property of Gentry's scheme allows for the evaluation of arbitrary circuits on the encrypted data. Given a function f and ciphertexts $c_1 = E_{pk}(m_1)$ and $c_2 = E_{pk}(m_2)$, the scheme enables the computation of $E_{pk}(f(m_1, m_2))$ without revealing m_1 or m_2.

Since Gentry's seminal work, significant progress has been made in improving the efficiency and practicality of FHE schemes, making them more suitable for resource-constrained environments such as TinyML. However, the computational cost and ciphertext expansion associated with FHE remain a challenge for its adoption in TinyML applications.

7.3.2 Partially and Fully Homomorphic Encryption Schemes

HE schemes can be classified into two main categories: partially HE (PHE) and FHE. PHE schemes support a limited set of operations on the encrypted data, typically either addition or multiplication, but not both. Examples of PHE schemes include

- **Paillier Cryptosystem [19]:** The Paillier cryptosystem is an additive HE scheme based on the decisional composite residuosity (DCR) assumption. Given two ciphertexts $c_1 = E_{pk}(m_1)$ and $c_2 = E_{pk}(m_2)$, the scheme allows for the computation of $E_{pk}(m_1 + m_2)$ as $c_1 \cdot c_2 \bmod n^2$, where n is the public key modulus.
- **ElGamal Cryptosystem [20]:** The ElGamal cryptosystem is a multiplicative HE scheme based on the decisional Diffie–Hellman (DDH) assumption. Given two ciphertexts $c_1 = \left(g^{r_1}, h^{r_1} \cdot m_1\right)$ and $c_2 = \left(g^{r_2}, h^{r_2} \cdot m_2\right)$, where g and h are public key components and r_1 and r_2 are random values, the scheme allows for the computation of $E_{pk}(m_1 \cdot m_2)$ as $\left(g^{r_1+r_2}, h^{r_1+r_2} \cdot m_1 \cdot m_2\right)$.

On the other hand, FHE schemes support an arbitrary number of both addition and multiplication operations on the encrypted data, enabling the evaluation of any computable function. Some notable FHE schemes include

- **Brakerski–Gentry–Vaikuntanathan (BGV) Scheme [21]:** The BGV scheme is an FHE scheme based on the ring learning with error (RLWE) problem. It introduces a technique called modulus switching to manage the noise growth in the ciphertext, allowing for the evaluation of deeper circuits compared to Gentry's original scheme.
- **Brakerski/Fan–Vercauteren (BFV) Scheme [22]:** The BFV scheme is another FHE scheme based on the RLWE problem. It employs a technique called scale-invariant HE, which enables the use of a fixed modulus throughout the evaluation process, leading to improved efficiency compared to the BGV scheme.
- **Cheon–Kim–Kim–Song (CKKS) Scheme [23]:** The CKKS scheme is an FHE scheme that supports approximate arithmetic on encrypted real numbers. It is based on the RLWE problem and uses a technique called rescaling to manage the precision loss during homomorphic operations, making it suitable for machine learning applications.

While FHE schemes offer greater flexibility and functionality compared to PHE schemes, they typically incur higher computational costs and ciphertext expansion. Therefore, the choice between PHE and FHE in TinyML applications depends on the specific requirements of the task and the available resources on the device.

Figure 7.1 provides a comprehensive overview of HE schemes, categorizing them into partially HE (PHE) and FHE.

The PHE schemes, including the Paillier and ElGamal cryptosystems, support either additive or multiplicative operations on encrypted data, respectively. In

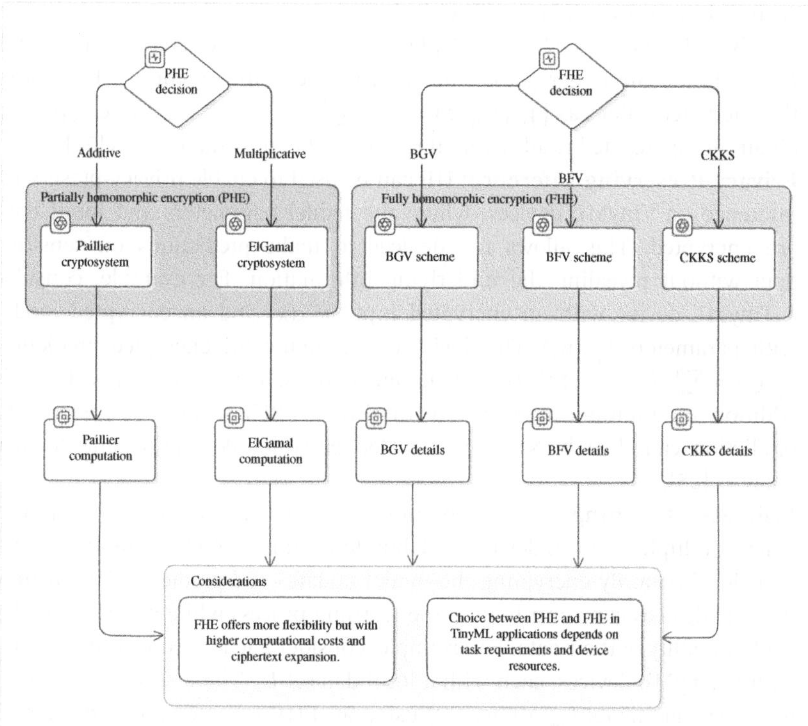

Figure 7.1 Homomorphic encryption schemes.

contrast, FHE schemes such as BGV, BFV, and CKKS enable both addition and multiplication, thus allowing the evaluation of any computable function on encrypted data. The diagram also highlights the trade-offs between these schemes, noting that FHE, despite its greater flexibility, incurs higher computational costs and ciphertext expansion, making the choice between PHE and FHE dependent on specific application requirements and available resources.

7.3.3 Applying Homomorphic Encryption to TinyML Tasks

HE has numerous applications in TinyML, particularly in scenarios where sensitive data need to be processed and analyzed on resource-constrained devices. Some key areas where HE can be applied in TinyML include

- **Secure Data Aggregation:** TinyML devices often need to aggregate data from multiple sources for analysis or decision-making. HE enables the devices to perform this aggregation on encrypted data, ensuring the privacy of the individual data points. For example, consider a scenario where n TinyML devices each hold a private value x_i. The devices can encrypt their values using an additive HE scheme, such as the Paillier cryptosystem, and send the ciphertexts $E_{pk}(x_i)$ to a central aggregator. The aggregator can then compute the sum of the encrypted values as $\prod_{i=1}^{n} E_{pk}(x_i) = E_{pk}\left(\sum_{i=1}^{n} x_i\right)$, which can be decrypted to obtain the aggregated result without revealing the individual values [24].

- **Privacy-Preserving Inference:** HE can be used to enable privacy-preserving inference on TinyML devices, where the model parameters and input data are encrypted. This allows the devices to make predictions on sensitive data without revealing the underlying information. For example, consider a TinyML device with an encrypted input $E_{pk}(x)$ and an encrypted model with parameters $E_{pk}(w_i)$. The device can compute the encrypted prediction $E_{pk}(y) = \sum_{i=1}^{n} E_{pk}(w_i) \otimes E_{pk}(x_i)$ using an FHE scheme that supports both addition and multiplication, such as the BGV or BFV scheme. The encrypted prediction can then be sent to a trusted party for decryption and further analysis [25].
- **Federated Learning:** HE can be employed in federated learning scenarios, where multiple TinyML devices collaborate to train a model without sharing their local data. By encrypting the model updates and gradients, HE ensures that the devices can contribute to the learning process while maintaining the confidentiality of their data. For example, consider a federated learning setting with n TinyML devices, each with a local dataset D_i. The devices can encrypt their local gradients $E_{pk}(\nabla L(w, D_i))$ using an FHE scheme and send them to a central server. The server can then compute the encrypted global gradient as $\sum_{i=1}^{n} E_{pk}(\nabla L(w, D_i))$, which can be decrypted and used to update the model parameters. This process is repeated iteratively until convergence, allowing the devices to collaboratively learn a model without revealing their local data [6].
- **Secure Multi-Party Computation:** HE is a key enabling technology for SMPC protocols, which allow multiple parties to jointly compute a function on their private inputs without revealing them to each other. In TinyML, HE-based MPC can be used for tasks such as privacy-preserving collaborative learning and secure decision-making. For example, consider a scenario where n TinyML devices, each with a private input x_i, want to compute a function $f(x_1, \ldots, x_n)$ without revealing their inputs to each other. The devices can encrypt their inputs using an FHE scheme and engage in an MPC protocol to compute the encrypted result $E_{pk}(f(x_1, \ldots, x_n))$, which can be decrypted by a trusted party. This approach ensures that the devices can collaborate on the computation while preserving the privacy of their individual inputs [26].

When applying HE to TinyML tasks, it is crucial to consider the specific requirements of the application, such as the desired level of security, the computational resources available on the devices, and the performance constraints. The choice

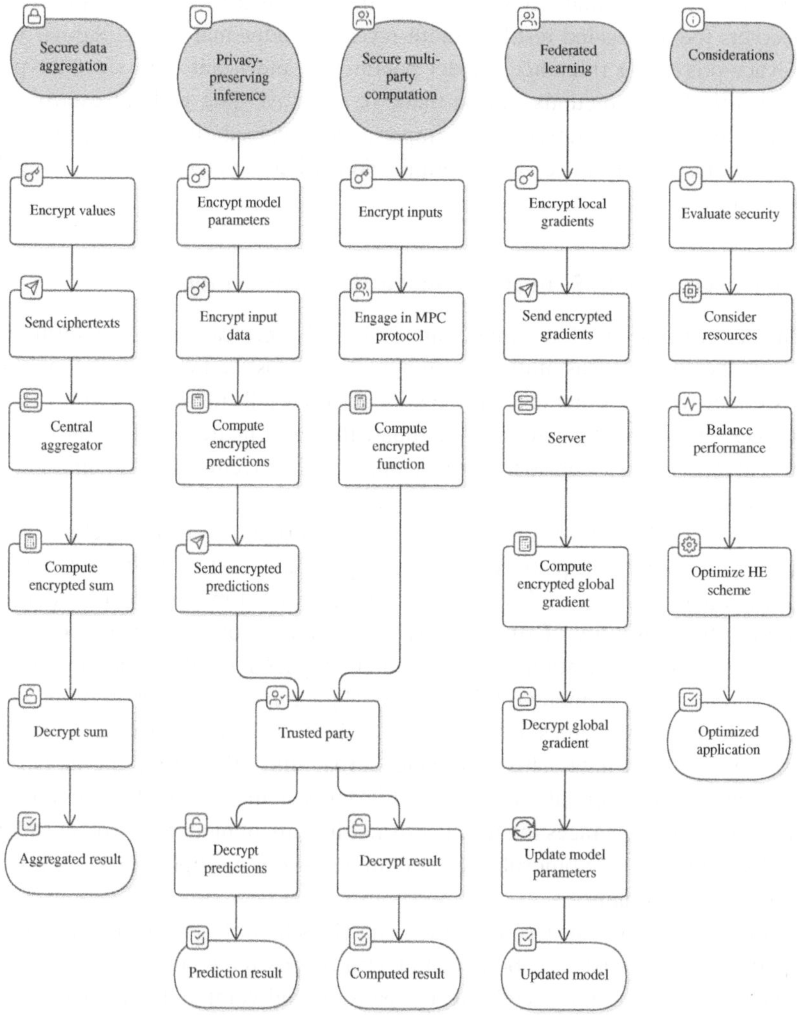

Figure 7.2 Applying homomorphic encryption to TinyML tasks.

of the HE scheme and its parameters should be carefully optimized to strike a balance between security, efficiency, and functionality.

Figure 7.2 illustrates the application of HE to various TinyML tasks, highlighting key areas such as secure data aggregation, privacy-preserving inference, federated learning, and SMPC.

Each section of the diagram details the specific steps involved in implementing HE for these tasks. For instance, in secure data aggregation, TinyML devices encrypt their data and send it to a central aggregator, which computes and decrypts the aggregated result without revealing individual values. Similarly, in privacy-preserving inference, model parameters and input data are encrypted to ensure that predictions can be made on sensitive data without exposing it. The diagram also underscores the importance of considering security, resource availability, performance, and optimization when applying HE to TinyML applications.

7.3.4 Performance Evaluation and Practical Considerations

The performance of HE in TinyML applications is a critical factor in determining its practicality and feasibility. The computational cost and ciphertext expansion associated with HE schemes can be significant, especially for resource-constrained devices. Therefore, it is essential to evaluate the performance of HE in the context of TinyML and to consider various optimization techniques to improve its efficiency.

Several studies have investigated the performance of HE schemes on microcontroller-based platforms. For example, Carpov et al. [27] implemented and evaluated the BGV scheme on an Arm Cortex-M4 microcontroller, demonstrating the feasibility of running FHE on resource-constrained devices. They also proposed optimizations, such as ciphertext packing and parallel processing, to improve the performance of the scheme. Ciphertext packing involves encoding multiple plaintext values into a single ciphertext, enabling batch operations on the encrypted data. This technique can significantly reduce the ciphertext expansion and computational overhead of HE schemes. Parallel processing, on the other hand, leverages the multi-core architecture of modern microcontrollers to distribute the homomorphic computations across multiple cores, leading to improved performance.

Other works have focused on optimizing HE schemes specifically for TinyML applications. For instance, Bourse et al. [16] proposed a compressed encryption scheme that reduces the ciphertext expansion and computational cost of HE, making it more suitable for resource-constrained environments. The scheme is based on the CKKS scheme and employs a technique called ciphertext compression, which reduces the size of the ciphertexts by discarding some of the least significant bits. This compression is achieved through a careful analysis of the noise growth during homomorphic operations, ensuring that the discarded bits do not affect the correctness of the computations.

Similarly, Jiang et al. [28] introduced a lightweight HE scheme based on the LWE problem, which achieves better performance compared to traditional HE schemes in terms of both computation time and memory consumption. The scheme employs a technique called dimension reduction, which reduces the size of the LWE problem instances while maintaining the same level of security. This is achieved by carefully selecting the parameters of the LWE problem based on the specific requirements of the TinyML application, such as the desired security level and the available computational resources.

In addition to algorithmic optimizations, the performance of HE in TinyML can be improved through hardware acceleration. Specialized hardware accelerators for HE have been proposed, such as the Homomorphic Encryption Acceleration (HEAX) accelerator [5], which achieves significant speedups compared to software implementations. The HEAX accelerator is designed as a co-processor for microcontrollers and implements the basic building blocks of HE schemes, such as modular arithmetic and polynomial operations, in hardware. By offloading these computationally intensive operations to the accelerator, the performance of HE in TinyML can be greatly improved, making it more practical for real-world applications.

When applying HE to TinyML tasks, it is essential to consider the practical aspects of the implementation, such as the memory footprint, power consumption, and latency. The choice of the HE scheme and its parameters should be carefully tuned to meet the specific requirements of the application while ensuring an acceptable level of performance. Furthermore, the integration of HE into existing TinyML frameworks and libraries should be seamless, providing developers with easy-to-use tools and interfaces for implementing privacy-preserving techniques in their applications.

In conclusion, HE is a powerful tool for enabling privacy-preserving computation in TinyML environments. By allowing computations to be performed directly on encrypted data, HE enables TinyML devices to process and analyze sensitive information while maintaining the confidentiality of the underlying data. The choice between partially and FHE schemes depends on the specific requirements of the application and the available resources on the devices. When applying HE to TinyML tasks, it is crucial to consider the performance implications and to employ various optimization techniques, at both the algorithmic and hardware levels, to ensure the practicality and feasibility of the approach. With ongoing research and advancements in HE schemes and their integration into TinyML frameworks, HE is poised to play a significant role in enabling secure and privacy-preserving TinyML applications in the future.

7.3.5 Comparison of Homomorphic Encryption Schemes

To determine the most suitable HE scheme for TinyML applications, it is essential to compare the various schemes in terms of their security, efficiency, and functionality. This section provides a comparative analysis of the key HE schemes discussed in this chapter, namely the Paillier, ElGamal, BGV, BFV, and CKKS schemes.

- **Security:** All the HE schemes considered in this comparison provide semantic security under appropriate hardness assumptions. The Paillier and ElGamal schemes are based on the DCR and DDH assumptions, respectively, while the BGV, BFV, and CKKS schemes are based on the RLWE assumption. The RLWE assumption is considered to be more secure against quantum attacks compared to the DCR and DDH assumptions, making the BGV, BFV, and CKKS schemes more suitable for long-term security in post-quantum scenarios.
- **Homomorphic Operations:** The Paillier and ElGamal schemes are partially homomorphic, supporting only additive and multiplicative operations, respectively. On the other hand, the BGV, BFV, and CKKS schemes are fully homomorphic, supporting an arbitrary number of both additive and multiplicative operations. This makes the fully homomorphic schemes more versatile and suitable for a wider range of TinyML applications that require complex computations on encrypted data.
- **Plaintext Space:** The Paillier and ElGamal schemes operate on integer plaintexts, while the BGV and BFV schemes support integer or polynomial plaintexts. The CKKS scheme, on the other hand, is designed specifically for real number plaintexts, making it more suitable for TinyML applications that involve floating-point arithmetic, such as deep learning and signal processing.
- **Ciphertext Size:** The ciphertext size of the HE schemes directly impacts the communication overhead and storage requirements in TinyML applications. The Paillier and ElGamal schemes have relatively small ciphertext sizes, typically twice the size of the plaintext. The BGV, BFV, and CKKS schemes have larger ciphertext sizes due to the use of noise to ensure security, with the ciphertext size growing linearly with the multiplicative depth of the homomorphic operations.
- **Computational Efficiency:** The computational efficiency of the HE schemes is crucial for their practical implementation in resource-constrained TinyML devices. The Paillier and ElGamal schemes have lower computational overhead compared to the fully homomorphic schemes, as they support only a single homomorphic operation. Among the fully homomorphic schemes, the BFV scheme is generally more efficient than the BGV scheme due to its use of a fixed modulus throughout the evaluation process. The CKKS scheme has a higher computational cost compared to the BGV and BFV schemes, as it involves additional operations for managing the precision loss during homomorphic operations.

Table 7.1 summarizes the key characteristics of the HE schemes considered in this comparison.

In conclusion, the choice of the most suitable HE scheme for a TinyML application depends on the specific requirements of the application, such as the desired security level, the types of homomorphic operations needed, the

Table 7.1 Comparison of homomorphic encryption schemes.

Scheme	Security assumption	Homomorphic operations	Plaintext space	Ciphertext size	Computational efficiency
Paillier	DCR	Additive	Integers	Small	High
ElGamal	DDH	Multiplicative	Integers	Small	High
BGV	RLWE	Fully homomorphic	Integers, polynomials	Large	Medium
BFV	RLWE	Fully homomorphic	Integers, polynomials	Large	Medium to high
CKKS	RLWE	Fully homomorphic	Real numbers	Large	Low to medium

plaintext space, and the available computational resources. For applications that prioritize computational efficiency and require only additive or multiplicative operations, the Paillier and ElGamal schemes may be preferred. For applications that demand fully homomorphic capabilities and can tolerate larger ciphertext sizes, the BGV and BFV schemes are suitable choices. If the application involves real number arithmetic and can accommodate the higher computational cost, the CKKS scheme is the most appropriate option. Ultimately, the selection of the HE scheme should be based on a careful analysis of the application's requirements and the trade-offs between security, efficiency, and functionality.

7.4 Differential Privacy for TinyML

DP has emerged as a strong standard for protecting the privacy of individuals' data in various applications, including machine learning [11, 29]. As TinyML devices often collect and process sensitive data, incorporating DP techniques into TinyML algorithms and systems is crucial to ensure that the privacy of users is protected [30, 31]. This section explores the principles of DP, the mechanisms for achieving it, and its application to TinyML algorithms. We also discuss the trade-offs between privacy and utility in TinyML applications and present techniques for balancing these competing objectives.

7.4.1 Principles of Differential Privacy

DP is a mathematical framework that provides strong privacy guarantees for individuals' data when performing statistical analysis or machine learning on a dataset [11]. The core principle of DP is that the output of a computation should not reveal too much information about any single individual in the dataset [29]. Formally, a randomized algorithm $\mathcal{A}$ is said to be (ε, δ)-differentially private if, for any two datasets D_1 and D_2 that differ in a single entry, and for any subset of outputs S, the following holds [11, 29]:

$$\Pr[\mathcal{A}(D_1) \in S] \leq e^{\varepsilon} \cdot \Pr[\mathcal{A}(D_2) \in S] + \delta.$$

Here, ε and δ are privacy parameters that quantify the level of privacy provided by the algorithm. Smaller values of ε and δ indicate stronger privacy guarantees. The parameter ε is often referred to as the privacy budget, as it represents the total amount of privacy loss that can be incurred during the analysis.

DP satisfies several important properties that make it a robust framework for privacy-preserving data analysis [11, 29]:

1) **Composition:** If two algorithms $\mathcal{A}_1$ and $\mathcal{A}_2$ are $(\varepsilon_1, \delta_1)$-differentially private and $(\varepsilon_2, \delta_2)$-differentially private, respectively, then their composition $\mathcal{A}_1 \circ \mathcal{A}_2$ is $(\varepsilon_1 + \varepsilon_2, \delta_1 + \delta_2)$-differentially private. This property allows for the analysis of multiple differentially private algorithms on the same dataset while still providing an overall privacy guarantee.
2) **Post-processing:** If an algorithm $\mathcal{A}$ is (ε, δ)-differentially private, then any post-processing of its output, denoted by $f(\mathcal{A}(D))$, is also (ε, δ)-differentially private. This property ensures that the privacy guarantees of a differentially private algorithm are preserved even if its output is subjected to further processing.
3) **Group Privacy:** If an algorithm $\mathcal{A}$ is (ε, δ)-differentially private, then it provides $(k\varepsilon, k\delta)$-DP for groups of k individuals. This property ensures that DP provides protection not only for individuals but also for groups of individuals whose data might be correlated.

These properties make DP a powerful tool for privacy-preserving data analysis, as they allow for the design of complex algorithms and systems that provide strong privacy guarantees under composition and post-processing.

7.4.2 Mechanisms for Achieving Differential Privacy

To achieve DP, algorithms must introduce carefully calibrated noise into the computation. This noise is typically drawn from a probability distribution, such as the Laplace or Gaussian distribution, and is added to the output of the algorithm to mask the contribution of any single individual in the dataset. The magnitude of the noise depends on the sensitivity of the computation, which measures how much the output can change when a single entry in the dataset is modified.

The two most common mechanisms for achieving DP are the Laplace mechanism and the Gaussian mechanism [11, 29]:

1) **Laplace Mechanism:** The Laplace mechanism adds noise drawn from the Laplace distribution to the output of a function f. The scale parameter of the Laplace distribution is determined by the sensitivity of f and the desired privacy budget ε. Formally, the Laplace mechanism is defined as

$$\mathcal{A}(D) = f(D) + \text{Lap}\left(\frac{\Delta f}{\varepsilon}\right).$$

Here, Δf is the sensitivity of the function f, which is defined as the maximum difference in the output of f when a single entry in the dataset is changed:

$$\Delta f = \max{}_{D_1,D_2} \|f(D_1) - f(D_2)\|_1.$$

The Laplace mechanism is (ε, 0)-differentially private, providing a strong privacy guarantee with no additional error probability.

2) **Gaussian Mechanism:** The Gaussian mechanism adds noise drawn from the Gaussian distribution to the output of a function f. The variance of the Gaussian distribution is determined by the sensitivity of f and the desired privacy parameters ε and δ. Formally, the Gaussian mechanism is defined as

$$\mathcal{A}(D) = f(D) + \mathcal{N}(0, \sigma^2).$$

Here, the variance σ^2 is chosen such that

$$\sigma \geq \frac{\sqrt{2\ln(1.25/\delta)}\Delta f}{\varepsilon}.$$

The Gaussian mechanism is (ε, δ)-differentially private, providing a slightly weaker privacy guarantee compared to the Laplace mechanism but allowing for a smaller amount of noise to be added to the output.

In addition to these basic mechanisms, several other techniques have been proposed for achieving DP, such as the exponential mechanism, the sparse vector technique, and the sample and aggregate framework. These techniques are designed to handle more complex data analysis tasks and provide DP guarantees under different settings.

7.4.3 Differentially Private TinyML Algorithms

Incorporating DP into TinyML algorithms is essential to protect the privacy of users' data when performing machine learning tasks on resource-constrained devices. This section presents several differentially private TinyML algorithms and discusses their implementation considerations.

1) **Differentially Private Stochastic Gradient Descent (DP-SGD):** SGD is a widely used optimization algorithm for training machine learning models. To make SGD differentially private, the gradients computed at each iteration are

clipped to a maximum norm and perturbed with Gaussian noise before being applied to update the model parameters. The amount of noise added to the gradients is determined by the sensitivity of the gradient computation and the desired privacy parameters. DP-SGD has been shown to provide strong privacy guarantees while achieving comparable performance to non-private SGD in various TinyML tasks, such as image classification and keyword spotting [30].

2) **Differentially Private Federated Learning:** Federated learning is a distributed machine learning approach that allows multiple parties to collaboratively train a model without sharing their raw data. By combining federated learning with DP, TinyML devices can participate in the training process while preserving the privacy of their local data. In differentially private federated learning, each device applies DP-SGD to its local data and sends the noisy gradients to a central server for aggregation. The server then updates the global model using the aggregated gradients and sends the updated model back to the devices. This process is repeated for multiple rounds until convergence. Differentially private federated learning has been successfully applied to various TinyML applications, such as human activity recognition and audio classification [32].
3) **Differentially Private Transfer Learning:** Transfer learning is a technique that allows TinyML devices to adapt pre-trained models to their local data, reducing the amount of data and computation required for training. To ensure the privacy of the local data during the adaptation process, differentially private transfer learning techniques have been proposed. These techniques typically involve fine-tuning the pre-trained model using DP-SGD on the local data, with the privacy budget allocated between the fine-tuning steps. Differentially private transfer learning has been shown to provide strong privacy guarantees while achieving good performance on various TinyML tasks, such as image classification and speech recognition [13].
4) **Differentially Private Distillation:** Model distillation is a technique that allows TinyML devices to compress large models into smaller ones, reducing the memory and computation requirements for deployment. To protect the privacy of the data used for distillation, differentially private distillation techniques have been proposed. These techniques involve training a differentially private teacher model on the sensitive data and then using the teacher model's outputs to train a smaller student model. The student model can then be deployed on the TinyML device without revealing information about the sensitive data used for training the teacher model. Differentially private distillation has been successfully applied to various TinyML applications, such as face recognition and object detection [33].

When implementing differentially private TinyML algorithms, several considerations should be taken into account. First, the choice of the privacy parameters ε and δ should be based on the desired level of privacy and the acceptable performance trade-off. Smaller values of ε and δ provide stronger privacy guarantees but may require more noise to be added to the computations, potentially degrading the performance of the algorithm.

Second, the sensitivity of the computations involved in the TinyML algorithm should be carefully analyzed to determine the appropriate amount of noise to be added. This may require adapting the algorithm to use differentially private building blocks, such as differentially private aggregation or differentially private gradient computation.

Third, the limited computational resources of TinyML devices should be considered when designing differentially private algorithms. This may involve optimizing the algorithms for efficiency, using techniques such as quantization or pruning to reduce the memory and computation requirements, or offloading some of the computations to more powerful devices or servers.

Finally, the composition of differentially private algorithms should be carefully managed to ensure that the overall privacy budget is not exceeded. This may require dividing the privacy budget among the different components of the TinyML system and using advanced composition techniques, such as the moment accountant, to track the privacy loss across multiple computations.

7.4.4 Balancing Privacy and Utility in TinyML Applications

While DP provides strong privacy guarantees for TinyML applications, it also introduces a trade-off between privacy and utility. Adding noise to the computations to achieve DP can degrade the performance of the TinyML algorithms, reducing the accuracy, throughput, or energy efficiency of the system. Therefore, balancing privacy and utility is a critical challenge in the design and deployment of differentially private TinyML applications.

Several techniques have been proposed to address this challenge and achieve a better balance between privacy and utility in TinyML:

1) **Privacy Budget allocation:** In TinyML applications that involve multiple computational stages or components, the privacy budget should be carefully allocated among these components to maximize the overall utility of the system. This can be achieved using optimization techniques that take into account the sensitivity and importance of each component, as well as the overall performance requirements of the application. For example, in a TinyML system that performs both data collection and inference, more of the privacy budget may be allocated to the inference component to ensure high

accuracy, while less budget is allocated to the data collection component to reduce the communication overhead [10, 13, 30].

2) **Adaptive Noise Calibration:** The amount of noise added to the computations to achieve DP can be adaptively calibrated based on the characteristics of the data and the performance requirements of the application. For example, in a TinyML system that performs real-time audio processing, the noise level can be adjusted based on the signal-to-noise ratio of the input data, adding more noise when the signal is strong and less noise when the signal is weak. This adaptive noise calibration can help maintain a good balance between privacy and utility across different operating conditions.
3) **Hybrid Privacy Mechanisms:** Combining different privacy mechanisms, such as the Laplace mechanism and the Gaussian mechanism, can help achieve a better trade-off between privacy and utility in TinyML applications. For example, in a federated learning scenario, the Gaussian mechanism can be used for the local computations on the TinyML devices, while the Laplace mechanism can be used for the global aggregation on the server [10]. This hybrid approach can provide strong privacy guarantees while reducing the overall noise added to the system, improving the performance of the learning process.
4) **Privacy-Preserving Compression:** Compressing the data or the models used in TinyML applications can help reduce the communication and storage overhead, while also providing some level of privacy protection. However, standard compression techniques may leak information about the underlying data. To address this issue, privacy-preserving compression techniques have been proposed, such as differentially private quantization [10, 13] and differentially private pruning [15]. These techniques combine compression with DP mechanisms to achieve a good balance between privacy, utility, and efficiency in TinyML applications.
5) **Privacy-Utility Trade-Off Analysis:** To make informed decisions about the balance between privacy and utility in TinyML applications, it is important to conduct a rigorous analysis of the privacy-utility trade-off. This can involve both theoretical analysis, using tools from DP and information theory, and empirical evaluation, using realistic datasets and performance metrics. The results of this analysis can help guide the design and configuration of the TinyML system, ensuring that the desired level of privacy is achieved while maximizing the utility of the application.

In conclusion, balancing privacy and utility is a critical challenge in the design and deployment of differentially private TinyML applications. By carefully allocating the privacy budget, adaptively calibrating the noise, combining different privacy mechanisms, employing privacy-preserving compression techniques, and conducting rigorous privacy-utility trade-off analysis, TinyML developers can achieve a good balance between protecting the privacy of users' data and maintaining the performance and efficiency of the system. As the field of TinyML continues to evolve, it is important to develop new techniques and tools for privacy-preserving machine learning that can address the unique challenges and constraints of resource-limited devices.

7.5 Secure Multi-Party Computation in TinyML

SMPC is a cryptographic technique that enables multiple parties to collaboratively compute a function on their private inputs without revealing the inputs to each other. In the context of TinyML, MPC can be used to enable distributed machine learning and privacy-preserving data aggregation among resource-constrained devices. This section presents the foundations of SMPC, the key building blocks of secret sharing and garbled circuits, and their applications to secure aggregation and efficiency optimization in TinyML.

7.5.1 Foundations of Secure Multi-Party Computation

SMPC was first introduced by Yao [34] in the context of the millionaires' problem, where two millionaires want to determine who is richer without revealing their actual wealth. Yao proposed a protocol based on garbled circuits that enables the millionaires to compute the comparison function on their private inputs without revealing the inputs to each other.

Since then, MPC has been extensively studied and generalized to support a wide range of functions and a number of parties. The goal of MPC is to enable n parties, each holding a private input x_i, to collaboratively compute a function $f(x_1, \ldots, x_n)$ without revealing their inputs to each other. The security of MPC is typically defined in terms of the ideal–/real-world paradigm [35], where the real-world execution of the protocol is compared to an ideal-world execution that involves a trusted third party. A protocol is considered secure if the real-world execution can be simulated in the ideal world, meaning that the protocol does not reveal any more information than what can be learned from the output of the function.

There are two main approaches to constructing MPC protocols: secret sharing-based approaches and garbled circuit-based approaches. Secret sharing-based approaches involve dividing the private inputs into shares and distributing them among the parties, who then perform computations on the shares to obtain the desired output. Garbled circuit–based approaches involve one party (the garbler) preparing an encrypted version of the function to be computed, which is then evaluated by the other parties (the evaluators) on their encrypted inputs.

Both approaches have been extensively studied and optimized for various settings, including dishonest majority, honest majority, and two-party computation.

In the dishonest majority setting, the majority of the parties may collude to try to learn the private inputs of the honest parties. In the honest majority setting, the majority of the parties are assumed to follow the protocol honestly. In the two-party setting, only two parties are involved in the computation.

MPC protocols have been designed for a wide range of functions, including arithmetic circuits, Boolean circuits, and more complex functions such as machine learning algorithms. The efficiency of MPC protocols is typically measured in terms of their communication complexity, which is the amount of data that need to be exchanged among the parties, and their computational complexity, which is the amount of computation that needs to be performed by each party.

7.5.2 Secret Sharing and Garbled Circuits

Secret sharing and garbled circuits are the two main building blocks of SMPC protocols.

Secret sharing is a technique that enables a secret value to be divided into shares and distributed among multiple parties, such that the secret can only be reconstructed when a sufficient number of shares are combined. The most common type of secret sharing is Shamir's secret sharing [36], which is based on polynomial interpolation. In Shamir's scheme, a secret value s is divided into n shares (x_1, y_1), …, (x_n, y_n), where x_i is a unique public value and y_i is a share of the secret. The shares are computed using a polynomial of degree t, where $t < n$, such that any subset of $t + 1$ shares can be used to interpolate the polynomial and recover the secret, while any subset of t or fewer shares reveals no information about the secret.

Secret sharing can be used to construct MPC protocols by having each party split their private input into shares and distribute them among the other parties. The parties can then perform computations on the shares, such as addition and multiplication, to obtain shares of the output. The shares of the output can then be combined to obtain the final result, without revealing the private inputs of the parties.

Garbled circuits, on the other hand, are a technique that enables a function to be evaluated on encrypted inputs without revealing the inputs or the intermediate values of the computation. The basic idea of garbled circuits is to represent the function as a Boolean circuit and to encrypt each wire of the circuit using a pair of random keys, one for each possible value of the wire (0 or 1). The encrypted wire values are then used to construct a garbled truth table for each gate in the circuit, which enables the output of the gate to be computed from the encrypted inputs without revealing the actual values.

In a typical garbled circuit protocol, one party (the garbler) prepares the garbled circuit and sends it to the other party (the evaluator), along with the keys corresponding to their own input. The evaluator then obtains the keys corresponding to their own input from the garbler through an oblivious transfer (OT) protocol [37], which enables the evaluator to obtain the keys without revealing their input to the garbler. The evaluator can then evaluate the garbled circuit on the encrypted inputs to obtain the encrypted output, which can be decrypted using the keys provided by the garbler.

Garbled circuits can be used to construct MPC protocols by having each party act as both a garbler and an evaluator. Each party prepares a garbled circuit for their own part of the function and sends it to the other parties, along with the keys corresponding to their own input. The parties then evaluate each other's garbled circuits on their encrypted inputs and combine the encrypted outputs to obtain the final result.

7.5.3 Secure Aggregation for Distributed TinyML

Secure aggregation is a common application of SMPC in TinyML, where multiple resource-constrained devices collaborate to perform machine learning tasks on their local data without revealing the data to each other or to a central server. Secure aggregation enables the devices to compute aggregate statistics or model updates on their local data, which can then be combined to obtain a global result without revealing the individual contributions.

One of the main challenges in secure aggregation for TinyML is the limited computational resources of the devices, which makes it difficult to implement complex MPC protocols. To address this challenge, several lightweight MPC protocols have been proposed for secure aggregation in TinyML, based on secret sharing and garbled circuits.

One approach is to use additive secret sharing to enable the devices to compute the sum of their local values without revealing the individual values [24]. In this approach, each device splits its local value into shares and sends one share to each of the other devices. The devices then compute the sum of the shares they received, which is equal to the sum of the original values. The devices can then send the sum of the shares back to a central server, which can combine the shares to obtain the final result.

Another approach is to use garbled circuits to enable the devices to compute more complex functions on their local data, such as machine learning models [26]. In this approach, each device prepares a garbled circuit for its local computation and sends it to the other devices, along with the keys corresponding to its own input. The devices then evaluate each other's garbled circuits on their encrypted inputs and send the encrypted outputs back to a central server, which can combine the outputs to obtain the final result.

Secure aggregation based on secret sharing and garbled circuits has been applied to various TinyML tasks, such as federated learning [6], where multiple devices collaborate to train a machine learning model on their local data without revealing the data to each other or to a central server. Secure aggregation can also be used for privacy-preserving data analysis [38], where multiple devices collaborate to compute aggregate statistics on their local data without revealing the individual values.

7.5.4 Efficiency Optimization for Secure Multi-Party Computation in TinyML

Efficiency is a critical concern in SMPC for TinyML, given the limited computational resources of the devices. Several optimization techniques have been proposed to improve the efficiency of MPC protocols for TinyML, based on the characteristics of the devices and the functions being computed.

One approach is to optimize the communication complexity of the MPC protocols by reducing the amount of data that need to be exchanged among the devices. This can be achieved by using more efficient secret sharing schemes, such as packed secret sharing [39], which enables multiple secrets to be packed into a single polynomial, reducing the number of shares that need to be exchanged. Another approach is to use compression techniques, such as arithmetic coding [40], to reduce the size of the messages exchanged among the devices.

Another approach is to optimize the computational complexity of the MPC protocols by reducing the amount of computation that needs to be performed by each device. This can be achieved by using more efficient garbled circuit constructions, such as half gates [41], which reduce the number of encryption operations required to prepare and evaluate the garbled circuits. Another approach is to use precomputation techniques, such as OT extension [42], which enable the devices to perform some of the computations in advance, reducing the online computational overhead.

In addition to these general optimization techniques, several specialized MPC protocols have been proposed for specific TinyML tasks, such as privacy-preserving inference [43] and privacy-preserving learning [44]. These protocols are designed to take advantage of the specific characteristics of the tasks, such as the structure of the machine learning models or the sparsity of the data, to reduce the communication and computational overhead of the MPC protocols.

Efficiency optimization for SMPC in TinyML is an active area of research, and new techniques are being developed to further improve the performance and scalability of MPC protocols for resource-constrained devices. Some promising directions include the use of trusted execution environments [45], such as Intel Software Guard Extensions (SGX), to reduce the overhead of MPC protocols, and the use of specialized hardware accelerators [46], such as field-programmable gate array (FPGAs), to speed up the computation of MPC protocols.

In conclusion, SMPC is a powerful technique for enabling privacy-preserving collaboration among resource-constrained devices in TinyML. Secret sharing and garbled circuits are the main building blocks of MPC protocols, and they have been applied to various TinyML tasks, such as secure aggregation and federated learning. Efficiency optimization is a critical concern in MPC for TinyML, and several techniques have been proposed to reduce the communication and computational overhead of MPC protocols, based on the characteristics of the devices and the functions being computed. As TinyML continues to evolve, MPC will play an increasingly important role in enabling privacy-preserving collaboration among resource-constrained devices, and new techniques will be developed to further improve the efficiency and scalability of MPC protocols for TinyML.

7.6 Case Studies and Applications of Privacy-Preserving TinyML

This section examines specific implementations and applications of privacy-preserving techniques in TinyML, providing insights into their effectiveness, performance, and practical implications.

7.6.1 Efficient Secure Multi-Party Computation for Privacy-Preserving Machine Learning

The work of Mohassel and Zhang [26, 47] on SecureML represents a significant advancement in privacy-preserving machine learning using SMPC. While not explicitly designed for TinyML, their techniques offer valuable insights and potential adaptations for privacy-preserving computation on resource-constrained devices.

Mohassel and Zhang introduce efficient protocols for privacy-preserving linear regression, logistic regression, and neural network training using SGD in a two-server model. Their approach significantly improves upon previous work in terms of computational efficiency and scalability.

Key contributions relevant to TinyML [47] include

1) A novel method for secure arithmetic on shared decimal numbers using truncation
2) MPC-friendly alternatives to nonlinear activation functions, crucial for the efficient implementation of complex models
3) Vectorized protocols that enhance computational efficiency
4) An offline/online protocol structure that allows for precomputation of data-independent operations

These techniques are particularly relevant for TinyML in IoT settings, where computational resources are limited, and efficiency is paramount. The authors' focus on minimizing both computation and communication overhead aligns well with the constraints of TinyML devices.

To evaluate the effectiveness of their approach, the authors employ several key metrics [47]:

1) Execution time (in seconds) for offline and online phases
2) Communication volume (in megabytes or gigabytes)
3) Accuracy of trained models (in percentages)
4) Speedup compared to previous works

The results of their experiments provide valuable insights into the potential of these techniques for TinyML applications. Table 7.2 presents the performance of their linear regression protocol for a dataset with 10,000 samples and 1,000 features, using a mini-batch size of 128 and 2 epochs [47].

In this table, "LHE-based" refers to the use of linear HE for generating multiplication triplets, "OT-based" refers to the use of OT for the same purpose, and "Client-aided" refers to a method where clients help generate multiplication triplets. "LAN" stands for local area network, representing a high-bandwidth, low-latency network setting.

The results demonstrate a trade-off between computation time and communication overhead. The LHE-based method has the highest offline computation time but the lowest communication overhead. The OT-based approach significantly reduces the offline computation time but at the cost of increased communication. The client-aided method offers the fastest offline phase but introduces changes to the security model and increases online overhead.

To contextualize the improvements achieved, Table 7.3 compares their approach with previous works for a dataset of 100,000 samples and 500 features [47].

The dramatic speedup achieved by Mohassel and Zhang's protocols (up to 1,270×) demonstrates the potential for practical implementation of privacy-preserving machine learning, even in resource-constrained environments. This comparison is particularly relevant for TinyML applications, where computational efficiency is crucial.

Table 7.2 Performance of linear regression (10,000 samples, 1,000 features, mini-batch size 128, and 2 epochs).

Method	Offline time (LAN)	Online time (LAN)	Communication
LHE-based	1,600.9 seconds	0.28 seconds	100 MB
OT-based	80.0 seconds	0.28 seconds	19 GB
Client-aided	4.0 seconds	1.0 seconds	687 MB

Table 7.3 Comparison with previous works (100,000 samples and 500 features).

Method	Total time (LAN)	Speedup
NWI + 13 [48]	~90,000 seconds	1×
GSB + 16 [49]	~30,000 seconds	3×
OT-based [26, 47]	70.8 seconds	1,270×
Client-aided [26, 47]	27.0 seconds	1,110×

In conclusion, the work of Mohassel and Zhang [26, 47] provides a solid foundation for developing efficient privacy-preserving techniques for TinyML. Their results demonstrate the feasibility of implementing secure machine learning protocols even in resource-constrained environments, paving the way for privacy-preserving IoT applications that can protect sensitive data while leveraging the power of machine learning at the edge.

7.6.2 A Generic Framework for Privacy-Preserving Deep Learning

Ryffel et al. [50, 51] propose a generic framework for privacy-preserving deep learning that integrates federated learning, SMPC, and DP into a unified system. Their work, implemented as an extension to PyTorch, aims to make privacy-preserving techniques more accessible to the broader deep learning community, including researchers working on TinyML applications.

Key contributions relevant to TinyML [50] include

1) A standardized protocol for communication between workers, enabling federated learning
2) A chain abstraction model for tensors, allowing efficient override of operations and encoding of new ones
3) Implementation of recent DP and MPC protocols within the framework

The authors introduce a novel abstraction called SyftTensor, which represents transformations or states of data that can be chained together. This chain structure, with a PyTorch tensor at its head, allows for seamless integration of privacy-preserving operations while maintaining a familiar PyTorch API for end users.

To evaluate their framework, the authors conducted experiments on two datasets [50]:

1) Boston Housing Dataset (regression task)
2) Pima Indian Diabetes Dataset (classification task)

Table 7.4 Training time comparison (Boston Housing Dataset, 10 epochs).

Training mode	Training time (seconds)
PySyft (Virtual)	10.1
PySyft (Socket)	14.6
PySyft (Virtual) + DP[a)]	15.3
Pure PyTorch	0.22

a) Equivalent time for the same number of batches processed for DP.

Table 7.5 Accuracy of differentially private federated learning.

(ε, δ) privacy	Boston housing MSE	Pima diabetes accuracy (%)
$(0.5, 10^{-5})$	29.4	60.6
$(1, 10^{-5})$	29.2	64.2
$(2, 10^{-5})$	28.5	66.1
$(4, 10^{-5})$	28.6	67.1
No privacy	23.7	70.3

Table 7.4 presents the training time for different configurations on the Boston Housing Dataset [50].

These results demonstrate that while the privacy-preserving framework introduces overhead compared to pure PyTorch, the performance impact is relatively modest, especially when considering the added privacy benefits.

The authors also evaluated the impact of DP on model accuracy. Table 7.5 shows the results for both datasets under different privacy budgets [50].

These results illustrate the trade-off between privacy and model performance. As the privacy budget (ε) increases, the model performance improves but at the cost of reduced privacy guarantees.

The framework's approach to DP in a federated learning context includes

1) Randomly choosing a worker and sampling from its data when batch sampling
2) Sanitizing gradients on remote workers to ensure data privacy

The authors note that their current implementation of SMPC based on the SPDZ protocol (named after Damgård, Pastro, Smart, and Zakarias) lacks a mechanism to ensure honest behavior from all participants. They suggest implementing media access control (MAC) authentication of secret shared values as a potential improvement.

Implications for TinyML:

1) The framework's ability to integrate federated learning, SMPC, and DP could be particularly valuable for TinyML applications where data privacy is crucial.
2) The modest performance overhead for virtual and socket-based workers suggests that the approach could be feasible for resource-constrained devices.
3) The trade-off between privacy and model performance demonstrated in the experiments is an important consideration for TinyML applications, where both resource efficiency and data protection are often critical.

Future work suggested by the authors includes

1) Improving the efficiency of the framework to reduce the current performance overhead
2) Implementing MAC authentication for SMPC to ensure honest behavior from all participants
3) Integrating additional DP approaches, such as Private Aggregation of Teacher Ensembles (PATE)

In conclusion, this work presents a promising approach to integrating privacy-preserving techniques into deep learning frameworks, with potential applications in TinyML. The framework's ability to combine federated learning, SMPC, and DP while maintaining a familiar API could significantly lower the barrier to entry for researchers and practitioners looking to implement privacy-preserving techniques in resource-constrained environments.

7.6.3 Maliciously Secure Coopetitive Learning for Linear Models

Zheng et al. [14, 52] present Helen, a system for maliciously secure coopetitive learning of linear models. Helen addresses a critical gap in privacy-preserving machine learning by providing strong security guarantees against malicious adversaries in a multi-party setting. This work is particularly relevant for TinyML applications where multiple parties with sensitive data wish to collaboratively train models while maintaining data privacy and integrity.

Key contributions relevant to TinyML [52] include

1) A maliciously secure protocol for training linear models in a multi-party setting
2) Efficient techniques for secure computation that scale to large datasets
3) A novel combination of SMPC and HE

Helen's architecture consists of several phases [52]:

1) **Initialization:** Parties compute threshold encryption parameters using a generic maliciously SMPC protocol.
2) **Input Preparation:** Each party precomputes and commits to summaries of their data.
3) **Model Compute:** Parties iteratively compute on encrypted data using the alternating direction method of multiplier (ADMM) algorithm.
4) **Model Release:** The final encrypted model is jointly decrypted and released to all parties.

To evaluate Helen's performance, the authors conducted experiments on both synthetic and real-world datasets. One of the key datasets used in this evaluation is the Gas Sensor Array Drift Dataset, referred to as the "gas sensor" dataset in the paper. This dataset is particularly relevant for assessing the system's performance in real-world scenarios and has several important characteristics [52]:

1) **Origin:** The dataset contains measurements from 16 chemical sensors used to analyze gas mixtures.
2) **Content:** The sensors record readings when mixing two types of gases at various concentration levels.
3) **Structure:** The dataset comprises 16 features (one for each sensor) and a large number of samples.
4) **Task:** The primary machine learning task for this dataset is regression to predict gas concentrations based on sensor readings.
5) **Size:** The authors used various sizes of this dataset, ranging from 1,000 to 1,000,000 samples per participant, to demonstrate their system's scalability.

The use of this gas sensor dataset is significant because it provides a realistic, complex, and scalable test case that is relevant to many IoT and industrial applications. It allows the authors to demonstrate Helen's performance on a practical problem while also enabling comparisons with other systems using the same publicly available dataset.

Table 7.6 presents a comparison of Helen's performance with a secure baseline implemented using SPDZ, a state-of-the-art maliciously SMPC framework, for the gas sensor dataset [52].

These results demonstrate Helen's significant performance improvements over the baseline, especially as the dataset size increases. For TinyML applications, this scalability is crucial when dealing with large numbers of IoT devices generating substantial amounts of data.

The authors also evaluated Helen's impact on model accuracy. Table 7.7 compares the L2 error and mean absolute error (MAE) of models trained using Helen and a plaintext sklearn implementation on the gas sensor dataset [52].

Table 7.6 Performance comparison on gas sensor dataset (4 parties and 16 features).

Samples per party	Helen (seconds)	Baseline (seconds)	Speedup
1,000	139	2,989	21.5×
10,000	140	29,890	213.5×
100,000	146	298,900	2,047×
1,000,000	193	3,989,000	20,689×

Table 7.7 Accuracy comparison on gas sensor dataset.

Samples per party	Helen L2 error	sklearn L2 error	Helen MAE	sklearn MAE
10,000	8,837.59	8,929.10	57.47	58.05
100,000	8,876.00	8,981.24	58.25	58.48
1,000,000	8,900.37	8,980.35	58.40	58.57

These results indicate that Helen achieves comparable or slightly better accuracy than the plaintext implementation, demonstrating that the privacy-preserving techniques do not significantly impact model quality.

Implications for TinyML [52]:

1) **Malicious Security:** Helen's strong security guarantees are particularly valuable in TinyML scenarios where devices may be physically compromised or operate in untrusted environments.
2) **Scalability:** The system's ability to handle large datasets efficiently is crucial for TinyML applications involving numerous IoT devices generating substantial data volumes.
3) **Accuracy Preservation:** Helen's ability to maintain model accuracy while providing privacy guarantees is essential for TinyML applications where model performance is critical.
4) **Resource Efficiency:** Although Helen focuses on a server-based setting, its techniques for efficient secure computation could be adapted for resource-constrained TinyML devices.
5) **Coopetitive Learning:** The framework enables collaborative learning among competing entities, which is applicable to TinyML scenarios where multiple organizations or devices need to jointly train models without sharing raw data.

Future research directions suggested by this work [52] include

1) Adapting Helen's techniques to more resource-constrained environments typical of TinyML devices
2) Extending the framework to support nonlinear models and deep learning architectures commonly used in TinyML applications
3) Investigating the energy efficiency implications of these privacy-preserving techniques for battery-powered TinyML devices
4) Developing specialized hardware accelerators for the cryptographic operations used in Helen to further improve performance on TinyML devices

In conclusion, Helen represents a significant advancement in privacy-preserving machine learning with strong security guarantees [52]. While not explicitly designed for TinyML, its techniques and insights provide a valuable foundation for developing secure and privacy-preserving TinyML systems in multi-party settings. The use of the gas sensor dataset in evaluating Helen demonstrates its potential for real-world IoT and industrial applications, which are closely aligned with many TinyML use cases.

7.6.4 Practical Secure Aggregation for Privacy-Preserving Machine Learning

Bonawitz et al. [24, 53] present a practical secure aggregation protocol for privacy-preserving machine learning in a federated learning context. Their work addresses the challenge of securely computing sums of vectors from multiple parties without revealing individual inputs, while also being robust to dropouts and efficient in terms of communication. This protocol is particularly relevant for TinyML applications, as it enables privacy-preserving collaborative learning among resource-constrained devices.

Key contributions relevant to TinyML [53] include

1) A communication-efficient secure aggregation protocol robust to user dropouts
2) A novel combination of secret sharing and key agreement techniques to achieve security and efficiency
3) Practical implementation and evaluation demonstrating scalability to large numbers of users

The protocol operates in four rounds [53]:

1) **Advertise Keys:** Users exchange public keys.
2) **Share Keys:** Users secret-share masks and encryption keys.
3) **Masked Input Collection:** Users submit masked inputs.
4) **Unmasking:** The server aggregates masked inputs and reconstructs the sum.

A key innovation in this protocol is the use of double-masking to protect security. Each user's input is masked with pairwise masks shared with other users, as well

Table 7.8 Wall-clock running time (seconds) as the number of clients increases.

Number of clients	Client time	Server time (0% dropouts)	Server time (10% dropouts)	Server time (30% dropouts)
100	0.85	1.02	3.12	7.31
500	4.25	10.09	311.20	716.95
1,000	8.50	24.44	904.38	2,068.84

Table 7.9 Communication costs per client.

Number of clients	Data vector size	Total data transfer	Expansion factor
100	100 K	1.60 MB	1.28×
1,000	100 K	2.30 MB	1.84×
100	1 M	12.21 MB	1.03×
1,000	1 M	12.91 MB	1.09×

as a self-mask. This approach prevents accidental leakage of individual inputs even if the server can reconstruct a user's pairwise masks.

To evaluate the protocol's performance, the authors conducted experiments with varying numbers of users and input vector sizes. Table 7.8 presents the wall-clock running times for clients and the server as the number of clients increases, with a fixed data vector size of 100 K entries [53].

These results demonstrate that the protocol scales well with the number of clients, with client-side computation remaining relatively low. The server-side computation increases significantly with the percentage of dropouts, primarily due to the cost of reconstructing the masks for dropped users.

The authors also evaluated the communication costs of their protocol. Table 7.9 shows the total data transfer per client and the expansion factor compared to sending raw data, for different numbers of clients and data vector sizes [53].

These results show that the protocol achieves low communication overhead, with the expansion factor decreasing as the data vector size increases. This efficiency is crucial for TinyML applications, where communication is often a significant constraint.

Implications for TinyML:

1) **Scalability:** The protocol's ability to handle large numbers of users efficiently is crucial for TinyML deployments involving numerous IoT devices.
2) **Dropout Resilience:** The robustness to user dropouts is particularly valuable in TinyML scenarios where devices may have unreliable connectivity.

3) **Communication Efficiency:** The low communication overhead is essential for resource-constrained TinyML devices with limited bandwidth.
4) **Privacy Preservation:** The protocol enables collaborative learning without revealing individual device data, addressing privacy concerns in TinyML applications.
5) **Practical Implementation:** The demonstrated feasibility of implementing this protocol provides a path for integrating secure aggregation into real-world TinyML systems.

Future research directions suggested by this work include

1) Adapting the protocol for even more resource-constrained devices typical in TinyML applications
2) Investigating the integration of DP techniques with secure aggregation for enhanced privacy guarantees in TinyML contexts
3) Exploring optimizations to reduce the computational cost of handling dropouts, which could be frequent in TinyML deployments
4) Developing techniques to verify the integrity of inputs from TinyML devices to prevent malicious manipulation of the aggregated result

In conclusion, the practical secure aggregation protocol presented by Bonawitz et al. [24, 53] offers a promising approach for enabling privacy-preserving collaborative learning in TinyML environments. Its efficiency, scalability, and robustness to dropouts make it well-suited for the challenges faced in deploying machine learning on resource-constrained IoT devices. As TinyML continues to evolve, techniques like this will be crucial in balancing the benefits of collaborative learning with the privacy and security needs of distributed, edge-based machine learning systems.

7.6.5 Falcon: Honest Majority Maliciously Secure Framework for Private Deep Learning

Wagh et al. [12, 54, 55] present Falcon, an efficient and expressive 3-party protocol for privacy-preserving training and inference of large machine learning models. Falcon addresses several key challenges in privacy-preserving machine learning, offering malicious security with honest majority, support for high-capacity networks, and efficient protocols for nonlinear operations. This work is particularly relevant for TinyML applications, as it demonstrates the feasibility of secure computation for complex neural network architectures on resource-constrained devices.

Key contributions relevant to TinyML include [12, 54]

1) Maliciously secure protocols for privacy-preserving training and inference with honest majority

2) Efficient implementations of nonlinear functions such as rectified linear unit (ReLU) and batch normalization
3) Support for high-capacity networks like AlexNet and VGG16
4) Comprehensive evaluation on large-scale datasets and networks

Falcon's architecture consists of four main phases [12, 54]:

1) **Initialization:** Parties compute threshold encryption parameters.
2) **Input Preparation:** Parties precompute and commit to summaries of their data.
3) **Model Compute:** Parties iteratively compute on encrypted data using the ADMM algorithm.
4) **Model Release:** The final encrypted model is jointly decrypted and released to all parties.

A key innovation in Falcon is the use of replicated secret sharing and novel protocol constructions for nonlinear operations. This approach allows for efficient computation of functions like ReLU and its derivative, which are typically expensive in SMPC protocols.

To evaluate Falcon's performance, the authors conducted extensive experiments on various network architectures and datasets. Table 7.10 presents a comparison of Falcon's inference time with prior work for Network-A on the MNIST dataset [12, 54].

These results demonstrate Falcon's significant performance improvements over prior work, especially considering its stronger security guarantees. For TinyML applications, this efficiency is crucial when dealing with resource-constrained devices.

The authors also evaluated Falcon's performance on larger networks and datasets. Table 7.11 shows the inference time for AlexNet and VGG16 on the Tiny ImageNet dataset [12, 54].

Table 7.10 Inference time comparison for Network-A (MNIST).

Framework	Threat model	LAN time (seconds)	WAN time (seconds)
SecureML	Semi-honest	4.88	4,336
MiniONN	Semi-honest	1.04	—
Gazelle	Semi-honest	0.09	—
SecureNN	Semi-honest	0.043	2.43
ABY3	Semi-honest	0.008	—
Falcon (SH)	Semi-honest	0.011	0.99
Falcon (Mal)	Malicious	0.021	2.33

Table 7.11 Inference time for large networks on Tiny ImageNet.

Network	Threat model	LAN time (seconds)	WAN time (seconds)
AlexNet	Semi-honest	1.81	2.43
AlexNet	Malicious	6.7	8.68
VGG16	Semi-honest	3.15	4.67
VGG16	Malicious	12.04	37.6

These results highlight Falcon's ability to handle complex network architectures efficiently, even under stronger security assumptions. This capability is particularly relevant for TinyML applications that require high-capacity models while maintaining privacy guarantees.

Implications for TinyML:

1) **Malicious Security:** Falcon's protocols provide strong security guarantees, which are crucial for TinyML applications operating in untrusted environments.
2) **Efficient Nonlinear Operations:** The novel protocols for ReLU and batch normalization enable efficient computation of complex neural networks on resource-constrained devices.
3) **Scalability:** Falcon's ability to handle large networks and datasets demonstrates its potential for scaling TinyML applications to more complex tasks.
4) **Performance Trade-offs:** The results show the trade-offs between semi-honest and malicious security, allowing TinyML developers to choose the appropriate security level based on their application requirements and resource constraints.
5) **Batch Normalization Support:** Falcon's efficient implementation of batch normalization enables the training of complex network architectures, which is critical for improving model convergence and accuracy in TinyML applications.

Future research directions suggested by this work include

1) Optimizing Falcon for specific TinyML hardware platforms to further improve performance
2) Investigating the integration of Falcon's protocols with quantization techniques to reduce the computational and memory requirements for TinyML devices
3) Exploring the application of Falcon's efficient nonlinear operation protocols to other TinyML tasks beyond image classification

4) Developing techniques to further reduce the communication overhead, which is particularly important for TinyML applications with limited network connectivity

In conclusion, Falcon represents a significant advancement in privacy-preserving machine learning, offering malicious security and support for complex neural network architectures [12, 54]. Its efficiency and scalability make it particularly relevant for TinyML applications, providing a foundation for secure and private computation on resource-constrained devices.

7.7 Discussion and Future Directions

The rapid advancement of TinyML has brought about significant opportunities for deploying machine learning models on resource-constrained devices. However, the privacy implications of processing potentially sensitive data on these devices remain a critical concern. This chapter has explored various privacy-preserving techniques specifically tailored for TinyML in IoT environments, addressing the unique challenges posed by limited computational resources, memory constraints, and energy efficiency requirements.

Throughout this chapter, we have examined several key approaches to privacy-preserving TinyML, including SMPC, HE, and DP. Each of these techniques offers distinct advantages and trade-offs in terms of security guarantees, computational efficiency, and communication overhead.

7.7.1 Secure Multi-party Computation for TinyML

The works of Mohassel and Zhang (SecureML), Wagh et al. (SecureNN), and Wagh et al. (Falcon) have demonstrated the feasibility of using MPC techniques for privacy-preserving machine learning in resource-constrained environments. These approaches have shown significant improvements in performance and scalability compared to earlier work.

SecureML introduced a two-party computation model for privacy-preserving machine learning, which laid the groundwork for subsequent research. SecureNN built upon this foundation, proposing a three-party computation model with improved efficiency for neural network training and inference. Falcon further advanced the state of the art by providing malicious security in a three-party setting, supporting high-capacity networks, and introducing efficient protocols for nonlinear operations such as ReLU and batch normalization.

The performance improvements achieved by these frameworks are substantial. For instance, Falcon demonstrated up to 6× faster training than SecureNN and

4.4× faster than ABY3. These advancements make privacy-preserving TinyML more practical for real-world applications.

However, challenges remain in adapting these techniques for extremely resource-constrained devices. Future research should focus on further optimizing these protocols for specific TinyML hardware platforms and investigating the integration of MPC techniques with model compression and quantization methods to reduce computational and memory requirements.

7.7.2 Homomorphic Encryption in TinyML

While not extensively covered in this chapter, HE offers another avenue for privacy-preserving TinyML. Works such as CryptoNets and their successors have shown the potential of HE for secure inference. However, the high computational overhead of FHE schemes remains a significant barrier to their adoption in TinyML applications.

Future research in this area should explore partial HE schemes or hybrid approaches that combine HE with other techniques to achieve a better balance between security and efficiency for TinyML devices.

7.7.3 Differential Privacy for TinyML

The integration of DP techniques with TinyML presents an opportunity to provide strong privacy guarantees while maintaining utility. The work of Ryffel et al. on a generic framework for privacy-preserving deep learning demonstrated how DP can be combined with secure aggregation to enhance privacy in federated learning settings.

For TinyML applications, adapting these techniques to account for the limited resources of edge devices is crucial. Future research should investigate efficient methods for generating and adding noise to model updates on resource-constrained devices, as well as exploring the trade-offs between privacy guarantees and model utility in the context of TinyML.

7.7.4 Practical Considerations and Future Directions

Several key considerations emerge from our examination of privacy-preserving techniques for TinyML:

1) **Performance-Security Trade-offs:** As demonstrated by the Falcon framework, there is often a trade-off between security guarantees and performance. TinyML developers must carefully consider the appropriate balance for their specific applications.

2) **Communication Overhead:** Many privacy-preserving protocols introduce significant communication overhead. Given the limited network connectivity of many IoT devices, optimizing communication efficiency is crucial for the practical deployment of privacy-preserving TinyML.
3) **Energy Efficiency:** The energy consumption of privacy-preserving protocols is a critical factor for battery-powered TinyML devices. Future research should focus on developing energy-aware privacy-preserving techniques.
4) **Model Accuracy:** Ensuring that privacy-preserving techniques do not significantly degrade model accuracy is essential. As shown in the evaluation of Falcon, careful implementation of secure protocols can maintain accuracy comparable to plaintext training and inference.
5) **Scalability:** As TinyML applications grow in complexity, privacy-preserving techniques must scale accordingly. The ability of frameworks like Falcon to handle large networks such as AlexNet and VGG16 is promising, but further work is needed to address the challenges of deploying such models on highly constrained devices.

Looking ahead, several promising research directions emerge:

1) **Hardware–Software Co-design:** Developing specialized hardware accelerators for privacy-preserving operations could significantly improve the efficiency of secure TinyML computations.
2) **Privacy-Preserving Federated Learning for TinyML:** Adapting and optimizing federated learning techniques for TinyML environments while preserving privacy is an important area for future research.
3) **Secure Model Updates:** Investigating secure methods for updating TinyML models in the field while preserving privacy and ensuring integrity is crucial for long-term deployment.
4) **Privacy-Preserving Transfer Learning:** Exploring techniques for securely transferring knowledge from larger models to TinyML models without compromising privacy could enable more powerful models on resource-constrained devices.
5) **Formal Verification:** Developing methods for formally verifying the security and privacy guarantees of TinyML systems is essential for building trust in these systems.

In conclusion, while significant progress has been made in developing privacy-preserving techniques for TinyML, numerous challenges and opportunities remain. As TinyML continues to evolve and find new applications, the development of efficient, secure, and privacy-preserving methods will be crucial in realizing its full potential while protecting user privacy [56, 57].

7.8 Conclusion

The integration of machine learning capabilities into resource-constrained IoT devices has opened new frontiers in edge computing, but it also presents significant privacy challenges. This chapter has provided a comprehensive examination of privacy-preserving techniques tailored for TinyML environments, focusing on SMPC, HE, and DP.

Our analysis of frameworks such as SecureML, SecureNN, and Falcon has demonstrated the rapid progress in this field, with each successive approach offering improved performance, stronger security guarantees, or support for more complex neural network architectures. These advancements have significantly reduced the performance gap between privacy-preserving and plaintext computations, making secure TinyML more viable for real-world applications.

However, several challenges remain. The trade-offs between security, computational efficiency, and communication overhead continue to be a critical consideration for TinyML deployments. Energy consumption and model accuracy in privacy-preserving settings also require careful attention, particularly for battery-powered devices with limited resources.

Looking forward, the field of privacy-preserving TinyML presents numerous exciting research opportunities. The development of specialized hardware accelerators for privacy-preserving operations, the adaptation of federated learning techniques for TinyML environments, and the exploration of secure transfer learning methods all hold promise for further advancing the field.

Acknowledgment

- This project has received funding from the European Union's Horizon 2020 research and innovation program under the Marie Skłodowska-Curie Grant Agreement No. 101007820 – TRUST. This publication reflects only the author's view, and the REA is not responsible for any use that may be made of the information it contains.
- This research was funded by the European Union – NextGenerationEU under the Italian Ministry of University and Research (MIUR), National Innovation Ecosystem Grant ECS00000041-VITALITY-CUP D83C22000710005.

References

1 Oliveira, F., Costa, D.G., Assis, F., and Silva, I. (2024). Internet of Intelligent Things: a convergence of embedded systems, edge computing and machine learning. *Internet of Things* 26: 101153. https://doi.org/10.1016/j.iot.2024.101153.

2 Saeed, M.M., Saeed, R.A., and Ahmed, Z.E. (2024). Chapter 9 – TinyML for 5G networks. In: *TinyML for Edge Intelligence in IoT and LPWAN Networks* (ed. B.S. Chaudhari, S.N. Ghorpade, M. Zennaro, and R. Paškauskas), 167–229. Academic Press. https://doi.org/10.1016/B978-0-44-322202-3.00014-2.

3 Patil, R.Y., Bhamare, M., Patil, Y.H., and Bannore, A. (2024). Chapter 13 – Securing TinyML in a connected world. In: *TinyML for Edge Intelligence in IoT and LPWAN Networks* (ed. B.S. Chaudhari, S.N. Ghorpade, M. Zennaro, and R. Paškauskas), 311–330. Academic Press. https://doi.org/10.1016/B978-0-44-322202-3.00018-X.

4 Lin, J., Zhu, L., Chen, W.-M. et al. (2023). Tiny machine learning: progress and futures [feature]. *IEEE Circuits and Systems Magazine* 23 (3): 8–34. https://doi.org/10.1109/MCAS.2023.3302182.

5 Al Badawi, A., Jin, C., Lin, J. et al. (2021). Towards the AlexNet moment for homomorphic encryption: HCNN, the first homomorphic CNN on encrypted data with GPUs. *IEEE Transactions on Emerging Topics in Computing* 9 (3): 1330–1343. https://doi.org/10.1109/TETC.2020.3014636.

6 Bonawitz, K., Eichner, H., Grieskamp, W., et al. (2019). Towards federated learning at scale: system design, March 22. *arXiv:1902.01046*. https://doi.org/10.48550/arXiv.1902.01046.

7 Ekenstedt, E., Ong, L., Liu, Y. et al. (2022). When differential privacy implies syntactic privacy. *IEEE Transactions on Information Forensics and Security* 17: 2110–2124. https://doi.org/10.1109/TIFS.2022.3177953.

8 Jiang, H., Pei, J., Yu, D. et al. (2023). Applications of differential privacy in social network analysis: a survey. *IEEE Transactions on Knowledge and Data Engineering* 35 (1): 108–127. https://doi.org/10.1109/TKDE.2021.3073062.

9 Kus, M.C. and Levi, A. (2022). Investigation and application of differential privacy in bitcoin. *IEEE Access* 10: 25534–25554. https://doi.org/10.1109/ACCESS.2022.3151784.

10 Mahawaga Arachchige, P.C., Bertok, P., Khalil, I. et al. (2020). Local differential privacy for deep learning. *IEEE Internet of Things Journal* 7 (7): 5827–5842. https://doi.org/10.1109/JIOT.2019.2952146.

11 Pan, K., Ong, Y.-S., Gong, M. et al. (2024). Differential privacy in deep learning: a literature survey. *Neurocomputing* 589: 127663. https://doi.org/10.1016/j.neucom.2024.127663.

12 Wagh, S., Tople, S., Benhamouda, F. et al. (2021). Falcon: honest-majority maliciously secure framework for private deep learning. *Proceedings on Privacy Enhancing Technologies* 2021 (1): 188–208. https://doi.org/10.2478/popets-2021-0011.

13 Zhao, Y., Zhao, J., Yang, M. et al. (2021). Local differential privacy-based federated learning for internet of things. *IEEE Internet of Things Journal* 8 (11): 8836–8853. https://doi.org/10.1109/JIOT.2020.3037194.

14 Zheng, W., Popa, R. A., Gonzalez, J. E., and Stoica, I. (2019). Helen: maliciously secure coopetitive learning for linear models. Presented at the 2019 IEEE Symposium on Security and Privacy (SP), IEEE Computer Society (May), pp. 724–738. https://doi.org/10.1109/SP.2019.00045.

15 Zhou, J., Wu, N., Wang, Y. et al. (2023). A differentially private federatedv learning model against poisoning attacks in edge computing. *IEEE Transactions on Dependable and Secure Computing* 20 (3): 1941–1958. https://doi.org/10.1109/TDSC.2022.3168556.

16 Bourse, F., Minelli, M., Minihold, M., and Paillier, P. (2018). Fast homomorphic evaluation of deep discretized neural networks. In: *Advances in Cryptology – CRYPTO 2018: 38th Annual International Cryptology Conference, Santa Barbara, CA, August 19–23, 2018, Proceedings, Part III*, 483–512. Berlin, Heidelberg: Springer-Verlag. https://doi.org/10.1007/978-3-319-96878-0_17.

17 Rivest, R. L. and Dertouzos, M. (1978). On data banks and privacy homomorphisms. [Online]. https://www.semanticscholar.org/paper/ON-DATA-BANKS-AND-PRIVACY-HOMOMORPHISMS-Rivest-Dertouzos/c365f01d330b2211e74069120e88cff37eacbcf5 (accessed 9 June 2024).

18 Gentry, C. (2009). Fully homomorphic encryption using ideal lattices. In: *Proceedings of the Forty-First Annual ACM Symposium on Theory of Computing in STOC'09*, 169–178. New York, NY: Association for Computing Machinery. https://doi.org/10.1145/1536414.1536440.

19 Paillier, P. (1999). Public-key cryptosystems based on composite degree residuosity classes. In: *Advances in Cryptology — EUROCRYPT'99*, Lecture Notes in Computer Science, vol. 1592 (ed. J. Stern), 223–238. Berlin, Heidelberg: Springer Berlin Heidelberg. https://doi.org/10.1007/3-540-48910-X_16.

20 Elgamal, T. (1985). A public key cryptosystem and a signature scheme based on discrete logarithms. *IEEE Transactions on Information Theory* 31 (4): 469–472. https://doi.org/10.1109/TIT.1985.1057074.

21 Brakerski, Z., Gentry, C., and Vaikuntanathan, V. (2011). Fully homomorphic encryption without bootstrapping, 2011/277. https://eprint.iacr.org/2011/277 (accessed 9 June 2024).

22 Fan, J. and Vercauteren, F. (2012). Somewhat practical fully homomorphic encryption, 2012/144. https://eprint.iacr.org/2012/144 (accessed 9 June 2024).

23 Cheon, J.H., Kim, A., Kim, M. et al. (2017). Homomorphic encryption for arithmetic of approximate numbers. In: *Advances in Cryptology – ASIACRYPT 2017* (ed. T. Takagi and T. Peyrin), 409–437. Cham: Springer International Publishing. https://doi.org/10.1007/978-3-319-70694-8_15.

24 Bonawitz, K., Ivanov, V., Kreuter, B. et al. (2017). Practical secure aggregation for privacy-preserving machine learning. In: *Proceedings of the 2017 ACM*

SIGSAC Conference on Computer and Communications Security in CCS'17 (October), 1175–1191. New York, NY: Association for Computing Machinery https://doi.org/10.1145/3133956.3133982.

25 Dowlin, N., Gilad-Bachrach, R., Laine, K. et al. (2016). CryptoNets: applying neural networks to encrypted data with high throughput and accuracy. In: *Proceedings of the 33rd International Conference on International Conference on Machine Learning - Volume 48, in ICML'16*, 201–210. New York, NY: JMLR.org.

26 Mohassel, P. and Zhang, Y. (2017). SecureML: A system for scalable privacy-preserving machine learning. *2017 IEEE Symposium on Security and Privacy (SP)*, May, pp. 19–38. https://doi.org/10.1109/SP.2017.12.

27 Carpov, S., Dubrulle, P., and Sirdey, R. (2015). Armadillo: A compilation chain for privacy preserving applications. In: *Proceedings of the 3rd International Workshop on Security in Cloud Computing in SCC'15*, 13–19. New York, NY: Association for Computing Machinery (April). doi: 10.1145/2732516.2732520.

28 Jiang, X., Kim, M., Lauter, K., and Song, Y. (2018). Secure outsourced matrix computation and application to neural networks. *ACM Conference on Computer and Communications Security* 2018: 1209–1222. https://doi.org/10.1145/3243734.3243837.

29 Yan, H., Yin, M., Yan, C., and Liang, W. (2024). A survey of privacy preserving methods based on differential privacy for medical data. *2024 7th World Conference on Computing and Communication Technologies (WCCCT)* (April), pp. 104–108. https://doi.org/10.1109/WCCCT60665.2024.10541778.

30 Abadi, M., Chu, A., Goodfellow, I. et al. (2016). Deep learning with differential privacy. In: *Proceedings of the 2016 ACM SIGSAC Conference on Computer and Communications Security, in CCS'16*, 308–318. New York, NY: Association for Computing Machinery (October) https://doi.org/10.1145/2976749.2978318.

31 Amiri, S., Belloum, A., Klous, S., and Gommans, L. (2021). Compressive differentially-private federated learning through universal vector quantization. https://www.semanticscholar.org/paper/Compressive-Differentially-Private-Federated-Vector-Amiri-Belloum/c7709d94baad53768f2ddf67da7b0c1f3049cfe5 (accessed 9 June 2024).

32 McMahan, H. B., Moore, E., Ramage, D., et al. (2023). Communication-efficient learning of deep networks from decentralized data, January 26, *arXiv:1602.05629*. https://doi.org/10.48550/arXiv.1602.05629.

33 Wang, Z., Song, M., Zhang, Z., et al. (2019). Beyond inferring class representatives: user-level privacy leakage from federated learning. *IEEE INFOCOM* 2019 – *IEEE Conference on Computer Communications* (April), pp. 2512–2520. https://doi.org/10.1109/INFOCOM.2019.8737416.

34 Yao, A. C.-C. (1986). How to generate and exchange secrets. *27th Annual Symposium on Foundations of Computer Science (SFCS 1986)*, October, pp. 162–167. 10.1109/SFCS.1986.25.

35 Oded, G. (2009). *Foundations of Cryptography: Volume 2, Basic Applications*, 1e. Cambridge University Press.

36 Shamir, A. (1979). How to share a secret. *Communications of the ACM* 22 (11): 612–613. https://doi.org/10.1145/359168.359176.

37 Rabin, M., (2005). How to exchange secrets with oblivious transfer. *IACR Cryptol. ePrint Arch.*, https://www.semanticscholar.org/paper/How-To-Exchange-Secrets-with-Oblivious-Transfer-Rabin/772cdcc8a67cc878b39409230cbf2488a1117e62 (accessed 9 June 2024).

38 Pettai, M. and Laud, P. (2015). Combining differential privacy and secure multiparty computation. In: *Proceedings of the 31st Annual Computer Security Applications Conference, in ACSAC'15*, 421–430. New York, NY: Association for Computing Machinery (December). https://doi.org/10.1145/2818000.2818027.

39 Franklin, M. and Yung, M. (1992). Communication complexity of secure computation (extended abstract). In: *Proceedings of the Twenty-Fourth Annual ACM Symposium on Theory of Computing, in STOC'92*, 699–710. New York, NY: Association for Computing Machinery (July). https://doi.org/10.1145/129712.129780.

40 Damgard, I., Geisler, M., and Kroigard, M. (2008). Homomorphic encryption and secure comparison. *International Journal of Applied Cryptography* 1 (1): 22–31. https://doi.org/10.1504/IJACT.2008.017048.

41 Zahur, S., Rosulek, M., Evans, D. et al. (2015). Two halves make a whole. In: *Advances in Cryptology – EUROCRYPT 2015* (ed. E. Oswald and M. Fischlin), 220–250. Berlin, Heidelberg: Springer. https://doi.org/10.1007/978-3-662-46803-6_8.

42 Asharov, G., Lindell, Y., Schneider, T., and Zohner, M. (2017). More efficient oblivious transfer extensions. *Journal of Cryptology* 30 (3): 805–858. https://doi.org/10.1007/s00145-016-9236-6.

43 Juvekar, C., Vaikuntanathan, V., and Chandrakasan, A., (2018). {GAZELLE}: a low latency framework for secure neural network inference. Presented at the 27th USENIX Security Symposium (USENIX Security 18), pp. 1651–1669. https://www.usenix.org/conference/usenixsecurity18/presentation/juvekar (accessed 9 June 2024).

44 Mohassel, P. and Rindal, P. (2018). ABY3: A mixed protocol framework for machine learning. In: *Proceedings of the 2018 ACM SIGSAC Conference on Computer and Communications Security, in CCS'18*, 35–52. New York, NY: Association for Computing Machinery. https://doi.org/10.1145/3243734.3243760.

45 Ohrimenko, O., Schuster, F., Fournet, C., et al. (2016). Oblivious {Multi-Party} Machine Learning on Trusted Processors. Presented at the 25th USENIX Security Symposium (USENIX Security 16), pp. 619–636. https://www.usenix.org/conference/usenixsecurity16/technical-sessions/presentation/ohrimenko (accessed 9 June 2024).

46 Riazi, M.S., Weinert, C., Tkachenko, O. et al. (2018). Chameleon: a hybrid secure computation framework for machine learning applications. In: *Proceedings of the 2018 on Asia Conference on Computer and Communications Security in ASIACCS'18*, 707–721. New York, NY: Association for Computing Machinery (May). https://doi.org/10.1145/3196494.3196522.

47 Mohassel, P. and Zhang, Y. (2017). SecureML: a system for scalable privacy-preserving machine learning, 2017/396. https://eprint.iacr.org/2017/396 (accessed 28 July 2024).

48 Nikolaenko, V., Weinsberg, U., Ioannidis, S., et al. (2013). Privacy-preserving ridge regression on hundreds of millions of records. *2013 IEEE Symposium on Security and Privacy* (May), pp. 334–348. https://doi.org/10.1109/SP.2013.30.

49 Gascón, A., Schoppmann, P., Balle, B., et al. (2016). Privacy-preserving distributed linear regression on high-dimensional data, 2016/892. https://eprint.iacr.org/2016/892 (accessed 28 July 2024).

50 Ryffel, T., Trask, A., Dahl, M., et al. (2018). A generic framework for privacy preserving deep learning. November, *ArXiv*. https://www.semanticscholar.org/paper/A-generic-framework-for-privacy-preserving-deep-Ryffel-Trask/76c6d39edecdb943ce0f68f5a44e6608db96e383 (accessed 28 July 2024).

51 Ryffel, T., Trask, A., Dahl, M., et al. (2018). A generic framework for privacy preserving deep learning, November 13. *arXiv:1811.04017*. https://doi.org/10.48550/arXiv.1811.04017.

52 Zheng, W., Popa, R. A., Gonzalez, J. E., and Stoica, I. (2019). Helen: maliciously secure coopetitive learning for linear models, September 03, *arXiv:1907.07212*. https://doi.org/10.48550/arXiv.1907.07212.

53 Bonawitz, K., Ivanov, V., Kreuter, B., et al. (2017). Practical secure aggregation for privacy preserving machine learning, 2017/281. https://eprint.iacr.org/2017/281 (accessed 28 July 2024).

54 Wagh, S., Tople, S., Benhamouda, F., et al. (2020). FALCON: honest-majority maliciously secure framework for private deep learning, September 07, *arXiv:2004.02229*. https://doi.org/10.48550/arXiv.2004.02229.

55 Wagh, S. (2024). *snwagh/falcon-public*. C++, July 17. https://github.com/snwagh/falcon-public (accessed 28 July 2024).

56 Brumnik, R., Klebanova, T., Guryanova, L. et al. (2014). Simulation of territorial development based on fiscal policy tools. *Mathematical Problems in Engineering* 2014 (1): 843976. https://doi.org/10.1155/2014/843976.

57 Kalashnikov, V., Dempe, S., Mordukhovich, B., and Kavun, S.V. (2017). Bilevel optimal control, equilibrium, and combinatorial problems with applications to engineering. *Mathematical Problems in Engineering* 2017 (1): 7190763. https://doi.org/10.1155/2017/7190763.

8

Enhancing Cybersecurity in TinyML with Lightweight Cryptographic Algorithms

Oleksandr Kuznetsov[1,2], Roman Minailenko[3], Aigul Shaikhanova[4], Yelyzaveta Kuznetsova[2], and Agbotiname Lucky Imoize[5,6]

[1] *Department of Theoretical and Applied Sciences, eCampus University, Novedrate, Italy*
[2] *Department of Information and Communication Systems Security, V. N. Karazin Kharkiv National University, Kharkiv, Ukraine*
[3] *Department of Cyber Security and Software, Central Ukrainian National Technical University, Kropyvnytskyi, Ukraine*
[4] *Department of Information Security, L.N. Gumilyov Eurasian National University, Astana, Republic of Kazakhstan*
[5] *Department of Electrical and Electronics Engineering, Faculty of Engineering, University of Lagos, Lagos, Nigeria*
[6] *Department of Electrical Engineering and Information Technology, Institute of Digital Communication, Ruhr University, Bochum, Germany*

8.1 Introduction

The convergence of machine learning and edge computing has given rise to tiny machine learning (TinyML), a paradigm that enables the deployment of intelligent algorithms on highly resource-constrained devices [1, 2]. As TinyML applications proliferate across various domains, from healthcare to industrial monitoring, the need for robust security measures becomes increasingly critical [3–5]. This chapter explores the intersection of TinyML and cryptography, focusing on the application of lightweight cryptographic algorithms to enhance data security in microcontroller-based systems.

The unique challenges posed by TinyML environments stem from severe limitations in computational power, memory, and energy resources. Traditional cryptographic algorithms, designed for more powerful systems, often prove impractical in these constrained settings [6, 7]. Consequently, the development and implementation of lightweight cryptographic primitives have become essential for securing TinyML applications without compromising their efficiency or functionality.

Tiny Machine Learning: Design Principles and Applications, First Edition.
Edited by Agbotiname Lucky Imoize, Dinh-Thuan Do, and Houbing Herbert Song.

This chapter begins by examining the security landscape of TinyML, identifying potential vulnerabilities and attack vectors specific to edge-deployed machine learning models. We then delve into the state of the art in lightweight cryptography, focusing on three primary categories: block ciphers, hash functions, and stream ciphers. Each category is analyzed in depth, with particular attention given to algorithms such as PRESENT [8, 9], GIFT [10, 11], PHOTON [12–14], SPONGENT [15–17], Trivium [18–20], and Grain [18, 20, 21], which have shown promise in TinyML contexts.

A comparative analysis of these lightweight cryptographic algorithms provides insights into their security levels, computational demands, and energy consumption. This analysis is supported by simulation results and practical implementations, offering a comprehensive view of their performance in real-world TinyML scenarios. The trade-offs between security strength and resource utilization are carefully examined, providing guidance for algorithm selection based on specific application requirements.

The chapter further addresses the crucial aspect of integrating lightweight cryptographic tools into existing TinyML frameworks. We explore implementation strategies for popular platforms such as TensorFlow Lite for microcontrollers, Edge Impulse, and uTensor, demonstrating how cryptographic primitives can be seamlessly incorporated into the TinyML development pipeline. Case studies of secure TinyML deployments across various domains illustrate the practical applicability and effectiveness of these integrated solutions.

Our contributions to the field are multifaceted. Firstly, we provide a comprehensive and up-to-date survey of lightweight cryptographic algorithms specifically tailored for TinyML applications, offering a unique perspective on their applicability in highly constrained environments. Secondly, we present novel comparative analyses of these algorithms, considering not only their security properties but also their performance characteristics in the context of TinyML deployment. Thirdly, we offer insights into the integration of these cryptographic primitives with popular TinyML frameworks, bridging the gap between theoretical cryptography and practical implementation. Finally, we present case studies that demonstrate the real-world applicability of our findings, providing valuable guidance for researchers and practitioners in the field.

The structure of this chapter is as follows: Section 8.2 provides a comprehensive literature review, examining the current state of research in lightweight cryptography for TinyML. Section 8.3 delves into detailed analyses of lightweight block ciphers, hash functions, and stream ciphers, respectively. Sections 8.4–8.6 offer a comparative analysis of these algorithms, while Section 8.7 explores implementation strategies for lightweight cryptography in TinyML environments. Finally, Section 8.8 concludes the chapter with a summary of key findings and directions for future research.

8.2 Literature Review

The integration of lightweight cryptography with TinyML has gained significant attention in recent years, driven by the growing need for secure and efficient machine learning applications on resource-constrained devices. This section provides a comprehensive review of the current state of research in this rapidly evolving field.

The proliferation of Internet of Things (IoT) devices and their applications in various domains, including healthcare, industrial automation, and smart cities, has necessitated the development of secure and efficient communication protocols [22]. As Kamath and Renuka [23] highlight, there is an increasing demand for lightweight deep learning-based object detection models that can perform efficiently on edge devices with limited resources.

In the context of TinyML security, Patil et al. [24] provide an extensive analysis of security issues and potential solutions for TinyML devices in a networked environment. They discuss various security vulnerabilities, including adversarial attacks, malware attacks, privacy concerns, and supply chain attacks, emphasizing the importance of developing robust security measures for TinyML applications.

The development of lightweight cryptographic algorithms has been crucial in addressing the security challenges of resource-constrained devices. Singh et al. [25] discuss the importance of hardware implementation of lightweight cryptographic algorithms for IoT devices, radio-frequency identification (RFID) tags, and wireless sensor network nodes. They propose three architectures of the KLEIN lightweight cryptographic algorithm and evaluate their field-programmable gate array (FPGA) implementation.

Several researchers have focused on developing and analyzing specific lightweight cryptographic algorithms. Thabit et al. [26, 27] propose a new lightweight cryptographic algorithm for enhancing data security in cloud computing and provide a comprehensive security analysis and performance evaluation of their proposed algorithm. Similarly, Thakor et al. [28] present a novel 5-bit S-box design for lightweight cryptography algorithms, focusing on maintaining trade-offs among cost, performance, and security for resource-constrained IoT devices.

In the realm of hash functions, Windarta et al. [17] introduce two new lightweight cryptographic hash functions, Alit-Hash and Tjuilik-Hash, based on the Saturnin block cipher and the Beetle mode of operation. Their work demonstrates the potential of these hash functions for ensuring data integrity in IoT networks. Furthermore, Windarta et al. [7] provide a comprehensive review of lightweight cryptographic hash functions, discussing design trends, cryptanalytic attacks, and comparative analysis of different implementations.

The integration of cryptographic techniques with emerging technologies has also been explored. Krishnan et al. [29] present an embedded Subscriber Identity Module (eSIM) and blockchain-integrated secure zero-touch provisioning system for autonomous cellular IoTs in 5G networks, highlighting the potential of combining multiple technologies to enhance security in IoT ecosystems.

As the field of quantum computing advances, researchers are also considering its impact on cryptographic systems. Paramasivam et al. [30] propose quantum circuit implementations of two lightweight block ciphers, PUFFIN and PRINT, and analyze their resistance to quantum key recovery attacks using Grover's algorithm.

The application of machine learning techniques to enhance cybersecurity has been explored by SumanPrakash et al. [31], who propose a learning-driven continuous diagnostics and mitigation program for secure edge management through zero-trust architecture. Their work demonstrates the potential of using recurrent neural networks to diagnose and mitigate security issues autonomously.

In the context of privacy-preserving artificial intelligence (AI) in healthcare, Khalid et al. [32] summarize state-of-the-art approaches for preserving privacy in AI-based healthcare applications, discussing techniques such as federated learning and hybrid approaches.

The intersection of TinyML and 5G networks is explored by Saeed et al. [33], who provide a detailed overview of TinyML applications in the context of 5G networks, including autonomous vehicles, smart cities, and industrial automation.

As the field of lightweight cryptography for TinyML continues to evolve, researchers are focusing on developing algorithms that balance security, performance, and resource efficiency. Lo'ai et al. [34] propose a new and efficient lightweight cryptography algorithm for mobile and web applications, while Ramakrishna et al. [6] provide an analysis of lightweight cryptographic algorithms for IoT gateways, comparing various encryption algorithms using real-time data.

This literature review highlights the multifaceted nature of research in lightweight cryptography for TinyML, encompassing algorithm design, security analysis, implementation strategies, and integration with emerging technologies. The ongoing challenge lies in developing cryptographic solutions that provide robust security while maintaining the efficiency required for deployment on resource-constrained TinyML devices.

8.3 Lightweight Cryptographic Algorithms for TinyML

In the realm of TinyML, where computational resources are severely constrained, the implementation of robust cryptographic measures presents a significant challenge. This section explores a suite of lightweight cryptographic algorithms specifically designed to operate efficiently within the limitations of microcontroller-based systems. We examine block ciphers, hash functions, and stream ciphers that have been optimized for minimal resource consumption while maintaining a high level of security.

8.3.1 Block Ciphers

Block ciphers form the backbone of many cryptographic protocols, offering a balance between security and efficiency. In TinyML applications, where every byte of memory and every clock cycle counts, lightweight block ciphers play a crucial role. This section delves into three prominent lightweight block ciphers: PRESENT, SIMON/SPECK, and GIFT.

8.3.1.1 PRESENT

PRESENT, introduced by Bogdanov et al. [35], is an ultra-lightweight block cipher designed specifically for constrained environments. With a block size of 64 bits and key sizes of 80 or 128 bits, PRESENT offers a good trade-off between security and efficiency.

Structure: PRESENT employs a substitution-permutation network (SPN) with 31 rounds. Each round consists of three layers [35]:

1) **addRoundKey:** A simple bitwise XOR of the round key with the current state
2) **sBoxLayer:** Substitution using a 4-bit S-box applied 16 times in parallel
3) **pLayer:** A bit permutation for diffusion

The S-box is defined as follows [35]:

x 0 1 2 3 4 5 6 7 8 9 A B C D E F S[x] C 5 6 B 9 0 A D 3 E F 8 4 7 1 2

The permutation layer is described by the following rule:

$$P(i) = \{16i \bmod 63, \quad \text{for } 0 \leq i \leq 62 \quad 63, \quad \text{for } i = 63$$

Key Schedule: PRESENT uses a relatively simple key schedule. For the 80-bit key version [35]:

1) The key is stored in a key register K and represented as $k_{79}k_{78}...k_0$.
2) The 64-bit round key is extracted from the leftmost bits of K.
3) The key register is then updated as follows:
   a) $[k_{79}k_{78}...k_1k_0] = [k_{18}k_{17}...k_{20}k_{19}]$
   b) $[k_{79}k_{78}k_{77}k_{76}] = S[k_{79}k_{78}k_{77}k_{76}]$
   c) $[k_{19}k_{18}k_{17}k_{16}k_{15}] = [k_{19}k_{18}k_{17}k_{16}k_{15}] \oplus roundcounter$

Security and Performance: PRESENT has withstood extensive cryptanalysis, with the best-known attacks being marginal improvements over brute force. Its hardware implementation requires only 1,570 gate equivalents (GEs), making it one of the most compact block ciphers available.

8.3.1.2 SIMON and SPECK

SIMON and SPECK, developed by the U.S. National Security Agency (NSA) in 2013 [36, 37], are families of lightweight block ciphers optimized for hardware and software implementations, respectively.

SIMON [36, 37]:

- **Structure:** Feistel network
- **Block Sizes:** 32, 48, 64, 96, or 128 bits
- **Key Sizes:** 64, 72, 96, 128, 144, 192, or 256 bits
- **Number of Rounds:** Varies from 32 to 72, depending on block and key size

The round function of SIMON is defined as

$$R_k(x, y) = (y \oplus f(x) \oplus k, x)$$

where $f(x) = (x \lll 1) \otimes (x \lll 8) \oplus (x \lll 2)$.

SPECK [36, 37]:

- **Structure:** Add–rotate–XOR (ARX)
- **Block Sizes:** 32, 48, 64, 96, or 128 bits
- **Key Sizes:** 64, 72, 96, 128, 144, 192, or 256 bits
- **Number of Rounds:** Varies from 22 to 34, depending on block and key size

The round function of SPECK is defined as

$$R_k(x, y) = ((S^{-\alpha}(x) + y) \oplus k, S^{\beta}(y) \oplus (S^{-\alpha}(x) + y) \oplus k)$$

where S^j denotes a left circular shift by j bits and α and β are rotation constants that depend on the block size.

Both SIMON and SPECK offer excellent performance in resource-constrained environments. SIMON requires as few as 958 GE for the smallest variant, while SPECK can be implemented in as few as 45 bytes of random access memory (RAM) on 8-bit microcontrollers.

8.3.1.3 GIFT

GIFT, introduced by Banik et al. [38, 39], is a lightweight block cipher designed as an improvement over PRESENT. It maintains a similar structure but offers better security margins and efficiency.

Structure: GIFT uses a 64-bit block size and supports 128-bit keys. It employs an SPN structure with 28 rounds. Each round [38, 39] consists of

1) **SubCells:** A 4-bit S-box applied to each nibble of the state
2) **PermBits:** A bit permutation for diffusion
3) **AddRoundKey:** XOR with the round key and round constant

The S-box [38, 39] is defined as

x 0 1 2 3 4 5 6 7 8 9 A B C D E F S[x] 1 A 4 C 6 F 3 9 2 D B 7 5 0 8 E

The bit permutation is more efficient than that of PRESENT, requiring only six operations on 16-bit words [38, 39]:

$$P(16i) = 4i, \quad P(16i+1) = 4i+1, \quad P(16i+2) = 4i+2, \quad P(16i+3) = 4i+3$$

$$P(16i+4) = 4i+1, \quad P(16i+5) = 4i+2, \quad P(16i+6) = 4i+3, \quad P(16i+7) = 4i$$

$$P(16i+8) = 4i+2, \quad P(16i+9) = 4i+3, \quad P(16i+10) = 4i, \quad P(16i+11) = 4i+1$$

$$P(16i+12) = 4i+3, \quad P(16i+13) = 4i, \quad P(16i+14) = 4i+1, \quad P(16i+15) = 4i+2$$

Key Schedule: GIFT uses a simple key schedule where the 128-bit key is divided into two 64-bit halves. In each round, one half is used as the round key, and both halves are then updated using rotations and a round constant.

Security and Performance [38, 39]: GIFT offers improved security margins compared to PRESENT while maintaining similar efficiency. It requires approximately 1,345 GE for hardware implementation, making it highly suitable for resource-constrained TinyML applications.

In conclusion, these lightweight block ciphers provide robust security solutions for TinyML environments, balancing the need for strong encryption with the severe resource limitations of microcontroller-based systems. Their optimized designs enable efficient implementation in both hardware and software, making them ideal choices for securing TinyML applications against potential threats.

8.3.2 Hash Functions

Hash functions play a crucial role in cryptography, providing data integrity and serving as building blocks for various security protocols. In the context of TinyML, where computational resources are scarce, lightweight hash functions become indispensable. This section examines three state-of-the-art lightweight hash functions: PHOTON, SPONGENT, and QUARK, each designed to operate efficiently in constrained environments while maintaining robust security properties.

8.3.2.1 PHOTON

PHOTON, introduced by Guo et al. [40, 41], is a lightweight hash function family based on the sponge construction. It offers five variants with different security levels and hash sizes, making it versatile for various TinyML applications.

Structure: PHOTON uses an internal permutation function P, which operates on a state of $t = d \times d$ elements, where each element is s bits long. The state is represented as a $d \times d$ matrix. The five variants are [40, 41] as follows:

- PHOTON-80/20/16
- PHOTON-128/16/16
- PHOTON-160/36/36
- PHOTON-224/32/32
- PHOTON-256/32/32

The naming convention is PHOTON-$n/r/r'$, where n is the hash size, r is the input bitrate, and r' is the output bitrate.

Internal Permutation: The permutation P consists of 12 rounds, each comprising four operations [40, 41]:

1) **AddConstants:** XOR round-dependent constants to the first column of the state.
2) **SubCells:** Apply an s-bit S-box to each cell of the state.
3) **ShiftRows:** Cyclically shift each row of the state.
4) **MixColumnsSerial:** Apply a maximum distance separable (MDS) matrix to each column.

The S-box used depends on the variant, with the 4-bit S-box being identical to the one used in PRESENT.

Sponge Construction: PHOTON follows the sponge construction with a capacity $c = t - r$ bits. The input message is padded and divided into r-bit blocks. Each block is XORed into the first r bits of the state, followed by applying the permutation P. The output is extracted r' bits at a time, interleaved with applications of P.

Security and Performance: PHOTON offers collision resistance of $n/2$ bits and preimage resistance of n bits. Its hardware implementation requires between 865 and 2,177 GE, depending on the variant, making it one of the most compact hash functions available for TinyML applications.

8.3.2.2 SPONGENT

SPONGENT, proposed by Bogdanov et al. [42, 43], is another lightweight hash function family based on the sponge construction. It is designed to be particularly efficient in hardware implementations.

Structure: SPONGENT offers various configurations with hash sizes ranging from 88 to 256 bits. The internal state size varies from 88 to 272 bits. The naming convention is SPONGENT-$n/c/r$, where n is the hash size, c is the capacity, and r is the rate.

Internal Permutation: The permutation consists of b rounds, where b is the width of the permutation. Each round [42, 43] comprises

1) **sBoxLayer:** Apply a 4-bit S-box to each nibble of the state.
2) **pLayer:** Perform a bit permutation.
3) **addConstantsR:** XOR round constants.

4) **addConstantsC:** XOR constants derived from an linear feedback shift register (LFSR).

The S-box used in SPONGENT [42, 43] is

x 0 1 2 3 4 5 6 7 8 9 A B C D E F S[x] E D B 0 2 1 4 F 7 A 8 5 9 C 3 6

The bit permutation [42, 43] is defined as

$$P(i) = \{i \times b/4 \bmod b - 1, \ \text{for}\ 0 \leq i \leq b - 2 \ \ b - 1, \ \text{for}\ i = b - 1$$

where b is the permutation width.

Sponge Construction: SPONGENT follows the sponge construction similar to PHOTON, with the message being padded and absorbed r bits at a time, followed by the squeezing phase to produce the hash output.

Security and Performance: SPONGENT provides collision resistance of $n/2$ bits and preimage resistance of n bits. Its hardware implementation requires between 738 and 1,950 GE, depending on the variant, making it highly suitable for resource-constrained TinyML devices.

8.3.2.3 QUARK

QUARK, introduced by Aumasson et al. [44, 45], is a family of lightweight hash functions designed for extremely constrained environments. It combines design principles from cryptographic hash functions and stream ciphers.

Structure: QUARK offers three variants [44, 45]:

- **U-QUARK:** 136-bit state, 8-bit rate, and 128-bit capacity
- **D-QUARK:** 176-bit state, 16-bit rate, and 160-bit capacity
- **S-QUARK:** 256-bit state, 32-bit rate, and 224-bit capacity

Internal Permutation: QUARK's permutation function P is based on a non-LFSR (NLFSR) design. The state is divided into three registers: X, Y, and L. The permutation consists of the following steps [44, 45]:

1) Update X using a nonlinear Boolean function f.
2) Update Y using a nonlinear Boolean function g.
3) Update L using a linear feedback function.
4) Perform an output filter h to produce the output bit.

The Boolean functions f, g, and h are specific to each QUARK variant and are designed to provide nonlinearity and algebraic complexity.

For example, in U-QUARK [44, 45]:

$$\begin{aligned} f(X) = {} & x_0 + x_9 + x_{14} + x_{21} + x_{28} + x_{33} + x_{37} + x_{45} \\ & + x_{50} + x_{52} + x_{55} + x_0x_{28} + x_9x_{33} + x_{21}x_{55} + x_{28}x_{45} \\ & + x_{33}x_{37} + x_{45}x_{52} + x_{50}x_{55} + x_0x_5x_9 + x_{21}x_{28}x_{33} + x_{37}x_{45}x_{50} \end{aligned}$$

Sponge Construction: QUARK follows the sponge construction, with the message being padded and absorbed r bits at a time, followed by the squeezing phase to produce the hash output.

Security and Performance: QUARK provides collision resistance of $n/2$ bits and preimage resistance of n bits, where n is the hash size. Its hardware implementation requires between 1,379 and 2,296 GE, depending on the variant. While slightly larger than PHOTON and SPONGENT, QUARK offers excellent security properties and is particularly suited for applications requiring high security in extremely constrained environments.

In conclusion, these lightweight hash functions provide efficient and secure options for TinyML applications. Their optimized designs allow for implementation in resource-constrained devices while maintaining strong cryptographic properties. The choice between PHOTON, SPONGENT, and QUARK depends on the specific requirements of the TinyML application, balancing factors such as hash size, security level, and available resources.

8.3.3 Stream Ciphers

Stream ciphers are fundamental cryptographic primitives that generate a pseudorandom keystream, which is then combined with the plaintext to produce the ciphertext. In the context of TinyML, where resources are severely constrained, lightweight stream ciphers offer an attractive solution for securing data streams with minimal computational overhead. This section examines three prominent lightweight stream ciphers: Trivium, Grain, and MICKEY, each designed to provide robust security within the limitations of resource-constrained environments.

8.3.3.1 Trivium

Trivium, designed by De Cannière and Preneel [46, 47], is a synchronous stream cipher that stands out for its simplicity and efficiency in hardware implementations. It was selected as part of the eSTREAM portfolio for hardware-oriented stream ciphers.

Structure: Trivium operates on a 288-bit internal state, divided into three interconnected NLFSRs of lengths 93, 84, and 111 bits. It uses an 80-bit key and an 80-bit initialization vector (IV).

Initialization: The internal state is initialized as follows [46]:

$$(s_1, s_2, \dots, s_{93}) \leftarrow (K_0, \dots, K_{79}, 0, \dots, 0)$$

$$(s_{94}, s_{95}, \dots, s_{177}) \leftarrow (IV_0, \dots, IV_{79}, 0, \dots, 0)$$

$$(s_{178}, s_{179}, \dots, s_{288}) \leftarrow (0, \dots, 0, 1, 1, 1)$$

The state is then updated $4 \times 288 = 1{,}152$ times without producing output.

Keystream Generation: For each iteration, Trivium generates one bit of keystream and updates its internal state as follows [46]:

$$t_1 \leftarrow s_{66} \oplus s_{93}$$

$$t_2 \leftarrow s_{162} \oplus s_{177}$$

$$t_3 \leftarrow s_{243} \oplus s_{288}$$

$$z_i \leftarrow t_1 \oplus t_2 \oplus t_3$$

$$t_1 \leftarrow t_1 \oplus s_{91} \cdot s_{92} \oplus s_{171}$$

$$t_2 \leftarrow t_2 \oplus s_{175} \cdot s_{176} \oplus s_{264}$$

$$t_3 \leftarrow t_3 \oplus s_{286} \cdot s_{287} \oplus s_{69}$$

$$(s_1, s_2, \dots, s_{93}) \leftarrow (t_3, s_1, \dots, s_{92})$$

$$(s_{94}, s_{95}, \dots, s_{177}) \leftarrow (t_1, s_{94}, \dots, s_{176})$$

$$(s_{178}, s_{179}, \dots, s_{288}) \leftarrow (t_2, s_{178}, \dots, s_{287})$$

Security and Performance: Trivium offers 80-bit security against key recovery attacks. Its compact design requires only about 2,800 GEs in hardware, making it highly suitable for TinyML applications. The simplicity of its operations allows for efficient implementation on various platforms, including 8-bit microcontrollers commonly used in IoT devices.

8.3.3.2 Grain

Grain, proposed by Hell, Johansson, and Meier [47, 48], is another lightweight stream cipher from the eSTREAM portfolio. It exists in two variants: Grain-128 and Grain-128a, with the latter being an improved version offering authentication.

Structure: We will focus on Grain-128a, which uses a 128-bit key and a 96-bit IV. The cipher consists of three main components [48]:

1) A 128-bit LFSR
2) A 128-bit NLFSR
3) A nonlinear output function

Initialization: The LFSR is initialized with the 96-bit IV followed by 32 ones. The NLFSR is loaded with the 128-bit key. The cipher is then clocked 256 times without producing output, feeding the output back into both registers.

Keystream Generation: Let s_i and b_i denote the ith state bits of the LFSR and NLFSR, respectively. The update functions for each clock [48] are

LFSR:

$$s_{i+128} = s_i \oplus s_{i+7} \oplus s_{i+38} \oplus s_{i+70} \oplus s_{i+81} \oplus s_{i+96}$$

NLFSR:

$$b_{i+128} = s_i \oplus b_i \oplus b_{i+26} \oplus b_{i+56} \oplus b_{i+91} \oplus b_{i+96} \oplus b_{i+3}b_{i+67}$$
$$\oplus\, b_{i+11}b_{i+13} \oplus b_{i+17}b_{i+18} \oplus b_{i+27}b_{i+59} \oplus b_{i+40}b_{i+48} \oplus b_{i+61}b_{i+65} \oplus b_{i+68}b_{i+84}$$

The output function y_i is defined as [48]

$$y_i = h(x) \oplus s_{i+93} \oplus \bigoplus_{j \in A} b_{i+j}$$

where $A = 2, 15, 36, 45, 64, 73, 89$ and $h(x)$ is a nonlinear function of five state bits.

Security and Performance: Grain-128a offers 128-bit security against key recovery attacks. Its hardware implementation requires approximately 3,000 GE, making it suitable for resource-constrained TinyML devices. The design allows for efficient bit-sliced software implementations, which can be advantageous in certain microcontroller architectures.

8.3.3.3 MICKEY

Mutual Irregular Clocking KEYstream generator (MICKEY), designed by Babbage and Dodd [49], is a stream cipher that employs a novel clocking mechanism to enhance security. Like Trivium and Grain, it is part of the eSTREAM portfolio for hardware-oriented stream ciphers.

Structure: MICKEY 2.0, the improved version, uses an 80-bit key and an 80-bit IV. It consists of two 100-bit registers [49]:

1) **R:** A linear feedback shift register
2) **S:** A nonlinear feedback shift register

Both registers are irregularly clocked based on their own states.

Initialization: The registers are initialized to zero. The IV is then loaded into R, followed by the key being mixed into both registers over 80 clock cycles. Finally, the cipher is clocked 100 times without producing output.

Keystream Generation: The core of MICKEY's operation is the irregular clocking of its registers. For each clock cycle [49],

1) Compute input bit $i = R_{99} \oplus S_{99}$
2) Compute control bits $COMP_0 = R_0 \oplus S_0$ and $COMP_1 = R_0 \oplus S_0 \oplus 1$
3) Clock register R: If $COMP_0 = 0$: $R \leftarrow R \ggg 1$. If $COMP_0 = 1$: $R \leftarrow (R \ggg 1) \oplus FBLIN$
4) Clock register S: If $COMP_1 = 0$: $S \leftarrow S \ggg 1$. If $COMP_1 = 1$: $S \leftarrow (S \ggg 1) \oplus FBNL(S)$
5) Mix in input bits: $R_0 \leftarrow R_0 \oplus i$ and $S_0 \leftarrow S_0 \oplus i$

Here, $\ggg$ denotes rightward shift, FBLIN is a fixed linear feedback mask, and FBNL is a nonlinear feedback function.

The output keystream bit is $R_0 \oplus S_0$.

Security and Performance: MICKEY 2.0 offers 80-bit security against key recovery attacks. Its hardware implementation requires approximately 3,700 GE, slightly more than Trivium or Grain, but still suitable for many TinyML applications. The irregular clocking mechanism provides resistance against various cryptanalytic attacks, including algebraic and correlation attacks.

In conclusion, these lightweight stream ciphers offer efficient and secure options for encrypting data streams in TinyML environments. Their optimized designs allow for implementation in highly resource-constrained devices while maintaining strong cryptographic properties. The choice among Trivium, Grain, and MICKEY depends on the specific requirements of the TinyML application, balancing factors such as security level, implementation efficiency, and available resources.

8.4 Comparative Analysis of Lightweight Block Ciphers

This section provides a comprehensive comparison of the lightweight block ciphers PRESENT, SIMON/SPECK, and GIFT, evaluating their performance across various metrics crucial for TinyML applications. We analyze their security levels, computational efficiency, energy consumption, memory footprint, and latency/throughput characteristics to offer insights into their suitability for resource-constrained environments.

8.4.1 Security Levels and Resistance to Known Attacks

The security of lightweight block ciphers is paramount, even in resource-constrained environments. Table 8.1 summarizes the security levels and known attacks for each cipher.

Commentary: PRESENT demonstrates robust security, with the best attacks falling short of full key recovery. SIMON and SPECK, with their flexible key sizes,

Table 8.1 Security analysis of lightweight block ciphers.

Cipher	Key size (bits)	Security level	Best-known attack	Attack complexity
PRESENT	80	80 bits	Linear hull attack	$2^{75.4}$
	128	128 bits	Biclique attack	2^{126}
SIMON	64–256	Key size	Differential cryptanalysis	$2^{(n-1)}$ to $2^{(n-4)}$
SPECK	64–256	Key size	Differential–linear attack	$2^{(n-2)}$ to $2^{(n-5)}$
GIFT	128	128 bits	Differential cryptanalysis	$2^{113.16}$

Table 8.2 Computational efficiency and resource utilization.

Cipher	Implementation	Gate equivalents (GEs)	Cycles per block	Throughput at 100 kHz (kbps)
PRESENT	Hardware	1,570	32	200
SIMON	Hardware	958–2,064	44–72	88.9–145.5
SPECK	Software (8-bit)	N/A	210–384	16.7–30.5
GIFT	Hardware	1,345	28	228.6

offer security levels commensurate with their key lengths, though some variants show slight vulnerabilities to advanced cryptanalytic techniques. GIFT, designed as an improvement over PRESENT, exhibits strong resistance to known attacks, with the best attack requiring computational complexity close to exhaustive search.

8.4.2 Computational Efficiency and Resource Utilization

In TinyML contexts, the efficiency of cipher implementations is crucial. Table 8.2 compares the computational efficiency and resource utilization of these ciphers.

Commentary: PRESENT and GIFT show excellent hardware efficiency, with GIFT slightly outperforming PRESENT in both area and speed. SIMON's hardware implementation is notably compact for smaller variants but scales up for larger key sizes. SPECK, optimized for software, demonstrates competitive performance on 8-bit microcontrollers common in TinyML devices.

8.4.3 Energy Consumption Profiles

Energy efficiency is critical in battery-powered TinyML devices. Table 8.3 presents the energy consumption characteristics of these ciphers.

Table 8.3 Energy consumption analysis.

Cipher	Implementation	Energy per bit (pJ/bit)	Energy per block (nJ)
PRESENT	Hardware (130 nm)	2.6	166.4
SIMON	Hardware (130 nm)	2.1–4.5	134.4–288
SPECK	Software (8-bit AVR)	11.5–13.8	736–883.2
GIFT	Hardware (130 nm)	2.3	147.2

Commentary: In hardware implementations, SIMON, PRESENT, and GIFT all exhibit excellent energy efficiency. SIMON's energy consumption varies with key size, offering flexibility for different security requirements. SPECK's software implementation, while less energy-efficient than hardware solutions, remains competitive for software-based encryption in TinyML scenarios.

8.4.4 Memory Footprint Analysis

Memory constraints are often severe in TinyML environments. Table 8.4 compares the memory requirements of these ciphers.

Commentary: SIMON and SPECK demonstrate exceptionally low memory footprints, making them highly suitable for extremely constrained devices. PRESENT and GIFT, while requiring more read only memory (ROM), still maintain very low RAM usage, crucial for many microcontroller-based TinyML applications.

8.4.5 Latency and Throughput Considerations

Latency and throughput can be critical in real-time TinyML applications. Table 8.5 presents these metrics for the ciphers under consideration.

Commentary: In hardware implementations, GIFT offers the lowest latency and highest throughput, closely followed by PRESENT. SIMON's performance varies with key size but remains competitive. SPECK's software implementation, while slower, provides a good balance for software-only scenarios in TinyML devices.

Table 8.4 Memory footprint comparison.

Cipher	Implementation	ROM (bytes)	RAM (bytes)
PRESENT	Software (8-bit)	1,046	18
SIMON	Software (8-bit)	290–576	24–32
SPECK	Software (8-bit)	234–510	32–64
GIFT	Software (8-bit)	970	18

Table 8.5 Latency and throughput analysis.

Cipher	Implementation	Latency (cycles)	Max throughput (Mbps)
PRESENT	Hardware	32	200 (at 100 MHz)
SIMON	Hardware	44–72	145.5 (at 100 MHz)
SPECK	Software (8-bit)	210–384	3.05 (at 16 MHz)
GIFT	Hardware	28	228.6 (at 100 MHz)

In conclusion, this comparative analysis reveals that each of these lightweight block ciphers offers unique advantages for TinyML applications. PRESENT and GIFT excel in hardware implementations, offering a good balance of security, efficiency, and energy consumption. SIMON provides flexibility with its range of key sizes and performs well in both hardware and software. SPECK stands out for its software efficiency, making it an excellent choice for TinyML devices without dedicated cryptographic hardware. The selection of the most appropriate cipher depends on the specific constraints and requirements of the TinyML application at hand.

8.5 Comparative Analysis of Lightweight Hash Functions

This section provides a detailed comparison of the lightweight hash functions PHOTON, SPONGENT, and QUARK, evaluating their performance across various metrics crucial for TinyML applications. We analyze their security levels, computational efficiency, energy consumption, memory footprint, and latency/throughput characteristics to offer insights into their suitability for resource-constrained environments.

8.5.1 Security Levels and Resistance to Known Attacks

The security of hash functions is critical, even in resource-constrained environments. Table 8.6 summarizes the security levels and known attacks for each function.

Commentary: All three hash functions demonstrate strong security properties, with collision and preimage resistance levels appropriate for their output sizes. PHOTON and SPONGENT currently have no known attacks better than generic attacks on their full versions. QUARK variants have shown some vulnerability to conditional cube attacks, but the attack complexities remain close to the ideal security levels.

8.5.2 Computational Efficiency and Resource Utilization

In TinyML contexts, the efficiency of hash function implementations is crucial. Table 8.7 compares the computational efficiency and resource utilization of these functions.

Commentary: SPONGENT demonstrates the lowest hardware footprint across its variants, making it particularly suitable for extremely constrained devices. PHOTON offers a good balance between area efficiency and throughput, with its

Table 8.6 Security analysis of lightweight hash functions.

Function	Variants	Output size (bits)	Collision resistance	Preimage resistance	Best-known attack
PHOTON	80/20/16	80	2^{40}	2^{80}	Generic
	128/16/16	128	2^{64}	2^{128}	Generic
	160/36/36	160	2^{80}	2^{160}	Generic
	224/32/32	224	2^{112}	2^{224}	Generic
	256/32/32	256	2^{128}	2^{256}	Generic
SPONGENT	88/80/8	88	2^{44}	2^{88}	Generic
	128/128/8	128	2^{64}	2^{128}	Generic
	160/160/16	160	2^{80}	2^{160}	Generic
	224/224/16	224	2^{112}	2^{224}	Generic
	256/256/16	256	2^{128}	2^{256}	Generic
QUARK	U-QUARK	136	2^{68}	2^{136}	Conditional cube attack (2^{74})
	D-QUARK	176	2^{88}	2^{176}	Conditional cube attack (2^{94})
	S-QUARK	256	2^{128}	2^{256}	Conditional cube attack (2^{134})

larger variants showing competitive performance. QUARK, while requiring more gates, provides higher security margins and may be preferred in applications demanding the utmost security.

8.5.3 Energy Consumption Profiles

Energy efficiency is critical in battery-powered TinyML devices. Table 8.8 presents the energy consumption characteristics of these hash functions.

Commentary: SPONGENT demonstrates excellent energy efficiency across its variants, particularly for smaller output sizes. PHOTON shows competitive energy consumption, with its larger variants being particularly efficient. QUARK, while generally consuming more energy, offers a good balance for applications requiring higher security margins.

8.5.4 Memory Footprint Analysis

Memory constraints are often severe in TinyML environments. Table 8.9 compares the memory requirements of these hash functions in software implementations.

Table 8.7 Computational efficiency and resource utilization.

Function	Variant	Gate equivalents (GEs)	Throughput (bps) at 100 kHz	Cycles per byte
PHOTON	80/20/16	865	2,000	400
	128/16/16	1,122	1,600	500
	160/36/36	1,396	3,600	222
	224/32/32	1,736	3,200	250
	256/32/32	2,177	3,200	250
SPONGENT	88/80/8	738	800	1,000
	128/128/8	1,060	800	1,000
	160/160/16	1,329	1,600	500
	224/224/16	1,728	1,600	500
	256/256/16	1,950	1,600	500
QUARK	U-QUARK	1,379	800	1,000
	D-QUARK	1,702	1,600	500
	S-QUARK	2,296	3,200	250

Commentary: PHOTON and SPONGENT demonstrate very low memory footprints, making them highly suitable for microcontroller-based TinyML applications. QUARK requires more memory, reflecting its focus on higher security margins at the cost of increased resource usage.

8.5.5 Latency and Throughput Considerations

Latency and throughput can be critical in real-time TinyML applications. Table 8.10 presents these metrics for the hash functions under consideration.

Commentary: QUARK demonstrates the highest throughput across its variants, making it suitable for applications requiring faster hashing. PHOTON offers a good balance between latency and throughput, with performance improving for larger variants. SPONGENT, while having higher latency, still provides acceptable throughput for many TinyML applications.

In conclusion, this comparative analysis reveals that each of these lightweight hash functions offers unique advantages for TinyML applications. SPONGENT excels in hardware efficiency and low energy consumption, making it ideal for extremely constrained devices. PHOTON provides a well-rounded performance across various metrics, offering flexibility for different application requirements. QUARK, while more resource-intensive, offers higher security margins and better

Table 8.8 Energy consumption analysis.

Function	Variant	Energy per bit (pJ/bit)	Energy per block (nJ)
PHOTON	80/20/16	2.8	44.8
	128/16/16	3.7	59.2
	160/36/36	2.6	93.6
	224/32/32	2.9	92.8
	256/32/32	3.6	115.2
SPONGENT	88/80/8	3.1	24.8
	128/128/8	3.5	44.8
	160/160/16	2.2	35.2
	224/224/16	2.9	64.96
	256/256/16	3.2	81.92
QUARK	U-QUARK	3.9	31.2
	D-QUARK	2.9	46.4
	S-QUARK	2.4	76.8

Table 8.9 Memory footprint comparison.

Function	Variant	ROM (bytes)	RAM (bytes)
PHOTON	80/20/16	364	20
	128/16/16	480	32
	160/36/36	596	45
	224/32/32	748	64
	256/32/32	924	80
SPONGENT	88/80/8	364	22
	128/128/8	456	32
	160/160/16	548	40
	224/224/16	732	56
	256/256/16	824	64
QUARK	U-QUARK	1,107	68
	D-QUARK	1,472	88
	S-QUARK	2,086	136

Table 8.10 Latency and throughput analysis.

Function	Variant	Latency (cycles per hash)	Throughput (Mbps) at 100 MHz
PHOTON	80/20/16	708	11.3
	128/16/16	924	13.9
	160/36/36	1, 224	13.1
	224/32/32	1, 716	13.1
	256/32/32	1, 716	14.9
SPONGENT	88/80/8	990	8.9
	128/128/8	1, 650	7.8
	160/160/16	1, 925	8.3
	224/224/16	2, 800	8.0
	256/256/16	3, 300	7.8
QUARK	U-QUARK	544	25.0
	D-QUARK	704	25.0
	S-QUARK	1, 024	25.0

throughput, suitable for TinyML applications where security is paramount. The selection of the most appropriate hash function depends on the specific constraints and security requirements of the TinyML application at hand.

8.6 Comparative Analysis of Lightweight Stream Ciphers

This section provides a comprehensive comparison of the lightweight stream ciphers Trivium, Grain, and MICKEY, evaluating their performance across various metrics crucial for TinyML applications. We analyze their security levels, computational efficiency, energy consumption, memory footprint, and latency/throughput characteristics to offer insights into their suitability for resource-constrained environments.

8.6.1 Security Levels and Resistance to Known Attacks

The security of stream ciphers is paramount, even in resource-constrained environments. Table 8.11 summarizes the security levels and known attacks for each cipher.

Table 8.11 Security analysis of lightweight stream ciphers.

Cipher	Key size (bits)	IV size (bits)	Security level	Best-known attack	Attack complexity
Trivium	80	80	80 bits	Cube attack	$2^{79.7}$
Grain	128	96	128 bits	Dynamic cube attack	2^{125}
MICKEY	80	80	80 bits	Guess and determine	2^{77}

Table 8.12 Computational efficiency and resource utilization.

Cipher	Implementation	Gate equivalents (GEs)	Throughput (bits/cycle)	Power at 100 kHz (μW)
Trivium	Hardware	2,580	1	1.92
Grain	Hardware	1,857	1	3.3
MICKEY	Hardware	3,188	1	2.8

Commentary: All three ciphers demonstrate strong security properties, with attack complexities close to their key sizes. Trivium shows remarkable resistance to attacks, with the best-known attack barely improving over exhaustive search. Grain-128a, with its larger key size, offers a higher security margin suitable for long-term security needs. MICKEY, while slightly more vulnerable to guess-and-determine attacks, still maintains a high security level appropriate for most TinyML applications.

8.6.2 Computational Efficiency and Resource Utilization

In TinyML contexts, the efficiency of cipher implementations is crucial. Table 8.12 compares the computational efficiency and resource utilization of these ciphers.

Commentary: Grain demonstrates the most compact hardware implementation, making it particularly suitable for extremely constrained devices. Trivium offers a good balance between area efficiency and power consumption. MICKEY, while requiring more gates, provides additional security features that may be beneficial in certain TinyML scenarios.

8.6.3 Energy Consumption Profiles

Energy efficiency is critical in battery-powered TinyML devices. Table 8.13 presents the energy consumption characteristics of these ciphers.

Table 8.13 Energy consumption analysis.

Cipher	Implementation	Energy per bit (pJ/bit)	Energy for 1 KB encryption (nJ)
Trivium	Hardware (65 nm)	2.5	20.48
Grain	Hardware (65 nm)	3.7	30.72
MICKEY	Hardware (65 nm)	3.2	26.21

Table 8.14 Memory footprint comparison.

Cipher	Implementation	ROM (bytes)	RAM (bytes)
Trivium	8-bit AVR	428	36
Grain	8-bit AVR	674	32
MICKEY	8-bit AVR	3, 192	27

Commentary: Trivium demonstrates excellent energy efficiency, consuming the least energy per bit of encryption. Grain and MICKEY show comparable energy consumption, with Grain being slightly more energy-intensive due to its more complex structure. All three ciphers are suitable for energy-constrained TinyML applications, with the choice depending on the specific energy budget and security requirements.

8.6.4 Memory Footprint Analysis

Memory constraints are often severe in TinyML environments. Table 8.14 compares the memory requirements of these ciphers in software implementations.

Commentary: Trivium and Grain demonstrate very low memory footprints, making them highly suitable for microcontroller-based TinyML applications. MICKEY requires significantly more ROM, reflecting its more complex structure, but uses slightly less RAM. The choice between these ciphers may depend on whether ROM or RAM is the more constrained resource in a given TinyML device.

8.6.5 Latency and Throughput Considerations

Latency and throughput can be critical in real-time TinyML applications. Table 8.15 presents these metrics for the ciphers under consideration.

Table 8.15 Latency and throughput analysis.

Cipher	Implementation	Initialization (cycles)	Throughput (Mbps) at 100 MHz
Trivium	Hardware	1,152	100
Grain	Hardware	160	100
MICKEY	Hardware	180	100

Commentary: All three ciphers offer high throughput in hardware implementations, capable of generating one bit per clock cycle. Grain and MICKEY demonstrate lower initialization latency compared to Trivium, which may be advantageous in applications requiring frequent re-keying. However, Trivium's longer initialization phase contributes to its security and may be preferable in scenarios where key agility is less critical.

In conclusion, this comparative analysis reveals that each of these lightweight stream ciphers offers unique advantages for TinyML applications. Trivium excels in energy efficiency and offers a good balance of security and performance, making it suitable for a wide range of TinyML scenarios. Grain provides the most compact hardware implementation and low memory footprint, ideal for extremely resource-constrained devices. MICKEY, while more resource-intensive, offers additional security features that may be beneficial in certain applications.

The selection of the most appropriate stream cipher depends on the specific constraints and requirements of the TinyML application at hand. Factors such as available hardware resources, energy budget, security requirements, and the need for key agility should all be considered when choosing between these ciphers. Additionally, the ongoing research in cryptanalysis may reveal new insights into the security of these ciphers, emphasizing the importance of staying updated with the latest developments in the field of lightweight cryptography for TinyML applications.

8.7 Implementation Strategies for Lightweight Cryptography in TinyML

The implementation of lightweight cryptographic algorithms in TinyML environments presents unique challenges due to severe resource constraints. This section explores various strategies to optimize the performance and efficiency of cryptographic operations in TinyML systems, focusing on hardware acceleration techniques, software optimization methods, and hybrid approaches that leverage both hardware and software optimizations.

8.7.1 Hardware Acceleration Techniques

Hardware acceleration plays a crucial role in enhancing the performance and energy efficiency of cryptographic operations in TinyML devices. Several techniques have been developed to optimize hardware implementations:

- **Bit-Sliced Implementation:** This technique involves parallelizing operations by processing multiple bits simultaneously.
- **Pipelining:** By breaking down cryptographic operations into smaller stages and processing them concurrently, pipelining can significantly increase throughput.
- **Reconfigurable Logic:** FPGAs offer flexibility in implementing cryptographic algorithms.
- **Application-Specific Integrated Circuits (ASICs):** Custom-designed ASICs can provide optimal performance for specific cryptographic algorithms.

8.7.2 Software Optimization Methods

Software optimizations are essential for TinyML devices that lack dedicated cryptographic hardware. Several techniques can enhance the performance of software implementations:

- **Loop Unrolling:** This technique reduces loop overhead and enables better instruction pipelining.
- **Table-Based Implementations:** Precomputed tables can replace complex operations, trading memory for speed.
- **Instruction Set Extensions:** Leveraging processor-specific instructions can significantly accelerate cryptographic operations.
- **Memory-Efficient Implementations:** Techniques such as in-place operations and careful memory management can reduce RAM usage.

8.7.3 Hybrid Approaches Leveraging Both Hardware and Software Optimizations

Hybrid approaches combine hardware and software optimizations to achieve optimal performance in TinyML systems:

- **Crypto-Coprocessors:** Lightweight cryptographic coprocessors can offload computationally intensive operations from the main processor.
- **Partial Hardware Acceleration:** This approach accelerates critical portions of cryptographic algorithms in hardware while implementing less intensive operations in software.
- **Dynamic Algorithm Selection:** Systems can dynamically switch between hardware and software implementations based on available resources and security requirements.

- **Instruction Set Architecture (ISA) Extensions:** Custom instructions tailored for cryptographic operations can bridge the gap between hardware and software implementations.

In conclusion, the implementation of lightweight cryptography in TinyML systems requires a multifaceted approach that considers the unique constraints of each application. Hardware acceleration techniques provide significant performance benefits but may not be feasible in all scenarios due to cost or power constraints. Software optimizations offer flexibility and can achieve remarkable efficiency on general-purpose microcontrollers. Hybrid approaches, leveraging the strengths of both hardware and software optimizations, often provide the best balance of performance, energy efficiency, and flexibility for TinyML applications. As the field of TinyML continues to evolve, ongoing research in lightweight cryptography implementation strategies will play a crucial role in enabling secure and efficient machine learning on highly constrained devices.

8.8 Conclusion

The integration of lightweight cryptographic algorithms with TinyML applications represents a critical advancement in the field of edge computing and IoT security. This chapter has explored the intricate balance between security, performance, and resource utilization in highly constrained environments, highlighting the importance of tailored cryptographic solutions for TinyML deployments.

Our comprehensive analysis of lightweight block ciphers, hash functions, and stream ciphers has revealed that no single algorithm universally outperforms others across all metrics. Instead, the selection of an appropriate cryptographic primitive depends heavily on the specific requirements and constraints of the TinyML application at hand. PRESENT and GIFT have shown remarkable efficiency in hardware implementations, while SPECK demonstrates superior performance in software. In the realm of hash functions, SPONGENT offers an excellent balance between security and resource utilization, whereas PHOTON provides flexibility across various security levels. Among stream ciphers, Trivium stands out for its energy efficiency, while Grain offers the most compact hardware implementation.

The implementation strategies discussed herein underscore the importance of a multifaceted approach to optimizing cryptographic operations in TinyML systems. Hardware acceleration techniques, such as bit-sliced implementations and pipelining, can significantly enhance performance but may not be feasible in all scenarios due to cost or power constraints. Software optimizations, including loop unrolling and memory-efficient implementations, offer flexibility and can achieve remarkable efficiency on general-purpose microcontrollers. Hybrid approaches, leveraging both hardware and software optimizations, often provide the best balance of performance, energy efficiency, and flexibility for TinyML applications.

The integration of lightweight cryptography with popular TinyML frameworks such as TensorFlow Lite for microcontrollers, Edge Impulse, and uTensor has demonstrated the practical feasibility of securing machine learning models and data on resource-constrained devices. Case studies in healthcare monitoring, industrial anomaly detection, and voice authentication have showcased the real-world applicability and benefits of these integrated solutions.

Looking forward, several key areas warrant further research and development:

1) **Post-quantum Lightweight Cryptography:** As quantum computing advances, developing and integrating quantum-resistant lightweight algorithms for TinyML will become increasingly important.
2) **Dynamic Security Adaptation:** Future TinyML systems could benefit from the ability to dynamically adjust their security levels based on real-time threat assessments and available resources.
3) **Homomorphic Encryption for TinyML:** Exploring lightweight partially homomorphic encryption schemes could enable secure collaborative learning and inference in distributed TinyML networks.
4) **Side-Channel Attack Resilience:** Developing lightweight cryptographic implementations that are inherently resistant to power analysis and other side-channel attacks is crucial for securing TinyML devices in physically accessible environments.
5) **Standardization Efforts:** Establishing standards for lightweight cryptography in TinyML applications will facilitate interoperability and promote wider adoption of secure TinyML solutions.

In conclusion, the field of lightweight cryptography for TinyML is rapidly evolving, driven by the increasing demand for secure and efficient machine learning capabilities on edge devices [50]. As TinyML applications continue to proliferate across various domains, the development and integration of optimized cryptographic solutions will play a pivotal role in ensuring the security, privacy, and trustworthiness of these systems. By continuing to advance the state of the art in lightweight cryptography and its integration with TinyML frameworks, researchers and practitioners can pave the way for a new generation of secure [51], intelligent edge devices that can operate reliably in resource-constrained environments while maintaining robust security guarantees.

Acknowledgment

- This project has received funding from the European Union's Horizon 2020 research and innovation program under the Marie Skłodowska-Curie Grant

Agreement No. 101007820 – TRUST. This publication reflects only the author's view, and the REA is not responsible for any use that may be made of the information it contains.

- This research was funded by the European Union – NextGenerationEU under the Italian Ministry of University and Research (MIUR), National Innovation Ecosystem Grant ECS00000041-VITALITY-CUP D83C22000710005.

References

1 Oliveira, F., Costa, D.G., Assis, F., and Silva, I. (2024). Internet of Intelligent Things: a convergence of embedded systems, edge computing and machine learning. *Internet of Things* 26: 101153. https://doi.org/10.1016/j.iot.2024.101153.

2 Lin, J., Zhu, L., Chen, W.-M. et al. (2023). Tiny machine learning: progress and futures [feature]. *IEEE Circuits and Systems Magazine* 23 (3): 8–34. https://doi.org/10.1109/MCAS.2023.3302182.

3 Ahmed, Z.E., Hashim, A.A., Saeed, R.A., and Saeed, M.M. (2024). Chapter 18 - TinyML network applications for smart cities. In: *TinyML for Edge Intelligence in IoT and LPWAN Networks* (ed. B.S. Chaudhari, S.N. Ghorpade, M. Zennaro, and R. Paškauskas), 423–451. Academic Press https://doi.org/10.1016/B978-0-44-322202-3.00023-3.

4 Bhamare, M., Kulkarni, P.V., Rane, R. et al. (2024). Chapter 14 - TinyML applications and use cases for healthcare. In: *TinyML for Edge Intelligence in IoT and LPWAN Networks* (ed. B.S. Chaudhari, S.N. Ghorpade, M. Zennaro, and R. Paškauskas), 331–353. Academic Press https://doi.org/10.1016/B978-0-44-322202-3.00019-1.

5 Chaudhari, B.S., Ghorpade, S.N., Zennaro, M., and Paškauskas, R. (2024). Chapter 1 - TinyML for low-power Internet of Things. In: *TinyML for Edge Intelligence in IoT and LPWAN Networks* (ed. B.S. Chaudhari, S.N. Ghorpade, M. Zennaro, and R. Paškauskas), 1–12. Academic Press https://doi.org/10.1016/B978-0-44-322202-3.00006-3.

6 Ramakrishna, C.J., Reddy, D.B.K., Priya, B.K. et al. (2024). Analysis of lightweight cryptographic algorithms for IoT gateways. *Procedia Computer Science* 233: 235–242. https://doi.org/10.1016/j.procs.2024.03.213.

7 Windarta, S., Suryadi, S., Ramli, K. et al. (2022). Lightweight cryptographic hash functions: design trends, comparative study, and future directions. *IEEE Access* 10: 82272–82294. https://doi.org/10.1109/ACCESS.2022.3195572.

8 Rashidi, B. (2019). High-throughput and lightweight hardware structures of HIGHT and PRESENT block ciphers. *Microelectronics Journal* 90: 232–252. https://doi.org/10.1016/j.mejo.2019.06.012.

9 Zhong, Y. and Gu, J. (2024). Lightweight block ciphers for resource-constrained environments: a comprehensive survey. *Future Generation Computer Systems* 157: 288–302. https://doi.org/10.1016/j.future.2024.03.054.

10 Syed, I., Nazish, M., Sultan, I., and Banday, M.T. (2022). Implementation techniques for GIFT block cypher: a real-time performance comparison. In: *2022 Smart Technologies, Communication and Robotics (STCR)*, 1–5. IEEE https://doi.org/10.1109/STCR55312.2022.10009581.

11 Jangir, A., Shekhawat, D., and Pandey, J.G. (2021). An FPGA prototyping of the GIFT cipher for image security applications. In: *2021 4th International Conference on Security and Privacy (ISEA-ISAP)*, 1–6. IEEE https://doi.org/10.1109/ISEA-ISAP54304.2021.9689764.

12 Mangole, T., Helberg, A.S.J., and Nair, K.K. (2022). Resource usage evaluation of the PHOTON hash function. In: *2022 Conference on Information Communications Technology and Society (ICTAS)*, 1–6. IEEE https://doi.org/10.1109/ICTAS53252.2022.9744686.

13 Al-Shatari, M.O.A., Hussin, F.A., Aziz, A.A. et al. (2020). FPGA-based lightweight hardware architecture of the PHOTON hash function for IoT edge devices. *IEEE Access* 8: 207610–207618. https://doi.org/10.1109/ACCESS.2020.3038219.

14 Al-Shatari, M., Hussin, F.A., Aziz, A.A. et al. (2022). Composite lightweight authenticated encryption based on LED block cipher and PHOTON hash function for IoT devices. In: *2022 IEEE 15th International Symposium on Embedded Multicore/Many-core Systems-on-Chip (MCSoC)*, 134–139. IEEE https://doi.org/10.1109/MCSoC57363.2022.00030.

15 Apriani, M. and Sari, R.F. (2021). Performance comparison of spongent and photon hashing algorithms in ethereum-based blockchain system. In: *2021 7th International Conference on Electrical, Electronics and Information Engineering (ICEEIE)*, 564–569. IEEE https://doi.org/10.1109/ICEEIE52663.2021.9616831.

16 Rashidi, B. (2021). Efficient full data-path width and serialized hardware structures of SPONGENT lightweight hash function. *Microelectronics Journal* 115: 105167. https://doi.org/10.1016/j.mejo.2021.105167.

17 Windarta, S., Suryadi, S., Ramli, K. et al. (2023). Two new lightweight cryptographic hash functions based on saturnin and beetle for the Internet of Things. *IEEE Access* 11: 84074–84090. https://doi.org/10.1109/ACCESS.2023.3301128.

18 Li, B., Liu, M., and Lin, D. (2020). FPGA implementations of Grain v1, Mickey 2.0, Trivium, Lizard and Plantlet. *Microprocessors and Microsystems* 78: 103210. https://doi.org/10.1016/j.micpro.2020.103210.

19 Kitsos, P., Sklavos, N., Provelengios, G., and Skodras, A.N. (2013). FPGA-based performance analysis of stream ciphers ZUC, Snow3g, Grain V1, Mickey V2, Trivium and E0. *Microprocessors and Microsystems* 37 (2): 235–245. https://doi.org/10.1016/j.micpro.2012.09.007.

20 Jafarpour, A., Mahdlo, A., Akbari, A., and Kianfar, K. (2011). Grain and Trivium ciphers implementation algorithm in FPGA chip and AVR micro controller. In: *2011 IEEE International Conference on Computer Applications and Industrial Electronics (ICCAIE)*, 656–658. IEEE https://doi.org/10.1109/ICCAIE.2011.6162215.

21 de Ree, M., Mantas, G., and Rodriguez, J. (2023). Grain-128PLE: generic physical-layer encryption for IoT networks. In: *2023 IEEE Globecom Workshops (GC Wkshps)*, 1844–1849. IEEE https://doi.org/10.1109/GCWkshps58843.2023.10464732.

22 Awad, A.I., Fouda, M.M., Khashaba, M.M. et al. (2023). Utilization of mobile edge computing on the Internet of Medical Things: a survey. *ICT Express* 9 (3): 473–485. https://doi.org/10.1016/j.icte.2022.05.006.

23 Kamath, V. and Renuka, A. (2023). Deep learning based object detection for resource constrained devices: systematic review, future trends and challenges ahead. *Neurocomputing* 531: 34–60. https://doi.org/10.1016/j.neucom.2023.02.006.

24 Patil, R.Y., Bhamare, M., Patil, Y.H., and Bannore, A. (2024). Chapter 13 - Securing TinyML in a connected world. In: *TinyML for Edge Intelligence in IoT and LPWAN Networks* (ed. B.S. Chaudhari, S.N. Ghorpade, M. Zennaro, and R. Paškauskas), 311–330. Academic Press https://doi.org/10.1016/B978-0-44-322202-3.00018-X.

25 Singh, P., Acharya, B., and Chaurasiya, R.K. (2021). Chapter 8 - Lightweight cryptographic algorithms for resource-constrained IoT devices and sensor networks. In: *Security and Privacy Issues in IoT Devices and Sensor Networks*, Advances in ubiquitous sensing applications for healthcare (ed. S.K. Sharma, B. Bhushan, and N.C. Debnath), 153–185. Academic Press https://doi.org/10.1016/B978-0-12-821255-4.00008-0.

26 Thabit, F., Alhomdy, A.P.S., Al-Ahdal, A.H.A., and Jagtap, P.D.S. (2021). A new lightweight cryptographic algorithm for enhancing data security in cloud computing. *Global Transitions Proceedings* 2 (1): 91–99. https://doi.org/10.1016/j.gltp.2021.01.013.

27 Thabit, F., Alhomdy, S., and Jagtap, S. (2021). Security analysis and performance evaluation of a new lightweight cryptographic algorithm for cloud computing. *Global Transitions Proceedings* 2 (1): 100–110. https://doi.org/10.1016/j.gltp.2021.01.014.

28 Thakor, V.A., Razzaque, M.A., Darji, A.D., and Patel, A.R. (2023). A novel 5-bit S-box design for lightweight cryptography algorithms. *Journal of Information Security and Applications* 73: 103444. https://doi.org/10.1016/j.jisa.2023.103444.

29 Krishnan, P., Jain, K., Poojara, S.R. et al. (2024). eSIM and blockchain integrated secure zero-touch provisioning for autonomous cellular-IoTs in 5G networks. *Computer Communications* 216: 324–345. https://doi.org/10.1016/j.comcom.2023.12.023.

30 Paramasivam, S., Jenitha, J., Sanjana, S., and Haghparast, M. (2023). Compact quantum circuit design of PUFFIN and PRINT lightweight ciphers for quantum key recovery attack. *IEEE Access* 11: 66767–66776. https://doi.org/10.1109/ACCESS.2023.3289764.

31 SumanPrakash, P., Ramana, K.S., CosmePecho, R.D. et al. (2024). Learning-driven Continuous Diagnostics and Mitigation program for secure edge management through Zero-Trust Architecture. *Computer Communications* 220: 94–107. https://doi.org/10.1016/j.comcom.2024.04.007.

32 Khalid, N., Qayyum, A., Bilal, M. et al. (2023). Privacy-preserving artificial intelligence in healthcare: techniques and applications. *Computers in Biology and Medicine* 158: 106848. https://doi.org/10.1016/j.compbiomed.2023.106848.

33 Saeed, M.M., Saeed, R.A., and Ahmed, Z.E. (2024). Chapter 9 - TinyML for 5G networks. In: *TinyML for Edge Intelligence in IoT and LPWAN Networks* (ed. B.S. Chaudhari, S.N. Ghorpade, M. Zennaro, and R. Paškauskas), 167–229. Academic Press https://doi.org/10.1016/B978-0-44-322202-3.00014-2.

34 Lo'ai, T., Alicea, M., and Alsmadi, I. (2022). New and efficient lightweight cryptography algorithm for mobile and web applications. *Procedia Computer Science* 203: 111–118. https://doi.org/10.1016/j.procs.2022.07.016.

35 Bogdanov, A., Knudsen, L.R., Leander, G. et al. (2007). PRESENT: an ultra-lightweight block cipher. In: *Cryptographic Hardware and Embedded Systems - CHES 2007* (ed. P. Paillier and I. Verbauwhede), 450–466. Berlin, Heidelberg: Springer https://doi.org/10.1007/978-3-540-74735-2_31.

36 Ashur, T. and Luykx, A. (2021). An account of the ISO/IEC standardization of the simon and speck block cipher families. In: *Security of Ubiquitous Computing Systems: Selected Topics* (ed. G. Avoine and J. Hernandez-Castro), 63–78. Cham: Springer International Publishing https://doi.org/10.1007/978-3-030-10591-4_4.

37 Beaulieu, R., Shors, D., Smith, J. et al. (2015). SIMON and SPECK: block ciphers for the Internet of things. 2015/585 [Online]. https://eprint.iacr.org/2015/585 (accessed 28 July 2024).

38 Banik, S., Pandey, S.K., Peyrin, T. et al. (2017). GIFT: a small present. 2017/622 [Online]. https://eprint.iacr.org/2017/622 (accessed 28 July 2024).

39 Banik, S., Pandey, S.K., Peyrin, T. et al. (2017). GIFT: a small present. In: *Cryptographic hardware and embedded systems – CHES 2017* (ed. W. Fischer and N. Homma), 321–345. Cham: Springer International Publishing https://doi.org/10.1007/978-3-319-66787-4_16.

40 Guo, J., Peyrin, T., Poschmann, A., and Rogaway, P. (2011). The PHOTON family of lightweight hash functions. In: *Advances in Cryptology – CRYPTO 2011* (ed. P. Rogaway), 222–239. Berlin, Heidelberg: Springer https://doi.org/10.1007/978-3-642-22792-9_13.

41 Guo, J., Peyrin, T., and Poschmann, A. (2011). The PHOTON family of lightweight hash functions. 2011/609 [Online]. https://eprint.iacr.org/2011/609 (accessed 28 July 2024).

42 Bogdanov, A., Knežević, M., Leander, G. et al. (2011). spongent: a lightweight hash function. In: *Cryptographic hardware and embedded systems – CHES 2011* (ed. B. Preneel and T. Takagi), 312–325. Berlin, Heidelberg: Springer https://doi.org/10.1007/978-3-642-23951-9_21.

43 Bogdanov, A., Knezevic, M., Leander, G. et al. (2011). Spongent: the design space of lightweight cryptographic hashing. 2011/697 [Online]. https://eprint.iacr.org/2011/697 (accessed 28 July 2024).

44 Aumasson, J.-P., Henzen, L., Meier, W., and Naya-Plasencia, M. (2010). Quark: a lightweight hash. In: *Cryptographic Hardware and Embedded Systems, CHES 2010* (ed. S. Mangard and F.-X. Standaert), 1–15. Berlin, Heidelberg: Springer https://doi.org/10.1007/978-3-642-15031-9_1.

45 Aumasson, J.-P., Henzen, L., Meier, W., and Naya-Plasencia, M. (2013). Quark: a lightweight hash. *Journal of Cryptology* 26 (2): 313–339. https://doi.org/10.1007/s00145-012-9125-6.

46 De Cannière, C. (2006). Trivium: a stream cipher construction inspired by block cipher design principles. In: *Information Security* (ed. S.K. Katsikas, J. López, M. Backes, et al.), 171–186. Berlin, Heidelberg: Springer https://doi.org/10.1007/11836810_13.

47 The eSTREAM portfolio page [Online]. https://web.archive.org/web/20120813111955/http://www.ecrypt.eu.org/stream/portfolio_revision1.pdf (accessed 28 July 2024).

48 Hell, M., Johansson, T., Maximov, A. et al. (2008). The grain family of stream ciphers. In: *New Stream Cipher Designs: The eSTREAM Finalists* (ed. M. Robshaw and O. Billet), 179–190. Berlin, Heidelberg: Springer https://doi.org/10.1007/978-3-540-68351-3_14.

49 Babbage, S. and Dodd, M. (2008). The MICKEY stream ciphers. In: *New Stream Cipher Designs: The eSTREAM Finalists*, Lecture Notes in Computer Science, vol. 4986 (ed. M. Robshaw and O. Billet), 191–209. Berlin, Heidelberg: Springer https://doi.org/10.1007/978-3-540-68351-3_15.

50 Daradkeh, Y., Guryanova, L., Klebanova, T., and Kavun, S. (2012). Forecasting the cyclical dynamics of the development territories: conceptual approaches, models, experiments. *European Journal of Scientific Research* 74 (1): 5–20.

51 Brumnik, R., Klebanova, T., Guryanova, L. et al. (2014). Simulation of territorial development based on fiscal policy tools. *Mathematical Problems in Engineering* 2014 (1): 843976. https://doi.org/10.1155/2014/843976.

9

Tiny Machine Learning for Enhanced Edge Intelligence

Emmanuel Alozie[1], Agbotiname Lucky Imoize[2], Hawau I. Olagunju[3], Nasir Faruk[4,5], Salisu Garba[6], and Ayobami P. Olatunji[7]

[1] *Department of Information Technology, Sule Lamido University, Kafin Hausa, Jigawa, Nigeria*
[2] *Department of Electrical and Electronics Engineering, Faculty of Engineering, University of Lagos, Akoka, Lagos, Nigeria*
[3] *Department of Information Technology, Sule Lamido University, Kafin Hausa, Jigawa, Nigeria*
[4] *Department of Information Technology, Sule Lamido University, Kafin Hausa, Jigawa, Nigeria*
[5] *Directorate of Communication and Information Technology, Sule Lamido University, Kafin Hausa, Jigawa, Nigeria*
[6] *Department of Software Engineering, Sule Lamido University, Kafin Hausa, Jigawa, Nigeria*
[7] *Department of Computer Science, Western Illinois University, Macomb, IL, USA*

9.1 Introduction

Edge devices, comprising the Internet of Things (IoT) and embedded devices, are fast proliferating in every aspect of human activity, finding useful applications in areas such as smart farming in agriculture, wearable systems in healthcare, predictive maintenance in manufacturing, and autonomous vehicles in transportation, and enabling a vast network of devices to interact for real-time data collection and processing [1]. A major concern for deploying these edge devices is the need for autonomous intelligent decision-making with minimal reliance on centralized cloud infrastructure, especially when these devices operate in environments with unreliable network connectivity, thus requiring on-device learning and inference [2, 3]. However, deploying large ML models on these resource-constrained devices is challenging due to their limited processing power, memory, and energy.

The ML-as-a-Service architecture at the edge was proposed to address this challenge [4]. This architecture supports on-device learning and inference by enabling collaboration between resource-constrained edge IoT devices and gateways where the IoT node handles simple machine learning (ML) tasks while the gateway, with higher processing capabilities, handles more complex models and tasks. This approach, therefore, optimizes the gateway resources while

Tiny Machine Learning: Design Principles and Applications, First Edition.
Edited by Agbotiname Lucky Imoize, Dinh-Thuan Do, and Houbing Herbert Song.

preserving the energy of the edge devices. Despite the benefits of this approach, it also introduces some challenges and trade-offs, particularly in terms of increased complexity and device demands [5]. Tiny Machine Learning (TinyML) was proposed to address these challenges.

TinyML is an emerging domain that facilitates the deployment of large ML models on resource-constrained embedded edge devices, enabling decentralization, on-device learning, and inference. Although still in its early stages, TinyML has already attracted significant attention from both academia and industry. For example, the works [6–8] provided extensive reviews on the definition, challenges, and applications of TinyML across various sectors, while the role of TinyML in edge intelligence was explored in [9–11]. However, none of these works provided a case study to illustrate how large ML can be implemented on edge devices, particularly considering the TinyML-as-a-Service (TMLaaS) architecture. While TMLaaS for edge intelligence was presented in [5], including a case study, it did not provide the key TinyML optimization techniques for enhanced edge intelligence and only utilized two datasets to evaluate the performance of different ML classifiers, with only one deep learning (DL) model.

In this chapter, we provide an overview of TinyML for edge intelligence, examining its implementation workflow using TensorFlow Lite Micro (TF-Lite Micro) and key optimization techniques. The chapter also introduces the TMLaaS architecture with a focus on its design trade-offs, its integration with machine learning operations (MLOps), and validating its implementation using a case study involving different ML and DL models as well as three large datasets obtained under different scenarios for on-device edge learning and inference, providing practical insights for deploying intelligent models on edge IoT devices.

9.1.1 Key Contributions

The following are the significant contributions of this chapter:

i) The chapter examines the critical challenges of deploying large ML models on edge devices.
ii) TinyML overview, its workflow implementation using TF-Lite Micro, and techniques for enhanced edge intelligence.
iii) Introduction of TMLaaS, its design trade-offs, and validation of its ease of development and deployment on edge devices through a case study.
iv) Open research challenges in TinyML for edge intelligence have been identified and discussed, and future research directions have been provided.

9.1.2 Chapter Organization

Section 9.2 provides an overview of TinyML, the challenges of deploying large ML models in edge intelligence, TinyML workflow implementation and TFLite

Micro design principles, and different TinyML techniques for enhanced edge intelligence. Section 9.3 introduces TMLaaS, including its design trade-offs and its integration with MLOps, and presents the validation of TMLaaS architecture on edge devices through a case study. Section 9.3 presents the open challenges of TinyML for edge intelligence and proffers further research directions. Finally, Section 9.4 concludes the chapter.

9.2 Overview of Tiny Machine Learning (TinyML)

TinyML is a novel domain that facilitates the deployment of ML models on extremely resource-constrained devices, such as microcontroller units (MCUs), sensors, and edge devices, with lower processing power, memory, and energy capacity, typically operating at milliwatt power consumption levels [8]. This transformation opens up new frontiers in IoT, wearable technology, industrial automation, and other fields where real-time, on-device intelligence is crucial for low-latency, low-power applications.

9.2.1 Challenges of Large Machine Learning in Edge Intelligence

Deploying ML and DL models directly on edge devices for local learning, inference, and decision-making is edge intelligence [12]. Leveraging this integration, edge devices such as sensors, cameras, and MCUs can perform real-time data analysis and pattern recognition to make quality decisions while reducing latency, bandwidth usage, and improving privacy. For example, sensors on a self-driving car should be able to perform local learning and inference to be able to identify different kinds of obstacles and make decisions in real-time with low latency [2]. Similarly, gyroscopes and accelerometers on the MCU of an emergency alerting bracelet for the elderly should be able to differentiate between a fall and normal laying down in real-time with low latency. Figure 9.1 compares the centralized and edge intelligence architecture for deploying ML models in edge IoT devices. In centralized intelligence, raw data obtained from edge devices are uploaded to the cloud for inference and decision-making, while in edge intelligence, models are partially or fully deployed on edge devices, enabling local processing and minimal reliance on cloud infrastructure. However, deploying ML models on these edge devices is not without significant challenges. This subsection highlights and discusses these challenges.

9.2.1.1 Resource Constraints in Edge Devices

Resource constraint is one of the primary challenges in edge intelligence, as these edge devices, such as MCUs, sensors, and embedded systems, typically have only kilobytes (KB) to a few megabytes (MB) of RAM and storage with

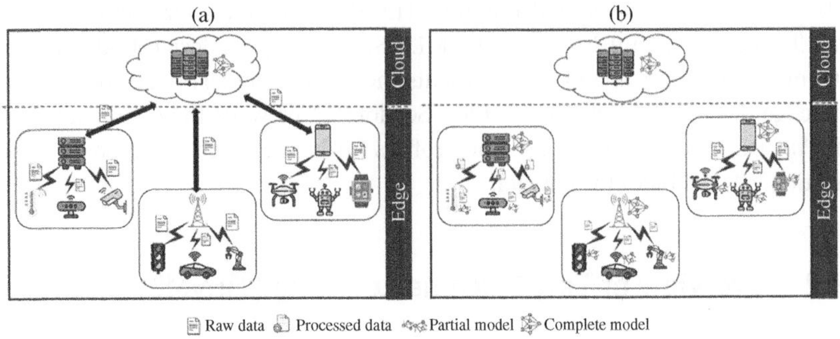

Figure 9.1 Comparison of central intelligence and edge intelligence architectures in IoT environments. (a) Central intelligence and (b) edge intelligence.

minimal energy consumption, functioning on small batteries [13]. Deploying a traditional ML model, especially DL models, on these devices can be difficult as these models require substantial RAM and storage that are not available, thus leading to memory overflow in edge devices and offloading computations to external servers, which increases latency and compromises real-time analysis and decision-making. Furthermore, these traditional ML models are energy-intensive and can drain the battery life of these devices, resulting in costly device maintenance and frequent battery replacements [14]. This necessitates the need to develop and deploy optimized models on edge devices.

9.2.1.2 Latency and Real-Time Inference Requirements

With the advancement of technology, there is an increase in the widespread adoption of edge devices in sectors where low-latency decision-making is paramount, for example, sensors on self-driving cars and wearable devices for real-time healthcare monitoring [15]. Even a minor delay in data processing can result in critical failures such as safety breaches or the loss of life. For example, sensors on a self-driving car failing to identify obstacles with minimal delay can result in catastrophic damage to property or loss of life. Deploying a traditional model on these devices, as mentioned earlier, can result in memory overflow and offloading processing to the cloud infrastructure, which can increase latency. Additionally, the balance between ML model complexity and inference speed is another fundamental challenge in edge intelligence, where complex models may achieve higher predictive accuracy but require substantial resources, thus increasing inference time, while optimized models might achieve lower accuracy, which can affect the reliability of the system [16]. Therefore, optimizing this balance while developing and deploying ML models on edge devices is critical.

9.2.1.3 Security and Privacy Concerns

A primary security and privacy concern in edge systems is data transmission vulnerability, especially when these devices are handling sensitive data, such as healthcare records and financial data [3]. Utilizing the centralized architecture, these devices are required to transmit the data to a cloud infrastructure or external server for processing and inference, thus exposing the system to data breaches and other security attacks [17]. Edge intelligence facilitates on-device learning and inference without the need to transmit sensitive data, thus preserving privacy and enhancing security. Furthermore, on-device security is another critical concern, as these edge devices, depending on where and how they operate, are vulnerable to both physical and cyber threats. Optimized ML can be utilized to detect and defend against these threats and enhance security [18, 19]; however, if not optimized, it can significantly degrade the performance of the system.

9.2.1.4 Heterogeneity in Edge Devices and Scalability

The heterogeneity of edge devices, in terms of varying levels of processing power, memory, and connectivity, is one of the significant challenges in edge intelligence systems as it complicates the development and deployment of unified ML models across such environments [12, 20]. Optimized ML models for one device may be incompatible with another, requiring extensive optimization or even reengineering to ensure they are compatible with a wide range of hardware configurations. This dampens the overall scalability of the system. Additionally, managing model deployment, updates, and operations while ensuring data synchronization across a vast network can result in operational bottlenecks and high network bandwidth demands. Therefore, maintaining model consistency and performance across all edge devices while preventing network overloads is crucial.

TinyML has been proposed to address these challenges, enabling the development and deployment of efficient ML/DL models for on-device learning and inference on edge devices, thus minimizing the need for offloading computations, reducing latency, and ensuring scalability in a heterogeneous edge environment.

9.2.2 TinyML Workflow Implementation

TinyML workflow implementation is categorized into three phases, which include the training, optimization, and deployment phases [9]. Figure 9.2 presents a pictorial representation of this workflow, including the three phases. The workflow begins with the training phase, where large datasets are obtained, often using sensors or any other hardware, and the datasets will then be utilized to develop the model using the TensorFlow library. This training phase can be conducted on high-end local workstations or cloud-based platforms such as Google Colab or the Microsoft Azure Machine Learning platform. The trained model is then

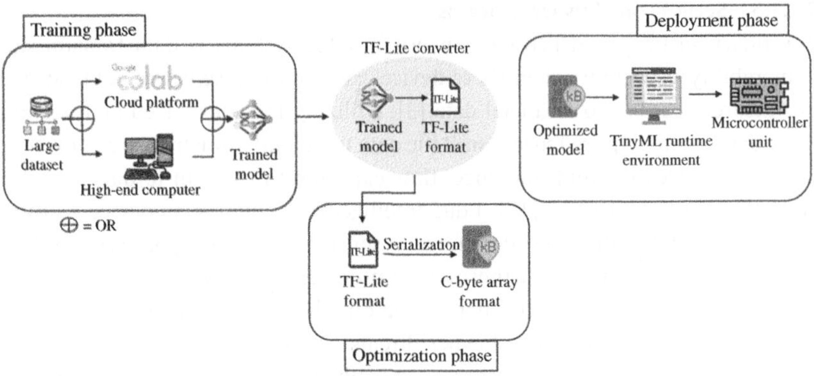

Figure 9.2 TinyML workflow implementation.

converted to a TF-Lite Micro format using the TF-Lite converter. This conversion step is very crucial, as it transforms the highly complex model into a lightweight version optimized for deployment on embedded edge devices. Post-conversion, the TF-Lite model is serialized into a C-byte array format, a compact representation of the model suitable for embedding within the firmware of edge devices. For the deployment phase, the optimized model is executed on the MCU using the TinyML runtime environment, which is typically facilitated by the TF-Lite Micro interpreter.

This limited operation set is managed by the OpResolver class, which allows the developer to explicitly specify the operators required for the ML model's inference. By linking only the necessary operators, the OpResolver significantly reduces the final binary size of the compiled firmware, optimizing for both memory footprint and execution speed. Despite the vast potential of TinyML, two critical challenges hinder its widespread adoption: interoperability and performance characterization [5].

- **Interoperability:** Edge devices are characterized by significant heterogeneity. For example, numerous MCU vendors offer a wide range of architectures, such as ARM Cortex-M, RISC-V, and ESP32, resulting in platform-specific TinyML implementation approaches. This diversity hinders the scalability of TinyML solutions across different hardware platforms. Each deployment often requires manual adjustments, vendor-tailored optimizations, or hardware-dependent code generation, which increases development complexity and restricts portability across devices.
- **Performance Characterization:** Due to the lack of standardization and uniform framework, it is difficult to evaluate the hardware performance of TinyML solutions in a neutral, non-vendor-based manner. When enhanced

performance is achieved, it is difficult to distinguish whether it is software- or hardware-based, complicating cross-platform comparisons and generalizing results.

To address these challenges, the TF-Lite Micro was developed, a framework designed specifically for running ML models on embedded edge devices with very low computational resources [21].

9.2.3 TinyML Implementation and TF-Lite Micro Design Principles

TF-Lite Micro, a lightweight inference engine, was designed specifically to support efficient on-device inference while minimizing resource overhead in terms of memory footprint, computational load, and energy consumption [22]. It assumes that models, input data, and output arrays are already stored within the device's limited memory, enabling it to perform computations directly on these arrays. This direct memory operation is crucial in environments where microcontrollers typically have only a few kilobytes of RAM available for all processing tasks. A key architectural distinction between TF-Lite Micro and traditional code generation approaches is its interpreter-based execution model, which dynamically handles data flow and the ML model computation graph, executing tensor operations during inference at runtime. Interpreter-based code generation architecture has several key advantages over static code generation, including the following:

1) **Model Update Flexibility:** Using the interpreter-based architecture, ML models can be updated or replaced without the need to rebuild the entire system because the architecture allows the ML models to operate independently from the main firmware running on the device.
2) **Cross-Platform Portability:** The interpreter-based architecture also facilitates the deployment of the same optimized ML model across different hardware platforms without the need for device-specific code generation, thus enhancing scalability and portability.
3) **Faster Development Cycles:** The interpreter-based architecture makes the development process faster because, with every change made, it does not require the entire system to be rebuilt, thus allowing swift prototyping and experimentation of new models or updates.
4) **Modular Model Optimization:** The interpreter-based architecture allows selective optimization techniques, such as model quantization or pruning, to be applied without needing to rebuild the system from scratch, thus facilitating the fine-tuning or partial optimization of models based on the specific requirements and providing more control over the balance between models' predictive accuracy and performance.

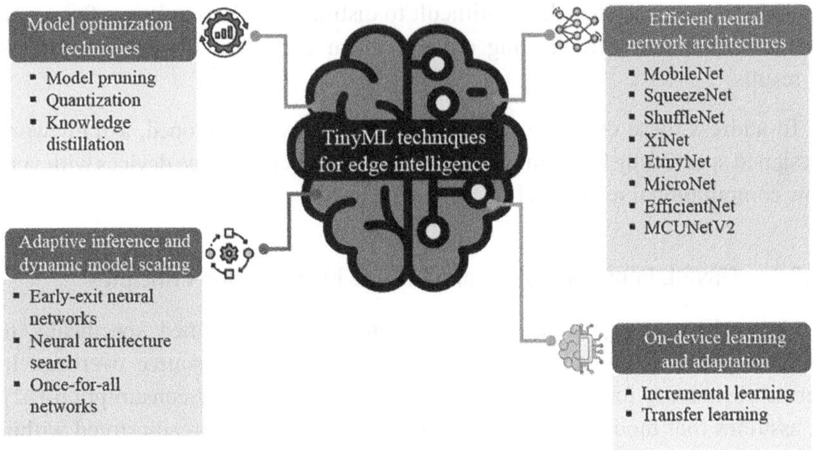

Figure 9.3 TinyML techniques for enhanced edge intelligence.

Additionally, the TF-Lite Micro architecture facilitates computational efficiency and minimizes the need for data movement across different memory hierarchies by operating directly on the in-memory tensors, ensuring that the overall system is optimized for resource-constrained embedded edge devices.

9.2.4 TinyML Techniques for Enhanced Edge Intelligence

To achieve efficient and enhanced edge intelligence, TinyML leverages a set of specialized techniques aimed at optimizing model performance, minimizing memory and power usage, and enabling adaptability at the edge. Figure 9.3 presents TinyML techniques for edge intelligence. This section further discusses these techniques that can be utilized to address the aforementioned challenges.

9.2.4.1 Model Optimization Techniques

One of the major challenges in edge intelligence, as mentioned earlier, is resource constraints, and optimizing ML models for these resource-constrained environments is central to the success of TinyML. This is achieved using different techniques, including model pruning, quantization, and knowledge distillation [9].

Model Pruning Model pruning is a technique that reduces the number of parameters in a neural network by removing unnecessary connections, neurons, or entire layers, which minimally contribute to the overall model accuracy [23]. This approach has enabled the deployment of neural networks in resource-constrained environments, such as embedded systems, by reducing the computational complexity of the models. Pruning techniques have been widely studied and classified in the literature. Vadera and Ameen [24] categorized pruning methods into eight groups based on their underlying mechanisms: magnitude-based pruning, similarity and clustering methods, sensitivity analysis, knowledge distillation, low-rank methods, quantization, architectural design, and hybrid approaches. Liang et al. [23] further categorized modern pruning techniques based on structure (structured vs. unstructured), target (neuron vs. connection pruning), and implementation (static vs. dynamic pruning).

Several studies have explored the application of model pruning techniques specifically for TinyML. For instance, depth pruning for TinyML was presented in [25], where an auxiliary network was used as a new head for the pruned model. This method significantly reduced model size with minimal accuracy loss, validated through on-device experiments. Similarly, a differentiable structured pruning method for convolutional neural networks (CNNs) was introduced in [26], integrating microcontroller-specific resource usage with parameter importance feedback to obtain highly compressed yet accurate models. The method outperformed existing MCU-specific compression techniques, proving effective in resource-constrained environments. He et al. [27] proposed an AutoML-based framework for model compression (AMC) that uses reinforcement learning to automate the search for optimal network architectures within hardware constraints, demonstrating significant compression without sacrificing accuracy.

Another advanced pruning method, unified DNAS for compressible (UDC) neural networks, was introduced in [28]. It searches both architecture choices and applies aggressive per-layer quantization and pruning, achieving superior performance compared to traditional methods. Specialized libraries like Deploying TinyML Models (DTMM) have been developed for deploying TinyML models on low-end IoT devices such as MCUs, leveraging pruning techniques to enhance performance while maintaining low memory usage [29]. These advanced pruning methods have proven particularly useful for deploying TinyML models on microcontrollers, allowing efficient image recognition and other tasks without the need for extensive retraining.

Quantization Quantization reduces the precision of model parameters, typically converting 32-bit floating-point values to lower-precision 8-bit integers, leading to significant savings in both memory and computational requirements [30]. Techniques such as post-training quantization and quantization-aware training (QAT) are employed to minimize performance loss. In post-training quantization, the model is first trained in full precision and then quantized after training is completed. QAT, on the other hand, simulates quantization effects during training, allowing the model to learn representations robust to precision loss [31].

Other quantization techniques have been proposed in the literature, particularly for TinyML. For example, [32] proposed a quantization-guided training (QGT) method to guide deep neural network (DNN) training toward optimized low-bit-precision targets, achieving extreme compression levels below 8-bit precision. Unlike standard QAT, QGT uses customized regularization to encourage weight distributions that maximize accuracy while reducing quantization errors. Similarly, [33] investigated seven quantization schemes, categorized into mixed-precision and uniform-precision approaches, on various neural network architectures, including MobileNetV1, ResNet-50, Xception, VGG16, and three custom-built CNNs. Mixed-precision schemes applied different bit widths to different network sections, while uniform schemes used fixed bit widths (8-bit, 4-bit, 2-bit). Using CIFAR-10 and QAT through QKeras, the study found that all schemes, except uniform 2-bit, significantly reduced memory with minimal performance loss.

Building on the challenge of deploying complex neural network models on memory-constrained devices [34], explored optimizing DL models within the TinyML framework. This study examined the effectiveness of four quantization techniques – Dynamic Range Quantization (DRQ), Full Integer Quantization (FIQ), Float16 Quantization, and Integer 16/8 Quantization – on popular architectures such as ResNetV1/V2 and MobileNetV1/V2, along with a custom-designed model. Using CIFAR-10 and MNIST datasets, the authors demonstrated that quantization plays a crucial role in reducing model size while maintaining acceptable performance, offering valuable insights for deploying DL models in resource-limited environments.

Knowledge Distillation Knowledge distillation is another technique that has proven to be effective in TinyML deployments. In this approach, a large, high-accuracy (teacher) model trains a smaller (student) model, transferring knowledge to it through a softened output probability distribution [35]. The smaller model retains much of the performance of the larger model but is computationally more efficient, making it suitable for edge inference. Knowledge distillation was explored in [36] for on-device training, focusing on single- or few-shot learning in resource-limited environments. The study employed histogram analysis and dimensionality reduction for dataset distillation, allowing decision-making at the edge. It was applied to a Radio Frequency Identification (RFID) dataset, where it achieved good performance compared to other state-of-the-art models, while reducing computational and memory usage, indicating its practical utilization for edge deployment.

Building on the principle of knowledge distillation, Jiao et al. [37] introduced transformer distillation, a method utilized for compressing BERT-based models by transferring knowledge from a large teacher BERT model to TinyBERT, a smaller student model, through a two-stage learning framework that comprised both general and task-specific knowledge. Analysis showed that TinyBERT, with 4 layers, achieved 96.8% of large-BERT's performance based on the GLUE benchmark while being 7.5 times smaller and 9.4 times faster, indicating its high efficiency for inference on resource-limited devices. This optimization technique provides a balance between TinyML model performance and resource efficiency, thus making it possible to deploy resource-intensive models, such as DL models, on embedded edge devices.

9.2.4.2 Efficient Neural Network Architectures

Efficient neural network architectures are highly optimized neural networks specifically tailored for TinyML applications. They are already optimized for memory, low latency, and computational efficiency, facilitating real-time inference on resource-limited devices.

MobileNet is one of these highly optimized neural network architectures, which reduces the number of parameters and operations per convolutional layer using depthwise separable convolutions [38]. It provides a good balance between model complexity and accuracy, making it well-suited for image classification and object detection tasks on both mobile and embedded devices. Enhancing its performance and efficiency further, the MobileNetV2 and MobileNetV3 versions introduced features such as lightweight attention modules and inverted residual blocks, where MobileNetV3 has proven very effective in real-world TinyML applications such as smartphone-based vision systems and smart camera systems [39–41]. TinyML models, SqueezeNet, AlexNet, CNN, MobileNetV2, and MobileNetV3, were applied for the detection of driver drowsiness in [42], where different quantization techniques were utilized, including QAT, FIQ, and DRQ. Results showed that MobileNetV2 achieved the highest accuracy of 99.64% while CNN achieved the lowest size of 50 kb.

SqueezeNet, another commonly utilized optimized neural network architecture, was introduced by Iandola et al. [43]. Compared to larger architectures such as AlexNet, SqueezeNet reduces memory requirements while maintaining high accuracy because it minimizes the number of parameters using "squeeze" and "expand" layers instead of conventional convolutions, making it well-suited for resource-limited devices. SqueezeNet achieves a higher accuracy of 99% compared to MobileNetV2 in [44] when utilized for real-time product quality inspection running at 5 and 2 frames-per-second on edge devices. ShuffleNet, introduced by Zhang et al. [45], is another commonly utilized efficient architecture that can be employed for group convolutions and channel shuffling to reduce computational complexity while maintaining high accuracy. It is well-suited for real-time video analytics on edge devices [46]. EfficientNet [47] and MCUNetV2 [48] are also commonly utilized architectures.

Several other efficient neural network architectures have been proposed in the literature. For example, XiNet, a novel efficient architecture, was proposed in [49] for edge devices, which enhances performance by optimizing parameters, operations, and memory usage. The architecture achieved 70% energy reduction while maintaining good performance on object detection benchmarks. EtinyNet, another novel architecture, was proposed in [50] for low-cost, resource-limited edge devices, where analysis showed that it achieved 57% ImageNet accuracy and 56.4% mean average precision on Pascal Visual Object Classes (VOC) with a 340 kb model size. The architecture also achieved 5.6 times greater power efficiency on a Field-Programmable Gate Array (FPGA) compared to previous designs. MicroNet, proposed in [51], was optimized using neural architecture search (NAS) to improve latency and energy efficiency for TinyML tasks such as visual wake words and keyword spotting.

9.2.4.3 On-Device Learning and Adaptation

Traditional ML systems use pretrained models deployed on edge devices. Recent advancements enable on-device learning and adaptation, allowing these devices to update their behavior in real time based on new data. This capability is crucial for dynamic environments where conditions change over time. On-device training not only provides personalized services but also protects user privacy by keeping data local [6]. Two primary approaches – incremental learning and transfer learning (TL) – have enabled significant advancements in on-device learning.

Incremental Learning Incremental learning allows a model to continuously learn from new data while retaining previously acquired knowledge [52]. This is essential for edge devices operating in environments with evolving data patterns. For example, a TinyML system deployed in a smart home for anomaly detection might encounter new types of events or conditions as the environment changes. A major challenge in incremental learning is catastrophic forgetting, where a model forgets previously learned information when adapted to new tasks or data distributions [53]. Recent research has introduced solutions to address catastrophic forgetting in TinyML. The work [34] developed an algorithm for TinyML incremental learning, utilizing knowledge distillation to prevent forgetting while dynamically adjusting the model size for efficiency. Evaluated on five different datasets, it achieved only a 1% accuracy drop while using just 43% of the floating-point operations per second (FLOPS) and 1% of the original data. Similarly, [54] proposed a method for onboard Class Incremental Learning (CIL) in TinyML, using Latent Replays and a Genetic Algorithm (GA) to handle new class data and prevent catastrophic forgetting. Evaluated on a gesture recognition task, the approach effectively adapted deep learning models on resource-limited devices, enhancing CIL performance.

PhiNets for class-incremental continual learning on edge devices were evaluated in [55], focusing on adapting models to dynamic environments. Evaluated on CORe50 and CIFAR-10, PhiNet outperformed MobileNet, achieving 4.47% higher accuracy on CIFAR-10 while using just 0.012% of the computation, demonstrating both efficiency and effectiveness for real-world applications. The work [56] proposed an incremental learning algorithm for TinyML that enables on-device learning using TL and k-nearest neighbor (KNN). It allows embedded systems and IoT devices to learn from the field, adapting to new conditions. Tested on image/audio benchmarks, the approach proved feasible and effective on two hardware platforms. The work [57] presented Train++, an incremental training algorithm for memory-constrained MCU-based IoT devices. Train++ enables local, on-device learning for high-dimensional data, addressing concept drift and improving accuracy by 5.15–7.3%. It reduced training time (10–226 seconds) and achieved real-time inference (1 ms), even on devices with limited memory.

Transfer Learning (TL) TL enables a pretrained model to be fine-tuned for a specific task or domain with minimal additional training. Unlike incremental learning, which continuously learns from new data while retaining previously acquired knowledge, TL utilizes previously acquired knowledge to help learn new knowledge and focuses on improving the performance on the new knowledge while neglecting the performance of the previous knowledge [52]. In TinyML, TL is particularly valuable because it allows large-scale models trained on powerful servers to be adapted to smaller, task-specific models that can run on resource-constrained edge devices. For instance, the work [58] introduced a CNN-LSTM-TL architecture to integrate TinyML with TL for Human Activity Recognition (HAR), utilizing pretrained models fine-tuned on edge devices. The approach reduced training time and memory usage using datasets like MotionSense and UCI. In [59], TL was applied to CNNs in TinyML environments, retraining both fully connected and convolutional layers, which improved accuracy and convergence speed, although with a slight energy cost.

Similarly, [60] presented a TinyML-based smart farming framework using UAVs and IoT sensors, combining DNNs, LSTMs, and TL for efficient soil moisture prediction. This highlighted the feasibility of TL on ultralow-power devices. In [61], TL was applied to machine fault diagnosis with a CNN architecture that improved domain generalization by fine-tuning convolutional layers and transferring dense layer parameters, achieving high accuracy even with minimal target data. Memory-efficient retraining was achieved by fine-tuning only biases. MicroTL, a TL framework for low-power IoT devices, was introduced in [62], which allowed on-device retraining of DNNs, reducing energy consumption and cloud dependency. The study demonstrated that MicroTL enhanced efficiency and privacy in resource-constrained environments.

9.2.4.4 Adaptive Inference and Dynamic Model Scaling

Two critical techniques in TinyML that can ensure optimal performance and efficiency in resource-limited devices are adaptive inference and dynamic model scaling. Leveraging these methods, edge devices can adjust computation loads based on the varying available resources and task requirements, therefore balancing model accuracy and power utilization.

Adaptive Inference In TinyML, an adaptive inference technique can be utilized to dynamically regulate the model complexity or the number of layers used for inference based on specific conditions of the task or the current state of power on the device, thus allowing models to run efficiently [63]. The early-exit neural network is a widely known adaptive inference method, where models can stop computations early if the inputs are detected to be adequately processed to make an accurate prediction. T-RecX, an early-exit classifier for TinyML models, was proposed in [64], where models were allowed to exit early when confident, thus reducing computation time. The proposed method was evaluated on CNNs from the MLPerf Tiny benchmark, where it outperformed previous methods by improving accuracy and reducing FLOPS by 31.58%, with only 1% accuracy loss. Similarly, ClassyNet, another early-exit architecture, was proposed in [65] for edge devices, prioritizing critical classes for faster inference and reduced computation demands while maintaining accuracy. When compared to existing models, ClassyNet not only improved speed and efficiency but also optimized resource use by facilitating class-specific exits, making it well-suited for time-sensitive edge applications. QUTE, a lightweight ensemble architecture for uncertainty quantification in TinyML, was proposed in [66], where analysis showed that, using an early-exit neural network and knowledge, it improved uncertainty estimates by 6% and reduced model size by 3.1 times compared to previous methods, while effectively detecting out-of-distribution inputs.

Dynamic Model Scaling Unlike the adaptive inference methods, dynamic model scaling enables the model structure to be changed during runtime to accommodate fluctuating computational resources and task complexity. This technique provides a balance between model accuracy and computational efficiency, considering different power and latency constraints. To achieve dynamic model scaling, different NAS techniques have been proposed, such as EfficientNet [47, 67] and Proxyless-NAS [68], which employ scaling mechanisms across width, depth, and resolution, creating highly adaptable models that can dynamically adjust to changing resource environments. These techniques are particularly effective in energy-constrained environments where maximizing efficiency is critical while maintaining accuracy. However, these techniques, despite their benefits, have challenges such as lack of adaptability, scalability issues, and high computational cost, which led to the development of the Once-for-All (OFA) networks.

The OFA network, introduced in [69], addressed the challenges of NAS techniques by training a single model to support multiple subnetworks with varying depths, widths, and kernel sizes. It facilitates efficient inference on edge devices by separating training and search, thus allowing them to be selected without the need for retraining. This technique allows for real-time optimization at inference time, allowing a TinyML device to select the most suitable subnetwork based on latency and resource limitations, thus improving inference speed and energy efficiency for edge scenarios.

Achieving adaptive inference dynamically can not only improve real-time performance but also enhance the longevity of battery-operated devices in a resource-limited setting.

9.3 TinyML as a Service (TMLaaS) Architecture

The TMLaaS architecture is a dynamic and flexible approach to deploying ML models on resource-limited environments. This architecture extends the standalone TinyML paradigm by distributing ML capabilities across both edge devices and gateway systems. Unlike the standalone TinyML, which focuses solely on the inference performance on resource-limited edge devices, TMLaaS provides a more hybridized approach that leverages both local and remote computations to improve performance and enhance energy efficiency. It has diverse applications for enabling edge intelligence, providing real-time and localized ML capabilities, across different sectors including industrial IoT (IIoT), healthcare and wearable devices, smart homes and buildings, security and surveillance, agriculture and precision farming, autonomous vehicles and drones, smart cities and urban infrastructure, as well as energy management and smart grids as depicted in Figure 9.4.

In TMLaaS architecture, computational capacity is divided between the edge device and a more powerful gateway, where the edge device handles low-latency and granular inference tasks to ensure real-time decision-making such as detecting basic anomalies or monitoring thresholds. The gateway, on the other hand, handles more complex tasks due to its high processing capabilities and available resources, including aggregation of data from multiple devices as well as retraining models and deploying them on edge devices.

9.3.1 Architectural Considerations in TMLaaS

Designing and implementing a TMLaaS-based system requires careful consideration of the division of computational tasks between the edge device and the gateway. One of the key design factors is the type of data being processed. In a scenario where the edge devices provide unique feature subsets for the ML

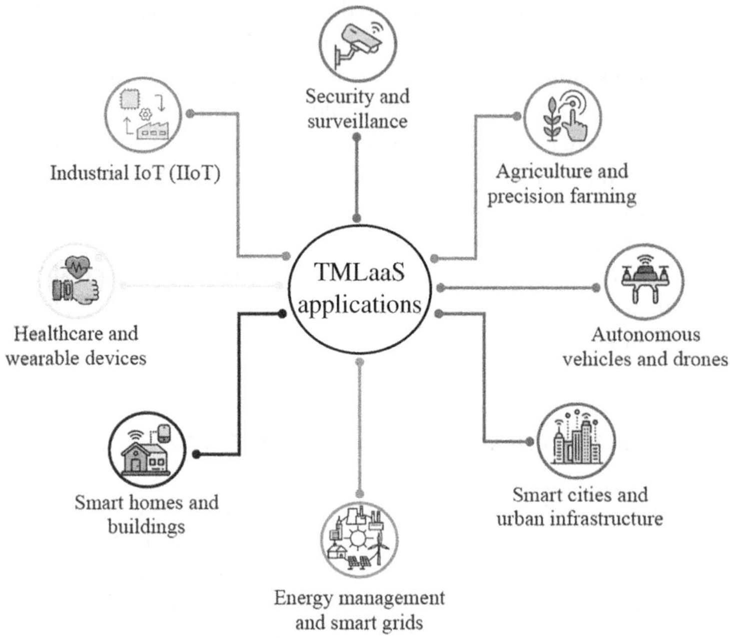

Figure 9.4 Application of TMLaaS.

model, it may be more efficient to offload most of the inference to the gateway. For example, applications that use heterogeneous sensors or multichannel data inputs can benefit from centralized inference on the gateway because of the computation-intensive operations involved. In a case where the application requires real-time, low-latency decision-making, it may be more efficient to perform inference on the device itself rather than offloading computation to the gateway, which may increase latency. However, this approach can result in higher energy consumption on the edge device due to increased computational load, which can be detrimental in a battery-powered environment, hence the need for careful consideration and management.

9.3.2 Energy and Computational Trade-Offs

Balancing on-device and wireless communication is one of the critical design trade-offs in a TMLaaS architecture. Executing ML models on-device requires computational energy, whereas offloading data to the gateway for processing requires energy in wireless transmission, which varies depending on the ML model complexity, transmitted data size, and the communication protocol utilized, such as Wi-Fi, Bluetooth Low Energy (BLE), or Low-Power Wide-Area

Network (LPWAN) technologies. If the gateway, however, is powered by a constant power supply, it is more energy-efficient to offload ML inference and computations to the gateway, especially for complex models or large datasets.

9.3.3 Security and Privacy Trade-Offs

Security and privacy considerations are other trade-offs in a TMLaaS-based system, as there is a crucial need for data transmitted between the edge device and gateway, as well as between the gateway and the cloud, to be secured, preventing unauthorized access or tampering. Choosing where to perform inference, whether on-device, on the gateway, or both, depends on the data sensitivity and the security of the communication channels. For instance, if the data involves multiple users or if the devices provide critical side information that may compromise privacy, the system architecture must account for these constraints, possibly requiring encryption, anonymization, or the use of privacy-preserving ML techniques

9.3.4 Modality of Communication

Maintaining and updating ML models within a TMLaaS architecture introduces its own set of challenges. Edge devices, especially those operating on unidirectional communication protocols such as SigFox or BLE beacons, may not support frequent firmware updates, making it difficult to deploy new models or configurations. Gateways, however, often have bidirectional communication with the cloud, facilitating more frequent updates and testing of ML models. This flexibility allows for the continuous deployment of model improvements, including bug fixes, accuracy enhancements, or even entirely new models, without requiring direct access to the end devices. Proper management of these updates and the overall TMLaaS applications lifecycle is handled by MLOps.

9.3.5 MLOps in TMLaaS Architecture

MLOps, a machine learning counterpart of DevOps, is a critical framework for managing the TinyML model lifecycle in embedded edge applications. It comprises a set of practices or stages to ensure seamless integration, deployment, testing, and monitoring of ML systems. Key stages of MLOps for TinyML are as follows:

Data Sampling and Analysis Data, in TinyML applications, can be obtained from diverse edge hardware, including sensors, cameras, and other sources, and must be processed rapidly to ensure real-time, low-latency inferences. Collecting data samples at accurate and consistent rates is crucial, as effective sampling ensures that the data used to train models to obtain inference is representative of the real-world conditions in which the model will operate. Once the data is accurately collected, analysis is conducted to classify and understand its structure, which often includes metadata analysis.

Data Structuring and Model Training Before training the models, the data must be formatted based on the requirements of the ML model. This often involves partitioning the data into training, validation, and test sets, ensuring that the model is trained and evaluated on diverse inputs and can generalize well to unseen data. Essentially, training TinyML models involves utilizing the structured data to adjust the parameters of the models so that it can make accurate predictions. This is done either on a high-end workstation or on a cloud platform, with the final model being optimized and deployed on edge devices.

Model Evaluation, Validation, and Deployment The model, after training, is evaluated against test data to measure its performance, which involves comparing the output of the model with expected results and measuring performance metrics such as accuracy, prediction latency, precision, and recall. For TinyML models, memory and processing constraints of the target hardware where the model is expected to operate within should be factored in. That is, after evaluating the performance of the model, it should be converted to TF-Lite Micro and deployed onto the edge device. The deployment process may also involve setting up infrastructure to monitor the performance of the model in a real-world setting and manage updates as needed.

Continuous Integration, Development, and Testing Continuous integration, in an MLOps framework, involves updating the system with new features, bug fixes, or performance improvements regularly. For the TinyML application, this may include firmware updates, model changes, or data processing adjustments to ensure the system is up to date. Continuous development, on the other hand, involves ensuring that the system is adaptable to evolving requirements, user needs, or environmental conditions, enabling new models to be optimized for battery-powered devices with power limitations or fluctuating network connectivity. Furthermore, continuous testing rigorously verifies that every update maintains or improves the device's performance, particularly in terms of prediction latency, accuracy, and energy efficiency, ensuring that no model update or changes negatively impact the device's performance.

9.3.6 Case Study

The development and deployment of the TMLaaS architecture are demonstrated through a simple case study. This case study, using three publicly available datasets, compares the performance of commonly used ML classifiers and DL models for on-device learning based on memory usage, prediction latency, and accuracy.

Scenario and Dataset Utilized Two main scenarios are considered for the case study: HAR and image recognition. Three different publicly available datasets – the UCI HAR, Wild-SHARD, and MNIST datasets – were selected to assess the performance of the models.

- **UCI HAR Dataset:** This dataset is a popular benchmark dataset used for recognizing human activities based on data collected from smartphones. The dataset contains accelerometer and gyroscope sensor readings from 30 participants who performed six different activities: walking, walking upstairs, walking downstairs, sitting, standing, and lying down. These readings are captured along the X, Y, and Z axes, and the dataset includes both raw sensor signals and preprocessed features such as time and frequency domain variables [70].
- **Wild-SHARD Dataset:** This dataset is also a HAR dataset collected in real-world, uncontrolled environments to overcome the limitations of existing simulated datasets. It captures time-series data of Activities of Daily Living (ADLs) on 40 adult subjects of varying ages, genders, weights, and heights. Using multiple smartphone models, such as Samsung Galaxy and OnePlus devices, data was gathered through accelerometer and gyroscope sensors, worn in the front pockets of participants. The activities recorded include sitting, walking, standing, running, going upstairs, and going downstairs, collected at a sampling frequency of 100 Hz [71].
- **MNIST Dataset:** This dataset is widely used in computer vision, especially for image classification tasks. It contains 70,000 grayscale images of handwritten digits (0–9), with 60,000 images in the training set and 10,000 in the test set. Each image is 28×28 pixels, and the task is to classify the digit represented in each image. MNIST serves as a standard benchmark for evaluating the performance of ML algorithms, particularly neural networks and DL models, due to its simplicity and the relatively low complexity of the images.

These datasets were selected based on their widespread use, real-world applicability, and suitability for evaluating different ML and DL models on diverse tasks.

Implementation The implementation of the case study followed a step-by-step approach, as illustrated in Figure 9.2. The datasets were obtained from their respective repositories, and based on their features and labels, different models were trained and tested offline using the Python programming language on the Google Colab platform.

The different ML classifiers considered, which included Support Vector Machine (SVM), Logistic Regression (LR), Light Gradient Boosting Machine (LightGBM), and Random Forest (RF), were trained using the Scikit-learn (Sklearn) library. Each dataset was divided into 80% training and 20% testing, and the GridSearchCV hyperparameter tuning technique was applied with fivefold cross-validation to obtain the optimal parameters for each ML model. The trained and tested models were then saved in Pickle (.pkl) format because there is currently no direct conversion of Sklearn-trained models to TF-Lite Micro (.tflite) format.

The DL models, which included DNN, MicroCNN, and Multilayer Perceptron (MLP), were trained directly using the TensorFlow library. For the DNN model, two hidden dense layers, 128x64, were utilized with the Rectified Linear Unit (ReLU) activation function and Adam optimizer. Similarly, the MLP model utilized two hidden dense layers, 256×128, with the ReLU activation function and Adam optimizer. However, for MicroCNN, two convolutional layers with 32 and 64 filters, respectively, were utilized with ReLU activation and Adam optimizer. MaxPooling layers with a 2×2 pool size are applied after each convolutional layer to reduce spatial dimensions and computational complexity. Since these DL models are TensorFlow-trained models, it was easily converted and saved in the TF-Lite Micro format.

The Edge Impulse platform [20] was employed in this case study as the gateway service, simulating the management of model deployment to edge devices and offloading certain computational tasks such as model conversion, optimization, and performance monitoring. After obtaining the trained models, both ML and DL models, they were uploaded to the Edge Impulse platform, which handled the conversion of non-TF-Lite ML models into TF-Lite Micro format using the JAX function, enabling their use in embedded devices. The platform enables further model training, optimization, and testing, ensuring the model is well-suited for deployment onto the target hardware. For this case study, the Cortex M4F 80 MHz microcontroller was utilized as the target hardware, where all trained models were deployed and evaluated to obtain key performance metrics such as prediction latency, model accuracy, and memory footprint (RAM and Flash usage). These metrics were monitored and recorded, allowing for a comparison of model efficiency and effectiveness on edge devices.

9.4 Results and Discussion

Figure 9.5 presents the performance of different DL models deployed on the Cortex M4F 80 MHz microcontroller. The DL models demonstrated a clear trade-off between latency, accuracy, and memory efficiency. For instance,

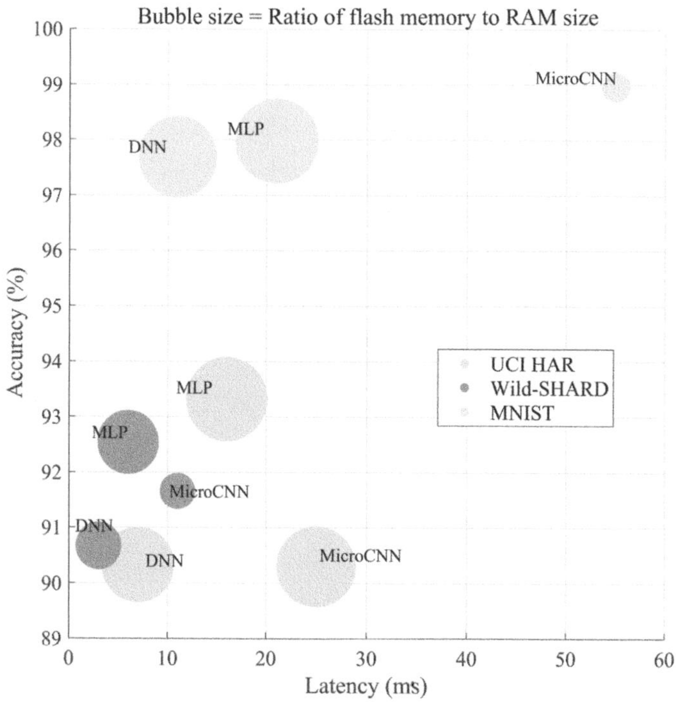

Figure 9.5 DL model comparison on Cortex M4F 80 MHz microcontroller.

MicroCNN, which achieved the highest accuracy for the MNIST dataset, showed significantly higher latency and required a larger memory footprint. This result indicates that while MicroCNN showed good feature extraction capabilities, its computational complexity resulted in increased inference time and higher memory demands, making it less suited for real-time applications on devices with strict power and memory constraints. In contrast, the MLP and DNN models achieved lower latency across all datasets, with latencies under 20 ms, which indicates faster inference with considerable accuracy levels, especially for the MNIST dataset. In addition, the reduced Flash-to-RAM memory ratio in these models further indicates their suitability for edge deployment.

Similar results and trade-offs between latency, accuracy, and memory efficiency were obtained for the ML classifiers deployed on the Cortex M4F 80 MHz microcontroller as presented in Figure 9.6. Results showed that all models, except LR (MNIST), achieved high accuracy above 80%. LightGBM and SVM showed the highest accuracy for MNIST. However, they showed significantly higher latencies of 127 and 82 ms, respectively, making them ideal for applications that prioritize

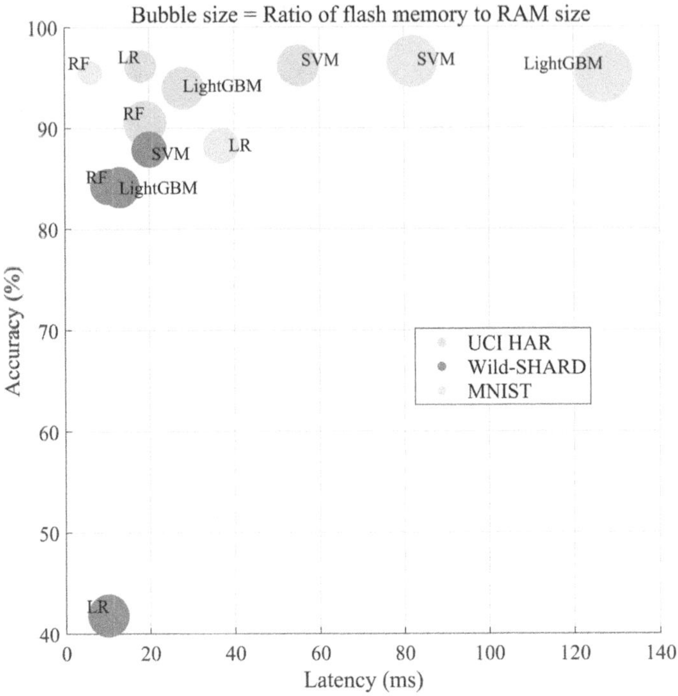

Figure 9.6 ML models comparison on the Cortex M4F 80 MHz microcontroller.

accuracy but tolerate increased inference speeds, such as smart agriculture for plant disease detection. The RF model showed considerable accuracy across all datasets with very low latency and memory footprint, making it ideal for edge applications that prioritize strict inference time and memory constraints. LR, on the other hand, achieved the lowest accuracy of 41.81% for the Wild-SHARD dataset, even after achieving a very high accuracy of 96.13% for the UCI HAR dataset. This is because the LR model is a very simple and linear model that failed to capture the complexities or nonlinearity of the Wild-SHARD dataset, which was obtained in uncontrolled environments. Thus, it is not well-suited for edge applications that deal with unstructured and highly varied data, such as audio, images, or sensor readings collected in dynamic, real-world settings.

Overall, the choice of model for edge deployment depends heavily on the specific application requirements. DNN, MLP, and Random Forest, with their low resource footprints and fast inference times, are particularly well-suited for environments with strict constraints on memory and processing power, while models such as LightGBM, SVM, and MicroCNN are better suited for accuracy-critical tasks where latency demands are not critical.

9.5 Open Challenges and Further Research Directions

TinyML offers several benefits and significant potential to further enhance edge intelligence; however, several open challenges need to be addressed for its widespread adoption and scalability. This section highlights and discusses these challenges and proffers further research directions.

9.5.1 TinyML for Transfer Learning

The computational intensity required for training models on large datasets is one of the major issues in deploying TinyML at scale. Training on traditional high-performance computing platforms is often not feasible due to resource constraints in edge devices. TL presents a solution by enabling edge devices to leverage pretrained models from centralized repositories, such as the TensorFlow Hub, which can then be adapted to specific tasks at the edge. However, the integration of TL in TinyML workflows remains underdeveloped. Key issues include the lack of efficient mechanisms for downloading and deploying pre-compiled models on edge devices and the inability to retrain or fine-tune models on-device due to limited computational power. Future research needs to explore how to optimize pretrained model deployment for real-time inference and enhance the adaptation of models to new, but related, tasks at the edge. Additionally, creating efficient workflows for on-the-fly model adaptation is a major research gap.

9.5.2 TinyML for Federated Learning

While TinyML enables edge devices to execute pre-compiled models, it cannot continuously improve these models through learning from other edge devices. Federated learning (FL) addresses this by distributing training across devices while keeping data localized, thus preserving privacy. However, combining FL with TinyML is still in its early stages and poses numerous challenges, such as efficiently aggregating locally computed models across devices with nonidentical and nonindependent data distributions. The computational overhead involved in performing local training on resource-limited devices and the communication cost of aggregating models across devices also need to be addressed. Further research should design and develop scalable FL algorithms that are energy efficient and can integrate with TinyML, enabling local device learning with minimal resource usage while preserving model accuracy.

9.5.3 Performance and Scalability

Optimizing the performance and scalability of TinyML models in real-world deployments is a critical challenge because achieving optimal trade-offs between energy utilization, latency, and model accuracy is difficult with constrained memory and processing power. Improving the scalability of TinyML models for large-scale edge deployments is crucial, particularly in TMLaaS architecture, where multiple embedded edge devices may rely on shared resources. Therefore, future research should focus on improving TinyML scalability and reducing the overhead associated with developing and deploying complex models on edge devices.

9.5.4 Trustworthiness of Pretrained Models in Transfer Learning

The trustworthiness of pretrained models is one of the key challenges in TL for TinyML, as edge devices often utilize pretrained models from external sources. Therefore, there is a need for methods that can assess the quality and integrity of these models, ensuring they are free from biases or adversarial inputs. Future research should aim at developing mechanisms that will enable edge devices to evaluate the credibility of pretrained models, including ensuring their source datasets are representative and unbiased before utilizing them for inference.

9.5.5 Secure Model-Sharing Protocols

The need for good security schemes or protocols cannot be overemphasized for any given system. Similarly, in TL for TinyML, models are often shared across multiple edge devices, which can be vulnerable to several threats and attacks. Thus, there is a crucial need to design secure model-sharing protocols that can ensure that models can be shared or updated across edge devices securely and reliably. Future research should aim to propose, design, and develop secure protocols for sharing or updating models without introducing vulnerabilities into the system.

9.6 Conclusion

This chapter provided a comprehensive overview of TinyML for enhanced edge intelligence. We addressed the challenges of deploying large models on edge devices using TinyML techniques such as model compression, quantization, and efficient neural networks. Using TF-Lite Micro, we provided a detailed workflow implementation for TinyML deployment on edge devices. The chapter introduced and validated the TMLaaS architecture through a case study that utilized three datasets to evaluate the on-device performance of different ML and DL models, where results showed clear trade-offs between latency, accuracy, and memory efficiency. Critical open challenges of TinyML were highlighted and discussed, including its integration with FL and TL, as well as proffered further research directions. This chapter can be utilized as a reference to further enhance TinyML for edge intelligence.

References

1 Shehu Yalli, J., Hilmi Hasan, M., and Abubakar Badawi, A. (2024). Internet of Things (IoT): Origins, embedded technologies, smart applications, and its growth in the last decade. *IEEE Access* 12: 91357–91382. https://doi.org/10.1109/ACCESS.2024.3418995.

2 Hua, H., Li, Y., Wang, T. et al. (2023). Edge computing with artificial intelligence: A machine learning perspective. *ACM Computing Surveys* 55 (9): 1–35. https://doi.org/10.1145/3555802.

3 Xu, D., Li, T., Li, Y., et al. (2020) Edge intelligence: Architectures, challenges, and applications. *arXiv Prepr. arXiv2003.12172*, March [Online]. http://arxiv.org/abs/2003.12172

4 Mira, J., Moreno, I., Bardisbanian, H., and Gorroñogoitia, J. (2023). Machine Learning (ML) as a Service (MLaas): Enhancing IoT with intelligence, adaptive online deep and reinforcement learning, model sharing, and zero-knowledge model verification. In: *Shaping the Future of IoT with Edge Intelligence* (ed. R.C. Sofia and J. Soldatos), 63–93. New York: River Publishers https://doi.org/10.1201/9781032632407-6.

5 Zaidi, S.A.R., Hayajneh, A.M., Hafeez, M., and Ahmed, Q.Z. (2022). Unlocking edge intelligence through Tiny machine learning (TinyML). *IEEE Access* 10: 100867–100877. https://doi.org/10.1109/ACCESS.2022.3207200.

6 Lin, J., Zhu, L., Chen, W.-M. et al. (2023). Tiny machine learning: Progress and futures. *IEEE Circuits and Systems Magazine* 23 (3): 8–34. https://doi.org/10.1109/MCAS.2023.3302182.

7 Ray, P.P. (2022). A review on TinyML: State-of-the-art and prospects. *Journal of King Saud University, Computer and Information Sciences* 34 (4): 1595–1623. https://doi.org/10.1016/j.jksuci.2021.11.019.

8 Abadade, Y., Temouden, A., Bamoumen, H. et al. (2023). A comprehensive survey on TinyML. *IEEE Access* 11: 96892–96922. https://doi.org/10.1109/ACCESS.2023.3294111.

9 Rajapakse, V., Karunanayake, I., and Ahmed, N. (2023). Intelligence at the extreme edge: A survey on reformable TinyML. *ACM Computing Surveys* 55 (13s): 1–30. https://doi.org/10.1145/3583683.

10 Sanchez-Iborra, R., Zoubir, A., Hamdouchi, A. et al. (2023). Intelligent and efficient IoT through the cooperation of TinyML and edge computing. *Informatica* 34 (1): 147–168. https://doi.org/10.15388/22-INFOR505.

11 Soro, S. (2021). TinyML for ubiquitous edge AI. *arXiv Prepr. arXiv2102.01255*, February [Online]. http://arxiv.org/abs/2102.01255

12 Mendez, J., Bierzynski, K., Cuéllar, M.P., and Morales, D.P. (2022). Edge intelligence: Concepts, architectures, applications, and future directions. *ACM Transactions on Embedded Computing Systems* 21 (5): 1–41. https://doi.org/10.1145/3486674.

13 Khalifeh, A., Mazunga, F., Nechibvute, A., and Nyambo, B.M. (2022). Microcontroller unit-based wireless sensor network nodes: A review. *Sensors* 22 (22): 8937. https://doi.org/10.3390/s22228937.

14 Jiang, C., Fan, T., Gao, H. et al. (2020). Energy-aware edge computing: A survey. *Computer Communications* 151: 556–580. https://doi.org/10.1016/j.comcom.2020.01.004.

15 Biswas, A. and Wang, H.-C. (2023). Autonomous vehicles enabled by the integration of IoT, edge intelligence, 5G, and blockchain. *Sensors* 23 (4): 1963. https://doi.org/10.3390/s23041963.

16 Liu, D., Kong, H., Luo, X. et al. (2022). Bringing AI to edge: From deep learning's perspective. *Neurocomputing* 485: 297–320. https://doi.org/10.1016/j.neucom.2021.04.141.

17 Alwakeel, A.M. (2021). An overview of fog computing and edge computing security and privacy issues. *Sensors* 21 (24): 8226. https://doi.org/10.3390/s21248226.

18 Laguduva, V.R., Katkoori, S., and Karam, R. (2020). Machine learning attacks and countermeasures for PUF-based IoT edge node security. *SN Computer Science* 1 (5): 282. https://doi.org/10.1007/s42979-020-00303-y.

19 Unal, D., Bennbaia, S., and Catak, F.O. (2022). Machine learning for the security of healthcare systems based on Internet of Things and edge computing. In: *Cybersecurity and Cognitive Science* (ed. A.A. Moustafa), 299–320. Elsevier https://doi.org/10.1016/B978-0-323-90570-1.00007-3.

20 Hymel, S., Banbury, C., Situnayake, D. et al. (2022). Edge impulse: An MLOps platform for tiny machine learning. In: *Proceedings of Machine Learning and Systems* (ed. D. Song, M. Carbin, and T. Chen), 254–268. Curan [Online]. http://arxiv.org/abs/2212.03332.

21 David, R., Duke, J., Jain, A. et al. (2020). TensorFlow lite micro: Embedded machine learning on TinyML systems. *Proceedings of Machine Learning and Systems* 3: 800–811. [Online]. http://arxiv.org/abs/2010.08678.

22 Shuvo, M.M.H., Islam, S.K., Cheng, J., and Morshed, B.I. (2023). Efficient acceleration of deep learning inference on resource-constrained edge devices: A review. *Proceedings of the IEEE* 111 (1): 42–91. https://doi.org/10.1109/JPROC.2022.3226481.

23 Liang, T., Glossner, J., Wang, L. et al. (2021). Pruning and quantization for deep neural network acceleration: A survey. *Neurocomputing* 461: 370–403. https://doi.org/10.1016/j.neucom.2021.07.045.

24 Vadera, S. and Ameen, S. (2022). Methods for pruning deep neural networks. *IEEE Access* 10: 63280–63300. https://doi.org/10.1109/ACCESS.2022.3182659.

25 De Leon, J. D. and Atienza, R. (2022). Depth pruning with auxiliary networks for TinyML. *ICASSP 2022–2022 IEEE International Conference on Acoustics, Speech and Signal Processing (ICASSP)*, IEEE, May, 3963–3967. https://doi.org/10.1109/ICASSP43922.2022.9746843.0.

26 Liberis, E. and Lane, N.D. (2022). Differentiable neural network pruning to enable smart applications on microcontrollers. *Proceedings of the ACM on Interactive, Mobile, Wearable and Ubiquitous Technologies* 6 (4): 1–19. https://doi.org/10.1145/3569468.

27 He, Y., Lin, J., Liu, Z., et al. (2018). AMC: AutoML for model compression and acceleration on mobile devices. *Lecture Notes in Computer Science (including subseries Lecture Notes in Artificial Intelligence and Lecture Notes in Bioinformatics)*, vol. 11211 LNCS, 815–832. https://doi.org/10.1007/978-3-030-01234-2_48.

28 Fedorov, I., Matas, R., Tann, H. et al. (2022). UDC: Unified DNAS for compressible TinyML Models for neural processing units. *Advances in Neural Information Processing Systems* 35.

29 Han, L., Xiao, Z., and Li, Z. (2024). DTMM: Deploying TinyML models on extremely weak IoT devices with pruning. *IEEE INFOCOM 2024 – IEEE Conference on Computer Communications*, IEEE, May, 1999–2008. https://doi.org/10.1109/INFOCOM52122.2024.10621387.

30 Siang, Y.Y., Ridzuan Ahamd, M., Shafinaz, M., and Abidin, Z. (2021). Anomaly detection based on tiny machine learning: A review. *Open International Journal of Informatics* 9 (Special Issue 2): 67–78. [Online]. https://oiji.utm.my/index.php/oiji/article/view/148.

31 Kirtas, M., Passalis, N., Oikonomou, A., et al. (2023). Quantization-aware training for mixed precision photonic neural networks. *2023 IEEE 33rd International Workshop on Machine Learning for Signal Processing (MLSP)*, IEEE, September, 1–6. https://doi.org/10.1109/MLSP55844.2023.10285966.

32 Ghamari, S., Ozcan, K., Dinh, T., et al. (2021). Quantization-guided training for compact TinyML models. *arXiv Prepr. arXiv2103.06231*, March [Online]. http://arxiv.org/abs/2103.06231

33 Moriello, C. (2023). Optimizing neural networks for TinyML: A study on quantization schemes. Politecnico Milano 1863, [Online]. https://www.politesi.polimi.it/bitstream/10589/222539/1/2024_07_Moriello.pdf

34 Shabir, M.Y., Torta, G., and Damiani, F. (2024). Edge AI on constrained IoT devices: Quantization strategies for model optimization. In: *Lecture Notes in Networks and Systems, vol. 1066 LNNS* (ed. K. Arai), 556–574. Cham: Springer Nature Switzerland https://doi.org/10.1007/978-3-031-66428-1_35.

35 Park, W., Kim, D., Lu, Y., and Cho, M. (2019). Relational knowledge distillation. *2019 IEEE/CVF Conference on Computer Vision and Pattern Recognition (CVPR)*, IEEE, June, 3962–3971. https://doi.org/10.1109/CVPR.2019.00409.

36 Accettola, A. G. and Merenda, M. (2023). Dataset distillation as an enabling technique for on-device training in TinyML for IoT: an RFID use case. *2023 8th International Conference on Smart and Sustainable Technologies (SpliTech)*, IEEE, June, 1–4. https://doi.org/10.23919/SpliTech58164.2023.10193138.

37 X. Jiao, Y. Yin, L. Shang, et al. (2020). TinyBERT: Distilling BERT for natural language understanding. *Findings of the Association for Computational Linguistics: EMNLP 2020*, Stroudsburg, PA: Association for Computational Linguistics, 4163–4174. https://doi.org/10.18653/v1/2020.findings-emnlp.372.

38 Howard, A. G., Zhu, M., Chen, B., et al. (2017). MobileNets: Efficient convolutional neural networks for mobile vision applications. *arXiv.1704.04861*, April. https://doi.org/10.48550/arXiv.1704.04861.

39 Dong, K., Zhou, C., Ruan, Y., and Li, Y. (2020). MobileNetV2 model for image classification. *2020 2nd International Conference on Information Technology and Computer Application (ITCA)*, IEEE, December, 476–480. https://doi.org/10.1109/ITCA52113.2020.00106.

40 Qian, S., Ning, C., and Hu, Y. (2021). MobileNetV3 for Image Classification. *2021 IEEE 2nd International Conference on Big Data, Artificial Intelligence and Internet of Things Engineering (ICBAIE)*, IEEE, March, 490–497. https://doi.org/10.1109/ICBAIE52039.2021.9389905.

41 Sandler, M., Howard, A., Zhu, M., et al. (2018). MobileNetV2: Inverted residuals and linear bottlenecks. *2018 IEEE/CVF Conference on Computer Vision and Pattern Recognition*, IEEE, June, 4510–4520. https://doi.org/10.1109/CVPR.2018.00474.

42 Alajlan, N.N. and Ibrahim, D.M. (2023). DDD TinyML: A TinyML-based driver drowsiness detection model using deep learning. *Sensors* 23 (12): 5696. https://doi.org/10.3390/s23125696.

43 Iandola, F. N., Han, S., Moskewicz, M. W., et al. (2016). SqueezeNet: AlexNet-level accuracy with 50x fewer parameters and <0.5MB model size. *arXiv Prepr. arXiv1602.07360*, February [Online]. http://arxiv.org/abs/1602.07360

44 Albanese, A., Nardello, M., Fiacco, G., and Brunelli, D. (2023). Tiny machine learning for high accuracy product quality inspection. *IEEE Sensors Journal* 23 (2): 1575–1583. https://doi.org/10.1109/JSEN.2022.3225227.

45 Zhang, X., Zhou, X., Lin, M., and Sun, J. (2018). ShuffleNet: An extremely efficient convolutional neural network for mobile devices. *2018 IEEE/CVF Conference on Computer Vision and Pattern Recognition*, IEEE, June, 6848–6856. https://doi.org/10.1109/CVPR.2018.00716.

46 Ma, N., Zhang, X., Zheng, H.-T., and Sun, J. (2018). ShuffleNet V2: Practical guidelines for efficient CNN architecture design. *Lecture Notes in Computer Science (including subseries Lecture Notes in Artificial Intelligence and Lecture Notes in Bioinformatics)*, vol. 11218 LNCS, 122–138. https://doi.org/10.1007/978-3-030-01264-9_8.

47 Tan, M. and Le, Q. V. (2019). EfficientNet: Rethinking model scaling for convolutional neural networks. *36th International Conference of Machine Learning ICML 2019*, vol. 2019-June, 10691–10700, May [Online]. http://arxiv.org/abs/1905.11946

48 Lin, J., Chen, W.-M., Cai, H. et al. (2021). MCUNetV2: Memory-efficient patch-based inference for tiny deep learning. *Advances in Neural Information Processing Systems* 4: 2346–2358. [Online]. http://arxiv.org/abs/2110.15352.

49 Ancilotto, A., Paissan, F., and Farella, E. (2023). XiNet: Efficient neural networks for TinyML. *2023 IEEE/CVF International Conference on Computer Vision (ICCV)*, IEEE, October, 16922–16931. https://doi.org/10.1109/ICCV51070.2023.01556.

50 Xu, K., Li, Y., Zhang, H., et al. (2022). EtinyNet: Extremely tiny network for TinyML. *Proceedings of the 36th AAAI Conference of Artificial Intelligence AAAI 2022*, vol. 36, 4628–4636, https://doi.org/10.1609/aaai.v36i4.20387.

51 Banbury, C., Zhou, C., Fedorov, I. et al. (2020). MicroNets: Neural network architectures for deploying TinyML applications on commodity microcontrollers. *Proceedings of Machine Learning and Systems* 3: 517–532. [Online]. http://arxiv.org/abs/2010.11267.

52 Luo, Y., Yin, L., Bai, W., and Mao, K. (2020). An appraisal of incremental learning methods. *Entropy* 22 (11): 1190. https://doi.org/10.3390/e22111190.

53 Ostrovan, E. (2021). TinyML on-device neural network training. Politecnico Milano 1863 [Online]. https://www.politesi.polimi.it/bitstream/10589/187690/6/TinyML_on_device_neural_network_training.fin.pdf

54 Pandey, S. K. and Nair, S. B. (2024). Onboard class incremental learning for resource-constrained scenarios using genetic algorithm and TinyML. *Proceedings of the Genetic and Evolutionary Computation Conference Companion, in GECCO'24 Companion*. New York, NY: ACM, July 299–302. https://doi.org/10.1145/3638530.3654392.

55 Tremonti, M., Pezze, D. D., Paissan, F., et al. (2024). An empirical evaluation of tinyML architectures for class-incremental continual learning. *2024 IEEE International Conference on Pervasive Computing and Communications Workshops and other Affiliated Events (PerCom Workshops)*, IEEE, March 690–695. https://doi.org/10.1109/PerComWorkshops59983.2024.10503398.

56 Disabato, S. and Roveri, M. (2020). Incremental on-device tiny machine learning. *Proceedings of the 2nd International Workshop on Challenges in*

Artificial Intelligence and Machine Learning for Internet of Things, in AIChallengeIoT'20. New York, NY: ACM, November 7–13. https://doi.org/10.1145/3417313.3429378.

57 Sudharsan, B., Yadav, P., Breslin, J. G., and Intizar Ali, M. (2021). Train++: An incremental ML model training algorithm to create self-learning IoT devices. *2021 IEEE SmartWorld, Ubiquitous Intelligence & Computing, Advanced & Trusted Computing, Scalable Computing & Communications, Internet of People and Smart City Innovation (SmartWorld/SCALCOM/UIC/ATC/IOP/SCI)*, IEEE, October, 97–106. https://doi.org/10.1109/SWC50871.2021.00023.

58 Hayajneh, A.M., Hafeez, M., Zaidi, S.A.R., and McLernon, D. (2024). TinyML empowered transfer learning on the edge. *IEEE Open Journal of the Communications Society* 5: 1656–1672. https://doi.org/10.1109/OJCOMS.2024.3373177.

59 Fragkou, E., Lygnos, V., and Katsaros, D. (2022). Transfer learning for convolutional neural networks in tiny deep learning environments. *ACM International Conference Proceeding Series*, 145–150. https://doi.org/10.1145/3575879.3575984.

60 Hayajneh, A.M., Aldalahmeh, S.A., Alasali, F. et al. (2024). Tiny machine learning on the edge: A framework for transfer learning empowered unmanned aerial vehicle assisted smart farming. *IET Smart Cities* 6 (1): 10–26. https://doi.org/10.1049/smc2.12072.

61 Asutkar, S., Chalke, C., Shivgan, K., and Tallur, S. (2023). TinyML-enabled edge implementation of transfer learning framework for domain generalization in machine fault diagnosis. *Expert Systems with Applications* 213: 119016. https://doi.org/10.1016/j.eswa.2022.119016.

62 Profentzas, C., Almgren, M., and Landsiedel, O. (2022). MicroTL: Transfer learning on low-power IoT devices. *Proceedings – Conference on Local Computer Networks, LCN*, IEEE, 1–8. https://doi.org/10.1109/LCN53696.2022.9843735.

63 Alajlan, N.N. and Ibrahim, D.M. (2022). TinyML: Enabling of inference deep learning models on ultra-low-power IoT Edge devices for AI applications. *Micromachines* 13 (6): 851. https://doi.org/10.3390/mi13060851.

64 Ghanathe, N. P. and Wilton, S. (2023). T-RecX: Tiny-resource efficient convolutional neural networks with early-eXit. *Proceedings of the 20th ACM International Conference on Computing Frontiers 2023, CF 2023*, 123–133. https://doi.org/10.1145/3587135.3592204.

65 Ayyat, M., Nadeem, T., and Krawczyk, B. (2024). ClassyNet: Class-aware early-exit neural networks for edge devices. *IEEE Internet of Things Journal* 11 (9): 15113–15127. https://doi.org/10.1109/JIOT.2023.3344120.

66 Ghanathe, N. P. and Wilton, S. (2024). QUTE: Quantifying uncertainty in TinyML models with early-exit-assisted ensembles. *arXiv Prepr. arXiv2404.12599*, Apr. 2024, [Online]. http://arxiv.org/abs/2404.12599

67 Koonce, B. and Koonce, B. (2021). EfficientNet. In: *Convolutional Neural Networks with Swift for Tensorflow: Image Recognition and Dataset Categorization* (ed. B. Koonce), 109–123. Apress https://doi.org/10.1007/978-1-4842-6168-2_10.

68 Cai, H., Zhu, L., and Han, S. (2018). ProxylessNAS: Direct neural architecture search on target task and hardware. *7th Int. Conf. Learn. Represent. ICLR 2019*, December [Online]. http://arxiv.org/abs/1812.00332

69 Cai, H., Gan, C., Wang, T., et al. (2019). Once-for-all: Train one network and specialize it for efficient deployment. *8th Int. Conf. Learn. Represent. ICLR 2020*, August [Online]. http://arxiv.org/abs/1908.09791

70 Reyes-Ortiz, J., Anguita, D., Ghio, A., et al. (2013). Human activity recognition using smartphones. https://doi.org/10.24432/C54S4K.

71 Soni, N. A. C. B. (2024). Wild-SHARD: Smartphone sensor-based human activity recognition dataset in wild. *IEEE Dataport*. 10.0.82.235/dc2h-a428.

10

Advanced Security Schemes for TinyML Devices

Wasswa Shafik[1] and Mumin Adam[2]

[1] *Dig Connectivity Research Laboratory (DCRLab), Kampala, Uganda*
[2] *Department of Computer Engineering, King Fahd University of Petroleum and Minerals (KFUPM) Dhahran, Saudi Arabia*

10.1 Introduction

Broadly speaking, security[1] and privacy[2] have been misinterpreted and misrepresented depending on the application, context, and discipline. Security in the realm of infotech describes a comprehensive collection of confidentiality, integrity, and availability[3] (CIA) aimed at safeguarding data, systems, and networks from unauthorized access, attacks, and prospective damages [1]. Furthermore, most people would be reluctant to use these devices in safety-critical scenarios. A car missing a stop command might change a pedestrian's life forever. Yet despite these applications' promise, TinyML developers have made security and privacy a secondary concern at best. A trivial search on existing systems finds naive models just listening for triggers in unprotected memory. As more effort is applied to close these kinds of vulnerabilities, privacy and adversarial attack vulnerabilities remain under-addressed in general. Many TinyML systems are devoid of security and privacy countermeasures [2].

TinyML is a branch of machine learning (ML) that enables intelligent processing at the edge. Contrary to traditional cloud computing applications, which feature centralized deployment of computing resources, TinyML devices rely on the deployment of intelligence along with the sensor or data source. The niche filled by TinyML devices is further amplified by the inclusion of additional constraints such as severe resource constraints including lower accuracy, limited

1 https://www.techtarget.com/searchsecurity/definition/security.
2 https://iapp.org/about/what-is-privacy.
3 https://www.fortinet.com/resources/cyberglossary/cia-triad.

Tiny Machine Learning: Design Principles and Applications, First Edition.
Edited by Agbotiname Lucky Imoize, Dinh-Thuan Do, and Houbing Herbert Song.

memory, storage, computation, low power, low complexity for better power efficiency, or increased latency [3]. As a result, extensions of ML at the edge are based on the concept of small ML models. TinyML applications are extensive. For instance, they can be employed in smart devices such as cameras, microphones, gas sensors, and accelerometers that are either worn, home-installed, or used in other everyday settings. Many Internet of Things (IoT) devices are also equipped with sensors capable of capturing environmental data that can serve myriad purposes, including safety monitoring, energy efficiency, and user comfort [4].

A basic TinyML deployment consists of a sensor, an ML algorithm, and a hardware platform for implementation. The sensor monitors and captures real-world data that is relayed to the TinyML hardware, which runs the ML. The processed data is then used to make intelligent decisions and actions. As these three individual components can present vulnerabilities, the whole TinyML system could be attacked and compromised. Traditional TinyML has shown resilience to noise, minor environmental changes, falls in an arbitrary position, or day-night variation. The main focus of this text is the vulnerabilities and security issues related to the implementation and deployment of these TinyML models on the available embedded hardware platforms [5]. The application of TinyML is largely the intersection of popular ML, Embedded Systems, and applications.

Security and privacy are paramount in TinyML. TinyML technology is mission-critical and is often used in systems where security and safety are paramount. A significant part of the data processed by embedded systems is personal and confidential. Consequently, these systems are prime targets for malicious attacks. A successful attack on any of these devices could lead to the exposure of sensor data or unauthorized entry. As time went on, the capabilities of embedded systems were upgraded, and organizations started to depend more heavily on their outputs. Naturally, this leads to higher expectations when it comes to performance, and an ecosystem of edge devices and cloud services started to be integrated. While these ecosystems offer various services for both residential and commercial users, consumers are hesitant to use them due to privacy and security concerns [6].

There have been countless efforts dedicated to enhancing the security of IoT systems. However, devices have some unique characteristics that necessitate changes in how these systems are approached with security in mind. As embracing new applications signifies continually extending the area of operation, one of the key difficulties we are considering here is finding a balance between protecting and maintaining TinyML functionality while acknowledging its resource limitations. One reason for the intrinsic security challenge in IoT is the size of its deployments. Because of the large number of deployed devices, even a small proportion of vulnerable or compromised devices produce a large networking footprint for an attacker [7].

10.1.1 The Chapter Contribution

This chapter presents the following contributions.

- The chapter defines TinyML, its applications, and its advantages, setting the foundation for understanding its context and significance.
- The chapter identifies and discusses the primary security threats and vulnerabilities that TinyML devices face.
- The chapter examines the key privacy concerns related to data collection, storage, and transmission in TinyML applications.
- The chapter presents various techniques and strategies, such as encryption and secure boot, to enhance the security of TinyML devices.
- The chapter explores methods to protect user privacy, including data minimization, federated learning (FL), and differential privacy (DP).
- The chapter further presents the sampled implementation of novel security schemes for TinyML applications.
- The chapter discusses emerging trends, research opportunities, and technological advancements in the field of TinyML security and privacy.
- Finally, the chapter summarizes the critical insights and emphasizes the importance of robust security and privacy protections for the successful deployment of TinyML devices.

10.1.2 The Chapter Organization

Section 10.2 exhibits, relates, and examines the overview of security threats and vulnerabilities that TinyML devices face. Section 10.3 examines the key privacy concerns related to data collection, storage, and transmission in TinyML applications. Section 10.4 describes various techniques and strategies to enhance the security of TinyML devices, such as encryption and secure boot. Section 10.5 explores methods to protect user privacy, including data minimization, federated learning (FL), and DP. Section 10.6 presents privacy-enhancing techniques for TinyML. Section 10.7 discusses emerging trends, research opportunities, and technological advancements in TinyML security and privacy. Lastly, Section 10.8 presents the lessons learned from the chapter and the conclusion.

10.2 Fundamentals of TinyML

TinyML refers to any deployment of ML technologies conducted on Cortex-M class microcontrollers and their connected sensor and actuator peripherals. This new field is an intersection of research in ML, embedded systems, low-power design, hardware efficiency, communications, and real-time systems. Fundamentally, the issue that distinguishes TinyML from traditional ML applications is that the latter typically assumes the presence of an internet connection and sufficient computational power to process the data, while also having access to extensive storage capacities for large-scale model weights. In contrast, TinyML devices must manage with minute fractions of computational time and energy, restricted data storage and network bandwidth, enabling fast decision-making, often on a microsecond timescale [8].

Elements that enable deployment in this space include model-efficient algorithmic techniques and compact hardware nodes featuring local processing capabilities alongside energy-optimized computation and memory elements. As a result, tasks such as decision trees, support vector machines, and k-nearest neighbors are generally favored over dense, deep learning models. Enabling technologies also include specialized hardware, such as custom digital or mixed-signal integration of ML cores onto mixed-signal sensors or processors, and optimized peripheral operation. To enable interoperability across sensors and actuators from different manufacturers, the TinyML community also jointly develops standards for operation, representation, and transmission [9]. These elements are stated here to distinguish the key technical drivers of the TinyML domain and not as required knowledge for this review. The means by which raw data is translated to signals for the learning system that motivate the constraints detailed above should be apparent from the outset as a clear, robust deployment with TinyML.

TinyML is defined as developing ML models for resource-constrained edge devices such as microcontrollers. These machines generally lack dedicated hardware to accelerate mathematical operations, a minimum and fixed amount of random-access memory (RAM) and read-only memory (ROM), and a battery-connected power source with a tiny capacity. TinyML employs specialized model optimization to optimize ML models so that they are light enough to deploy and run on resource-constrained edge devices. In traditional ML, after data collection and model training on a server or the cloud, we need to deploy models on edge devices "close to sensors and actuators," usually for real-time data interpretation and device actuation based on the outcome [10]. In other words, only the preprocessed or sensor data are transferred to the model, and the model typically takes only a few seconds to produce an outcome.

TinyML powers today to make edge devices "embedded with intelligence," as reflected by smart home, energy, building automation, industrial monitoring, predictive maintenance, predictive farming, diagnostics, and healthcare monitoring. Edge devices can use TinyML chips to interpret raw data from the sensors, actuators, and nearby assets in real time, give commands, leverage on-device intelligence to tag raw data instead of streaming everything to the cloud, and make inferences and suggest possible solutions. The hardware components of a TinyML system typically include microcontrollers, sensors (on-chip and off-chip components), and memory. Microcontrollers' memory footprints vary significantly from kilobytes to a few megabytes, limiting the size of the model and the capacity of the learning algorithm due to memory constraints [11]. These microcontrollers implement at least one general-purpose processing core capable of running multiple threads with ultralow standby energy via fast entry-exit capabilities. The processing cores are optimized for low-power operation.

Likewise, software components designed to operate in remote settings or battery-powered nodes consume less energy during operation, providing computational operations compared to their server-based or cloud variants. The ML community is in the process of designing TinyML algorithms that have low memory footprint and low power consumption, especially as the tiny operations need to run many times in battery-powered systems. These tiny operations take several shapes and sizes but focus on tiny learning operations with a heavy emphasis on executing predictions with less computational processing, reduced memory access, and thus minimal energy for execution. The software tools, comprising both ML algorithms optimized for low-power consumption and tools to convert these learning algorithms into lightweight firmware that can operate on tiny microcontrollers, form the essentials of TinyML. Recognizing the potential of edge computing, many software framework players have launched lightweight embedded ML inferencing engines to perform inference directly on microcontrollers by deploying learning models [12]. Some prominent open-source ML deployment frameworks include lightweight versions optimized for operation analysis and are key to understanding the operational dynamics.

10.2.1 Security Threats in TinyML Devices

The proliferation of TinyML has been encouraging, though TinyML devices bring certain threats that can be a risk for use in commercial settings. For instance, an electronic device, no matter how tiny, is not immune to physical damage or subversion. One of the significant constraints for the security of small electronic devices is anonymity from the rest of the world, such as wireless connectivity. If two or more devices do not communicate with the outside world, the security threat these devices face is minimal. When it comes to security, gaining a digital interface is like opening a can of worms.

Physical attacks, software attacks, and side-channel attacks are significant threats. Physical attacks center around unauthorized access to both hardware devices and physical systems; if operators can prevent access, the integrity of the device is likely to remain intact. Software attacks usually come in the form of errors or vulnerabilities in operating systems and application software; they aim to use defective code to cause the application to crash or to compromise the device and hardware [13]. Therefore, side-channel attacks purport to empower attackers with knowledge about the physical properties in use or governing the design of a cryptographic algorithm; for example, repeated outputs may enable an attacker to guess if two messages use the same encryption key.

10.2.1.1 Physical Attacks

Physical attacks are performed directly on the TinyML device, and attackers can interact with the hardware as well. Unlike software-based attacks, physical security threats should be prevented at the device level to ensure that every functional component is adequately secured. Some common physical attacks include tampering, where attackers try to modify the device configuration or hardware design parameters to unlock protected features, and probing or reverse engineering, where adversaries try to extract sensitive information to access functionalities that are otherwise protected and prohibit unauthorized access. Most of the time, physical security measures are overlooked in a design due to ignorance of possible threats or the added costs involved in designing tamper-resistant devices [14]. However, with the growing number of secure installations in congested cyberspace, the cost per hack has increased, encouraging both attackers and organizations to invest in technology to follow attackers and defend their infrastructure.

A commercial off-the-shelf microcontroller can be physically attacked by a side-channel analysis that reveals the device's secret decryption key. A COTS flash memory chip could be cleaned out entirely with a laser-based side-channel attack. An attacker used chemical stripping, a microscope repair operation, and a reverse engineer's expertise to extract read-only memory physical unclonable feature secrets from a microcontroller. A successful ROM physically unclonable function (PUF) secret recovery entails the complete functionality invasion of the system, namely, having full access to its flash memory. Once this attack is accomplished, the ability to release the attacks in the future is a matter of reverse engineering efforts and analysis on the COTS microcontroller. Hostile ROM PUF secrets can now be calculated using the broadcast behavior of COTS flash memory storing adversarial prefetched instruction operations and PUF challenges, coupled with programming errors in the soft reconfigurable state [15].

10.2.1.2 Software Attacks

The main threat to TinyML devices is software attacks, which exploit flaws in the operating system or application software. Since most of these devices run an embedded operating system and execute some applications to collect and analyze data, software attacks may lead to unauthorized access, data manipulation, or malfunctioning of the device. Software vulnerabilities can be exploited in various ways, such as by infecting the operating system with malware, by infecting the application software, or simply by denying that service. Common attacks include malicious software infections such as ransomware, denial-of-service attacks, and code injections [16].

TinyML devices have limited processing capability, which may make them more prone to software attacks. Unauthorized access to these devices may result in data theft, data confidentiality compromise, and user privacy violations. Execution of malicious applications may result in data manipulation, data loss, and performance degradation in the devices. A denial-of-service attack may render the TinyML device unavailable to users. Given the vulnerabilities and potential consequences of software security breaches, rigorous software security practices such as secure design, secure coding, and software testing and validation are necessary.

Furthermore, access control, secure communication, and data monitoring can also contribute to thwarting malware and malicious code execution, as well as preventing unauthorized access to and data exfiltration from TinyML devices [17]. Recently, there has been a significant push to run sanitizers, static analysis, and formal verification tools on modern IoT devices, and similar methods can likely be used to detect potential security risks when building a TinyML model for deployment. For security-related issues where verification of code is standard, it should be clear that future software technologies provided for general-purpose devices, like those embedding TinyML, may lack these same constraints. The conclusion should be that more certain, simplistic, and transparent TinyML constructions may be necessary on these platforms in order to protect sensitive data stored in a deep learning classifier.

10.2.1.3 Side-Channel Attacks

Side-channel attacks exploit information made available by a system while it performs its regular functions. The information leak usually occurs in unintended ways via the physical implementation of the system, such as power consumption, timing information, and electromagnetic emanations. Such leakage might enable attackers to recover secrets or other sensitive information, including cryptographic keys or the direct output of the TinyML model. One of the most promising paths of side-channel attacks is ML models, which have been shown to suffer from privacy leakage via inversion attacks and backdoor attacks. Hardware side channels, captured through microarchitectural probes such as data caches and micro-op caches, can lead to the extraction of full cryptographic keys in the presence of known plaintext attacks and to plaintext exfiltration on encrypted channels even if the attacker only controls the network stack.

In the context of TinyML, powerful side-channel attacks have been developed to extract the output of binarized, quantized, and pruned models and leverage this information to backdoor ML retraining [17]. These types of attacks can provide backdoor reprogramming success rates of up to 99.375% under strong evasion attack scenarios. The adaptive countermeasures to mitigate such attacks can lead to the extraction of TinyML model output with a deviation reduction of up to 76.46% compared to non-mitigated output extraction. Such attacks, when extended to TinyML, would totally undermine trust in autonomous smart devices. Security analysts have come up with interesting case studies of passive side-channel attacks against IoT devices. For privacy leakage, the models are recovered by considering physical access to the training data, showing a successful inversion attack scenario. Finally, conducting side-channel resistance measurements allows the designer to ensure that their proposed design contains appropriate mechanisms to protect against passive observation attacks.

10.3 Privacy Concerns in TinyML

The introduction of processors with ML capabilities has led to the development of low-powered TinyML devices, which are gaining extensive interest from the engineering and scientific community. These tiny devices will, most of the time, be used close to the end user. Being the major or sole source of private information being dealt with, the use of such data marks a need for attention. Most applications of TinyML are centered on data from physical sensors or smartphones, and this data is frequently sensitive and personal. Neither the consumer nor the end user knows the extent to which information can be crafted based on the data. A primary concern regarding the use of TinyML is the potential misuse of user data by employers and governments. ML models created using user information might be used to compromise the privacy of the user's data or to block freedom of speech [17].

The definition of privacy is multifaceted and spans various domains, including data acquisition, usage limitations, data sharing, and data deletion, making it essential to define and consider data privacy separately. Consumer confidence is essential when launching new TinyML technologies, and this is an essential issue for those trying to develop secure TinyML devices. User confidentiality and privacy preferences should be taken into account during the TinyML system design, while user experience considerations should be taken into account during the actual implementation of the system. As more and more of such implementations occur, concern for data privacy and security has raised a wave of laws and regulations worldwide [18]. Though ML applications are increasingly important, ethics and privacy remain pressing issues.

10.3.1 Data Privacy and Protection

The first and foremost issue that needs to be addressed, from the security and privacy viewpoint, while considering a data-driven application, is how to protect the privacy of the data that is collected by the application. Data privacy is involved in all data lifecycle aspects, such as collection, processing, storage, and sharing. Therefore, data collection, the first step in the data lifecycle, is critical since once data are collected, the privacy of the owner of the data is at risk. As a result, it is crucial to protect data and ensure that they are only accessible by authorized users. To achieve this, privacy-preserving techniques must be put in place at the data collection stage as well as across the rest of the data lifecycle processes to safeguard against unauthorized access. This is critical in TinyML, where devices are built to interact with each other and share data, and so challenges to the privacy of data are increased [18].

This motivates the urgent need for methodologies to protect the data privacy of systems for TinyML. For example, data anonymization techniques can help in hiding the identity of data owners, and encryption can protect sensitive data. In addition, FL and DP can also be utilized to enhance the privacy of the system. Also, users have the right to control the way that the service or entity provider processes their data, such as providing their consent to accept or deny usage of their data for specific purposes or accessing their data. Service providers have a legal obligation to comply with regulations that state the right of users to have full access to and full control of their data in the process. Data privacy, however, doesn't merely refer to protecting the originator's personally identifiable information; it encompasses the protection of all individual data, such as aggregated data [18]. Therefore, what is necessary in the context of TinyML applications is the development of frameworks that address the privacy of data across the entire data lifecycle.

10.3.2 User Privacy and Consent Issues

It is important to acknowledge that the way we collect data can impact the autonomy and possibly trust between users and the organizations deploying such sensing mechanisms. Engaging in data collection without obtaining consent from the users to collect details of a potentially sensitive nature could raise ethical and possibly regulatory issues. The role of transparency is essential. Informed, explicit consent is likely under the control of regulatory frameworks, and there are challenges that the right to give informed consent could impose on embedded devices. What would be even more difficult for an embedded learning scenario is if the inferred properties of an individual were derived as a result of data sharing using cloud platforms. Data sharing and interoperability mandates providing adequate information to allow an individual to make an informed choice prior to consent [18].

One of the overarching issues with a lot of data usage and exchange practices is often the adequacy of transparency and informed consent in this space. There are regulatory frameworks that enforce some level of user rights to understand and question their data sharing. There must be prior consent to transfer personal information to a third party if the jurisdiction is different from the one in which the data was collected and the receiving country does not have similar privacy protections. The consent of the data subject is acknowledged as the legitimate basis for transferring data between jurisdictions with different privacy protections. Specific requirements for consent include that it must be freely given, specific, informed, and explicit. However, managing informed consent is challenging in the space where there is a shortage of capabilities with some of the emerging edge ML [19]. It is essential to ensure that the interests and rights of the individual are prioritized in the operation and design of technologies.

10.4 Security and Privacy Solutions

Developing effective security and privacy solutions that are compatible with the constraints of TinyML is necessary. However, the diminished technical and economic incentives present in the edge and IoT markets mean that many device manufacturers will not undertake security and privacy-enhancing efforts unless there are external pressures to do so. The dissemination of security and privacy knowledge is therefore necessary so that consumers or regulatory bodies will have the requisite information to qualitatively differentiate these sensor deployments based on their designs or the manufacturer's security posture [10, 19].

There are two primary features of the device design and development process that will have a significant impact on the fundamentals of device security and privacy posture: hardware and software. A range of hardware-based protections also exists in the data protection and encryption space, but many of these are designed for larger general-purpose computers and have performance-speed-accuracy trade-offs that make them prohibitive for a small microcontroller. Examples of hardware-based protections for larger computers include the use of secure enclaves and tamper-resistant hardware designs. These ensure that if an adversary were to compromise the device and extract the computational secrets physically, it would be infeasible to use the results of that computation as substitutes in a new operating device or to learn the details of what transpired [11].

Practices that can enhance software security on these devices include defect mitigation, for example, Software Development Life Cycle (SDLC) and secure coding practices, ensuring that the software (and hardware, when possible) is constructed in a manner that prohibits whole classes of attacks on the system. This could include performing vulnerability assessments to enumerate the potential attack surfaces on a device and then hardening those surfaces to resist attacks. Protection of data in transit and at rest is necessary to ensure the integrity of the data and the privacy of the user. Combine this with ensuring that network traffic associated with these systems can be authenticated to prevent spoofing and depersonalization via encryption so that intermediate parties on the network, with potentially intrusive insight into user traffic, will not have personalized user data that violates user privacy [12].

10.4.1 Hardware-Based Solutions

Integrating hardware solutions in TinyML devices is essential to enhancing their security. Key hardware security technologies include secure chips and hardware security modules, which can be used to provide, for instance, encryption functionalities and secure key storage. These modules can also be leveraged to perform relevant security functionalities, including securely identifying a device, secure data storage, secure firmware updates, secure remote access, code and data protection, secure boot procedures, and secure on-board cryptographic operations. A secure chip can be used for, among other things, establishing a trusted execution and data processing environment, establishing a secure link with the device, and enabling secure registration and verification of the devices [13]. Security-aware training of TinyML Models is presented in Table 10.1.

For embedded devices in general, and TinyML devices in particular, the combined use of hardware-based cryptographic operations, hardware-based encryption tamper protection, and physical security features can be precious in resisting physical attacks. Hardware-based encryption typically requires a secret or confidential key to be stored securely in a key storage memory or dedicated hardware security module inside or connected to a microcontroller or processor. During the decryption process, minimal or no sensitive data is made available to the software or becomes directly accessible as unencrypted data, reducing the risk of unauthorized data access. A hardware security module can improve the security of key and data operations by integrating functionalities resistant to external active and semi-invasive physical attack techniques [14]. Relevant industry standards and guidelines exist, defining requirements and checking methods for secure cryptographic devices.

The importance of hardware-level security features is also clearly depicted in real-world use cases and case studies. For example, a modern approach to secure code storage and execution on a microcontroller in the cloud utilizes the microcontroller's architecture-based components to make this process as

Table 10.1 Security-aware training of TinyML models.

References	Security-aware training	Description	Technology used	Devices	Loophole
[7]	Data provenance and integrity	Tracks and verifies the origin and integrity of training data to ensure it is reliable and secure	Blockchain, data integrity verification	Agricultural sensors, industrial IoT devices	Data provenance systems may be bypassed or compromised, leading to trust issues
[8]	Access control mechanisms	Implements strict access controls to manage who can view or modify the model and its parameters	Role-based access control (RBAC), multi-factor authentication	Data servers, cloud-based platforms	Access controls can be bypassed if not properly configured or if credentials are compromised
[9]	Model encryption	Encrypts the model itself to prevent unauthorized access and reverse engineering	Model encryption, obfuscation	Embedded systems, smart appliances	Encryption can be broken if cryptographic standards are weak or outdated
[10]	Privacy-preserving techniques	Implements methods like differential privacy and federated learning to protect sensitive data	Differential privacy, federated learning	Healthcare devices, smart home systems	Privacy methods might not fully prevent data inference attacks
[11]	Secure update mechanisms	Uses cryptographic techniques to verify and apply model updates securely	Digital signatures, hash functions	Smart devices, medical equipment	Secure update processes may be compromised if not properly implemented
[12]	Robust model evaluation	Involves rigorous testing to identify and address potential vulnerabilities	Penetration testing, stress testing	Security cameras, industrial IoT Devices	The evaluation may miss novel or sophisticated attacks

[13]	Quantization and secure deployment	Reduces model size and computational needs while ensuring secure deployment practices	Quantization techniques (post-training quantization), secure boot	Mobile devices, consumer electronics	Quantization may introduce precision errors, and deployment may be vulnerable if not secured
[14]	Model pruning with security in mind	Reduces model size while preserving critical features to avoid introducing vulnerabilities	Model optimization, sparse training	Edge devices, industrial sensors	Over-pruning may lead to loss of critical features, affecting model performance
[15]	Secure data handling	Encrypts and anonymizes training data to protect it from unauthorized access	Encryption algorithms (RSA,[a) AES[b)]), data masking	IoT sensors, medical devices	Encryption keys may be exposed, or data may be improperly anonymized
[16]	Adversarial training	Exposes models to adversarial examples to improve resilience against attacks	Adversarial ML, perturbation techniques	Smart home devices, healthcare wearables	May not cover all possible adversarial examples, leading to potential blind spots

RSA, Rivest–Shamir–Adleman; AES, advanced encryption standard.

a) https://www.encryptionconsulting.com/education-center/what-is-rsa.

b) https://www.techtarget.com/searchsecurity/definition/Advanced-Encryption-Standard.

efficient as possible. The solution was also compatible with a growing set of security features that extend over a trust boundary from the silicon to the cloud. The secure chips used in practice acted as a hardware root of trust. Combined with a typically used method for the secure operation of the main processor in the industry, it allowed remote attestation and secure condition detection processes in their protocol. Secure chips were also used to store RSA and symmetric secret keys in a project, focusing on a low-power IoT-based facial authentication and identification system [15]. This was one of the key security features of the facial recognition solution, respecting the identification standard for driver's licenses and electronic visa systems. Each module was given a unique, secure chip, ensuring that the symmetric keys stored in them always satisfied the requirements provided in the guidelines. This unique hardware feature was verified to give the system software and developers knowledge of whether the module functions under the correct purchased quality level. According to the guidelines, the setup was tested using an obtainment setup that would generate a unique, non-containing backdoor, data hiding, and a feature for financially identifying the chip [16]. A secure reference firmware was created for the security check. Development of the main module system firmware continues.

10.4.2 Software-Based Solutions

To bolster the security posture of TinyML devices, companies can also aim to address software-based issues on both the device and server sides. This includes: Secure Coding Practices: The main focus should be to follow a secure development lifecycle during the development of software for TinyML devices. During the design phase, identify various types of nodes for identification purposes. Frequent Vulnerability Assessment: It is quite likely that companies can easily find or assess vulnerabilities within the TinyML software stack. These identified vulnerabilities can be reported to the developer for fixing , which can be provided to the user through regular software updates. Regular Software Updates: Supply the TinyML software updates in a phased manner to address the known vulnerabilities on a priority basis (where unknown vulnerabilities are not fixed once the device is under deployment) [17].

By developing software, identifying weaknesses, and updating software with added security solutions, ML can be used for real-time threat detection. Vulnerabilities will continue to exist in the environment from which solutions are derived, and the emphasis will be on addressing these vulnerabilities as quickly as possible. Manufacturers of TinyML devices will be forthcoming in implementing control measures to address the identified vulnerabilities. By implementing software-based solutions, organizations can demonstrate a prudent level of control to identify and address weaknesses in software through a consolidated known vulnerability document. In this way, developers can fix the known issues well in advance during the development lifecycle.

Organizations that develop TinyML embedded devices must use internal or external known threat intelligence services to find any embedded malware in their solution that resides on the device itself. By providing malware-free solutions that monitor device memory and constantly look for any embedded malware in the device or solution, manufacturers can win the trust and satisfaction of the consumer. Developers can start providing such exceptions at design time when the device is being developed and functions are being provided [18]. This needs to be built in during the initial device design and not as an add-on feature at the end.

10.4.3 Encryption and Authentication Techniques

Encryption is a critical component for security on TinyML devices. Data transmitted from these devices to a remote server or database and stored on the device can be intercepted or otherwise breached if left unencrypted. As such, one of the primary tenets of encryption is its ability to prevent unauthorized access or exposure of sensitive (and often private) data. Systems or devices that utilize cryptography will want to be sure that the encryption standards used are intense, challenging to break, and provide protection from techniques such as cryptanalysis.

Moreover, these standards should be able to ensure the integrity and confidentiality of the data being transmitted or analyzed. Some suggested algorithms and standards include the Advanced Encryption Standard, elliptic-curve cryptography, and protocols such as RSA and hash-based message authentication code – Secure Hash Algorithm 1 (HMAC-SHA1) [19]. The prime concepts in protecting data include two significant ways of encryption. Authentication, such as encryption, is of equal importance in maintaining a safeguard on devices and data; a valid user identification or device recognition must be enforced before any data is potentially stored or transmitted further.

For these TinyML devices to be secure, robust methods must be used to verify that data is encrypted and that a user's data is held confidential. This often takes the form of some physical device in the user's possession. One of the most common forms of protecting data on a device is the use of cryptographic keys, typically a pair consisting of a public key and a private key. Key management is an essential aspect of security that must be taken into account when developing these systems [20]. For that reason, it is necessary to guarantee that privacy policies are transparent, carefully implemented, and routinely updated to deal with arising personal privacy worries and governing needs. Table 10.2 summarizes the privacy-aware training of TinyML models.

Table 10.2 Privacy-aware training of TinyML models.

References	Security-aware training	Description	Technology	Devices	Loophole
[9]	Anonymized data synthesis	Synthesizes data that is anonymized to prevent reidentification of individuals while preserving data utility	Anonymization algorithms, synthetic data generators	Data analytics platforms, research tools	Anonymization methods might not be foolproof; synthetic data may still be inferable
[10]	Encrypted Computation	Performs computations on encrypted data without decrypting it, protecting privacy during processing	Homomorphic Encryption, Secure Computation Protocols	Secure Cloud Platforms, Encrypted Data Servers	Computational overhead requires robust encryption methods
[11]	Privacy-aware hyperparameter tuning	Adjusts model hyperparameters with privacy considerations to minimize leakage of sensitive information	Hyperparameter optimization techniques, privacy-enhanced algorithms	Training platforms, cloud-based services	Hyperparameter adjustments may compromise privacy if not appropriately managed
[12]	Synthetic data validation	Validates synthetic data to ensure it accurately represents the real data while protecting privacy	Data validation tools, statistical analysis	Data generators, testing platforms	Synthetic data might still reveal patterns; validation must be thorough
[13]	Secure model deployment	Ensures that deployed models are protected against tampering and unauthorized access	Model encryption, secure boot, access controls	Edge devices, cloud services	Attacks can target deployment environments; encryption must be robust
[14]	Transparent privacy policies	Clearly communicates data collection, usage, and protection practices to users	Policy management tools, consent management systems	Platforms and applications	Users may not fully understand policies; enforcement must be rigorous

[15]	Privacy-enhancing technologies	Uses encryption, anonymization, and access controls to protect data privacy	RSA and AES, encryption, anonymization techniques, access controls	Data storage systems, cloud platforms	Performance overhead: vulnerabilities if not implemented correctly
[16]	Data minimization	Collects only the data necessary for model training and operation, reducing risk of exposure	Data filtering, selection algorithms	Sensors, smartphones	May limit model performance; essential data might be excluded
[17]	Privacy-preserving data augmentation	Generates synthetic data to train models without exposing real personal data	Generative adversarial networks, data anonymization	Data collection devices, edge devices	Synthetic data may impact model performance; anonymization effectiveness varies
[18]	Secure multiparty computation	Enables joint model training across parties without sharing raw data using cryptographic techniques	Homomorphic encryption, secret sharing	Collaborative research platforms, distributed IoT networks	Complex to implement; cryptographic protocols must be robust
[19]	Federated learning	Trains models across decentralized devices with only model updates shared, keeping data local	Federated learning frameworks, secure aggregation	Mobile phones, IoT devices	Vulnerable to model poisoning and inference attacks; aggregation must be secure
[20]	Differential privacy	Adds controlled noise to data or model outputs to obscure individual data points, ensuring privacy	Differential privacy algorithms, noise functions	Healthcare wearables, smart home devices	Noise can degrade model accuracy; managing the privacy budget is crucial

IoT, Internet of Things.

10.5 The Implementation of Novel Security Schemes for TinyML Applications

TinyML represents the execution of ML models on ultralow-power edge devices, including microcontrollers, wearables, and IoT sensors. The use of TinyML is proliferating, particularly in domains like healthcare, autonomous systems, and smart homes. Given the constrained nature of these devices, ensuring security poses significant challenges. Developing and implementing novel security schemes tailored specifically for TinyML applications is critical to protect against cyber threats. This section explores several advanced security techniques specifically adapted to the needs and limitations of TinyML systems.

10.5.1 Intrusion Detection System (IDS)

As IoT resource-constrained systems expand, the complexity of cyber threats grows accordingly, making the need for effective threat mitigation more critical. Intrusions, unauthorized activities that compromise the confidentiality, integrity, and availability of these systems, must be detected and countered [21]. Intrusion Detection Systems[4] (IDSs) are essential tools in this regard, as they monitor and analyze network traffic or system activities to detect unauthorized access, misuse, or potential breaches [22]. IDSs generally fall into two categories: Signature-based Intrusion Detection Systems (SIDSs) and Anomaly-based Intrusion Detection Systems (AIDSs) [23].

The SIDS functions by matching network activity to a database of known attack signatures, leveraging supervised ML algorithms to enhance its ability to detect familiar threats. However, these systems face limitations due to their dependence on labeled datasets, which makes it difficult to detect novel or previously unseen threats. Additionally, maintaining the large databases required for SIDSs can be resource-intensive [24]. Traditional SIDS architectures, often centralized, may also introduce vulnerabilities, such as single points of failure and privacy concerns [25].

In contrast, the AIDS uses ML and statistical techniques to establish a baseline of normal system behavior, flagging deviations as potential security threats [26]. These systems, particularly those utilizing unsupervised learning, are capable of detecting anomalies and new patterns without relying on predefined criteria. This flexibility enables AIDSs to support sophisticated malware detection, spam classification, and intrusion detection frameworks. To achieve this, AIDSs typically rely on powerful servers with high computational capacity to analyze vast datasets and identify vulnerabilities [27]. However, AIDSs, or behavior-based

4 https://www.ibm.com/topics/intrusion-detection-system.

IDSs, are often plagued by a high False Alarm Rate (FAR), which can lead to unnecessary disruptions, wasted resources, and system downtime while verifying alerts. Furthermore, the growing threat of zero-day attacks – exploits targeting unknown vulnerabilities poses a significant challenge to AIDSs. These attacks require rapid adaptation, a difficult task given the unpredictable nature of such threats [24].

Several studies have investigated various ML methods for IDSs. For example, authors of [28] present an IDS that employs a deep auto-encoder along with multiple deep decoders for unsupervised classification. This system processes raw, unlabeled network data, compressing it into a simpler form using the auto-encoder, while the decoders work to reconstruct the original data. Through this reconstruction process, the IDS learns to identify anomalies in network traffic that could signify potential threats. Also, [29] introduces an IDS using XGBoost, an ML technique renowned for tree-boosting capabilities. In this system, decision trees are constructed sequentially, correcting errors identified by previous trees. The primary strength of XGBoost lies in its ability to strike a balance between bias and variance, allowing the system to make accurate predictions while minimizing the risk of overfitting. In a different context, the authors of [30] explore the use of several ML classifiers for IDSs in electric vehicle charging stations (EVCSs).

The importance of IDS in electric vehicle charging infrastructure (EVCI) is critical to maintaining both cybersecurity and operational integrity. Authors in [31] demonstrate the effectiveness of ML classifiers in detecting distributed-denial-of-service (DDoS) attacks within EVCS networks by analyzing IoT datasets to detect malicious traffic patterns specific to EVCS environments. In wearable devices, Fyntanidou et al. [32] developed a wrist-worn device that tracks vital signs such as body temperature, respiratory patterns, and blood oxygen levels to assist in prioritizing COVID-19 patients in emergency settings.

Utilizing TinyML technology, the device's neural network processes respiratory data locally, eliminating the need to transfer information to the cloud and ensuring the privacy of sensitive patient data. Moving beyond traditional ML approaches, TinyML emerges as a game changer for IDS, particularly in resource-constrained environments. TinyML enhances both SIDS and AIDS by optimizing ML models for edge devices, enabling real-time, resource-efficient anomaly detection. Its adaptive capabilities minimize latency, reduce network dependence, and strengthen cybersecurity resilience, allowing systems to adapt to evolving threats.

10.5.2 Trusted Execution Environment (TEE)

Trusted execution environment (TEE) refers to a secure, isolated environment within a device that ensures sensitive computations, such as those involving ML models or private data, are performed securely. TEEs are designed to protect data and code integrity from unauthorized access or tampering, even if the primary operating system is compromised. TEEs are becoming more commonly available on modern embedded processors [33]. One notable example of such embedded processors is ARM, where a TrustZone security feature is integrated into ARM-based processors that create a secure execution environment [34]. It enables the separation of trusted and untrusted resources on a system, effectively dividing the hardware into two modes: secure and normal. The Secure mode handles sensitive tasks like encryption and authentication with access to both secure and nonsecure resources, while the normal mode runs regular applications with restricted access to ensure the protection of critical operations. Because of its TEE capability in providing a secure environment, it has gained significant attention in the development of speech-based applications.

VoiceGuard [35], for example, was one of the pioneering efforts to implement Intel SGX, a well-established TEE solution for automatic speech recognition (ASR) systems. Similarly, Bayerl et al. [36] introduced the Offline Model Guard (OMG), a TEE-based architecture designed to support keyword spotting (KWS) tasks on ARM-powered mobile devices. Furthermore, iFlask [34] introduced a security system that isolates the Flask security subsystem from unsafe environments using ARM TrustZone, by placing the security server in a protected enclave. They address mobile runtime security issues and mitigate threats from malicious environments. Implemented on SELinux, which underpins SEAndroid, their prototype successfully protects against all known vulnerabilities, according to experimental results. Similarly, authors in [37] developed a system to prevent sensitive information leakages, like contact lists or private photos, using a fine-grained access control scheme based on Ciphertext-Policy Attribute-Based Encryption (CPABE), and TEE. Their approach supports secure access during critical operations, outperforming traditional access-control methods.

Moreover, the work [38] proposed PrivateRS, a private recommender system using a client/server framework. Assuming an untrustworthy server, the client randomizes user ratings with DP techniques and replaces them with private symbols, allowing the server to predict item ratings without revealing sensitive data. Their experiments showed that the system can generate accurate recommendations while maintaining privacy. The works [39, 40] propose two notable approaches for ensuring the security of on-device learning models: ShadowNet and MirrorNet. ShadowNet, designed around convolutional neural networks (CNNs), modifies the linear layer weights before sending them to the untrusted environment and subsequently restores them within the TEE. This enables secure outsourcing of the model's heavy linear layers without exposing the weights.

Alternatively, MirrorNet connects a lightweight mirror network to the base deep neural network (DNN) model, with the mirror network residing in a secure environment. This solution effectively prevents model extraction and addresses the computational and storage constraints of TEEs. Finally, [41] presents STML, a system and algorithm codesign aimed at securing the intellectual property (IP) of TinyML on MCUs equipped with ARM TrustZone, focusing on resource-constrained devices. The design optimizes both memory utilization and latency while also ensuring the security and accuracy of emerging models. A prototype was implemented and benchmarked using seven models, showing that STML significantly reduces model protection runtime overhead.

10.5.3 Lightweight Cryptography

TinyML devices typically operate with limited memory, processing power, and battery life, making the use of conventional cryptographic algorithms infeasible. Lightweight cryptography has emerged as an essential approach for securing TinyML applications, offering data protection without overburdening the system's resources. Generally, lightweight cryptography is broadly categorized into symmetric and asymmetric algorithms. Symmetric algorithms are further broken down into two types: Lightweight Block Ciphers (LWBC) and Lightweight Stream Ciphers (LWSC). In contrast, elliptic curve cryptography (ECC) is a form of asymmetric cryptography. The performance of lightweight cryptographic techniques is assessed based on several parameters, including key size, block size, the number of rounds, and the underlying algorithmic structure. In this direction, authors [42] introduced the LWHC system, combining LED and PRESENT ciphers with the SPECK key scheduling algorithm, enhancing security through the RECTANGLE S-Box. Although resistant to key attacks, the system leaves other vulnerabilities unaddressed. Noura et al. [43] proposed a dynamic lightweight stream cipher (LSC) optimized for IoT, combining CR4, PRNG, and LFSR with high resistance to various attacks but remaining vulnerable to disclosure attacks. Also, [44] developed SAT_Jo, a lightweight block cipher for IoT, balancing security and resource efficiency through a substitution-permutation network (SPN). Authors of [45] created a hybrid cryptographic model using PRESENT and Salsa20 with a chaotic logistic map for key generation, providing both speed and complexity.

Notably, Noura et al. [46] presented the One Round Cipher (ORC), a dynamic lightweight encryption model offering high randomness and resistance to critical attacks. Authors in [47] introduced G-TBSA for energy-efficient encryption in wireless sensor networks, though it is restricted to sensor devices. The framework in [48] enhanced the PRESENT cipher by reducing rounds, and improving tiny can significantly enhance privacy and security. Register security, though battery consumption remains untested. The KDF scheme [49] introduced dynamic key generation with real-time applications, ensuring resistance to several attacks while maintaining low computational complexity. Furthermore, authors in [50] proposed Lightweight CA (LCC), a cipher that outperforms DES and 3DES with strong randomness and efficiency.

Zhao et al. [51] developed a modified QARMA architecture, enhancing speed and gate area efficiency but requiring further optimization. Similarly, reference [52] presented MBCT, a block cipher with matrix rotation and XOR functions that reduced memory use and improved speed. Mishra et al. [53] introduced lightweight cryptographic primitives, combining differential logic and random key generation to reduce memory and time while increasing complexity. Authors in [54] developed chaos-based PRNG encryption for patient data security, demonstrating strong real-time performance and resistance to attacks. Gyamfi et al. [55] enhanced ECC by integrating Diffie-Hellman key exchange and AES, making it suitable for resource-limited IoT systems. Khan et al. [56] proposed IECC, an elliptic curve-based security framework for medical IoT sensors, offering faster encryption and lower overheads. Mohandas et al. [57] introduced A4, a stream cipher using LFSR and FCSR, with strong resistance to algebraic and brute-force attacks. Finally, authors of [58] designed NLSC, a chaos-based stream cipher suitable for resource-constrained devices, demonstrating resistance to statistical attacks.

10.6 Privacy-Enhancing Techniques for TinyML

TinyML, with its ability to perform on-device ML on edge devices such as IoT sensors and wearables, introduces privacy concerns, especially when dealing with sensitive data in healthcare, smart homes, and autonomous vehicles. To address these privacy concerns, several techniques are employed, each carefully designed to align with the resource constraints and computational limitations of TinyML environments.

10.6.1 Federated Learning

Data privacy remains a critical concern for individuals, particularly when it involves the transmission and storage of their personal information. FL can address this by enabling TinyML models to be trained locally on devices while data, such as medical records, must remain local, eliminating the need to transfer sensitive raw data to the cloud server [59]. Hence, incorporating FL into TinyML can significantly enhance privacy and security. For instance, authors in [60] have introduced TinyFedTL, a decentralized learning mechanism. This approach keeps data processing on the device itself, thereby avoiding the centralization of sensitive information and strengthening security. Federated transfer learning provides a privacy-respecting solution for the TinyML community, reducing communication overhead and accommodating the hardware limitations of microcontrollers like the Arduino Nano 33 BLE Sense.

Similarly, Mathur et al. [61] demonstrated FL on TinyML edge devices such as Android phones, Raspberry Pi, and NVIDIA Jetson using the Flower framework. They implemented the clients in Java and Python, allowing on-device training with TensorFlow Lite. This approach differs from resource-limited systems, such as the Arduino Nano 33 BLE, where models must be trained off-device and later transferred to the device for inference [62]. Moreover, the authors of [63] review key challenges in TinyML, such as power consumption, limited memory, and computational constraints, while exploring solutions like energy harvesting, optimization techniques, and privacy-preserving FL. They also examine TinyML's role in enhancing cybersecurity for EVCIs through an experimental case study. This study shows that TinyML reduces delay and memory usage compared to traditional ML, with a minor accuracy trade-off, and includes a practical demonstration using the ESP32 microcontroller.

Although FL keeps the raw data safe, it is still vulnerable to various privacy and security threats, mainly because sensitive information can still be inferred from shared model updates [63]. Similarly, in [64], the authors propose FedLVR, a privacy-preserving FL framework optimized for fine-grained vehicle detection in edge-assisted intelligent transportation systems. This framework deploys a lightweight vehicle recognition model on edge nodes, achieving high accuracy while notably reducing model complexity, as demonstrated in experimental evaluations.

Rahim et al. [65] present the EET-SAGIN algorithm, leveraging the TinyFDRL framework in a four-layer aerial system combining ground nodes, low- and high-altitude UAVs, and satellites. Integrating FL, deep reinforcement learning, and TinyML enables distributed UAV applications across diverse environments. Simulations show that while energy consumption increases with more UAVs, TinyFDRL consistently achieves the lowest consumption with 20 drones due to optimized parameter learning.

10.6.2 Differential Privacy (DP)

This has emerged as a critical privacy-preserving technique for TinyML, ensuring that sensitive information in local datasets is protected while allowing for collaborative model training and inference. In TinyML, where models are deployed on resource-constrained edge devices [66], DP can prevent the leakage of private data during model updates and inference by adding controlled noise to the data or model gradients. DP was initially introduced by Dwork [67]; it is a statistical model aimed at preserving data privacy. It works by applying carefully calculated random noise to the data through various mathematical techniques to ensure anonymity. There are many types of DP. Global DP (GDP) adds noise after data aggregation, while Local DP (LDP) applies noise directly to individual data, ensuring privacy at the source. Zero-concentrated DP (zCDP) and Rényi DP (RDP) offer fine control over privacy budgets, particularly in iterative ML processes. Approximate DP (ε, δ)-DP allows minor privacy breaches for practical use. Group DP extends privacy to groups, and PATE applies DP in model aggregation for sensitive data. Each type balances privacy and utility based on application needs [67, 68].

Conventional DP relies on the assumption of a trusted central authority, a requirement that is frequently impractical [69]. Therefore, LDP [70] has been adopted in FL to eliminate the need for a trusted central party, but it struggles with deep learning on heterogeneous IoT data due to the trade-off between privacy and utility [71]. Key challenges include the degradation of model performance from fixed-range gradient clipping and the rapid accumulation of the privacy budget during iterative training. Sun et al. [72] proposed adaptive weight clipping, though it only adapts at the layer level, and the noise introduced by LDP continues to impair accuracy. For TinyML devices, authors of [73] integrated several LDP mechanisms, including PM-OPT and PM-SUB, with FedSGD to form the LDP-FedSGD algorithm, enabling privacy-preserving traffic prediction in vehicular crowdsourcing by transmitting noisy gradients.

Experiments show superior performance, with plans to extend this to deep learning tasks. Similarly, an LDP-based framework for clustered FL on heterogeneous IoT data is proposed in [74]. To reduce LDP noise impact, an adaptive layer-wise perturbation method, along with parameter shuffling and compression, is used to minimize privacy budget issues and communication overhead. This approach effectively balances privacy and utility in FL models. DP is also applied in centralized speech applications to obscure speaker voiceprints, as demonstrated in VoiceMask [75]. In contrast, another approach ensures privacy by introducing dummy words into the output transcripts [76]. While these techniques focus on centralized training, there is a growing trend of incorporating DP within FL. For example, DP has been applied to mitigate property inference attacks in the training of FL-based speech-emotion recognition models [77].

10.6.3 Homomorphic Encryption (HE)

This is a cryptographic technique enabling computation on encrypted data, ensuring privacy by never revealing the underlying data to the processing entity. For TinyML, which involves ML models deployed on resource-constrained edge devices, homomorphic encryption (HE) has emerged as a promising solution for privacy-preserving computations. However, the computational overhead and memory usage associated with HE pose significant challenges in the context of TinyML. Recent innovations focus on reducing this overhead and making HE feasible for real-time TinyML applications.

Hybrid homomorphic encryption (HHE), which combines the efficiency of symmetric cryptography with HE, has been a key advancement. HHE reduces the computational burden by offloading intensive cryptographic operations to the cloud, allowing edge devices to perform lightweight encryption tasks while maintaining robust privacy [78]. In this direction, Xu et al. [79] developed the OpenFHE library, an open-source tool designed to enhance the performance of fully homomorphic encryption (FHE). The library integrates several hardware acceleration techniques, such as GPU and FPGA support, to make HE more practical for edge computing scenarios like TinyML. This work focused on the bootstrapping mechanism, which is critical for reducing noise and enabling deeper computations on encrypted data.

A HHE framework enables privacy-preserving ML on edge devices. This framework demonstrated its effectiveness in reducing computational costs by selectively applying encryption only to sensitive parts of the data [78]. The proposed solution was tested on a heart disease prediction model, showing that HHE could enable efficient and secure model inference on TinyML devices. Shabir et al. [80] presented an aggregation scheme for FL using HE, designed specifically for TinyML devices. Their study addressed the communication bottlenecks in secure FL environments, showing how efficient homomorphic operations could significantly reduce both computation and communication overhead while maintaining privacy.

Lou and Jiang [81] introduced HEMET, a HE-friendly privacy-preserving mobile neural network architecture. This work addressed the challenge of using traditional mobile network architectures for privacy-preserving neural networks (PPNNs). They found that simply reducing parameters did not lower inference latency due to the increased multiplicative depth in HE. HEMET optimized this architecture, resulting in a 59.3–61.2% reduction in inference latency while maintaining accuracy, which makes it particularly valuable for real-time TinyML applications. Also, the approach integrates FHE with deep learning, specifically for edge devices. They demonstrated the application of FHE in time-series analysis using a one-dimensional convolutional neural network (1D CNN) for predicting dairy herd performance [82]. This work highlighted the practical challenges of integrating FHE in real-world neural network models and laid the groundwork for further optimizations in edge computing scenarios.

10.7 Future Directions of Security and Privacy in TinyML Devices

As TinyML continues to expand its applications across various industries, the security and privacy of these devices have become critical concerns. TinyML operates at the edge, often processing sensitive information directly on-device, which presents unique challenges related to the protection of data and model integrity. The resource-constrained nature of TinyML devices, limited in computing power, memory, and energy, adds further complexity to the implementation of robust security measures. Moving forward, research and development in the security and privacy of TinyML devices are expected to focus on several key areas.

10.7.1 Secure Model Inference and Execution

Ensuring the security of model inference is a significant challenge for TinyML devices, as adversarial attacks can target both the model's input and output. Future advancements will focus on developing secure inference protocols that protect against adversarial examples, deliberately manipulated inputs designed to deceive the model into making incorrect predictions. Additionally, techniques like HE and secure multiparty computation (SMPC) will evolve to enable secure model inference without revealing sensitive data to external entities, albeit with more efficient and lightweight adaptations suitable for TinyML devices.

10.7.2 Adversarial Attack and Defense Mechanisms

As TinyML devices become more prevalent, they will become targets for various adversarial attacks, including data poisoning, model inversion, and evasion attacks. Future research will focus on creating robust adversarial defense mechanisms that can be deployed on resource-constrained TinyML platforms. This includes developing lightweight adversarial training methods to improve model robustness, as well as designing detection systems to identify and mitigate adversarial attacks in real time. Additionally, the concept of "secure by design" will become more prominent, where TinyML models are designed with inherent resistance to adversarial manipulation.

10.7.3 Federated Learning (FL) with Differential Privacy Security

FL is widely recognized as a privacy-preserving method of distributed model training, but it faces unique security challenges in the context of TinyML. In the future, secure FL for TinyML devices will focus on mitigating threats like model poisoning, where malicious devices can corrupt the global model. Innovations such as Byzantine-resilient aggregation techniques, secure model updates, and DP-integrated FL will be essential to ensure the integrity of the model without relying on computationally expensive operations. Moreover, advancements in hierarchical and decentralized FL architectures will enable more secure, efficient, and scalable deployment of TinyML systems in environments with a large number of devices.

10.7.4 Tamper-Resistant Hardware for TinyML

Physical tampering poses a serious risk to TinyML devices, particularly when deployed in untrusted environments, such as industrial settings or remote locations. Future research will focus on developing tamper-resistant hardware architectures that protect TinyML devices against physical attacks. Secure hardware enclaves, TEEs, and PUFs will become more prevalent in TinyML devices. These hardware-based security mechanisms will ensure that sensitive data, model parameters, and cryptographic keys are protected even if an attacker gains physical access to the device.

10.7.5 Energy-Efficient Security Protocols

One of the critical challenges in securing TinyML devices is balancing the energy efficiency requirements with the computational demands of security protocols. Future directions will emphasize the development of energy-efficient security solutions specifically designed for TinyML environments. This includes lightweight authentication and authorization protocols, as well as context-aware security solutions that adapt the security level based on the device's current energy state and threat level [83]. The rise of energy-harvesting TinyML devices, which rely on ambient energy sources like solar or kinetic energy, will also drive the need for ultralow-power security techniques.

10.7.6 Real-Time Security Monitoring and Anomaly Detection

TinyML devices often operate in dynamic environments where new threats can emerge at any time. Real-time security monitoring and anomaly detection will be critical to protecting TinyML devices from attacks or abnormal behavior. Future advancements will include lightweight ML-based anomaly detection systems that can run directly on TinyML devices, enabling them to detect and respond to security threats in real time. These systems will leverage behavior profiling, context awareness, and unsupervised learning techniques to identify deviations from normal operations without the need for extensive computational resources.

10.8 Lessons Learned

In the exploration of security and privacy for TinyML devices, several critical lessons have emerged that highlight the importance of adapting security mechanisms to the unique characteristics of resource-constrained edge devices. Key lessons learned include:

- TinyML devices have limited computational power, memory, and energy, so security mechanisms must be lightweight yet practical. Traditional cryptographic methods and security protocols must be reengineered to operate within TinyML's constraints without compromising the device's performance.
- FL has emerged as a promising solution for decentralized model training on TinyML devices. However, FL introduces new attack vectors, such as model poisoning and data leakage, emphasizing the need for secure aggregation methods and privacy-preserving techniques like DP to ensure both data integrity and user privacy.
- TinyML systems are particularly vulnerable to adversarial attacks, where small input manipulations can lead to incorrect model predictions. This underscores the need for lightweight adversarial defense mechanisms tailored to TinyML devices, capable of detecting and mitigating such attacks without imposing excessive computational burdens.
- Tamper-resistant hardware, secure enclaves, and TEEs are essential to safeguarding TinyML devices from physical attacks. Security needs to be implemented at the hardware and software levels to ensure holistic protection against adversarial threats.
- Many TinyML devices operate in environments where power sources are limited, making energy-efficient security protocols essential. Techniques like lightweight cryptography, context-aware security solutions, and efficient key management are necessary to maintain robust security without draining device power.
- While DP and HE show promise, their implementation on TinyML devices remains a challenge due to their resource demands. More efficient versions of these techniques are needed to maintain privacy while enabling real-time, on-device inference and learning.
- As TinyML evolves to support on-device learning, new security challenges arise. Securely updating models and preventing adversarial tampering during training requires new protocols and mechanisms that can operate within the limited resources of TinyML systems.

10.9 Limitations of this Study

This chapter provides a comprehensive exploration of privacy-enhancing techniques for TinyML; however, certain limitations persist, and some have been identified. The implementation complexity and computational demands of methods such as SMPC and DP can be prohibitive for resource-constrained TinyML devices. Moreover, FL, while effective in decentralizing data, remains vulnerable to challenges such as model poisoning and inference attacks, which are not fully mitigated by existing solutions. Privacy-preserving data augmentation techniques may also affect model accuracy due to the synthetic nature of generated data, potentially limiting the robustness of the models. While methods such as encryption and anonymization are promising, they often involve trade-offs between security and computational efficiency, which may restrict their practicality for real-time TinyML applications. Besides the detailed analysis of security and privacy, this study did not consider other aspects of TinyML devices, which are also critical to fostering acceptable technology usage and application, such as ethical and legal considerations, explainability and interpretability, data quality and management, edge device heterogeneity, real-time processing and latency, deployment and integration challenges, model optimization and compression and energy efficiency.

10.10 Conclusion

This chapter explored privacy-aware training methods to boost the discretion of data processed by TinyML models. It begins by exploring differential personal privacy, which adds statistical sound to version outcomes of unknown individual data payments. This makes personal privacy more robust while balancing the compromise between personal privacy and design precision. Federated understanding is looked at for its ability to keep information centered by sharing only model updates and mitigating threats of information breaches, yet encountering difficulties such as version poisoning and reasoning assaults. It was identified that SMPC is scrutinized as a technique that enables collective design training without raw information exchange. It depends on cryptographic methods to preserve information confidentiality, though its complexity and computational demands can be high. Privacy-preserving data augmentation, which creates artificial information to simulate actual information while protecting personal privacy, is highlighted, noting its possible effect on design performance and the requirement for adequate anonymization. Information minimization, which entails gathering vital information, is a simple strategy to reduce direct exposure risks. Simultaneously, privacy-enhancing innovations such as encryption and anonymization are reviewed for their duties in safeguarding information. The chapter addresses added methods such as safe and secure design implementation, synthetic information validation, privacy-aware hyperparameter tuning, encrypted calculation, and anonymized information synthesis, each contributing to a detailed structure for privacy-aware TinyML version training.

References

1 Patil, R.Y., Bhamare, M., Patil, Y.H., and Bannore, A. (2024). Securing TinyML in a connected world. In: *TinyML for Edge Intelligence in IoT and LPWAN Networks* (ed. B.S. Chaudhari, S.N. Ghorpade, M. Zennaro, and R. Paškauskas), 311–333. Academic Press https://doi.org/10.1016/B978-0-44-322202-3.00018-X.

2 Mwaura, J., Araki, S., and Kakizaki, K. I. (2024). A study on DDoS attacks detection on IoT devices using machine learning for microcontrollers. *IEEE International Conference on Consumer Electronics (ICCE),* Las Vegas, NV, USA, 1–4. IEEE. https://doi.org/10.1109/ICCE59016.2024.10444344.

3 Ficco, M., Guerriero, A., Milite, E. et al. (2024). Federated learning for IoT devices: enhancing TinyML with on-board training. *Information Fusion* 104: 102189. https://doi.org/10.1016/j.inffus.2023.102189.

4 Bhamare, M., Kulkarni, P.V., Rane, R. et al. (2024). TinyML applications and use cases for healthcare. In: *TinyML for Edge Intelligence in IoT and LPWAN Networks* (ed. B.S. Chaudhari, S.N. Ghorpade, M. Zennaro, and R. Paškauskas), 331–353. Academic Press.

5 Szydlo, T. and Nagy, M. (2024). Management of TinyML enabled internet of things devices. *IEEE Micro* 45: 1–7. https://doi.org/10.1109/MM.2024.3382488.

6 Chaudhari, B.S., Ghorpade, S.N., Zennaro, M., and Paškauskas, R. (2024). TinyML for low-power Internet of Things. In: *TinyML for Edge Intelligence in IoT and LPWAN Networks* (ed. B.S. Chaudhari, S.N. Ghorpade, M. Zennaro, and R. Paškauskas), 1–12. Academic Press https://doi.org/10.1016/B978-0-44-322202-3.00006-3.

7 Tsoukas, V., Gkogkidis, A., Boumpa, E., and Kakarountas, A. (2024). A review on the emerging technology of TinyML. *ACM Computing Surveys* 56 (10): 1–37. https://doi.org/10.1145/3661820.

8 Porsia, A., Ruospo, A., and Sanchez, E. (2024). Model theft attack against a tinyML application running on an Ultra-Low-Power Open-Source SoC. *Proceedings of the 21st ACM International Conference on Computing Frontiers: Workshops and Special Sessions*, 63–66. https://doi.org/10.1145/3637543.3652877.

9 Ancilotto, A., Paissan, F., and Farella, E. (2024). Ximswap: many-to-many face swapping for tinyml. *ACM Transactions on Embedded Computing Systems* 23 (3): 1–6. https://doi.org/10.1145/3603173.

10 Kulkarni, V. and Jujare, V. (2024). TinyML using neural networks for resource-constrained devices. In: *TinyML for Edge Intelligence in IoT and LPWAN Networks* (ed. B.S. Chaudhari, S.N. Ghorpade, M. Zennaro, and R. Paškauskas), 87–101. Academic Press https://doi.org/10.1016/B978-0-44-322202-3.00010-5.

11 Hayajneh, A.M., Hafeez, M., Zaidi, S.A., and McLernon, D. (2024). TinyML empowered transfer learning on the edge. *IEEE Open Journal of the Communications Society* 5: 1656–1672. https://doi.org/10.1109/OJCOMS.2024.3373177.

12 Zheng, T., Qiu, Y., Zheng, Y. et al. (2024). Enhancing TinyML-based container escape detectors with systemcall semantic association in UAVs networks. *IEEE Internet of Things Journal* 11: 21158–21169. https://doi.org/10.1109/JIOT.2024.3361452.

13 Fusco, P., Rimoli, G.P., and Ficco, M. (2024). TinyIDS-an IoT intrusion detection system by tiny machine learning. *International Conference on Computational Science and Its Applications*, 71–82. Cham: Springer Nature Switzerland. https://doi.org/10.1007/978-3-031-65223-3_5.

14 Siddique, T. and Mahmud, M.S. (2024). Physics-enhanced TinyML for real-time detection of ground magnetic anomalies. *IEEE Access* 12: 25372–25384. https://doi.org/10.1109/ACCESS.2024.3362346.

15 Huang, Z., Zandberg, K., Schleiser, K., and Baccelli, E. (2024). RIOT-ML: toolkit for over-the-air secure updates and performance evaluation of TinyML models. *Annals of Telecommunications* 80: 283–297. https://doi.org/10.1007/s12243-024-01041-5.

16 Shafik, W. (2024). Artificial intelligence and machine learning with cyber ethics for the future world. In: *Future Communication Systems Using Artificial Intelligence, Internet of Things and Data Science* (ed. I. Ullah, I.U. Khan, M. Ouaissa, et al.), 110–130. CRC Press https://doi.org/10.1201/9781032648309-9.

17 Saeed, M.M., Saeed, R.A., and Ahmed, Z.E. (2024). TinyML for 5G networks. In: *TinyML for edge intelligence in IoT and LPWAN networks* (ed. B.S. Chaudhari, S.N. Ghorpade, M. Zennaro, and R. Paškauskas), 167–229. Academic Press https://doi.org/10.1016/b978-0-44-322202-3.00014-2.

18 Kallimani, R., Pai, K., Raghuwanshi, P. et al. (2024). TinyML: tools, applications, challenges, and future research directions. *Multimedia Tools and Applications* 83 (10): 29015–29045. https://doi.org/10.1007/s11042-023-16740-9.

19 Shafik, W. (2024). Machine learning for advanced wireless communication: applications, challenges, problems, and open research questions. In: *Microwave Devices and Circuits for Advanced Wireless Communication: Design and Analysis*, vol. 1 (ed. D.K. Choudhary, N. Mishra, I. Singh, et al.), 225–278. CRC Press.

20 Valencia-Arias, A., González-Ruiz, J.D., Verde Flores, L. et al. (2024). Machine learning and blockchain: a bibliometric study on security and privacy. *Information* 15 (1): 65. https://doi.org/10.3390/info15010065.

21 Ghorpade, S.N., Chougule, S., Chaudhari, B.S., and Zennaro, M. (2024). TinyML: principles and algorithms. In: *TinyML for Edge Intelligence in IoT and LPWAN Networks* (ed. B.S. Chaudhari, S.N. Ghorpade, M. Zennaro, and

R. Paškauskas), 65–86. Academic Press https://doi.org/10.1016/B978-0-44-322202-3.00009-9.

22 Adam, M., Hammoudeh, M., Alrawashdeh, R., and Alsulaimy, B. (2024). A survey on security, privacy, trust, and architectural challenges in IoT systems. *IEEE Access* 12: 57128–57149. https://doi.org/10.1109/ACCESS.2024.3382709.

23 Thakkar, A. and Lohiya, R. (2022). A survey on intrusion detection system: feature selection, model, performance measures, application perspective, challenges, and future research directions. *Artificial Intelligence Review* 55 (1): 453–563. https://doi.org/10.1007/s10462-021-10037-9.

24 Abd-Alhussain hadi al-zubaidy, Q., Alfoudi, A.S., and Mahdi, A.M. (2023). IoT cybersecurity threats and detection mechanisms: a review. *Wasit Journal for Pure Sciences* 2 (2): 231–250. https://doi.org/10.31185/wjps.136.

25 Z. Pourmirza, "*Cybersecurity in Centralised vs Decentralised Energy Systems*," 2023.

26 Khraisat, A. and Alazab, A. (2021). A critical review of intrusion detection systems in the internet of things: techniques, deployment strategy, validation strategy, attacks, public datasets and challenges. *Cybersecurity* 4: 1–27.

27 Albanese, A., Nardello, M., Fiacco, G., and Brunelli, D. (2022). Tiny machine learning for high accuracy product quality inspection. *IEEE Sensors Journal* 23 (2): 1575–1583. https://doi.org/10.1007/978-3-031-48121-5_17.

28 Dutta, V., Pawlicki, M., Kozik, R., and Choraś, M. (2022). Unsupervised network traffic anomaly detection with deep autoencoders. *Logic Journal of the IGPL* 30 (6): 912–925.

29 Bhati, B.S., Chugh, G., Al-Turjman, F., and Bhati, N.S. (2021). An improved ensemble based intrusion detection technique using XGBoost. *Transactions on Emerging Telecommunications Technologies* 32 (6): e4076.

30 ElKashlan, M., Elsayed, M.S., Jurcut, A.D., and Azer, M. (2023). A machine learning-based intrusion detection system for IoT electric vehicle charging stations (evcss). *Electronics* 12 (4): 1044.

31 ElKashlan, M., Aslan, H., Said Elsayed, M. et al. (2023). Intrusion detection for electric vehicle charging systems (EVCS). *Algorithms* 16 (2): 75.

32 Fyntanidou, B., Zouka, M., Apostolopoulou, A. et al. (2020). IoT-based smart triage of Covid-19 suspicious cases in the Emergency Department. *2020 IEEE Globecom Workshops (GC Wkshps)*, 1–6. IEEE.

33 Gangal, A., Ye, M., and Wei, S. (2020). Hybridtee: secure mobile DNN execution using hybrid trusted execution environment. *2020 Asian Hardware Oriented Security and Trust Symposium (AsianHOST)*, 1–6. IEEE.

34 Zhang, D. and You, S. (2020). iFlask: isolate flask security system from dangerous execution environment by using ARM TrustZone. *Future Generation Computer Systems* 109: 531–537.

35 Brasser, F., Frassetto, T., Riedhammer, K. et al. (2018). VoiceGuard: secure and private speech processing. *Interspeech* 18: 1303–1307.

36 Bayerl, S.P., Frassetto, T., Jauernig, P. et al. (2020). Offline model guard: secure and private ML on mobile devices. *2020 Design, Automation & Test in Europe Conference & Exhibition (DATE)*, 460–465. IEEE.

37 Fan, Y., Liu, S., Tan, G., and Qiao, F. (2020). Fine-grained access control based on trusted execution environment. *Future Generation Computer Systems* 109: 551–561.

38 Xiong, P., Zhang, L., Zhu, T. et al. (2020). Private collaborative filtering under untrusted recommender server. *Future Generation Computer Systems* 109: 511–520.

39 Sun, Z., Sun, R., Liu, C. et al. (2023). Shadownet: a secure and efficient on-device model inference system for convolutional neural networks. *2023 IEEE Symposium on Security and Privacy (SP)*, 1596–1612. IEEE.

40 Liu, Z., Luo, Y., Duan, S. et al. (2023). MirrorNet: a TEE-friendly framework for secure on-device DNN inference. *2023 IEEE/ACM International Conference on Computer Aided Design (ICCAD)*, 1–9. IEEE.

41 Wang, J., Wu, Y., Liu, H. et al. (2023). IP protection in TinyML. *2023 60th ACM/IEEE Design Automation Conference (DAC)*, 1–6. IEEE.

42 Prakash, V., Singh, A.V., and Kumar Khatri, S. (2019). A new model of light weight hybrid cryptography for internet of things. *2019 3rd International Conference on Electronics, Communication and Aerospace Technology (ICECA)*, Coimbatore, India, 282–285. https://doi.org/10.1109/ICECA.2019.8821924.

43 Noura, H., Couturier, R., Pham, C., and Chehab, A. (2019). Lightweight stream cipher scheme for resource-constrained IoT devices. *2019 International Conference on Wireless and Mobile Computing, Networking and Communications (WiMob)*, 1–8. IEEE.

44 Shantha, M.J.R. and Arockiam, L. (2018). SAT_Jo: an enhanced lightweight block cipher for the internet of things. *2018 Second International Conference on Intelligent Computing and Control Systems (ICICCS)*, 1146–1150. IEEE.

45 Kubba, Z.M.J. and Hoomod, H.K. (2020). A hybrid modified lightweight algorithm combined of two cryptography algorithms PRESENT and Salsa20 using chaotic system. *2019 First International Conference of Computer and Applied Sciences (CAS)*, 199–203. IEEE.

46 Noura, H., Chehab, A., Sleem, L. et al. (2018). One round cipher algorithm for multimedia IoT devices. *Multimedia Tools and Applications* 77: 18383–18413.

47 Ahmed, S.F., Islam, M.R., Nath, T. D. et al. (2019). G-TBSA: A generalized lightweight security algorithm for IoT. *2019 4th International Conference on Electrical Information and Communication Technology (EICT)*, 1–6. IEEE.

48 Chatterjee, R. and Chakraborty, R. (2020). A modified lightweight PRESENT cipher for IoT security. *2020 International Conference on Computer Science, Engineering and Applications (ICCSEA)*, 1–6. IEEE.

49 Noura, H., Chehab, A., and Couturier, R. (2019). Lightweight dynamic key-dependent and flexible cipher scheme for IoT devices. *2019 IEEE Wireless Communications and Networking Conference (WCNC)*, 1–8. IEEE.

50 Roy, S., Rawat, U., and Karjee, J. (2019). A lightweight cellular automata based encryption technique for IoT applications. *IEEE Access* 7: 39782–39793.

51 Zhao, C., Yan, Y., and Li, W. (2019). An efficient ASIC implementation of QARMA lightweight algorithm. *2019 IEEE 13th International Conference on ASIC (ASICON)*, 1–4. IEEE.

52 Chaudhary, R.R.K. and Chatterjee, K. (2020). An efficient lightweight cryptographic technique for IoT based E-healthcare system. *2020 7th International Conference on Signal Processing and Integrated Networks (SPIN)*, 991–995. IEEE.

53 Mishra, A.K., Tripathy, A.K., Puthal, D., and Yang, L.T. (2018). Analytical model for sybil attack phases in internet of things. *IEEE Internet of Things Journal* 6 (1): 379–387.

54 Hamza, R., Yan, Z., Muhammad, K. et al. (2020). A privacy-preserving cryptosystem for IoT E-healthcare. *Information Sciences* 527: 493–510.

55 Gyamfi, E., Ansere, J.A., and Xu, L. (2019). ECC based lightweight cybersecurity solution for IoT networks utilising multi-access mobile edge computing. *2019 Fourth International Conference on Fog and Mobile Edge Computing (FMEC)*, 149–154. IEEE.

56 Khan, M.A., Quasim, M.T., Alghamdi, N.S., and Khan, M.Y. (2020). A secure framework for authentication and encryption using improved ECC for IoT-based medical sensor data. *IEEE Access* 8: 52018–52027.

57 Mohandas, N. A., Swathi, A., Abhijith, R. et al. (2020). A4: A lightweight stream cipher. *2020 5th International Conference on Communication and Electronics Systems (ICCES)*, 573–577. IEEE.

58 Ding, L., Liu, C., Zhang, Y., and Ding, Q. (2019). A new lightweight stream cipher based on chaos. *Symmetry* 11 (7): 853.

59 Choudhury, O., Gkoulalas-Divanis, A., Salonidis, T. et al. (2019). Differential privacy-enabled federated learning for sensitive health data. arXiv preprint arXiv: 1910.02578.

60 Kopparapu, K. and Lin, E. (2021). TinyFedTL: federated transfer learning on tiny devices. arXiv preprint arXiv: 2110.01107.

61 Mathur, A., Beutel, D.J., de Gusmao, P.P.B. et al. (2021). On-device federated learning with flower. arXiv preprint arXiv:2104.03042.

62 Kwon, J. and Park, D. (2021). Toward data-adaptable tinyML using model partial replacement for resource frugal edge device. *The International Conference on High Performance Computing in Asia-Pacific Region*, 133–135.

63 Dehrouyeh, F., Yang, L., Ajaei, F. B. and Shami, A. (2024). On TinyML and cybersecurity: electric vehicle charging infrastructure use case. arXiv preprint arXiv:2404.16894.

64 Bagdasaryan, E., Veit, A., Hua, Y. et al. (2020). How to backdoor federated learning. *International Conference on Artificial Intelligence and Statistics*, 2938–2948. PMLR.

65 Zeng, J., Zhang, K., Wang, L., and Li, J. (2023). FedLVR: a federated learning-based fine-grained vehicle recognition scheme in intelligent traffic system. *Multimedia Tools and Applications* 82 (24): 37431–37452.

66 Rahim, S., Peng, L., and Ho, P.-H. (2024). TinyFDRL-enhanced energy-efficient trajectory design for integrated space–air–ground networks. *IEEE Internet of Things Journal* 11 (12): 21391–21401. https://doi.org/10.1109/JIOT.2024.3361394.

67 Dwork, C. (2008). Differential privacy: a survey of results. *International Conference on Theory and Applications of Models of Computation*, 1–19. Berlin, Heidelberg: Springer Berlin Heidelberg.

68 Long, Y., Zhang, J., Wang, G., and Fu, J. (2023). Hierarchical federated learning with global differential privacy. *Electronic Research Archive* 31 (7): 3741–3758.

69 Demelius, L., Kern, R., and Trügler, A. (2023). Recent advances of differential privacy in centralized deep learning: a systematic survey. arXiv preprint arXiv:2309.16398.

70 He, Z., Wang, L., and Cai, Z. (2023). Clustered federated learning with adaptive local differential privacy on heterogeneous IoT data. *IEEE Internet of Things Journal.* 11: 137–146.

71 Wang, N., Xiao, X., Yang, Y. et al. (2019). Collecting and analyzing multidimensional data with local differential privacy. *2019 IEEE 35th International Conference on Data Engineering (ICDE)*, 638–649. IEEE.

72 Cai, Z., He, Z., Guan, X., and Li, Y. (2016). Collective data-sanitization for preventing sensitive information inference attacks in social networks. *IEEE Transactions on Dependable and Secure Computing* 15 (4): 577–590.

73 Sun, L., Qian, J., and Chen, X. (2020). LDP-FL: practical private aggregation in federated learning with local differential privacy. arXiv preprint arXiv:2007.15789.

74 Qian, J., Du, H., Hou, J. et al. (2018). Hidebehind: Enjoy voice input with voiceprint unclonability and anonymity. *Proceedings of the 16th ACM Conference on Embedded Networked Sensor Systems* (82–94).

75 Ahmed, S., Chowdhury, A.R., Fawaz, K., and Ramanathan, P. (2020). Preech: a system for privacy-preserving speech transcription. *29th USENIX Security Symposium (USENIX Security 20)*, 2703–2720.

76 Feng, T., Peri, R. and Narayanan, S. (2022). User-level differential privacy against attribute inference attack of speech emotion recognition in federated learning. arXiv preprint arXiv:2204.02500.

77 Shoemate, M., Jett, K., Cowan, E. et al. (2022). Sotto voce: federated speech recognition with differential privacy guarantees. arXiv preprint arXiv:2207.07816.

78 Nguyen, K., Budzys, M., Frimpong, E. et al. (2024). A pervasive, efficient and private future: realizing privacy-preserving machine learning through hybrid homomorphic encryption. arXiv preprint arXiv:2409.06422.

79 Al Badawi, A., Bates, J., Bergamaschi, F. et al. (2022). Openfhe: Open-source fully homomorphic encryption library. *Proceedings of the 10th Workshop on Encrypted Computing & Applied Homomorphic Cryptography*, 53–63. https://ia.cr/2022/915.

80 Shabir, M.Y., Torta, G., Basso, A., and Damiani, F. (2023). Toward secure TinyML on a standardized AI architecture. *Device-Edge-Cloud Continuum: Paradigms, Architectures and Applications*, 121–139. Cham: Springer Nature Switzerland. https://doi.org/10.1007/978-3-031-42194-5_7.

81 Lou, Q. and Jiang, L. (2021). Hemet: a homomorphic-encryption-friendly privacy-preserving mobile neural network architecture. *International Conference on Machine Learning*, 7102–7110. PMLR.

82 Onoufriou, G., Mayfield, P., and Leontidis, G. (2021). Fully homomorphically encrypted deep learning as a service. *Machine Learning and Knowledge Extraction* 3 (4): 819–834. https://doi.org/10.3390/make3040041.

83 Shafik, W. (2025). Wireless sensor network-assisted fuzzy sink based model. *Energy* 333: 137512. https://doi.org/10.1016/j.energy.2025.137512.

11

Robust Ground Truth Data Mining for Enhanced Privacy and Accuracy in Noisy TinyML Environments*

Yuichi Sei[1] *and Agbotiname Lucky Imoize*[2,3]

[1] *Graduate School of Informatics and Engineering, The University of Electro-Communications, Chōfu, Tokyo, Japan*
[2] *Department of Electrical and Electronics Engineering, Faculty of Engineering, University of Lagos, Lagos, Nigeria*
[3] *Department of Electrical Engineering and Information Technology, Institute of Digital Communication, Ruhr University, Bochum, Germany*

11.1 Introduction

Recently, tiny machine leasrning (TinyML) has emerged as a groundbreaking technology that enables machine learning (ML) algorithms to operate on microcontrollers and other resource-constrained devices [2]. By integrating the Internet of Things (IoT) with ML, TinyML facilitates intelligent, data-driven decisions at the edge without relying on cloud-based processing. This capability is crucial for applications that require low latency, reduced bandwidth, and enhanced privacy. TinyML considerably influences the IoT sector, where numerous devices generate vast amounts of data daily [3, 4]. These devices, ranging from environmental sensors to wearable health monitors, operate with limited computational power, memory, and energy [5–7]. Despite these constraints, they must perform reliably and accurately to support applications such as predictive maintenance, anomaly detection, and real-time monitoring. TinyML brings advanced ML capabilities to edge devices that traditionally could not handle such complex computations. The benefits of TinyML include the following aspects.

First, local data processing minimizes delays associated with cloud data transmission, which is crucial for time-sensitive applications such as autonomous driving and industrial automation [8, 9]. Second, local data processing keeps sensitive information on edge devices, reducing the risk of data breaches [10].

* An earlier version of the chapter appeared in Sei and Ohsuga [1].

Tiny Machine Learning: Design Principles and Applications, First Edition.
Edited by Agbotiname Lucky Imoize, Dinh-Thuan Do, and Houbing Herbert Song.

Third, processing data at edges limits the data sent to the cloud, leading to lower bandwidth usage and costs [11]. Fourth, TinyML models are designed to be lightweight and efficient, thereby extending the battery life of portable and remote devices.

Despite its potential, TinyML faces challenges due to resource constraints, necessitating innovative approaches for designing, training, and deploying TinyML models. Techniques like model quantization and pruning have been developed to reduce the size and complexity of ML algorithms without compromising accuracy, making them suitable for edge devices. Integrating ML with IoT also requires robust frameworks for seamless deployment and operation. Frameworks such as TensorFlow Lite for Microcontrollers and Edge Impulse provide necessary tools and libraries for efficient TinyML model development, training, and deployment. TinyML applications are extensive. In health care, TinyML-powered wearables monitor vital signs and detect anomalies in real-time, providing timely alerts to patients and healthcare providers. In agriculture, TinyML-based sensors monitor soil conditions and crop health, optimizing irrigation and fertilization practices. In smart homes, TinyML enhances security, optimizes energy consumption, and offers personalized user experiences. As TinyML continues to evolve, future developments will likely focus on enhancing edge devices by improving algorithms, hardware support, and data processing efficiency. Integrating TinyML with emerging technologies like 5G and edge computing will further expand its potential, enabling previously unimaginable innovative applications.

Nevertheless, sensing data related to individuals or their surroundings can contain personally identifiable information [12], risking privacy breaches. To address this issue, data can be anonymized using ϵ-differential privacy (DP) [13, 14], a well-established privacy framework. In ϵ-DP, the privacy budget ϵ balances data utility and privacy preservation. This approach has been widely used in various studies [15–17] and is crucial for data analysis.

Local DP (LDP) is a variant of DP designed for data collection from individuals. In this context, "DP" refers to "local DP." To achieve DP, numerical values are typically protected by adding Laplace noise [13]. For categorical values, category identifiers (IDs) can be disguised with a certain probability to ensure privacy preservation [18, 19]. While these methods ensure ϵ-DP, they do not address errors in the data.

This chapter introduces a novel privacy model, inferred ground truth DP (GDP), which distinguishes between "true" values (original, error free, and possibly unknown to the data owner or anonymizer) and "measured" values (potentially containing errors). Unlike traditional DP, which focuses on sensor data protection, GDP aims to safeguard the ground truth values. We acknowledge that ground truth values are often unknown and assume they follow a specific

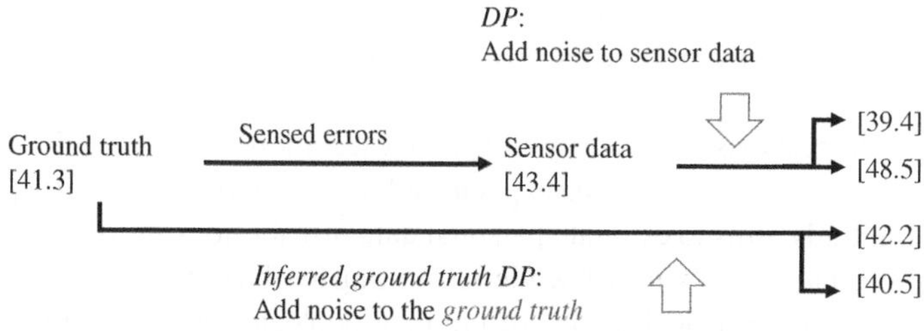

Figure 11.1 Inferred ground truth differential privacy (GDP). The goal of GDP is to add noise to the ground truth values rather than to the sensor data.

probability distribution, such as a normal distribution (ND), based on the knowledge or error theory of the data owner or anonymizer [20, 21].

By leveraging GDP, the noise added to sensor data can be reduced because GDP accounts for the inherent errors in the sensor data, thereby requiring less additional noise for privacy preservation. The effectiveness of GDP relies on the anonymizer's ability to estimate the measurement error distribution (Figure 11.1). However, even with inaccurate estimates, GDP remains valid provided the errors are not overestimated. Therefore, GDP can still be applied with uncertain error distributions by conservatively estimating the error amount. When no error is anticipated, GDP defaults to traditional DP.

In scenarios with no information about the error distribution available, the proposed method cannot be applied. However, we believe that in several real-world scenarios, reasonable error estimates can be made, enabling the GDP to be used effectively by underestimating the error amount. A more detailed discussion on this issue is available in Section 11.6.

In our evaluation presented in Section 11.5, we assessed the effectiveness of our proposed method using a combination of three synthetic and three real-world datasets. We compared its performance against three existing methods for numerical data and four methods for categorical data. Results demonstrated an approximate 20% reduction in noise.

Recently, several novel LDP methods have been developed, primarily focusing on estimating the distribution of user data and federated learning [22–25]. They employ techniques such as adding noise to data values or randomly changing category values to preserve privacy. The underlying objective is to mitigate the adverse effects of noise on statistical analysis. However, these methods do not explicitly account for the inherent sensing errors prevalent in IoT environments.

We address a scenario where data are collected on an untrusted server, which is one of the two main scenarios identified by Husnoo et al. [26] in their categorization of DP techniques in IoT environments. This scenario is particularly challenging due to the potential for considerable cumulative noise added to each data point, thereby compromising data utility. Considering measurement errors, the proposed method effectively reduces this cumulative noise, as evidenced by experimental results.

Further, although most existing security and privacy techniques in IoT overlook sensing noise issue, the proposed method explicitly addresses this factor in privacy preservation. This study aims to estimate personal data distributions in IoT environments and protect user data with DP while accounting for the inherent sensing noise. By doing so, we fill a critical gap in current methodologies that often introduce excessive privacy noise into sensor data due to neglecting sensing errors.

The proposed method, referred to as GDP, derives the probability distribution of the true values based on the known probability distribution of noise and noisy measurement. Noise is then added according to this derived distribution to achieve DP.

The major contributions of this study are summarized as follows:

- We introduce GDP, a novel DP concept that explicitly incorporates sensing noise.
- We develop anonymization algorithms for both numerical and categorical data within the GDP framework.
- We provide rigorous proofs that our algorithms ensure GDP.
- Through experiments with synthetic and real datasets, we showcase that the proposed method reduces DP noise.
- We further demonstrate that the proposed method improves the accuracy of estimated personal data distributions when applied to the same datasets.

The remainder of this chapter is structured as follows. We discuss related methods in Section 11.2. Section 11.3 outlines the application scenarios, underlying assumptions, and formal privacy definitions. Section 11.4 details our design and mechanisms for achieving GDP. In Section 11.5, we present simulation results obtained using synthetic real-world datasets. We address design considerations specific to the proposed method in Section 11.6. Finally, we conclude this study in Section 11.7.

11.2 Related Research Work

This section provides an overview of related research in privacy-preserving methods within the context of TinyML. It begins with Section 11.2.1, which reviews various DP methods applied to TinyML, highlighting the key advancements and limitations. Section 11.2.2 discusses anonymized data collection techniques, emphasizing their relevance to TinyML and the unique challenges they address. Section 11.2.3 explores approaches that incorporate sensing noise into privacy-preserving methods, examining how they handle errors in data collection.

Finally, Section 11.2.4 focuses on DP methods specifically designed for dirty data, showcasing their effectiveness in ensuring privacy while dealing with inconsistencies in large datasets. Collectively, these subsections illustrate the breadth of research in privacy preservation and its implications for TinyML offering a comprehensive understanding of the current landscape of this research area.

11.2.1 DP and TinyML

Numerous DP methods have been proposed to address privacy concerns in various contexts [27, 28]. Waheed et al. [29] designed a DP-based system to safeguard data privacy in smart homes by applying LDP techniques, specifically using randomized response methods before data transmission to an aggregator. They further enhanced privacy protection by incorporating hidden Markov model technology at the aggregator level, achieving a dual-layer privacy shield. Yang and Al-Masri [30] proposed a user-centric LDP optimization method that empowers users to manage their privacy preferences more effectively. This method uses multicriteria decision-making to simplify the management of privacy preferences for users. Evaluation results indicate that this method successfully strikes an optimal balance between data accuracy and privacy protection. He et al. [31] proposed an adaptive LDP method to train clustered federated learning models on heterogeneous IoT data, aiming to reduce LDP noise and communication overhead while balancing between privacy and utility through utilizing adaptive clipping, weight compression, and parameter shuffling.

Additionally, several studies have explored ML techniques, including deep neural networks, for analyzing IoT sensing with DP. Shi et al. [32] introduced a reinforcement learning technique for transportation network companies using passenger data. Xu et al. [33] focused on mobile data analysis in edge computing. Guan et al. [34] applied ML to the Internet of Medical Things. However, these studies did not incorporate the proposed GDP. Incorporating GDP could enhance their accuracy while maintaining robust privacy preservation.

11.2.2 Anonymized Data Collection for TinyML

Many studies have explored anonymized data collection techniques. Wang et al. [35] proposed a method to identify the top-k most frequent new terms by collecting usage data under DP. Kim et al. [36] used DP to derive population statistics from indoor positioning data. Encryption-based approaches for anonymized data collection have also been investigated [37, 38]. These approaches focus on aggregating data values and do not address individual data retrieval or consider value errors. In contrast, the proposed method aims to achieve high accuracy in individual data values, which is crucial for applications such as recommender systems that rely on detailed personal attributes.

Table 11.1 Comparison of different works.

	DP	Noises	For TinyML	For local
[29–31]	√	—	√	√
[39, 40]	—	√	√	√
[41–44]	√	√	√	—
This work	√	√	√	√

11.2.3 Privacy-Preserving Methods Accounting for Sensing Noises

Abul et al. [39] and Sei and Ohsuga [40] developed location anonymization methods that account for location errors, achieving k-anonymity, a basic privacy metric. However, these methods are not applicable to ϵ-DP.

11.2.4 DP Methods Considering Dirty Data

Ge et al. [41], Krishnan et al. [42], and Blass et al. [43] proposed methods for privately cleaning "dirty data" using DP as a privacy metric. These methods focused on resolving inconsistencies in large databases under the assumption that all database values were accurate and applying the Laplace mechanism without accounting for value errors.

Table 11.1 shows a comparison of different methods.

11.3 Models

This section describes the application model of this study, including the assumptions, attack model, and privacy metric.

11.3.1 Application Model

IoT devices are capable of collecting and inferring various personal attributes, such as location, heart rate, health conditions, age, and movement patterns. This data can be used to provide personalized services, including recommender systems. The data collection entity can anonymize this information before transmitting it to the intended recipient.

The collected attribute data can be classified into two types: numerical attributes (e.g., heart rate in beats per minute) and categorical attributes (disease diagnosis like COVID-19).

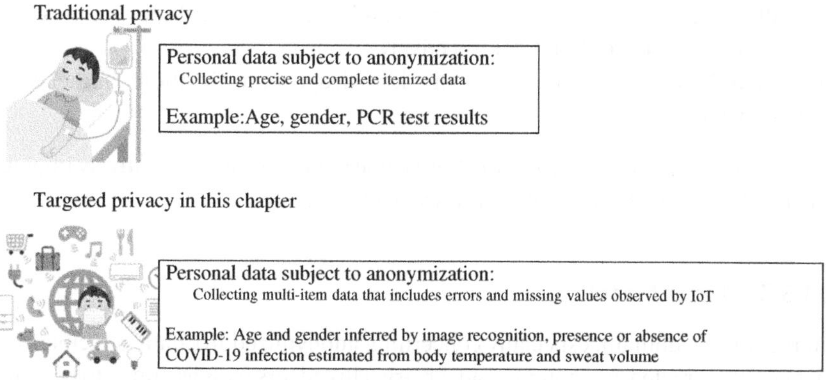

Figure 11.2 Application model discussed in this chapter.

Notably, the collected attribute data often contains some degree of sensing errors due to the inherent difficulty associated with accurately sensing and estimating personal attributes. In extreme cases, some data may be entirely missing. While techniques like multiple imputations or regression-based predictions can address missing data [45–47], the estimates may still carry notable errors. Figure 11.2 illustrates the application model discussed in this chapter.

11.3.2 Assumptions

Anonymizers may not possess the exact true attribute values but can make estimations. However, these estimations are prone to errors. For numerical attributes, anonymizers can estimate the error distribution, often modeled as an ND due to the common occurrence of measurement errors following this pattern [48]. This distribution is characterized by the standard deviation parameter (σ). Notably, the concept of GDP is adaptable to other error models as well.

For categorical attributes, the probability of misclassification, denoted as $p_{i \to j}$, represents the likelihood that the true category ID is i while the anonymizer assumes it to be j.

In this study, the parameters σ and $p_{i \to j}$ for all i, j are referred to as "error parameters."

We consider the following three possible scenarios:

Scenario 1: The anonymizer has precise knowledge of the error parameters.

Scenario 2: The anonymizer lacks precise knowledge of the error parameters. The estimated parameters may deviate from the actual parameters, but the anonymizer is not overly optimistic about the degree of error. Formal mathematical definitions for numerical and categorical attributes are provided in Sections 11.4.1 and 11.4.2, respectively.

Scenario 3: The anonymizer has no knowledge or estimation of the error parameters.

This chapter focuses on Scenario 2, as Scenario 1 is considered relatively impractical and Scenario 3 is not within the scope of this study.

11.3.3 Attack Model

In the privacy domain, especially concerning anonymized IoT data, a recipient of data is often treated as a potential adversary. This adversary is assumed to be semi-honest, meaning that while one may adhere to established protocols for collecting anonymized IoT data, one might also attempt to infer personal information from that anonymized data. This assumption is crucial for understanding the potential privacy risks associated with data sharing.

An anonymized dataset is typically not an exact replica of the original data but is modified to mask the true values. This modification includes errors from two primary sources: original sensing errors and intentionally added noise. Sensing errors are inherent inaccuracies that occur during data collection due to limitations in sensor accuracy, environmental factors, or other technical constraints. In addition to these inherent inaccuracies, additional noise is intentionally introduced during the anonymization process to further obfuscate the data and protect individual privacy. This intentional noise addition is often guided by principles such as privacy, ensuring that the data cannot be easily traced back to an individual.

The concept of DP is fundamental in this context, as it involves adding calibrated noise to the data to obscure whether a specific individual's information is included in the dataset. This technique provides a mathematical guarantee of privacy. However, for IoT data, which already contains measurement errors, adding excessive noise can degrade the data utility, making it less useful for analytical purposes.

Given the presence of both sensor errors and intentionally added noise, even if an attacker adheres to the data collection protocols, the inaccuracies embedded in the anonymized data prevent them from accurately identifying individuals' true data values. Instead, the attacker can only estimate the probability distribution of these values that accounts for the errors and noise. This results in a range of possible values rather than a single precise value, adding a layer of security by masking the true data within a range of plausible values, thereby complicating the attacker's ability to determine the exact original data.

This approach aligns with advanced privacy-preserving techniques, such as the GDP proposed in this chapter. GDP extends traditional DP by focusing on the "ground truth" – the true values that are often unknown to both the data owner

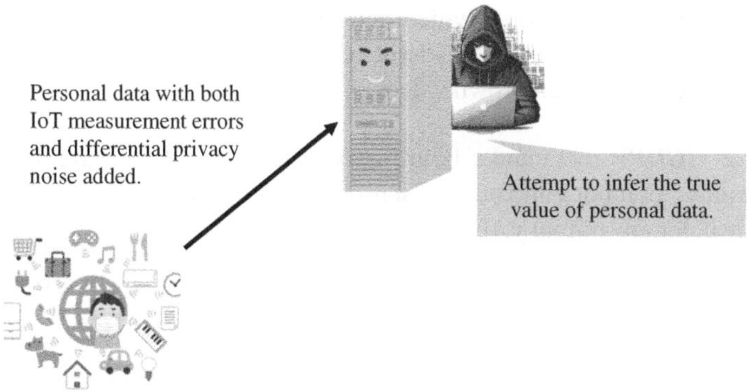

Figure 11.3 Attack model presented in this chapter.

and the anonymizer. By accounting for the inherent sensor data errors, GDP reduces the amount of additional noise needed, thereby preserving the data utility while still safeguarding privacy. This method ensures that the noise added for privacy preservation is optimized by considering existing inaccuracies, thereby improving the balance between data utility and privacy preservation. Figure 11.3 shows the attack model discussed in this chapter.

In summary, treating the recipient of anonymized data as a semihonest adversary emphasizes the need for robust privacy-preserving mechanisms. The dual errors – original sensor inherent inaccuracies and intentionally added noise – play a critical role in protecting the true data values. Advanced methods like GDP further refine this protection, ensuring that the data remains useful for analysis while protecting individual privacy against potential inference attacks.

11.3.4 Privacy Metric

In privacy-preserving data mining, DP [13] serves as a crucial metric. Originally developed for query–response databases, DP has been applied in anonymized data collection.

Consider a scenario where an individual's attribute value is anonymized, either by the individual or by an anonymizer. Let ϵ be a positive real number. LDP is defined as follows:

Definition 11.1 ***(LDP)*** Let x and x' be arbitrary values within the target domain. A mechanism $\mathcal{A}$ satisfies ϵ-LDP if and only if, for any output y, the following equation holds:

$$\frac{P(\mathcal{A}(x) = y)}{P(\mathcal{A}(x') = y)} \leq e^{\epsilon} \tag{11.1}$$

It is assumed that the attribute value x may include sensing errors. Thus, the focus shifts to the ground truth of x, which may be unknown even to the data owner or an anonymizer. To address this issue, GDP is proposed as a means of managing the privacy of these unknown values.

Definition 11.2 *(GDP)* Let x and x' be arbitrary (true) values. A measurement function $\mathcal{M}$ acquires an input x and outputs sensor data. A mechanism $\mathcal{A}$ satisfies GDP if and only if, for any output y, the following equation holds:

$$\frac{P(\mathcal{A}(\mathcal{M}(x)) = y)}{P(\mathcal{A}(\mathcal{M}(x')) = y)} \leq e^{\epsilon} \tag{11.2}$$

11.4 GDP

Previous studies have typically defined variables x and x' in Definition 11.1 as the values obtained directly from measurements. However, we deviate from this convention and redefine these variables as the underlying ground truths, devoid of any measurement errors. This shift in perspective necessitates the development of novel anonymization mechanisms for both numerical and categorical attributes, which are detailed in this chapter.

11.4.1 Numerical Value Anonymization

For numerical attributes, a well-known method to add noise is the Laplace mechanism [13], which uses a Laplace distribution (LD). However, this mechanism does not inherently account for sensing errors. In the traditional approach, often referred to as the **baseline approach** for numerical attributes, noise following an ND is first added to the ground truth to simulate sensing errors. Subsequently, additional noise from the Laplace mechanism is introduced into this already noisy value. The proposed method invariably adds Laplace noise, regardless of the underlying data characteristics. The resulting probability density function (PDF), which describes the probability of the distance between the final noisy value and ground truth, can be obtained by convolving the ND and LD.

To express this mathematically, let $\mathcal{N}(x; \sigma^2)$ represent the PDF of the ND with standard deviation σ and $\mathcal{L}(x; b)$ represent the PDF of the LD with scale parameter b. For simplicity and without loss of generality, we focus on centered distributions that peak at zero. Let $\mathcal{V}(x; \sigma^2, b)$ represent the convolution of $\mathcal{N}(x; \sigma^2)$ and $\mathcal{L}(x; b)$.

In Scenario 2, note that the value of σ used by the anonymizer does not have to be perfectly accurate, provided it is not overly optimistic (i.e., underestimating the true error). Let σ_t denote the true standard deviation of the error distribution and

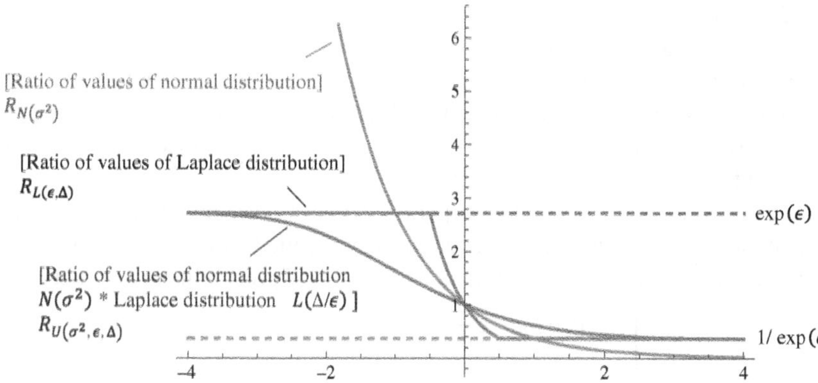

Figure 11.4 Ratio of probability density function (PDF) values.

σ represent the standard deviation that the anonymizer believes to be true. In this context, being overly optimistic means that

$$\sigma > \sigma_t \tag{11.3}$$

Figure 11.4 shows the ratio of PDF values separated by a distance of Δ for the ND $\mathcal{N}(x;\sigma^2)$, the LD $\mathcal{L}(x;\Delta/\epsilon)$, and their convolution $\mathcal{V}(x;\sigma^2,\Delta/\epsilon)$ with ϵ and σ set to 1. For the ND, the ratio of PDF values separated by Δ is computed as follows:

$$R_{\mathcal{N}(x;\sigma^2)} = \frac{\mathcal{N}(x+\Delta/2;\sigma^2)}{\mathcal{N}(x-\Delta/2;\sigma^2)} = e^{-\frac{\Delta x}{\sigma^2}} \tag{11.4}$$

As shown in equation 11.4, $R_{\mathcal{N}(x;\sigma^2)}$ tends to infinity as x approaches negative infinity. This implies that even with a large σ, additional noise must be introduced to satisfy the ϵ-DP requirement.

In Figure 11.4, $R_{\mathcal{L}(x;\epsilon,\Delta)}$ and $R_{\mathcal{V}(x;\sigma^2,\epsilon,\Delta)}$ represent the ratios of PDF values separated by Δ for the LD $\mathcal{L}(x;\Delta/\epsilon)$ and convolution $\mathcal{V}(x;\sigma^2,\Delta/\epsilon)$, respectively.

To adhere to the definition of ϵ-DP, the ratio of PDF values separated by Δ should fall within the bounds of $\exp(\epsilon)$ and $1/\exp(\epsilon)$. Figure 11.4 demonstrates that both $R_{\mathcal{L}(x;\epsilon,\Delta)}$ and $R_{\mathcal{V}(x;\sigma^2,\epsilon,\Delta)}$ satisfy this condition, confirming that the mechanisms $\mathcal{L}(x;\Delta/\epsilon)$ and $\mathcal{V}(x;\sigma^2,\Delta/\epsilon)$ achieve ϵ-DP (where $\sigma = \Delta = \epsilon = 1$ in the figure). However, although $R_{\mathcal{V}(x;\sigma^2,\Delta/\epsilon)}$ approaches $\exp(\epsilon)$ (or $1/\exp(\epsilon)$) as the absolute value of x increases, it does so slower than $R_{\mathcal{L}(x;\Delta/\epsilon)}$. This slower convergence leads to the addition of more noise than is strictly necessary for privacy preservation.

To address this issue, we propose a simple yet effective algorithm that selectively adds Laplace noise. Specifically, Laplace noise is only added when the magnitude of the generated noise exceeds a predefined threshold w. This approach is expected to reduce the overall noise and allow the ratio of PDF values separated by Δ to converge more rapidly to $\exp(\epsilon)$ and $1/\exp(\epsilon)$.

Determining the optimal value for the threshold w is a complex task. Setting w too high would result in a failure to achieve either traditional ϵ-DP or GDP. Conversely, setting w too low would lead to the inclusion of unnecessary noise in the output.

We can express the PDF that incorporates Laplace noise only when the noise magnitude $|x|$ is greater than or equal to a predefined threshold w as follows (notably, this is technically a combination of a PDF and probability mass function):

$$\hat{\mathcal{L}}(x; b, w) = \begin{cases} \int_{-w}^{w} \mathcal{L}(t; b)dt & x = 0 \\ \frac{e^{-x/b}}{2b} & x \geq w \\ \frac{e^{x/b}}{2b} & x \leq -w \\ 0 & otherwise \end{cases} \tag{11.5}$$

Consequently, we can express the PDF resulting from the original sensing error and Laplace noise defined in equation 11.5 as follows:

$$\begin{aligned} \mathcal{V}(x; \sigma^2, b, w) &= \int_{-\infty}^{\infty} \mathcal{N}(t; \sigma^2)\hat{\mathcal{L}}(x - t; b, w)dt \\ &+ \mathcal{N}(x; \sigma^2) \int_{-w}^{w} \mathcal{L}(t; b)dt \\ &= \frac{e^{-\frac{w+x}{b} - \frac{x^2}{2\sigma^2}}}{4b\sigma} \times \left\{ \sigma e^{\frac{1}{2}\left(\frac{2bw+\sigma^2}{b^2} + \frac{x^2}{\sigma^2}\right)} \left[\operatorname{erfc}\left(\frac{b(w-x)+\sigma^2}{\sqrt{2}b\sigma} \right) \right. \right. \\ &\left. \left. + e^{\frac{2x}{b}} \operatorname{erfc}\left(\frac{b(w+x)+\sigma^2}{\sqrt{2}b\sigma} \right) \right] + 2\sqrt{\frac{2}{\pi}} b \left(e^{\frac{w}{b}} - 1 \right) e^{\frac{x}{b}} \right\} \end{aligned} \tag{11.6}$$

The ratio of PDF values separated by a distance Δ for the proposed algorithm is expressed as follows:

$$R_{\mathcal{V}(x;\sigma^2,\epsilon,\Delta,w)} = \frac{\mathcal{V}(x + \Delta/2; \sigma^2, \Delta/\epsilon, w)}{\mathcal{V}(x - \Delta/2; \sigma^2, \Delta/\epsilon, w)} \tag{11.7}$$

Our goal is to determine an appropriate value for w such that $R_{\mathcal{V}}$ closely approximates $\exp(\epsilon)$ without exceeding the bounds of $\exp(\epsilon)$ or $1/\exp(\epsilon)$.

To achieve this, we propose the following theorem (illustrated in Figure 11.5):

Theorem 11.1 As w approaches infinity, the value of $R_{\mathcal{V}}$ converges to the value of $R_{\mathcal{N}}$. Conversely, as w approaches zero, $R_{\mathcal{V}}$ converges to $R_{\mathcal{U}}$.

Proof: By calculating the limit of $\mathcal{V}(x; \sigma^2, \Delta/\epsilon, w)$ (equation 11.6) as w tends to zero and infinity, we obtain $\mathcal{U}(x; \sigma^2, \Delta/\epsilon)$ and $\mathcal{N}(x; \sigma^2)$ (Equation 11.4), respectively.

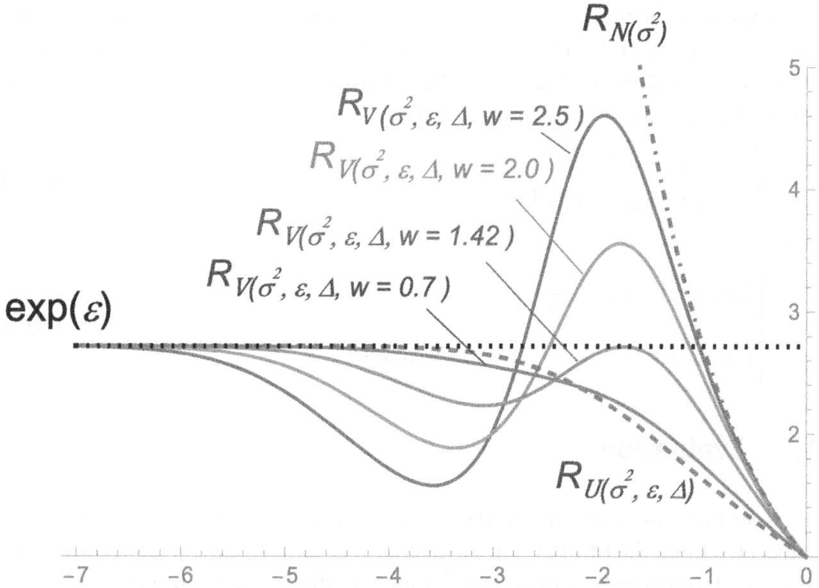

Figure 11.5 Analysis of the metric $R_{\mathcal{V}}$ for different values of w. Results show that an excessively large w fails to meet the differential privacy (DP) requirement, while an excessively small w introduces more noise than necessary.

In this study, we define the ratio of $x + \Delta/2$ to $x - \Delta/2$. Therefore, we can examine the range $-w - \Delta/2 < x < 0$ to determine whether the maximum ratio exceeds $\exp(\epsilon)$. Notably, owing to the symmetry of $\mathcal{V}$ with respect to the point $(x, y) = (0, 1)$, where y denotes the ratio of PDF values separated by Δ, it is sufficient to check for $x < 0$.

11.4.2 Categorical Attributes

First, consider Scenario I. Assume that an attacker obtains category ID γ as the anonymized version of a categorical attribute. Let $P(v_a = \gamma | v_t = i)$ represent the probability that the anonymized category ID is γ when the true category ID is i. The proposed mechanism ensures that

$$P(v_a = \gamma | v_t = i) = \begin{cases} \frac{e^{\epsilon}}{M-1+e^{\epsilon}} & (i = \gamma) \\ \frac{1-\frac{e^{\epsilon}}{M-1+e^{\epsilon}}}{M-1} & (otherwise) \end{cases} \tag{11.8}$$

The ratio of the two cases in equation 11.8 is e^{ϵ} or $1/e^{\epsilon}$. Therefore, Definition 11.2 holds. Based on the post-processing property of DP, the resulting values also satisfy GDP.

Next, consider Scenario II. Assume that the anonymizer's knowledge of sensing errors is incorrect but not pessimistic. Let $x_{i \to j,t}$ and $x_{i \to j}$ denote the disguising probabilities based on the true and believed error parameters, respectively. If the error parameters are not pessimistic, then

$$\begin{cases} x_{i \to j,t} \geq x_{i \to j} & (i = j) \\ x_{i \to j,t} \leq x_{i \to j} & (otherwise) \end{cases} \tag{11.9}$$

Therefore,

$$\begin{cases} P(v_a = \gamma | v_t = i) \leq \frac{e^{\epsilon}}{M-1+e^{\epsilon}} & (i = \gamma) \\ P(v_a = \gamma | v_t = i) \geq \frac{1-\frac{e^{\epsilon}}{M-1+e^{\epsilon}}}{M-1} & (otherwise) \end{cases} \tag{11.10}$$

11.5 Evaluation

In this section, we explain the parameter values used in the experiments in Section 11.5.1 and the utility metric in Section 11.5.2. Then, we present the results of the real datasets in Section 11.5.3. Finally, we discuss the computation cost of the proposed method in Section 11.5.4.

11.5.1 Parameter Settings

Realistic values need to be set for ϵ, σ, and $\boldsymbol{p}_i$ for all i.

11.5.1.1 Value of ϵ

Apple's deployment ensures that ϵ is set to either 1 or 2 per datum [49]. In evaluations by Apple's DP team, ϵ values of 2, 4, and 8 were used [50]. The paper proposing Randomized Aggregatable Privacy-Preserving Ordinal Response (RAPPOR) by Google [51] used $\epsilon = \log(3)$ as a primary setting. Based on these references, ϵ is chosen to be in the range of 1–10. For this range, if Δ is set to 100, the average noise added by the Laplace mechanism ranges from 5 to 50, indicating adequate privacy protection. Notably, if Δ is scaled by a factor of a, the average noise is also scaled by a. For categorical attributes with two categories ($M = 2$), the retention probability value ranges between 73.11% and 99.996% when ϵ is between 1 and 10. An ϵ value of 10 suggests very low privacy protection when M is small, emphasizing the relevance of evaluating performance at small ϵ values for categorical attributes.

11.5.1.2 Value of σ

The average sensing error, $ASE(\sigma)$, is defined as follows:

$$ASE(\sigma) = \int_{x=-\infty}^{\infty} |x| \cdot \mathcal{N}(x; \sigma^2) dx = \sqrt{\frac{\pi}{2}} \sigma \tag{11.11}$$

If σ is set to 1/40 of Δ (the range of possible values), $ASE(\sigma)$ will be 2.0% of Δ, indicating high accuracy of the corresponding IoT device. Similarly, setting σ to 1/10, 1/4, and 1/2 of Δ yields $ASE(\sigma)$ values of 8.0% (relatively high accuracy), 20% (relatively low accuracy), and 40% (low accuracy) of Δ, respectively. Based on this, σ is set between 1/40 and 1/2 of Δ.

11.5.1.3 Value of p_i

For all i, $p_{i\to i}$ is set to a common value τ within the range of 0.3–0.9. This reflects the probability that an IoT device correctly identifies an attribute category, from low (0.3) to high (0.9) accuracy. For all $i, j (i \neq j)$, $p_{i\to j}$ is defined as follows:

$$p_{i\to j} = \frac{1-\tau}{M-1}. \tag{11.12}$$

11.5.2 Utility Metric

The goal is to use anonymized data for various services; thus, the estimated values should be close to the ground truth. Let N be the number of individuals whose attribute values are collected, v_i is the ground truth for person i, and $\widetilde{v_i}$ is the corresponding anonymized value.

For numerical attributes, utility is defined as follows:

$$U_n = \frac{1}{N}\sum_{i=1}^{N}\left(1 - \frac{|v_i - \widetilde{v_i}|}{\Delta}\right) \tag{11.13}$$

For categorical attributes, utility is defined as follows:

$$U_c = \frac{1}{N}\sum_{i=1}^{N}\delta_{v_i,\widetilde{v_i}} \tag{11.14}$$

where $\delta_{i,j}$ is the Kronecker delta:

$$\delta_{i,j} = \begin{cases} 1 & (i = j) \\ 0 & (i \neq j) \end{cases} \tag{11.15}$$

Higher utility values indicate better performance.

Some methods estimate statistical values or generate cross-tabulations from collected data. For cross-tabulations, the total loss comparing true and generated cross-tabulations is used. However, this study focuses on personalized data applications in which analyzing each individual's attribute value is crucial, such as health monitoring, context-aware recommendation, and navigation (Figure 11.6).

11.5.3 Real Dataset Results

Simulations were performed using the Adult dataset [55], a common benchmark in privacy-preserving data mining, which includes six numerical and nine categorical attributes, with 30,162 records after removing unknown values.

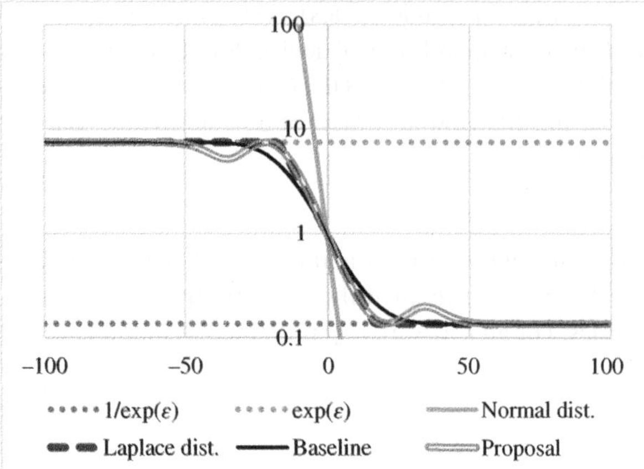

Figure 11.6 An example result for numerical values.

Table 11.2 Results of numerical attributes of adult.

	Age	fnlwgt	Education	Gain	Loss	Hours
Δ	73	1,470,936	15	99,999	4,356	98
Baseline	0.85	0.85	0.85	0.85	0.85	0.85
Proposal	**0.92**	**0.92**	**0.92**	**0.92**	**0.92**	**0.92**

Each value in the Adult dataset is assumed true, with IoT devices estimating age, sex, race, and nationality using various methods [56–58]. For numerical attributes, σ was set to 0.1 of Δ and ϵ was set to 8. For categorical attributes, τ was set to 0.6 and ϵ to 2. Simulation results, including attribute names with their respective Δ and M, are summarized in Tables 11.2 and 11.3. Compared with the baseline, the proposed mechanism increased U_n to approximately 92% from 85% for all numerical attributes and increased U_c by up to 20% for categorical attributes. These results demonstrate the proposed mechanism's ability to enhance utility (i.e., reduce total loss) on real datasets.

Additional simulations were performed using other real datasets using the same parameter settings.

The AReM dataset [59], containing 42,239 instances of six numerical attributes, was used for the numerical attributes.

For categorical attributes, the Activity Recognition Data (ADL) dataset [52], Radio Frequency Identification (RFID) dataset [53], and Localization dataset [54] with 741, 75,128, and 164,860 instances, respectively, were used. The simulation

Table 11.3 Results of categorical attributes of adult.

	Workclass	Edu.	Marital	Occ.	Rel.	Race	Sex	Country	Salary
M	7	16	7	14	6	5	2	41	2
Baseline	0.36	0.22	0.36	0.24	0.40	0.42	0.58	0.10	0.58
Proposal	**0.55**	**0.33**	**0.55**	**0.36**	**0.60**	**0.60**	**0.60**	**0.16**	**0.60**

Table 11.4 Results of numerical attributes of AReM.

	avg_rss12	var_rss12	avg_rss13	var_rss13	avg_rss23	var_rss23
Δ	56.25	17.24	35	11.42	40.33	13.61
Baseline	0.85	0.85	0.85	0.85	0.85	0.85
Proposal	**0.92**	**0.92**	**0.92**	**0.92**	**0.92**	**0.92**

Table 11.5 Results of categorical attributes of three datasets.

	ADL dataset [52]	RFID dataset [53]	Localization dataset [54]
M	10	4	11
Baseline	0.30	0.47	0.28
Proposal	**0.45**	**0.60**	**0.43**

results are summarized in Tables 11.4 and 11.5, indicating that the proposed mechanism outperforms the baseline approach on all considered datasets.

11.5.4 Computation Cost

The proposed anonymizing algorithm is executed on local devices. We tested them on a laptop with 8 GB of memory and a Core i5 10210U central processing unit (CPU), where each dataset was processed in less than a second. User data distribution estimation was performed on a server with 128 GB memory and an Intel Xeon W-2295 CPU, taking an average of 7.8 seconds for the AReM dataset with 42,239 values. The proposed method's computation time is proportional to the number of users and attributes; thus, it increases with more users. However, because data collection itself takes time, a few minutes of computation is practical for estimating data distribution, although the comparison methods have shorter computation times due to their simpler algorithms.

11.5.5 Overhead Cost

In Section 11.5.4, we discussed computation costs. Here, we address the overhead cost, specifically the network cost between the server and the user. Our proposed method adds differential privacy noise to the IoT measurements, while accounting for the measurement errors of the IoT devices. The number of bits required to represent the IoT measurement remains the same before and after applying differential privacy noise. In addition, encryption of the communication channel between the server and the user is beyond the scope of this chapter. If encryption is necessary to prevent external attackers from eavesdropping on the communication channel, then encryption should be implemented. The encryption process, similar to conventional differential privacy techniques, does not affect our proposed scheme in any way. In summary, there is no additional overhead in implementing our proposed scheme.

11.6 Discussion

We discuss the relationships between TinyML and IoT data in Section 11.6.1, followed by the connection between TinyML and DP in Section 11.6.2. Our future research directions are outlined in Section 11.6.3.

11.6.1 TinyML with Noisy IoT Data

Several studies have developed ML models from noisy IoT data, achieving high accuracy [60, 61]. However, DP studies [32–34] have shown that small privacy budgets, which correspond to high noise levels, can substantially degrade model accuracy.

This study demonstrates that accounting for observation noise in IoT data enables a reduction in the privacy preservation noise required. By integrating our method with existing DP techniques, we can enhance the accuracy of ML models while preserving privacy.

11.6.2 TinyML and DP

This section discusses the technical and applied aspects of integrating TinyML and DP, exploring how these technologies can synergistically enhance privacy-preserving ML at edges.

11.6.2.1 Implementing DP on Edge Devices

Implementing DP on resource-constrained edge devices presents unique challenges due to their computational constraints and limited resources. Traditional

DP algorithms, known for their computational intensity, must be optimized to match the capabilities of edge devices. This includes exploring efficient noise addition techniques and decentralized privacy-preserving algorithms.

Efficient Noise Addition Techniques Noise addition is a cornerstone of DP. On edge devices, it is critical to use computationally efficient noise generation algorithms. Techniques such as optimizing the Gaussian/Laplacian noise generation, or using precomputed noise values, can help achieve this efficiency.

Decentralized Privacy-Preserving Algorithms To ensure privacy across an entire network of edge devices, decentralized DP algorithms are employed. These algorithms allow each device to locally process data while collectively ensuring privacy. Integrating this approach with federated learning eliminates the need for centralized data collection, thereby further enhancing privacy.

11.6.2.2 Use Cases and Applications

The integration of TinyML and DP opens up a wide range of applications, including real-time medical data analysis, IoT device data processing, and smart home technologies. Each application presents unique benefits and challenges.

Real-Time Medical Data Analysis In the healthcare sector, there is a growing demand for real-time data analysis while preserving patient privacy. TinyML enables the real-time processing of data from wearable devices and medical equipment, while DP ensures that patient confidentiality is maintained. This combination supports more accurate and privacy-conscious diagnostic processes.

IoT Device Data Processing IoT devices, prevalent in smart homes and cities, generate massive data. Privacy concerns require careful handling of this data. By integrating TinyML with DP, data can be locally processed on these devices, enabling valuable insights without compromising user privacy.

Smart Home Applications Smart homes collect data on user behaviors and habits, which are inherently personal and requires robust privacy protection. TinyML enables on-device data processing, and when combined with DP, it ensures that users benefit from smart home technology while preserving their privacy.

11.6.2.3 Technical Challenges and Solutions

The integration of TinyML and DP presents several technical challenges. Major challenges include resource limitations, privacy–utility trade-offs, and data validation.

Resource Constraints Implementing DP on edge devices with limited processing power and memory poses considerable challenges. To address this issue, researchers are actively developing lightweight algorithms and optimizing hardware.

Privacy–Utility Trade-Off Achieving an optimal balance between strong privacy preservation and data utility is critical. Excessive noise can render the data meaningless, making it essential to carefully calibrate the amount of noise added.

Data Validation Ensuring data accuracy and consistency when applying DP is crucial. Establishing robust validation methods for edge devices is essential for reliable data processing.

11.6.2.4 Future Directions

The convergence of TinyML and DP is poised to considerably influence the future of edge computing. Ongoing research and development efforts are expected to yield more efficient and secure algorithms, as well as practical implementations across diverse domains. The establishment of standards and guidelines will also be crucial in promoting widespread adoption and increasing the trustworthiness of these technologies.

This section provides a comprehensive overview of both the technical and applied aspects of integrating TinyML with DP. We have explored the implementation of DP on edge devices, highlighted real-world use cases, addressed technical challenges, and provided a glimpse into the future prospects of this exciting field. The findings presented in this study aim to drive further advances in edge computing and privacy preservation, paving the way for more secure and effective technologies.

11.6.3 Future Work

There are various mechanisms for achieving DP beyond the Laplace and randomized response mechanisms. For instance, Geng and Viswanath [62] proposed the staircase mechanism for numerical values, which proves to be more effective when ϵ is large. Additionally, Andrés et al. [63] introduced a geo-indistinguishable mechanism for location information. We believe that GDP can be applied to other privacy metrics that manage attribute value probability distributions, such as pk-anonymity [64, 65] and t-closeness [66].

Future research should investigate the application of GDP to these and other privacy mechanisms. Although our focus has been on ND errors, exploring other error distributions could lead to the development of new algorithms.

Furthermore, our study has concentrated on individual anonymized data collection. Another crucial area of privacy research is privacy-preserving data publishing, where a data holder anonymizes and publishes personal data. GDP could be applied to ϵ-DP in this context as well, and we plan to address this aspect in future work.

11.7 Conclusions

DP is a key technique for protecting user privacy by adding noise to sensor values. However, in the context of IoT, where sensor data often contains inherent errors, existing solutions fail to account for this inherent sensing noise. Instead, they focus on protecting the already noisy detected noisy values rather than the ground truth. Our approach addresses this issue by modifying the added noise based on the sensor noise model, thereby reducing the DP noise by approximately 20%. We introduce GDP, a novel privacy metamodel that applies to DP and acknowledges that the ground truth might be unknown to both the data owner and the anonymizer. We validate our results with both synthetic and real datasets, marking the first application of GDP.

TinyML has revolutionized the deployment of ML models on resource-constrained devices, remarkably advancing various domains, such as IoT, health care, and smart environments. With the rapid growth of IoT devices that generate massive data, ensuring data privacy while maintaining high accuracy is of paramount importance. Due to their inherent limitations in computational power, memory, and energy, TinyML necessitates innovative privacy-preservation approaches that do not compromise performance.

The proposed GDP framework perfectly aligns with the objectives of TinyML by optimizing the balance between privacy and accuracy. By considering the inherent sensing errors in IoT data, GDP reduces the need for excessive noise addition, which is crucial for maintaining the utility of data processed on TinyML devices. This balance is particularly relevant in applications such as health monitoring, in which accurate real-time data analysis can be life-saving, and in smart homes, where user behavior data must be protected without degrading the service quality.

The integration of TinyML with DP opens several exciting avenues for future research and development.

These include further refining GDP and other privacy-preserving algorithms to enhance their efficiency and effectiveness on TinyML platforms. This could involve exploring alternative noise mechanisms and error models that are better suited for the constraints of TinyML devices. Additionally, developing specialized hardware accelerators that can efficiently handle privacy-preserving computations on TinyML devices, possibly through designing new microcontroller architectures optimized for differential privacy tasks, presents a great opportunity. Leveraging the synergy between TinyML and edge computing to create robust, privacy-preserving systems that can process and analyze data locally while offloading more complex tasks to edge servers is another promising direction. This hybrid approach can remarkably enhance the scalability and flexibility of IoT deployments. Furthermore, tailoring privacy-preserving techniques to specific TinyML applications such as environmental monitoring, industrial automation, and personalized health care, is essential. Each application has unique privacy and accuracy requirements that must be addressed through customized solutions. Collaboration with regulatory bodies and industry stakeholders will also be crucial for developing standardized guidelines and policies for implementing DP in TinyML applications. Ensuring that privacy-preserving practices are consistently applied and trusted by users will help to foster confidence and broader adoption.

The integration of GDP with TinyML represents notable advancements in the field of privacy-preserving data analysis on resource-constrained devices. By addressing the inherent sensor errors in IoT data and optimizing the noise addition process, GDP enhances the utility of data processed on TinyML-based devices while maintaining robust privacy protection. As TinyML continues to expand its influence across various industries, incorporating advanced privacy-preserving techniques such as GDP will be crucial for building secure, efficient, and reliable IoT systems. Future research and development in this area will unlock new potentials and applications, driving the next wave of innovation in smart, connected devices.

References

1 Sei, Y. and Ohsuga, A. (2022). Private true data mining: differential privacy featuring errors to manage Internet-of-Things data. *IEEE Access* 10 (1): 8738–8757.

2 Kallimani, R., Pai, K., Raghuwanshi, P. et al. (2024). TinyML: tools, applications, challenges, and future research directions. *Multimedia Tools and Applications* 83 (10): 29015–29045.

3 Ficco, M., Guerriero, A., Milite, E. et al. (2024). Federated learning for IoT devices: enhancing TinyML with on-board training. *Information Fusion* 104: 102189.

4 Bhattacharya, S. and Pandey, M. (2024). Deploying an energy efficient, secure & high-speed sidechain-based TinyML model for soil quality monitoring and management in agriculture. *Expert Systems with Applications* 242: 122735.

5 Zhang, Z., He, J., Kumar, A., and Rahman, S. (2024). Ai-based space occupancy estimation using environmental sensor data. *2024 IEEE Power & Energy Society Innovative Smart Grid Technologies Conference (ISGT)*, 1–5. IEEE.

6 Putra, K.T., Arrayyan, A.Z., Hayati, N. et al. (2024). A review on the application of internet of medical things in wearable personal health monitoring: a cloud-edge artificial intelligence approach. *IEEE Access* 12: 21437–21452.

7 Andrew, J. and Karthikeyan, J. Privacy-preserving Internet of Things: techniques and applications. *International Journal of Engineering and Advanced Technology* 8 (6): 3229–3234.

8 Saeed, M.M., Saeed, R.A., and Ahmed, Z.E. (2024). TinyML for 5G networks. In: *TinyML for Edge Intelligence in IoT and LPWAN Networks*, 167–229. Elsevier.

9 Pradhan, N., Chaudhari, B.S., and Zennaro, M. (2024). 6TiSCH adaptive scheduling for Industrial Internet of Things. In: *TinyML for Edge Intelligence in IoT and LPWAN Networks*, 283–309. Elsevier.

10 Patil, R.Y., Bhamare, M., Patil, Y.H., and Bannore, A. (2024). Securing TinyML in a connected world. In: *TinyML for Edge Intelligence in IoT and LPWAN Networks*, 311–330. Elsevier.

11 Andrade, P., Silva, I., Diniz, M. et al. (2024). Online processing of vehicular data on the edge through an unsupervised TinyML regression technique. *ACM Transactions on Embedded Computing Systems* 23 (3): 1–28.

12 Peng, T., Zhong, W., Wang, G. et al. (2024). Privacy-preserving truth discovery based on secure multi-party computation in vehicle-based mobile crowdsensing. *IEEE Transactions on Intelligent Transportation Systems* 25 (7): 7767–7779.

13 Dwork, C., McSherry, F., Nissim, K., and Smith, A. (2006). Calibrating noise to sensitivity in private data analysis. *Proceedings of Theory of Cryptography (TCC)*, 265–284.

14 Dwork, C. and Roth, A. (2014). The algorithmic foundations of differential privacy. *Foundations and Trends in Theoretical Computer Science* 9 (3-4): 211–407.

15 Liu, F. (2019). Generalized Gaussian mechanism for differential privacy. *IEEE Transactions on Knowledge and Data Engineering* 31 (4): 747–756.

16 Isawa, R., Sei, Y., Tahara, Y., and Ohsuga, A. (2024). Minimizing noise in location privacy protection through equipment error consideration. *International Journal of Electrical and Computer Engineering Systems* 15 (3): 285–296.

17 Wang, M., Wang, Q.-Y., Zhang, Y.-H. et al. (2024). Preserving differential privacy in neural networks for foreign object detection with heterogeneity-based noising among distributed devices. *The Journal of Supercomputing* 80 (14): 21447–21474.

18 Warner, S.L. (1965). Randomized response: a survey technique for eliminating evasive answer bias. *American Statistical Association* 60 (309): 63–69.

19 Huang, Z. and Du, W. (2008). OptRR: optimizing randomized response schemes for privacy-preserving data mining. *Proceedings of IEEE ICDE*, 705–714.

20 Airy, G.B. (2007). *On the Algebraical and Numerical Theory of Errors of Observations and the Combination of Observations*. Kessinger Publishing, LLC.

21 Taylor, J.R. (1997). *Introduction to Error Analysis: The Study of Uncertainties in Physical Measurements*, 2e. University Science Books.

22 Murakami, T. and Kawamoto, Y. (2019). Utility-optimized local differential privacy mechanisms for distribution estimation. *Proceedings of USENIX Security Symposium*, 1877–1894.

23 Gu, X., Li, M., Xiong, L., and Cao, Y. (2020). Providing input-discriminative protection for local differential privacy. *Proceedings of ICDE*, vol. 2020-April, 505–516.

24 Li, Z., Wang, T., Lopuhaä-Zwakenberg, M. et al. (2020). Estimating numerical distributions under local differential privacy. *Proceedings of ACM SIGMOD*, 621–635.

25 Zhao, Y., Zhao, J., Yang, M. et al. (2021). Local differential privacy-based federated learning for Internet of Things. *IEEE Internet of Things Journal* 8 (11): 8836–8853.

26 Husnoo, M.A., Anwar, A., Chakrabortty, R.K. et al. (2021, 2021). Differential privacy for IoT-enabled critical infrastructure: a comprehensive survey. *IEEE Access* 9: 153276–153304.

27 Sei, Y. and Ohsuga, A. (2017). Differential private data collection and analysis based on randomized multiple dummies for untrusted mobile crowdsensing. *IEEE Transactions on Information Forensics and Security* 12 (4): 926–939.

28 Sei, Y., Onesimu, J.A., Okumura, H., and Ohsuga, A. (2023). Privacy-preserving collaborative data collection and analysis with many missing values. *IEEE Transactions on Dependable and Secure Computing* 20 (3): 2158–2173.

29 Waheed, N., Khan, F., Mastorakis, S. et al. (2023). Privacy-enhanced living: a local differential privacy approach to secure smart home data. *2023 IEEE International Conference on Omni-layer Intelligent Systems (COINS)*, 1–6. IEEE.

30 Yang, W. and Al-Masri, E. (2024). ULDP: a user-centric local differential privacy optimization method. *2024 IEEE World AI IoT Congress (AIIoT)*, 0316–0322. IEEE.

31 He, Z., Wang, L., and Cai, Z. (2023). Clustered federated learning with adaptive local differential privacy on heterogeneous IoT data. *IEEE Internet of Things Journal* 11 (1): 137–146.

32 Shi, D., Ding, J., Errapotu, S.M. et al. (2019). Deep Q-network based route scheduling for TNC vehicles with passengers' location differential privacy. *IEEE Internet of Things Journal* 6 (5): 7681–7692.

33 Xu, C., Ren, J., She, L. et al. (2019). EdgeSanitizer: locally differentially private deep inference at the edge for mobile data analytics. *IEEE Internet of Things Journal* 6 (3): 5140–5151.

34 Guan, Z., Lv, Z., Du, X. et al. (2019). Achieving data utility-privacy tradeoff in Internet of Medical Things: a machine learning approach. *Future Generation Computer Systems* 98: 60–68.

35 Wang, N., Xiao, X., Yang, Y. et al. (2018). PrivTrie: effective frequent term discovery under local differential privacy. *Proceedings of IEEE ICDE*, 821–832.

36 Kim, J.W., Kim, D.-H., and Jang, B. (2018). Application of local differential privacy to collection of indoor positioning data. *IEEE Access* 6: 4276–4286.

37 Tonyali, S., Akkaya, K., Saputro, N. et al. (2018). Privacy-preserving protocols for secure and reliable data aggregation in IoT-enabled smart metering systems. *Future Generation Computer Systems* 78: 547–557.

38 Liu, Y., Guo, W., Fan, C.-I. et al. (2019). A practical privacy-preserving data aggregation (3PDA) scheme for smart grid. *IEEE Transactions on Industrial Informatics* 15 (3): 1767–1774.

39 Abul, O., Bonchi, F., and Nanni, M. (2008). Never walk alone: uncertainty for anonymity in moving objects databases. *Proceedings of IEEE ICDE*, 376–385.

40 Sei, Y. and Ohsuga, A. (2017). Location anonymization with considering errors and existence probability. *IEEE Transactions on Systems, Man, and Cybernetics: Systems* 47 (12): 3207–3218.

41 Ge, C., Ilyas, I.F., He, X., and Machanavajjhala, A. (2017). Private exploration primitives for data cleaning. *arXiv preprint arXiv:1712.10266*, 1–17.

42 Krishnan, S., Wang, J., Franklin, M.J. et al. (2016). PrivateClean: data cleaning and differential privacy. *Proceedings of ACM SIGMOD*, 937–951.

43 Blass, E.-O. and Kerschbaum, F. (2023). Private collaborative data cleaning via non-equi PSI. *2023 IEEE Symposium on Security and Privacy (SP)*, 1419–1434. IEEE.

44 Ma, L., Pei, Q., Zhou, L. et al. (2020). Federated data cleaning: collaborative and privacy-preserving data cleaning for edge intelligence. *IEEE Internet of Things Journal* 8 (8): 6757–6770.

45 Curnow, E., Carpenter, J.R., Heron, J.E. et al. (2023). Multiple imputation of missing data under missing at random: compatible imputation models are not sufficient to avoid bias if they are mis-specified. *Journal of Clinical Epidemiology* 160: 100–109.

46 Jia, F. and Wu, W. (2023). A comparison of multiple imputation strategies to deal with missing nonnormal data in structural equation modeling. *Behavior Research Methods* 55 (6): 3100–3119.

47 Zahid, F.M. and Heumann, C. (2019). Multiple imputation with sequential penalized regression. *Statistical Methods in Medical Research* 28 (5): 1311–1327.

48 Lyon, A. (2014). Why are normal distributions normal? *British Journal for the Philosophy of Science* 65 (3): 621–649.

49 Tang, J., Korolova, A., Bai, X. et al. (2017). Privacy loss in apple's implementation of differential privacy on MacOS 10.12. *arXiv:1709.02753*, 1–12.

50 Differential Privacy Team Apple (2017). Learning with privacy at scale. *Apple Machine Learning Journal* 1 (8): 1–25.

51 Erlingsson, Ú., Pihur, V., and Korolova, A. (2014). RAPPOR: randomized aggregatable privacy-preserving ordinal response. *Proceedings of ACM CCS*, 1054–1067.

52 Ordóez, F.J., de Toledo, P., Sanchis, A. et al. (2013). Activity recognition using hybrid generative/discriminative models on home environments using binary sensors. *Sensors* 13 (5): 5460–5477.

53 Torres, R.L.S., Ranasinghe, D.C., Shi, Q., and Sample, A.P. (2013). Sensor enabled wearable RFID technology for mitigating the risk of falls near beds. *Proceedings of IEEE RFID*, 191–198.

54 Kaluža, B., Mirchevska, V., Dovgan, E. et al. (2010). An agent-based approach to care in independent living. *Proceedings of International Joint Conference on Ambient Intelligence*, 177–186.

55 Dua, D. and Graff, C. (2019). UCI machine learning repository. http://archive.ics.uci.edu/ml (accessed 23 September 2025).

56 Wang, X., Ly, V., Lu, G., and Kambhamettu, C. (2013). Can we minimize the influence due to gender and race in age estimation?*Proceedings of ICMLA*, 309–314.

57 Choi, S.E., Lee, Y.J., Lee, S.J. et al. (2011). Age estimation using a hierarchical classifier based on global and local facial features. *Pattern Recognition* 44 (6): 1262–1281.

58 Han, H., Otto, C., Liu, X., and Jain, A.K. (2015). Demographic estimation from face images: human vs. machine performance. *IEEE Transactions on Pattern Analysis and Machine Intelligence* 37 (6): 1148–1161.

59 Palumbo, F., Gallicchio, C., Pucci, R., and Micheli, A. (2016). Human activity recognition using multisensor data fusion based on reservoir computing. *Journal of Ambient Intelligence and Smart Environments* 8 (2): 87–107.

60 Li, M., Xie, L., Lv, Z. et al. (2020). Multistep deep system for multimodal emotion detection with invalid data in the Internet of Things. *IEEE Access* 8: 187208–187221.

61 Lee, C.H., Lin, J.W., Chen, P.H., and Chang, Y.C. (2019). Deep learning-constructed joint transmission-recognition for Internet of Things. *IEEE Access* 7: 76547–76561.

62 Geng, Q. and Viswanath, P. (2016). The optimal noise-adding mechanism in differential privacy. *IEEE Transactions on Information Theory* 62 (2): 925–951.

63 Andrés, M.E., Bordenabe, N.E., Chatzikokolakis, K., and Palamidessi, C. (2013). Geo-indistinguishability: differential privacy for location-based systems. *Proceedings of ACM CCS*, 901–914.

64 Kimura, E., Chida, K., Ikarashi, D. et al. (2012). Statistical disclosure limitation of health data based on Pk-anonymity. *Studies in Health Technology and Informatics* 180: 1117–1119.

65 Kakizawa, M., Watanabe, C., Furukawa, R., and Takahashi, T. (2014). Improvement of Pk-anonymization. *Proceedings of IEEE SRDS Workshop*, 82–87.

66 Li, N., Li, T., and Venkatasubramanian, S. (2007). t-closeness: privacy beyond k-anonymity and l-diversity. *Proceedings of IEEE ICDE*, 106–115.

12

Security and Privacy of TinyML Devices

Eftychia Mistillioglou, Evangelia Konstantopoulou, Nicolas Sklavos, and Andronikos Kyriakou

SCYTALE Group, Computer Engineering and Informatics Department, University of Patras, Patras, Greece

12.1 Introduction

As tiny machine learning (TinyML) technology continues to advance and integrate into numerous applications, ranging from healthcare to industrial monitoring and smart cities, a critical issue that often goes unaddressed is security. TinyML models enable sophisticated machine learning (ML) directly on resource-constrained devices, often deployed on microcontrollers and low-power hardware. While this capability opens new opportunities, it also raises critical concerns regarding data integrity and device reliability in sensitive applications. Threats such as adversarial attacks, physical tampering, data exfiltration, and side-channel attacks can severely compromise model functionality and data privacy. Ensuring robust security for TinyML within the limitations of small-scale devices is essential to maintain trust and reliability across such applications.

The resource limitations of TinyML devices present unique challenges for implementing robust security measures. Unlike traditional Internet of Things (IoT) or edge devices, which often possess megabytes of random access memory (RAM) and processors running at gigahertz speeds, TinyML devices typically operate with kilobytes of memory and processors functioning at megahertz speeds [1]. Because of these strict limitations, traditional security measures are not feasible. For example, deploying strong cryptographic protocols or sophisticated authentication methods is often infeasible. Given these issues, the proprietary nature of many TinyML models makes them attractive targets for theft or manipulation, further emphasizing the need for tailored security solutions.

Tiny Machine Learning: Design Principles and Applications, First Edition.
Edited by Agbotiname Lucky Imoize, Dinh-Thuan Do, and Houbing Herbert Song.

The limitations of traditional ML methods and cloud-based systems further complicate the security landscape for TinyML. While ML has proven effective in addressing certain cyberthreats, its computational complexity, high energy consumption, and reliance on extensive hardware resources make it unsuitable for TinyML's resource-constrained environment [2]. Similarly, cloud-based ML systems face challenges such as high-power usage, security vulnerabilities, privacy risks, and latency issues, which are incompatible with the real-time, low-power requirements of TinyML applications. Additionally, the deployment of TinyML on IoT devices raises sustainability concerns, as battery life becomes critical in scenarios where regular maintenance or replacement is not feasible [2]. These factors highlight the urgency of developing security mechanisms that are lightweight, efficient, and adaptive to TinyML's specific constraints.

To meet these challenges, the development of lightweight, adaptive security strategies tailored to TinyML ecosystems is essential. Traditional security mechanisms, such as complex encryption schemes or advanced adversarial defenses, are often infeasible due to the computational and energy constraints inherent in TinyML devices. Instead, innovative approaches that balance security with operational efficiency must be prioritized. By addressing vulnerabilities through adaptive measures that consider the unique characteristics of TinyML, it becomes possible to ensure robust security and privacy without compromising the performance and functionality of these critical systems [1]. The solutions addressed in this work encompass a combination of hardware-based protections, federated learning (FL) frameworks, adversarial defenses, and real-time anomaly detection methods that jointly address the multifaceted security requirements of TinyML.

12.1.1 Main Contributions

The key contributions of this chapter are as follows:

- **Lightweight Security for TinyML Devices**: Development of security strategies tailored to the unique constraints of TinyML devices.
- **Federated Learning Frameworks**: Privacy-aware training methods, including federated meta-learning for TinyML (TinyMetaFed) and distributed federated learning (D-FedAvg), that ensure secure and decentralized model updates.
- **Enhanced Adversarial Robustness**: Adoption of adversarial training techniques to defend against manipulated inputs.
- **Hardware-Based Security Solutions**: Implementation of physically unclonable functions (PUFs) for secure authentication and deployment of trusted execution environments (TEEs) to isolate sensitive computations.
- **Real-Time Threat Detection**: Integration of D-score methodologies for low-latency, real-time detection of security breaches.

12.1.2 Chapter Organization

The organization of this chapter begins with an introduction to the security and privacy challenges faced in deploying TinyML models on constrained devices. Following this, Section 12.2 reviews related work in TinyML security, identifying existing approaches and highlighting gaps in securing TinyML applications. Section 12.3, titled "Secure and Privacy-Aware Training of TinyML Models," delves into privacy-preserving training methodologies, such as FL, intellectual property (IP) protection, and adversarial robustness strategies. Section 12.4, "Implementation of Novel Security Schemes for TinyML Applications," discusses advanced security schemes, including hardware-based protections like PUFs, TEEs, and anti-tampering mechanisms, as well as techniques for anomaly detection and adaptive learning. Section 12.5 describes not only the lessons learned from this study but also the limitations of the presented methods. The chapter concludes with Section 12.6, where a summary of key insights and implications for secure TinyML deployment in critical tiny applications is provided.

12.2 Related Work

As TinyML becomes an integral part of many ecosystems, addressing its security and privacy challenges has grown increasingly urgent. TinyML's limited computational resources, small memory, and constrained power capacities present unique obstacles that render conventional cybersecurity measures impractical. To mitigate these limitations, research in TinyML security has focused on several areas, including privacy-preserving training, adversarial robustness, hardware-based protections, and secure deployment frameworks. These approaches collectively represent significant strides toward securing TinyML applications, though critical gaps remain.

Privacy-preserving training is essential for TinyML models, especially in applications where sensitive data is processed directly on the device, such as in healthcare or finance. FL has been widely adopted as a solution, allowing devices to collaboratively train models while retaining data locally. FedAvg, a foundational FL algorithm, has been adapted into frameworks like D-FedAvg and TinyMetaFed, optimized for TinyML's constraints by reducing the communication load and computational demand. However, while FL enhances privacy by keeping raw data decentralized, studies indicate that it is still susceptible to model inversion and data reconstruction attacks, where attackers infer sensitive information from model updates. To mitigate these vulnerabilities, differential privacy (DP) has been explored as a supplement to FL; however, DP's computational demands limit its effectiveness for TinyML. Addressing this gap, the current study incorporates lightweight encryption techniques and proposes privacy-enhanced FL algorithms that are customized for TinyML's constrained resources, reducing the risk of data inference attacks.

Another significant concern in TinyML security is adversarial robustness. Adversarial attacks, where inputs are manipulated to cause model misclassifications, pose substantial risks, particularly in sensitive applications like healthcare and industrial IoT. Conventional defenses, such as adversarial training, aim to expose models to perturbed data to increase resilience but are often computationally intensive. For TinyML, defensive distillation offers a lighter alternative by simplifying model outputs to obscure exploitable information, yet current defenses are limited in adaptability. Most adversarial defenses remain static, unable to counter dynamic attacks that adjust strategies based on model feedback. This study addresses this gap by implementing a multifaceted adversarial robustness framework that combines real-time anomaly detection with adaptive model updates, equipping TinyML models to detect and respond to adaptive attacks more effectively.

Hardware-based security measures have also been extensively explored to protect TinyML from both physical and logical threats. TEEs and PUFs are two widely studied solutions in this area. TEEs, such as ARM's TrustZone, create secure, isolated environments within microcontrollers, protecting critical model computations from tampering. PUFs, on the other hand, leverage device-specific physical characteristics as a lightweight alternative to cryptographic keys, enabling secure device authentication. However, both TEEs and PUFs come with certain limitations. TEEs may impose a memory overhead and cannot fully prevent side-channel attacks, such as power analysis, which can compromise data integrity. PUFs, while energy-efficient, may produce inconsistent responses in different environmental conditions, affecting reliability. Recent literature addresses these challenges with additional protections, such as anti-tampering sensors, signal obfuscation techniques, and fault-resilient architectures, which help ensure that TinyML devices remain secure and operational even in volatile conditions. These approaches demonstrate the potential for multilayered hardware-based security to enhance TinyML resilience against both physical and logical threats.

Secure deployment frameworks are another area of critical importance for TinyML applications, particularly when data must be transmitted across IoT networks. Frameworks such as TinyML as a Service (TMLaaS) have been proposed to support secure, scalable inference by enabling on-device processing and reducing dependence on cloud infrastructure. TMLaaS leverages lightweight encryption algorithms, including Advanced Encryption Standard (AES) with reduced rounds and Diffie–Hellman key exchange, which provide essential data protection with minimal computational burden. Lightweight cryptography has been a well-researched area in the past decade, with extremely low-area hardware implementations proposed for resource-constrained devices [3]. However, many TinyML devices still face challenges in supporting even lightweight encryption schemes in real-time applications. Moreover, current deployment frameworks often lack robust multi-device coordination and network-level anomaly detection, leaving TinyML applications susceptible to network-based attacks. To address these limitations, various studies have explored solutions such as secure edge-to-cloud communication protocols and multi-device coordination frameworks specifically tailored to TinyML's constrained computational resources, underscoring the ongoing need for advancements in data security and scalability to ensure dependable TinyML deployments.

Real-time anomaly detection is essential for safeguarding TinyML systems operating in dynamic IoT environments, where data patterns can shift rapidly. TinyML online learning (TinyOL) is one approach that enables TinyML models to perform incremental learning directly on microcontrollers, helping models retain accuracy despite concept drift. This capability is especially important in applications like industrial IoT, where sudden changes in data may signal equipment malfunctions or security threats. However, TinyOL and similar anomaly detection frameworks are constrained by limited computational resources, which can restrict their ability to continuously adapt. To address these limitations, various studies have proposed efficient, resource-aware anomaly detection algorithms that enable TinyML models to respond to emerging threats with minimal power consumption. These advancements contribute to TinyML's adaptability and security, making it more resilient across diverse applications.

The current study addresses several identified gaps in the general study of TinyML security and privacy. For privacy-preserving training, vulnerabilities in model inversion and data interference attacks are addressed with the introduction of lightweight encryption and privacy-enhanced FL, tailored to balance computational constraints and robust privacy protection. For adversarial defense, the limited adaptability of existing approaches to dynamic and adaptive adversarial attacks is mitigated through a multifaceted robustness framework incorporating real-time anomaly detection, adaptive model updates, and preprocessing techniques. In hardware-based security, challenges such as memory overhead in TEEs and the sensitivity of PUFs to environmental factors are resolved by implementing advanced solutions, including anti-tampering sensors, signal obfuscation, and fault-resilient architectures for consistent device functionality. Regarding deployment frameworks, the gaps in multi-device coordination and the lack of network-level anomaly detection are bridged by designing secure edge-to-cloud communication protocols and multi-device coordination frameworks to ensure scalable and secure IoT deployments. Lastly, in anomaly detection, the energy constraints that limit continuous real-time adaptation are addressed through resource-efficient techniques enabling robust, real-time anomaly detection

Table 12.1 A summary of related works and contributions.

Area	Key approaches	Identified gaps	Contributions of current study
Privacy-preserving training	Federated learning (FL), D-FedAvg, TinyMetaFed	Vulnerability to model inversion and data inference attacks	Lightweight encryption and privacy-enhanced FL tailored for TinyML
Adversarial defense	Adversarial training, defensive distillation	Limited adaptability to dynamic/adaptive adversarial attacks	Multifaceted adversarial robustness framework with real-time adaptation
Hardware-based security	TEEs, PUFs	Memory overhead in TEEs; PUFs sensitive to environmental factors	Anti-tampering sensors, signal obfuscation, and fault-resilient architecture
Deployment frameworks	TMLaaS, AES encryption	Limited multi-device coordination; lack of network-level anomaly detection	Secure edge-to-cloud communication and multi-device coordination
Anomaly detection	TinyOL	Energy constraints limit continuous real-time adaptation	Resource-efficient, real-time anomaly detection for dynamic environments

in dynamic environments without excessive energy consumption. Table 12.1 outlines key approaches, identified gaps, and the contributions of the current study, highlighting how each of the identified challenges is addressed to advance TinyML security.

Despite advancements in privacy-preserving methods, adversarial defenses, hardware-based protections, secure deployment, and anomaly detection, significant gaps remain in TinyML security. Privacy-preserving frameworks often lack robust encryption suited to TinyML's limited resources, and adversarial defenses require further development to adapt dynamically to evolving threats. Similarly, hardware protections face limitations in handling side-channel vulnerabilities, while deployment frameworks and anomaly detection methods need optimization for multi-device coordination and real-time adaptability. By addressing these gaps, the current study contributes a unified, resource-efficient security framework for TinyML, enabling secure and resilient deployments across diverse IoT applications.

12.3 Secure and Privacy-Aware Training of TinyML Models

Training TinyML models securely and with privacy awareness is a necessity for devices operating in increasingly sensitive and distributed environments, particularly those embedded within IoT networks. These models, designed for deployment on resource-constrained devices like microcontrollers, demand special considerations for data protection, model integrity, and efficiency. TinyML's utility across various sectors, including healthcare, smart cities, and industrial IoT, heightens the importance of robust security and privacy strategies that can withstand a range of cyberthreats. Traditional ML security measures often exceed the computational, memory, and energy capacities of TinyML devices. Therefore, novel approaches to FL, IP protection, and adversarial robustness have been adapted to meet the specific needs of these constrained environments.

12.3.1 Federated Learning for Privacy-Preserving Training in TinyML

FL is an innovative approach for training ML models across decentralized devices without aggregating raw data centrally, a crucial feature for preserving privacy. Instead, the model is trained locally on each device, and only the updates (parameter adjustments) are sent to a central server, which aggregates these updates to create a refined global model. This decentralization is especially relevant for TinyML, as these devices often handle sensitive data but lack the resources for advanced local security measures. TinyMetaFed is an optimized FL framework designed specifically for TinyML, incorporating privacy-conscious techniques like partial local reconstruction, top-P% selective communication, and online learning [4].

In FL, consider a network of N devices, each with a local dataset D_i and a model parameterized by θ. Instead of aggregating datasets, FL trains a model on each device individually, where each device computes an update, $\Delta\theta_i$, based on its data. The global model parameters θ are then updated at the server using a weighted average, as explained in (12.1), where η is the learning rate and w_i denotes the weight of each device's contribution, often determined by the dataset size on each device [4]:

$$\theta = \theta - \eta \sum_{i=1}^{N} w_i \Delta\theta_i \tag{12.1}$$

This averaging process reduces data exposure risks by retaining raw data on devices, making it essential for privacy-sensitive environments.

TinyMetaFed's partial local reconstruction enhances privacy by splitting the model parameters θ into global parameters, θ_g, and local parameters, θ_l. Only the global parameters are shared with the server, while local parameters are retained on each device. This distinction allows for device-specific learning, wherein local parameters capture unique aspects of each device's data, making the model more personalized without risking privacy. For instance, in a healthcare application with wearable devices, local parameters may learn individual-specific health metrics, which remain private, while global parameters contribute to generalizable health insights.

Mathematically, this approach is represented by dividing θ as defined in (12.2) that

$$\theta = \{\theta_g, \theta_l\} \tag{12.2}$$

It can, also, be seen from (12.3) that during each training round, only $\Delta\theta_g$, the gradient of the global parameters, is sent to the server:

$$\Delta\theta_g = average(\Delta\theta_{g,1}, \Delta\theta_{g,2}, \dots, \Delta\theta_{g,N}) \tag{12.3}$$

This averaging process minimizes data exposure and reinforces privacy, as each device retains its unique data-specific local parameters. In use cases like distributed health monitoring, this approach enables general insights without exposing device-specific data to external servers.

Top-P% selective communication is another critical enhancement in TinyMetaFed, designed to optimize both communication efficiency and data privacy. Instead of transmitting all global weights after every training iteration, each device selects only the top-P% of weights with the largest updates, transmitting only these values to the central server. This process reduces bandwidth and energy usage while minimizing data exposure by limiting the amount of information transmitted.

The top-P% updates are then transmitted to the server, which aggregates them. This technique is especially advantageous in settings with constrained bandwidth, such as environmental monitoring systems where devices are remotely located. By sending only the most impactful updates, top-P% selective communication helps conserve energy and limits potential data leakage.

TinyMetaFed's online learning capability enables continuous, real-time model adaptation. Devices update their models incrementally as new data arrives, without needing to store historical data, which reduces memory usage and enhances privacy. Online learning is essential in TinyML, where applications often operate in dynamic environments and must respond to evolving data patterns. For instance, a TinyML-powered device in industrial monitoring might adapt to changing operational conditions without retaining large datasets.

In online learning, TinyMetaFed enables each TinyML device to update its model parameters continuously as new data arrives. Unlike traditional batch learning – which trains models using large, stored datasets – online learning processes each data point individually and immediately, discarding it once processed.

This approach reduces memory requirements, making it ideal for TinyML's limited storage capacities.

As each new data instance arrives, the model on the device adjusts its parameters to better reflect the most recent data, allowing the model to stay relevant to real-time conditions. For example, in a sensor network monitoring temperature, a model might adapt continuously to changing weather without retaining prior measurements. By updating in small, incremental steps, TinyMetaFed ensures that the model gradually improves without retaining sensitive historical data, preserving privacy while remaining adaptive to dynamic environments.

The incorporation of online learning within TinyMetaFed (Figure 12.1) allows TinyML devices to process incoming data in real-time, discarding old data as new information arrives. This incremental learning approach enables models to remain current while avoiding the privacy risks associated with storing large quantities of historical data on device. Given the memory constraints of TinyML devices, this technique not only conserves resources but also aligns well with privacy requirements, as it reduces the likelihood of data breaches through stored information. Online learning ensures that models continually adapt to their environments while adhering to privacy constraints, making it an essential feature for TinyML in dynamic applications.

The incorporation of online learning within TinyMetaFed allows TinyML devices to process incoming data in real-time, discarding old data as new

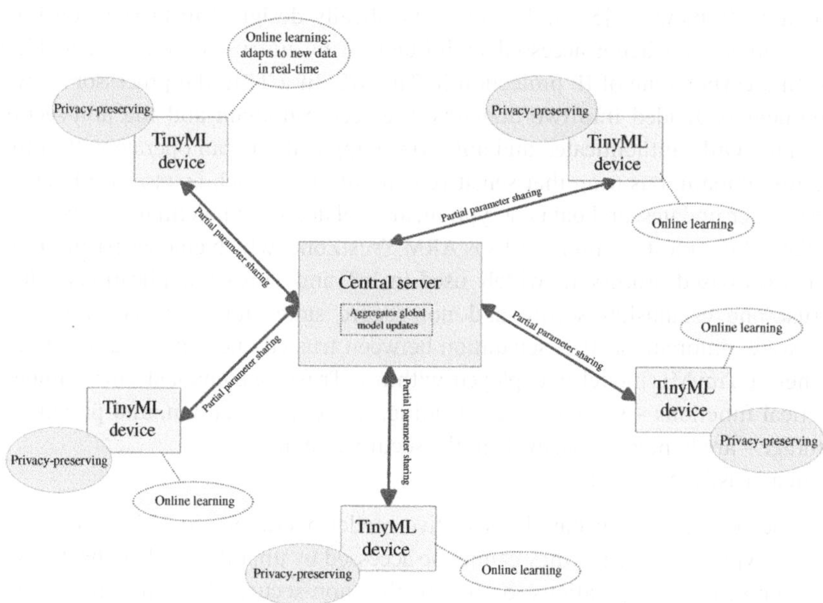

Figure 12.1 Federated learning architecture for TinyML using TinyMetaFed.

information arrives. This incremental learning approach enables models to remain current while avoiding the privacy risks associated with storing large quantities of historical data on device. Given the memory constraints of TinyML devices, this technique not only conserves resources but also aligns well with privacy requirements, as it reduces the likelihood of data breaches through stored information. Online learning ensures that models continually adapt to their environments while adhering to privacy constraints, making it an essential feature for TinyML in dynamic applications.

12.3.2 Intellectual Property Protection in TinyML

Protecting the IP of TinyML models is essential, as these models represent significant investments in development, computational resources, and proprietary knowledge. TinyML models often encode unique features, optimizations, and decision-making processes that offer a competitive advantage to their developers. As a result, unauthorized access or reverse engineering can lead to considerable financial loss and compromise competitive positioning. In the constrained environments where TinyML operates – such as microcontrollers and low-power embedded devices – these risks are heightened due to limited physical security and lower processing power. This makes dedicated IP protection strategies crucial to prevent reverse engineering, unauthorized model extraction, and tampering.

A TEE is a secure area within the processor that provides a hardware-based isolation framework [5, 6]. TEEs are specifically designed to protect sensitive operations from being accessed or influenced by untrusted software, making them a cornerstone of IP protection in TinyML. In a TEE, the processor's environment is divided into two domains: the secure domain and the non-secure domain. Only authenticated and authorized applications can operate within the secure domain, ensuring that sensitive computations, such as model inference, parameter updates, and data encryption, are isolated from potential threats.

One of the most common TEEs is ARM TrustZone, which enables partitioning on ARM-based processors, widely used in IoT and TinyML applications. ARM TrustZone establishes secure and non-secure states for both hardware and software, enforcing a strict separation between trusted and untrusted resources. When a TinyML model is deployed within a TrustZone-enabled environment, critical functions – such as model loading, inference execution, and parameter storage – are handled exclusively in the secure domain.

In a TrustZone setup:

1) The secure domain handles sensitive model operations (e.g., inference and encryption), ensuring they cannot be accessed by unauthorized applications.
2) Non-sensitive operations occur in the non-secure domain, minimizing resource strain on the secure environment.

During execution, model parameters and computations remain protected within the secure domain. For example, in an IoT healthcare device that uses TinyML for patient monitoring, TrustZone ensures that patient data and inference tasks are isolated from general system operations, maintaining data privacy and preventing unauthorized access. This strict compartmentalization safeguards the IP embedded within the TinyML model, ensuring that even if the device's non-secure domain is compromised, the secure domain remains protected.

Memory isolation and encryption offer additional layers of security to protect the TinyML model IP. While the TEE ensures that sensitive computations are secure, memory isolation and encryption protect stored model parameters from physical or software-based attacks. In TinyML, where devices are often resource-limited, lightweight encryption techniques like AES are particularly effective.

AES, particularly in its 128-bit format, is a popular encryption method for TinyML devices due to its balance of strong security and computational efficiency. AES encryption secures sensitive model parameters by transforming them into encrypted data blocks, which are only decrypted within the secure environment of the TEE when needed. For instance, parameters that define a model's specific decision-making thresholds or feature weights are encrypted before being stored in the device's external memory. This approach ensures that if an attacker gains physical access to the device's memory, they cannot access the model's sensitive parameters without the decryption key, which is securely stored within the TEE.

Encryption is performed as follows:

1) **Key Generation and Storage**: A secure key is generated and stored exclusively within the TEE. This key is inaccessible to external applications and remains protected throughout the model's lifecycle.
2) **Parameter Encryption**: Before storing model parameters in memory, each parameter is encrypted using AES-128, resulting in ciphertext that is unintelligible to unauthorized users.
3) **On-Demand Decryption**: During inference or parameter updating, the TEE decrypts the necessary parameters, using the stored key. Once the operation is complete, the parameters are re-encrypted or cleared from memory.

In a TinyML device, monitoring sensitive industrial equipment, for example, ensures that only encrypted parameters are stored in external memory (see Figure 12.2). If a malicious actor attempts to access the device's memory directly, they encounter only encrypted data, preserving the model's integrity and IP.

Given the memory limitations of TinyML devices, dynamic memory allocation and secure memory swapping play an essential role in IP protection. These techniques ensure that sensitive data is stored securely and efficiently without overwhelming the device's limited memory resources. In dynamic memory management, memory is allocated for secure operations only when necessary and

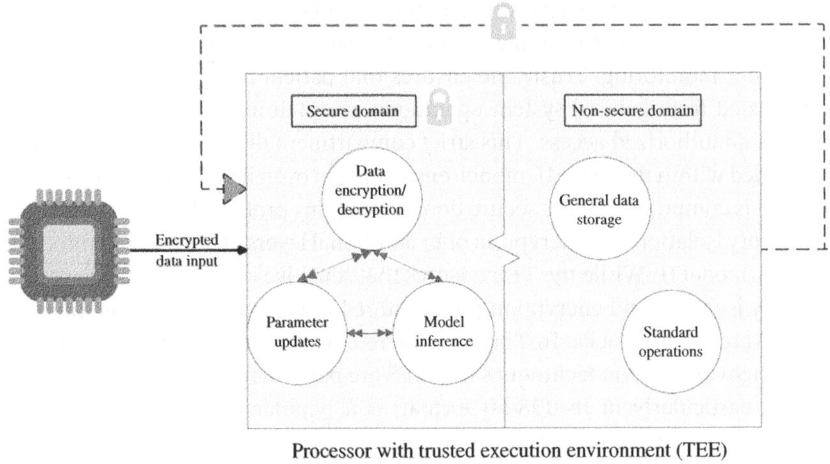

Figure 12.2 Trusted execution environment (TEE) for IP protection.

is deallocated immediately after use, reducing the potential exposure of sensitive data. More specifically:

1) **Dynamic Memory Allocation**: Dynamic memory allocation allows TinyML devices to manage memory more flexibly by dedicating secure memory spaces only when performing sensitive computations. For example, during model inference, the device allocates secure memory to temporarily store decrypted parameters. Once inference completes, the allocated memory is cleared, preventing any residual data from being accessed by unauthorized applications.
2) **Secure Memory Swapping**: Secure memory swapping extends dynamic memory allocation by moving sensitive data between secure and non-secure domains as needed. For instance, during non-sensitive operations, memory can be allocated for general-purpose tasks. When switching to a sensitive operation, such as parameter decryption or inference, memory is temporarily shifted to the secure domain. Once the operation concludes, data is immediately cleared or encrypted before releasing the memory back to the non-secure domain. This swapping mechanism ensures efficient memory use without compromising data integrity or model confidentiality.

To further protect the IP of TinyML models, anti-tampering techniques are employed to detect and respond to any unauthorized physical or logical access attempts. Anti-tampering techniques are especially important in scenarios where TinyML devices are deployed in remote or high-risk environments, such as industrial monitoring and critical infrastructure.

Anti-tampering measures can include the following:

1) **Tamper-Resistant Hardware Design**: Devices can be designed with physical barriers, such as conductive mesh covers or specialized epoxy coatings, which disrupt operations if tampered with. For example, any attempt to open the device physically may sever these barriers, effectively shutting down the device.
2) **Tamper Detection Sensors**: Sensors within the device can detect physical intrusions, such as pressure or light changes, which may indicate tampering. If such tampering is detected, the device initiates data-wiping protocols, ensuring that sensitive model parameters are erased before unauthorized access occurs.
3) **Physically Unclonable Functions (PUFs)**: PUFs exploit physical variations in the device's hardware to generate unique cryptographic keys. Each PUF-generated key is unique to a specific device and can serve as a robust identifier for IP protection. For TinyML, PUFs can be used to encrypt model parameters, ensuring that the model can only operate on authorized devices with matching hardware characteristics.

For instance, in a TinyML device used for environmental monitoring in a sensitive government facility, anti-tampering measures would protect the model's IP from being accessed by unauthorized personnel. If tampering is detected, the device can trigger a self-erasure protocol, deleting sensitive model data and rendering the device unusable, thereby preserving IP confidentiality. To illustrate how these IP protection techniques function in real-world applications, here are two use cases where IP protection is critical for TinyML deployments:

1) **Smart Home Security Systems**: In smart home applications, TinyML models are embedded in devices like security cameras, motion sensors, and access control systems. These models often contain proprietary detection algorithms that differentiate between genuine users and potential intruders. By deploying TEEs, memory encryption, and dynamic memory management, these devices ensure that sensitive parameters related to motion detection or face recognition remain secure. If a device is stolen, memory encryption and anti-tampering measures prevent unauthorized access to the model, protecting both the IP and the privacy of users.
2) **Wearable Health Monitoring Devices**: Wearable devices using TinyML models monitor vital health metrics such as heart rate, oxygen saturation, and electrocardiograms. These models are trained to detect health anomalies, often incorporating proprietary algorithms developed by medical device companies. By encrypting model parameters and storing them within the TEE, the wearable device ensures that only authorized software can access and process

these sensitive computations. Furthermore, dynamic memory allocation helps conserve battery life, while anti-tampering features protect IP by detecting physical tampering attempts and wiping data if necessary.

12.3.3 Adversarial Robustness in TinyML Models

In addition to IP protection, ensuring robustness against adversarial attacks is crucial for TinyML models, especially given the rise of adversarial threats in critical applications. TinyML models are vulnerable to adversarial tactics that exploit model weaknesses, either through subtle manipulations of input data – known as evasion attacks – or by querying the model to replicate its functionality, known as model extraction attacks [6].

Evasion attacks present another serious threat, where attackers alter input data just enough to trigger misclassification or erroneous predictions. TinyML models, due to their constrained resources, have limited capacity for filtering out such manipulated inputs. As a result, adversarial training has emerged as an effective defense mechanism. In adversarial training, models are exposed to both standard and adversarial examples during training, enabling them to recognize and withstand inputs that are intentionally altered. This exposure improves the model's resilience in real-world settings, as it learns to detect and mitigate subtle input manipulations. By integrating adversarial training, TinyML devices become more adept at detecting adversarial perturbations, which is critical for applications requiring high accuracy and reliability.

The fast gradient sign method (FGSM) is often used in adversarial training to generate adversarial examples [7]. For a given input x with a true label y, the adversarial example x' is computed according to (12.4):

$$x' = x + \varepsilon \cdot sign(\nabla_x \ell(f(x;\theta), y)) \tag{12.4}$$

where

- $f(x;\theta)$ is the model's prediction function,
- $\ell(f(x;\theta), y)$ is the loss function (e.g., cross-entropy) used to calculate prediction error,
- $\nabla_x \ell$ represents the gradient of the loss with respect to x_i, and
- ε is a small constant that controls the perturbation size.

By training TinyML models on these adversarial examples, they learn to better detect and resist minor adversarial changes in inputs. This method (illustrated in Figure 12.3) is particularly advantageous in critical applications such as healthcare monitoring, where models are trained to detect naturally fluctuating signals rather than artificially manipulated patterns.

Additionally, input preprocessing techniques, such as feature squeezing and randomization, are lightweight yet effective defenses for TinyML. Feature

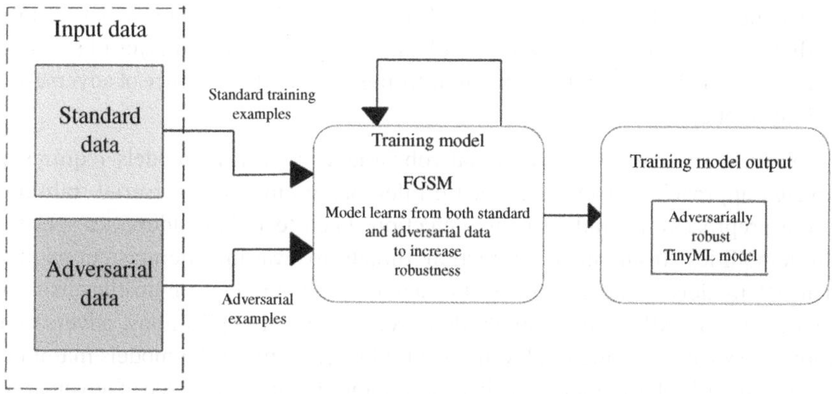

Figure 12.3 Adversarial training workflow for TinyML.

squeezing simplifies input data by reducing unnecessary detail, making it more challenging for adversarial perturbations to succeed. For example, reducing color depth in image classification tasks can limit the effectiveness of pixel-level perturbations. In resource-constrained TinyML environments, feature squeezing is computationally efficient and helps prevent minor adversarial changes from impacting model predictions.

Anomaly detection systems provide a reactive defense by identifying and flagging inputs that deviate significantly from expected distributions. Simple statistical anomaly detection, such as Z-score analysis or moving averages, can be implemented on TinyML devices with minimal computational load. For instance, in an industrial TinyML application, an anomaly detection algorithm could flag unusual sensor readings that deviate from typical operational ranges, potentially identifying tampered or manipulated data in real-time. This approach enhances model resiliency against evolving adversarial tactics, providing a secondary defense mechanism that complements adversarial training.

Consider a practical application of adversarial robustness techniques in a wearable health monitor powered by TinyML, which continuously tracks vital signs such as heart rate and oxygen saturation. In such a context, adversarial attacks that alter sensor readings could lead to incorrect health assessments, potentially endangering the user.

To secure this device:

- Adversarial training incorporates perturbed versions of vital sign readings, enabling the model to differentiate between natural health patterns and adversarially altered signals.
- Feature squeezing rounds input values to reduce precision, which helps filter out small adversarial manipulations in sensor readings.

- Anomaly detection systems track real-time inputs, flagging unusual data points that deviate from the user's normal health range. This added layer of security ensures reliable, accurate health monitoring, even in the presence of adversarial interference.

In summary, ensuring adversarial robustness in TinyML models requires a layered approach that includes secure inference protocols, adversarial training, input preprocessing, and anomaly detection. Each technique addresses specific vulnerabilities, from model extraction threats to evasion attempts, providing TinyML models with the resilience needed to function reliably in hostile environments. As TinyML is increasingly deployed in critical applications, adversarial robustness will continue to play an essential role, ensuring the models maintain high standards of accuracy, security, and reliability even in the face of adversarial threats.

12.3.4 Distributed Federated Learning (D-FedAvg) for Enhanced Privacy

D-FedAvg represents a further innovation in FL for TinyML, offering additional privacy and security enhancements [8]. D-FedAvg minimizes communication between devices and central servers by focusing on local model updates rather than raw data, preserving the decentralization that is central to privacy-conscious FL. By processing updates in parallel across client devices, D-FedAvg reduces the frequency of communication rounds, a particularly advantageous feature for TinyML deployments where devices may have intermittent connectivity or restricted bandwidth. This model further conserves device resources and aligns well with privacy requirements, allowing for efficient, decentralized learning that maintains data confidentiality.

Equation (12.5) shows how the global model in D-FedAvg is updated by aggregating the local updates from each device i, where each device's contribution is weighted by its dataset size, ensuring that updates represent each device proportionally:

$$\theta = \theta - \eta \sum_{i=1}^{N} w_i \Delta\theta_i \tag{12.5}$$

This decentralization allows efficient, privacy-respecting training across a diverse network of devices.

D-FedAvg also implements privacy-preserving mechanisms, such as secure aggregation, which aggregates model updates rather than individual data points, ensuring that no single device's data can be isolated or analyzed independently. This approach is particularly well-suited for sensitive environments where data regulations restrict centralized data processing, and the device-level privacy guaranteed by D-FedAvg makes it a valuable framework for maintaining data integrity and user privacy in TinyML applications. By distributing the model update load across devices and focusing on client-level parallelism, D-FedAvg achieves a balance between efficiency, data protection, and performance, making it a cornerstone technique in privacy-aware TinyML model training.

12.3.5 Security-Aware Quantization in TinyML

Quantization is a widely used technique in TinyML, primarily for reducing model size and improving computational efficiency on resource-constrained devices. It operates by mapping high-precision model parameters, such as 32-bit floating-point values, to lower-precision formats, like 8-bit integers. This reduction in precision leads to significant savings in memory and processing requirements, allowing complex models to run on low-power devices. However, while quantization offers substantial efficiency gains, it also introduces vulnerabilities, particularly in the context of adversarial robustness.

In adversarially trained models, high-precision representations help capture subtle variations in input data that are useful for distinguishing genuine inputs from adversarially crafted ones. When a model is quantized, its ability to detect these subtle variations diminishes, as the reduced precision blurs the fine-grained information that adversarial defenses rely upon. Specifically, quantizing to integer-only operations can cause a degradation in the model's robustness, as quantization increases the sensitivity of the model to adversarial perturbations. This is particularly problematic for models used in security-sensitive TinyML applications, where reliability is paramount.

To understand why, consider an adversarial example x' crafted from an input x with a perturbation δ designed to exploit the model's decision boundary. In high-precision models, the gradient-based defenses in adversarial training help the model recognize and ignore these perturbations. However, in a quantized model, the reduced numerical range may result in the model interpreting x' similarly to x, since the fine-grained distinction is lost. Thus, the quantized model might fail to distinguish adversarial examples from genuine ones, making it susceptible to attacks.

Security-aware quantization techniques are designed to mitigate the adversarial vulnerabilities introduced by standard quantization processes [9]. These techniques modify the quantization process itself or introduce pre- and post-processing steps to ensure that the model's robustness is preserved, even with reduced precision. Key security-aware quantization methods include differential quantization, stochastic rounding, and adversarially aware quantization.

1) **Differential Quantization**: Differential quantization selectively applies different levels of quantization precision based on the sensitivity of each layer

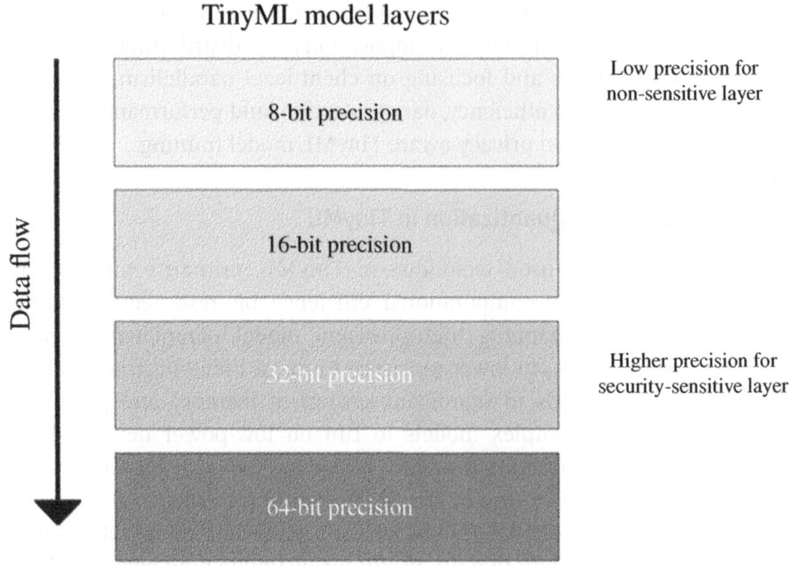

Figure 12.4 Differential quantization for TinyML model layers.

or parameter to adversarial attacks (see Figure 12.4). For instance, layers that are more prone to adversarial influence, such as early layers capturing feature representations, can retain higher precision, while less sensitive layers are quantized to lower precision. This selective approach maintains computational efficiency without compromising security. Differential quantization is explained in (12.6), where $\|\nabla\ell(x)\|$ is the magnitude of the gradient, indicating layer sensitivity to adversarial perturbations [10]:

$$Q(x) = \begin{cases} quantize_{high}(x), & if\,\|\nabla\ell(x)\|\ is\ large \\ quantize_{low}(x), & otherwise \end{cases} \tag{12.6}$$

2) **Stochastic Rounding**: In stochastic rounding [11], each value is rounded probabilistically to either the nearest or next-nearest quantization level, with the probability proportional to the distance from each level. This randomness helps reduce the deterministic pathways that adversarial examples often exploit, making the quantized model less predictable and thus more resistant to adversarial manipulation. This stochastic approach is particularly useful in high-stakes applications, such as financial TinyML models, where adversarial robustness is critical.
3) **Adversarially Aware Quantization**: It incorporates adversarial training specifically for the quantized model. During training, the model is subjected to adversarial examples at lower precision, ensuring that it can identify and

mitigate adversarial patterns even after quantization. This method improves the model's robustness by making it familiar with adversarial perturbations in its final quantized state, enabling it to resist such inputs after deployment.

12.3.6 Federated Transformed Learning and Model Resiliency

FL enables secure, privacy-preserving training by distributing the model learning process across multiple devices, each of which keeps its data locally. However, in adversarial settings, FL alone is often insufficient to safeguard models from attacks, especially when data or model updates can be adversarially manipulated. To address this, federated transformed learning introduces transformation layers within the FL framework, enhancing robustness by restructuring model representations to make them less vulnerable to adversarial attacks [12].

Federated transformed learning applies transformation operations to the model's intermediate representations during training. These transformations can include operations such as manifold defense and topological data analysis (TDA). The goal is to create a model architecture that can identify and neutralize adversarial inputs by recognizing atypical data structures or patterns indicative of adversarial manipulation. In TinyML, where resources are limited, these transformations are applied selectively to ensure efficiency while maintaining robustness.

1) **Manifold Defense**: Manifold-based defenses enhance model robustness by leveraging the natural structure of data, which often lies on a lower-dimensional "manifold" within a high-dimensional space. The idea is that authentic data points (like images and sensor readings) form patterns or clusters within this space. In contrast, adversarial examples, which are maliciously altered data designed to deceive the model, tend to fall outside of these clusters or deviate from these natural structures (see Figure 12.5).

   In a manifold-based defense, the model first projects each new input onto this data manifold before making predictions. This step acts as a filter, removing or minimizing any adversarial noise that does not conform to the natural patterns of the data. By focusing on this intrinsic data structure, the model becomes better at distinguishing genuine inputs from those that have been subtly manipulated. This defense approach helps maintain accuracy and security, as the model is less likely to be tricked by small but malicious alterations to input data, a significant advantage in security-sensitive TinyML applications.
2) **Topological Data Analysis (TDA)**: TDA is a technique that examines the "shape" of the data to identify anomalies or adversarial patterns. TDA constructs a topological representation, such as a persistence diagram or a simplicial complex, that captures the structural properties of the data. Adversarial examples that disrupt these structural properties are identified through

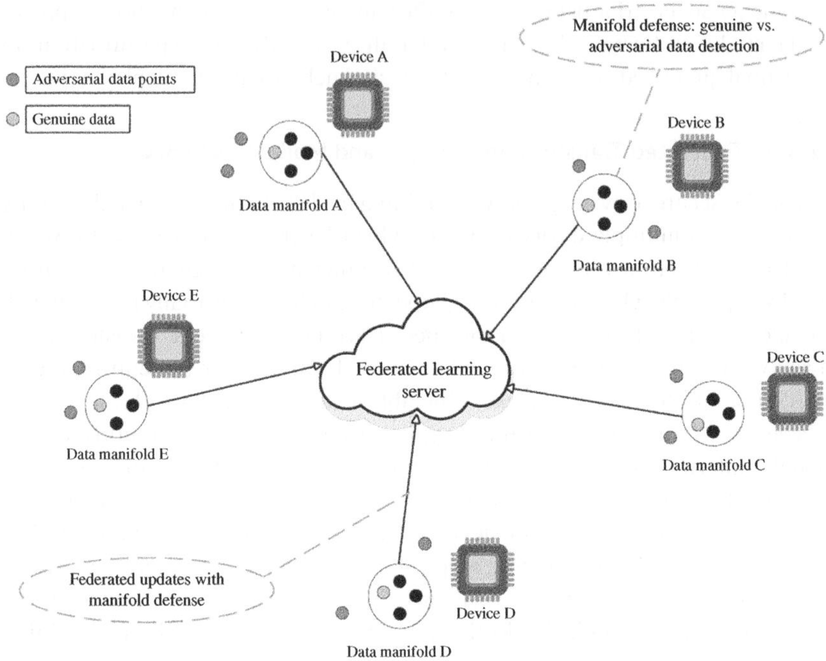

Figure 12.5 Federated transformed learning with manifold defense.

discrepancies in the topological representation. TDA is computationally efficient and well-suited for TinyML, as it reduces the input data to a simplified topological form, making it easier to detect deviations without complex computations.

In the context of federated transformed learning, each device in the network applies these transformations locally, generating intermediate representations that are inherently more robust against adversarial attacks. These transformed representations are then aggregated at a central server, which assembles a global model that benefits from both the diversity of device-specific data and the resilience of transformed representations.

Federated transformed learning is particularly valuable in TinyML applications deployed in sensitive environments where adversarial robustness is critical. Here are two practical use cases:

1) **Smart Agriculture**: In smart agriculture, TinyML devices monitor environmental factors like soil moisture, temperature, and light to optimize crop growth. These devices operate in remote and potentially hostile environments

where adversarial attacks could manipulate data to disrupt agricultural operations. By applying federated transformed learning with manifold defense, each device's sensor data is mapped to a manifold representing typical environmental conditions. If an input deviates significantly from the manifold, it is flagged as potentially adversarial, ensuring reliable monitoring despite adversarial attempts.

2) **Distributed Health Monitoring**: In a federated health monitoring network, TinyML devices measure and transmit vital signs to predict health events. Federated transformed learning with TDA enables each device to examine the "shape" of incoming data, detecting adversarial perturbations that might affect diagnostics. If a model receives an input that distorts the topological pattern typical of genuine health data, it can respond by flagging or disregarding the input, thereby maintaining patient safety even in the presence of adversarial inputs.

Security-aware quantization and federated transformed learning provide crucial defenses against adversarial threats in TinyML models. By ensuring that quantization preserves adversarial robustness and applying federated transformations that enhance model resilience, TinyML models achieve both computational efficiency and security. These approaches are fundamental to deploying TinyML in high-stakes, resource-constrained environments, ensuring reliable performance even when faced with adversarial manipulation.

12.3.7 Use Cases

12.3.7.1 Smart Healthcare: Private and Secure Patient Monitoring

Imagine a healthcare environment where wearable devices monitor patients' vital signs in real-time. Equipped with TinyML models, these wearables can detect early signs of potential health risks – abnormal heart rhythms, sudden drops in oxygen saturation, or other unusual patterns [13]. Given the personal nature of this data, privacy is paramount. Here, FL shines, allowing each device to update a global model without ever sharing patient data directly.

Using TinyMetaFed, each wearable learns its user's unique health patterns, while only general updates are shared with the central server. For instance, a device might learn a specific user's heart rate trends, like higher rates during their morning run or a lower baseline during rest, without uploading this detailed personal information. This approach keeps each person's data private while enriching the global model with broader health insights that can help refine early-warning systems for doctors and researchers alike.

In this setup, D-FedAvg adds an additional layer of privacy by ensuring only aggregated updates are sent, safeguarding user-specific details in the process.

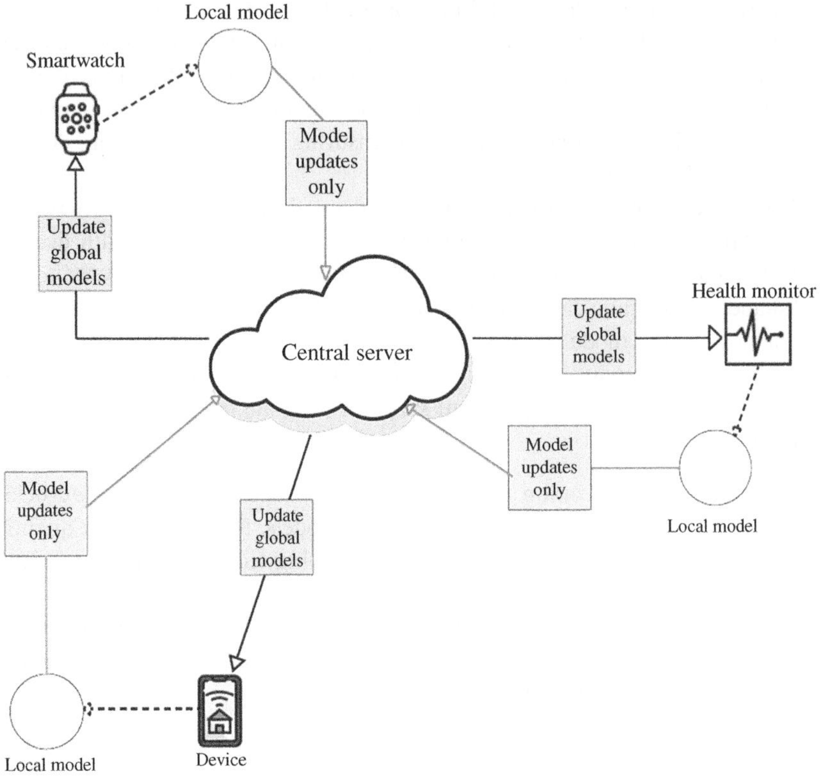

Figure 12.6 Privacy-aware training use case in healthcare.

This model lets healthcare providers monitor trends across patient populations without compromising individual privacy. And with TEEs handling sensitive operations like inference, any unauthorized attempt to access patient data is blocked, securing sensitive computations. Should a wearable be lost, the TEE's isolation mechanism ensures that the device's memory remains inaccessible, keeping patient data safe.

This use of FL in healthcare not only respects individual privacy but also enhances the predictive power of health monitoring devices, making healthcare more proactive and personalized. This use case is visually presented in Figure 12.6.

12.3.7.2 Industrial IoT Monitoring: Secure Predictive Maintenance

Industrial environments, from manufacturing floors to energy plants, rely on complex equipment that must function smoothly to avoid costly downtime. TinyML devices monitor critical components – think turbines, motors, or robotic arms – tracking subtle changes in vibration, temperature, or sound to identify signs of wear before they escalate into failures. However, such operational data is sensitive, often proprietary, and vital for preventing disruptions.

Here, TEEs play a central role, isolating and protecting essential data processing tasks within the secure domain of each device. For example, in a power plant, TEEs help analyze equipment status data securely, ensuring only authorized personnel have access to insights about the machinery's health. This isolation ensures data privacy and secures the predictive maintenance model, which is often a valuable proprietary asset for the company.

Given the real-time demands, security-aware quantization is critical, as it allows the model to function efficiently without being vulnerable to attacks [14]. Here, differential quantization selectively applies precision to model layers critical for identifying anomalies, reducing memory usage while ensuring accuracy remains intact. For instance, quantizing layers that detect vibration anomalies with higher precision makes the system resilient to adversarial noise, ensuring accurate predictions in high-stakes environments.

Moreover, federated transformed learning enhances robustness by adding transformation layers that recognize tampering attempts. By ensuring each model's representation is inherently secure, federated transformed learning protects operational data across devices, making predictive maintenance a more secure, scalable option for industrial IoT.

12.3.7.3 Smart Agriculture: Privacy-Conscious Environmental Monitoring

Consider a farm where TinyML-enabled sensor nodes are spread across acres of land to monitor essential conditions like soil moisture, temperature, and crop health. These devices help farmers optimize irrigation, predict pest outbreaks, and manage crop yield, all without relying on constant Internet connectivity or large-scale data aggregation. But here is the catch: Agricultural data is often proprietary, with insights into crop health, soil quality, and irrigation schedules holding significant value.

Using TinyMetaFed, each sensor adapts to its location's microclimate and soil properties without sharing this data directly. Only generalized updates reach the server, enabling a regional-level model to emerge that assists farmers without revealing specific farm data. For instance, a sensor in a low-lying part of the field might adapt to high-moisture readings during early mornings, learning the area's unique conditions and adjusting irrigation recommendations accordingly.

When multiple farms use the same network, D-FedAvg ensures that each farm's data remains private. Each sensor transmits only aggregated updates, so even if one farm's network node were compromised, it would not expose other farms' data. Finally, to ensure efficient, secure operation, stochastic rounding in quantization introduces minor, random variations that make it difficult for adversarial actors to tamper with sensor readings.

These privacy-protecting techniques empower farmers to improve yield predictability while ensuring the confidentiality of their operations, making smart agriculture both efficient and secure.

12.3.7.4 Smart City Infrastructure: Privacy-Aware Traffic and Pollution Monitoring

In a smart city [15], sensors embedded in streetlights and buildings collect data on air quality, traffic flow, and noise levels. Aggregating this data enables city planners to make informed infrastructure decisions, address pollution hotspots, and manage traffic flow. However, the data collected often includes sensitive details, such as real-time traffic patterns and pedestrian densities, which makes privacy protection a high priority.

In this setting, FL ensures that raw data stays local. Each street-level sensor analyzes its environment and sends only model updates to a central server, which combines them to identify larger trends. For example, a sensor may track air quality spikes during rush hour, adjusting its predictions and contributing insights to the central model without sharing specific timestamps or traffic counts.

To prevent adversarial disruptions, federated transformed learning applies TDA, which detects structural inconsistencies that may signal tampering. For instance, if an attacker tried to manipulate a sensor's pollution readings to produce a false alarm, the TDA-based transformations could identify unusual data structures and flag the input as anomalous. This detection strategy safeguards the integrity of city data, ensuring that critical services and emergency response systems function reliably.

Additionally, TEEs secure sensitive computations like pollution assessment and traffic density estimation, allowing only authorized city agencies access to the insights. This combination of privacy-preserving techniques ensures that smart city planners can access valuable data without compromising individual privacy or system reliability.

12.3.7.5 Consumer Electronics: Privacy-Aware Voice Assistants and Smart Home Devices

In consumer electronics, TinyML models are embedded in smart home devices and voice assistants, handling user commands, adjusting preferences, and learning from individual interactions. These devices interact with highly personal information, which calls for robust privacy protections, to both safeguard user data and protect the proprietary value of the models themselves.

Through FL combined with D-FedAvg, each device personalizes its TinyML model based on user interactions – such as frequently used commands or room-specific preferences – without transferring personal data to the cloud. A voice assistant, for instance, may learn to recognize a specific user's accent and adjust its response patterns accordingly. By sending only aggregated updates to a central model, the device improves performance while keeping user data private.

Given the low-power requirements of these devices, security-aware quantization is crucial for maintaining performance without sacrificing robustness. Stochastic rounding helps the model remain accurate even in the presence of minor background noise, such as television (TV) sounds or cross-conversations. This randomization limits adversarial opportunities by making it difficult to inject manipulative noise patterns into the model's input.

Finally, TEEs protect the most privacy-sensitive operations, like processing voice commands or managing device settings. When a user issues a command, the TEE ensures that both the request and the response remain isolated from other applications, protecting user interactions and model IP alike. This secure setup allows smart home devices to adapt to users' preferences without risking data breaches or compromising proprietary model assets.

12.4 Implementation of Novel Security Schemes for TinyML Applications

With the rapid advancement and deployment of TinyML within resource-constrained environments like IoT, ensuring robust security for these applications has become critical. TinyML's constrained hardware presents unique security challenges, especially as it becomes embedded within increasingly sensitive environments, from healthcare to industrial IoT. This section explores practical, state-of-the-art security schemes tailored for TinyML applications, addressing the unique challenges posed by cyber- and physical threats.

12.4.1 Detectability Assessment for IoT-Related Cyber Attacks: Adapting the D-Score Methodology

In TinyML applications deployed within IoT ecosystems, it is essential to ensure that the system can identify and respond to cyberthreats effectively and with minimal latency. Cyberattacks targeting IoT systems, such as data exfiltration, distributed denial of service (DDoS) attacks, and command-and-control (C&C) infiltration, exploit the limited computational and storage resources of TinyML-enabled devices, making conventional cybersecurity approaches infeasible. The D-score methodology, initially developed to assess detectability in IoT, provides a quantifiable framework for evaluating the detectability of various attack types in TinyML applications.

The D-score methodology systematically evaluates the detectability of cyberattacks based on device-specific network behaviors and threat characteristics, generating a detectability score that quantifies how well a system can identify abnormal activities [16]. By defining baseline behaviors for TinyML devices – such as predictable packet flow rates, latency, and signal strength – one can establish an expected range of activity patterns. These baselines are critical because TinyML systems on microcontroller units (MCUs) tend to have fixed operational patterns and data flow characteristics, which are typically simpler and less dynamic compared to full-scale IoT devices.

Applying D-score principles in TinyML systems involves monitoring deviations from established baseline behaviors to detect potential attacks. For instance, DDoS attacks, which typically manifest as an increase in low-data packet frequency, can be identified by anomalously high packet rates or unusual congestion within the network. Detecting C&C attacks, on the other hand, may involve identifying anomalous interactions that deviate from expected device communication patterns. This detectability is enhanced through the compact nature of TinyML device interactions, which allows for more precise anomaly detection algorithms tailored to limited network activities.

The D-score methodology provides quantitative metrics for evaluating detectability based on factors such as signal-to-noise ratio (SNR), frequency of abnormal activity, and response time. For TinyML applications, these metrics are refined to align with microcontroller capabilities. Detectability can be expressed as a function, as shown in (12.7):

$$D = f(\alpha \times F + \beta \times R + \gamma \times T) \tag{12.7}$$

where

- F represents the frequency of anomalies,
- R denotes the response effectiveness upon detection,
- T captures the time delay between attack initiation and detection, and
- α, β, γ are the weighting factors assigned based on the device's security priorities and operational constraints.

To improve robustness, FL can be integrated within TinyML systems to enable cooperative learning across multiple devices without centralizing sensitive data. By aggregating device-specific anomaly scores in a FL framework, TinyML applications can develop a distributed model that identifies broader patterns in potential attacks across a network of devices. This approach not only strengthens individual device detectability but also enhances collective detectability by sharing insights from similar TinyML deployments without compromising data privacy.

A significant benefit of the D-score methodology for TinyML systems is its adaptability for continuous threat assessment. As new attack types emerge and

TinyML applications evolve, updating D-score baselines and recalibrating weights α, β, and γ allow for ongoing optimization of detection capabilities. Additionally, incorporating ML-driven feedback into the D-score framework enables the system to self-adjust baselines and improve anomaly detection sensitivity over time.

Continuous monitoring, powered by real-time feedback loops, enables TinyML systems to adaptively recalibrate detection thresholds and response protocols to respond to emerging threats effectively. This adaptability makes D-score particularly valuable in high-risk environments, where TinyML devices face dynamic and evolving attack vectors.

12.4.2 Hardware-Based Protection for TinyML Against Physical and Logical Threats

In resource-constrained TinyML applications, hardware-based protection against both physical and logical threats is critical [17]. These threats encompass both invasive attacks (e.g., micro-probing and laser etching) and noninvasive methods (e.g., side-channel attacks like power and electromagnetic analysis). Such attacks exploit the physical characteristics and signal emissions of TinyML systems to infer sensitive data or interfere with device functionality, presenting a significant risk, especially in high-stakes applications.

To fortify TinyML devices, four main anti-tampering strategies – resistance, detection, response, and evidence – are implemented:

1) Resistance focuses on increasing the difficulty of physical tampering through protective coatings and shielding techniques, which add layers to prevent direct access.
2) Detection involves incorporating sensors or meshes to monitor for tampering attempts. Conductive mesh-based sensors, for example, detect when a physical breach is attempted by observing disruptions in the electrical field.
3) Response mechanisms initiate protective actions (such as erasing sensitive data or disabling functionalities) upon tamper detection, limiting the attacker's ability to exploit any accessed data.
4) Evidence methods securely log tampering incidents, enabling forensic analysis.

In Figure 12.7, the hardware-based protection mechanism, as described above, for a TinyML device is presented.

12.4.2.1 Noninvasive Attack Countermeasures Using Signal Obfuscation

TinyML devices are particularly susceptible to noninvasive side-channel attacks due to their constrained hardware, which limits the range of protections available. Power analysis, for example, enables attackers to extract cryptographic

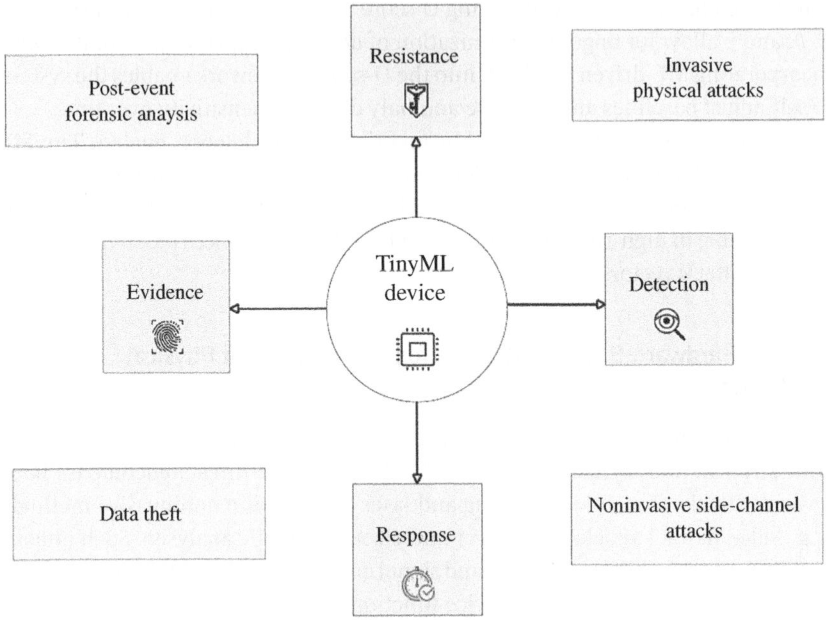

Figure 12.7 Hardware-based protection mechanism for TinyML.

keys by analyzing power consumption variations during operations. To counteract this, signal obfuscation techniques are applied. Letting E_{emit} represent emitted electromagnetic signals, TinyML systems can employ methods such as randomizing processing patterns or adding noise to reduce the SNR, as shown in equation (12.8):

$$SNR = \frac{Signal\ Power\ (S)}{Noise\ Power\ (N)} \tag{12.8}$$

A lower SNR value reduces the effectiveness of side-channel attacks, as the randomized emissions obscure patterns that attackers typically exploit. Introducing adaptive power regulation and dynamic clock adjustments can further reduce the predictability of signal patterns, enhancing resilience against power and electromagnetic analysis.

12.4.2.2 Physically Unclonable Functions (PUFs) for Secure Authentication

PUFs are hardware-based authentication mechanisms that generate unique signatures from the inherent manufacturing variability of microcontroller circuits. For TinyML applications, PUFs provide a lightweight yet effective means of ensuring device authenticity without the need for traditional cryptographic keys, which are often impractical in constrained environments.

The response R generated by a PUF is a function of the device's physical properties and an input challenge C, as explained in (12.9), where the unique characteristics of each device produce a distinctive response:

$$R = PUF(C) \tag{12.9}$$

This response R can be used to authenticate the device or verify its integrity, as any attempt to physically alter the circuit would disrupt the consistency of the response. By incorporating PUF-based verification, TinyML systems can enforce device authentication with minimal computational overhead, reinforcing security against both invasive and cloning attacks.

12.4.2.3 Hardware Countermeasures Available on Commodity Microcontrollers

TinyML is deployed on very resource-constrained devices such as MCUs. The strict resource limitation of such devices defines which hardware-based security solutions can be used on TinyML class MCUs without violating their size and performance limitations. In this section, countermeasures that are readily available on commodity MCUs, and thus viable for TinyML devices, are summarized.

Secure boot is an essential feature to prevent malicious code injection and application spoofing by establishing a hardware root of trust. This is done by authenticating device firmware before booting, usually using a manufacturer's public key [1]. Thus, the integrity of firmware is ensured by using cryptographic signatures and hashing, preventing unauthorized or malicious loading. It is widely available on commodity MCUs and does not impose substantial overhead.

Memory protection mechanisms like readout protection (RDP), execute-only firmware (PCROP), and write protection (WRP) are already present on STM32 devices [1]. These security mechanisms protect device interfaces from being debugged by an adversary to gain access to the memory, also known as memory extraction attacks. Although these would offer some level of protection, they are not a comprehensive solution, especially on lightweight MCUs, often leaving large sections of memory open to attack.

In addition to memory protection, various countermeasures are available on commodity STM32 MCUs to counter fault injection attacks (FIAs), such as error code correction (ECC) mechanisms and programmable voltage detectors (PVDs), which are not foolproof solutions [1]. ECC is only a supplementary protection feature added in the design; it is not highly robust against FIAs. PVDs provide limited reliability by detecting voltage drops below the threshold and cleaning memory, but they are only applicable to voltage glitching attacks. Generally speaking, these approaches represent the only feasible solutions for deeply embedded devices due to the high overhead introduced by more robust solutions.

12.4.2.4 Implementation of Lightweight Encryption for Secure Memory Access

In TinyML, encrypted memory access is a key component of secure data storage. Lightweight encryption schemes can be employed to secure sensitive data without imposing significant processing burdens [18]. The AES with reduced rounds (AES-128) in particular is often used for TinyML. Given that TinyML devices often operate with limited memory and processing capabilities, the encryption–decryption time T_{enc} can be minimized as shown in (12.10), where k is a constant representing the computational efficiency of the encryption algorithm, n is the size of the data block, and f is the device's clock frequency:

$$T_{enc} \approx k \cdot \frac{n}{f} \tag{12.10}$$

This model allows system designers to adjust the encryption parameters, balancing security with processing time and power consumption. By deploying AES-based encryption within TEEs on TinyML devices, sensitive data remains protected from unauthorized access while ensuring rapid, secure access when needed.

12.4.3 Intelligent Edge-Enabled Security and Blockchain Integration for TinyML Applications

The intelligent secure edge computing (ISEC) model, designed for efficient data routing in green IoT environments [19], aligns with TinyML's role in IoT by combining edge computing with secure, distributed models. By leveraging blockchain and distributed hashing tables, TinyML applications can enhance data integrity and secure key management across IoT networks. Blockchain technology, with its decentralized nature, supports TinyML in environments requiring transparent and tamper-proof data handling.

For lightweight, resource-efficient encryption, symmetric key cryptosystems based on the Diffie–Hellman protocol offer a viable solution. It is a reasonable security measure to implement on TinyML devices due to the small code size it requires and an execution time on the order of milliseconds [1]. Therefore, this key exchange method suits TinyML's constrained computational resources, ensuring that data privacy and security can be maintained even on low-power devices. By integrating blockchain and symmetric encryption, TinyML systems can protect data confidentiality across distributed IoT networks, particularly where decentralized secure data storage and transmission are necessary.

12.4.4 Comprehensive Security Strategies for TinyML in IoT Environments

TinyML's deployment in IoT networks brings unique security challenges, especially concerning data storage, authentication, and rogue node detection.

In environments where devices are exposed to both digital and physical vulnerabilities, implementing anti-tampering mechanisms [20] is crucial for protecting data stored on TinyML devices. Access control frameworks and tamper-resistant designs enhance security by establishing boundaries against unauthorized access, protecting the integrity of both data and devices. These measures ensure that TinyML applications, despite their constrained resources, remain resilient to modifications by unauthorized entities.

To fortify TinyML systems against threats to data integrity, particularly in environments where rogue devices may be present, advanced authentication mechanisms are indispensable. Secure device identification methods, including multifactor authentication (MFA), enable stricter access control, which is especially critical in sensitive domains like healthcare, where patient confidentiality is paramount. By validating the authenticity of each device, TinyML deployments in privacy-focused applications can mitigate risks associated with impersonation or unauthorized access. In Figure 12.8, we present the security measures that need to be taken in every layer of a TinyML device.

Integrating hardware and software security measures through cross-layer security optimization further enhances TinyML's resilience in resource-constrained environments. TinyML applications frequently balance security, performance, and power efficiency; hence, techniques such as model compression, pruning, and quantization are applied to optimize memory use while safeguarding against threats. By synchronizing hardware and software protections across layers, TinyML systems can effectively address risks such as side-channel attacks,

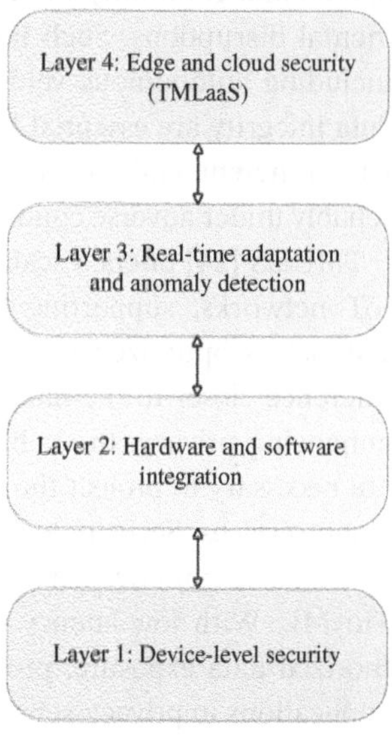

Figure 12.8 Layered security framework for TinyML in IoT.

data poisoning, and unauthorized intrusions. Additionally, cross-layer security strategies enhance energy efficiency, which is crucial for mobile and low-power applications. Approaches like adaptive neural architecture search (NAS) and fault-tolerant mapping offer a sustainable, fault-resilient framework for TinyML, ensuring consistent performance even under challenging conditions in IoT settings.

To address the dynamic nature of many IoT environments, where data patterns evolve over time, TinyML applications require adaptive capabilities to recognize and respond to new patterns or threats. TinyOL [21] facilitates real-time learning and adaptation directly on microcontrollers, allowing models to incrementally update and maintain accuracy in the face of concept drift. This is particularly valuable in sectors like industrial monitoring and healthcare diagnostics, where data shifts rapidly, and anomalies or unusual behaviors must be identified promptly. TinyOL's architecture supports both real-time anomaly detection and classification, making it an effective layer for dynamic security. As TinyML devices quickly adapt to new data, they maintain an agile security posture suitable for dynamic and resource-limited IoT environments.

For TinyML applications deployed at the edge, particularly in artificial intelligence (AI)-driven contexts, it is essential to adopt security frameworks that are both energy-efficient and fault-resilient [22]. Techniques like model pruning, quantization, and NAS reduce the computational load on TinyML devices, allowing them to sustain security without excessive power consumption. Integrating fault-aware training and fault-tolerant mapping within these frameworks ensures operational continuity even in the event of hardware malfunctions or environmental disruptions. Such fault tolerance is crucial in high-stakes applications, including autonomous vehicles and medical diagnostics, where reliability and data integrity are essential for safe operation [23]. By designing TinyML models with inherent fault resilience, these systems are better equipped to function reliably under adverse conditions.

TMLaaS [24] offers a scalable solution for deploying TinyML across extensive IoT networks, supporting both on-device and gateway-based ML inference. TMLaaS is optimized for data privacy and power efficiency by enabling local inference closer to the data source, thereby minimizing cloud dependency and enhancing privacy. In such implementations, secure communication channels are necessary to protect the data exchanged between IoT devices, gateways, and cloud services. Security techniques like model quantization, selective inference, and lightweight encryption support a scalable, privacy-focused architecture for TinyML. With low-latency edge inference, TMLaaS reduces the risk of unauthorized data exposure, providing a robust and secure framework for TinyML applications in privacy-sensitive areas.

In summary, securing TinyML applications in IoT environments requires a layered approach that spans access control, adaptive anomaly detection, and scalable deployment frameworks. By integrating cross-layer security, real-time learning, and energy-efficient, fault-resilient architectures, TinyML devices can operate securely even under challenging conditions and evolving threat landscapes. Each of these strategies enables TinyML to meet the stringent demands of IoT networks, safeguarding data integrity and device functionality across diverse and high-stakes applications – from healthcare to industrial IoT. As TinyML continues to expand within IoT, the need for sophisticated, adaptable security schemes will remain paramount to ensure both resilience and user trust in connected, low-power devices.

12.5 Lessons, Challenges, and Future Directions

This work has demonstrated how efficient, secure, and privacy-aware training can be achieved for TinyML applications and has pointed out the implementation of various new security schemes that are customized for the limitations of such systems. Some of the privacy-aware training methods, like TinyMetaFed and D-FedAvg, allow for decentralized model updates with sensitive data staying on local devices and never needing to be centralized in raw form. This would provide privacy but would still have the advantages of FL. Also, top-P% selective communication minimizes resource utilization due to the need to transfer only the most important model updates, thus drastically reducing bandwidth and energy use. Enhancements include the following: These techniques are very suitable for TinyML devices because of the strict resources of such devices. Capabilities for online learning further enhance efficiency by letting models adapt to new data in real-time without storing any historical information, thus reducing memory usage while preserving privacy.

Implementations of novel security schemes in TinyML are based on multilayered hardware and software protections. PUFs allow for light device authentication based on unique physical properties inherent in the underlying hardware. These functions produce authentication keys in a highly secured way, with no need to execute complex cryptographic operations that will suit the limited resources in TinyML. Similarly, the use of TEEs isolates the zones around sensitive computations and guarantees safety against critical tasks such as model inference and data encryption in the event of compromising a system component. Consequently, sensitive operations are protected against possible threats from outside.

Anti-tampering mechanisms, such as tamper-resistant coatings, physical barriers, and detection sensors, add to the physical security of devices. These are especially useful in high-risk deployments where physical access by attackers is a significant threat. Real-time anomaly detection further strengthens security by enabling devices to identify suspicious behavior or malicious inputs proactively. Efficient techniques like feature squeezing and statistical anomaly detection make it possible for TinyML devices to monitor and respond to threats without overburdening their computational capacity.

Despite these advances, challenges and limitations remain. While privacy-aware training methods are effective, they still remain vulnerable to risks such as model inversion and data inference attacks. Existing solutions, such as lightweight encryption and selective communication, mitigate these risks but often introduce additional energy and computational demands. The trade-offs between privacy, efficiency, and device constraints remain a key area for future work. Scalability is also challenging for FL, whereby the coordination of heterogeneous devices with different capabilities and different levels of connectivity to networks complicates the operation; more adaptive frameworks must be developed to support such diversity.

While hardware security measures are generally robust, this is not to say there are no limitations at all. PUFs tend to be sensitive to a wide variety of environmental impacts. These would lower its dependability on less highly regulated settings. TEE, though very secure, requires excess additional memory and computational capabilities well beyond those of ultra-low-power devices. Finding their solutions will be quite significant for ensuring the wider applicability of the proposed security schemes.

In all, this work presents how privacy-aware training and innovative security schemes could help TinyML overcome some unique challenges. While the developed methods represent significant advances, further refinements are essential for their efficiency and adaptability in a wide range of applications. As these challenges continue to be overcome, TinyML will be able to achieve robust security and privacy that enable applications in increasingly complex and sensitive environments.

12.6 Conclusions and Outlook

The deployment of TinyML models in IoT applications offers transformative potential across sectors like healthcare, industrial monitoring, and smart cities. However, the integration of TinyML into such critical applications also brings with it a complex interplay of security and privacy challenges that must be addressed for the full realization of its transformative potential.

One of the biggest challenges is providing data integrity and privacy within a TinyML application's whole lifecycle, especially given the limitations of resource-constrained devices. This chapter has highlighted an array of adaptive, lightweight security measures tailored to TinyML's unique environment, including FL frameworks like TinyMetaFed and D-FedAvg, hardware-based defenses such as PUFs and TEEs, and adversarial resilience through techniques like adversarial training and defensive distillation.

These strategies collectively address both cyber- and physical threats, enhancing the resilience and operational integrity of TinyML applications while maintaining efficiency. Privacy-preserving model updates, robust adversarial defenses, and real-time threat detection mechanisms create a scalable, secure foundation for TinyML deployment in sensitive and high-stakes IoT contexts. As TinyML technology continues to expand, these security frameworks will play a vital role in supporting secure, privacy-conscious, and resilient operations, ensuring TinyML's suitability for diverse applications that demand both low-power consumption and robust protection.

References

1 Huckelberry, J., Zhang, Y., Sansone, A. et al. (2024). TinyML security: exploring vulnerabilities in resource-constrained machine learning systems. arxiv, 11 November 2024. https://doi.org/10.48550/arXiv.2411.07114 (accessed 7 December 2024).

2 Dehrouyeh, F., Yang, L., Ajaei, F., and Shami, A. (2024). On TinyML and cybersecurity: electric vehicle charging infrastructure use case. arXiv, 25 April 2024. https://doi.org/10.48550/arXiv.2404.16894 (accessed 7 December 2024).

3 Konstantopoulou, E., Athanasiou, G.S., and Sklavos, N. (2024). Minimizing area footprint of UAV communication security using FPGAs. *2024 IEEE International Conference on Cyber Security and Resilience (CSR)*, London, United Kingdom. https://doi.org/10.1109/CSR61664.2024.10679502.

4 Ren, H., Li, X., Anicic, D., and Runkler, T. (2023). TinyMetaFed: efficient federated meta-learning for TinyML. arXiv, 13 Jul 2023. https://doi.org/10.48550/arXiv.2307.06822 (accessed 25 October 2024).

5 Vidaković, M. and Vinko, D. (2023). Hardware-based methods for electronic device protection against invasive and non-invasive attacks. *Electronics* 12 (21): 4507. https://doi.org/10.3390/electronics12214507.

6 Wang, J., Wu, Y., Liu, H. et al. (2023). IP protection in TinyML. *2023 60th ACM/IEEE Design Automation Conference (DAC)*, San Francisco, CA, USA. https://doi.org/10.1109/DAC56929.2023.10247898.

7 Shah, P., Govindarajulu, Y., Kulkarni, P., and Parmar, M. (2024). Enhancing TinyML security: study of adversarial attack transferability. arXiv, July 2024. https://doi.org/10.48550/arXiv.2407.11599 (accessed 31 October 2024).

8 Papadopoulos, P., Thornewill von Essen, O., Pitropakis, N. et al. (2021). Launching Adversarial Attacks against Network Intrusion Detection Systems for IoT. *Journal of Cybersecurity and Privacy* 1 (2): 252–273. https://doi.org/10.3390/jcp1020014.

9 Ghimire, A., Asiri, A.N., Hildebrand, B., and Amsaad, F. (2023). Implementation of secure and privacy-aware AI hardware using distributed federated learning. in *2023 IEEE 16th Dallas Circuits and Systems Conference (DCAS)*, Denton, TX, USA. https://doi.org/10.1109/DCAS57389.2023.10130231.

10 Preuveneers, D., Verheyen, W., Joos, S., and Joosen, W. (2023). On the adversarial robustness of full integer quantized TinyML models at the edge. *Proceedings of the 2nd International Workshop on Middleware for the Edge (MiddleWEdge '23)*, New York, NY, USA. https://doi.org/10.1145/3630180.3631201.

11 Lin, C.-Y., Kostina, V., and Hassibi, B. (2022). Differentially quantized gradient methods. *IEEE Transactions on Information Theory* 68 (9): https://doi.org/10.1109/TIT.2022.3171173.

12 Croci, M., Fasi, M., Higham, N.J. et al. (2022). Stochastic rounding: implementation, error analysis and applications. *Royal Society Open Science* 9: 211631.

13 Guo, W., Sun, S., Li, B., and Blakeman, S. (2023). Federated transformed learning for a circular, secure, and tiny AI. arXiv, 24 November 2023. https://doi.org/10.48550/arXiv.2311.14371 (accessed 31 October 2024).

14 Boumpa, E., Tsoukas, V., Gkogkidis, A. et al. (2022). Security and privacy concerns for healthcare wearable devices and emerging alternative approaches. *Wireless Mobile Communication and Healthcare. MobiHealth 2021. Lecture Notes of the Institute for Computer Sciences,* Social Informatics and Telecommunications Engineering, vol. 440. https://doi.org/10.1007/978-3-031-06368-8_2.

15 Atif Butt, M., Qayyum, A., Ali, H. et al. (2023). Towards secure private and trustworthy human-centric embedded machine learning: an emotion-aware facial recognition case study. *Computers & Security* 123: 103058. https://doi.org/10.1016/j.cose.2022.103058.

16 Konstantopoulou, E., Sklavos, N., and Ognjanovic, I. (2024). Securing public safety mission-critical 5G communications of smart cities. *Internet of Everything for Smart City and Smart Healthcare Applications*, Springer, Cham. https://doi.org/10.1007/978-3-031-34601-9_4.

17 Meidan, Y., Benatar, D., Bitton, R. et al. (2023). D-Score: an expert-based method for assessing the detectability of IoT-related cyber-attacks. *Computers & Security* 126: 103073. https://doi.org/10.1016/j.cose.2022.103073.

18 Konstantopoulou, E. and Sklavos, N. (2022). Design and implementation of a lightweight cryptographic module, for wireless 5G communications and beyond. *IEEE 2022 International Conference on Internet of Things and Intelligence Systems (IOTAIS'22)*, Bali. https://doi.org/10.1109/IoTaIS56727.2022.9975849.

19 Haseeb, K., Ud Din, I., Almogren, A. et al. (2021). Intelligent and secure edge-enabled computing model for sustainable cities using green internet of things. *Sustainable Cities and Society* 68: 102779. https://doi.org/10.1016/j.scs.2021.102779.

20 Adlakha, R. and Kabbar, E. (2024). The challenges of TinyML implementation: a literature review. *CITRENZ 2023 Conference*, Auckland. https://doi.org/10.34074/proc.240120.

21 Ren, H., Anicic, D., and Runkler, T.A. (2021). TinyOL: TinyML with online-learning on microcontrollers. *2021 International Joint Conference on Neural Networks (IJCNN)*, Shenzhen, China. https://doi.org/10.1109/IJCNN52387.2021.9533927.

22 Shafique, M., Theocharides, T., Reddy, V.J., and Murmann, B. (2021). TinyML: current progress, research challenges, and future roadmap. *2021 58th ACM/IEEE Design Automation Conference (DAC)*, San Francisco, CA, USA. https://doi.org/10.1109/DAC18074.2021.9586232.

23 Shafique, M., Marchisio, A., Wicaksana Putra, R.V., and Hanif, M. A. (2021). Towards energy-efficient and secure edge AI: a cross-layer framework. *2021 IEEE/ACM International Conference On Computer Aided Design (ICCAD)*, Munich, Germany. https://doi.org/10.1109/ICCAD51958.2021.9643539.

24 Zaidi, S.A.R., Hayajneh, A.M., Hafeez, M., and Ahmed, Q.Z. (2022). Unlocking edge intelligence through Tiny machine learning (TinyML). *IEEE Access* 10: 100867–100877. https://doi.org/10.1109/ACCESS.2022.3207200.

13

Semantic Management of TinyML for Industrial Application

Kinzah Noor[1], Hasnain Ahmad[2], and Agbotiname Lucky Imoize[3]

[1] *Office of Research Innovation and Commercialization, University of Management and Technology, Lahore, Pakistan*
[2] *Department of Computer Science, COMSATS University Islamabad-Lahore Campus, Lahore, Pakistan*
[3] *Department of Electrical and Electronics Engineering, Faculty of Engineering, University of Lagos, Lagos, Nigeria*

13.1 Introduction

The Internet of Things (IoT) [1] has gained significant attention due to its diverse range of smart applications. More recently, advancements in the Industrial Internet of Things (IIoT) [2] have enabled sophisticated real-time processing of vast amounts of data. This development has facilitated progress in areas like anomaly detection, smart manufacturing, and predictive maintenance, demonstrating the transformative potential of IoT technologies across various industries.

Onboard-embedded technological intelligence refers to learning algorithms that enable devices to make reliable decisions based on collected data. However, implementing machine learning (ML) on small devices is challenging due to significant constraints in architecture, energy, and latency [3]. This is particularly crucial in safety-critical systems, such as industrial robotics, autonomous vehicles, and medical devices, where delays can have serious consequences [4].

The term tiny machine learning (TinyML), introduced in 2018, represents a paradigm that enables ML on edge devices with minimal processing power and memory requirements. Typically, these systems consume only a few milliwatts of power [5], and their memory footprint ranges between 64 and 256 KB, often not exceeding 1 MB. The computing units in such devices generally operate at frequencies between 40 and 400 MHz. Rapid advancements in TinyML have enabled ML models to operate on low-power microcontrollers (MCUs) found in IoT devices. The deployment of ML directly on these devices enhances latency and

Tiny Machine Learning: Design Principles and Applications, First Edition.
Edited by Agbotiname Lucky Imoize, Dinh-Thuan Do, and Houbing Herbert Song.

energy efficiency, strengthens data privacy, and reduces reliance on continuous Internet connectivity. With demand continuing to rise, over 25 billion MCUs were shipped in 2019 [6]. By integrating ML into these low-power embedded devices, the intelligence of IoT applications can be substantially improved. Additionally, significant progress has been made in the development of TinyML algorithms and the optimization of low-power IoT devices.

The heterogeneity of the TinyML ecosystem poses significant challenges to effective management and large-scale implementation, particularly in IIoT networks comprising hundreds or thousands of devices. Embedded systems are often customized for specific use cases, resulting in a diverse range of computational and energy efficiency capabilities. These devices incorporate various sensors, processing power, runtimes, and sizes of random access memory (RAM) and flash memory. Developing a TinyML model entails numerous decisions, including selecting the neural network (NN) type, layer combinations, and neuron sizes to optimize performance under different conditions. Furthermore, the preprocessing and postprocessing for each model must be tailored, as meaningful outputs depend on accurately formatted input data. Additionally, the lack of comprehensive documentation accompanying TinyML models contributes to a fragmented ecosystem. TinyML practitioners face numerous challenges, particularly as some of the most advanced technologies, such as contemporary NNs, now require billions of parameters [7].

While larger networks tend to yield better results and broader applicability, their energy consumption increases with size, rendering this trend energetically unsustainable at larger scales [8]. This highlights the importance of pursuing TinyML as a vital area of research. In this context, TinyML empowers developers to implement sophisticated algorithms on constrained devices that are portable, cost-effective, and energy-efficient, making them suitable for a variety of applications. Recent market trends further support this approach, emphasizing the demand for less power-hungry technology; indeed, there is a pressing need for TinyML to become even more compact.

There are two conventional approaches to developing a TinyML solution: the hardware (HW)-oriented and ML-oriented approaches, along with a more recent co-design methodology. In traditional approaches, the development of ML models and HW implementation are treated as separate processes. In the ML-oriented approach, experts focus on designing, training, and refining the model before deploying it onto suitable HW [9]. The HW-oriented approach, on the other hand, aims to design optimized HW using simplified models and techniques without prior selection of the HW [10]. The co-design workflow derives its name from the collaborative efforts of HW engineers and ML specialists, who work together from the outset, actively sharing knowledge throughout the design process [11].

The management of TinyML systems in the industry is significantly complicated by the rapid evolution and diversity of TinyML components, including embedded devices and ML models. A primary challenge is identifying IoT devices in the field that possess the necessary sensors and capabilities for specific ML models without requiring a thorough examination of each device. Additionally, it is essential to locate existing NN models that are compatible with particular embedded devices to avoid unnecessary redundancy when developing new models. Moreover, effective management and repurposing of TinyML components at scale is a critical concern, alongside the need to integrate discovery results into coherent IoT applications.

To address these challenges, a scalable approach is proposed in this chapter for managing TinyML systems utilizing Semantic Web technologies. In this research, semantics encompass HW requirements, information, and NN representations for ML models in TinyML. Additionally, it is recommended to employ the standardized Thing Description (TD) [12] semantics in conjunction with the World Wide Web Consortium's (W3C) Web of Things (WoT) architecture [13] to enhance IoT devices and facilitate the specification of semantic HW. Integrating the proposed TinyML model semantics with the TD semantics enables the alignment of model requirements with IoT device capabilities. This unified knowledge representation, hosted in a central knowledge graph (KG), supports vendor-agnostic information management and reasoning. Moreover, the use of SPARQL Protocol and RDF Query Language (SPARQL) enhances system administration and interoperability across diverse sectors by enabling the discovery and interaction of TinyML components in various contexts.

This chapter aims to support the community in developing, maintaining, and utilizing KG for TinyML. Examples and code are provided to create semantic representations for NN models in the ".tflite" format, which is compatible with TensorFlow Lite for MCUs. Although the focus of the research has not been on IoT device semantics due to the established TD semantics, the proposed repository includes six semantic descriptions. Moreover, the distribution of models to devices is enabled through knowledge matching, allowing for the orchestration of TinyML components. Additionally, the use of low-code platforms simplifies TinyML workflows for engineers who lack expertise in SPARQL queries or semantic technologies, facilitating more efficient system management.

13.1.1 Key Contributions of the Chapter

The following are the significant contributions of this chapter:

i) This chapter comprehensively introduces TinyML, discussing its core principles and the challenges surrounding its broader adoption. It further examines recent advancements in TinyML, including essential workflows, HW, and software for efficient implementation.
ii) A scalable approach is discussed for managing TinyML systems on NNs using Semantic Web technologies. The approach enhances interoperability by integrating model and device requirements into a unified KG, enabling

efficient discovery and management of TinyML components across diverse IoT environments.

iii) This chapter showcases a semantic management system using a KG that encompasses six IoT devices and 23 TinyML models. Three case studies illustrate the advantages of this approach, while the performance of TinyEngine is rigorously assessed through comparisons of speed, memory usage, and binary size across frameworks.

13.1.2 Chapter Organization

The remaining part of this chapter is organized as follows. Section 13.2 introduces TinyML, outlining its core concepts and the challenges associated with its widespread adoption. Section 13.3 delves into recent advancements in TinyML, including the required workflow, HW, and software for effective implementation. Section 13.4 provides a detailed methodology on NNs, examining TinyML integration with IoT through three in-depth case studies. Section 13.5 presents the results and discussion of the TinyEngine, offering a performance comparison based on memory usage and binary size. Finally, Section 13.6 concludes the chapter by exploring future directions and potential advancements in TinyML.

13.2 Introduction to TinyML

The field of TinyML is experiencing rapid growth within both industrial and academic communities. The research [14] outlined the challenges and advancements in the development of embedded ML. Improved algorithms [15] have been designed to optimize deployment in real-world scenarios, enhance model performance, and increase resource efficiency. Furthermore, the development of ultra-low-power artificial intelligence (AI) circuits and accelerators [16] has been proposed to enable continuous, always on ML capabilities, even when operating on battery power for extended periods.

Since TinyML delivers ML solutions to resource-constrained devices, a holistic approach combining HW and algorithm [17] design is essential to maximize performance. Frameworks for evaluating ML deployment at the edge [18] play a critical role in systematically addressing potential challenges. Several platforms and tools have been developed to accelerate TinyML, such as Google's TensorFlow Lite for MCUs [19], STM's X-CUBE-AI, Apache TVM, and various open online courses. Moreover, notable applications of TinyML are emerging across multiple domains [20].

Effective management of the TinyML ecosystem necessitates enhanced transparency and interoperability to address its inherent heterogeneity. TensorFlow Lite metadata provides a structured framework for documenting model characteristics, including input and output data, while the Open Neural Network Exchange (ONNX) serves as an open-source initiative for sharing ML models across various frameworks. Emphasizing efficiency and limitations, Google's model card [21] establishes a standard for characterizing ML models. However, these systems typically rely on isolated, manually created documentation that lacks interchangeability and interoperability. Within the context of TinyML, the main focus of research is on enhancing the reusability of ML model formats with IoT devices rather than on converting models between different platforms.

The management of ML models has garnered increased attention in recent years. While research [22] introduced a database for tracking ML models, research [23] highlighted the challenges associated with managing models across diverse ML use cases. A KG called FAIRnets aggregates and facilitates the interchange of information regarding publicly available NNs. However, despite their expressiveness, these approaches are insufficient for TinyML, as the growing complexity demands a thorough examination of both HW specifications and ML models.

Semantic Web technologies have been effectively applied within the IoT domain and are particularly adept at representing the diverse knowledge pertaining to TinyML components in a manner comprehensible to both humans and machines [24]. The WoT approach aims to make various IoT devices accessible as web resources. Additionally, the W3C WoT interest group has proposed a formal information schema known as the TD semantic, which semantically characterizes the interfaces and capabilities of these devices.

Despite recent advancements, a disconnect remains between ML models and IoT devices, particularly in the large-scale management of TinyML components. Key challenges include identifying suitable combinations and initiating extensive deployments. In research [25], a semantics-based approach for managing on-device applications across distributed IoT devices was proposed. This approach is adopted by developing an NN semantics tailored for TinyML in conjunction with the TD semantic. To enhance interoperability, research leveraged established semantic vocabularies: the semantic sensor network (SSN) schema, which encompasses sensors and their interactions, incorporates the sensor, observation, sample, and actuator semantics. Additionally, the semantic smart sensor network (S3N) schema extends SSN to provide insights into smart sensors and their characteristics. Another relevant semantic model is iot.schema.org, an extension of schema.org designed to enrich structured data across the Internet.

Finally, TinyML as a service [26] facilitates the deployment of ML models across various embedded devices by creating an abstraction layer over multiple HW compilers and automatically generating compiled ML models for specific HW platforms. This service has the potential to enhance the deployment phase of the presented workflow significantly. Table 13.1 presents a summary of the semantic management of TinyML.

Table 13.1 Semantic management of TinyML.

Emphasis	Approach	Significant outcome	Unaddressed areas	Contribution of research	Future directions
IoT semantic management	Semantic Web technologies	Suggested a structure for governing IoT applications	Insufficient attention to on-device handling	Framework is extended to TinyML	Establish shared vocabulary among IoT devices and systems
Effective implementation of ML models on TinyML devices with limited resources	Applied ML to embedded systems by TensorFlow Lite Micro Framework	ML model deployment on embedded devices was shown to be effective	Lacks connectivity and support for various IoT ecosystems	TensorFlow Lite Micro's operation on devices with limited resources was demonstrated	Significant resource limitations and missing program functionalities
Using sensors like Wi-Fi and IMU fusion for positioning indoors	Fusion algorithm that fused Wi-Fi, IMU, and floor plan data for indoor localization	Increased precision in positioning within indoor spaces	No direct interaction with IoT or TinyML apps	Improved the accuracy of positioning in indoor settings	The creation of a real-time indoor GPS using multi-modal sensor fusion
Detecting objects in real-time on inexpensive FPGAs	FPGA optimization for object recognition	Allowed for effective real-time object identification on FPGAs with limited resources	No use of TinyML integration or IoT	An FPGA is used to construct a hardware accelerator for CNN that is economical and resource-efficient	Real-time applications, throughput, and FPS were further increased by sophisticated quantization
Investigated TinyML's dynamic updates for online learning	Microcontroller-based online learning environment	Dynamic ML upgrades on embedded devices were demonstrated	Real-time deployment has not been thoroughly investigated	Incorporates strategies for real-time application administration	Methods to combine TinyOL with IoT sensors and platforms to enhance data gathering

Semantic administration of IoT applications was discussed	Semantic models suggested for on-device application administration	Semantic models to improve IoT apps' discoverability and reusability	Extensive use of IoT devices for TinyML implementation	Enhance TD by semantic modeling for on-device application details	Establish benchmarks and criteria to evaluate the effectiveness of semantic management
Examined weaknesses in accelerators for neural inference	Analysis of vulnerabilities in restricted accelerators	Identified security threats and solutions for accelerators operating in restricted settings	No particular emphasis on TinyML security	Found possible security flaws in environmental limitations	Extending investigation to include inference accelerators and neural designs
Ecosystem for ML model deployment on IoT devices	TinyMLaaS framework automating deployment abstracting hardware differences	Hardware abstraction to install ML models on a variety of IoT devices	Does not prioritize large-scale management and orchestration	TinyML models are automatically deployed on a variety of devices	Deployed models' version management and real-time upgrades guarantee peak performance
IoT device interoperability through semantic Web	Semantic technologies to improve the integration of IoT devices	Framework for semantic web-based IoT device management	The implementation of TinyML in the IoT ecosystem is not specifically addressed	Created semantic framework for IoT device management	Investigating to combine AI-powered systems with semantic IoT

(Continued)

Table 13.1 (Continued)

Emphasis	Approach	Significant outcome	Unaddressed areas	Contribution of research	Future directions
Dynamic, adaptable edge-based TinyML is required in settings with low resources	Examining real-world applications, classifying architectures, and surveying reformable TinyML systems	Overview of proposed architecture taxonomy, reformable TinyML systems, IoT, healthcare, smart city applications	Insufficient practical implementations and investigation of scalability in dynamic reconfiguration methods	The taxonomy of reformable TinyML architectures emphasizes flexibility, and dynamic reconfiguration is suggested	Examine self-adaptation, security, scalability, real-time learning, and reformable TinyML in industrial settings
Optimize neural networks for TinyML, maximizing speed and minimizing energy	Proposed hardware-aware NAS with advanced quantization boosts efficiency in real-time applications	Created framework optimizing TinyML neural networks to enhance performance in resource-limited settings	Limited focus on adaptability to diverse edge hardware real-time implementation	Achieved 25% faster inference and 40% lower power use; offers open-source tools for TinyML optimization	Support various edge hardware platforms enhancing energy efficiency for complex neural networks

13.3 Recent Advances in TinyML

ML has significantly transformed various domains, such as language, audio processing, and computer vision. Nonetheless, advanced models often require substantial memory and processing resources, which can lead to high deployment costs. To overcome these challenges, researchers are focusing on developing efficient algorithms, methods, and HW tailored to diverse applications. Efficient ML [27] is primarily divided into two categories, as shown in Figure 13.1. CloudML aims to enhance latency and throughput in cloud environments, and EdgeML emphasizes improving energy efficiency, reducing latency, and ensuring privacy on edge devices. These two areas intersect in several ways, including over-the-air updates, hybrid inference, and federated learning.

A notable advancement in this field is TinyML, which allows ML on ultra-low-power devices that require as little as a few hundred kilobytes of memory. As the number of IoT devices continues to rise, there is an increasing demand for low-power, on-device AI that enhances responsiveness and privacy while minimizing the costs associated with wireless communication. This capability is especially critical for real-time applications, such as autonomous vehicles. Moreover, efforts to develop TinyML enable on-device training, strengthening privacy by allowing IoT devices to adjust their models locally rather than relying on cloud servers. This approach can enhance various applications, such as email clients that refine their predictive capabilities using personalized language models and home security cameras that recognize unfamiliar faces. Additionally, it facilitates IoT solutions in environments lacking Internet connectivity, including applications in precision agriculture and ocean monitoring.

13.3.1 Latest Hardware Trends Empowering TinyML

TinyML practitioners face significant challenges, primarily due to the high parameter requirements of modern NNs, which can reach billions. While larger

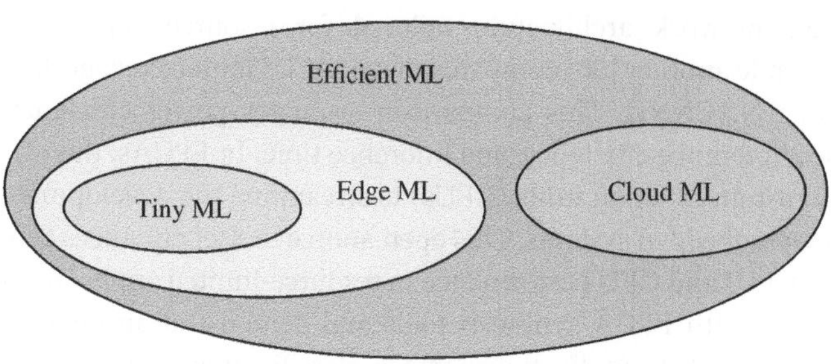

Figure 13.1 Domains of efficient ML: EdgeML and CloudML.

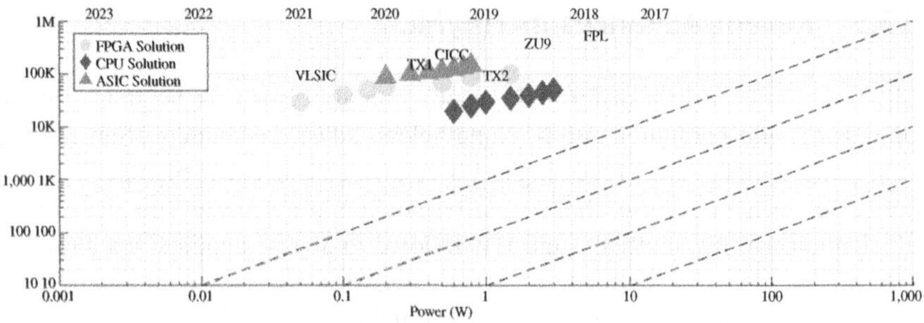

Figure 13.2 Latest TinyML hardware trends.

networks offer improved performance and versatility, their energy demands make large-scale deployment unsustainable. Consequently, TinyML emerges as a crucial field that enables developers to implement advanced algorithms on compact, cost-effective, and energy-efficient devices suitable for diverse applications. Recent market trends further emphasize the need for low-power solutions, underscoring the importance of making TinyML even more efficient and compact. One noteworthy trend in the most recent HW design for TinyML is that the emphasis is on reducing power consumption. This suggests that the main focus is on making ML compatible with low-resource devices, as shown in Figure 13.2. TinyML solutions primarily utilize three HW technologies: application-specific integrated circuits (ASICs), MCUs, and field-programmable gate arrays (FPGAs). Notably, FPGAs tend to have significantly higher power consumption than ASICs.

ASICs have gained popularity among researchers and industry practitioners, particularly the RISC-V architecture. Selecting the most suitable RISC-V processors, architectures, and ML software frameworks can be challenging. A survey [28] examines various RISC-V HW implementations and software frameworks for system-on-chip development, highlighting newer frameworks that support open HW integration for ML applications. The research [29] focusing on TinyML frameworks for MCUs emphasizes the integration of ML algorithms. It introduces a multi-radio access network architecture tailored for resource-constrained networked devices, while another [30] compares seven MCUs for face recognition using convolutional NN (CNNs). This comparison evaluates power efficiency, energy per inference, inference efficiency, and inference time. In FPGAs, the [31] framework utilizes custom function units (CFUs) to streamline the development of ML accelerators for embedded systems. This open-source toolset enhances central processing unit (CPU) and CFU performance in resource-limited applications by combining software with FPGA synthesis tools and demonstrating iterative improvements in two TinyML applications: image classification and keyword detection.

Figure 13.3 ML-focused process for TinyML-based system.

13.3.2 TinyML Workflows

The main steps for developing TinyML systems focus on the integration of HW platforms with ML models [32]. While working on both elements simultaneously can yield a more unified and efficient outcome, developers typically start with the area they know best. Industrial limitations give rise to two primary approaches: an ML-oriented workflow that centers on model design, adaptation, training, and evaluation with predetermined HW and HW-oriented workflow that starts with HW selection to optimize resource use. Co-design, a more holistic approach, combines both elements from the outset for improved performance.

13.3.2.1 ML-Focused Process

The ML-oriented workflow in Figure 13.3 emphasizes developing, training, and assessing ML models, often with HW platform choices restricted by industry requirements. For example, adapting NN models [33] to embedded devices demands rigorous testing to ensure efficient use of power, latency, and memory. These devices are more limited than in cloud systems. There are five key stages of the workflow. The first stage is the model framework, where models are created and validated with a focus on generalization. The next step is model optimization, balancing efficiency with performance. After this, direct evaluation, assessment, and adjustment of the model will be performed. The next stage is the deployment destination, applying device-specific optimizations. Finally, the destination evaluation confirms the effectiveness of the model in real-world use.

13.3.2.2 Hardware (HW)-Focused Process

In the HW-oriented approach in Figure 13.4, developers focus on creating advanced HW platforms optimized for embedded applications to support current and future ML algorithms. This process often involves identifying bottlenecks

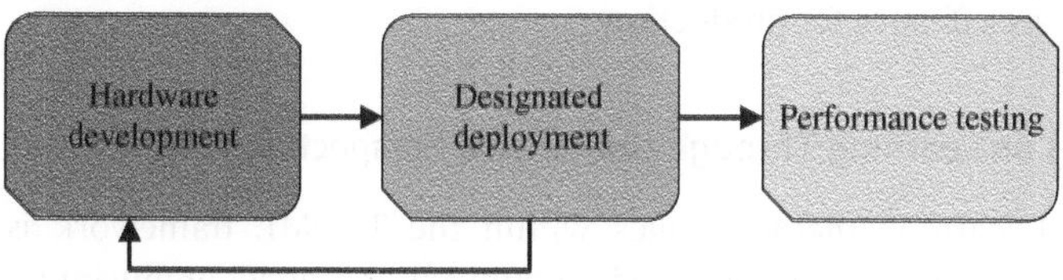

Figure 13.4 HW-focused process for TinyML-based system.

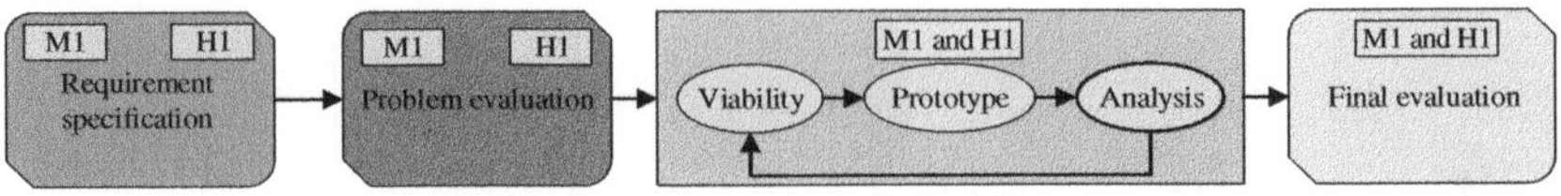

Figure 13.5 Co-design process for TinyML-based system.

in existing architectures, particularly in ML computations like NNs [34], and designing HW accelerators to boost throughput and reduce energy use, for example, by simplifying convolutional computations or developing efficient, low-power perceptrons. In some cases, new platforms are built with integrated digital signal processing, requiring optimized computing libraries to enhance performance. The HW-oriented workflow typically includes three stages. The first stage is HW development, where developers build architectures or accelerator modules to improve performance for specific computational tasks. The next step is designated deployment, which assesses the optimized HW in simulated environments, with a redesign as needed. Finally, performance testing involves the production and testing of the final HW.

13.3.2.3 Co-design Process

In the integrated design workflow in Figure 13.5, HW design and model optimization are developed together from the start to enhance performance and resource efficiency. Unlike previous workflows that address these as separate stages, this approach focuses on co-optimization. This may include using HW-informed training to support NN computations on analog computation-in-memory accelerators or creating specialized architectures for certain ML algorithms on FPGAs [35]. The key steps in this workflow are as follows:

a) **Requirements Specification**: Two teams – one of ML specialists (M1) and the other of HW specialists (H1) – define the device's requirements and capabilities.
b) **Problem Evaluation**: The teams review various options and select the most suitable architecture.
c) **Co-design**: A collaborative design process selects specific HW and software components with an overall system perspective.
d) **Final Evaluation**: A final assessment ensures that model- and device-specific optimizations are effective for production.

13.3.3 Insights on Learning Through the TinyML Perspective

The taxonomy of learning methodologies within the TinyML framework is divided into three primary categories: model optimization, model architecture, and modeling algorithms, as illustrated in Figure 13.6. This framework provides a

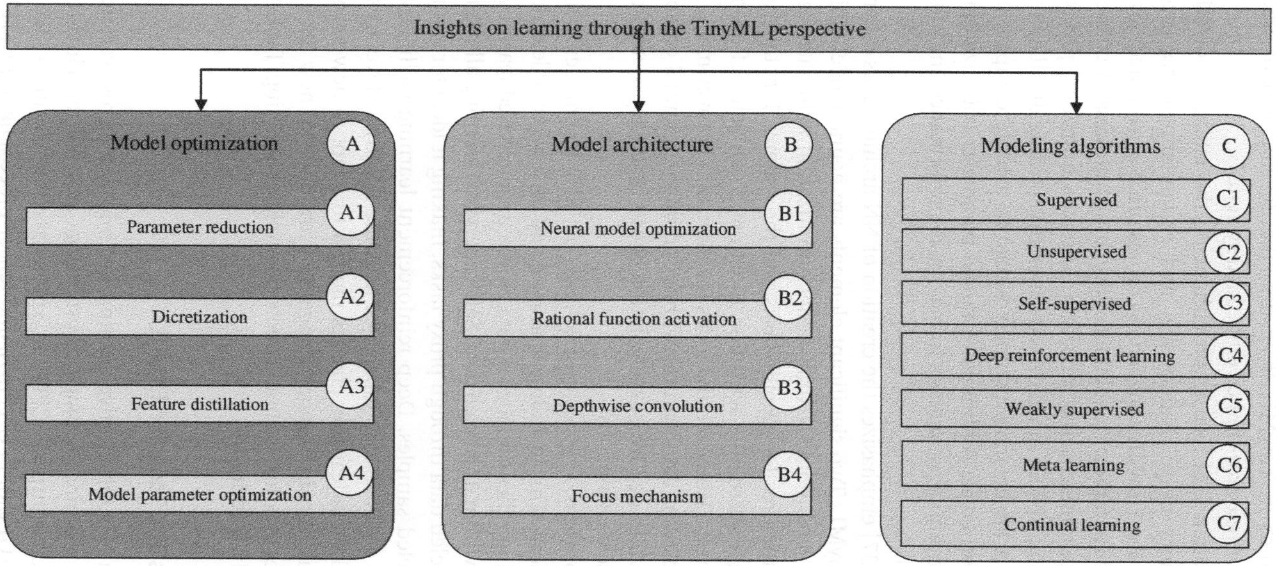

Figure 13.6 Insights on learning through ML, including three key areas.

comprehensive overview of strategies for creating, optimizing, and implementing modeling algorithms in TinyML. It ensures the development of efficient ML models suitable for resource-limited embedded systems.

Model optimization [36] focuses on enhancing the efficiency of ML models concerning memory usage, battery life, and computational resource factors that are essential for TinyML applications on resource-constrained devices. Key techniques include parameter reduction, eliminating unnecessary connections by reducing model parameters. Discretization decreases numerical precision from 32-bit floating-point to 8-bit integer to save memory. Feature distillation transfers knowledge from a larger, more complex model to a more compact one while maintaining performance. Additionally, model parameter optimization seeks to identify the best training parameters to maximize performance and efficiency in embedded systems.

Model design [37] emphasizes the creation of NN architectures tailored to the constraints of TinyML. Two significant elements are neural model optimization and depthwise convolutions. Neural model optimization automates the discovery of optimal structures for specific tasks and the rational function activation, which implements alternative activation functions to enhance efficiency while conserving resources. Moreover, the focus mechanism improves model accuracy by focusing on relevant data features without adding substantial computational overhead. Meanwhile, depthwise convolution boosts efficiency by separating spatial and depthwise convolutions.

Modeling algorithms [38] categorize various ML techniques suitable for TinyML applications. This includes supervised and unsupervised learning, where supervised learning relies on labeled data, while unsupervised learning seeks to identify patterns within unlabeled data. Self-supervised learning allows models to learn from unlabeled data through proxy tasks, making it ideal for environments with limited labeled samples. Deep reinforcement learning facilitates learning through trial and error. Weakly supervised learning effectively leverages partially labeled data, and meta-learning enables models to adapt to new tasks quickly. Lastly, continuous learning supports ongoing learning from new data, allowing low-power devices to adjust without needing complete retraining. Table 13.2 gives an overview of ML algorithms.

13.3.4 Devices and Tools for TinyML

TinyML relies on HW for effective training and inference across various devices. The [29] review examines a range of processors suitable for TinyML applications, from general-purpose CPUs to specialized architectures like graphics processing units (GPUs), FPGAs [39], and tensor processing units (TPUs). Each type presents flexibility, programmability, and efficiency trade-offs based on

Table 13.2 Overview of ML algorithms.

Algorithm	Overview	Core methods	Applications	Advantages
Supervised learning	Learns to predict outcomes from labeled input–output pairs	Neural networks, K-nearest neighbors, SVM, decision trees, linear regression	Spam detection, image recognition, predictive analytics	High precision with enough data that is labeled
Unsupervised learning	Discovers hidden structures in data by identifying unlabeled patterns	PCA, clustering K-means, association analysis	Recommendation systems, customer segmentation, anomaly detection	Useful when labeled data is unavailable
Semi-supervised learning	It uses both labeled and unlabeled data to improve model performance	Graph-based models, semi-supervised SVM	Web content classification, medical image classification	The hassle of labeling, lowers the expense
Weakly supervised learning	Reduces labeling costs by learning from sparse, noisy, or partially labeled data	Weak supervision frameworks, inexact, inaccurate, or incomplete labeling	Sentiment analysis, medical diagnosis, object detection	Effective with faulty labels
Self-supervised learning	Uses proxy tasks to infer hidden data properties for feature learning from unlabeled data	Predictive coding, contrastive learning, autoencoders	Representation learning, NLP, computer vision	Useful for extracting representations from huge unlabeled datasets
Reinforcement learning	Maximizes reward through trial and error by interacting with the environment to optimize actions	Actor-critic, Q-learning, deep Q-networks, policy gradient	Game playing, robotics, autonomous vehicles	Learns the best practices for challenging projects

(Continued)

Table 13.2 (Continued)

Algorithm	Overview	Core methods	Applications	Advantages
Meta-learning	Learning to learn enables models to leverage past knowledge for quick adaptation to new challenges	Few-shot learning, model-agnostic meta-learning (MAML)	Few-shot classification, personalized recommendations, rapid model adaptation	High capacity for task adaptation
Continual learning	Enables models to learn and adapt over time without forgetting previous knowledge	Memory-based models, incremental learning, elastic weight consolidation (EWC)	Lifelong learning applications robotics, adaptive systems	Prevents disastrous forgetfulness
Deep learning	Captures complex data patterns and high-level abstractions using multilayer neural networks	Transformers, CNNs, RNNs	Language translation, image classification, text generation	High-dimensional data in Excel
Transfer learning	Improves learning efficiency by transferring knowledge from one model to a similar task	Pre-trained models, fine-tuning	Object detection, medical imaging, NLP	Cuts down on training time for associated duties
Ensemble learning	Combines multiple models to improve feature selection, reduce bias, or decrease variance	Stacking, random forest, gradient boosting, bagging	Customer churn prediction, fraud detection, financial forecasting	Improves the accuracy and resilience of the model

Table 13.3 Comparison of hardware options for TinyML.

Hardware	Benefits	Power consumption	Application	Drawbacks
CPU	It has a large memory capacity and is appropriate for general tasks	Low to moderate	IoT devices, basic ML tasks	Reduced processing speed and limited parallelism
GPU	High processing speed, perfect for sophisticated structures	High	Deep learning, image recognition	Expensively and energy-intensive
FPGA	Adaptable and energy-efficient	Moderate to low	Edge devices, custom ML applications	Limited software support and difficulty in using
TPU	Significantly increase the pace of model inference	Moderate	AI applications, high-performance inference	Costly and difficult to create for

application requirements. While CPUs, particularly the Arm Cortex-M, are efficient, they may not fully exploit parallelism. In contrast, GPUs excel at parallel processing for ML, and TPUs developed by Google are optimized for tensor operations, offering exceptional performance and scalability for TinyML applications. Table 13.3 shows a comparison of HW options for TinyML.

13.3.5 Software Platforms

The rising demand for HW-based ML emphasizes the critical role of the software layer in developing TinyML systems. Key frameworks include uTensor, which converts ML models into optimized C++ files for easier integration with arm architectures, and TensorFlow Lite Micro [40], enabling ML model execution on MCUs without an operating system. Edge impulse supports real-time data processing and cloud-based training for edge device model development. Additionally, X-CUBE-AI automates AI algorithm conversion for STM32 products, while the Embedded Learning Library (ELL) serves as a cross-compiler for low-resource platforms. Furthermore, MinUn addresses memory management, bandwidth optimization, and number representation issues in TinyML applications, and uTVM enhances deep learning workloads across various HW. Together, these frameworks significantly advance TinyML research.

13.3.6 TinyML Constraints

Due to the stringent constraints of devices like MCUs, TinyML applications cannot afford the processing resources necessary for the success of deep learning

Table 13.4 Comparison of memory and storage for AI models across mobile platforms.

	Cloud AI (NVIDIA V100)	Mobile AI (iPhone 11)	TinyAI (STM32F746)	ResNet-50	MobileNetV2	MobileNetV2 (int8)
Memory	16 GB	4 GB	320 kB	gap → 7.2 MB	6.8 MB	1.7 MB
Storage	TB ~ PB	>64GB	1 MB	gap → 102 MB	13.6 MB	3.4 MB

models. Training and implementing AI models on MCUs is challenging because of their significant memory limitations, such as read-only flash memory (FLASH) and static random-access memory (SRAM) being less than 256 kB, lack of operating systems, and absence of dynamic random-access memory (DRAM). Compared to mobile platforms, they have far fewer available resources. Prior studies have either focused on optimizing operator kernels without improving network design or aimed to reduce model parameters without addressing the critical issue of activation bottlenecks. Neither strategy adopts a co-design perspective, leading to suboptimal solutions for TinyML. Table 13.4 presents a comparative analysis of memory and storage for AI models across mobile platforms.

13.3.6.1 Mobile-Optimized Models Are Unsuitable for TinyML

Because mobile devices typically have ample memory, optimization efforts for mobile deep learning models are required. MobileNets and ShuffleNet [41] have largely focused on reducing parameters, floating point operations per second (FLOPs), and latency rather than addressing peak memory usage. Although MobileNetV2-1.4 has a model size that is 4.2× smaller than ResNet-50, it requires 2.3× more peak memory, further straining the SRAM limitations observed with ResNet, as shown in Figure 13.7. This highlights the need to rethink model design strategies specifically for TinyML applications.

13.3.6.2 Challenges in Adapting Inference Models for Tiny Training

Backpropagation requires storing intermediate activations, making training on resource-limited systems like MCUs challenging, with memory usage increasing by 6.9 times compared to inference. As shown in Figure 13.7, MobileNet's training memory usage is only 10% lower than ResNet, which remains inadequate for devices with limited SRAM, such as 256 KB. Despite prior efforts to reduce FLOPs or optimize inference memory, even efficient models like MCUNet [42] are not suitable for training on tiny devices. New backpropagation techniques and learning algorithms are needed to reduce activation memory requirements, enabling efficient and accurate training on these platforms.

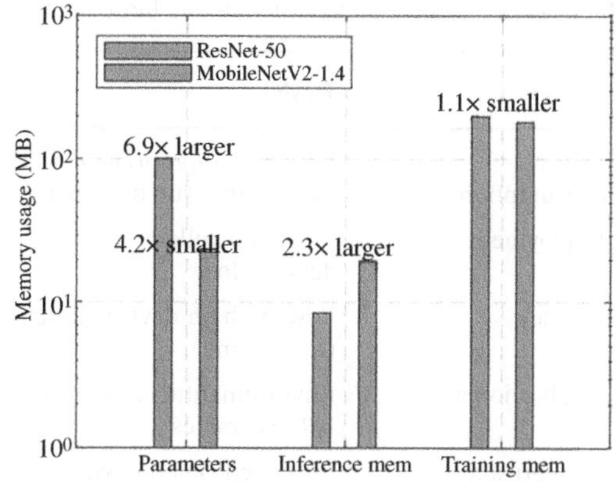

Figure 13.7 Memory usage comparison: parameters, inference, and training.

13.3.7 TinyML Application

The review [43] focuses on the applications of TinyML across various fields, including IoT, environmental challenges, predictive maintenance, anomaly detection, and healthcare. It highlights the benefits of TinyML, such as reduced latency and enhanced data privacy, and examines its integration with low-power wide-area network (LPWAN) and 5G technologies. The survey [44] discusses the role of TinyML in healthcare, particularly with wearable devices. It provides insights into optimizing predictive maintenance systems and detecting anomalies, culminating in a taxonomy of existing TinyML techniques across these domains.

13.3.8 TinyML and Large ML Comparison

Building effective models under constraints such as memory, computing, engineering resources, and data is essential for both TinyML [45] and large ML. Although TinyML focuses on optimizing models for small devices, it involves techniques like quantization and sparse learning. It proves valuable for large-scale applications in cloud environments, making these efficiency methods broadly applicable.

TinyML continues to evolve alongside HW advancements. For example, ResNet-50 [46], introduced in 2016 with 25 million parameters, was initially considered a large model. Today, it can perform inference under a millisecond on a smartphone digital signal processor (DSP) such as Qualcomm Snapdragon 8 Gen1. As HW progresses, models once deemed “large” may become “tiny,” highlighting the need for TinyML to expand and adapt over time. Table 13.5 compares TinyML and large ML features.

Table 13.5 Comparison of TinyML and large ML features.

Feature	TinyML	Large ML
Power consumption	Very low, appropriate for gadgets that run on batteries	High, requiring GPUs/TPUs and strong servers
Deployment	Microcontrollers, edge devices, IoT	Data centers, cloud servers
Latency	Low, with on-device inference in real-time	Higher, affected by network and processing delays
Application areas	Environmental monitoring, IoT, wearables	Large-scale analytics, image recognition, NLP
Data privacy	Improved via on-device processing	Reliant on cloud security and data storage
Model complexity	Lightweight models, compact	Deep, intricate models with lots of parameters
Resource requirements	Minimal computation and memory	High-end hardware and memory
Recent developments	Model compression and effective algorithms	Instruction on specialized hardware (such as TPUs) and distributed systems

13.4 Methodology

The study [47] provides a detailed explanation of the semantic schema along with essential guidelines for establishing the semantics of neural network (NN) models tailored to TinyML. The semantics for TinyML models was designed to integrate with TD, facilitating the expression of IoT device capabilities. To illustrate the semantic management system in action, KG encompassing six IoT devices and 23 publicly accessible TinyML models is presented. Finally, three case studies were included to showcase the benefits of this approach.

13.4.1 Designing a Semantics for TinyML Models

The scope of the TinyML model semantics [48] and the level of detail are defined by establishing five essential requirements. First, it was crucial to develop a method for describing ML models that run on embedded devices, ensuring the semantics could capture the unique attributes of models optimized for embedded applications. Second, the requirement was identifying the necessary metadata for each model, including elements such as the originator, unique identifiers, and references. This metadata provides important context for understanding and effectively managing the models. Third, it was crucial to outline the standard features of embedded ML models, covering aspects like layer configurations, quantization methods, and input/output specifications, which are essential for categorizing and organizing TinyML models in a standardized manner.

The fourth one is information about HW, particularly memory and sensor requirements, which was another essential factor in the design of semantics. Including these specifics allows the semantics to accurately capture the unique demands and limitations of embedded environments. Finally, a developed ontology vocabulary aligned with TD to facilitate seamless integration and compatibility with other frameworks, which is vital for effective data sharing across IoT systems. It emphasized that reusing existing ontologies enhances interoperability and avoids unnecessary redundancy. In line with this approach, the S3N ontology is adopted as the foundational layer, which builds upon the sensor, observation, sample, and actuator ontology (SOSA), semantic sensor network ontology (SSN), and W3C WoT TD ontologies. This choice of S3N enhances compatibility with current web standards and strengthens semantic alignment between ML models and HW, particularly regarding TD and the proposed NN ontology. This decision effectively addresses the requirements for both interoperability and vocabulary alignment. The prefixes and the namespaces of the used semantics are available in [48].

13.4.1.1 Semantic Framework for TinyML Neural Networks

Figure 13.8 presents the semantic schema designed for TinyML models, where an NN is categorized as a subclass of both "s3n" and "schema." Metadata attributes enrich each ML model, enhancing its functionality and description. For example, a citation attribute allows linkage to a reference uniform resource locator (URL), such as a research article or code repository, while an identifier serves as a unique model identity (ID). Other metadata features include the repository location, details about the training dataset, creation date, author, description, and additional relevant characteristics. These attributes enable effective searching across various criteria for the model's data. A NN consists of three main types of layers: input, output, and hidden or middle layers. Each layer stores specific details such as input and output dimensions as well as quantization details. The type of layer – such as fully connected, quantized, or depthwise Conv2D – defines additional type-specific properties, allowing for a detailed model representation. Layer indexing further illustrates the structure of the ML model.

The semantics also includes model-specific information that links it to the HW, providing insights into the characteristics of the model. For instance, accuracy metrics can evaluate model performance on a given training dataset.

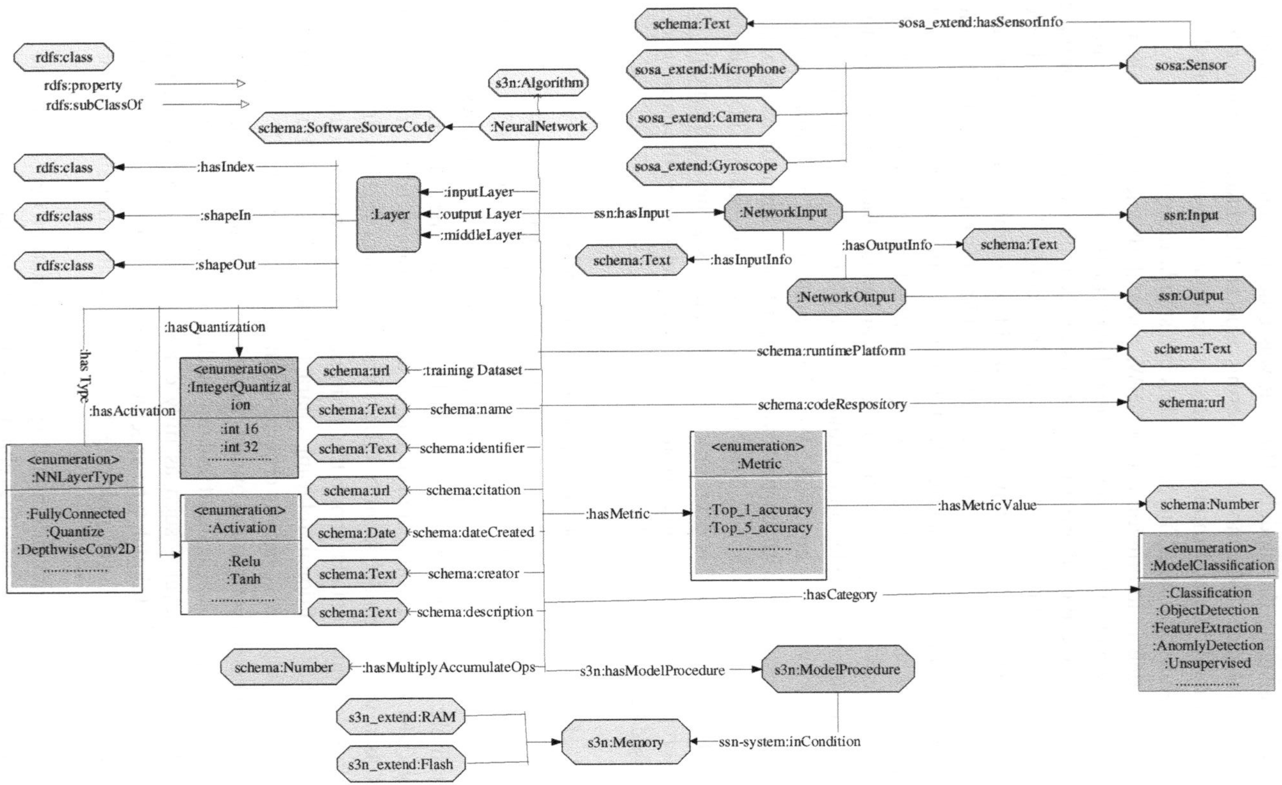

Figure 13.8 TinyML neural network model semantics with layer and metadata relationship.

However, accuracy is only one factor to consider when selecting a TinyML model. Model latency, for example, can be estimated based on the count of multiple MAC operations, as latency scales linearly with the multiply–accumulate operation (MAC) count. To align with TD standards, the RAM and flash memory requirements – critical for evaluating the compatibility of the model with IoT devices – are classified as subclasses of "s3n." The runtime platform attribute specifies the model format and compatible frameworks. The input source for the model is determined by the types of sensors associated with it, such as a camera, microphone, or accelerometer. Input/output specifications and category attributes provide additional information on the application of the model. Finally, the input and output components of the ML model are represented as subclasses of "ssn."

13.4.2 Semantic Knowledge Graph (KG) for TinyML Model

The design of the ML model ontology accounted for compatibility with the TD standard, providing a structured approach to describing embedded devices in IoT applications. Leveraging Semantic Web technologies, this semantics enables the representation of ML models and IoT devices within a unified KG by semantically modeling them according to a shared semantic. To evaluate the semantics and demonstrate its potential, a KG is developed for featuring six different IoT boards and 23 publicly available NN models, encompassing a broad range of devices within the TinyML community. These IoT devices typically include various sensors, limited flash memory, and a modest amount of RAM. The provided TinyML models were trained using the TensorFlow Lite Microcontroller library, an open-source framework that is widely adopted and frequently updated. Each model is under 1 MB, which enhances both the scalability and acceptance of the presented approach within the TinyML community.

The repository also includes scripts and examples for automating the knowledge-building and information-extraction process using the semantics. This setup requires minimal manual input for details such as sensor specifications and citation information, streamlining the integration and application of these models in diverse IoT environments. GraphDB and similar tools can be used to visually explore the KG itself, as shown in Figure 13.9. The illustration displays six embedded devices on the right and 23 ML models on the left, with one model ("Model Activity") highlighted in the center. Due to space constraints, some features and connections are omitted. Graph databases offer valuable flexibility in visualizing relationships within complex TinyML systems, facilitating the efficient maintenance and expansion of the KG as needed. This flexibility ensures that the KG can seamlessly accommodate more complex scenarios and evolving semantic schemas in the future.

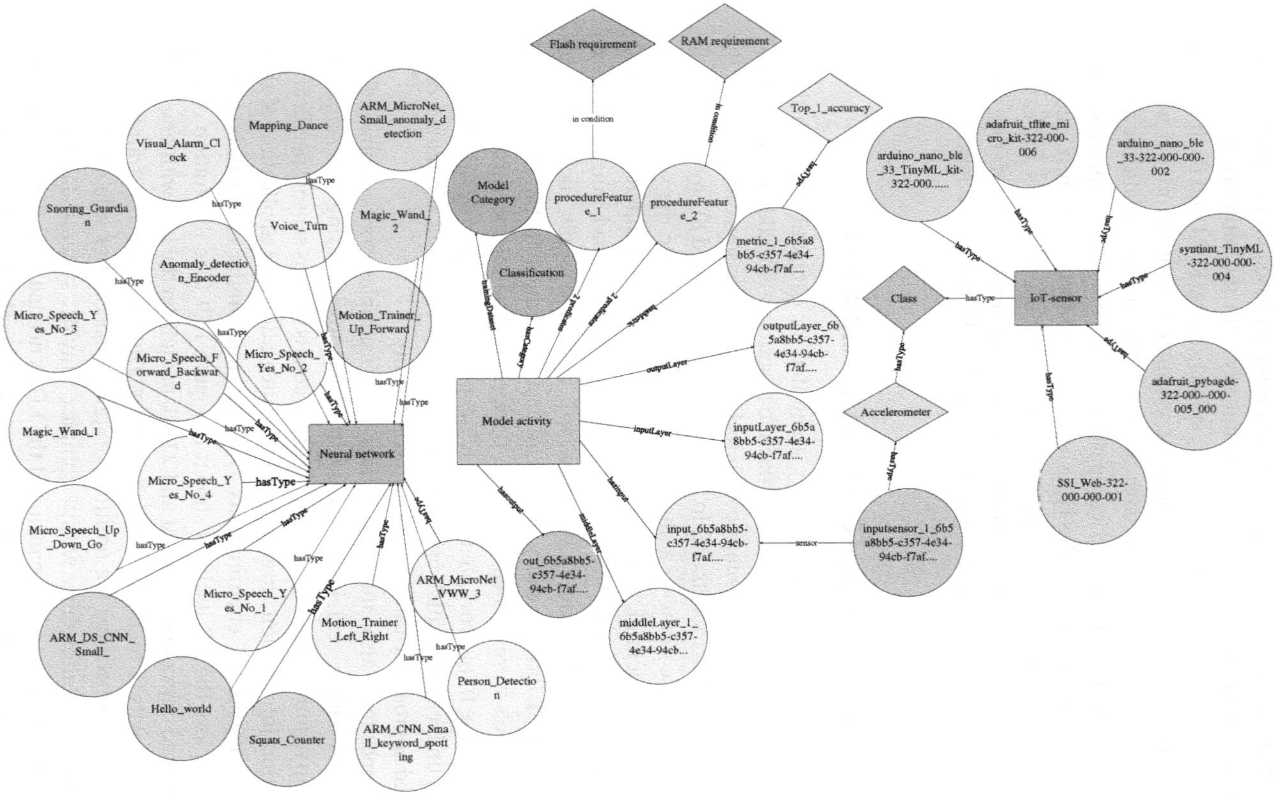

Figure 13.9 Simplified KG for TinyML, showing 23 ML models and six IoT devices.

13.4.3 Case Study 1

Suppose accelerometer and gyroscope data are used to train a TinyML model for a motion classification task. Based on the parsed results, the model requires a minimum of 128 KB of RAM and 556 KB of flash memory. In the field, there are numerous IoT boards, each with different memory capacities and integrated sensors. Instead of manually reviewing each IoT device, a more efficient way to identify which boards can run the model is through a structured query. As shown below, Algorithm 13.1 produces an organized result that pinpoints compatible boards based on the specified memory and sensor requirements.

Algorithm 13.1 Classification of a motion task for TinyML

SELECT ?Device ?MemoryRAM ?MemoryFlash

```
WHERE {
?Device a s3n:IoTSensor ;
ssn:hasSubSystem ?subsystem1 ;
ssn:hasSubSystem ?subsystem2 ;
ssn:hasSubSystem ?subsystem3 .
?subsystem1 a sosa_extend:Accelerometer .
?subsystem2 a sosa_extend:Gyroscope .
?subsystem3 a s3n:MicroController ;
s3n:hasSystemCapability ?capability .
?capability ssn-system:hasSystemProperty ?property1, ?property2 .
?property1 a s3n_extend:RAM ;
schema:value ?MemoryRAM ;
schema:unitCode om:kilobyte .
?property2 a s3n_extend:Flash ;
schema:value ?MemoryFlash ;
schema:unitCode om:kilobyte .
FILTER (?MemoryRAM >= 128)
FILTER (?MemoryFlash >= 556)
}
Result:
Board: 000; RAM: 172 kb; Flash: 628 kb.
Board: 001; RAM: 187 kb; Flash: 785 kb.
```

13.4.4 Case Study 2

The objective is to enable ML capabilities on a customized IoT board. The board is equipped with a camera sensor and has 164 KB of RAM and 596 KB of flash memory available, based on the report of the operating system. Instead of investing significant effort in developing a new model from scratch, it can explore the model zoo to find suitable pre-trained ML models that can operate on this board. Algorithm 13.2 can help identify compatible models from the model zoo that meet the board's specifications.

Algorithm 13.2 Empower a specified IoT board with ML functionality.

SELECT ?id ?MACs ?MemoryRAM ?MemoryFlash

```
WHERE {
?nn a nnet:NeuralNetwork ;
schema:identifier ?id ;
ssn:hasInput ?inputData ;
nnet:hasMultiplyAccumulateOps ?MACs ;
s3n:hasProcedureFeature ?feature1, ?feature2 .
?feature1 ssn-system:inCondition ?condition1 .
?feature2 ssn-system:inCondition ?condition2 .
?condition1 a s3n_extend:RAM ;
schema:minValue ?MemoryRAM ;
schema:unitCode om:kilobyte .
?condition2 a s3n_extend:Flash ;
schema:minValue ?MemoryFlash ;
schema:unitCode om:kilobyte .
?sensor ssn_extend:provideInput ?inputData ;
a sosa_extend:Camera .
FILTER (?MemoryRAM <= 164)
FILTER (?MemoryFlash <= 596)
}
Result:
uuid: e3... ; MACs: 7387976; RAM: 126 kb; Flash: 541 kb.
uuid: 10... ; MACs: 7158144; RAM: 84 kb; Flash: 246 kb.
```

13.4.5 Case Study 3

Suppose you are to explore available ML models for a specific task, such as recognizing "yes/no" commands in speech. At the same time, the aim is to compare the options based on memory requirements, processing delay, and accuracy on a defined dataset. The structure for this scenario is given in Algorithm 13.3 as follows.

Algorithm 13.3 Browser available ML models for a specific task.

SELECT ?id ?Category ?Accuracy ?MACs ?RequiredRAM ?RequiredFlash

```
WHERE {
?nn a nnet:NeuralNetwork ;
schema:identifier ?id ;
schema:description ?desc ;
ssn:hasInput ?inputSignal ;
nnet:hasCategory ?Category ;
nnet:hasMetric ?metric ;
nnet:hasMultiplyAccumulateOps ?MACs ;
nnet:trainingDataset ?dataset ;
s3n:hasProcedureFeature ?feature1, ?feature2 .
?metric a nnet:Top_1_accuracy ;
nnet:hasMetricValue ?Accuracy .
?feature1 ssn-system:inCondition ?condition1 .
?feature2 ssn-system:inCondition ?condition2 .
?condition1 a s3n_extend:RAM ;
schema:minValue ?RequiredRAM ;
schema:unitCode om:kilobyte .
?condition2 a s3n_extend:Flash ;
schema:minValue ?RequiredFlash ;
schema:unitCode om:kilobyte .
?sensor ssn_extend:provideInput ?inputSignal ;
a sosa_extend:Microphone .
FILTER regex(?desc, "yes/no", "i")
FILTER regex(str(?dataset), "speech_commands", "i")
}ORDER BY ?Accuracy
Result:
uuid: 3e...; Category: nnet:Classification; Acc: 0.82;
MACs: 381824; Min_RAM: 9.7 kb; Min_Flash: 9.9 kb.
... (Additional results have been excluded for brevity)
```

13.5 Results and Discussion

13.5.1 TinyEngine: A Compact and Memory-Optimized Inference Library

Previously, it was believed that variations in deep learning frameworks only influenced inference speed without affecting accuracy. However, in TinyML, the choice

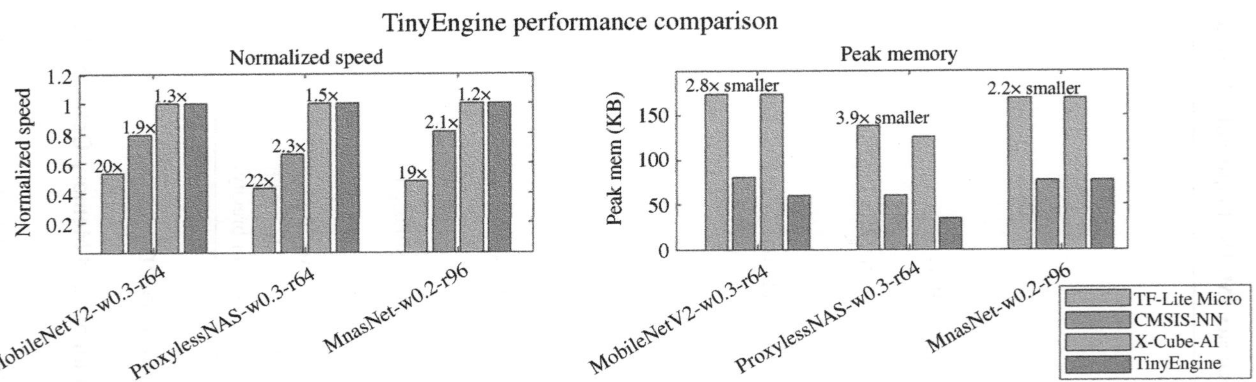

Figure 13.10 TinyEngine achieves higher inference efficiency and reduced memory usage.

of inference library plays a crucial role in both accuracy and latency. Effective frameworks like TinyEngine [49] maximize the limited resources of MCUs, minimize memory usage, and expand the design search space, enabling the creation of higher-accuracy models. Unlike interpreter-based libraries such as TF-Lite Micro and CMSIS-NN, which require additional memory for storing meta-information, TinyEngine adopts a code generation approach. This eliminates the need for runtime interpretation, freeing up memory and enhancing performance. Compared to CMSIS-NN, TinyEngine boosts inference efficiency by 22% and reduces memory usage by 2.1×, as shown in Figure 13.10.

TinyEngine's compact binary size ensures exceptional memory efficiency for MCUs. Popular deep learning programming languages such as Python and Cython often produce models whose dependencies and package sizes exceed the flash memory capacity of MCUs, making direct execution impractical. In contrast, interpreter-based frameworks like TF-Lite Micro introduce significant redundancy by preparing code for every operation, such as convolution and softmax, to enable cross-model inference, even for unused operations. TinyEngine takes a different approach by compiling only the specific operations a given model utilizes into the binary. This leverages the inherent advantages of compilation over interpretation and introduces model-specific optimization and specialization. As illustrated in Figure 13.11, this adaptive, model-focused compilation strategy reduces code size by up to 4.5× compared to TF-Lite Micro and up to 5.0× compared to CMSIS-NN.

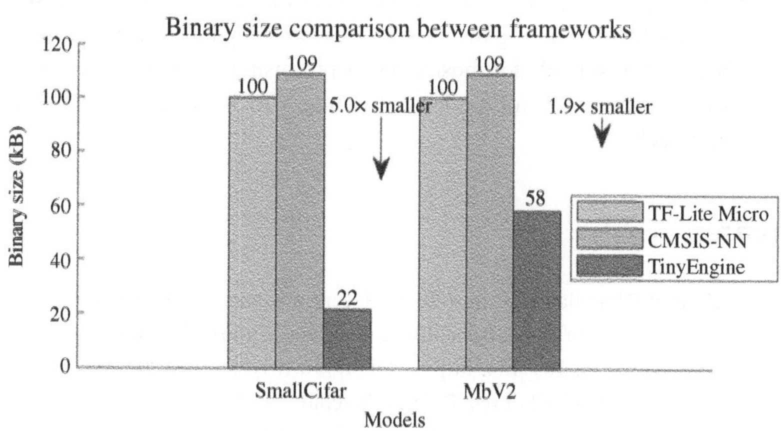

Figure 13.11 Binary size comparison of frameworks and models.

13.6 Conclusions and Future Scope

TinyML enables private, customized AI applications directly at the data source, bringing deep learning capabilities to resource-limited devices. Key challenges include redesigning model architectures, adapting learning algorithms for lightweight training, and co-optimizing models with HW. Recent advancements in TinyML allow AI applications on IoT systems to perform continuous learning and efficient inference. Despite challenges in co-managing ML models and IoT devices, TinyML brings intelligence to embedded systems. TinyML will require ongoing efforts to enhance energy efficiency and performance as a rapidly evolving field. This chapter introduces a scalable, interpretable, and shareable knowledge framework crucial for advancing the IoT by integrating TinyML with Semantic Web technology. A tailored semantic schema for TinyML NN models, aligned with the W3C WoT TD semantic schema, provides a unified information model for TinyML systems. Tools and techniques for constructing and maintaining a TinyML KG were validated on six devices and 23 ML models and further assessed in three case studies. The framework also supports a low-code platform for scalable and accessible TinyML management. Future work will focus on incorporating sensor-specific settings for more precise knowledge management. The framework will be extended to support a broader range of TinyML libraries and enable TinyML as a service for seamless model deployment across platforms. Efforts will also be made to ensure the toolchain and framework are accessible to all users. Additionally, the integration of graph NNs with TinyML will be explored to enhance IoT systems, particularly for large-scale anomaly detection. Accurately labeling data for on-device learning presents a significant challenge, particularly for applications where supervision is costly, such as in vision tasks. While techniques like predicting the next input word may be effective in some contexts, they are not suitable for most tasks. A promising solution is the use of self-supervised learning tasks for on-device training, which has gained attention as a viable approach.

References

1 Sinha, B.B. and Dhanalakshmi, R. (2022). Recent advancements and challenges of Internet of Things in smart agriculture: A survey. *Future Generation Computer Systems* 126: 169–184. https://doi.org/10.1016/j.future.2021.08.006.

2 Liao, Y., de Freitas Rocha Loures, E., and Deschamps, F. (2018). Industrial Internet of Things: A systematic literature review and insights. *IEEE Internet of Things Journal* 5 (6): 4515–4525. https://doi.org/10.1109/JIOT.2018.2834151.

3 Branco, S., Ferreira, A.G., and Cabral, J. (2019). Machine learning in resource-scarce embedded systems, FPGAs, and end-devices: A survey. *Electronics* 8 (11): 1289. https://doi.org/10.3390/electronics8111289.

4 Lin, J., Chen, W.-M., Lin, Y. et al. (2020). MCUNet: tiny deep learning on IoT devices. *Proceedings of the Advances in Neural Information Processing Systems(NeurIPS)*, NY, USA,vol. 33, pp. 11711–11722. https://doi.org/10.48550/arXiv.2007.10319

5 Warden, P. and Situnayake, D. (2019). *TinyML: Machine Learning With TensorFlow Lite on Arduino and Ultra-Low-Power Microcontrollers.* O'Reilly Media.

6 Statista. (2021). Microcontroller unit (MCU) shipments worldwide from 2015 to 2023.

7 Pramod, A., Naicker, H.S., and Tyagi, A.K. (2021). Machine learning and deep learning: open issues and future research directions for the next 10 years. In: *Computational Analysis and Deep Learning Medical Care: Principles, Methods, and Application* (ed. A.K. Tyagi), 463–490. Wiley https://doi.org/10.1002/9781119785750.ch18.

8 Thompson, N.C., Greenewald, K., Lee, K., and Manso, G. F. (2022). The computational limits of deep learning. arXiv: 2007.05558, July. https://doi.org/10.48550/arXiv.2007.05558.

9 Dutta, L. and Bharali, S. (2021). TinyML meets IoT: a comprehensive survey. *Internet of Things* 16: 100461. https://doi.org/10.1016/j.iot.2021.100461.

10 Rajapakse, V., Karunanayake, I., and Ahmed, N. (2023). Intelligence at the extreme edge: A survey on reformable TinyML. *ACM Computing Surveys* 55 (13s): 1–30. https://doi.org/10.1145/3583683.

11 Bringmann, O., Ecker, W., Feldner, I. et al. (2021). Automated HW/SW co-design for edge AI: State, challenges and steps ahead. *Proceedings of the International Conference on Hardware/Software Co-Design and System Synthesis*, Austin, TX, USA, pp. 11–20.

12 Charpenay, V., Kabisch, S., and Kosch, H. (2016). Introducing thing descriptions and interactions: an ontology for the web of things. *Proceedings of the International Semantic Web Conference*, Kobe, Japan, pp. 55–66.

13 W3C. (2021). W3C Web of Thing Ontology.

14 Shafique, M., Theocharides, T., Reddy, V.J., and Murmann, B. (2021). TinyML: current progress, research challenges, and future roadmap. *Proceedings of the 58th ACM/IEEE Design Automation Conference (DAC)*, San Francisco, CA, USA, pp. 1303–1306. https://doi.org/10.1109/DAC18074.2021.9586232.

15 Ren, H., Anicic, D., and Runkler, T.A. (2021). TinyOL: TinyML with online-learning on microcontrollers. *Proceedings of the International Joint Conference on Neural Networks(IJCNN)*, Shenzhen, China, pp.1–8. https://doi.org/10.1109/ijcnn52387.2021.9533927.

16 Jiao, B., Zhang, J., Xie, Y. et al. (2021). A 0.57-GOPS/DSP object detection PIM accelerator on FPGA. *Proceedings of the 26th Asia and South Pacific Design Automation Conference (ASPDAC '21)*, Tokyo, Japan, pp. 13–14. https://doi.org/10.1145/3394885.3431659.

17 Jain, V., Jadhav, N., and Verhelst, M. (2022). Enabling real-time object detection on low cost FPGAs. *Journal of Real-Time Image Processing* 19 (1): 217–229. https://doi.org/10.1007/s11554-021-01177-w.

18 Corneliou, P., Nikolaou, P., Michael, M., and Theocharides, T. (2021). Fine-grained vulnerability analysis of resource constrained neural inference accelerators. *Proceedings of the IEEE International Symposium on Defect and Fault Tolerance in VLSI and Nanotechnology Systems (DFT)*, Athens, Greece, pp. 1–6. https://doi.org/10.1109/dft52944.2021.9568281.

19 David, R., Duke, J., Jain, A. et al. (2021). TensorFlow lite micro: embedded machine learning on TinyML systems. *Proceedings of Machine Learning and Systems* 3: 800–811. https://doi.org/10.48550/arXiv.2010.08678.

20 Herath, S., Irandoust, S., Chen, B. et al. (2021). Fusion-DHL: WiFi, IMU, and floorplan fusion for dense history of locations in indoor environments. *Proceedings of the IEEE International Conference on Robotics and Automation (ICRA)*, Xi'an, China, pp. 5677–5683. https://doi.org/10.1109/ICRA48506.2021.9561115.

21 Mitchell, M., Wu, S., Zaldivar, A. et al. (2019). Model cards for model reporting. *Proceedings of the Conference on Fairness, Accountability, and Transparency (FAT*'19), Atlanta, GA, USA, pp. 220–229. https://doi.org/10.1145/3287560.3287596.

22 Sudharsan, B., Yadav, P., Breslin, J.G., and Intizar Ali, M. (2021). An SRAM optimized approach for constant memory consumption and ultra-fast execution of ML classifiers on TinyML hardware. *Proceedings of the IEEE International Conference on Services Computing (SCC)*, Chicago, IL, USA, pp. 319–328. https://doi.org/10.1109/SCC53864.2021.00045.

23 Schelter, S., Biessmann, F., Januschowski, T. et al. (2018). On challenges in machine learning model management. *IEEE Data Engineering Bulletin* 41: 5–15. https://doi.org/10.5555/3375993.3375995.

24 Pandey, R. (2021). *Semantic IoT: Theory and Applications*. Springer Nature.

25 Ren, H., Anicic, D., and Runkler, T.A. (2022). Towards semantic management of on-device applications in industrial IoT. *ACM Transactions on Internet Technology* 22 (4): 1–30. https://doi.org/10.1145/3510820.

26 Doyu, H., Morabito, R., and Brachmann, M. (2021). A TinyMLaaS ecosystem for machine learning in IoT: overview and research challenges. *Proceedings of the International Symposium on VLSI Design, Automation and Test (VLSI-DAT)*, IEEE, Hsinchu, Taiwan, pp. 1–5. https://doi.org/10.1109/vlsi-dat52063.2021.9427352.

27 Liu, Z., Zhang, X., Chen, Y., and Wu, J. (2020). Data mix: efficient privacy-preserving edge-cloud inference. *Proceedings of the European Conference on Computer Vision (ECCV)*, Glasgow, Scotland, pp. 5–20. https://doi.org/10.1007/978-3-030-58621-8_34.

28 Kalapothas, S., Galetakis, M., Flamis, G. et al. (2023). A survey on RISC-V-based machine learning ecosystem. *Information* 14 (2): 64. https://doi.org/10.3390/info14020064.

29 Sanchez-Iborra, R. and Skarmeta, A.F. (2020). TinyML-enabled frugal smart objects: challenges and opportunities. *IEEE Circuits and Systems Magazine* 20 (3): 4–18. https://doi.org/10.1109/MCAS.2020.3005467.

30 Giordano, M., Piccinelli, L., and Magno, M. (2022). Survey and comparison of milliwatts microcontrollers for tiny machine learning at the edge. *Proceedings of the IEEE 4th International Conference Artificial Intelligence Circuits and Systems (AICAS)*, Incheon, Korea Republic, pp. 94–97. https://doi.org/10.1109/AICAS54282.2022.9870017.

31 Prakash, S., Callahan, T., Bushagour, J. et al. (2023). CFU playground: full-stack open-source framework for Tiny machine learning (TinyML) acceleration on FPGAs. *Proceedings of the IEEE International Symposium Performance Analysis of Systems and Software (ISPASS)*, Raleigh, NC, USA, pp. 157–167. https://doi.org/10.1109/ISPASS57527.2023.00024.

32 Capogrosso, L., Cunico, F., Cheng, D.S. et al. (2024). A machine learning-oriented survey on tiny machine learning. *IEEE Access* 12: 23406–23426. https://doi.org/10.1109/ACCESS.2024.3365349.

33 Ren, H., Anicic, D., and Runkler, T.A. (2023). TinyReptile: TinyML with federated meta-learning. arXiv preprint arXiv:2304.05201.

34 Garofalo, A., Rusci, M., Conti, F. et al. (2020). PULP-NN: accelerating quantized neural networks on parallel ultra-low-power RISC-V processors. *Philosophical Transactions of the Royal Society A* 378 (2164): 20190155. https://doi.org/10.1098/rsta.2019.0155.

35 Zhou, C., Redondo, F.G., Büchel, J. et al. (2022). ML-HW co-design of noise-robust TinyML models and always-on analog compute-in-memory edge accelerator. *IEEE Micro* 42 (6): 76–87. https://doi.org/10.1109/MM.2022.3198321.

36 Zhao, H., Sun, X., Dong, J. et al. (2022). Highlight every step: knowledge distillation via collaborative teaching. *IEEE Transactions on Cybernetics* 52 (4): 2070–2081. https://doi.org/10.1109/TCYB.2020.3007506.

37 Liu, Y., Sun, Y., Xue, B. et al. (2023). A survey on evolutionary neural architecture search. *IEEE Transactions on Neural Networks Learning Systems* 34 (2): 550–570. https://doi.org/10.1109/TNNLS.2021.3100554.

38 Abououf, M., Mizouni, R., Singh, S. et al. (2022). Self-supervised online and lightweight anomaly and event detection for IoT devices. *IEEE Internet of Things Journal* 9 (24): 25285–25299. https://doi.org/10.1109/JIOT.2022.3196049.

39 Shawahna, A., Sait, S.M., and El-Maleh, A. (2019). FPGA-based accelerators of deep learning networks for learning and classification: a review. *IEEE Access* 7: 7823–7859. https://doi.org/10.1109/ACCESS.2018.2890150.

40 David, R., Duke, J., Jain, A. et al. (2021). TensorFlow lite micro: embedded machine learning for TinyML systems. *Proceedings of Machine Learning and Systems* 3: 800–811.

41 Sandler, E., Hu, M., Panda, N. et al. (2018). MobileNetV2: inverted residuals and linear bottlenecks. *Proceedings of the IEEE/CVF Conference on Computer Vision and Pattern Recognition*, Salt Lake City, UT, USA, pp. 4510–4520. https://doi.org/10.1109/CVPR.2018.00474.

42 Lin, J., Chen, W.-M., Cai, H. et al. (2021). MCUNetV2: memory-efficient patch-based inference for tiny deep learning. *Proceedings Neural Information Processing Systems*, USA.

43 Schizas, N., Karras, A., Karras, C., and Sioutas, S. (2022). TinyML for ultra-low power AI and large scale IoT deployments: A systematic review. *Future Internet* 14 (12): 363. https://doi.org/10.3390/fi14120363.

44 Abadade, Y., Temouden, A., Bamoumen, H. et al. (2023). A comprehensive survey on TinyML. *IEEE Access* 11: 96892–96922. https://doi.org/10.1109/ACCESS.2023.3294111.

45 Lin, J., Zhang, Y., Liu, H. et al. (2020). MCUNet: tiny deep learning on IoT devices. *Proceedings of the Conference Neural Information Processing Systems (NeurIPS)*, Vancouver, BC, Canada, pp. 1–12. https://doi.org/10.48550/arXiv.2007.10319

46 He, K., Zhang, X., Ren, S., and Sun, J. (2016). Deep residual learning for image recognition. *Proceedings of the IEEE Conference on Computer Vision and Pattern Recognition (CVPR)*, Las Vegas, NV, USA, pp. 770–778. https://doi.org/10.1109/CVPR.2016.90.

47 Ren, H., Anicic, D., and Runkler, T.A. (2022). How to manage tiny machine learning at scale: an industrial perspective. ArXiv, abs/2202.09113. https://doi.org/10.48550/arXiv.2202.09113.

48 Noy, N.F. and McGuinness, D.L. (2001). Ontology development 101: a guide to creating your first ontology. *Knowledge Systems Laboratory* 32 (1): 1.

49 Lin, J., Zhu, L., Chen, W.-M. et al. (2023). Tiny machine learning: progress and futures. *IEEE Circuits and Systems Magazine* 23 (3): 8–34. https://doi.org/10.1109/MCAS.2023.3302182.

14

Fight Poison with Poison: Tiny Machine Learning Resilience Against Poisoning Attacks*

Tomoki Chiba[1], Yasuyuki Tahara[1], Akihiko Ohsuga[1], Agbotiname Lucky Imoize[2,3], and Yuichi Sei[1]

[1]Graduate School of Informatics and Engineering, The University of Electro-Communications, Chōfu, Tokyo, Japan
[2]Department of Electrical and Electronics Engineering, Faculty of Engineering, University of Lagos, Lagos, Nigeria
[3]Department of Electrical Engineering and Information Technology, Institute of Digital Communication, Ruhr University, Bochum, Germany

14.1 Introduction

Machine learning technology has emerged as an auxiliary tool for human decision-making processes. Machine learning infers patterns and features from large datasets. Systems that use machine learning have become commonplace and are currently used in many fields, such as autonomous driving and health monitoring [2–4]. Thus, machine learning is increasingly being integrated into many systems, assuming a role akin to social infrastructure. Moreover, the use of machine learning technology to solve cybersecurity problems has been proposed, and such solutions are already in use [5, 6]. Therefore, the safety of the systems and services that use machine learning technology (hereinafter referred to as "machine learning systems") has become extremely important. If there are vulnerabilities in machine learning systems, other systems that have implemented these machine learning systems could be adversely affected.

Considering these circumstances, as the popularity of machine learning systems grows, the value of attackers who target machine learning systems is increasing. The methods of attack mainly include targeted attacks, which generate mislabeled data and mix it into the training data, and indiscriminate attacks, which cause general misclassification and lead to denial of service, among other things.

* An earlier version of the chapter appeared in Chiba et al. [1].

Tiny Machine Learning: Design Principles and Applications, First Edition.
Edited by Agbotiname Lucky Imoize, Dinh-Thuan Do, and Houbing Herbert Song.

The methods of attack on machine learning systems are diverse, and some are difficult to visually detect (Section 14.3.1). Therefore, in the current era of smart cities, if machine learning systems are introduced without countermeasures against such attacks, there is a chance that large-scale accidents or incidents causing significant financial damage and even loss of life could occur.

Extensive research has focused on attacks on and defenses for machine learning systems and on disseminating knowledge about the vulnerabilities of machine learning systems [7–9]. Various methods of attack on machine learning systems have been reported [10, 11]. The current study focuses on poisoning attacks. A poisoning attack is an attack that degrades the predictive accuracy of a model by injecting harmful data (poisonous data) into the training data [12, 13].

This study aims to mitigate the degradation of the accuracy of machine learning models by poisoning attacks. Specifically, the goals of this study are as follows:

- To reduce the degradation of model accuracy due to poisoning attacks.
- To stabilize the accuracy of detecting the harmful data used in poisoning attacks.

In this study, we propose a defense method to achieve these goals, and we verify the effectiveness of the proposed method through evaluation experiments.

The contributions of this study are as follows:

- We propose a novel approach that reverses the use of poisoning attacks to defend the model.
- We reduce the false detection of actual data by prioritizing the removal of data.
- We demonstrate the effectiveness of the proposed method against poisoning attacks through validation experiments in the assumed environment.

The introduction of tiny machine learning (TinyML), which deploys machine learning models on resource-constrained devices, introduces additional layers of complexity and also potential vulnerabilities to the security landscape. As TinyML systems become integral to various applications, from smart home devices to wearable health monitors, ensuring the security and robustness of these systems is also becoming paramount [14].

TinyML systems operate, by design, under extreme constraints in terms of computational power, memory, and energy consumption [15]. These constraints make necessary the use of highly optimized and often simplified machine learning models, which can inadvertently introduce new attack surfaces. For instance, the low complexity of models in TinyML systems may make them more susceptible to certain types of poisoning attacks that exploit these simplifications.

Furthermore, the deployment environments for TinyML systems are often less controlled than those of traditional machine learning systems, as these devices are common and users interact with them directly. This increases the potential number of attack vectors and the ease with which malicious actors can introduce poisonous data. Therefore, the security strategies employed must be lightweight yet effective and must balance the need for robust defenses with the limited resources available.

In this context, our focus on developing efficient detection and mitigation strategies for poisoning attacks is quite relevant. The proposed defense method, which aims to minimize model degradation and accurately identify harmful data, is particularly suited for the TinyML paradigm. By ensuring that the defensive mechanisms of TinyML systems are computationally efficient and do not overly burden the limited resources of TinyML devices, we can enhance the security of these systems without compromising their performance or functionality.

Additionally, the practical implications of ensuring the security of TinyML systems are significant. As these devices are often connected to critical infrastructure and personal environments, any successful attack could have far-reaching consequences and impact not just individual users but entire communities. Hence, our efforts to improve the resilience of machine learning systems against poisoning attacks are crucial for the broader adoption of and trust in TinyML technologies.

In conclusion, the integration of robust, lightweight security measures into TinyML systems is essential to safeguarding the next generation of smart devices. Our proposed method contributes to this goal by addressing the unique challenges posed by the resource constraints and deployment scenarios of TinyML and ensures that these innovative systems are powerful and secure.

The structure of this chapter is as follows. In Section 14.2, we present the assumed environment and threat model. Section 14.3 introduces related research and explores the differences between this study and previous studies. The proposed method is described in Section 14.4, and the experimental setup and results are presented in Section 14.5. Section 14.6 discusses the proposed method. Finally, the conclusion to this study is summarized in Section 14.7.

14.2 Problem Definition

In this section, we discuss the assumed environment and the threat model.

14.2.1 Assumed Environment

Machine learning systems are utilized in various scenarios. The use cases are particularly numerous in IoT environments, such as autonomous driving and automated robot control. In these contexts, if a machine learning system is attacked as described in Section 14.1, the damage could be considerable.

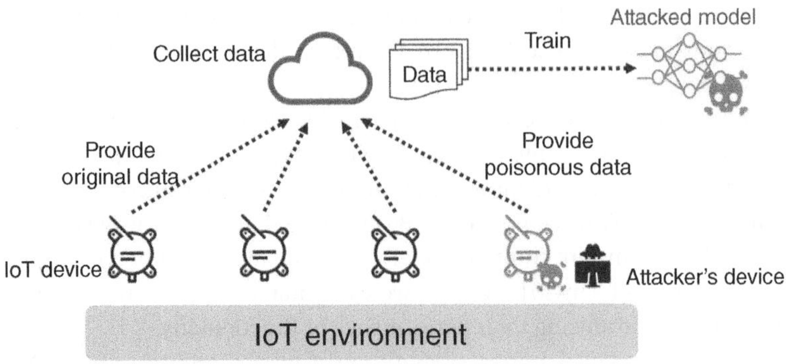

Figure 14.1 Assumed environment.

In this study, we assume an IoT environment in which the data used for model training is obtained from multiple sources. Our assumed environment is illustrated in Figure 14.1. As an example, we consider data obtained from multiple IoT devices. Here, we envision a context in which a company builds a road sign recognition model for autonomous vehicles. This recognition model is installed in the autonomous vehicle to identify road signs, and the vehicle's speed is adjusted according to the recognition results. In this assumed environment, the attacker executes a poisoning attack by intervening in the data collection process.

Furthermore, this study assumes there exist multiple entities or IoT devices with access to the dataset used for training, and some of them have malicious intent. In other words, some data providers supply harmful data, thereby launching a poisoning attack on the model being constructed by the data collector. Hereinafter, the data providers will be referred to as "devices."

There are no limitations on the kinds of machine learning systems that we consider to be the target of attacks and defenses. Due to the mechanism of poisoning attacks, the attacker must access the training data before the model is trained and used. Therefore, machine learning systems that learn online easily serve as targets. However, the types of scenarios in which attackers are able to access training data differ. For instance, if the attacker has access to a provider who builds and offers the model as a service, poisoning attacks can occur even if the machine learning system does not learn online.

14.2.2 Threat Model

The attacker assumed in this study intends to perform a poisoning attack, which involves injecting harmful data into the training dataset to undermine the accuracy of the model. The objective of the assumed poisoning attack is not

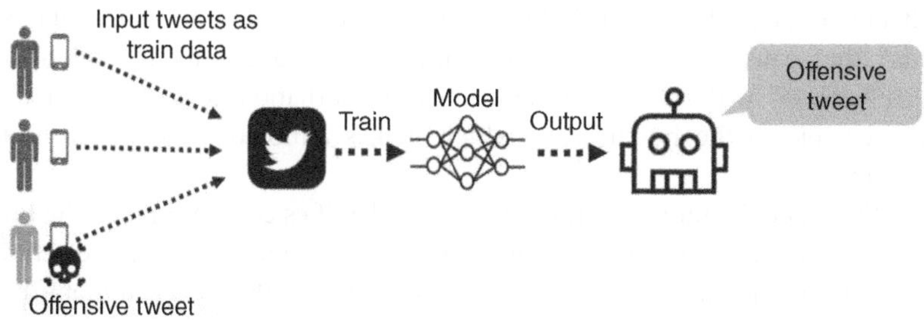

Figure 14.2 Example of damage by poisoning attack in the assumed environment.

the intentional misclassification of specific data by the attacker but an overall degradation of accuracy. This objective is based on the attacker's intention to cause significant events, such as accidents involving autonomous vehicles, in the assumed environment.

Moreover, the attacker is presumed to have background knowledge about the learning algorithm and to be able to draw or generate data from the distribution of the data used for training. Although this setting may seem unrealistic in general, using proxy training data from the same distribution would be feasible in a real-world setting [16].

An example of a poisoning attack in a scenario where training data is obtained from multiple devices is illustrated with the case of the Microsoft chatbot "Tay" in Figure 14.2. Tay was a program designed to generate human-like tweets by learning from conversations with humans. However, some individuals intentionally taught Tay offensive expressions, which led Tay to tweet similarly offensive content. Consequently, Microsoft decided to terminate the service approximately 16 hours after its launch. In our study, we assume data is collected from multiple devices, but the attacker is able to access only a subset of these devices.

14.3 Related Work

We introduce related work in this section.

14.3.1 Poisoning Attacks

As shown in Section 14.1, poisoning attacks involve the insertion of harmful data (poisonous data) into a training dataset to degrade the predictive accuracy of a machine learning model [17]. In this context, harmful data is data whose features have been deliberately modified to cause the model to misclassify normal input data. An example of data that has been intentionally modified is shown in

Goodfellow et al. [18]. Here, a slight perturbation is added to an image labeled "panda" that causes a normally functioning model to classify the modified data as "gibbon" with 99.3% confidence. The perturbation is so subtle that it is undetectable to the human eye, which makes annotation-based detection challenging.

After the model is trained on this data, it misclassifies certain inputs. The targets of poisoning attacks mainly include spam detection and malware detection systems, but online recommendation systems have recently also been identified as targets [19].

For poisoning attacks that occur during model training, the attacker must have access to the training data. For example, online tools for detecting malware in PDFs use feedback from end users to verify the detection results for the provided data [20]. Some systems update classifiers based on these verification results [21]. In this context, end users have access to the training data for the classification system. Another example is web-based honeypots, which collect malware data for training purposes [17]. Here, too, although it is not possible to obtain different data, arbitrary data can be added to the training dataset. Poisoning attacks have also occurred in scenarios where data is collected via crowdsensing [22], and methods of accessing training data may vary depending on the environment. An example of a poisoning attack is shown in Figure 14.3.

Another variant of the poisoning attack is the backdoor attack. This attack method inserts hidden triggers (backdoor keys) into the model during training, which do not affect the normal behavior of the model. The model is then manipulated so that it performs unexpected actions only when the backdoor key is added to the input data. Unlike regular poisoning attacks, backdoor attacks do not cause the model to misclassify specific backdoor-keyed data [23]. Therefore, the approaches to defending against poisoning attacks and backdoor attacks generally differ.

As mentioned in Section 14.1, poisoning attacks are classified mainly as two types: targeted attacks and indiscriminate attacks. The objectives of indiscriminate attacks are to undermine the integrity of the target system and to reduce its

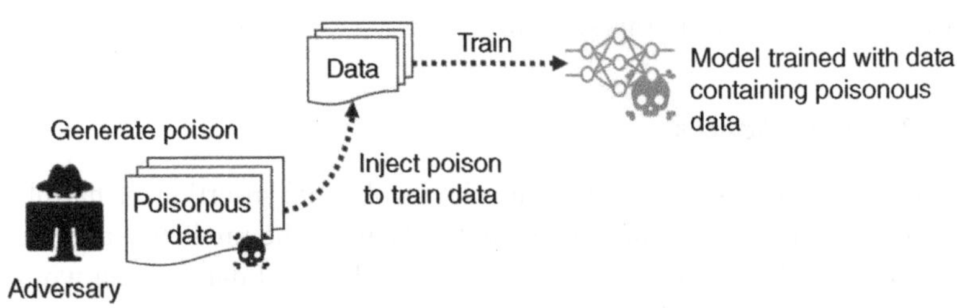

Figure 14.3 Example of a poisoning attack.

availability. Targeted attacks seek to maintain normal operation while causing the misclassification of specific samples [24].

Indiscriminate attacks, on the other hand, are intended to reduce the overall predictive accuracy of the model.

The transferability of poisoning attacks has also been studied [25]. Shafahi et al. demonstrated that even if an attacker cannot access the target model, they may still create an attack method in a white-box environment for use on a self-generated model that exhibits the same behavior as the target model that would be effective against the target model as well [26].

The expected damage from poisoning attacks depends heavily on the task executed by the target model. There are challenges related to the feasibility of the attacks, such as the right of access to the training data and the difficulty for the model creators and users to verify the training data, which indicates that the definition of feasibility is still unclear.

14.3.2 Defensive Methods Against Poisoning Attacks

Various approaches to defending against poisoning attacks have been studied [27]. The primary objective of existing methods for defense against poisoning attacks is to detect data that negatively impacts the model's accuracy [23]. Other approaches include sanitizing outliers [28], but several problems have been noted [29]. Therefore, the defense method proposed in this paper is intended to mitigate the degradation of model accuracy by detecting data that adversely affects the accuracy and the data sources that provide such data, which is similar to the former approach.

Van et al. proposed a robust training approach that mitigated the effects of data poisoning attacks on machine learning models [30]. They introduced the Healthy Influential Noise-Based Training (HINT) method, which leverages influence functions to generate "healthy noise" that hardens the model against adversarial data without significantly affecting its generalization ability. They demonstrated the effectiveness of HINT using extensive evaluations on two image datasets, and they showed that the method efficiently protects models from both untargeted and targeted poisoning attacks. Wu et al. investigated the vulnerabilities of reward model learning to malicious preference poisoning attacks [31]. The authors proposed two algorithmic approaches – a gradient-based framework and a rank-by-distance method – and demonstrated their effectiveness in three application domains: autonomous control, recommendation systems, and textual prompt-response learning. Their findings revealed significant vulnerabilities, where some attacks achieved a success rate of up to 100% with minimal poisoned data.

Although our scenario assumes that training data is collected from multiple devices, defensive methods against poisoning attacks under similar assumptions in IoT environments have been studied [32]. Baracaldo et al. envisioned sensing data from multiple IoT devices aggregated in the cloud for use as training data for models. Among these IoT devices, some carry malicious intent and send these data to the cloud to execute a poisoning attack. Baracaldo et al.'s study demonstrated that the information that links IoT devices to data (provenance data) is robust against tampering [33], and they validated each IoT device by partitioning the training data according to the provenance data. Given the similarity of the assumed environments, Baracaldo et al.'s method is presented as an existing method in this study. As a foundation to Baracaldo et al.'s defensive method, we mention Nelson et al.'s reject on negative impact (RONI) method [16, 34]. Their study proposed an approach that removed from devices data that negatively impacted spam filter defenses. However, Baracaldo et al. compared the defense performance of their method with that of RONI and demonstrated that their method outperformed RONI. Therefore, our study uses only Baracaldo et al.'s method for comparison. There are many studies that address poisoning attacks; however, the capabilities of poisoning attacks are continually improving, which leads to a cat-and-mouse game between attacks and defense methods. As a result, existing methods may not always be effective against new poisoning attacks. Our proposed method employs the most recent poisoning attack techniques that are available at the time of use to detect such attacks. This novel approach has the advantage that it is adaptable to new attack methods, provided that the attack algorithms are known.

In TinyML, privacy protection is also essential. When information for machine learning is collected from individuals, differential privacy techniques are sometimes used to protect privacy [35]. Differential privacy is a crucial technology that has recently been used by companies such as Google and Apple [36]. However, it has the drawback of being vulnerable to poisoning attacks. Cao et al. introduced the first comprehensive study of poisoning attacks against key-value data with local differential privacy (LDP) protocols [37]. They formulated the attacks as two-objective optimization problems to maximize both the estimated frequency and mean values of the target keys. Their approach, which was called maximal gain attack (MGA), demonstrated good effectiveness in compromising the privacy and integrity of the key-value data. They also proposed and evaluated defenses against these attacks and found them to be somewhat effective, thereby highlighting the need for protective measures that are more robust in LDP protocols. Wu et al. explored the susceptibility of LDP protocols for key-value data to poisoning attacks [38]. They developed new poisoning strategies that targeted the frequency and mean estimation of keys, which proved the efficiency of their attacks through both theoretical and empirical analyses. Their findings suggest that although some defenses can mitigate the impact of these attacks under certain conditions, there remains a significant need to develop stronger

Table 14.1 Comparison between proposed method and existing methods.

	TinyML scenario	Adaptable to latest attack methods	Applicable beyond LDP scenarios
[16, 30–32, 34]	√	—	√
[37–39]	√	—	—
This work	√	√	√

and more adaptive defenses to protect the integrity of LDP-based systems. Sun et al. investigated the vulnerability of LDP protocols to poisoning attacks, and they focused specifically on recovery mechanisms and their robustness [39]. The authors proposed various attack strategies for manipulating the data aggregated by LDP protocols, and they showcased how adversaries could inject fake users to significantly alter the analytic outcomes of the data. They highlighted the challenges and limitations of existing defenses and called for improved methods to detect and mitigate poisoning attempts by fake users.

Table 14.1 shows a comparison of our work with existing methods.

14.3.3 Other Attacks on TinyML

Traditional machine learning algorithms do not consider attacks of an adversarial nature, and their performance degrades significantly when they are used in adversarial settings [20, 40].

In addition to the poisoning attacks mentioned in Section 14.3.1, attacks on machine learning models include evasion attacks [41] and examples of adversarial attacks [42]. These attacks target machine learning systems in a way similar to poisoning attacks. Evasion attacks manipulate the machine learning model so that it fails to detect events that should be detected, particularly in applications like anomaly detection.

Attacks that infer model parameters after training have also been studied [43]. By inferring the parameters of models used in existing services, attackers are able to generate models that respond to inputs similarly to the target model. This technique can be used to generate adversarial examples that cause the target model to reject large amounts of input. Another attack on machine learning systems that is related to privacy invasion is the membership inference attack [44].

Some attacks in TinyML environments generate a large burst of data to overwhelm the system. In IoT environments in particular, where individual device performance is often insufficient, it is critical to identify such data sources quickly and efficiently [45]. As there are a variety of attacks on TinyML, this chapter focuses on attacks that degrade the accuracy of machine learning models through poisoning.

14.4 Proposed Method

We propose our new method in this section.

14.4.1 Overview

As described in Section 14.1, our proposed method aims to mitigate the degradation of model accuracy, caused by poisoning attacks, in a multi-device scenario. To achieve this objective, we partition the data according to device, and we compare the accuracies of the model before and after the data of each device is filtered to define the impact of each device on the model's accuracy. In this context, filtering is defined as the process of excluding from the training data the data obtained from a specific device. Subsequently, a method is proposed for removing data from devices deemed harmful by comparing the defined metrics for each device with their impact on the model. The flowchart for the proposed method is shown in Figure 14.4.

A schematic diagram of the proposed method is shown in Figure 14.5. As described in Section 14.2.1, this study assumes a model built in an environment where data is obtained from multiple IoT devices, and some of these devices

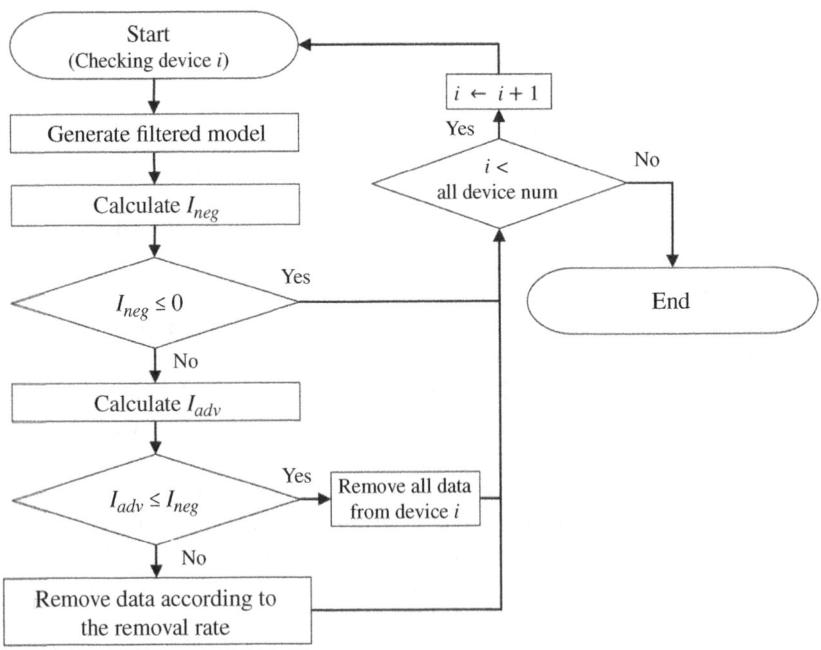

Figure 14.4 Flowchart for the proposed method.

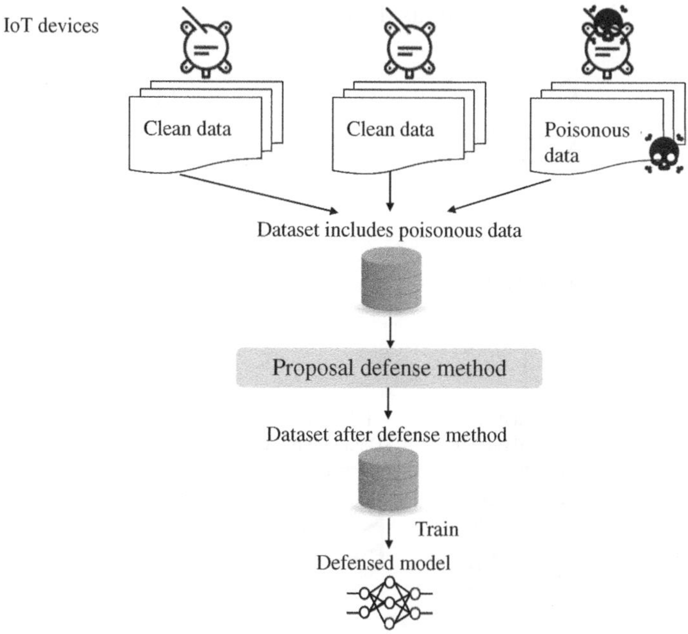

Figure 14.5 Defense flow for the proposed method.

launch poisoning attacks by providing harmful data. In other words, the training data that is used to build the model includes harmful data (the dataset includes poisonous data). The defense system of the proposed method removes the harmful data from the training data. The data that is obtained through this process is referred to as "the dataset" after the defense method, and the model trained on these data is referred to as the "defense model."

The proposed method can be broadly divided into the following three steps:

1) Filter the data from the target device to create filtered training and validation datasets (hereinafter referred to as "filtered train set" and "filtered validation set"). Train a model on the filtered train set to create a filtered model (Figure 14.6). Compare, on the filtered validation set, the accuracies of the filtered model and the unfiltered model to evaluate the impact of the device on the model's accuracy.
2) Create datasets where each device's data is replaced with harmful data; generate adversarial models that are trained on these datasets. Replace the unfiltered model used in step (1) with the adversarial model and calculate the "adversarial impact" by comparing the accuracy of the filtered model.
3) Based on the values defined in steps (1) and (2), derive the confidence level for use in judging the target device as harmful and determine the proportion of

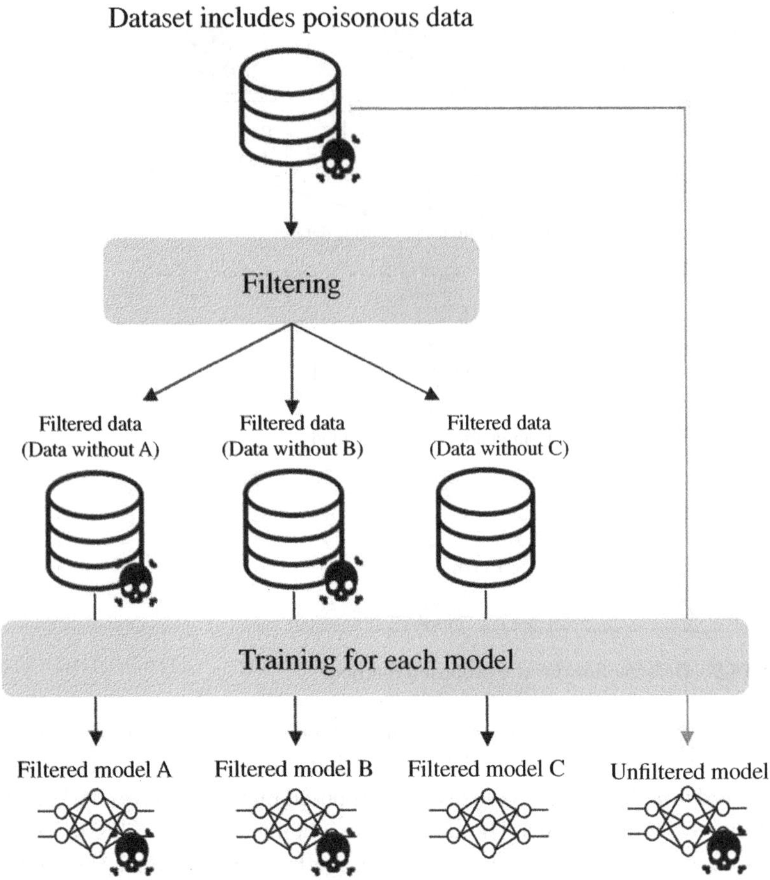

Figure 14.6 Workflow for generating the filtered model.

data to be removed. If this proportion is neither 0% nor 100%, use a data removal priority algorithm to decide which data to remove considering the priority and the amount of data to be removed (Figure 14.7).

The above steps are repeated for each device. The data that remains after removal is considered to be the data after defense, and the model that is trained on these data is the defense model proposed in this method.

The algorithm of the proposed method is presented as Algorithm 14.1. The definitions for the symbols used in this algorithm are shown in Table 14.2. The proposed method receives D, *clf*, and *device_info* as inputs. *device_info* is a list of device identifiers that represent the source for each data sample. The original dataset D is first divided into training and validation data (D_{tr} and D_{val}). The

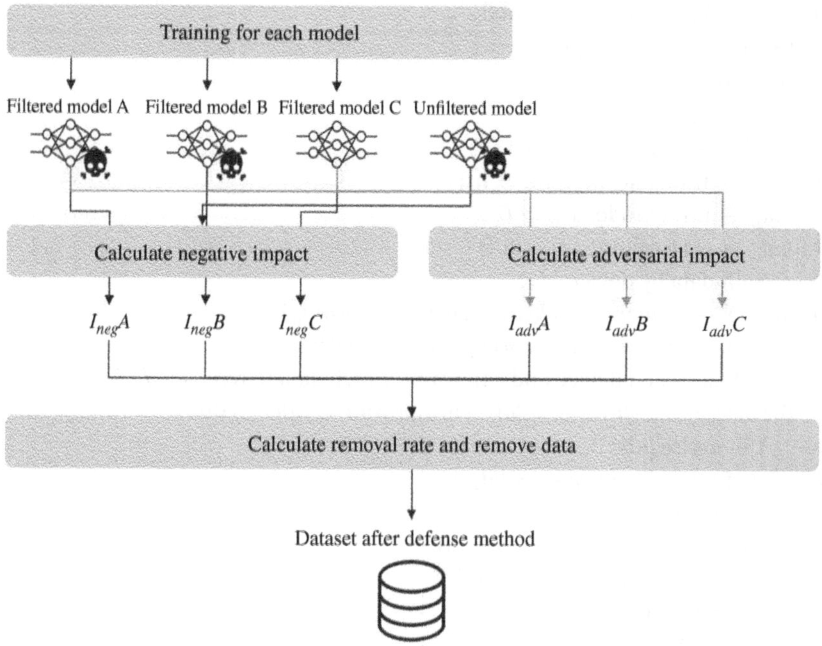

Figure 14.7 Filtering and data generation.

Table 14.2 Notation.

Notation	Description
D	Dataset before protecting
clf	Classifier before training
device_info	Identifier for the device that corresponds to the data
$D_{protected}$	Dataset after protecting by the proposed method

validation data generated here is used as test data to verify the accuracy of the model within the method. Ind_{tr} and Ind_{val} are lists of indices that represent the correspondence between D_{tr} and D_{val} within D, respectively. Next, D_{tr} and D_{val} are divided into data for each device using *device_info* along with Ind_{tr} and Ind_{val}. As the proposed method evaluates the validity of data based on device rather than the data itself, processing the data by device is necessary. The processing for each device then begins. *train_s* and *valid_s* refer to the training and validation data, respectively, that are provided by device i. First, the filtering process

Algorithm 14.1 Proposed method

```
Input: D, clf, device_info
Output: D_protected
M_u, M_f, M_a ← clf.copy()
D_tr, D_val ← train_valid_s(D)
/* Store train and valid index list */
Ind_tr ← List of index D_tr in D
Ind_val ← List of index D_val in D
/* Split data by device */
train_s ← segment_by_class(D_tr, device_info[Ind_tr])
valid_s ← segment_by_class(D_val, device_info[Ind_val])
/* Validate each device and determine which data to remove */
for device_idx, (train_s, valid_s) in zip(train_s, valid_s) do
    i ← device_idx
    n_data ← |train_s| + |valid_s|
    /* Generate filtered data */
    train_f_i ← filtering(D_tr, train_s)
    valid_f_i ← filtering(D_val, valid_s)
    /* Train each model */
    M_u.fit(D_tr)
    M_f.fit(train_f_i)
    /* Calculate I_neg and I_adv */
    I_neg ← calc_neg(valid_f_i, M_u, M_f)
    I_adv ← calc_adv(train_s, train_f_i, valid_f_i, M_f, M_a)
    /* Remove poisonous data */
    D_tr, D_val ← remove_data(D_tr, D_val, train_f_i, valid_f_i, I_neg, I_adv, i, n_data)
end for
D_protected ← append(D_tr, D_val)
return D_protected
```

filtering() defines *filtered data*. The filtering process removes from D_{tr} and D_{val} the data belonging to device *i*. This processing corresponds to the steps shown in Figure 14.6.

In the for loop, the *negative impact* introduced in Section 14.4.2 and the *adversarial impact* described in Section 14.4.3 are calculated. Using these impact values, the data that is selected to be removed is removed using the function *remove_data*(). The function *remove_data*() is described in Section 14.4.4. During this process, the aim is to improve detection accuracy by defining the priority for removing data from among the data deemed to be harmful. By removing data that is deemed to be harmful from the D_{tr} and D_{val} for each device, D_{tr} and D_{val}

ultimately become the datasets after defense by the proposed method. Finally, the two datasets are concatenated and $D_{protected}$ is output. These processing steps are shown in Figure 14.7.

14.4.2 Negative Impact

As indicated in Section 14.3.1, poisoning attacks degrade the model accuracy by injecting harmful data into the training dataset. Therefore, when we search for and remove harmful data from a dataset, one of the criteria to consider is the "impact on the original model's accuracy."

This method uses the RONI [16] concept to evaluate the effect of each device's training data on the accuracy of a model. The effect I_{neg} on model accuracy is defined as the difference between the accuracy of the unfiltered model on the filtered validation set (*unfiltered score*) and the accuracy of the filtered model on the filtered validation set (*filtered score*).

$$I_{neg} = \textit{filtered score} - \textit{unfiltered score} \tag{14.1}$$

The higher the value of I_{neg}, the more negatively it affects the model accuracy. Thus, with the proposed method, the higher the value of I_{neg}, the greater the likelihood that the data is harmful. To prevent the harmful data from having an effect on the validation, the filtered validation set is used to confirm *unfiltered score* and *filtered score*.

If the value of equation 14.1 is less than or equal to 0, the impact on the model accuracy is considered, at the least, not to be negative. In such cases, the device is judged to be normal by the proposed method regardless of the *filtered impact* I_{adv} in Section 14.4.3. The data obtained from devices that are judged not to be normal is not subject to removal; thus, no further steps are taken. Conversely, if I_{neg} is greater than 0, then the data is considered to negatively affect the model accuracy, and the data becomes candidates for removal. The formula for the number of data points to be removed is described in Section 14.4.4.

The function *calc_neg* for calculating I_{neg} is shown in Algorithm 14.2. This function requires the validation data $valid\_f_i$ that is used to calculate model accuracy and the models M_u and M_f. The algorithm simply calculates the accuracy of the two input models on the input validation data, and it stores the accuracy data in $score_u$ and $score_f$, respectively. The process for calculating accuracy corresponds to the function *score*() in the algorithm. The result of Equation 14.1 is then stored in I_{neg} and returned as the output of the function *calc_neg*.

14.4.3 Adversarial Impact

Unlike existing methods, the proposed method defines an adversarial impact for each device. By executing a poisoning attack on the defense side and injecting

Algorithm 14.2 *calc_neg*

1: **Input:** $valid\_f_i, M_u, M_f$
2: **Output:** I_{neg}
3: $score_u \leftarrow score(M_u, valid\_f_i)$
4: $score_f \leftarrow score(M_f, valid\_f_i)$
5: $I_{neg} \leftarrow score_f - score_u$
6: **return** I_{neg}

harmful data into the filtered dataset, the method is able to virtually output the model accuracy if the data being validated is harmful data (referred to as *adversarial score*). The difference between *adversarial score* and *filtered score* is defined as the threshold, as shown in equation 14.2:

$$I_{adv} = \textit{filtered score} - \textit{adversarial score} \tag{14.2}$$

Here, it is assumed that even if an attack is executed by the use of harmful data, the harm caused to the model remains. This assumption is based on our preliminary experimental results. When a normal device is inspected, an executed attack on that device's data renders the data harmful data, which increases the proportion of total data that is harmful data. Conversely, when a harmful device is inspected, an attack on its data does not increase the total amount of harmful data according to the aforementioned assumption. Consequently, the value of *adversarial score* is expected to be low for normal devices and high for harmful devices. Similarly, *filtered score* is expected to be higher for harmful devices than for normal devices. Therefore, the value of I_{adv} itself does not vary significantly between normal and harmful devices. However, compared with I_{neg}, which is a factor in normal/harmful judgments, the *adversarial score* for normal devices is expected to be lower than the *unfiltered score*. As a result, I_{adv} for normal devices does not fall below the value of I_{neg}, thereby preventing the false detection of normal devices. For harmful devices, the values of *adversarial score* and *unfiltered score* are expected to be similar, which makes harmful devices easier to detect than normal devices.

The function *calc_adv* for calculating I_{adv} is shown in Algorithm 14.3. The inputs required by the function are the data for the target device *train_s*, the filtered training and validation data $train\_f_i$ and $valid\_f_i$, the model trained on $train\_f_i$ M_f, and the model before training M_a. Unlike the algorithm for calculating I_{neg}, the function *calc_adv* first needs to generate harmful data. Therefore, harmful data is generated using the function *generate_poison* based on *train_s*. The algorithm specifically for *generate_poison* depends on the attack method used, which is determined by the user of the defense system (the proposed method). In the experiments run for this study, the attack method proposed by Biggio et al. [17]

was used. The generated harmful data is stored in D^p, and then $train\_f_i$ is concatenated with D^p and stored in *adversarial_data*. This way, *adversarial_data* virtually recreates the unfiltered data while assuming device i is a harmful device. Subsequently, the accuracy values $score_f$ and *adversarial_score* are calculated using the function *score*(). The calculated accuracies are then used to compute the value of I_{adv} according to equation 14.2, and the result is stored in I_{adv}, which is returned as the output of function *calc_adv*.

Algorithm 14.3 *calc_adv*

```
1: Input: train_s, train_f_i, valid_f_i, M_f, M_a
2: Output: I_adv
3: D^p ← generate_poison(train_s)
4: adversarial_data ← append(train_f_i, D^p)
5: M_a.fit(adversarial_data)
6: score_f ← score(M_f, valid_f_i)
7: adversarial_score ← score(M_a, valid_f_i)
8: I_adv ← score_f − adversarial_score
9: return I_adv
```

14.4.4 Data Removal

Finally, the proportion of data to be removed is calculated by using the relationship between the impact on model accuracy I_{neg} defined in Section 14.4.2 and I_{adv} defined in Section 14.4.3. The equation for the data removal rate is shown as equation 14.3:

$$removal\ rate = \begin{cases} 0.0 & if\ I_{neg} \leq 0 \\ \frac{I_{neg}}{I_{adv}} & if\ 0 < I_{neg} < I_{adv} \\ 1.0 & else \end{cases} \tag{14.3}$$

In the proposed method, if the value of the negative impact I_{neg} is less than or equal to 0, the device is judged to be normal, and the corresponding data is not removed. This policy is based on the definition of negative impact, where data from a device with a value less than or equal to 0 does not adversely affect (degrade) the accuracy of the other data. When I_{neg} is greater than 0 but less than I_{adv}, the removal rate is defined as the ratio of I_{neg} to I_{adv}. If I_{neg} exceeds 0, it adversely affects the model accuracy and thus becomes a candidate for data removal. However, in this study, we calculate an independent removal rate and apply it to each device instead of removing all the data from a device marked for removal. The value of I_{adv} represents the I_{neg} value that assumes that all data from the target device is harmful. The closer I_{neg} is to I_{adv}, the higher the likelihood that the data from the device is harmful. Therefore, the removal rate is determined from the proximity of I_{neg} to I_{adv}. If I_{neg} exceeds I_{adv}, it is considered to be worse than the value that would be assumed if all the data were harmful, and the removal rate is set to 1.0, which indicates that all the data from the target device will be removed.

After the data removal process, the removed data is excluded from all subsequent data that is used for validating the next device. That is, no data that is deemed harmful exists in the data used after the judgment. Therefore, the data judged as harmful does not affect the judgment of subsequent devices.

The algorithm that is used to calculate the data removal rate is presented as Algorithm 14.4. The function *remove_data* takes the training and validation data D_{tr} and D_{val}, the filtered data $train\_f_i$ and $valid\_f_i$, and the negative impact I_{neg} and adversarial impact I_{adv} used for calculating the removal rate, as well as the number of data points from the target device *n_data* as inputs.

Algorithm 14.4 *remove_data*

Input: D_{tr}, D_{val}, $train\_f_i$, $valid\_f_i$, I_{neg}, I_{adv}, n_data
Output: D_{tr}, D_{val}
if $I_{neg} > 0 \land I_{neg} \geq I_{adv}$ **then**
    /* Remove all data obtained from device i */
    $D_{tr} \leftarrow filtered\_data_i$
    $D_{val} \leftarrow valid\_f_i$
else if $I_{neg} > 0$ **then**
    /* Remove some of the data obtained from device i */
    $removal\_rate \leftarrow I_{neg}/I_{adv}$
    $n\_delete \leftarrow n\_data * removal\_rate$
    $n\_delete\_train, n\_delete\_valid \leftarrow n\_delete/2$
    $pri\_train, pri\_valid \leftarrow det\_priority(removal\_rate)$
    $D_{tr} \leftarrow delete(D_{tr}, n\_delete\_train, pri\_train)$
    $D_{val} \leftarrow delete(D_{val}, n\_delete\_valid, pri\_valid)$
end if
return D_{tr}, D_{val}

This function processes the conditions according to Equation 14.3 to determine the removal rate within those conditions. The first conditional branch of the algorithm checks whether I_{neg} exceeds the value of I_{adv}, which corresponds to the last case in equation 14.3. If this condition is met, the data removal rate is 100%, and the training and validation data is set to the filtered data, which effectively removes the data from the target device. The second conditional branch checks whether I_{neg}

is greater than 0 but does not exceed I_{adv}, which corresponds to the middle case in equation 14.3. If this condition is met, the removal rate is calculated by dividing I_{neg} by I_{adv}. After the removal rate is calculated, the number of data points to be removed is determined using *n_data* on both training and validation data. Then, the function for determining the removal priority is called to store the priority lists in *pri_train* and *pri_valid*. Finally, the function *delete*() is called with the original data to be removed, the number of data points to be removed, and the priority lists to perform data removal according to the priority. After data removal is carried out, the training and validation data D_{tr} and D_{val} are returned. If neither condition is met, then I_{neg} is less than or equal to 0, and no data is removed as the device is judged to be normal.

14.4.5 Removal Priority

Harmful devices can be classified into two types: those that provide only harmful data (all data is harmful) and those that provide both harmful and non-harmful data (some data is harmful). This section describes the scenario where only some of the provided data is harmful.

This functionality is called only when the algorithm in Section 14.4.4 determines that part of the data should be removed. The priority for removing data is set highest for data whose features are furthest away from the opposite side of the Support Vector Machine (SVM) decision boundary. In other words, data that deviates significantly from the SVM prediction is preferentially removed. First, the data from the target device is compressed into two dimensions using principal component analysis (PCA). Then, the distance between the SVM and the mapped data is calculated in the two-dimensional space. The data whose SVM prediction is incorrect is prioritized for removal according to the calculated distance, starting from the largest. If the number of data points that are removed by this method is less than the number of data points to be removed as determined by the removal rate, the data points whose SVM prediction is correct but that are close to the SVM boundary are prioritized for removal.

The algorithm for this functionality is shown in Algorithm 14.5. The function *det_priority* takes as inputs the training and validation data D_{tr} and D_{val}, the device identifier i, the data from the target device *train_*s_i and *valid_*s_i, and the filtered data *train_f* and *valid_f*. Because the removal priority is determined within a two-dimensional plane, dimensionality reduction is needed for data whose dimensionality exceeds two. In this study, PCA is used to reduce the dimensionality of *train_*s_i, *valid_*s_i, and D_{tr}. Next, to determine whether the data is on the opposite side of the SVM decision boundary, an array mesh grid is generated to represent the two-dimensional plane. The model boundary can

Algorithm 14.5 *det_priority*

```
Input: D_{tr}, D_{val}, i, train_s_i, valid_s_i, train_f, valid_f
Output: pri_train, pri_valid
priority_array ← ϕ
clf ← SVC
if D_{tr}.feature > 2 then
    /* Compress features into two dimensions i */
    decomp ← PCA(n_components = 2)
    decomp_tr ← decomp.fit_transform(train_s_i)
    decomp_val ← decomp.fit_transform(valid_s_i)
    decomp_unfil ← decomp.fit_transform(D_{tr})
else
    decomp_tr ← copy(train_s_i)
    decomp_val ← copy(valid_s_i)
    decomp_unfil ← copy(D_{tr})
end if
decomp_segment ← append(decomp_tr, decomp_val)
/* Generate mechgrid */
xx, yy ← meshgrid
clf.fit(decomp_unfil)
Z ← clf.predict(xx, yy)
Z ← Z.reshape(xx.shape)
for j < |train_s_i| + |valid_s_i| do
    /* Define priority */
    distance ← calc_distance(boundary, decomp_segment[j])
    location ← is_opposite(Z, xx, yy, decomp_segment[j])
    priority_array ← append(priority_array, array(location, distance))
    j ← j + 1
end for
pri_train, pri_valid ← train_val(priority_array)
return pri_train, pri_valid
```

be understood by predicting the components of this mesh grid. The prediction results are stored in Z. Then, for each data point, the following are calculated:

1) Distance from the decision boundary.
2) Whether it is located on the opposite side of the boundary.

These values are stored in *distance* and *location*, respectively. After all data points are processed, *priority_array* is split into training and validation and stored in *pri_train* and *pri_valid*, respectively. These two lists are returned as the output of the function *det_priority*.

Although this algorithm assumes a binary classification, in the case of multi-class classification, a single data label value can be assumed to be binary against other labels. By generating a linear SVM as in Algorithm 14.5 and calculating the distance to the target data, the priority can be determined in the same manner as in binary classification.

14.5 Evaluation Experiment

This section presents the evaluation results.

14.5.1 Overview

We conducted experiments to verify the effectiveness of the proposed defense method against the poisoning attacks presented in this paper. The accuracies of the following four models were compared in the experiments:

- Model trained with data containing no harmful data (Perfect defense)
- Model trained with data containing harmful data (No defense)
- Model defended by the proposed method (Proposal)
- Model defended by the existing method (Existing)

The accuracies of these models were evaluated using test data independent of the experimental environment.

Furthermore, the attack method and existing method used in this study were implemented according to the open-source code provided in Nicolae et al. [46]. To shorten the duration of the experiment, the harmful data generated by the attack method was prepared and used in advance. We varied the number of malicious devices and compared the accuracy of each defended model against the proportion of harmful data. Additionally, the threshold for determining harmful devices in the existing method was set to values of 0.0 and 0.2 (the default value set in the open-source code by Nicolae et al. [46]) for comparison. The experimental results reported in the present study are the values averaged over 10 runs.

The following datasets were used for experiments:

- Modified National Institute of Standards and Technology (MNIST)
- Synthetic data
- Environmental sensor telemetry data

The experimental settings and the results obtained on each dataset are described in Sections 14.5.2–14.5.4.

14.5.2 MNIST

14.5.2.1 Experimental Setting

In this experiment, the MNIST dataset of handwritten digits was used. Although the original dataset has 10 classes labeled from 0 to 9, we set up a binary classification task for digits 0 and 4, similar to the experiment settings by Biggio et al. Harmful data was generated using the attack method proposed by Biggio et al. [17]. Their attack method is an algorithm that generates harmful data in a poisoning attack that targets SVM. The experiment was conducted in two ways: with and without determining the removal priority. For the case with the removal priority determination, the proportion of harmful data in malicious devices was also used as a parameter to verify the effectiveness of the removal priority described in Section 14.4.5. For the case without the removal priority determination, the proportion of harmful data in each malicious device was set to 100%, and the experiment was conducted in an environment in which all the data provided by the malicious devices was harmful.

14.5.2.2 Results

The graph depicting the accuracy of the defended models is shown in Figure 14.8a. The results in this figure do not represent the case that used the removal priority described in Section 14.4.5. As shown in Figure 14.8a, both the existing method and the proposed method significantly outperform the model with no defense up to a proportion of harmful data of 40%. However, when the proportion of harmful data exceeds this level, the accuracy deteriorates even compared with the model without defense. This is likely because an increase in the proportion of harmful data in the training data causes the I_{neg} values for both the existing method and the proposed method to be high for normal data, which indicates that normal data is judged to have a negative impact on model accuracy.

Furthermore, the F1 score and accuracy for the harmful data detection rate of the existing method and the proposed method are shown in Figures 14.8b and c, respectively. The proposed method with a threshold of 0.2 significantly outperforms the existing method on the F1 score. However, the existing method, with a threshold of 0.0, has nearly the same values as the proposed method. These results suggest that the definition of the removal rate in the proposed method (equation 14.3) causes it to produce results similar to the existing method with a threshold of 0.0.

Figures 14.9a and b show the precision and recall results for harmful data detection in the above experiment. From the results in (a), it is evident that the proposed method outperforms the existing methods on both threshold values of 0.0 and 0.2 up to a proportion of harmful data of 40%. In (b), however, at a threshold of 0.0, the results for the existing method generally surpass those of the proposed method.

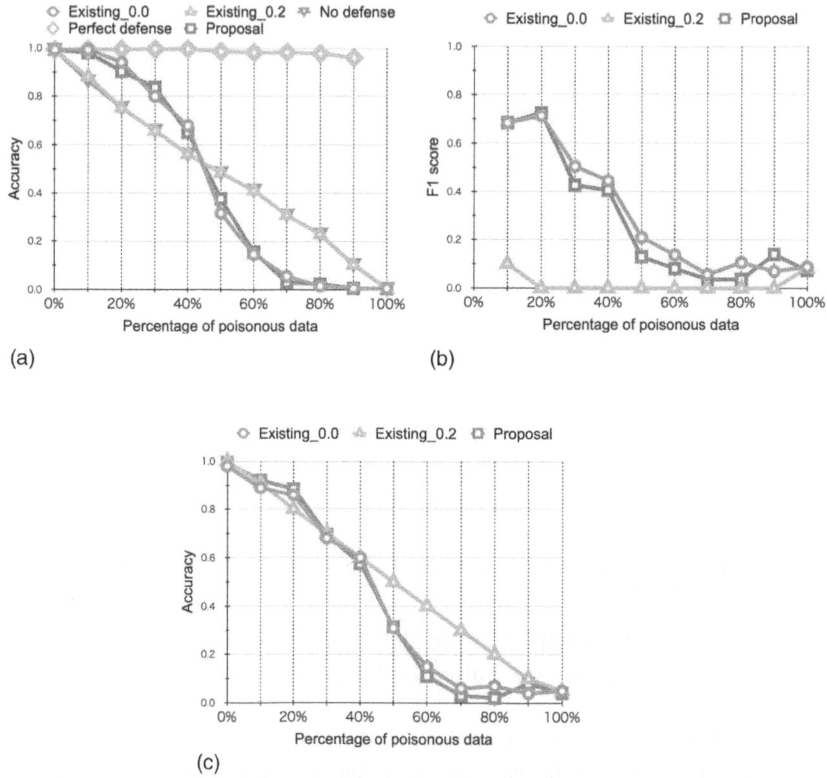

Figure 14.8 Results of model accuracy and harmful data detection accuracy (without removal priority). (a) Model accuracy, (b) detection accuracy F1 score, and (c) detection accuracy.

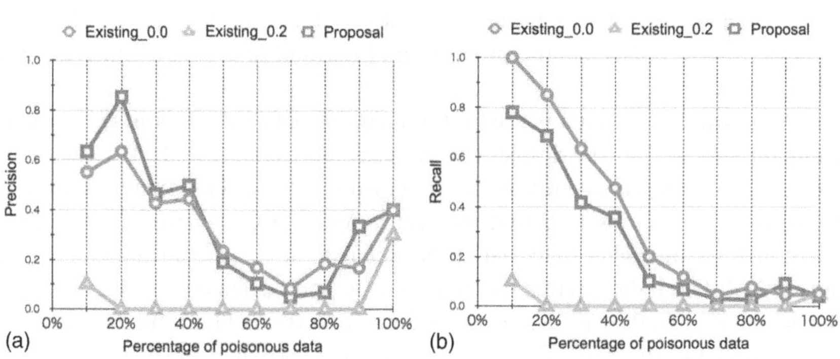

Figure 14.9 Detection accuracy results of each defense method (without removal priority) (a) Detection accuracy precision and (b) detection accuracy recall.

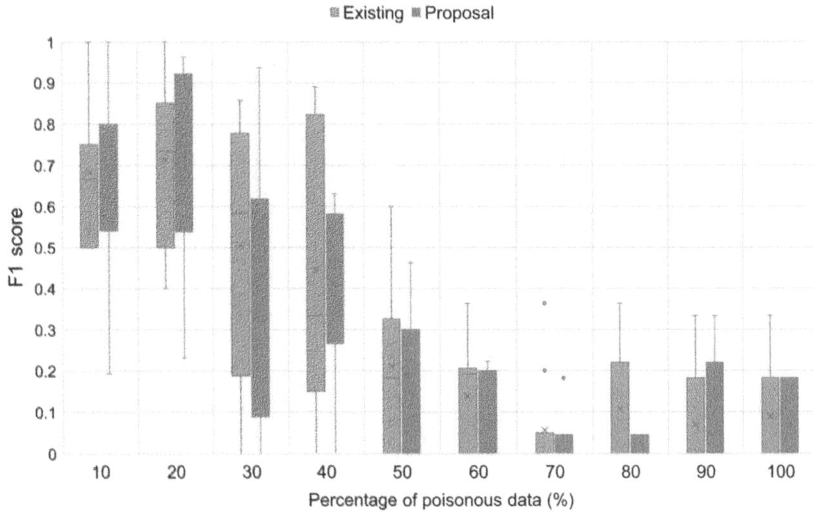

Figure 14.10 Box plot of detection accuracy (F1 score) results for each defense method (without removal priority).

These results indicate that the proposed method suppresses false positives with its removal rate formula, but that it also has difficulty detecting a large amount of harmful data.

Figure 14.10 shows, using a box plot, the distribution of F1 score values from the detection accuracy of the experiment in Figure 14.8. As can be seen, both the existing and proposed methods exhibit similar shapes, which indicates that neither has superiority in detection stability.

Figures 14.11 and 14.12 show the experimental results for the case when removal priority is determined. As mentioned in the previous section, the proportion of harmful data in each malicious device was set to 50%. Therefore, even if all of the 10 devices were malicious, the proportion of harmful data in the entire dataset would be only 50%.

Because the MNIST data is used for a binary classification task in this experiment, the task is relatively simple. Therefore, as shown in Figure 14.11a, the accuracies of the models are generally high after each of the defense methods is used. However, it can be observed that the accuracies of "Existing_0.0" and the proposed method are higher than that of "No defense" up to a proportion of harmful data of 25%. This indicates that the proposed method's defense effectively reduces the degradation in the model's accuracy for a proportion of harmful data of up to 25%. Figures 14.11b and c show the detection accuracy of harmful data in terms of F1 score and accuracy, respectively. These results show that the proposed method generally shows higher detection accuracy than the existing method.

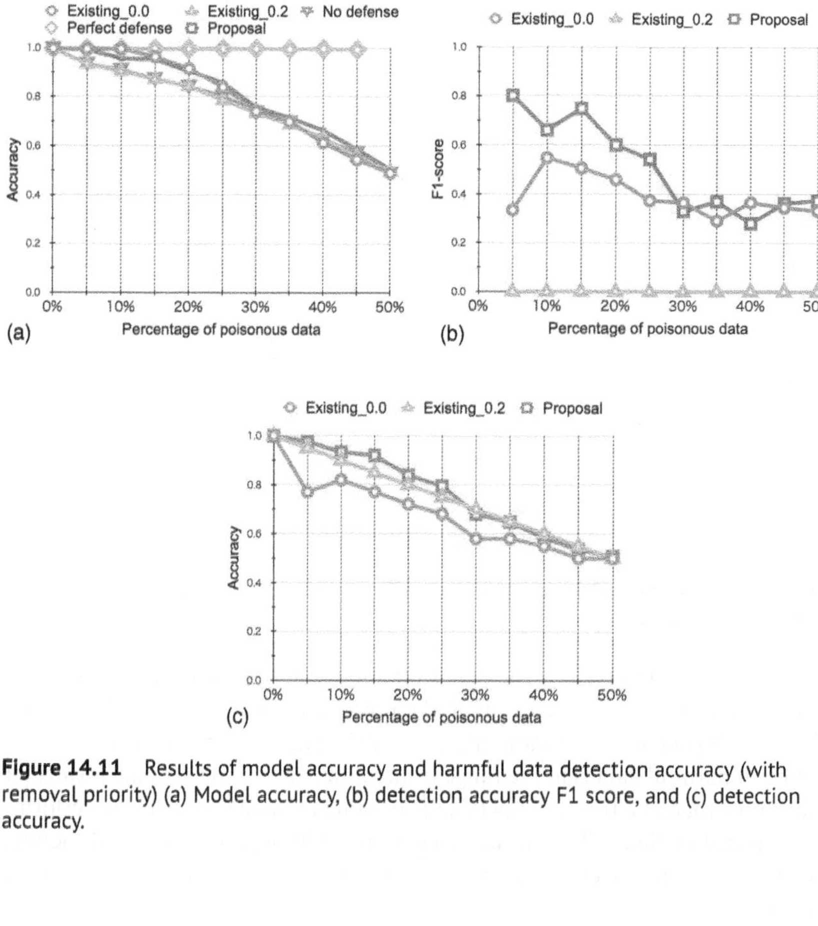

Figure 14.11 Results of model accuracy and harmful data detection accuracy (with removal priority) (a) Model accuracy, (b) detection accuracy F1 score, and (c) detection accuracy.

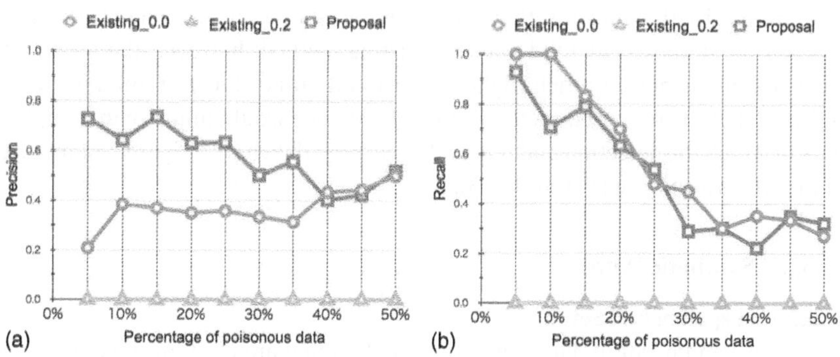

Figure 14.12 Detection accuracy results of each defense method (with removal priority) (a) Detection accuracy precision and (b) detection accuracy recall.

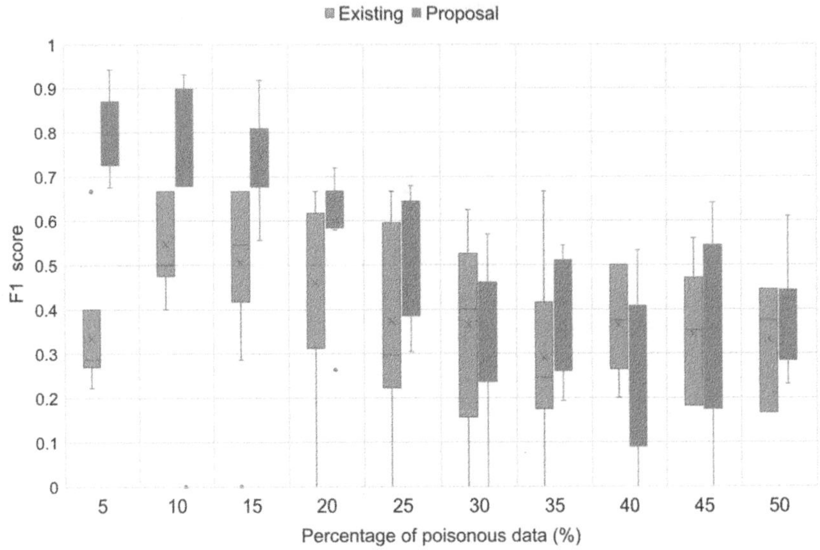

Figure 14.13 Box plot of detection accuracy (F1 score) results for each defense method (with removal priority).

Figures 14.12a and b show the precision and recall results for the detection of harmful data in the above experiment. For precision, the proposed method generally shows higher accuracy. Particularly, the difference is dramatic when the proportion of harmful data is in the range from 5% to 35%. In contrast, the results for recall show that, at a threshold of 0.0, the existing method generally outperforms the proposed method. These results suggest that although the proposed method suppresses false positives, it has greater difficulty detecting harmful data than the existing method.

To summarize these results, there is no significant difference in model accuracy between the proposed method and the existing method; however, the proposed method demonstrates a much higher detection accuracy for harmful data. This indicates that the proposed method's model accuracy would improve in complex tasks in which the detection and removal of harmful data have a significant impact on model performance (Figure 14.13).

14.5.3 Synthetic Data

14.5.3.1 Experimental Setting

The dataset and harmful data generation method that are used for the poisoning attack in this experiment are based on the method used in Zhou et al. [47]. Figure 14.14 shows a sample of the synthetic data. This attack algorithm specifies

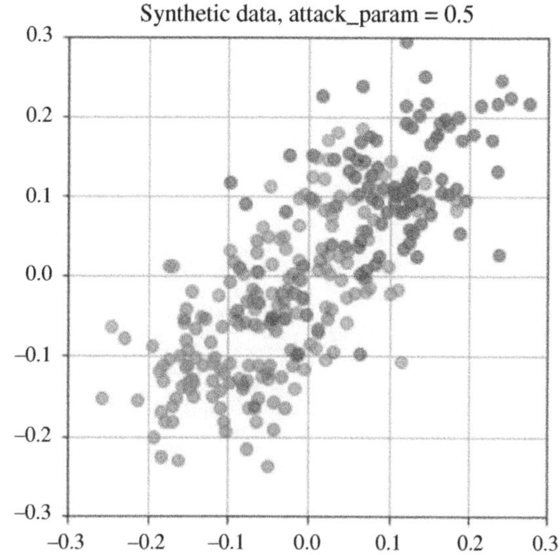

Figure 14.14 Synthetic data (*attack_parameter* = 0.5).

a parameter (*attack_parameter*) that represents the strength of the attack, with values ranging from 0 to 1. The strength of the attack indicates how much the generated harmful data deviates from the original distribution, where data that is farther from the original distribution shows greater deviation. As can be seen from Figure 14.14, the light gray and dark gray plots represent normal data, whereas the medium gray represents harmful data. With the attack parameter set to 0.5, the harmful data is distributed around the midpoint between the ight gray and dark gray normal data. An attack parameter value set close to 0 represents a weaker setting, whereas a value close to 1 represents more harmful data.

We conducted experiments with three different values for the attack parameter (0.3, 0.5, 0.7). As was done in the experiments in Section 14.5.2, the total number of devices was set to 10, where each device provided 100 data points. The thresholds for the existing method were set the same values as in Section 14.5.2.

14.5.3.2 Results

The results for the accuracy of each model after the defense methods were implemented are shown in Figures 14.15a, 14.16a, and 14.17a. The accuracy of the models that were trained without any defense decreases as the proportion of harmful data increases, whereas the existing and proposed methods maintain their accuracy relatively well. Additionally, with higher values of the attack parameter, the accuracy of the models for each defense method decreases.

The rate of detection of harmful data when the attack parameter value is set to 0.5 is shown in Figures 14.16b and c. These results indicate that the

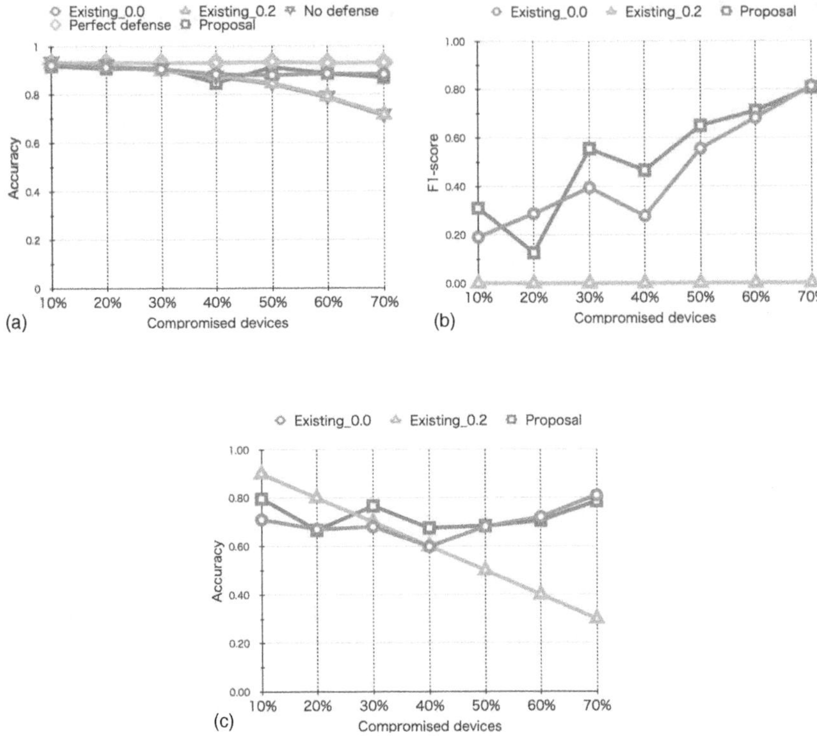

Figure 14.15 Results of model accuracy and harmful data detection rate (*attack_parameter* = 0.3) (a) Model accuracy, (b) detection accuracy F1 score, and (c) detection accuracy.

proposed method is slightly better at detecting harmful data than the existing method. Regarding the F1 score, it can be seen that the detection rate increases as the proportion of harmful data increases up to a proportion of 60%. This may be because the degree of influence of false positives decreases with an increase in the number of harmful data. However, the detection rate drops sharply when the proportion of harmful data reaches 70%. When the proportion of harmful data is 70%, the ratio of harmful data to the total data exceeds 50% for both the existing and proposed methods. Therefore, it is thought that the reason for this is the difficulty in judging whether harmful data has an adverse effect on the model. Figure 14.18 shows the negative impact values for a given ratio of harmful data to total data. As can be seen from this figure, the difference in negative impact between normal and harmful devices is small, which indicates that the lower detection rate may be due to the above reason.

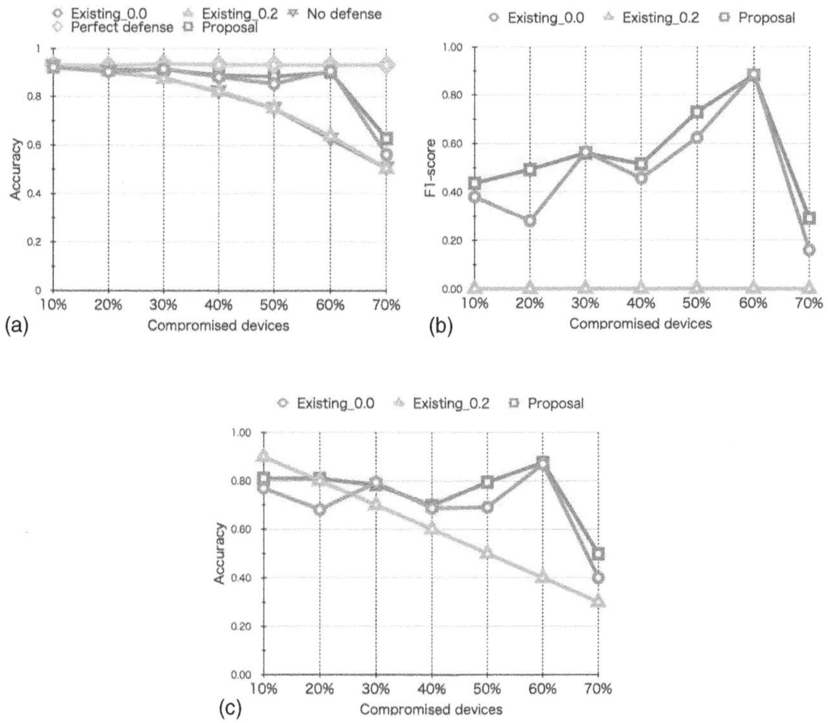

Figure 14.16 Results of model accuracy and harmful data detection rate (*attack_parameter* = 0.5) (a) Model accuracy, (b) detection accuracy F1 score, and (c) detection accuracy.

Similar to the accuracy results for each model, the rate of detecting harmful data decreases for higher values of the attack parameter. This result is likely due to the following two factors:

- The stronger effect of harmful data makes it easier to detect the proportion of harmful data up to a certain value.
- The stronger effect of harmful data makes it more difficult to detect a smaller amount of harmful data compared with weaker parameters.

Figure 14.19 shows, using box plots, the distribution of F1 scores for the detection accuracy from the experiment with *attack_parameter* = 0.5. Although there is no significant difference between these box plots and the box plots in Figure 14.10 for the MNIST experiment, a closer look reveals that except when the proportion of harmful data is 40% and 60%, the proposed method's box plots are generally higher and smaller.

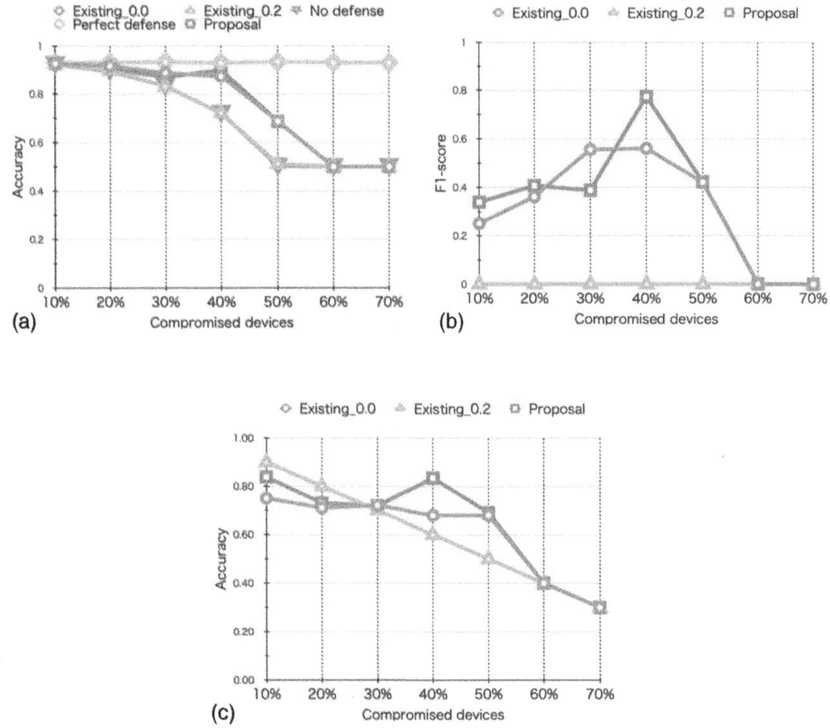

Figure 14.17 Results of model accuracy and harmful data detection rate (*attack_parameter* = 0.7) (a) Model accuracy, (b) detection accuracy F1 score, and (c) detection accuracy.

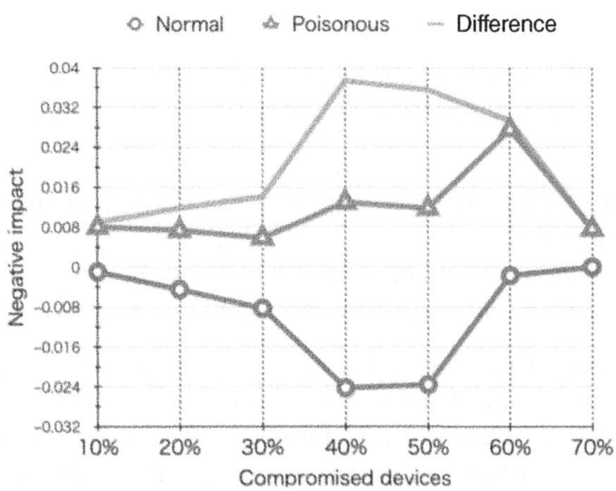

Figure 14.18 Results of I_{neg} values for different proportions of harmful data.

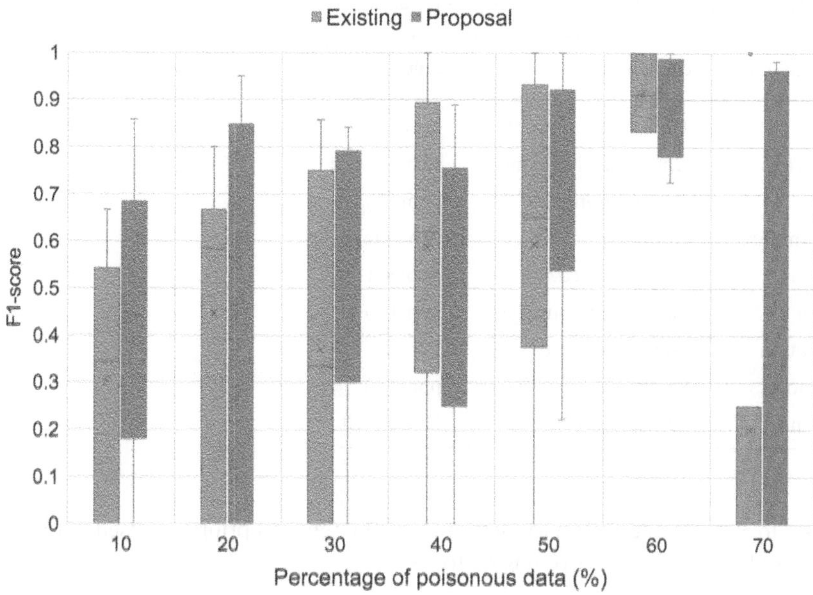

Figure 14.19 Box plot of detection accuracy (F1 score) results for each defense method.

14.5.4 Environmental Sensor Telemetry Data

14.5.4.1 Experimental Setting

The dataset used in this experiment was obtained from the data analysis platform Kaggle [48]. This sensing data was generated by an environmental sensor array [49]. The breadboard-based sensor array is connected to a Raspberry Pi single-board computer. These IoT devices are intentionally placed in physical locations under various environmental conditions, such as temperature and humidity conditions, to collect data. The attributes of these data are shown in Table 14.3.

In this experiment, we aim to predict the "light" attribute value using other sensed data except for "motion." The reason for excluding "motion" from the explanatory variables is to avoid making the task overly simple, as "motion" is a Boolean value.

The experimental settings are similar to those of the experiments in Sections 14.5.2 and 14.5.3, with the total number of devices set to 10. The number of data points for each device is set to 20 and 100, and we conduct experiments for both settings. The proportion of harmful data in the malicious devices is set to 50% and 100%, as was done in Section 14.5.2. When the proportion

Table 14.3 Environmental sensor telemetry data.

Notation	Description	Units
ts	Timestamp of event	epoch
device	Unique device name	string
co	Carbon monoxide	ppm (%)
humidity	Humidity	percentage
light	Light detected?	Boolean
lpg	Liquid petroleum gas	ppm (%)
motion	Motion detected?	Boolean
smoke	Smoke	ppm (%)
temp	Temperature	Fahrenheit

is set to 50%, the algorithm for determining the removal priority is used, and it is not used when the proportion is set to 100%.

14.5.4.2 Results

The results obtained when the number of data points per device is 20 and the proportion of harmful data in the malicious devices is 100% are shown in Figures 14.20 and 14.21. Both the existing method and the proposed method can detect harmful data up to a proportion of harmful data of 40%. However, when the proportion exceeds 40%, neither method detects harmful data effectively. Correspondingly, the model accuracy shown in Figure 14.20a is almost the same as that for the no defense for this proportion (Figure 14.22).

Table 14.4 shows the I_{neg} values for each device in the experiment with three to five malicious devices. The results, which are averaged over 10 runs, show that as soon as the number of malicious devices reaches four (i.e., 40% harmful data), the I_{neg} values for all malicious devices become 0. As was defined in equation 14.1 of Section 14.4.2, I_{neg} is the difference between the accuracy of the filtered model and that of the unfiltered model. A value of 0 indicates that both models have either endpoint accuracy values (0 or 1) or the same accuracy value. Because the accuracy values are not extreme, as is shown in Figure 14.20a, the accuracy of the filtered and unfiltered models is taken to be the same. This means that including or excluding the data from a single malicious device does not affect the model accuracy.

Figures 14.23 and 14.24 show the experimental results for the number of data points per device set to 100 and the proportion of harmful data in malicious devices set to 50%. In this experimental setting, the removal priority is defined, and the data

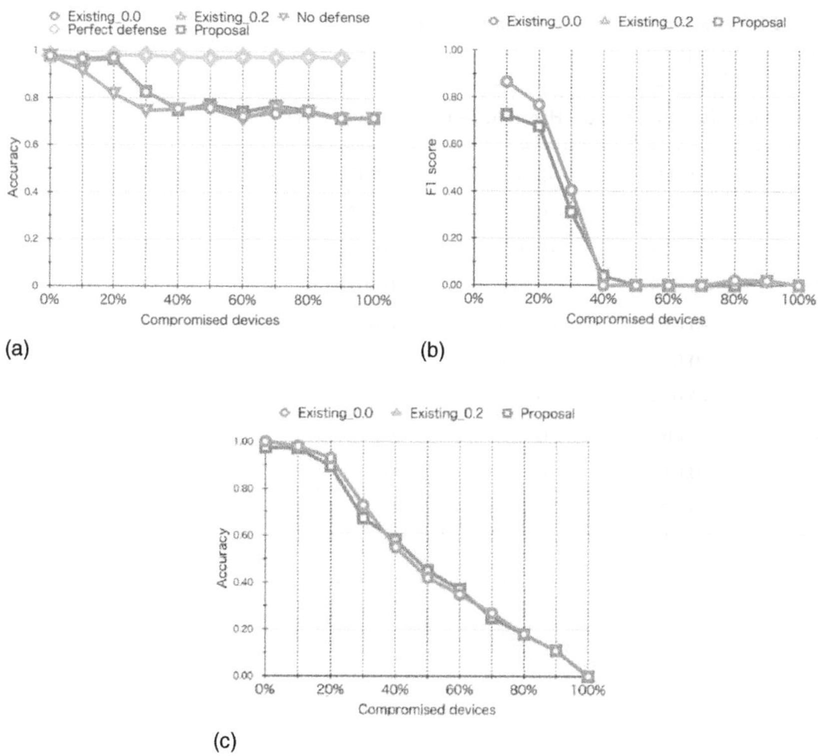

Figure 14.20 Results of model accuracy and harmful data detection accuracy (without removal priority) (a) Model accuracy, (b) detection accuracy F1 score, and (c) detection accuracy.

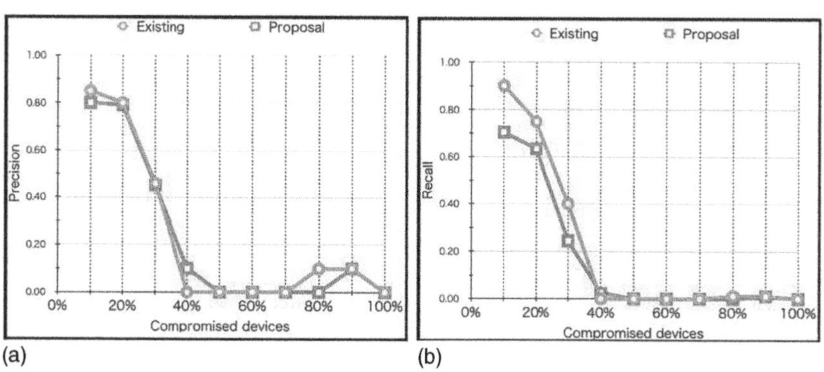

Figure 14.21 Detection accuracy results of each defense method (without removal priority) (a) Detection accuracy precision and (b) detection accuracy recall.

Table 14.4 Results of I_{neg} values for each device in Figure 14.20 experiment.

Device no.	Harmful: 8–10	Harmful: 7–10	Harmful: 6–10
1	0.0003	−0.002	0.004
2	0.012	−0.001	0.002
3	−0.018	−0.013	−0.0001
4	−0.025	−0.012	−0.002
5	−0.020	−0.008	0.014
6	−0.031	−0.011	0
7	−0.004	0	0
8	0.008	0	0
9	0.033	0	0
10	0.057	0	0

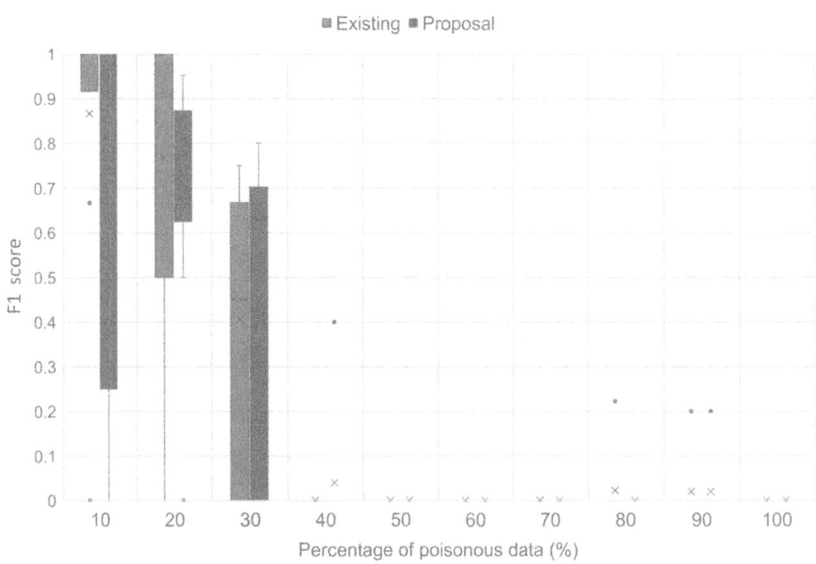

Figure 14.22 Box plot of detection accuracy (F1 score) results for each defense method.

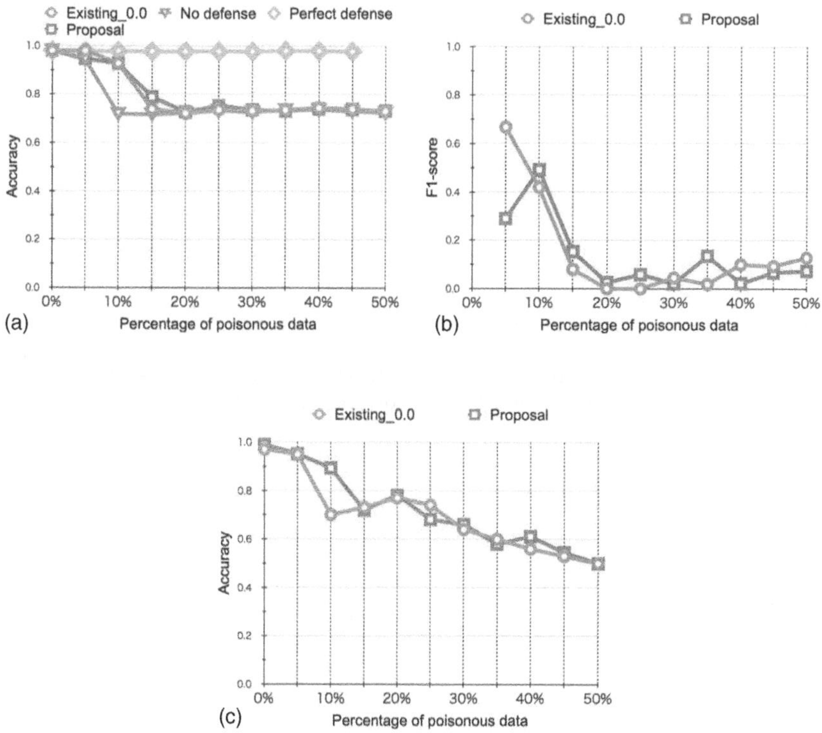

Figure 14.23 Results of model accuracy and harmful data detection accuracy (with removal priority) (a) Model accuracy, (b) detection accuracy F1 score, and (c) detection accuracy.

is removed accordingly. Figure 14.23a shows the results for the prediction accuracy of the model, and the accuracy did not differ significantly from the results for the previous experimental setting (Figure 14.20). The No defense model reaches its lower accuracy limit when the proportion of harmful data is 10%. In Figure 14.20a, the lower limit is reached at a proportion of harmful data of approximately 30%, but in this setting, it is 10%, which indicates that the proportion of harmful devices to the total number of devices has a more significant impact on model accuracy than does the proportion of harmful data alone. This means that the increase in the number of harmful devices has a greater negative impact on the accuracy of the model than does the proportion of harmful data within those devices.

Furthermore, Figure 14.25 shows the box plot of the F1 score values for the detection accuracy in the experimental setting shown in Figure 14.23. The results in Figure 14.24 indicate that in a setting where the proportion of harmful data is set to 5%, the existing method consistently detects the harmful data with high accuracy. However, the proposed method sometimes achieves higher detection rates

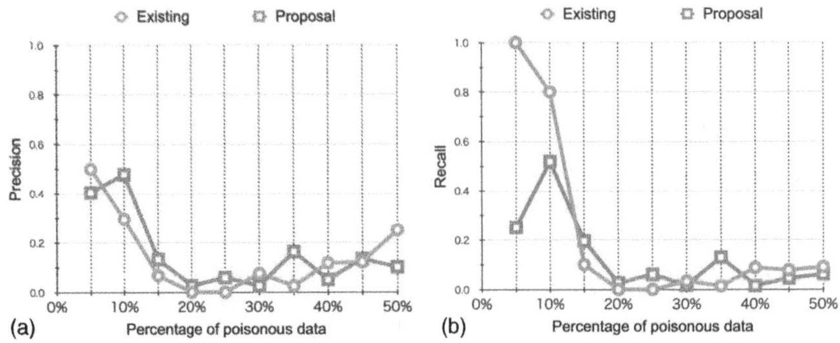

Figure 14.24 Detection accuracy results of each defense method (with removal priority) (a) Detection accuracy precision and (b) detection accuracy recall.

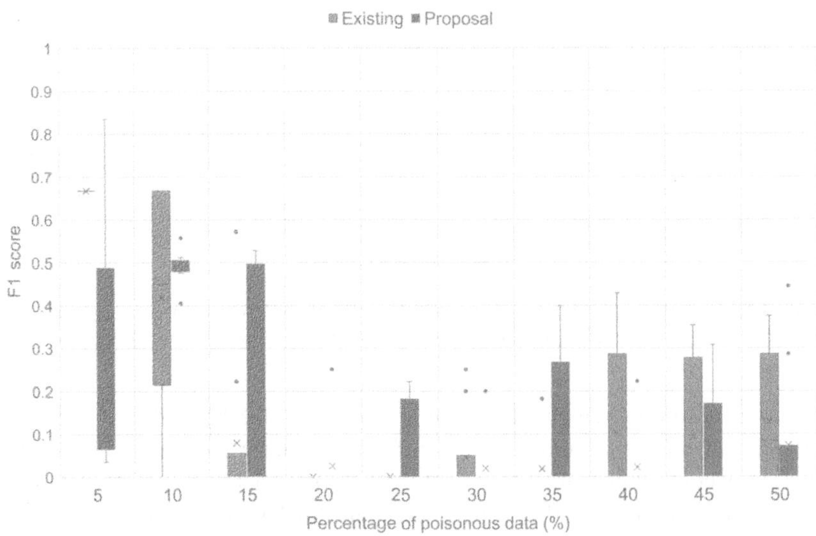

Figure 14.25 Box plot of detection accuracy (F1 score) results for each defense method.

than the existing method, but it generally has lower average detection rates. When the proportion of harmful data is set to 10% and 15%, the median values indicate that the proposed method obtains higher detection rates than the existing method. When the proportion of harmful data is set to 20% or higher, there is no significant difference in detection stability between the existing method and the proposed method. These results suggest that in the experimental settings for this section, when the proportion of harmful data in malicious devices is 50%, there is little difference in detection stability between the existing and proposed methods.

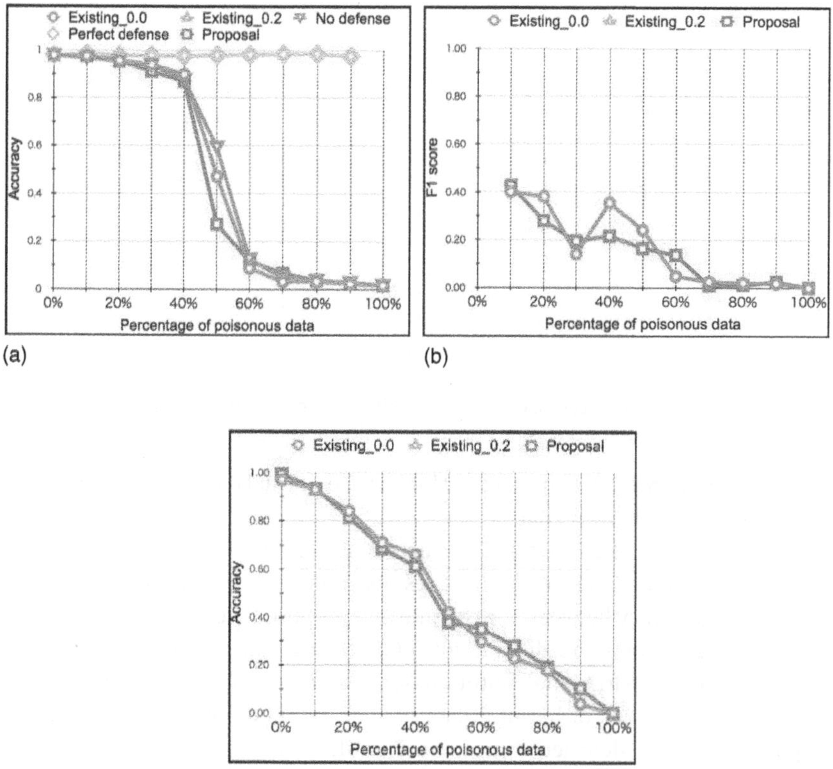

Figure 14.26 Results of model accuracy and harmful data detection accuracy (Attack method: Label flip) (a) Model accuracy, (b) detection accuracy F1 score, and (c) detection accuracy.

Figures 14.26 and 14.27 show the results of experiments conducted with the number of data points per device set to 20, the proportion of harmful data in malicious devices set to 100%, and the attack method changed to "label flip." "Label flip" is an attack in which the labels of the attack data are simply flipped in the binary data.

As shown in Figure 14.26a, the accuracy of the models after each defense method is applied drops to almost zero, except for the "perfect defense" model. Additionally, the accuracy of the models tends to rapidly decrease when the proportion of harmful data reaches 50%, and the accuracy takes values that are opposite to those for the settings where the proportion of harmful data is 0% to 40%. These results suggest that the increase in reversed data due to the label flip caused significant learning interference in the training data, which resulted in reversed prediction outcomes. This means that for the settings where the

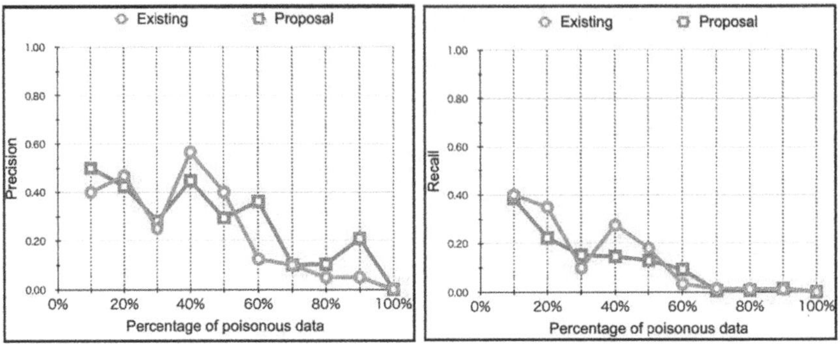

(a) (b)

Figure 14.27 Detection accuracy results of each defense method (Attack method: Label flip) (a) Detection accuracy precision and (b) detection accuracy recall.

proportion of harmful data was 60% or higher, the normal data became a minority and was treated as harmful data.

The results for the accuracy in detecting harmful data are shown in Figures 14.26b and c and 14.27. These results show there is almost no difference in detection accuracy between the existing and proposed methods.

Figure 14.28 shows the distribution of F1 score values from the detection accuracy of the experimental setting described above using a box plot. In the results for proportions of harmful data of 70% or higher, both the existing and proposed methods detected almost no harmful data, and this is consistent

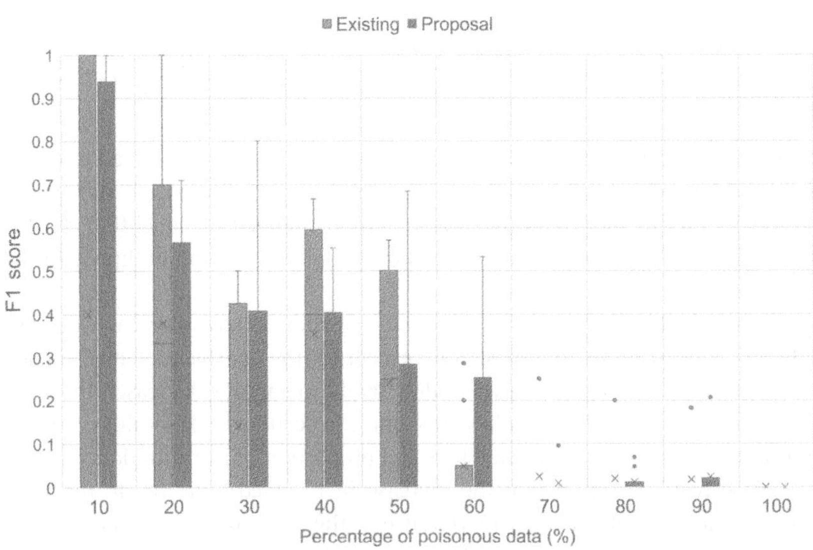

Figure 14.28 Box plot of detection accuracy (F1 score) results for each defense method.

with the results for model accuracy. For the settings where the proportion of harmful data is set to 10%–60%, both methods detect some harmful data, but the box plots cover a wide vertical margin, which indicates instability in detection.

14.6 Discussion

We discuss several issues related to our proposed method in this section.

14.6.1 Attack Methods Utilized

The method proposed in this study is a defense mechanism against poisoning attacks that leverages the harmful data generated by the attack algorithms. In the evaluation experiments presented in Section 14.5, both the attack and defense sides use the same poisoning attack method as in the setting. However, in real-world scenarios, it is unlikely that the defense would be able to use the same methods as the attackers. The proposed method takes advantage of the characteristic that harmful data does not revert to normal data, even when the attack algorithm is rerun. Therefore, it is necessary to verify in the future whether this premise holds when different attack algorithms are combined.

14.6.2 Adversarial Impact of Each Device

Baracaldo et al. [32] used a fixed threshold to determine the harmful or benign status of each device. In contrast, the method proposed in this study defines a unique threshold, I_{adv}, for each device. Table 14.5 shows the values of I_{adv} for the normal and harmful devices according to the proposed method. From these results, it can be seen that normal devices have values of I_{adv} that are relatively more likely to be judged as harmful. Intuitively, smaller values of I_{adv} for the harmful devices would be expected to improve detection accuracy. The results presented in Section 14.5 suggest that the accuracy of the proposed method is not inferior to that of the existing method for two reasons: Normal devices with an I_{neg} less than or equal to zero are judged as benign, regardless of I_{adv}, and harmful devices having I_{neg} values close to I_{adv} undergo higher data removal rates (Table 14.6).

14.6.3 When Harmful Data Constitutes the Majority

As described in Section 14.4.2, the proposed method adopts the strategy of judging all devices with a value of I_{neg} less than or equal to zero as normal and not

Table 14.5 Values of I_{adv} in the proposed method.

Harmful data inclusion rate	Normal device	Harmful device
8%	0.064	0.115
17%	0.041	0.078
25%	0.027	0.064
33%	0.004	0.024

Table 14.6 Values of I_{neg} in the proposed method.

Harmful data inclusion rate	Normal device	Harmful device
8%	−0.011	0.091
17%	−0.008	0.076
25%	−0.022	0.055
33%	−0.014	0.008

removing the corresponding data. However, when more than half of the dataset under protection consists of harmful data, it is conceivable that "normal data have an adverse effect on harmful data." In other words, the roles of normal and harmful data are essentially reversed. In such a case, the defense algorithm proposed in this paper may detect a greater number of normal devices than harmful devices because the I_{neg} of normal data would have a relatively high value.

14.6.4 Against Attackers Aware of the Defense Method's Internal Structure

As described in Section 14.2.1, this study assumes an environment in which data is collected from multiple devices in an IoT setting. However, the proposed method could be applied to many sources of data outside such environments. For instance, consider the scenario in which the annotation of training data is outsourced. In this case, data providers may be divided by company or employee. If an attacker is among these providers, the attacker can execute a poisoning attack. This scenario assumes that modifications to the data before and after the annotation are not detected by comparing hash values.

Additionally, it is conceivable that attackers could possess partial or complete knowledge about the algorithms of proposed method. In such cases, attackers may attempt to minimize I_{neg} to evade detection. Because the proposed method uses the impact on the accuracy of the model as a criterion for identifying harmful data, reducing I_{neg} would make detecting harmful data more challenging. However, reducing I_{neg} also reduces the effectiveness of an attack. Therefore, this method could potentially serve as a deterrent to attacks.

14.6.5 Limitations

Our proposed method determines whether the received data has been subject to a poisoning attack by applying known poisoning attacks to the data and then checking if the accuracy of the machine learning model subsequently decreases. If the server receives data that has been subject to an unknown poisoning attack, our proposed method may not correctly detect the attack. Existing defense methods for detecting poisoning attacks, other than our proposed method, may also be vulnerable to unknown attacks. Therefore, this drawback is not unique to our method. However, it is necessary to further evaluate how robust our method is against unknown attacks.

Additionally, the proposed method requires that the machine learning model be repeatedly trained, which results in significant computational cost. It will become necessary to explore ways to reduce this computational burden in the future.

14.7 Conclusion

The advent of TinyML, which focuses on deploying machine learning models on resource-constrained devices, introduces additional layers of complexity and potential vulnerabilities to the security landscape. As TinyML systems become integral to various applications, from smart home devices to wearable health monitors, ensuring the security and robustness of these systems becomes paramount. TinyML systems, by design, operate under extreme constraints in terms of computational power, memory, and energy consumption. These constraints necessitate highly optimized and often simplified machine learning models, which may inadvertently introduce new attack surfaces. For example, the reduced complexity of models in TinyML systems could make them more susceptible to certain types of poisoning attacks that exploit these simplifications.

Furthermore, the environments in which TinyML systems are deployed are often less controlled than those of traditional machine learning systems, as these devices are widely distributed and interact directly with the physical world. This increases the number of potential attack vectors and the ease with which malicious actors can introduce poisonous data. Therefore, the security strategies employed must be lightweight yet effective to balance the need for robust defenses with the limited resources available.

In this context, our study focuses on developing efficient detection and mitigation strategies for poisoning attacks. The proposed defense method, which is intended to minimize model degradation and accurately identify harmful data, is particularly suited to the TinyML paradigm. By ensuring that the defensive mechanisms are computationally efficient and do not overburden the limited resources of TinyML devices, we can enhance the security of these systems without compromising their performance or functionality.

Additionally, the practical implications of securing TinyML systems are considerable. As TinyML devices are often embedded in critical infrastructure and personal environments, any successful attack may have far-reaching consequences and impact not only individual users but also entire communities. Hence, our efforts to improve the resilience of machine learning systems against poisoning attacks are crucial for the broader adoption of and trust in TinyML technologies.

In conclusion, the integration of robust, lightweight security measures into TinyML systems is essential to safeguarding the next generation of smart devices. Our proposed method contributes to this goal by addressing the unique challenges posed by the resource constraints and deployment scenarios of TinyML, and it ensures that these innovative systems are both powerful and secure.

References

1 Chiba, T., Sei, Y., Tahara, Y., and Ohsuga, A. (2021). A countermeasure method using poisonous data against poisoning attacks on IoT machine learning. *International Journal of Semantic Computing* 15 (02): 215–240.

2 Medeiros, M., Flores, T., Silva, M., and Silva, I. (2024). A multi-layered methodology for driver behavior analysis using TinyML and edge computing. *2024 IEEE International Conference on Evolving and Adaptive Intelligent Systems (EAIS)*, 1–8. IEEE.

3 Kim, E., Kim, J., Park, J. et al. (2023). TinyML-based classification in an ECG monitoring embedded system. *Computers, Materials and Continua* 75 (1): 1751–1764.

4 Sun, B., Bayes, S., Abotaleb, A.M., and Hassan, M. (2023). The case for TinyML in healthcare: CNNs for real-time on-edge blood pressure estimation. *Proceedings of the 38th ACM/SIGAPP Symposium on Applied Computing*, 629–638.

5 Patil, R.Y., Bhamare, M., Patil, Y.H., and Bannore, A. (2024). Securing TinyML in a connected world. *TinyML for Edge Intelligence in IoT and LPWAN Networks*, 311–330. Elsevier.

6 Apruzzese, G., Laskov, P., Montes de Oca, E. et al. (2023). The role of machine learning in cybersecurity. *Digital Threats: Research and Practice* 4 (1): 1–38.

7 Pitropakis, N., Panaousis, E., Giannetsos, T. et al. (2019). A taxonomy and survey of attacks against machine learning. *Computer Science Review* 34: 100199.

8 Huang, L., Joseph, A.D., Nelson, B. et al. (2011). Adversarial machine learning. *Proceedings of the 4th ACM Workshop on Security and Artificial Intelligence*, ser. AISec'11, no. 16, 43–58.

9 Barreno, M., Nelson, B., Joseph, A.D., and Tygar, J.D. (2010). The security of machine learning. *Machine Learning* 81 (2): 121–148.

10 Chakraborty, A., Alam, M., Dey, V. et al. (2018). Adversarial attacks and defences: a survey. *arXiv preprint arXiv:1810.00069.*

11 Barreno, M., Nelson, B., Sears, R. et al. (2006). Can machine learning be secure? *Proceedings of the 2006 ACM Symposium on Information, Computer and Communications Security*, 16–25.

12 Zakariyya, I., Kalutarage, H., and Al-Kadri, M.O. (2023). Towards a robust, effective and resource efficient machine learning technique for IoT security monitoring. *Computers & Security* 133: 103388.

13 Zheng, J., Xu, J., Du, H. et al. (2024). Trust management of tiny federated learning in internet of unmanned aerial vehicles. *IEEE Internet of Things Journal* 11 (12): 21046–21060.

14 Rajapakse, V., Karunanayake, I., and Ahmed, N. (2023). Intelligence at the extreme edge: a survey on reformable TinyML. *ACM Computing Surveys* 55 (13s): 1–30.

15 Abadade, Y., Temouden, A., Bamoumen, H. et al. (2023). A comprehensive survey on TinyML. *IEEE Access* 11: 96892–96922.

16 Nelson, B., Barreno, M., Chi, F.J. et al. (2008). Exploiting machine learning to subvert your spam filter. *Proceedings of the 1st Usenix Workshop on Large-Scale Exploits and Emergent Threats*, ser. LEET'08.

17 Biggio, B., Nelson, B., and Laskov, P. (2012). Poisoning attacks against support vector machines. *Proceedings of the 29th International Conference on International Conference on Machine Learning*, ser. ICML'12, 2012, 1467–1474.

18 Goodfellow, I.J., Shlens, J., and Szegedy, C. (2015). Explaining and harnessing adversarial examples. *arXiv preprint arXiv:1412.6572.*

19 Zhang, H., Li, Y., Ding, B., and Gao, J. (2020). Practical data poisoning attack against next-item recommendation. *Proceedings of the Web Conference 2020*, ser. WWW '20, 2458–2464.

20 Biggio, B. and Roli, F. (2018). Wild patterns: ten years after the rise of adversarial machine learning. *Pattern Recognition* 84: 317–331.

21 Smutz, C. and Stavrou, A. (2012). Malicious PDF detection using metadata and structural features. *Proceedings of the 28th Annual Computer Security Applications Conference*, ser. ACSAC'12, 239–248.

22 Li, M., Sun, Y., Lu, H. et al. (2020). Deep reinforcement learning for partially observable data poisoning attack in crowdsensing systems. *IEEE Internet of Things Journal* 7 (7): 6266–6278.

23 Yao, Y., Li, H., Zheng, H., and Zhao, B.Y. (2019). Latent backdoor attacks on deep neural networks. *Proceedings of the 2019 ACM SIGSAC Conference on Computer and Communications Security*, ser. CCS'19, 2041–2055.

24 Jagielski, M., Oprea, A., Biggio, B. et al. (2018). Manipulating machine learning: poisoning attacks and countermeasures for regression learning. *2018 IEEE Symposium on Security and Privacy (SP)*, 19–35.

25 Demontis, A., Melis, M., Pintor, M. et al. (2019). Why do adversarial attacks transfer? Explaining transferability of evasion and poisoning attacks. *28th USENIX Security Symposium (USENIX Security 19)*, 321–338.

26 Shafahi, A., Huang, W.R., Najibi, M. et al. (2018). Poison frogs! targeted clean-label poisoning attacks on neural networks. *Advances in Neural Information Processing Systems 31*, 6103–6113.

27 Madry, A., Makelov, A., Schmidt, L. et al. (2018). Towards deep learning models resistant to adversarial attacks. *ICLR*.

28 Steinhardt, J., Koh, P.W.W., and Liang, P.S. (2017). Certified defenses for data poisoning attacks. *Advances in Neural Information Processing Systems 30*, 3517–3529.

29 Dunn, C., Moustafa, N., and Turnbull, B. (2020). Robustness evaluations of sustainable machine learning models against data poisoning attacks in the Internet of Things. *Sustainability* 12 (16): 1–17.

30 Van, M.-H., Carey, A.N., and Wu, X. (2023). Hint: healthy influential-noise based training to defend against data poisoning attacks. *2023 IEEE International Conference on Data Mining (ICDM)*, 608–617. IEEE.

31 Wu, J., Wang, J., Xiao, C. et al. (2024). Preference poisoning attacks on reward model learning. *arXiv preprint arXiv:2402.01920*.

32 Baracaldo, N., Chen, B., Ludwig, H. et al. (2018). Detecting poisoning attacks on machine learning in IoT environments. *2018 IEEE International Congress on Internet of Things (ICIOT)*, 57–64.

33 Baracaldo, N., Bathen, L.A.D., Ozugha, R.O. et al. (2016). Securing data provenance in Internet of Things (IoT) systems. *Service-Oriented Computing –ICSOC 2016 Workshops*, 92–98.

34 Nelson, B.A. (2010). Behavior of machine learning algorithms in adversarial environments. PhD dissertation, USA.

35 Dwork, C. (2006). Differential privacy. In: *International Colloquium on Automata, Languages, and Programming* (ed. M. Bugliesi, B. Preneel, V. Sassone, and I. Wegener), 1–12. Springer.

36 Isawa, R., Sei, Y., Tahara, Y., and Ohsuga, A. (2024). Minimizing noise in location privacy protection through equipment error consideration. *International Journal of Electrical and Computer Engineering Systems* 15 (3): 285–296.

37 Cao, X., Jia, J., and Gong, N.Z. (2021). Data poisoning attacks to local differential privacy protocols. *30th USENIX Security Symposium (USENIX Security 21)*, 947–964.

38 Wu, Y., Cao, X., Jia, J., and Gong, N.Z. (2022). Poisoning attacks to local differential privacy protocols for Key-Value data. *31st USENIX Security Symposium (USENIX Security 22)*, 519–536.

39 Sun, X., Ye, Q., Hu, H. et al. (2024). LDPRecover: recovering frequencies from poisoning attacks against local differential privacy. *IEEE 40th Annual International Conference on Data Engineering (ICDE)*, 1619–1631. IEEE.

40 Sei, Y. and Honiden, S. (2010). Distributed detection of node replication attacks resilient to many compromised nodes in wireless sensor networks. *4th International ICST Conference on Wireless Internet.*

41 Biggio, B., Corona, I., Maiorca, D. et al. (2013). Evasion attacks against machine learning at test time. *Machine Learning and Knowledge Discovery in Databases*, 387–402.

42 Szegedy, C., Zaremba, W., Sutskever, I. et al. (2014). Intriguing properties of neural networks. *ICLR.*

43 Wang, B. and Gong, N.Z. (2018). Stealing hyperparameters in machine learning. *2018 IEEE Symposium on Security and Privacy (SP)*, 36–52.

44 Shokri, R., Stronati, M., Song, C., and Shmatikov, V. (2017). Membership inference attacks against machine learning models. *2017 IEEE Symposium on Security and Privacy (SP)*, 3–18.

45 Sei, Y. and Ohsuga, A. (2022). False event message detection robust to burst attacks in wireless sensor networks. *IEEE Open Journal of the Communications Society* 3: 1630–1642.

46 Nicolae, M.-I., Sinn, M., Tran, M.N. et al. (2018). Adversarial robustness toolbox v1.2.0. *arXiv preprint arXiv:1807.01069.*

47 Zhou, Y., Kantarcioglu, M., Thuraisingham, B., and Xi, B. (2012). Adversarial support vector machine learning. *KDD'12*, 1059–1067.

48 Kaggle. https://www.kaggle.com/ (accessed 22 August 2025).

49 Towards Data Science (2020). Getting started with IoT analytics on AWS. https://towardsdatascience.com/getting-started-with-iot-analytics-on-aws-5f2093bcf704 (accessed 22 August 2025).

15

TinyML for Real-Time Medical Image Classification and Diagnosis

Jelil O. Agbo-Ajala[1], Lateef A. Akinyemi[1,2], Olufisayo S. Ekundayo[1], and Ernest Mnkandla[1]

[1] *Centre for Augmented Intelligence and Data Science, School of Computing, University of South Africa, Johannesburg, South Africa*
[2] *Department of Electrical and Smart Systems Engineering, School of Computing, University of South Africa, Johannesburg, South Africa*

15.1 Introduction

Medical image classification and diagnosis involve the analysis and interpretation of medical images, such as X-ray, computed tomography (CT), magnetic resonance imaging (MRI), and ultrasound images, to detect and classify various diseases, medical conditions, and abnormalities [1, 2]. This process classifies medical images into different conditions using their visual features and patterns. The diagnosis of medical images goes a step further by using classified images to make a diagnosis, identify the type and stage of the conditions, and detect the location and severity of injuries [3, 4].

Real-time medical image classification and diagnosis are critical in health care, with accurate and timely information to make reliable decisions and appropriate patient care [5]. Conventional methods like machine learning (ML) and deep learning (DL) have limitations, including requiring large amounts of training images, being computationally expensive and resource-intensive, and susceptibility to overfitting, especially with small data. However, these limitations led to delayed diagnoses and encouraged ongoing research in deep learning and real-time medical image classification and diagnosis, as well as alternative techniques to address these limitations [6, 7].

Recently, TinyML (Tiny Machine Learning) has proven to be a promising technology in medical image analysis, supporting real-time classification and diagnosis in resource-constrained devices. TinyML focuses on developing tiny and

Tiny Machine Learning: Design Principles and Applications, First Edition.
Edited by Agbotiname Lucky Imoize, Dinh-Thuan Do, and Houbing Herbert Song.

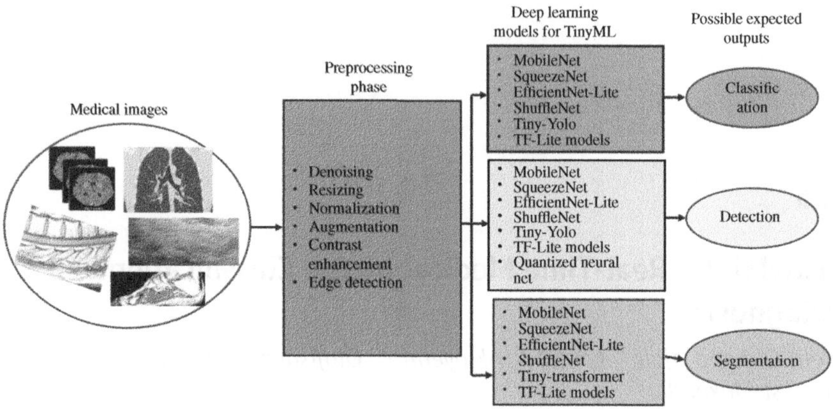

Figure 15.1 A typical TinyML architecture; the architecture consists of four major stages: input, pre-processing, model training and testing, and output. The input stage utilizes medical images from sources such as CT, X-ray, MRI, or PET scans. The pre-processing stage encompasses image denoising, resizing, augmentation, and other data preparation techniques. Next, the model training and testing phase. Finally, the output stage that produces the disease detection, classification, or segmentation results.

lightweight models that can run efficiently and effectively on devices with limited computational resources, such as edge devices, smartphones, and embedded systems. These benefits make TinyML suitable for real-time analysis and interpretation of medical images. TinyML empowers healthcare professionals to make more accurate and timely diagnoses that reduce healthcare costs, improve patient outcomes, and increase accessibility in remote and resource-constrained areas [8, 9]. This research aims to explore TinyML models for real-time medical image classification and diagnosis (Figure 15.1), achieving the following objectives:

1) We present a comprehensive review and detailed analysis of various methods used in medical image classification and diagnosis.
2) We conducted a comparative study between TinyML models and other machine learning methods, including traditional machine learning (ML) and deep learning (DL).
3) We also explore the advantages and benefits of TinyML models over other approaches, particularly in real-time classification and diagnosis performance across different datasets, and their potential to improve the field.
4) We also present a description of popular datasets and evaluation metrics used in medical image classification, especially breast cancer, brain tumor, and COVID-19.
5) We then examine future research and development directions for TinyML, in medical image classification and diagnosis.

This work serves as a comprehensive guide for researchers interested in employing TinyML for medical image classification and presents new and advanced methods utilized in this area that provide important insight into the current and existing works.

The rest of this work is arranged as follows: Section 15.2 presents a comprehensive review and analysis of medical image classification and diagnosis methods, with a thorough examination of various methods/models used for medical image classification and diagnosis. Section 15.3 presents the performance of conventional methods, including Machine Learning (ML) and Deep Learning (DL), as well as novel TinyML models. In Section 15.4, we present the results of our findings, highlighting the benefits of TinyML models while stating the limitations of other mechanisms. Finally, we summarize our contributions and highlight future research directions, which provide a conclusion to our work in Section 15.5.

15.2 Literature Review

This section presents a comprehensive review of existing work, including research focusing on machine learning and deep learning techniques for medical image analysis. Despite varying degrees of success, these studies have collectively advanced the field, highlighting key methodologies, achievements, and challenges.

15.2.1 Methods Employed in Medical Image Classification and Diagnosis

Medical imaging is a significant area of computer vision research, with many publications published over the last few decades [10]. Machine learning algorithms, particularly deep neural networks, have become popular in this field of image analysis. Therefore, this section presents a comprehensive and detailed analysis of the most recent and relevant work (see Tables 15.1 and 15.2), including TinyML, on the classification and diagnosis of three popular diseases: breast cancer, brain tumor, and COVID-19. We aim to present an overview of the current and recent methods, understand the complexities of the field, identify application areas for novel TinyML, and highlight areas that require further exploration.

Breast Cancer: In [11], the authors presented a novel channel attention model that employed "multiscale grid average pooling," for the segmentation of breast cancer on ultrasound images. This module incorporated both local and global spatial information, to enhance the segmentation task. The proposed network was trained and validated on popular dataset of breast cancer ultrasound

Table 15.1 Summary of methods employing TinyML models for medical imaging.

Authors	Year of publication	Classification task	Backbone model	Method	Datasets	Evaluating metrics
Polsinelli et al. [41]	2020	COVID-19	SqueezeNet	A light convolutional neural network	Italian and Zhao	Sensitivity; specificity; precision; F1 score; accuracy
Akpinar et al. [44]	2020	Chest diseases	SqueezeNet	Deep learning model	Chest X-ray14	Accuracy
Faith Özyurt [63]	2020	COVID-19	ShuffleNet	ShuffleNet with ReliefF and neighborhood component analysis	CT images; X-ray images	Accuracy; precision; geometric mean; F1-score
Owais et al. [40]	2021	COVID-19	MobileNet, ShuffleNet, and FCNet	Light weighted ensembled CNN + multilevel activation visualization	BIMCV COVID-19; COVID-CT; Cancer Archive; Shenzhen; Montgomery; CoronaHack	Accuracy; F1 score; average precision; average recall; AUC
Huang et al. [53]	2021	Pneumonia; COVID-19; lyme; tumor	EfficientNet-B1, MobileNet-V3-Small, ShuffleNet-V2, and PeleeNet	Batch similarity based triplet loss + light-weighted CNN	Chest 1; Chest 2; rash image; osteosarcoma histology images	Sensitivity; specificity; accuracy; AUC
Wang et al. [64]	2021 cumulative disease	ShuffleNet	Three-dimensional ShuffleNet3, PCANet, KCCA, and SVM	MRI images	Accuracy; precision; sensitivity; specificity; F1 score; AUC	

Hoang et al. [65]	2022	Skin lesion	ShuffleNet	New segmentation approach: entropy-based weighting first-order cumulative moment and wide-ShuffleNet	HAM10000; ISIC2019	Accuracy; precision; sensitivity; specificity; F1 score
Rasool et al. [32]	2022	Brain tumor	SqueezeNet	SqueezeNet model with SVM and finetuning	MRI	Accuracy: precision; recall; error rate; F1 score
Vaiyapuri et al. [5]	2022	Pancreatic tumor	MobileNet	Emperor penguin optimizer (EPO) with multilevel thresholding (EPO-MLT) technique with MobileNet	CT images	Accuracy; sensitivity; specificity; F1 score
Tsivgouls et al. [66]	2022	—	SqueezeNet	Improved SqueezeNet: SqueezeNodu-leNetV1 and SqueezeNodu-leNetV2	LUNA16	Sensitivity; specificity; accuracy; G-mean
Bourennane and Naimi [31]	2023	Brain tumor	Efficientnetb0	CNN models + SVM	Br35H	Sensitivity; specificity; precision; F1 score; accuracy

(Continued)

Table 15.1 (Continued)

Authors	Year of publication	Classification task	Backbone model	Method	Datasets	Evaluating metrics
Vijayan and Kuruvila [67]	2023	Lung cancer	SqueezeNet	Modified lightweight SqueezeNet architecture	LIDC data; 500 CT slides	Accuracy; recall; precision; specificity; AUC
Ullah et al. [42]	2023	COVID-19	ShuffleNet	ShuffleNet DL framework	ECG image of cardiac; COVID-19 patients; SARS-CoV-2 CT scan images; COVID-19 radiography	Accuracy; precision; sensitivity (recall); specificity; F1 score
Obaid et al. [68]	2023	Gallbladder disease type	VGG16, InceptionV3, ResNet152 and MobileNet	Deep convolution neural network	Ultrasound images	Accuracy; sensitivity; specificity; F1 score; AUC; PPV; NPV; time processing
Alghamdi et al. [43]	2023	COVID-19	VGG16, MobileNet, EfficientNetB0, NASNetMobile, and ResNet50V2	Deep convolution neural network	Chest X-ray images	Accuracy; sensitivity; specificity; precision
Parveen et al. [33]	2023	Brain tumor	ShuffleNet	Equilibrium optimization with deep learning	Br35H	Accuracy; precision; F1 score; AUC
Mehmood et al. [45]	2024	COVID-19 and pneumonia	SqueezeNet	Deep learning model	Chest X-ray images for pulmonary diseases	Accuracy; precision; recall; specificity; area under curve; F1 score

Table 15.2 Summary of methods employing other deep learning models for medical imaging.

Authors	Year of publication	Classification task	Backbone model	Method	Datasets	Evaluating metrics
Fradi et al. [69]	2020	Bone	Inception-V3; MobileNet; ameobaNet	Transfer deep learning method	Ultrasonic computed tomographic (USCT) images	Accuracy
Babukarthik et al. [38]	2020	COVID-19	Deep CNN	Genetic deep learning convolutional neural network	CT scan; chest X-ray	Accuracy; precision; sensitivity; specificity; recall; F1 score; log loss
Hasan et al. [70]	2021	COVID-19	VGG-16	Deep transfer learning model VGG-16	Chest X-ray images (other chest X-ray image RSNA pneumonia detection challenge)	Accuracy; sensitivity; specificity; precision; F score
Zazzaro et al. [37]	2021	COVID-19	Deep CNN	CNN and K-nearest neighbor	CT images	Accuracy; recall; precision; ROC
Jin et al. [34]	2021	COVID-19	AlexNet	Hybrid models (AlexNet, ReliefF, and SVM classifier)	NIH chest X-ray; COVID-19 CXRs	Accuracy; specificity; sensitivity; F1 score
Aljuaid et al. [17]	2022	Breast cancer	ResNet18, ShuffleNet, and Inception-V3Net	Transfer Learning and deep neural Networks	BreakHis; histopathology	Accuracy; precision; recall; specificity
Alharbi et al. [71]	2022	Pneumonia	ImgNet; SqueezeNet	Deep learning model (BoxENet)	Kaggle X-ray; pulmonary mask; PND pneumonia; Bel pnem; NIAPn pneumonia; RPNA	Accuracy; specificity; sensitivity; precision; dice

(Continued)

Table 15.2 (Continued)

Authors	Year of publication	Classification task	Backbone model	Method	Datasets	Evaluating metrics
Selma and Eltrass [3]	2022	Skin lesion	VGG16; ResNet50; ResNetX; InceptionV3; MobileNet	Deep learning model and SVM	ISIC 2017; HAM10000	AUC; accuracy; sensitivity; precision; F1 score; computational time
Balaha et al. [23]	2022	Breast cancer	MobileNet, MobileNetV2, NASNetMobile, VGG16, and VGG19	Transfer learning and aquila optimizer	MRI; mammographic; ultrasound images; histopathology slides	Loss; accuracy; F1 score; precision; recall; specificity; AUC; sensitivity; IoU; dice coef; cosine similarity
Simanjuntak et al. [72]	2022	Cataract	GoogleNet, MobileNet, and ResNet	Convolution neural network	Primary	Precision; recall; F1 score; accuracy
Mahmud et al. [73]	2023	Breast cancer	ResNet50	Pre-trained deep transfer learning models	Histopathology images	Accuracy; AUC; recall; loss
Bag et al. [25]	2023	Brain growth	VGG, ResNet, DenseNet, and MobileNet	Ensembles of VGG, ResNet, DenseNet, and MobileNet models	MRI scan images	Precision; recall; F1 score
Raza et al. [24]	2023	Breast cancer	Inception	Deep learning-based model (DeepBreast-CancerNet)	BC images	Accuracy; precision; recall; F1 score

Sufian et al. [35]	2023	COVID-19	InceptionV3, MobileNet, ResNet50, VGG16, ViTB16, and ViT	Histogram Equalization, contrast Limited Adaptive HistogramEqualization + deep learning	SARS-CoV-2; New_Data_CoV2	Precision; recall; F1 score; accuracy
Vaiyapuri et al. [26]	2023	Brain tumor	EfficientNet, DenseNet, and MobileNet	Ensemble learning (EfficientNet, DenseNet, and MobileNet), denoising autoencoder (DAE), and social spider optimization algorithm	BRATS 2015	Precision; recall; F1 score; accuracy; MCC; AUC
Canales-Fiscal and Tamez-Peña [74]	2023	Melanoma and glaucoma	CNN	Morphological and convolutional neural network (CNN), and random forest classifier	ISIC; ORIGA	ROC; AUC; accuracy; error
Güler and Namlı [30]	2024	Brain tumor	VGG, ResNet, DenseNet, and SqueezeNet	Ensemble of CNN + ensemble of machine learning algorithms	Brain tumor MRI dataset (Figshare, SARTAJ and Br35H)	Accuracy; precision; sensitivity; specificity
Lilhore et al. [75]	2024	Skin cancer	U-Net and MobileNet-V3	Hybrid U-Net and improved MobileNet-V3 with hyperparameter optimizer	HAM-10000, melanoma, skin cancer	Precision; sensitivity; accuracy; specificity

images. The experimental results demonstrated the ability of the proposed model to perform the segmentation task. In [12], the authors introduced a deep convolutional neural network (CNN) architecture for automatic breast cancer segmentation. The model was trained on a diverse, large, and evenly distributed dataset and evaluated on popular benchmark datasets. Robust training on these datasets enabled the model to achieve excellent generalization that demonstrated its capacity to accurately segment breast cancer. In [10], the authors presented a novel method for breast cancer segmentation in ultrasonography images that addressed the challenges posed by complex noise and artifacts. They introduced an integrated architecture that combined ultrasonography image pre-processing, "linear iterative clustering," and a U-Net model. This approach enhanced the accuracy of breast cancer diagnosis and provided medical professionals with a valuable tool for improved patient outcomes. Also, the authors in [13] developed an effective and intelligent method for breast cancer diagnosis and segmentation of breast cancer images. They proposed a novel pre-processing technique for pathological images and an image recognition technology for breast cancer. This intelligent architecture for breast cancer employed digital image processing technology and filtering techniques to improve the clarity of the image data. Moreau et al. [14] proposed a model that automated the segmentation of metastatic breast cancer lesions on positron emission tomography (PET) and CT images. The novel model was evaluated on benchmark datasets. The results showed that the proposed model can accurately segment lesions in images with metastatic breast cancer. Similarly, in [15], the authors developed an automatic method for evaluating histopathological cancer sections. Using the whole images of breast cancer collected from Norway, they proposed a convolutional but cascaded neural network model, H2G-Net, to segment the regions with breast cancer from histopathological images. The design also employed a patch-wise technique and a convolution autoencoder. To validate the proposed model, they investigated ablation study to analyze the importance of the chosen regions on cancer segmentation. In [16], the authors investigated a novel model that combined a multiscale CNN with a joint transformer, for segmenting breast cancer in "Dynamic Contrast-Enhanced MRI (DCE-MRI)." This two-stage segmentation method first employed a U-Net network model to delineate the breast area in MRI images. The model was then evaluated using standard evaluation metrics, where it achieved excellent results in the segmenting and classifying of breast cancer lesions. In [17], the authors proposed a breast cancer model that detected and classified cancer at its early stage. They developed a diagnosis and segmentation method for breast cancer using both deep neural networks and transfer learning that is validated on a publicly available dataset. The results showed that the proposed method achieved a high accuracy for both binary and multi-class classification of cancer cases. In [18], the authors presented a model for automatic diagnosis of high-resolution breast cancer of pathological images. The proposed model is based on a "dual-path feature extraction network" for breast pathology, which utilized dilated convolution networks to encode rich contextual information. Additionally, they introduced a "channel attention mechanism" and a feature attention and fusion module to selectively filter and combine features. This fused approach achieved excellent performance on the image data while reducing the computational complexity associated with existing models. Patra et al. [19] presented a novel approach for automatic breast cancer segmentation using "DCE-MRI." Their framework employed a multiple-level thresholding technique, with an optimization algorithm, to segment breast lesions in the DCE-MRI images. Additionally, they utilized an enhanced "simple metaheuristic algorithm (SMA)" with a "quasi-opposition-based learning mechanism," to determine the optimal threshold values. This approach showed superior performance compared to existing methods. In [20], Xianjun et al. employed an attention-learning approach for breast cancer segmentation of mammography images. The proposed model comprised a novel architecture, an attention-based scheme, and a method for learning image labels. This multi-task approach integrated multiscale feature fusion and an enhanced U-Net architecture that improves the segmentation capability. The model achieved significant success when evaluated on benchmark datasets. In [21], Liping et al. presented a novel breast cancer segmentation method that comprises a hybrid "AttendSeg architecture" and "gravitational clustering optimization" method. The proposed hybrid method was employed to segment breast cancer images into various regions such as normal, malignant, and benign. The model demonstrated great performance when evaluated using some evaluating metrics, such as recall, precision, Dice coefficient, and Jaccard index. Ahmed et al. [22] introduced a deep learning-based model for classifying, detecting, and segmenting cancerous regions in mammogram images. They introduced a pre-processing mechanism to remove noise, artifacts, and muscle regions, followed by converting the pre-processed image into 512 × 512 patches. The model was trained on two benchmark breast cancer datasets and evaluated on popular metrics where it achieved excellent performance in terms of area under the receiver operating curve (AUC-ROC), average precision, and accuracy. These results demonstrated the proposed model's potential to assist medical practitioners in breast cancer classification and segmentation. Balaha et al. [23] proposed a novel approach for diagnosing breast cancer at its early stage. They suggested a hybrid framework for breast cancer classification and segmentation, consisting of two phases: classification and segmentation. The classification phase employed five different CNN models, with the Aquila optimizer used to calculate the optimal hyperparameters for the transfer learning architectures. The framework was trained on common datasets, and the results demonstrated good performance in binary and multiclass classification, as well as in the segmentation task. Also, Raza et al. [24] proposed an efficient and robust deep-learning model, "DeepBreastCancerNet," for breast cancer detection and classification. The novel architecture employed "clipped rectified linear unit (ReLU) and leaky ReLU activation functions," cross-channel normalization, and batch normalization. Experimental results showed that the proposed model achieved comparable performance with other state-of-the-art models on classification accuracy.

Brain Tumor: In [25], Bag et al. proposed an advanced CNN technique for detecting abnormalities in MRI scans, aimed at supporting prompt decision-making by radiologists. The technique comprises three stages including pre-processing of brain MRI images, application of machine learning feature extraction methods to identify significant features from these images, and utilization of CNN models to classify the detected regions in the MRI images. The results showed the effectiveness of the automated abnormal brain growth detection model in real-world scenarios, showcasing its potential to save medical practitioners time and facilitate early diagnosis. In [26], Vaiyapuri et al. introduced an ensemble learning-driven diagnosis model that employed deep learning techniques and CNNs on MRI scans. The model featured a Gabor filter to enhance image quality and utilized popular CNNs like EfficientNet, MobileNet, and DenseNet for feature extraction. Hyperparameter tuning was performed using the "social spider optimization algorithm." The proposed model achieved significant success compared to other methods. Similarly, Dhakshnamurthy et al. [27] explored a algorithms for detecting and classifying brain tumors using transfer learning models, for medical imaging. They employed three CNN-based models: VGG16, AlexNet, and ResNet-50. The ensemble of VGG16 and ResNet-50 models showed exceptional performance on benchmark datasets, with remarkable results in specificity, sensitivity, accuracy, and F1 score. Also in [28], the authors proposed a deep transfer learning architecture for brain tumor diagnosis. They applied four different transfer learning architectures of DenseNet121, VGG19, InceptionV3, and MobileNet. They also employed images from three benchmark datasets of Br35H, Figshare, and SARTAJ, to validate the models. The experimental results showed that the MobileNet outperformed other existing models achieving great accuracy. Similarly, in [29], Noreen et al. proposed a method utilizing pre-trained models, where features are extracted from down layers. They employed a multilevel feature extraction and concatenation technique for early detection and diagnosis of brain tumors. They specifically used the pre-trained DenseNet201 to extract features from different DenseNet layers, which were then concatenated and transferred to a softmax classifier. The two models achieved excellent performance in detecting brain tumors when evaluated on a three-class brain tumor dataset. Güler and Namlı [30] presented a novel deep-learning model for detecting brain tumors on MRI images. Their automatic brain tumor diagnosis and classification system aim to assist radiologists by eliminating the need for manual diagnosis. The authors trained and evaluated their proposed model on a large dataset of 7022 brain MR images. Initially, they employed machine learning methods to classify the extracted features, followed by an ensemble learning approach. The model demonstrated outstanding results, confirming the effectiveness of the employed techniques and methods. In [31], Mohammed and Hilai developed a hybrid model that integrated CNN and support vector machine (SVM) kernels for brain tumor classification. They utilized a CNN-based model to identify regional variations and classify brain tumors, using already pre-trained CNN models, including VGG19, EfficientNetB0, and NASNetLarge. The extracted features were then classified using an SVM classifier, which demonstrated comparable performance to existing models. Rasool et al. [32] proposed a novel approach that combined SqueezeNet with SVM for brain tumor classification. The hybrid deep learning method utilized a pre-trained CNN for feature extraction phase and an SVM for pattern classification, where it achieved high accuracy on MRI scans. Parveen et al. [33] developed an automated brain tumor diagnosis and classification model using an equilibrium optimization method and deep learning. Their model employed a Wiener filtering technique to remove noise from images and a ShuffleNet model for feature extraction. The model was validated on the benchmark Br35H dataset, comprising 1500 tumor and 1500 non-tumor images, and demonstrated significant success over existing techniques.

COVID-19: In [34], Jin et al. proposed a hybrid model that included a feature extractor, feature selector, and classifier. They primarily pre-trained AlexNet CNN for the feature extraction phase before employing ReliefF algorithm for feature sorting and selected important features using a trial-and-error technique. The proposed model showed great performance compared to other state-of-the-art models. Sufian et al. [35] introduced a novel image processing technique for grayscale medical images where it employed "histogram equalization and contrast-limited adaptive histogram equalization," to produce a "three-channel image representation." The proposed representation technique improved the classification performance when evaluated on six popular pre-trained models. Al et al. [36] developed a deep transfer learning (VGG16) model for the diagnosis of COVID-19 infection in chest X-ray images. They developed six different classifiers for different chest X-ray image data before combining them into a final classifier, and this demonstrated an encouraging capability of the model in COVID-19 detection. Zazzaro et al. [37] proposed a deep learning and ensemble technique for automated COVID-19 detection on chest CT images. They employed transfer learning techniques of a CNN originally pre-trained on popular ImageNet dataset, to allow for feature extraction from CT images. Babukarthik et al. [38] also proposed a genetic deep learning and CNN model, for COVID-19 prediction. The model was trained from scratch, to extract features needed for the classification task. The model achieved a good performance on precision, accuracy, sensitivity, and specificity, in COVID-19 diagnosis. Ozyurt et al. [39] proposed an automatic COVID-19 detection method using transfer learning of a lightweight deep learning approach. They used a lightweight (ShuffleNet) architecture and a multilevel activation visualization layer, to detect COVID-19 pneumonia from large-scale chest radiographic images. Owais et al. [40] proposed a method for analyzing and diagnosing COVID-19 infection from large-scale radiographic image datasets. They used a lightweight network with an ensemble design of ShuffleNet, MobileNet, and FCNet, for the diagnosis of COVID-19, and it achieved good performance when compared to existing and conventional procedures. In [41], Polsinelli et al. developed a lightweight CNN-based model that used SqueezeNet architecture for the diagnosis of COVID-19 in CT images. The model achieved good accuracy and competitive results. Ullah et al. [42] employed a holistic approach using ShuffleNet CNN architecture, to detect and classify COVID-19 from CT scans, chest radiographs, and electrocardiogram (ECG) images. The extensive classification experiments demonstrated the ability of the proposed method on the classification task. Alghamdi et al. [43] introduced an enhanced deep learning approach for the multiclass classification of COVID-19 infection of chest X-rays. They utilized various deep CNN models, data normalization and augmentation, and approaches, to improve performance. The model successfully classified the images into five different cases, including normal, COVID-19, and other lung infections. Akpinar et al. [44] developed a CNN-based model using SqueezeNet for detecting abnormalities in chest X-ray images, and this achieved great success after experimental evaluations. Mehmood et al. [45] proposed an automated system (CP_DeepNet) for detecting COVID-19 infections in lung X-ray images. They utilized a pre-trained SqueezeNet model and convolution layers and employed data augmentation, to avoid overfitting.

In general, despite the significant advances in medical image classification and diagnosis presented in the literature, TinyML still outperforms state-of-the-art methods in classification and diagnosis of medical images, particularly in real-time applications.

15.2.2 TinyML Models and Architecture Used in Real-Time Medical Image Classification and Diagnosis

Medical image classification and diagnosis involve developing models and architectures to analyze and process medical images, such as computed tomography (CT), magnetic resonance imaging (MRI), X-rays, and ultrasound images, to support disease classification and diagnosis. These models and architectures are typically optimized for accuracy, memory constraints, and computational efficiency, making them suitable for deployment on resource-constrained devices, including smartphones, tablets, and embedded systems. In this section, we therefore present some popular TinyML models and architectures used in real-time medical image classification and diagnosis.

MobileNet: MobileNet is a renowned lightweight neural network model developed by Google researchers, specifically designed for embedded devices and devices that are memory-constrained. Its efficient design has made it a popular choice among computer vision researchers, who adapt it for various applications, including medical image classification, detection, and segmentation. MobileNet excels in managing computational resources and memory usage, making it an ideal choice for mobile devices, smartphones, and other resource-constrained devices. The architecture employs "depthwise separable convolutions" that significantly reduce the number of parameters and computational requirements. The MobileNet family includes several variants, such as MobileNet-v1, MobileNet-v2, MobileNet-v3, and MobileNet-v3-Small, which are commonly used to classify medical images, classify magnetic resonance scans, and detect breast cancer. These variants offer developers a variety of options to choose from, depending on their specific requirements and constraints [46].

ShuffleNet: ShuffleNet is a lightweight CNN architecture that has demonstrated exceptional performance in various medical image classification tasks, including breast cancer and lung nodule. The model was designed for mobile devices with limited computing power, and ShuffleNet was introduced in 2018 by researchers at Shanghai Jiao Tong University, China. Its architecture combines "depthwise separable convolutions," pointwise convolutions, and channel shuffle operations, to achieve efficient computation and memory usage. The network has several variants: ShuffleNet-v1, ShuffleNet-v2, and ShuffleNet-v3, which hold great promise for various real-world computer vision applications, such as detecting brain tumors, classifying MRI scans, diagnosing Alzheimer's diseases. ShuffleNet's efficiency and accuracy make it an attractive solution for resource-constrained devices and real-time applications [47].

SqueezeNet: SqueezeNet is a compact and lightweight neural network architecture widely used in medical image classification tasks, including cardiovascular disease detection, diabetic retinopathy detection, chest radiograph classification, and skin lesions. SqueezeNet was developed by DeepScale, specifically for deployment on resource-constrained mobile and embedded devices. Its unique architecture enables reduced parameters and computations while still maintaining accuracy. This is achieved through a combination of techniques, including fire modules, depthwise separable convolution, squeeze layers, and expand layers [48].

Tiny CNN: Tiny CNN is one of the common lightweight CNN architectures that have been used in medical image classification and the diagnosis of breast cancer and prostate cancer. The CNN architecture was specifically designed to be used on resource-constrained devices (smartphones, embedded systems, and other mobile devices). It is commonly used for detecting cardiovascular diseases, classifying ultrasound images, and classifying mammography images. Tiny CNN has several key benefits including low memory usage, fast inference times, power efficiency, and its ability to work on resource-constrained devices including mobile and embedded systems. It also has its usability in medical devices, Internet of Things (IoT) devices, and autonomous vehicles [49].

NASNet: NASNet (neural architecture search network) is a search-based CNN architecture that has shown promising applications in medical image classification, diagnosis, and segmentation. NASNet was designed to be highly efficient and has achieved state-of-the-art accuracy on the popular ImageNet dataset, with exceptional performance. With neural architecture search techniques, NASNet can automatically learn optimal network architectures, making it a powerful tool for various computer vision tasks, including medical image analysis [50].

ESPNet: ESPNet is another lightweight CNN architecture, designed for medical image classification and segmentation. It is particularly useful in classifying MRI, classifying CT imaging, and detecting prostate cancer. The model employed spatial pyramid architecture, channel shuffle operations, "depthwise separable convolutions," hierarchical feature fusion, and spatial attention techniques to reduce memory usage and computational complexity and achieve high accuracy and fast inference for efficient feature extraction [51].

EfficientNet: EfficientNet is a family of convolutional neural networks (CNNs) proposed by Mingxing Tan and Quoc VL. The model is renowned for achieving state-of-the-art accuracy in medical image classification with reduced computations and parameters when compared to other models. The EfficientNet model is specifically designed for classifying medical images, including CT scans, and detecting diabetic retinopathy. The EfficientNet family comprises various variants, such as EfficientNet-B0 and EfficientNet-B1, each offering a unique trade-off between model size and accuracy for deployment in different applications [52].

PeleeNet: PeleeNet is also a lightweight neural network architecture optimized for the efficient classification of medical images. The model was introduced by Robert Cheng et al. in their paper "Pelee: A Real-Time Object Detection System on Mobile Devices," and PeleeNet is a streamlined variant of the singleshot detector (SSD) network, specifically designed for mobile and embedded devices. It uses a simple architecture with reduced computation requirements and parameters that enable real-time medical image detection on resource-constrained devices, making it an ideal solution for applications where computational resources are limited [53].

In summary, TinyML models and architectures, such as MobileNet, ShuffleNet, SqueezeNet, and Tiny CNN, have demonstrated exceptional promise in medical image classification and diagnosis, achieving outstanding accuracy and efficiency. These models are particularly valuable in resource-constrained devices, including mobile devices and embedded systems, where computational resources and memory are limited. With techniques like model compression, quantization, and pruning, TinyML models can attain high accuracy with significantly reduced computational requirements, making them ideal for real-time classification and diagnosis of medical images. As medical imaging continues to advance, TinyML models and architectures are poised to play a vital role in enabling accurate and efficient image classification, ultimately improving patient care and outcomes.

15.2.3 Application Areas of TinyML in Medical Image Classification and Diagnosis

The rapid advancement of medical imaging technologies has led to an increase in the number of medical images generated daily [54]. However, a timely and accurate interpretation of these images was a challenge, particularly in real-time scenarios [55]. Recently, with the advent of TinyML, a solution has been provided for classification and diagnosis tasks. TinyML can run on resource-constrained devices while still maintaining its high accuracy and real-time processing abilities, making it a promising solution for timely diagnostics and remote healthcare services [6, 8, 56]. In this section, we therefore study a comprehensive overview of TinyML applications in medical image classification and diagnosis, exploring their applicability in various medical imaging modalities, including CT, MRI, X-rays, and ultrasound images. Some of those application areas include the following.

Detection of Diseases: TinyML can be used to detect several diseases, including cancer, Alzheimer's disease, Parkinson's disease, brain tumor, cardiovascular disease, diabetic retinopathy, pneumonia, skin, and lung, among others types. The model is utilized to analyze images to detect signs of infections in MRI, CT, X-rays, and ultrasound images [57, 58].

Classification of Medical Images: TinyML can also be used to classify medical images into different categories, such as normal or abnormal, tumor or non-tumor, and benign or malignant. TinyML has numerous applications in medical image classification, including skin disease classification, tumor disease classification, orthopedic disorder classification, and lung disease classification [57, 59].

Segmentation of Medical Images: TinyML is also used in segmenting medical images to identify specific structures or features such as organs, tumors, or tissues. The model has its applications in lesion segmentation, tissue segmentation, organ segmentation, and tumor segmentation. It can also segment different tissues such as muscle, bone, and fat [60, 61].

Denoising of Medical Images: TinyML can also be employed to remove noise from medical images, to improve their quality and diagnostic ability. It can be used to reduce noise in medical images, including CT, MRI, X-ray, ultrasound, mammography, and fluoroscopy [62].

TinyML therefore has varying and vast applications in medical imaging including medical image classification, segmentation, disease detection, diagnosis, and image denoising, and it therefore has the potential to improve medical imaging by enabling efficient, accurate, and personalized diagnosis and treatment. It helps to improve accuracy and efficiency that enhances patient care and outcomes, reducing computational resources and energy consumption, among others.

15.3 Methodology

In this section, we present a description of three state-of-the-art approaches in medical image classification and diagnosis. These methods include the following.

15.3.1 Machine Learning (ML)

Machine learning (ML) is an aspect of artificial intelligence (AI) technology that allows machines to learn from data, identify the patterns, and automatically make decisions without human intervention [76]. In medical image classification and diagnosis, ML models play a vital role because they can learn from medical images and adapt to them, identify the patterns in those images, and produce reliable results on new image data, among other key characteristics. ML is also useful in medical imaging to interpret medical images employing computer-aided design (CAD) and algorithms where it applies ML techniques, to CT scans, X-ray scans, MRI scans, etc. The conventional ML models rely on structured datasets with predefined techniques for each step. However, new algorithms can adapt to missing data, ensuring robustness in disease classification and diagnosis [77].

15.3.2 Deep Learning (DL)

Deep learning (DL) is also a branch of machine learning that has been widely applied in computer vision and image processing. It utilizes multiple processing layers or multiple layers of complex structures that included multiple nonlinear transformations. Recently, various medical image processing and image recognition tasks have adopted deep learning methods on CT images, MRI images, ultrasound images, X-ray images, pathological images, etc. Deep learning technology has applications in medical imaging to detect, classify, and segment infections. The key benefits of deep learning techniques lie in their ability to replace manual feature learning with unsupervised or semi-supervised feature learning and hierarchical feature extraction efficiency. Some of the popular deep learning models used in classification tasks include GoogLeNet, VGGNet, and ResNet. In medical imaging, deep learning has demonstrated exceptional capabilities in segmenting diseases, discriminating between diseases, and classifying diseases [64].

15.3.3 TinyML

TinyML utilizes intelligence on edge layers and mobile devices, such as smartphones and tablets. TinyML techniques are designed to perform effectively on mobile devices with limited computational resources and battery restrictions. This technique focuses on deploying compressed and optimized machine learning models on low-powered, tiny devices like micro-controllers and embedded systems. TinyML was inspired by mobile machine learning features, including resource constraints, low latency, and moderate cost. The model emerged from technical breakthroughs in the IoT field and can be applied in various areas, particularly in medical image classification. TinyML technology addresses the need to apply intelligence in a wide range of applications previously hindered by the high power and computing requirements of standard machine learning models [78].

15.4 Results and Discussion

This section contains information on the medical imaging data used for this study, the popular evaluation metrics used to evaluate TinyML models, comparative analysis, and presentation of the results of our findings.

15.4.1 Description of Datasets Employed in this Work

Medical image classification and diagnosis play an important role in the healthcare industry. To develop and evaluate accurate classification models, researchers rely on various datasets. These datasets consist of labeled medical images, annotated with essential information such as the presence of severity and other relevant features of the disease. This section reviews popular datasets used in medical image classification research, focusing on three popular applications: breast cancer, brain tumor, and COVID-19 diagnosis.

BreakHis (Breast Cancer on Histopathology Images): BreakHis dataset was collected to be used in medical image classification, for binary and multiclass problems. The dataset consists of 7,909 microscopic images of breast tumor samples collected from 82 patients with different magnification factors, including 10×, 100×, 200×, and 400× of histopathological images of BC. The dataset is divided into 5,429 malignant image samples and 2,480 benign image samples. Those image samples were pre-processed and resized to 224 × 224 pixels [79].

Brain Tumor MRI: Brain tumor MRI dataset consists of medical images of human brain MRI scans, consisting of information about the location, presence, and characteristics of the brain tumor. Those MRI images are classified into 4 classes, namely, pituitary, glioma, meningioma, and no tumor. This dataset is essential for training computer vision algorithms to automate brain tumor detection, to help with early diagnosis and treatment planning [80].

COVID-19 CXR Dataset: This dataset comprises COVID-19 positive and negative chest X-ray (CXR) images obtained from various sources, including the BIMCV-COVID19+ database, a German medical school, Eurorad, SIRM, Kaggle, GitHub, and other COVID-19 repositories. The images, representing both male and female subjects, were curated to facilitate detection of lung damage caused by COVID-19 pneumonia. Specifically, subjects with confirmed cases and clinical symptoms are likely to exhibit abnormal lung X-ray patterns, allowing researchers to develop accurate diagnostic models for early detection and treatment planning [81]. Some of the sample images are presented in Figures 15.2–15.4.

15.4.2 Model Evaluation Metrics

Evaluation metrics are essential in medical image classification where they help to evaluate the performance of machine learning models. In this work, we focus on common and benchmark evaluation metrics comprising true positive (TP), false positive (FP), false negative (FN), and true negative (TN). TP refers to correctly classified positive images, FP refers to incorrectly classified positive images, FN represents incorrectly classified negative images, and TN refers to correctly predicted negative images. We analyze our findings using five performance

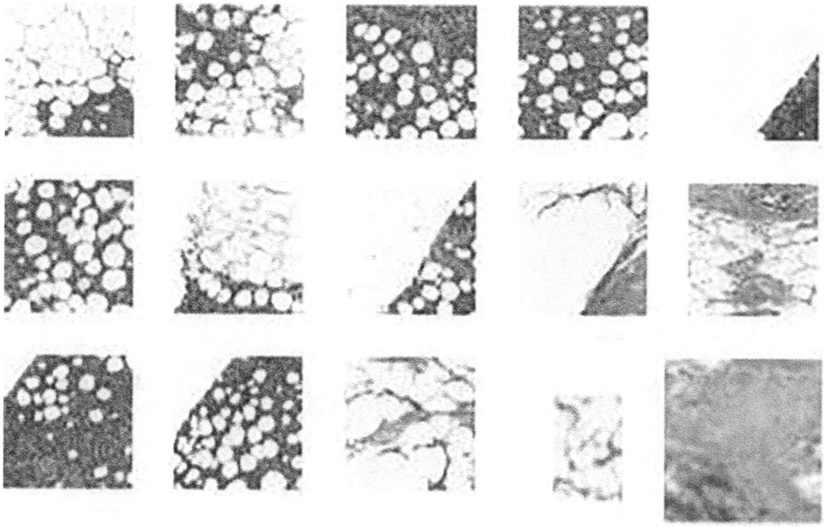

Figure 15.2 Sample images of breast cancer data.

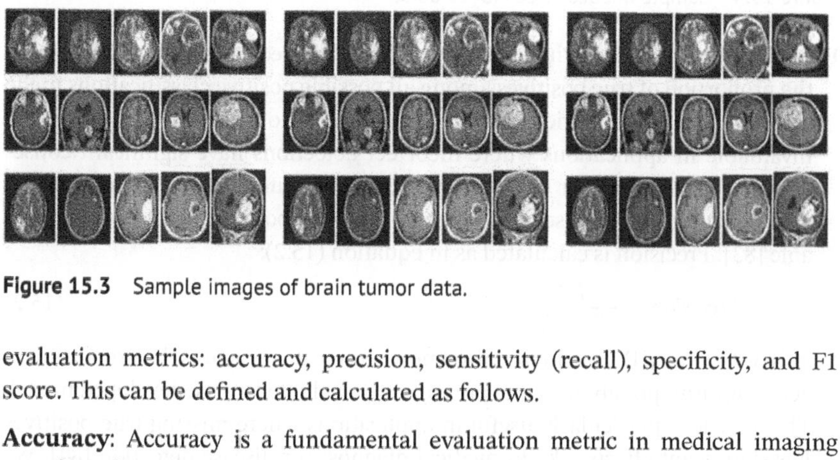

Figure 15.3 Sample images of brain tumor data.

evaluation metrics: accuracy, precision, sensitivity (recall), specificity, and F1 score. This can be defined and calculated as follows.

Accuracy: Accuracy is a fundamental evaluation metric in medical imaging tasks, where it measures the percentage of correctly classified images by the proposed model. The metric provides an overall assessment of the model performance. However, accuracy can be influenced by class imbalance in the datasets, potentially masking underlying issues. However, the higher the accuracy, the better the performance of the model [82]. Accuracy is therefore calculated as in equation (15.1):

$$Accuracy = \frac{TP + TN}{TP + TN + FP + FN} \tag{15.1}$$

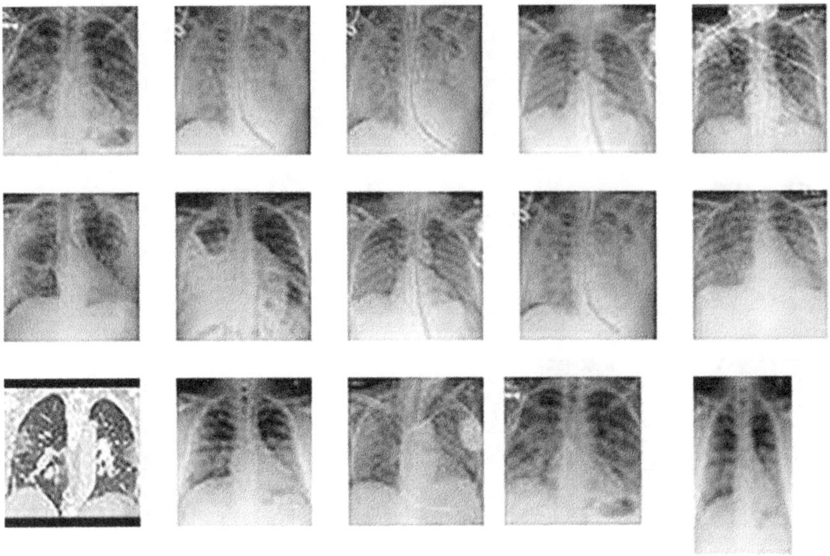

Figure 15.4 Sample images of COVID-19 data.

Precision: Precision is a vital evaluation metric in medical imaging, calculating the proportion of true positives among all possible positive classifications made by the model. This metric is particularly sensitive to false positives, making it invaluable in applications where incorrect detections have significant consequences, such as cancer diagnosis or disease monitoring. A higher precision value indicates fewer false positives, ensuring that positive predictions are reliable [83]. Precision is calculated as in equation (15.2):

$$Precision = \frac{TP}{TP + FP} \tag{15.2}$$

Sensitivity (Recall): Sensitivity, also known as recall, measures the proportion of actual positive possibilities that are correctly evaluated by the developed model. This metric is particularly useful in applications where missing true positives has significant effects, like in medical diagnosis or disease detection [84]. We can calculate sensitivity as in equation (15.3):

$$Sensitivity\ (Recall) = \frac{TP}{TP + FN} \tag{15.3}$$

Specificity: Specificity measures the value of actual negative occurrences that are correctly classified by the model, which provided insight into its ability to avoid false positives. This metric is crucial in situations where false positives can lead to unnecessary treatment, costly interventions, or adverse consequences. High specificity indicates that more actual negative instances are accurately identified, reducing the risk of unnecessary actions [85]. Specificity is calculated as in equation (15.4):

$$Specificity = \frac{TN}{FP + TN} \tag{15.4}$$

F1 Score: F1 score metric is a widely used measure of performance in computer vision, providing a balanced evaluation of recall and precision through its harmonic mean. A higher F1 score indicates a better balance between recall and precision that makes it a valuable metric in applications where both false positives and false negatives are important. Specifically, the F1 score is extensively utilized in medical image classification, detection, and segmentation [86]. The F1 score is calculated from the recall and precision of a prediction as in equation (15.5):

$$F1 = \frac{2 * Precision * Recall}{Precision + Recall} = \frac{2 * TP}{2 * TP + FP + FN} \tag{15.5}$$

15.4.3 Results of Key Findings

In this section, we present the performance evaluation results (Figures 15.5–15.7) of existing and recent state-of-the-art methods used in the classification and diagnosis tasks for breast cancer, brain tumors, and COVID-19 diseases.

Results of Breast Cancer Models: From Table 15.3, the accuracy results of the latest models are as follows: 99.87% on ResNet50, 99.86% on lightweight CNN, 99.28% on DenseNet, 99.15% on SVM and 98.95% on VGG19. The ResNet50 model was observed to achieve the best result; however, the lightweight and TinyML models achieved comparable results despite having lighter models with lower resources. Regarding precision metrics, custom and lightweight CNN achieved a perfect score of 100%, outperforming machine learning models like SVM (99%) and deep learning models like VGG19 (98.95%), ResNet50 (98.77%), and DenseNet (97%). For sensitivity, the SVM machine learning model achieved the best result (99.0%), while the DenseNet, VGG19, and ResNet50 deep learning models achieved comparable results but used more resources and were not suitable on devices limited in resources. The specificity results were 98.95% for VGG19, 89.69% for VGG, and 84% for deep belief networks. Furthermore, DenseNet achieved the best F1 score result (99.48%). In particular, TinyML achieved comparable results with significantly lower memory usage, making it suitable for real-time applications.

Results of Brain Tumor Models: Table 15.4 shows the results of brain tumor classification. The models achieved the accuracy scores of VGG16-ResNet50 (99.98%), EfficientNetb0 (99.78%), Lightweight CNN (99.74%), MobileNet

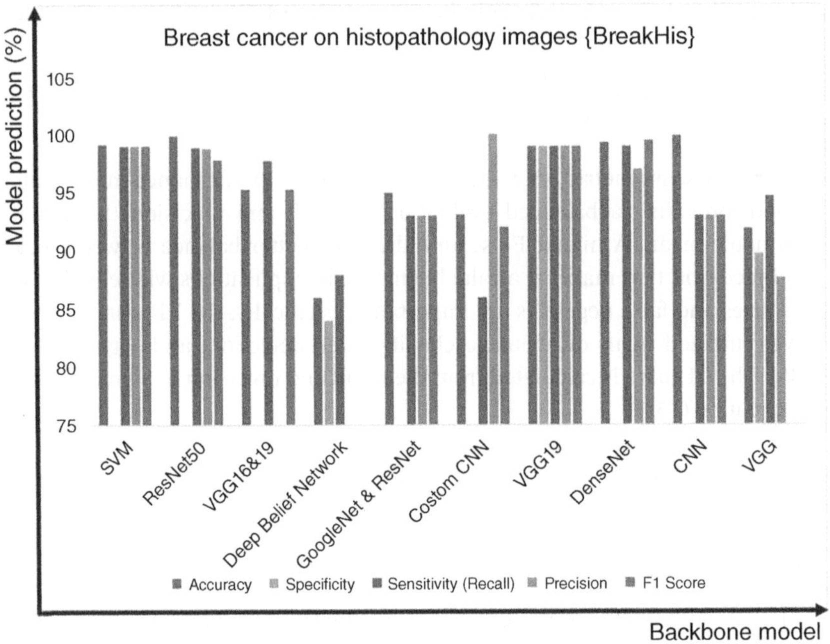

Figure 15.5 Results of the comparison of the TinyML model with other deep learning models for breast cancer detection.

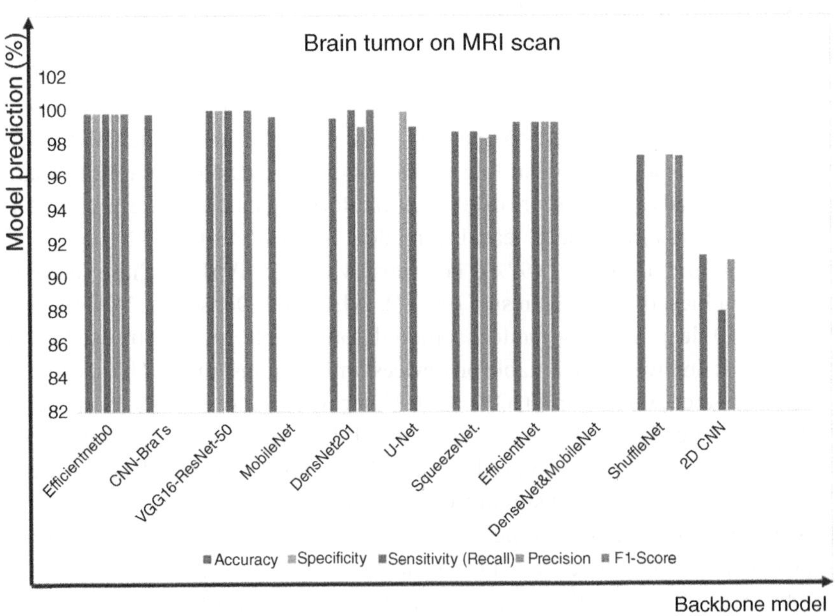

Figure 15.6 Results of the comparison of the TinyML model with other deep learning models for brain tumor detection.

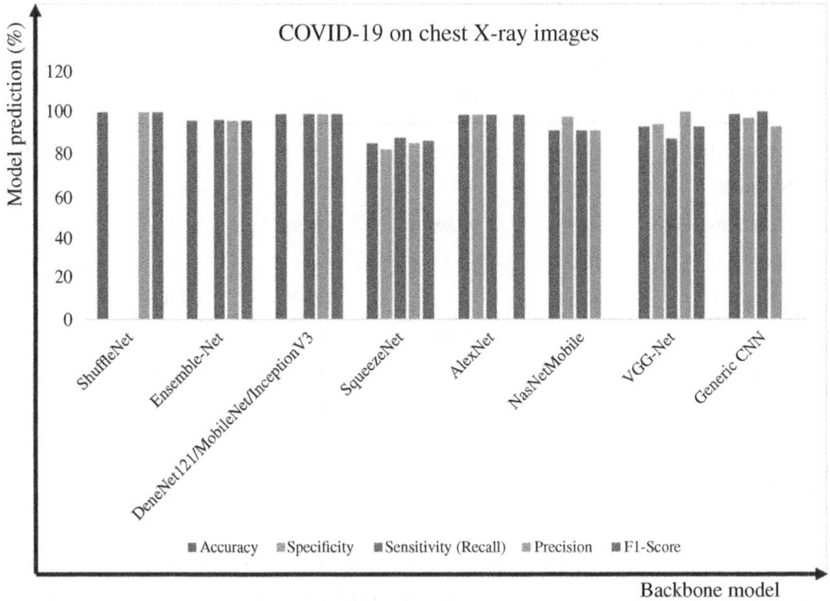

Figure 15.7 Results of the comparison of the TinyML model with other deep learning models for COVID-19 detection.

Table 15.3 Results of breast cancer performance on histopathology images (BreakHis).

Backbone model	Accuracy (%)	Specificity (%)	Sensitivity (%)	Precision (%)	F1 score (%)
SVM	99.15	—	99.00	99.00	99.00
ResNet50	99.87	—	98.87	98.77	97.83
VGG16&19	95.29	—	97.73	—	95.29
Deep belief network	86.00	84.00	87.90	—	—
GoogLeNet&ResNet	95.00	—	93.00	93.00	93.00
Custom CNN	93.13	—	86.00	100.00	92.00
VGG19	98.95	98.95	98.95	98.95	98.95
DenseNet	99.28	—	98.96	97.00	99.48
CNN	99.86	—	93.00	93.00	93.00
VGG	91.86	89.69	94.67	87.65	—

Table 15.4 Results of brain tumor performance on MRI scans.

Backbone model	Accuracy (%)	Specificity (%)	Sensitivity (%)	Precision (%)	F1 score (%)
Efficientnetb0	99.78	99.78	99.78	99.78	99.78
CNN-BraTs	99.74	—	—	—	—
VGG16&ResNet50	99.98	99.98	99.98	—	99.98
MobileNet	99.60	—	—	—	—
DensNet201	99.51	—	100.00	99.00	100.00
U-Net	—	99.90	98.99	—	—
SqueezeNet	98.70	—	98.70	98.30	98.50
EfficientNet	99.24	—	99.24	99.26	99.24
ShuffleNet	97.28	—	—	97.30	97.24
2D CNN	91.30	—	88.00	91.00	—

(99.60%), and DenseNet201 (99.51%). Deep learning models demonstrated the best results, while TinyML of EfficientNetb0, lightweight CNN, and MobileNet achieved comparable results with enhanced resource efficiency. On precision metric, EfficientNetb0(TinyML) achieved 99.78% (best result), fusion of EfficientNet, DenseNet, and MobileNet achieved 99.26%, DenseNet201 achieved 99.0%, SqueezeNet achieved 98.30%, and ShuffleNet achieved 97.30%. The result on sensitivity metric, DenseNet201, achieved 100% but EfficientNetb0 (TinyML) achieved a comparable result of 99.78%. Also, on the specificity metric, VGG16-ResNet50 achieved the best result (99.98%), and EfficientNetb0 (TinyML) achieved comparable result of 99.78%. On F1 score, DenseNet201 achieved 100%, while TinyML models such as EfficientNetb0 and SqueezeNet achieved comparable result of 99.78% and 98.50%, respectively.

Results of COVID-19 Models: As summarized in Table 15.5, TinyML-based models for COVID-19 detection achieved outstanding accuracy results, with ShuffleNet achieving 99.98% and SqueezeNet achieving 99.62%. Other models' results are also recorded in the table. In terms of precision, ShuffleNet (TinyML) excelled with a perfect score of 100.0%. Another ShuffleNet model achieved 99.98%, followed by SqueezeNet (99.78%) and DenseNet121, MobileNet, and InceptionV3 (99%). Regarding sensitivity, ShuffleNet also achieved a perfect score of 100%, with another ShuffleNet model achieving 99.98%. SqueezeNet attained 99.52%, while DenseNet121, MobileNet, and InceptionV3 produced 99%. AlexNet achieved 98.64%. For specificity, the results were 99.97% (ShuffleNet), 99.69% (SqueezeNet), 98.64% (AlexNet), 97.70% (NASNetMobile),

Table 15.5 Results of COVID-19 performance on chest X-ray images.

Baseline model	Accuracy (%)	Specificity (%)	Sensitivity (%)	Precision (%)	F1 score (%)
ShuffleNet	99.98	—	—	99.98	99.98
EnsembleNet	95.83	—	96.20	95.68	95.94
DenseNet121& MobileNet& InceptionV3	99.00	—	99.00	99.00	99.00
SqueezeNet	85.03	81.95	87.55	85.01	86.20
AlexNet	98.64	98.64	98.64	—	98.64
ShuffleNet	99.98	99.97	100.00	100.00	100.00
NASNetMobile	91.00	97.70	91.00	91.00	—
VGGNet	93.00	94.00	87.00	100.00	93.00
SqueezeNet	99.62	99.69	99.52	99.79	—
Generic CNN	98.84	97.00	100.00	93.00	—

and 97.00% (Generic CNN). Lastly, the F1 score results were 100% (ShuffleNet), 99.98% (ShuffleNet1), 99% (DenseNet121), 98.64% (AlexNet), and 95.94% (Ensemble-Net).

15.4.4 Discussion

This section presents a discussion of the significance and interpretation of key findings within the context of classification of medical imaging in real-time scenarios. Furthermore, we evaluate the results of the study highlighting areas for improvement by exploring the advantages and disadvantages of TinyML models compared to other approaches including the following:

Traditional Machine Learning (ML) Approaches: From our findings, we observed the following key differences between TinyML and traditional ML:

- TinyML is designed for resource-constrained devices, while traditional ML was designed for powerful hardware.
- TinyML models are compact, efficient, and optimized for inference, while traditional ML models are complex, large, and optimized for training.
- TinyML typically uses pre-processing techniques like pruning, while traditional ML typically uses techniques like deep learning and ensemble methods.
- TinyML focuses on real-time inference and low latency, while traditional ML focuses on high accuracy and large datasets.

- TinyML is used for edge AI applications, while traditional ML is often used for cloud-based applications.

Deep Learning (DL) Approaches: The following are some of the comparisons and key differences between TinyML and other deep learning approaches:

- TinyML employs lightweight models such as MobileNet, SqueezeNet, etc., while DL uses large, complex models like CNNs, recurrent neural networks (RNNs), etc.
- TinyML uses shallow model architectures, while DL uses deep model architectures with many layers.
- TinyML focuses on efficiency, interpretability, and generalization, while DL focuses on high accuracy, often at the cost of interpretability.
- TinyML requires a small amount of training data, while DL requires a large amount of training data.
- TinyML typically uses techniques like feature engineering, regularization, and early stopping, while DL typically uses techniques like batch normalization, dropout, and transfer learning.
- TinyML is designed for resource-constrained devices, while DL requires powerful hardware and large datasets.

We can therefore deduce that in medical image classification, TinyML offers several advantages over other approaches, including

- TinyML approach enables real-time medical image diagnosis, which allows for timely and accurate diagnoses, improved patient outcomes, and enhanced healthcare delivery.
- TinyML is also suitable for battery-powered devices and remote applications, which has led to significant reductions in power consumption and enables the development of efficient, portable, and wearable medical devices that can perform image classification tasks in real time.
- TinyML allows for efficient storage and transmission. The models are designed to be compact and efficient, with small model sizes that offer several advantages in reduced memory usage, and faster inference.
- TinyML models can achieve comparable accuracy when compared with larger models. They are designed to achieve comparable accuracy to larger models despite their smaller size and computational requirements.
- TinyML enables analysis on edge devices, which offers several benefits including reducing latency and dependence on cloud systems.
- TinyML enables medical image analysis in resource-constrained areas, achieving faster diagnosis and treatment, a cost-effective solution, that offers scalable and sustainable medical image analysis benefits.

- TinyML can update and refine models with new data, and this ensures the medical image analysis remains accurate and effective.
- TinyML enables autonomous medical devices, such as robots and drones, to perform remote healthcare tasks that expand access to medical care.

Limitations of the Study: Although the study on TinyML for real-time medical image classification and diagnosis showed significant potential, it also has several limitations, including the following:

- The compact nature and limited model size of typical TinyML models may compromise their ability to capture complex patterns in medical images.
- The limitation in terms of dataset size may not provide a comprehensive representation of all medical image diagnosis scenarios and may affect model training and performance.
- The memory and computational limitations of most mobile and edge devices reduce the processing power, which may affect the model's performance.
- Issues with a class imbalance in most medical imaging datasets result in the under-representation of certain diseases, which impacts the model's performance.

Future research in developing TinyML models should provide solutions to these challenges in order to further improve their accuracy and suitability of the model in medical image analysis.

Future research in developing TinyML models should focus on addressing these challenges to further enhance their accuracy, robustness, and suitability for medical image analysis applications.

15.4.5 Lessons Learned and Future Research Direction

Some of the key lessons learned from the study on TinyML for real-time medical image classification and diagnosis include the following:

- The novel TinyML models can achieve high accuracy in medical image classification and diagnosis despite computational resource constraints.
- Class imbalance issues must be considered when preparing and pre-processing datasets.
- The size and diversity of the dataset can affect the generalization of TinyML model.
- Using appropriate model optimization techniques can enhance model performance.
- Real-time medical image classification and diagnosis improve speed and diagnostic accuracy.
- TinyML-based models enable point-of-care and real-time diagnostics.

- Transfer learning techniques applied to pre-trained models can improve performance.
- Hyperparameter tuning is very important for optimal model performance.
- Combining models through ensemble methods improves model robustness.

Despite the existing challenges, the future research directions of TinyML in medical image classification and diagnosis are promising. More research in areas such as transfer learning, distributed learning, and the collection of larger and diverse datasets will continue to advance the field to achieve robust, faster, more accurate, and accessible diagnosis and treatment of various medical conditions.

15.5 Conclusion

In conclusion, TinyML has the potential to improve the diagnosis and classification of medical images in real-time scenarios, particularly on devices limited by resources, while maintaining an accurate and accessible diagnosis and treatment of various medical conditions. The model has various applications in the diagnosis of medical images, including tumor detection and classification, disease diagnosis, image segmentation, and denoising. Among these numerous benefits, its efficiency and effectiveness make it an invaluable tool in medical imaging.

References

1 Cai, L., Gao, J., and Zhao, D. (2020). A review of the application of deep learning in medical image classification and segmentation. *Annals of Translational Medicine* 8 (11): 713.
2 Dai, Y., Gao, Y., and Liu, F. (2021). TransMed: transformers advance multi-modal medical image classification. *Diagnostics* 11 (8): 1384.
3 Salma, W. and Eltrass, A.S. (2022). Automated deep learning approach for classification of malignant melanoma and benign skin lesions. *Multimedia Tools and Applications* 81 (22): 32643–32660.
4 Aloyayri, A. and Krzyżak, A. (2020). Breast cancer classification from histopathological images using transfer learning and deep neural networks. *Artificial Intelligence and Soft Computing: 19th International Conference, ICAISC 2020*, Zakopane, Poland (12–14 October 2020), *Proceedings, Part I 19*, 491–502. Springer.
5 Vaiyapuri, T., Dutta, A.K., Punithavathi, I.S.H. et al. (2022). Intelligent deep-learning-enabled decision-making medical system for pancreatic tumor classification on CT images. In: *Healthcare*, vol. 10 (ed. R. Shiri), 677. MDPI.

6 Dutta, L. and Bharali, S. (2021). TinyML meets IoT: a comprehensive survey. *Internet of Things* 16: 100461.

7 Abadade, Y., Temouden, A., Bamoumen, H. et al. (2023). A comprehensive survey on TinyML. *IEEE Access* 11: 96892–96922.

8 Ray, P.P. (2022). A review on TinyML: state-of-the-art and prospects. *Journal of King Saud University, Computer and Information Sciences* 34 (4): 1595–1623.

9 Tsoukas, V., Boumpa, E., Giannakas, G., and Kakarountas, A. (2021). A review of machine learning and TinyML in healthcare. *Proceedings of the 25th Pan-Hellenic Conference on Informatics*, 69–73.

10 Inan, M.S.K., Alam, F.I., and Hasan, R. (2021). Deep integrated pipeline of segmentation leading to classification for automated detection of breast cancer from breast ultrasound images. *arXiv preprint arXiv:2110.14013.*

11 Lee, H., Park, J., and Hwang, J.Y. (2020). Channel attention module with multiscale grid average pooling for breast cancer segmentation in an ultrasound image. *IEEE Transactions on Ultrasonics, Ferroelectrics, and Frequency Control* 67 (7): 1344–1353.

12 Hirsch, L., Huang, Y., Luo, S. et al. (2021). Radiologist-level performance by using deep learning for segmentation of breast cancers on MRI scans. *Radiology: Artificial Intelligence* 4 (1): e200231.

13 Tang, X., An, Y., and Li, C. (2021). Intelligent segmentation and recognition method of breast cancer based on digital image processing technology. *Wireless Communications and Mobile Computing* 2021 (1): 2256316.

14 Moreau, N., Rousseau, C., Fourcade, C. et al. (2021). Automatic segmentation of metastatic breast cancer lesions on ^{18}F-FDG PET/CT longitudinal acquisitions for treatment response assessment. *Cancers* 14 (1): 101.

15 Pedersen, A., Smistad, E., Rise, T.V. et al. (2022). H2G-Net: a multi-resolution refinement approach for segmentation of breast cancer region in gigapixel histopathological images. *Frontiers in Medicine* 9: 971873.

16 Qin, C., Wu, Y., Zeng, J. et al. (2022). Joint transformer and multi-scale CNN for DCE-MRI breast cancer segmentation. *Soft Computing* 26 (17): 8317–8334.

17 Aljuaid, H., Alturki, N., Alsubaie, N. et al. (2022). Computer-aided diagnosis for breast cancer classification using deep neural networks and transfer learning. *Computer Methods and Programs in Biomedicine* 223: 106951.

18 Lu, X. and Zhu, X. (2022). Automatic segmentation of breast cancer histological images based on dual-path feature extraction network. *Mathematical Biosciences and Engineering* 19 (11): 11137–11153.

19 Patra, D.K., Si, T., Mondal, S., and Mukherjee, P. (2023). Breast lesion detection from MRI images using quasi-oppositional slime mould algorithm. *Multimedia Tools and Applications* 82 (20): 30599–30641.

20 Fu, X., Cao, H., Hu, H. et al. (2023). Attention-based active learning framework for segmentation of breast cancer in mammograms. *Applied Sciences* 13 (2): 852.

21 Yu, L., Shaheema, S.B., Sunil, J. et al. (2023). Breast cancer segmentation using a hybrid AttendSeg architecture combined with a gravitational clustering optimization algorithm using mathematical modelling. *Open Physics* 21 (1): 20230105.

22 Ahmed, L., Iqbal, M.M., Aldabbas, H. et al. (2023). Images data practices for semantic segmentation of breast cancer using deep neural network. *Journal of Ambient Intelligence and Humanized Computing* 14 (11): 15227–15243.

23 Balaha, H.M., Antar, E.R., Saafan, M.M., and El-Gendy, E.M. (2023). A comprehensive framework towards segmenting and classifying breast cancer patients using deep learning and Aquila optimizer. *Journal of Ambient Intelligence and Humanized Computing* 14 (6): 7897–7917.

24 Raza, A., Ullah, N., Khan, J.A. et al. (2023). DeepBreastCancerNet: a novel deep learning model for breast cancer detection using ultrasound images. *Applied Sciences* 13 (4): 2082.

25 Bag, V.V., Patil, M.B., and Kendre, S.N. (2024). Frequent CNN based ensembling for MRI classification for abnormal brain growth detection. *Journal of Integrated Science and Technology* 12 (4): 785–785.

26 Vaiyapuri, T., Mahalingam, J., Ahmad, S. et al. (2023). Ensemble learning driven computer-aided diagnosis model for brain tumor classification on magnetic resonance imaging. *IEEE Access* 11: 91398–91406.

27 Dhakshnamurthy, V.K., Govindan, M., Sreerangan, K. et al. (2024). Brain tumor detection and classification using transfer learning models. *Engineering Proceedings* 62 (1): 1.

28 Islam, M.M., Barua, P., Rahman, M. et al. (2023). Transfer learning architectures with fine-tuning for brain tumor classification using magnetic resonance imaging. *Healthcare Analytics* 4: 100270.

29 Noreen, N., Palaniappan, S., Qayyum, A. et al. (2020). A deep learning model based on concatenation approach for the diagnosis of brain tumor. *IEEE Access* 8: 55135–55144.

30 Güler, M. and Namlı, E. (2024). Brain tumor detection with deep learning methods' classifier optimization using medical images. *Applied Sciences* 14 (2): 642.

31 Bourennane, M. and Naimi, H. (2023). An efficient hybrid model of CNNs and different kernels of SVM for brain tumor classification. *The Journal of Engineering and Exact Sciences* 9 (8): 16739–01e.

32 Rasool, M., Ismail, N.A., Al-Dhaqm, A. et al. (2022). A novel approach for classifying brain tumours combining a squeezenet model with SVM and fine-tuning. *Electronics* 12 (1): 149.

33 Nayak, D.R., Padhy, N., Mallick, P.K. et al. (2022). Brain tumour classification using noble deep learning approach with parametric optimization through metaheuristics approaches. *Computers* 11 (1): 10.

34 Jin, W., Dong, S., Dong, C., and Ye, X. (2021). Hybrid ensemble model for differential diagnosis between COVID-19 and common viral pneumonia by chest X-ray radiograph. *Computers in Biology and Medicine* 131: 104252.

35 Sufian, M.M., Moung, E.G., Hijazi, M.H.A. et al. (2023). COVID-19 classification through deep learning models with three-channel grayscale CT images. *Big Data and Cognitive Computing* 7 (1): 36.

36 Hasan, Md.A.M., Shin, J., and Parvin, F. (2021). Deep transfer learning based detection of COVID-19 from chest X-ray images. *Proceedings of the 2021 11th International Conference on Biomedical Engineering and Technology*, 64–70.

37 Zazzaro, G., Martone, F., Romano, G., and Pavone, L. (2021). A deep learning ensemble approach for automated COVID-19 detection from chest CT images. *Journal of Clinical Medicine* 10 (24): 5982.

38 Babukarthik, R.G., Adiga, V.A.K., Sambasivam, G. et al. (2020). Prediction of COVID-19 using genetic deep learning convolutional neural network (GDCNN). *IEEE Access* 8: 177647–177666.

39 Özyurt, F. (2021). Automatic detection of COVID-19 disease by using transfer learning of light weight deep learning model. *Traitement du Signal* 38 (1): 147.

40 Owais, M., Yoon, H.S., Mahmood, T. et al. (2021). Light-weighted ensemble network with multilevel activation visualization for robust diagnosis of COVID19 pneumonia from large-scale chest radiographic database. *Applied Soft Computing* 108: 107490.

41 Polsinelli, M., Cinque, L., and Placidi, G. (2020). A light CNN for detecting COVID-19 from CT scans of the chest. *Pattern Recognition Letters* 140: 95–100.

42 Ullah, N., Khan, J.A., El-Sappagh, S. et al. (2023). A holistic approach to identify and classify COVID-19 from chest radiographs, ECG, and CT-scan images using shufflenet convolutional neural network. *Diagnostics* 13 (1): 162.

43 Alghamdi, M.M.M., Dahab, M.Y.H., and Alazwary, N.H.A. (2023). Enhancing deep learning techniques for the diagnosis of the novel coronavirus (COVID-19) using X-ray images. *Cogent Engineering* 10 (1): 2181917.

44 Akpinar, K.N., Genc, S., and Karagol, S. (2020). Chest X-ray abnormality detection based on squeezenet. *2020 International Conference on Electrical, Communication, and Computer Engineering (ICECCE)*, 1–5. IEEE.

45 Mehmood, M.H., Hassan, F., Rahman, A.U. et al. (2024). CPDeepNet: a novel automated system for COVID-19 and pneumonia detection through lung X-rays. *Multimedia Tools and Applications* 83 (41): 88681–88698.

46 Chen, Y., Dai, X., Chen, D. et al. (2022). Mobile-former: bridging mobilenet and transformer. *Proceedings of the IEEE/CVF Conference on Computer Vision and Pattern Recognition*, 5270–5279.

47 Fang, S., Yang, J., Wang, M. et al. (2022). An improved image classification method for cervical precancerous lesions based on shufflenet. *Computational Intelligence and Neuroscience* 2022 (1): 9675628.

48 Naveen, A., Nadhana, R., Tarunika, K., and Munirathinam, R. (2024). Effective categorization of lung abnormalities using squeezenet features and machine learning models. *2024 2nd International Conference on Emerging Trends in Information Technology and Engineering (ICETITE)*, 1–6. IEEE.

49 Khan, T.M., Robles-Kelly, A., and Naqvi, S.S. (2022). T-Net: a resource-constrained tiny convolutional neural network for medical image segmentation. *Proceedings of the IEEE/CVF Winter Conference on Applications of Computer Vision*, 644–653.

50 Rahman, M.A., Bazgir, E., Hossain, S.M.S., and Maniruzzaman, Md. (2024). Skin cancer classification using NASNet. *International Journal of Science and Research Archive* 11 (1): 775–785.

51 Mehta, S., Nuechterlein, N., Mercan, E. et al. (2021). Applications of the ESP-Net architecture in medical imaging. In: *State of the Art in Neural Networks and their Applications* (ed. A.S. El-Baz and J.S. Suri), 117–131. Elsevier.

52 Marques, G., Agarwal, D., and De la Torre Díez, I. (2020). Automated medical diagnosis of COVID-19 through EfficientNet convolutional neural network. *Applied Soft Computing* 96: 106691.

53 Huang, Z., Zhou, Q., Zhu, X., and Zhang, X. (2021). Batch similarity based triplet loss assembled into light-weighted convolutional neural networks for medical image classification. *Sensors* 21 (3): 764.

54 Kurian, N.C., Sethi, A., Konduru, A.R. et al. (2021). A 2021 update on cancer image analytics with deep learning. *Wiley Interdisciplinary Reviews: Data Mining and Knowledge Discovery* 11 (4): e1410.

55 Pillai, A.S. (2021). Utilizing deep learning in medical image analysis for enhanced diagnostic accuracy and patient care: challenges, opportunities, and ethical implications. *Journal of Deep Learning in Genomic Data Analysis* 1 (1): 1–17.

56 Immonen, R. and Hämäläinen, T. (2022). Tiny machine learning for resource-constrained microcontrollers. *Journal of Sensors* 2022 (1): 7437023.

57 Kumar, M., Kumar, H.P.M., Sultana, H., and Shweta, J.S. (2023). Mobile-based classification and detection of diabetic retinopathy using TinyML. *2023 IEEE North Karnataka Subsection Flagship International Conference (NKCon)*, 1–6. IEEE.

58 Acharya, D., Eashwer, K., Kumar, S. et al. (2023). Multiple disease detection using machine learning techniques. *International Journal of Online & Biomedical Engineering* 19 (13): 120.

59 Ashwath, V.A., Sikha, O.K., and Benitez, R. (2023). TS-CNN: a three-tier self-interpretable CNN for multi-region medical image classification. *IEEE Access* 11: 78402–78418.

60 Nia, N.G., Kaplanoglu, E., and Nasab, A. (2023). Evaluation of artificial intelligence techniques in disease diagnosis and prediction. *Discover Artificial Intelligence* 3 (1): 5.

61 Widyaningrum, R., Sela, E.I., Pulungan, R., and Septiarini, A. (2023). Automatic segmentation of periapical radiograph using color histogram and machine learning for osteoporosis detection. *International Journal of Dentistry* 2023 (1): 6662911.

62 Hamza, M.A. (2023). Hyperparameter tuned deep hybrid denoising autoencoder breast cancer classification on digital mammograms. *Intelligent Automation & Soft Computing* 36 (3): 2879.

63 Tuncer, T., Ozyurt, F., Dogan, S., and Subasi, A. (2021). A novel COVID-19 and pneumonia classification method based on F-transform. *Chemometrics and Intelligent Laboratory Systems* 210: 104256.

64 Wang, W., Liang, D., Chen, Q. et al. (2020). Medical image classification using deep learning. In: *Deep Learning in Healthcare: Paradigms and Applications* (ed. Y.-W. Chen and L.C. Jain), 33–51. Cham: Springer International Publishing.

65 Hoang, L., Lee, S.-H., Lee, E.-J., and Kwon, K.-R. (2022). Multiclass skin lesion classification using a novel lightweight deep learning framework for smart healthcare. *Applied Sciences* 12 (5): 2677.

66 Tsivgoulis, M., Papastergiou, T., and Megalooikonomou, V. (2022). An improved SqueezeNet model for the diagnosis of lung cancer in CT scans. *Machine Learning with Applications* 10: 100399.

67 Vijayan, N. and Kuruvilla, J. (2022). The impact of transfer learning on lung cancer detection using various deep neural network architectures. *2022 IEEE 19th India Council International Conference (INDICON)*, 1–5. IEEE.

68 Obaid, A.M., Turki, A., Bellaaj, H. et al. (2023). Detection of gallbladder disease types using deep learning: an informative medical method. *Diagnostics* 13 (10): 1744.

69 Fradi, M., Afif, M., and Machhout, M. (2020). Deep learning based approach for bone diagnosis classification in ultrasonic computed tomographic images. *International Journal of Advanced Computer Science and Applications* 11 (12): 1–9.

70 Ibn Hasan, N. (2021). A hybrid method of COVID-19 patient detection from modified CT-scan/chest-X-ray images combining deep convolutional neural network and two-dimensional empirical mode decomposition. *Computer Methods and Programs in Biomedicine Update* 1: 100022.

71 Alshehri, A., Alharbi, B., and Alharbi, A. (2022). Pneumonia detection from chest X-ray images based on sequential model. *International Journal of Computer Science & Network Security* 22 (4): 53–58.

72 Simanjuntak, R.B.J., Fuâ, Y., Magdalena, R. et al. (2022). Cataract classification based on fundus images using convolutional neural network. *JOIV: International Journal on Informatics Visualization* 6 (1): 33–38.

73 Mahmud, Md.I., Mamun, M., and Abdelgawad, A. (2023). A deep analysis of transfer learning based breast cancer detection using histopathology images. *arXiv e-prints*, arXiv-2304.

74 Canales-Fiscal, M.R. and Tamez-Pea, J.G. (2023). Hybrid morphological-convolutional neural networks for computer-aided diagnosis. *Frontiers in Artificial Intelligence* 6: 1253183.

75 Lilhore, U.K., Simaiya, S., Sharma, Y.K. et al. (2024). A precise model for skin cancer diagnosis using hybrid U-Net and improved MobileNet-V3 with hyperparameters optimization. *Scientific Reports* 14 (1): 4299.

76 Liu, X., Gao, K., Liu, B. et al. (2021). Advances in deep learning-based medical image analysis. *Health Data Science* 2021: 8786793.

77 Barragán-Montero, A., Javaid, U., Valdés, G. et al. (2021). Artificial intelligence and machine learning for medical imaging: a technology review. *Physica Medica* 83: 242–256.

78 Alajlan, N.N. and Ibrahim, D.M. (2022). TinyML: enabling of inference deep learning models on ultra-low-power IoT edge devices for AI applications. *Micromachines* 13 (6): 851.

79 Benhammou, Y., Achchab, B., Herrera, F., and Tabik, S. (2020). BreakHis based breast cancer automatic diagnosis using deep learning: taxonomy, survey and insights. *Neurocomputing* 375: 9–24.

80 Nawaz, S.A., Khan, D.M., and Qadri, S. (2022). Brain tumor classification based on hybrid optimized multi-features analysis using magnetic resonance imaging dataset. *Applied Artificial Intelligence* 36 (1): 2031824.

81 Constantinou, M., Exarchos, T., Vrahatis, A.G., and Vlamos, P. (2023). COVID-19 classification on chest X-ray images using deep learning methods. *International Journal of Environmental Research and Public Health* 20 (3): 2035.

82 Aggarwal, R., Sounderajah, V., Martin, G. et al. (2021). Diagnostic accuracy of deep learning in medical imaging: a systematic review and meta-analysis. *npj Digital Medicine* 4 (1): 65.

83 Müller, D., Soto-Rey, I., and Kramer, F. (2022). Towards a guideline for evaluation metrics in medical image segmentation. *BMC Research Notes* 15 (1): 210.

84 Hicks, S.A., Strümke, I., Thambawita, V. et al. (2022). On evaluation metrics for medical applications of artificial intelligence. *Scientific Reports* 12 (1): 5979.

85 Varoquaux, G. and Cheplygina, V. (2022). Machine learning for medical imaging: methodological failures and recommendations for the future. *npj Digital Medicine* 5 (1): 48.

86 Chan, J.H. and Li, C. (2022). Learning from imbalanced COVID-19 chest X-ray (CXR) medical imaging data. *Methods* 202: 31–39.

16

Biometric Authentication in TinyML: Opportunities and Challenges

Oleksandr Kuznetsov[1,2], Emanuele Frontoni[3], Marco Arnesano[1], Oleksii Smirnov[4], and Boris Khruskov[5]

[1]*Department of Theoretical and Applied Sciences, eCampus University, Novedrate, Italy*
[2]*Department of Information and Communication Systems Security, V. N. Karazin Kharkiv National University, Kharkiv, Ukraine*
[3]*Department of Political Sciences, Communication and International Relations, University of Macerata, Macerata, Italy*
[4]*Department of Cyber Security and Software, Central Ukrainian National Technical University, Kropyvnytskyi, Ukraine*
[5]*Department of Cyber Security and Information Technology, University of Customs and Finance, Dnipro, Ukraine*

16.1 Introduction

In the era of ubiquitous computing and the Internet of Things (IoT), securing access to devices and protecting sensitive information has become a paramount concern [1, 2]. Traditional authentication methods, such as passwords and personal identification numbers (PINs), often fall short in providing the necessary level of security and user convenience required by the diverse range of IoT applications. Biometric authentication, which relies on the unique physiological or behavioral characteristics of individuals, has emerged as a promising solution to address these challenges [3].

The integration of biometric authentication techniques into tiny machine learning (TinyML) systems has gained significant attention in recent years [4]. TinyML refers to the deployment of machine learning algorithms on resource-constrained devices, such as microcontrollers and low-power IoT nodes [5, 6]. By combining the power of biometric authentication with the efficiency and scalability of TinyML, it becomes possible to develop secure and user-friendly authentication mechanisms for a wide range of IoT applications.

This chapter explores the opportunities and challenges associated with the integration of biometric authentication techniques into TinyML systems. It

Tiny Machine Learning: Design Principles and Applications, First Edition.
Edited by Agbotiname Lucky Imoize, Dinh-Thuan Do, and Houbing Herbert Song.

provides a comprehensive overview of various biometric modalities, including fingerprint recognition, facial recognition, iris scanning, voice recognition, gait recognition, and ear recognition [7]. The chapter delves into the fundamental concepts, algorithms, and performance metrics associated with each modality, highlighting their strengths and limitations in the context of resource-constrained environments.

One of the key challenges addressed in this chapter is the adaptation of biometric techniques for TinyML systems. The limited computational power, memory capacity, and energy budget of microcontrollers pose significant constraints on the implementation of biometric algorithms. The chapter discusses various strategies for developing resource-efficient biometric algorithms, such as lightweight feature extraction, template compression, efficient matching, and hardware acceleration. It also explores the hardware considerations for integrating biometric sensors into IoT devices, focusing on factors such as sensor selection, form factor, power consumption, and interface requirements.

The chapter presents a comparative analysis of different biometric authentication methods, evaluating their performance based on metrics such as accuracy, efficiency, and user acceptance. The analysis provides valuable insights into the trade-offs and suitability of each modality for specific TinyML applications. It also highlights the potential of multimodal biometric authentication, where multiple biometric traits are combined to enhance the overall security and reliability of the system.

Furthermore, the chapter discusses the privacy and ethical considerations associated with the deployment of biometric authentication in IoT devices. It emphasizes the importance of establishing clear privacy policies, implementing robust data protection measures, and adhering to relevant regulations and standards. The chapter also explores future research directions in the field of biometric authentication for TinyML systems, including the development of novel biometric modalities, optimization techniques for energy efficiency and real-time performance, and the integration of biometric authentication with other security mechanisms.

The key points covered in this chapter are

- Overview of biometric authentication techniques and their relevance to TinyML systems
- Comparative analysis of different biometric modalities, including fingerprint, facial, iris, voice, gait, and ear recognition
- Adaptation of biometric techniques for resource-constrained environments, focusing on resource-efficient algorithms and hardware considerations
- Future research directions and opportunities in the field of biometric authentication for TinyML systems

The chapter is organized as follows: Section 16.2 presents the related work. Section 16.3 provides an overview of various biometric authentication techniques, discussing their fundamental concepts, algorithms, and performance metrics. Section 16.4 presents a comparative analysis of these techniques, evaluating their strengths and limitations in the context of TinyML systems. Section 16.5 focuses on the adaptation of biometric techniques for resource-constrained environments, exploring resource-efficient algorithms and hardware considerations. Finally, Sections 16.6 and 16.7 conclude the chapter by summarizing the key insights and emphasizing the potential of biometric authentication in enhancing the security and user experience of TinyML systems.

16.2 Related Work

The integration of biometric authentication techniques into TinyML systems has garnered significant attention from the research community in recent years. This section provides an overview of the relevant literature, highlighting the key contributions and advancements in this field.

Agarwal and Joshi [8] proposed a secure and effective self-error-correctable physical unclonable function (PUF) for tiny IoT nodes. They employed systematic Golay code and decoding algorithms to improve the reliability of PUF-based authentication in IoT applications. Using an efficient error correction mechanism, their hardware implementation on an field-programmable gate array (FPGA) board demonstrated good performance with 45% uniqueness and 97% reliability.

Awad et al. [9] conducted a comprehensive review of artificial intelligence (AI)-powered biometrics for IoT security. They analyzed the interdependencies between AI, biometrics, and IoT domains, covering the applications of AI and biometrics in the context of IoT. The authors highlighted the current challenges and future research directions for the deployment of intelligent biometrics in various IoT application domains.

Coelho et al. [10] presented a security management approach for authentication that combines two convolutional neural networks (CNNs) for the biometric identification of users. Their approach relies on federated learning (FL) to support data management and privacy by training decentralized models collaboratively without effective data exchange. The authors combined photoplethysmography (PPG) and electrocardiogram signals to improve identification accuracy and establish multimodal authentication.

Cremonezi et al. [11] provided a comprehensive state-of-the-art review of identity management (IdM) systems, presenting the fundamental concepts and challenges when developing an IdM for IoT applications. They explored existing works that propose solutions to address at least one component of IdMs in response to the current challenges, including the scalability of identity provisioning, authentication for IoT devices, and concerns regarding the performance and management of the authorization processes.

Ding et al. [12] introduced a fuzzy-based temporal convolutional autoencoder (FTCAE) framework for gait recognition from inertial gait time series. They incorporated an interval type-2 fuzzy set (IT2FS) dense layer to handle uncertainties, imprecision, and noise of feature maps, enabling the learning of curious representations in fuzzy latent space. Their experimentation on public gait sensory datasets validated the efficiency of the proposed FTCAE with accuracies of 98.48% and 95.11% for authentication and identification, respectively.

Galasso et al. [13] focused on classifying subjects' physical activity levels using kinematic features provided by wearable wireless inertial measurement unit (IMU) sensors. They employed the neighborhood component analysis (NCA) algorithm to process the statistical feature space and select the most significant ones. Their approach resulted in a significant increase in accuracy for various classifiers, with an improvement of approximately 20%.

Kamruzzaman and Alruwaili [14] proposed a new methodology named optimizing computer vision with AI-enabled technology (OCV-AI) for solving image classification, object classification, and facial recognition problems in 6G wireless networks. They introduced deep learning 6G proposed frameworks (Dl-6G) to recognize pattern recognition and intelligent management systems and provide a driven methodology planned to be provisioned automatically.

Khalid et al. [15] summarized the state-of-the-art approaches for preserving privacy in AI-based healthcare applications. They elaborated on prominent privacy-preserving techniques such as FL and hybrid techniques, along with potential privacy attacks, security challenges, and future directions.

Oliveira et al. [4] comprehensively reviewed the emerging concept of the Internet of Intelligent Things (IoIT), adopting an integrated perspective centered on the areas of embedded systems, edge computing, and machine learning. They performed an extensive state-of-the-art survey to identify recent works that have implemented TinyML models on different off-the-shelf embedded devices, analyzing the development of practical solutions and discussing recent research trends and future perspectives.

Praveen Kumar and Priyanka [16] provided a comprehensive overview of the security issues related to various hardware-level attacks, memory attacks, hardware Trojans, and their countermeasures in different domains. They investigated the use of primitive hardware techniques, such as physically unclonable functions (PUFs), to generate unique keys for devices by preventing unauthorized access to the hardware during authentication for different domains.

Sodhro et al. [17] conducted a survey on intelligent authentication for securing IoT-5G devices deployed in the healthcare domain. They reviewed IoT-5G device authentication, radio-frequency fingerprinting, and mutual authentication and envisioned the use of AI to integrate the attributes of the physical layer and 5G networks to empower intelligent healthcare devices.

Wan et al. [18] proposed a novel approach for biometric authentication using PPG signals and deep learning techniques. They designed and trained a multi-scale feature fusion deep learning (MFFD) model that takes the relevant features extracted from PPG signals as input and learns to differentiate between individuals based on their unique PPG patterns. Their evaluation demonstrated the effectiveness of the PPG-based biometric authentication system, achieving high accuracy while addressing key security concerns.

These studies collectively contribute to the growing body of knowledge on biometric authentication techniques for TinyML systems. They encompass various aspects, including the development of resource-efficient algorithms, the integration of AI and deep learning techniques, the exploration of novel biometric modalities, and the consideration of privacy and security challenges. The insights gained from these works provide a solid foundation for further research and development in this field, paving the way for the widespread adoption of biometric authentication in IoT and TinyML applications.

Despite the significant advancements in biometric authentication techniques for TinyML systems, there is still a need for a comprehensive and systematic approach that addresses the unique challenges posed by resource-constrained environments. Our work aims to fill this gap by providing a holistic framework that encompasses the adaptation of biometric techniques, the development of resource-efficient algorithms, and the consideration of privacy and security aspects, enabling the effective deployment of biometric authentication in a wide range of TinyML applications.

16.3 Overview of Biometric Authentication Techniques

Biometric authentication has emerged as a robust and reliable means of verifying user identity in various domains, including access control, financial transactions, and personal device security [19, 20]. This chapter provides a comprehensive overview of the most prevalent biometric modalities, focusing on their fundamental principles, algorithmic foundations, and suitability for integration into TinyML systems. By examining the strengths and limitations of each technique, we aim to guide researchers and practitioners in selecting the most appropriate biometric solution for their specific TinyML application.

16.3.1 Fingerprint Recognition

Fingerprint recognition stands as one of the most widely adopted biometric authentication techniques, owing to its long-standing use in forensic

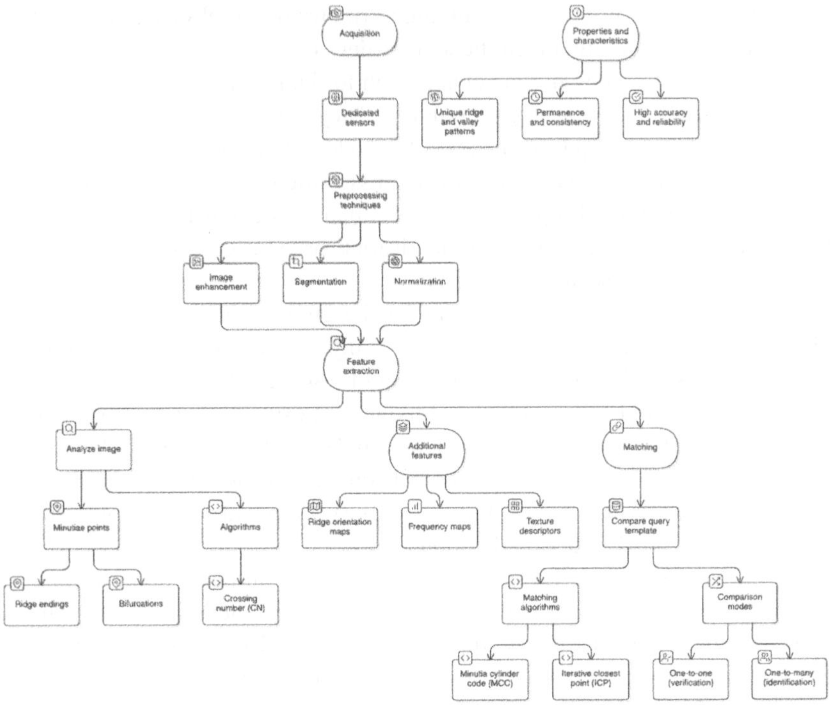

Figure 16.1 Fingerprint recognition process.

investigations and its high accuracy in identifying individuals [3, 21]. The unique patterns of ridges and valleys on the surface of a finger, formed during embryonic development, provide a distinctive and permanent signature that can be captured and analyzed for authentication purposes [21].

The flowchart in Figure 16.1 delineates the comprehensive process of fingerprint recognition, highlighting each critical stage from acquisition to matching. Initially, a high-resolution fingerprint image is obtained through dedicated sensors, followed by preprocessing techniques to enhance image quality. Subsequently, feature extraction algorithms identify unique characteristics such as minutiae points and additional features like ridge orientation and texture descriptors. The matching phase involves comparing the query template with stored database templates using sophisticated algorithms. This process is underscored by the inherent properties of fingerprints, including their unique patterns, permanence, and high reliability in identification.

At the core of fingerprint recognition lie two fundamental concepts: feature extraction and matching [22]. The first step involves acquiring a high-resolution image of the fingerprint using a dedicated sensor, such as an optical, capacitive, or ultrasonic scanner [9, 17]. Preprocessing techniques, including image enhancement, segmentation, and normalization, are then applied to improve the quality and consistency of the captured image.

Feature extraction algorithms subsequently analyze the preprocessed image to identify and encode the discriminative characteristics of the fingerprint. The most common features employed in fingerprint recognition are minutiae points, which represent the local discontinuities in the ridge patterns, such as ridge endings and bifurcations [22]. Minutiae-based methods, such as the widely used crossing number (CN) algorithm, detect and map these points, recording their spatial coordinates and orientation [9, 21].

In addition to minutiae, other feature sets, such as ridge orientation maps, frequency maps, and texture descriptors, can be extracted to complement the minutiae-based representation [3, 17]. These additional features capture the global and local structural properties of the fingerprint, enhancing the discriminative power of the biometric template.

Once the features are extracted, the matching process compares the query fingerprint template with the enrolled templates stored in a database. Matching algorithms, such as the minutia cylinder code (MCC) and the iterative closest point (ICP) algorithm, assess the similarity between two templates by aligning their feature sets and computing a matching score. The comparison can be performed in a one-to-one (verification) or one-to-many (identification) manner, depending on the application requirements.

Despite its widespread adoption and proven accuracy, fingerprint recognition faces challenges when applied to TinyML systems. The limited computational resources and memory constraints of microcontrollers necessitate the development of lightweight and efficient algorithms for feature extraction and matching. Furthermore, the form factor and power consumption of fingerprint sensors must be optimized to ensure seamless integration into IoT devices.

Ongoing research efforts in fingerprint recognition for TinyML focus on developing compact and energy-efficient sensing technologies, such as thin-film transistor (TFT) and microelectromechanical system (MEMS) sensors [3, 21]. Additionally, researchers are exploring the use of deep learning techniques, such as CNNs, to enhance the robustness and accuracy of fingerprint recognition algorithms in resource-constrained environments.

The flowchart in Figure 16.2 illustrates the challenges and research directions in fingerprint recognition for TinyML systems. It identifies the primary challenges, such as limited computational resources, the need for lightweight algorithms, and sensor optimization. Research directions focus on developing compact sensors, optimizing algorithms, applying deep learning techniques, and ensuring

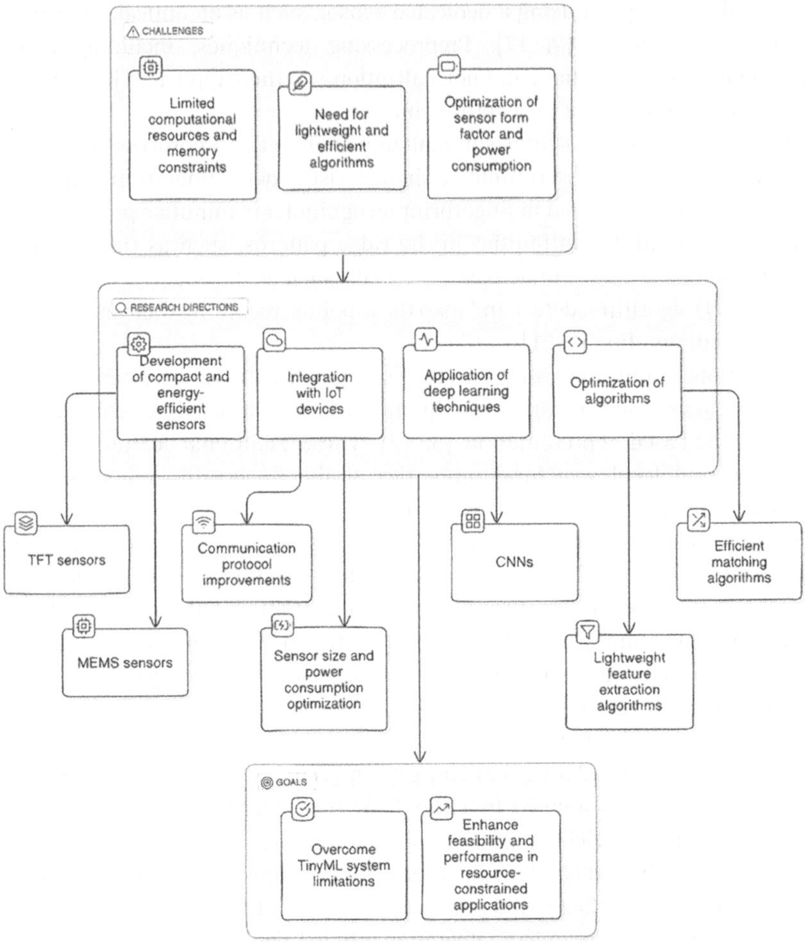

Figure 16.2 Challenges and research directions in fingerprint recognition for TinyML.

integration with IoT devices. These efforts aim to overcome TinyML system limitations and enhance performance in resource-constrained applications.

As the demand for secure and convenient authentication in IoT devices continues to grow, fingerprint recognition remains a promising biometric modality for TinyML systems. By addressing the challenges of resource efficiency and sensor miniaturization, researchers can unlock the full potential of fingerprint-based authentication in a wide range of TinyML applications, from smart homes and wearables to industrial IoT and beyond.

16.3.2 Facial Recognition

Facial recognition has gained significant attention as a biometric authentication technique due to its non-intrusive nature and the ubiquity of cameras in modern devices [9, 14]. Leveraging the unique features and geometry of an individual's face, this modality offers a natural and intuitive means of verifying identity in various applications, including access control, surveillance, and user authentication on mobile devices [23, 24].

The process of facial recognition involves several key stages: face detection, preprocessing, feature extraction, and matching [3, 25]. Face detection algorithms, such as the Viola–Jones framework, and deep learning-based methods, like Single Shot MultiBox Detector (SSD) and You Only Look Once (YOLO), localize and isolate the facial region from the input image or video frame. These algorithms exploit the distinctive characteristics of facial features, such as eyes, nose, and mouth, to rapidly identify faces in real time [3, 25].

Once the facial region is detected, preprocessing techniques are applied to normalize the image and mitigate the impact of variations in illumination, pose, and expression. Common preprocessing steps include face alignment, which involves rotating and scaling the face to a canonical pose, and illumination normalization, which compensates for uneven lighting conditions using techniques like histogram equalization and homomorphic filtering.

Feature extraction is a critical step in facial recognition, as it aims to capture the discriminative information that distinguishes one individual from another. Traditional approaches, such as principal component analysis (PCA) and linear discriminant analysis (LDA), project the high-dimensional facial data onto a lower-dimensional subspace, retaining the most salient features. These methods, often referred to as holistic or appearance-based techniques, treat the entire face as a single entity and learn a compact representation of facial variations [3, 25].

In recent years, deep learning techniques have revolutionized facial recognition by enabling the learning of hierarchical and abstract features directly from raw pixel data. CNNs, such as DeepFace, FaceNet, and ArcFace, have achieved remarkable performance in face verification and identification tasks [26–28]. These networks employ a series of convolutional and pooling layers to extract local and global facial features, followed by fully connected layers that learn a compact face embedding [3, 25].

The derived facial features, whether obtained through traditional methods or deep learning, are then compared during the matching stage to determine the identity of an individual [29, 30]. Matching algorithms compute the similarity between the query face and the enrolled face templates stored in a database. Common similarity measures include Euclidean distance, cosine similarity, and learned similarity metrics optimized by the deep learning models.

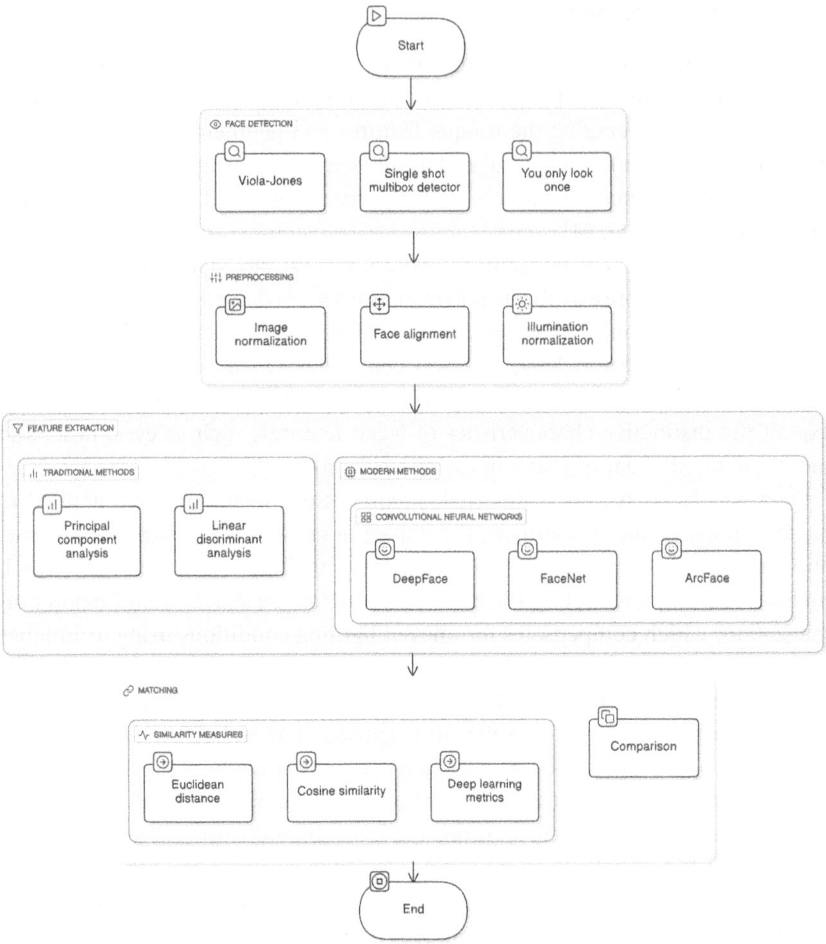

Figure 16.3 Facial recognition process.

The flowchart in Figure 16.3 illustrates the key stages and methodologies in the facial recognition process. It begins with face detection using algorithms like Viola–Jones, SSD, and YOLO, followed by preprocessing to normalize the image and adjust for illumination and pose variations. Feature extraction employs both traditional methods like PCA and LDA and modern CNN-based approaches, including DeepFace, FaceNet, and ArcFace. The final stage, matching, compares the extracted features with database templates using similarity measures such as Euclidean distance and cosine similarity. This structured approach highlights the integration of advanced techniques to enhance the accuracy and reliability of facial recognition systems.

Facial recognition systems face several challenges, particularly when deployed on resource-constrained TinyML platforms. The high dimensionality of facial data and the complexity of deep learning models pose significant computational and memory requirements. To address these challenges, researchers have explored techniques such as model compression, quantization, and pruning to reduce the size and computational cost of facial recognition models without compromising accuracy.

Another critical aspect of facial recognition in TinyML is the acquisition of high-quality face images under varying environmental conditions. Low-resolution cameras, limited processing power, and constrained energy budgets of IoT devices necessitate the development of efficient face detection and preprocessing algorithms that can operate in real time with minimal overhead.

Privacy concerns surrounding the collection and use of facial data have also spurred research into privacy-preserving facial recognition techniques. Approaches such as FL, homomorphic encryption, and differential privacy enable the training and deployment of facial recognition models without the need for centralized storage of sensitive biometric data.

16.3.3 Iris Scanning

Iris scanning has emerged as a highly accurate and reliable biometric authentication technique, leveraging the intricate patterns of the human iris for identity verification [3]. The iris, the colorful ring surrounding the pupil, contains a complex arrangement of features, including furrows, ridges, and crypts, which remain stable throughout an individual's lifetime [7, 31]. These unique characteristics, along with the iris's protected location behind the cornea, make it an ideal candidate for biometric authentication in various domains, including high-security access control and border control systems.

The process of iris recognition begins with image acquisition, where a specialized camera captures a high-resolution image of the eye under near-infrared (NIR) illumination. NIR light is preferred due to its ability to penetrate the surface of the iris and reveal the underlying texture details, while minimizing reflections and artifacts caused by ambient light [32, 33].

Preprocessing techniques are then applied to the captured image to isolate the iris region from surrounding structures, such as the pupil, sclera, and eyelids [32, 34]. This segmentation step typically involves a combination of edge detection, Hough transform, and active contour models to precisely localize the iris boundaries. Once segmented, the iris region undergoes normalization, which unwraps the circular iris into a rectangular block of fixed dimensions, accounting for variations in pupil size and image orientation.

Feature extraction is a crucial step in iris recognition, aiming to capture the discriminative information within the normalized iris texture. The most widely adopted approach is Daugman's algorithm, which applies 2D Gabor wavelets to the iris image to extract a binary feature vector known as the IrisCode. The Gabor wavelets, with their ability to capture both spatial and frequency information, effectively encode the local and global iris patterns into a compact representation [29, 35].

Other feature extraction techniques have also been explored, including the use of discrete cosine transform (DCT), wavelet packets, and deep learning-based methods. These approaches aim to capture complementary iris features and enhance the discriminative power of the extracted representation.

During the matching stage, the extracted iris features are compared against a database of enrolled iris templates to determine the identity of an individual. The comparison is typically performed using the Hamming distance, which measures the dissimilarity between two binary IrisCodes. A decision threshold is applied to the Hamming distance to classify the comparison as a match or a non-match.

Figure 16.4 illustrates the process of iris recognition, highlighting the sequential stages from image acquisition to matching. The process begins with capturing a high-resolution image of the eye under NIR illumination. Preprocessing techniques then isolate the iris, segment the image, and normalize it. Feature extraction is performed using methods such as Daugman's algorithm and other techniques, followed by the matching stage, which compares the extracted features with database templates using similarity measures like the Hamming distance.

Deploying iris recognition on TinyML platforms presents several challenges due to the resource constraints of microcontrollers. The high-resolution imaging requirements and the computational complexity of iris segmentation and feature extraction algorithms demand efficient implementations that can operate within the limited memory and processing power of TinyML devices.

Researchers have explored various techniques to optimize iris recognition for resource-constrained environments. These include the use of lightweight segmentation algorithms, such as the integro-differential operator and the Hough transform, which reduce the computational burden while maintaining accurate iris localization. Feature extraction methods based on compressed sensing and sparse representations have also been investigated to minimize the storage and processing requirements of iris templates.

Figure 16.5 outlines the properties, challenges, and research directions for iris recognition, particularly in the context of TinyML platforms. It underscores the high accuracy and reliability of iris patterns, their security advantages, and key applications. The challenges include limited resources and computational

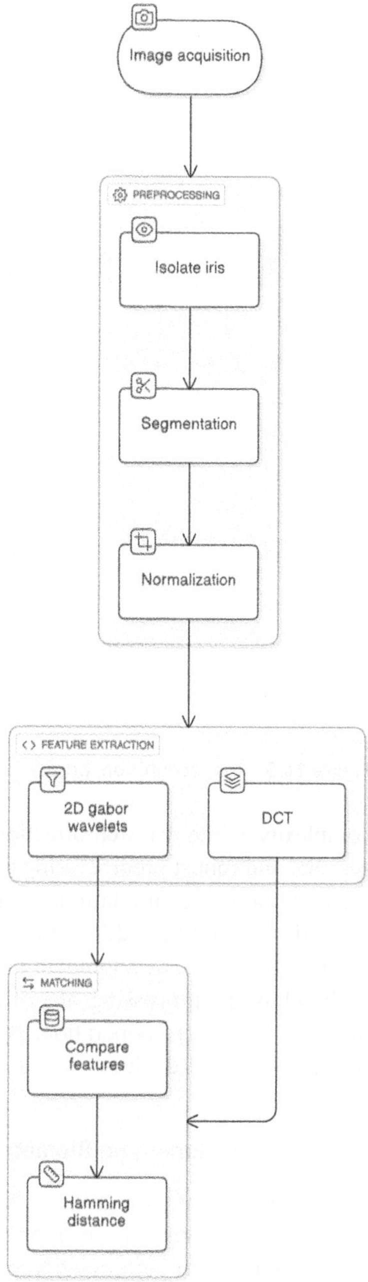

Figure 16.4 Iris recognition process.

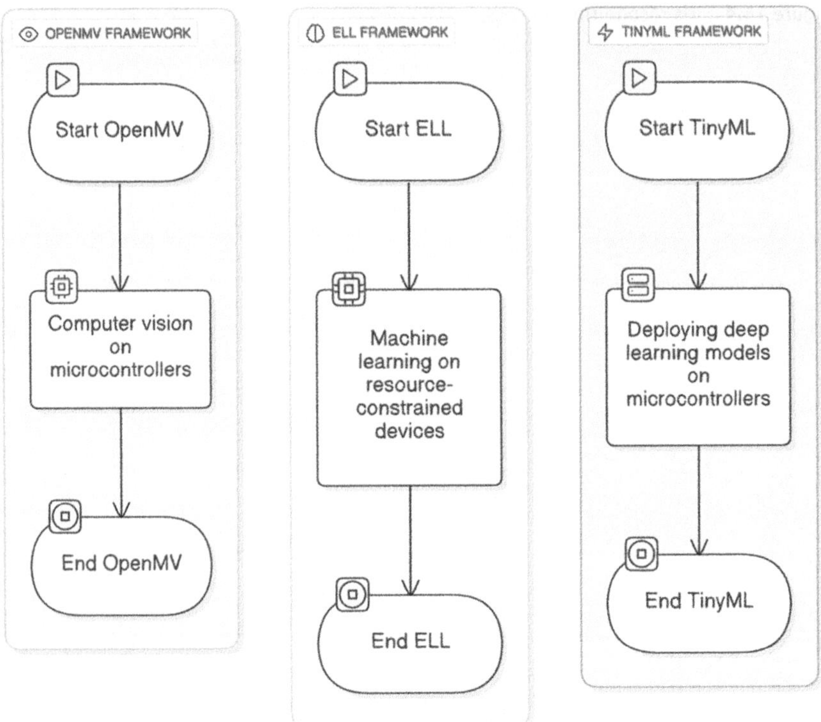

Figure 16.5 Iris recognition: properties, challenges, and research directions.

complexity, while research directions focus on algorithm optimization, compact sensors, and robust preprocessing techniques to handle varying conditions.

In addition to computational efficiency, the integration of iris scanning into TinyML systems must address the challenges of image acquisition in uncontrolled environments. The development of compact and low-power NIR cameras, along with robust preprocessing algorithms that can handle variations in lighting, distance, and user cooperation, is crucial for the practical deployment of iris recognition in IoT devices.

16.3.4 Other Emerging Biometric Modalities

While fingerprint recognition, facial recognition, and iris scanning have established themselves as the most widely adopted biometric authentication techniques, researchers continue to explore novel and emerging modalities to address the limitations of traditional approaches and cater to the diverse requirements of TinyML applications. This subsection delves into three promising biometric modalities: voice recognition, gait recognition, and ear recognition, discussing their principles, advantages, and challenges in the context of TinyML systems.

16.3.4.1 Voice Recognition

Voice recognition, also known as speaker recognition, leverages the unique acoustic characteristics of an individual's voice for biometric authentication [36]. The human voice contains a wealth of information, including the fundamental frequency, pitch, and timbre, which are determined by the physical structure of the vocal tract and learned speaking habits. Voice recognition systems aim to capture and analyze these distinctive features to verify or identify speakers.

The process of voice recognition typically involves two main stages: feature extraction and matching [3, 37]. During feature extraction, the speech signal is segmented into short frames, and various acoustic features are computed from each frame. Common feature extraction techniques include Mel-frequency cepstral coefficients (MFCCs), linear predictive coding (LPC), and perceptual linear prediction (PLP). These features capture the spectral and temporal characteristics of the speech signal, providing a compact representation of the speaker's voice.

Matching algorithms then compare the extracted features against a database of enrolled speaker templates to determine the identity of the speaker. Techniques such as Gaussian mixture models (GMMs), hidden Markov models (HMMs), and deep learning-based approaches like CNNs and long short-term memory (LSTM) networks have been successfully applied to voice recognition tasks [37, 38].

Figure 16.6 illustrates the voice recognition process, detailing the sequential stages from feature extraction to matching. Initially, the speech signal is segmented into short frames, and acoustic features such as MFCCs, LPC, and PLP are computed. These features capture the spectral and temporal characteristics of the voice. The extracted features are then compared with speaker templates in a database using techniques like GMMs, HMMs, CNNs, and LSTM networks [3, 37, 38].

The integration of voice recognition into TinyML systems offers several advantages. Voice-based authentication provides a hands-free and natural user experience, making it suitable for applications where physical contact with the device is inconvenient or impractical. Moreover, the ubiquity of microphones in IoT devices enables the seamless acquisition of speech data for biometric processing.

However, voice recognition in TinyML faces challenges related to resource constraints and environmental factors. The limited computational power and memory of microcontrollers necessitate the development of lightweight feature extraction and matching algorithms that can operate efficiently on TinyML platforms. Additionally, the presence of background noise, variations in recording

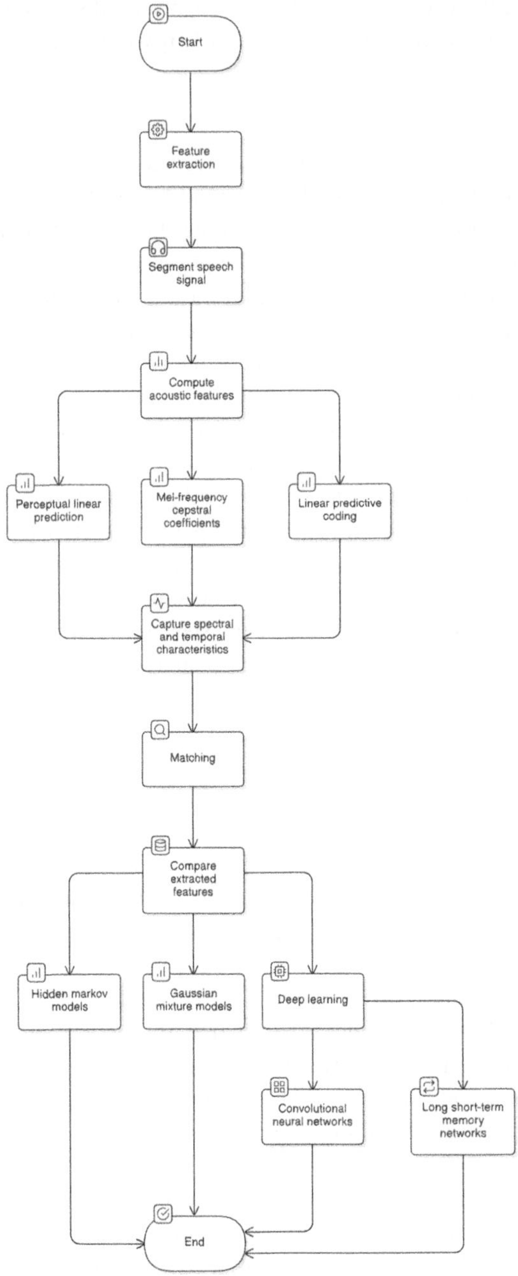

Figure 16.6 Voice recognition process.

conditions, and the impact of emotions and health conditions on an individual's voice can affect the accuracy and robustness of voice recognition systems.

Researchers have explored techniques to address these challenges, such as noise reduction algorithms, feature normalization methods, and domain adaptation techniques to improve the performance of voice recognition in diverse acoustic environments. The use of deep learning models, particularly those optimized for resource-constrained devices, has shown promise in enhancing the accuracy and generalization ability of voice recognition systems in TinyML applications.

16.3.4.2 Gait Recognition

Gait recognition is an emerging biometric modality that utilizes an individual's walking pattern for authentication [39, 40]. The human gait is a complex and idiosyncratic motion that involves the coordination of various body parts, including the legs, arms, and torso. Gait recognition systems capture and analyze the spatiotemporal characteristics of an individual's gait to extract unique features for biometric identification [41].

The process of gait recognition typically involves the following steps: data acquisition, preprocessing, feature extraction, and matching. Data acquisition is performed using various sensors, such as cameras, pressure sensors, or IMUs, which capture the motion dynamics of the individual. Preprocessing techniques, including background subtraction, silhouette extraction, and gait cycle detection, are applied to isolate the relevant gait information from the raw sensor data [13, 42].

Feature extraction methods aim to capture the discriminative characteristics of the gait pattern. Approaches like the gait energy image (GEI) and the gait entropy image (GEnI) generate compact 2D representations of the gait by averaging the silhouettes over a complete gait cycle. Other techniques, such as the use of spatial-temporal motion descriptors, kinematic features, and deep learning-based feature learning, have also been explored to capture the dynamics and variability of gait patterns [12, 42].

Matching algorithms compare the extracted gait features against a database of enrolled templates to determine the identity of an individual. Distance metrics, such as Euclidean distance and dynamic time warping (DTW), are commonly used to measure the similarity between gait features. Machine learning classifiers, including support vector machines (SVMs) and neural networks, have also been employed for gait recognition [13, 39].

Figure 16.7 illustrates the gait recognition process, detailing the key stages from data acquisition to matching. Data are acquired using various sensors, followed by preprocessing steps such as background subtraction, silhouette extraction, and gait cycle detection. Feature extraction methods like GEI, GEnI, spatial-temporal motion descriptors, and deep learning-based methods are used to capture the

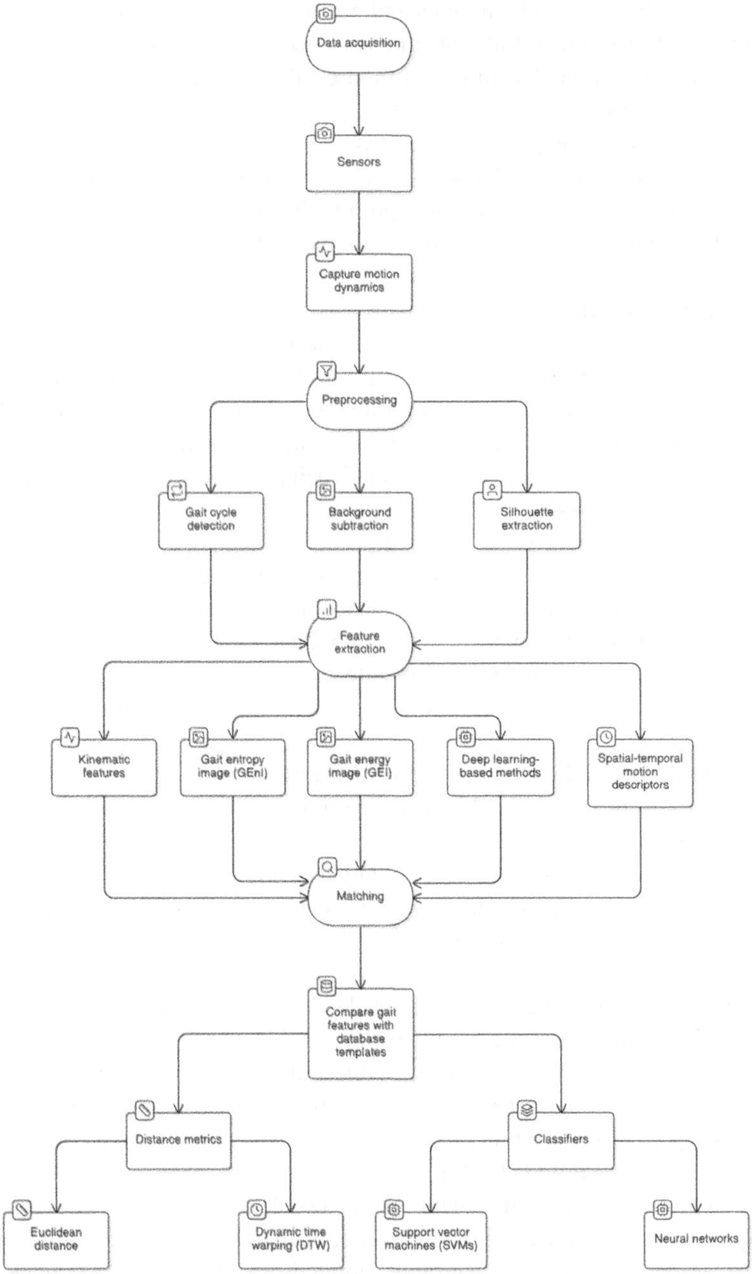

Figure 16.7 Gait recognition process.

unique gait characteristics. The extracted features are then matched with database templates using distance metrics and classifiers.

Gait recognition offers several advantages for TinyML systems. It enables unobtrusive and continuous authentication, as gait data can be collected passively without requiring explicit user interaction. Gait recognition can be particularly useful in applications like activity monitoring, access control, and user identification in smart environments.

However, gait recognition in TinyML faces challenges related to sensor limitations, environmental factors, and individual variations. Factors such as walking speed, surface conditions, footwear, and physical changes due to aging or injuries can affect its accuracy. The resource constraints of TinyML platforms also require the development of computationally efficient gait recognition algorithms that can operate with limited memory and processing power.

Figure 16.8 outlines the properties, challenges, and research directions for gait recognition, particularly in the context of TinyML. The properties emphasize the unique gait pattern and its advantages for unobtrusive and continuous authentication. Challenges include sensor limitations, environmental factors, individual variations, and resource constraints. Research directions focus on multimodal sensors, feature selection and dimensionality reduction, and domain adaptation techniques to enhance gait recognition performance.

Researchers have investigated techniques to improve the robustness and efficiency of gait recognition in TinyML systems. These include the use of

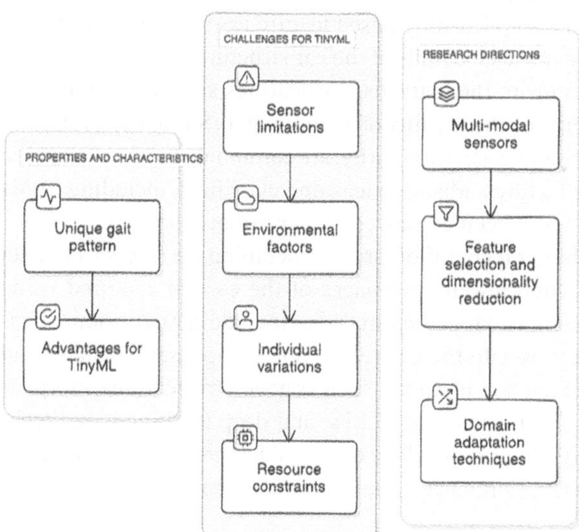

Figure 16.8 Gait recognition: properties, challenges, and research directions.

multimodal sensors to capture complementary gait information, feature selection and dimensionality reduction methods to optimize the gait representation, and domain adaptation techniques to handle variations across different walking conditions and individuals.

16.3.4.3 Ear Recognition

Ear recognition is a relatively new biometric modality that exploits the unique shape and structure of the human ear for authentication. The external ear, or pinna, exhibits a complex and stable pattern of curves, ridges, and valleys that remain relatively unchanged throughout an individual's lifetime. These distinctive characteristics make the ear a promising biometric trait for personal identification [3, 21, 43].

Ear recognition systems typically involve the following stages: image acquisition, preprocessing, feature extraction, and matching. Image acquisition is performed using cameras or specialized ear scanners that capture high-resolution images of the ear. Preprocessing techniques, such as ear detection, alignment, and normalization, are applied to isolate the ear region from the captured image and compensate for variations in pose and scale.

Feature extraction methods aim to capture the discriminative information present in the ear structure. Approaches like the use of geometric features, such as the shape and size of the ear, and the analysis of the ear's morphological characteristics, including the helix, antihelix, and tragus, have been explored for ear recognition. Other techniques, such as the use of wavelet transforms, local binary patterns (LBPs), and deep learning-based feature learning, have also been investigated to capture the intricate details of the ear structure [44, 45].

Matching algorithms compare the extracted ear features against a database of enrolled templates to determine the identity of an individual. Similarity measures, such as Euclidean distance and cosine similarity, are commonly used to assess the resemblance between ear features. Machine learning classifiers, including SVMs and neural networks, have also been employed for ear recognition [43].

Figure 16.9 illustrates the ear recognition process, detailing the stages from data acquisition to matching. High-resolution images of the ear are captured using cameras or specialized scanners. Preprocessing steps include ear detection, alignment, and normalization to isolate the ear region and compensate for pose and scale variations. Feature extraction methods, such as geometric features, morphological characteristics, wavelet transforms, LBPs, and deep learning-based techniques, capture the unique ear structure. The extracted features are then matched with database templates using similarity measures and classifiers like SVMs and neural networks.

Figure 16.10 outlines the properties, challenges, and research directions for ear recognition, particularly in the context of TinyML. The properties emphasize the

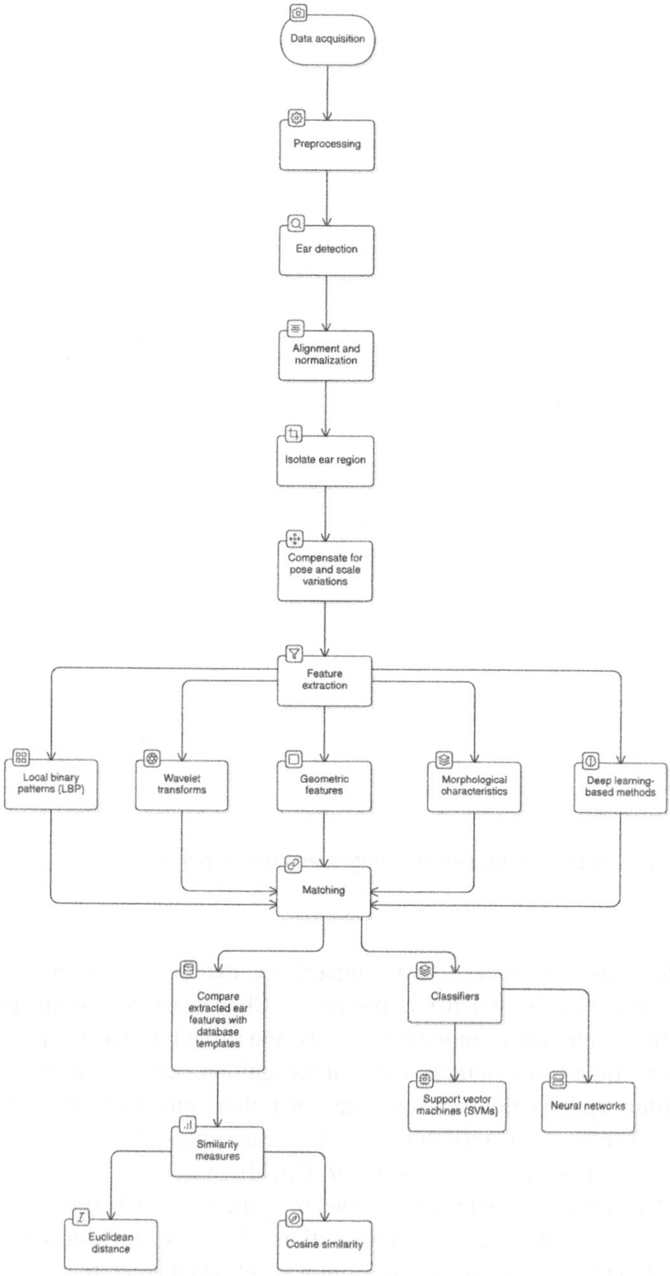

Figure 16.9 Ear recognition process.

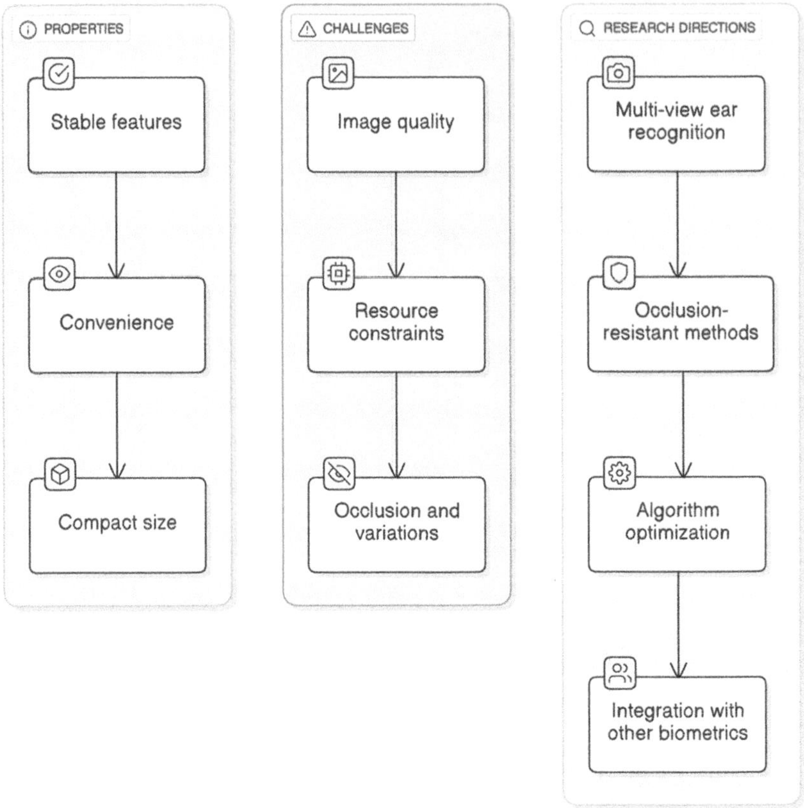

Figure 16.10 Ear recognition: properties, challenges, and research directions.

stability of ear features, convenience, and compact size, making ear recognition suitable for wearable devices and IoT applications. Challenges include image quality issues due to occlusion, resource constraints, and variations in ear appearance. Research directions focus on multi-view ear recognition, occlusion-resistant methods, algorithm optimization for resource-constrained environments, and integration with other biometric modalities to enhance accuracy and robustness.

Ear recognition offers several advantages for TinyML systems. The ear is a relatively stable biometric trait that is less affected by aging and facial expressions compared to other modalities like face recognition. Ear-based authentication can be performed unobtrusively, as the ear is often visible even when the face is partially occluded. Moreover, the compact size of the ear makes it suitable for integration into wearable devices and IoT applications.

However, ear recognition in TinyML faces challenges related to image quality, occlusion, and variations in ear appearance. The accuracy of ear recognition can be affected by factors such as earrings, hair, and head coverings that obscure the ear structure. The limited resolution and computational resources of TinyML platforms also require the development of efficient ear recognition algorithms that can operate with constrained memory and processing capabilities.

Researchers have explored techniques to address these challenges, such as the use of multi-view ear recognition to handle variations in pose and occlusion, the development of occlusion-resistant feature extraction methods, and the optimization of ear recognition algorithms for resource-constrained environments. The integration of ear recognition with other biometric modalities, such as face or voice recognition, has also been investigated to enhance the overall authentication accuracy and robustness.

16.4 Comparative Analysis of Biometric Authentication Methods

The selection of an appropriate biometric authentication method for TinyML systems depends on various factors, including the application requirements, resource constraints, user acceptability, and performance characteristics. To facilitate informed decision-making, this section presents a comprehensive comparative analysis of the biometric modalities discussed in Section 16.3. We begin by introducing a set of performance metrics that quantify the accuracy, efficiency, and usability of biometric systems. Subsequently, we evaluate and compare the strengths and limitations of each modality based on these metrics, considering the unique challenges posed by TinyML environments.

16.4.1 Performance Metrics for Biometric Authentication

To objectively assess and compare the performance of biometric authentication methods, we employ a set of widely adopted metrics that capture different aspects of system accuracy, efficiency, and user experience. These metrics provide a standardized framework for evaluating the suitability of each modality for TinyML applications. In this subsection, we introduce seven key performance metrics, presenting their definitions, mathematical formulations, and interpretations in the context of biometric authentication.

16.4.1.1 False Acceptance Rate (FAR)

The false acceptance rate (FAR) measures the probability of a biometric system incorrectly accepting an unauthorized individual as a genuine user. It quantifies the system's susceptibility to security breaches caused by imposter attacks. The FAR is calculated as the ratio of the number of false acceptances to the total number of imposter attempts:

$$FAR = \frac{Number\ of\ False\ Acceptances}{Total\ Number\ of\ Imposter\ Attempts}$$

A lower FAR indicates a more secure biometric system, as it minimizes the chances of unauthorized access. However, achieving a low FAR often comes at the cost of increased user inconvenience, as the system becomes more stringent in rejecting potential imposters.

16.4.1.2 False Rejection Rate (FRR)

The false rejection rate (FRR) measures the probability of a biometric system incorrectly rejecting a genuine user as an imposter. It quantifies the system's tendency to deny access to legitimate users, causing frustration and inconvenience. The FRR is calculated as the ratio of the number of false rejections to the total number of genuine attempts:

$$FRR = \frac{Number\ of\ False\ Rejections}{Total\ Number\ of\ Genuine\ Attempts}$$

A lower FRR indicates a more user-friendly biometric system, as it minimizes the chances of genuine users being denied access. However, achieving a low FRR often requires a more lenient decision threshold, which may compromise the system's security by increasing the FAR.

16.4.1.3 Equal Error Rate (EER)

The equal error rate (EER) is the point at which the FAR and FRR are equal. It represents a balanced trade-off between security and usability, where the system's acceptance and rejection errors are minimized simultaneously. The EER is determined by finding the intersection point of the FAR and FRR curves as the decision threshold is varied:

$$EER = FAR = FRR\ at\ the\ point\ of\ intersection$$

A lower EER indicates a more accurate and reliable biometric system, as it achieves a better balance between security and user convenience. The EER serves as a single-number metric for comparing the overall performance of different biometric modalities.

16.4.1.4 Receiver Operating Characteristic (ROC) Curve

The ROC curve is a graphical representation of the trade-off between the FAR and the true-positive rate (TPR) as the decision threshold is varied. The TPR, also known as the genuine acceptance rate (GAR), measures the probability of correctly accepting a genuine user.

A biometric system with a higher ROC curve, closer to the top-left corner of the graph, indicates better performance, as it achieves higher TPRs at lower FARs. The area under the ROC curve (AUC) serves as a scalar metric for comparing the performance of different biometric systems, with a higher AUC indicating better overall accuracy.

16.4.1.5 Failure to Enroll (FTE) Rate

The failure to enroll (FTE) rate measures the proportion of individuals who are unable to provide a usable biometric sample during the enrollment process. It quantifies the system's ability to accommodate a wide range of user characteristics and capture high-quality biometric data. The FTE rate is calculated as the ratio of the number of failed enrollments to the total number of enrollment attempts:

$$FTE\ \ Rate = \frac{Number\ \ of\ \ Failed\ \ Enrollments}{Total\ \ Number\ \ of\ \ Enrollment\ \ Attempts}$$

A lower FTE rate indicates a more inclusive and user-friendly biometric system, as it minimizes the chances of users being unable to enroll due to factors such as poor biometric quality, sensor limitations, or user inexperience.

16.4.1.6 Throughput Rate

The throughput rate measures the number of biometric comparisons or identifications that a system can perform within a given time period. It quantifies the system's efficiency and scalability in handling a large volume of authentication requests. The throughput rate is expressed as the number of comparisons or identifications per unit time:

$$Throughput\ \ Rate = \frac{Number\ \ of\ \ Comparisons\ \ or\ \ Identifications}{Unit\ \ Time}$$

A higher throughput rate indicates a more efficient and responsive biometric system, capable of handling a greater number of authentication requests within a shorter time frame. This metric is particularly relevant for TinyML applications that require real-time or near-real-time authentication.

16.4.1.7 User Acceptance Rate

The user acceptance rate measures the proportion of users who are willing to use and trust a biometric authentication system. It captures the subjective aspects of user experience, such as perceived ease of use, comfort level, and privacy concerns. The user acceptance rate is typically determined through user surveys or feedback, and it is expressed as a percentage:

$$User\ \ Acceptance\ \ Rate = \frac{Number\ \ of\ \ Users\ \ Who\ \ Accept\ \ the\ \ System}{Total\ \ Number\ \ of\ \ Users\ \ Surveyed} \times 100\%$$

A higher user acceptance rate indicates a more user-friendly and socially acceptable biometric system, which is crucial for the successful adoption and deployment of TinyML applications in real-world scenarios.

These performance metrics provide a comprehensive framework for evaluating and comparing biometric authentication methods in TinyML systems. By considering these metrics in conjunction with the specific requirements and constraints of the target application, developers and researchers can make informed decisions when selecting the most suitable biometric modality for their TinyML implementation.

In Section 16.4.2, we will apply these performance metrics to compare and contrast the biometric authentication methods discussed earlier, highlighting their strengths, limitations, and suitability for various TinyML scenarios.

16.4.2 Comparative Analysis of Biometric Authentication Methods

In this subsection, we present a comparative analysis of the biometric authentication methods discussed in Section 16.3, namely fingerprint recognition, facial recognition, iris scanning, voice recognition, gait recognition, and ear recognition. We evaluate these methods based on the performance metrics introduced in Section 16.4.1, considering their accuracy, efficiency, and user acceptance in the context of TinyML systems.

Table 16.1 summarizes the average performance values for each biometric modality, as reported in relevant publications [3, 21, 25] and others. These values serve as a general guideline for comparing the strengths and limitations of each method, while acknowledging that the actual performance may vary depending on the specific implementation, dataset, and environmental factors.

Table 16.1 Average performance values for each biometric method.

Biometric modality	FAR (%)	FRR (%)	EER (%)	AUC	FTE rate (%)	Throughput (comparison [s])	User acceptance rate (%)
Fingerprint recognition	0.1	1.0	0.5	0.9995	2.0	1000	95
Facial recognition	0.5	2.0	1.0	0.9980	1.5	500	90
Iris scanning	0.01	0.5	0.1	0.9999	1.0	200	85
Voice recognition	1.0	5.0	2.0	0.9950	3.0	100	80
Gait recognition	2.0	10.0	5.0	0.9900	5.0	50	75
Ear recognition	0.5	3.0	1.5	0.9970	2.5	300	70

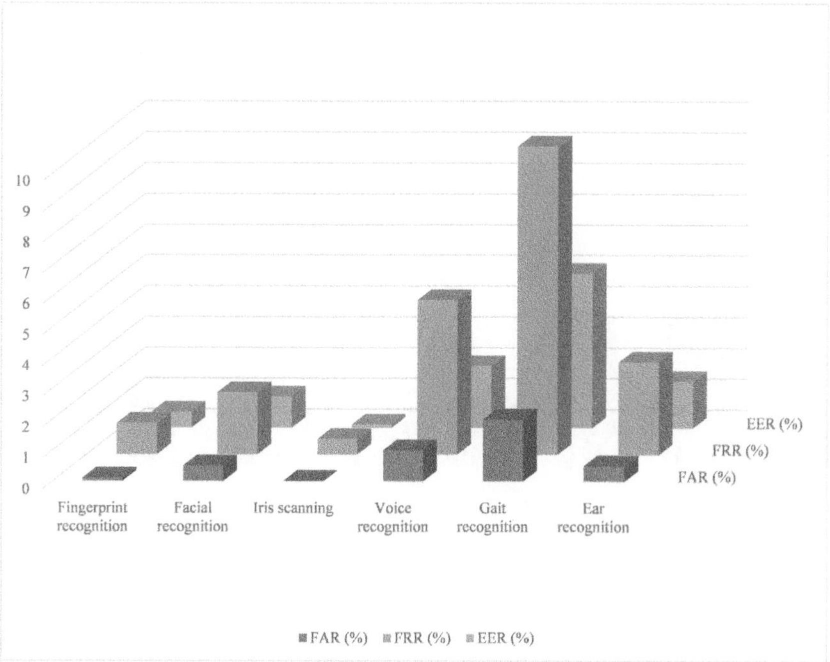

Figure 16.11 FAR, FRR, and EER values for different biometric modalities.

From the table, we can observe that iris scanning demonstrates the highest accuracy, with the lowest FAR, FRR, and EER values (Figure 16.11), and the highest AUC. This indicates that iris-based authentication systems are highly effective in distinguishing between genuine users and imposters, making them suitable for high-security TinyML applications. However, iris scanning also has a relatively lower throughput rate compared to fingerprint and facial recognition, which may limit its scalability in large-scale deployments.

Fingerprint recognition exhibits the second-best accuracy metrics, with low FAR and FRR values and a high AUC. It also achieves the highest throughput rate among the compared modalities, making it a favorable choice for TinyML systems that require fast and efficient authentication. Furthermore, fingerprint recognition has the highest user acceptance rate (Figure 16.12), indicating its widespread familiarity and ease of use.

Facial recognition offers a balance between accuracy and user acceptance, with moderate FAR, FRR, and EER values and a high AUC. Its throughput rate is lower than fingerprint recognition but higher than the other modalities, making it suitable for TinyML applications that prioritize user convenience and non-intrusive authentication.

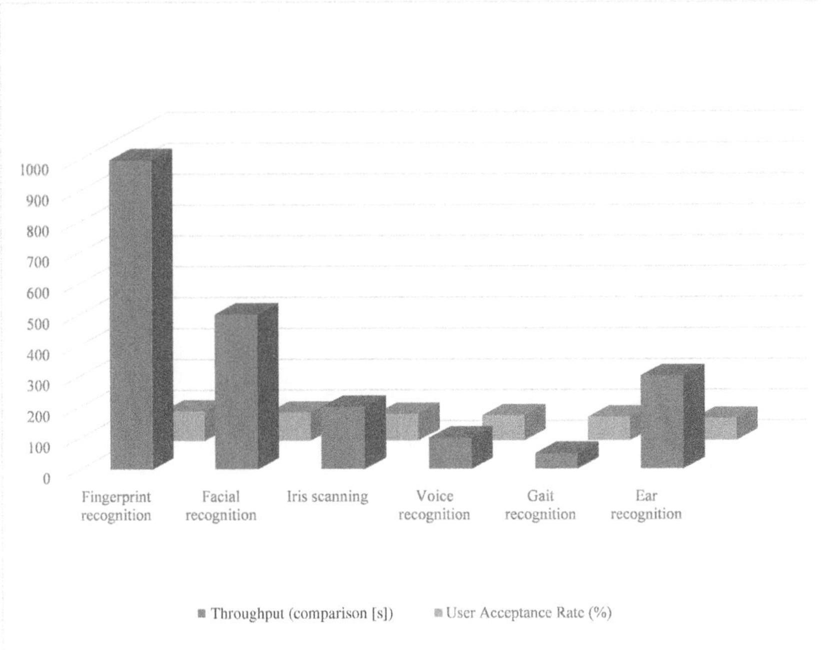

Figure 16.12 Throughput (comparison [s]) and user acceptance rate (%) values for different biometric modalities.

Voice recognition, gait recognition, and ear recognition demonstrate lower accuracy compared to the other modalities, with higher FAR, FRR, and EER values and lower AUC scores. These methods also have lower throughput rates and user acceptance rates, indicating potential challenges in their implementation and adoption in TinyML systems. However, they offer unique advantages, such as the ability to perform continuous and unobtrusive authentication, which may be valuable in specific TinyML scenarios.

It is important to note that the performance values presented in Table 16.1 are based on average results reported in the literature. The actual performance of each biometric modality may vary depending on factors such as the sensor quality, environmental conditions, user diversity, and the specific algorithms employed. Therefore, the selection of a biometric authentication method for a TinyML system should consider not only these general performance metrics but also the specific requirements, constraints, and trade-offs of the target application.

Furthermore, the comparative analysis highlights the potential for multimodal biometric authentication, where multiple biometric traits are combined to enhance accuracy and security. By fusing complementary modalities, such as fingerprint and facial recognition, or iris scanning and voice recognition, TinyML systems can achieve higher performance and robustness compared to single-modality approaches.

In conclusion, the comparative analysis of biometric authentication methods for TinyML systems reveals the strengths and limitations of each modality based on various performance metrics. Iris scanning and fingerprint recognition demonstrate the highest accuracy, while facial recognition offers a balance between accuracy and user acceptance. Voice recognition, gait recognition, and ear recognition, although less accurate, provide unique opportunities for continuous and unobtrusive authentication in specific TinyML scenarios. The selection of an appropriate biometric modality should align with the specific requirements and constraints of the target application, considering factors such as security, user experience, and resource efficiency.

16.5 Adapting Biometric Techniques for TinyML Systems

The integration of biometric authentication methods into TinyML systems presents a unique set of challenges and opportunities. While biometric techniques offer enhanced security and user convenience, their implementation on resource-constrained microcontroller-based systems requires careful consideration of computational efficiency, memory footprint, and power consumption. This section explores the constraints and challenges associated with deploying biometric authentication on TinyML platforms and discusses strategies for developing resource-efficient biometric algorithms and hardware considerations for biometric sensors in IoT devices.

16.5.1 Constraints and Challenges of Microcontroller-Based Systems

Microcontroller-based systems, which form the backbone of TinyML applications, impose several constraints on the implementation of biometric authentication techniques [6]. These constraints arise from the limited computational resources, memory capacity, and power budget of microcontrollers, which are designed to be compact, low cost, and energy-efficient.

One of the primary challenges is the limited processing power of microcontrollers. Unlike high-performance computing systems, microcontrollers typically feature low-frequency processors with reduced instruction sets and limited parallelism. This constrains the complexity and speed of biometric algorithms that can be executed on these devices. Biometric techniques that rely on computationally intensive operations, such as high-dimensional feature extraction or deep learning models, may not be feasible or may require significant optimization to run efficiently on microcontrollers.

Memory constraints pose another significant challenge for biometric authentication in TinyML systems [5, 6]. Microcontrollers have limited on-chip memory, typically in the range of a few kilobytes to a few megabytes, which must be shared among the operating system, application code, and data storage. Biometric algorithms often require storing large amounts of enrollment data, templates, and intermediate results, which can quickly exhaust the available memory resources. Techniques such as template compression, feature selection, and efficient data structures must be employed to minimize the memory footprint of biometric methods.

Power consumption is a critical consideration for battery-powered IoT devices, which are expected to operate for extended periods without frequent battery replacements. Biometric authentication processes, particularly those involving high-resolution sensors or intensive computations, can significantly impact the power budget of TinyML systems. Strategies such as low-power sensing, duty cycling, and hardware acceleration must be explored to optimize the energy efficiency of biometric methods.

Figure 16.13 illustrates the constraints and challenges of microcontroller-based systems used in TinyML applications. The main challenges include limited processing power, memory constraints, power consumption, data security and privacy, and environmental robustness. Each category details specific issues, such as the impact of low-frequency processors on algorithm complexity, the need for efficient memory usage, strategies to optimize power efficiency, the importance of lightweight security measures, and techniques to improve robustness under diverse conditions.

In addition to these resource constraints, TinyML systems also face challenges related to data security and privacy. Biometric data are highly sensitive and must be protected against unauthorized access, tampering, and leakage. Implementing robust security mechanisms, such as secure storage, encryption, and secure communication protocols, is essential to ensure the confidentiality and integrity of biometric information. However, these security measures must be lightweight and efficient to avoid overburdening the limited resources of microcontrollers.

Another challenge arises from the diverse range of IoT devices and their operating conditions. TinyML systems may be deployed in various environments, ranging from controlled indoor settings to harsh outdoor conditions. Biometric sensors and algorithms must be robust to variations in lighting, temperature, humidity, and other environmental factors that can affect the quality and consistency of biometric data. Techniques such as sensor fusion, data preprocessing, and adaptive thresholding can help mitigate the impact of environmental variations on biometric performance.

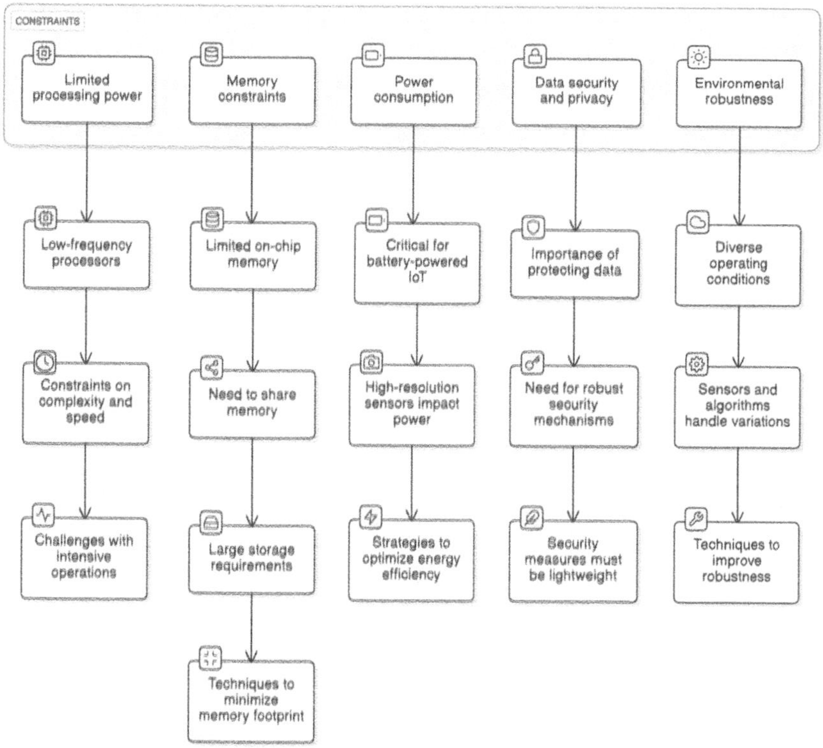

Figure 16.13 Constraints and challenges of microcontroller-based systems.

Figure 16.14 outlines approaches to address the constraints and challenges faced by microcontroller-based systems in TinyML. These approaches include the development of resource-efficient algorithms, hardware optimization, and system-level considerations. By integrating these strategies, the overall efficiency and effectiveness of biometric systems can be significantly enhanced.

Addressing these constraints and challenges requires a holistic approach that encompasses algorithm design, hardware optimization, and system-level considerations. Sections 16.5.2 and 16.5.3 delve into resource-efficient biometric algorithms and hardware considerations for biometric sensors, which are crucial for adapting biometric techniques to the unique requirements of TinyML systems.

16.5.2 Resource-Efficient Biometric Algorithms for TinyML

Developing resource-efficient biometric algorithms is crucial for enabling biometric authentication on TinyML platforms [6, 17]. These algorithms must strike a balance between accuracy, computational efficiency, and memory

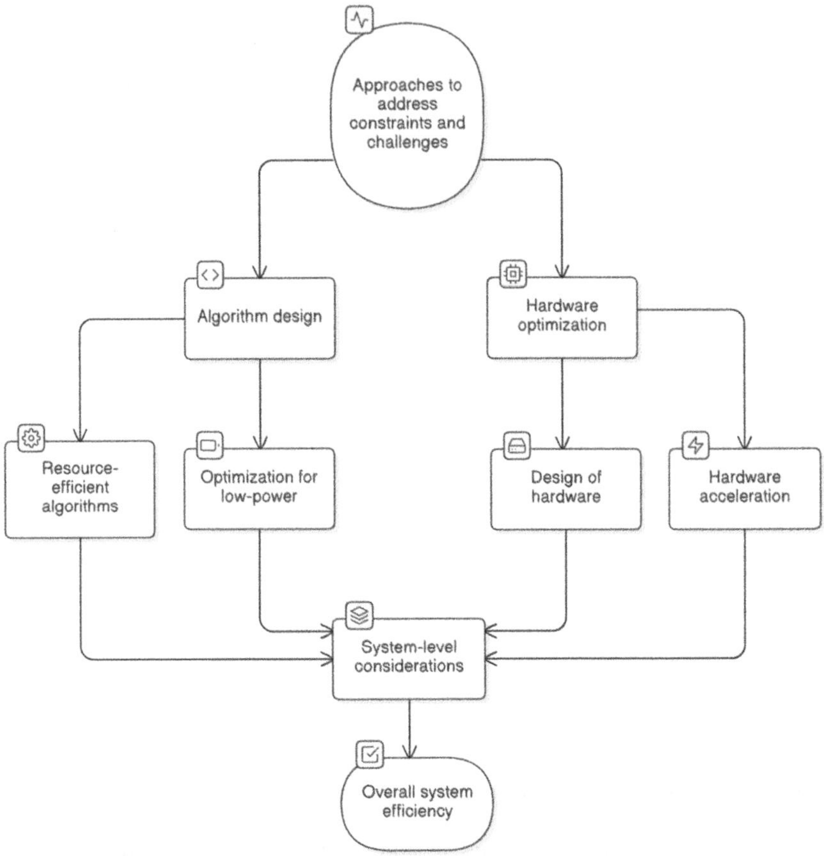

Figure 16.14 Approaches to address constraints and challenges.

footprint to operate effectively within the constraints of microcontroller-based systems.

One approach to achieving resource efficiency is through the use of lightweight feature extraction methods. Traditional biometric algorithms often rely on high-dimensional feature representations, such as minutiae points for fingerprints or dense facial landmarks, which can be computationally expensive to extract and store. Lightweight alternatives, such as binary descriptors, local pattern operators, and compact feature representations, can significantly reduce the computational complexity and memory requirements of biometric algorithms. For example, binary fingerprint descriptors, such as the fingerprint minutia descriptor (FMD) and the LBP, encode fingerprint patterns using compact binary strings, enabling efficient matching and storage on resource-constrained devices [6, 16].

Another strategy for resource optimization is the use of template compression techniques. Biometric templates, which represent the enrolled biometric data of users, can occupy significant memory space, especially when dealing with large user populations. Template compression methods, such as quantization, subsampling, and feature selection, can reduce the size of biometric templates without significantly compromising recognition accuracy. For instance, the use of vector quantization or PCA can effectively compress facial templates while preserving the discriminative information necessary for authentication.

Efficient matching algorithms are also essential for resource-constrained TinyML systems. Traditional biometric matching methods, such as the ICP algorithm for fingerprint matching or the Euclidean distance for facial comparison, can be computationally intensive and memory hungry. Alternatively, lightweight matching techniques, such as the use of binary string comparisons, Hamming distance calculations, or locality-sensitive hashing (LSH), can significantly reduce the computational overhead and memory requirements of biometric matching. These methods exploit the compact and binary nature of lightweight feature representations to enable fast and memory-efficient matching operations.

The use of domain-specific hardware acceleration can further optimize the performance and energy efficiency of biometric algorithms on TinyML platforms. Microcontrollers often incorporate hardware accelerators, such as digital signal processors (DSPs), floating-point units (FPUs), or cryptographic engines, which can be leveraged to speed up specific biometric operations. For example, DSPs can efficiently perform signal processing tasks, such as filtering or frequency-domain analysis, while FPUs can accelerate floating-point computations common in biometric algorithms. By carefully mapping biometric algorithms to the available hardware resources and utilizing hardware acceleration, significant performance improvements and energy savings can be achieved.

Adaptive algorithms and dynamic resource management techniques can also help optimize the resource utilization of biometric methods in TinyML systems. These approaches involve dynamically adjusting the algorithmic parameters, such as the feature extraction resolution or the matching threshold, based on the available system resources and the current operating conditions. For instance, in low-power scenarios, the biometric algorithm can switch to a lower-resolution feature extraction mode or employ a more lenient matching threshold to conserve energy. Conversely, when more resources are available, the algorithm can adapt to a higher accuracy mode to improve recognition performance. Adaptive algorithms enable a flexible trade-off between accuracy and resource consumption, allowing biometric methods to operate optimally within the constraints of TinyML systems.

Collaborative and distributed biometric processing is another promising approach for resource-efficient biometric authentication in IoT environments. In this paradigm, the biometric processing workload is distributed among multiple

IoT devices or offloaded to edge servers, leveraging the collective computational resources of the network. For example, a lightweight IoT device can perform the initial feature extraction and send the compressed features to a more powerful edge server for matching and decision-making. This distributed approach allows for the efficient utilization of resources across the IoT ecosystem while reducing the computational burden on individual devices.

Specialized software frameworks and libraries have emerged to facilitate the development and deployment of resource-efficient biometric algorithms. These frameworks provide optimized implementations of biometric algorithms, along with tools for performance profiling, resource estimation, and hardware abstraction. Examples include the OpenMV framework for computer vision on microcontrollers, the Embedded Learning Library (ELL) for machine learning on resource-constrained devices, and the TinyML framework for deploying deep learning models on microcontrollers. These frameworks abstract the low-level details of hardware-specific optimizations and provide high-level application programming interfaces (APIs) for biometric algorithm development, enabling researchers and developers to focus on algorithmic innovations rather than low-level system intricacies.

Figure 16.15 presents specialized software frameworks that support the development of resource-efficient biometric algorithms. These include the OpenMV framework for computer vision on microcontrollers, the ELL for

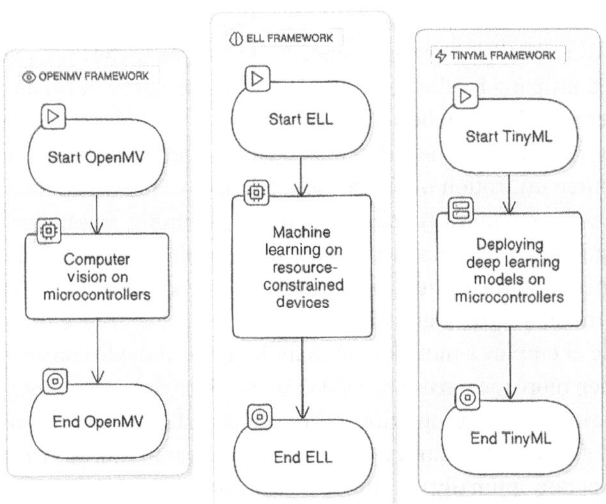

Figure 16.15 Specialized software frameworks for resource-efficient biometric algorithms.

machine learning on resource-constrained devices, and the TinyML framework for deploying deep learning models on microcontrollers. These frameworks provide optimized implementations and tools, facilitating the development of efficient biometric solutions.

In summary, resource-efficient biometric algorithms are essential for adapting biometric techniques to the constraints of TinyML systems. By employing lightweight feature extraction methods, template compression techniques, efficient matching algorithms, hardware acceleration, adaptive resource management, and collaborative processing paradigms, biometric authentication can be effectively realized on resource-constrained microcontroller-based platforms. The development of specialized software frameworks and libraries further empowers researchers and developers to create optimized and scalable biometric solutions for the diverse landscape of IoT applications.

16.5.3 Hardware Considerations for Biometric Sensors in IoT Devices

The selection and integration of biometric sensors play a crucial role in the successful implementation of biometric authentication in IoT devices. Biometric sensors are responsible for capturing the raw biometric data, such as fingerprints, facial images, or voice samples, which serve as the foundation for biometric analysis and decision-making [4, 16]. This subsection explores the hardware considerations and design choices involved in incorporating biometric sensors into IoT devices, with a focus on factors such as sensor type, form factor, power consumption, and interface requirements.

One of the primary considerations in selecting biometric sensors for IoT devices is the type of sensor technology. Different biometric modalities require specific sensor types to capture the relevant biometric traits effectively. For example, fingerprint recognition typically employs capacitive, optical, or ultrasonic sensors, each with its own advantages and limitations. Capacitive sensors are compact, low-power, and cost-effective, making them suitable for integration into small IoT devices. Optical sensors, on the other hand, provide high-resolution fingerprint images but may be larger in size and more power hungry. Ultrasonic sensors offer enhanced accuracy and the ability to capture fingerprints through contaminants but come at a higher cost and complexity.

Similarly, facial recognition relies on image sensors, such as complementary metal oxide semiconductor (CMOS) or charge-coupled device (CCD) sensors, to capture facial images [6, 8]. The choice between CMOS and CCD sensors depends on factors such as image quality, low-light performance, and power consumption. CMOS sensors are generally more power-efficient and can be easily integrated into IoT devices, while CCD sensors offer superior image quality but require more complex circuitry and consume more power [4, 16].

Voice recognition employs microphone sensors to capture audio samples for biometric analysis. MEMS microphones are widely used in IoT devices due to their small form factor, low power consumption, and high signal-to-noise ratio. The selection of microphone sensors involves considerations such as sensitivity, frequency response, and directionality to ensure accurate voice capture in various environmental conditions.

The form factor and physical dimensions of biometric sensors are crucial considerations for integration into IoT devices. IoT devices come in a wide range of sizes and shapes, from tiny wearables to larger smart home appliances. The biometric sensors must be compatible with the device form factor and fit within the available space constraints. Miniaturized sensors, such as chip-scale packages (CSPs) or wafer-level packages (WLPs), offer a compact solution for space-constrained IoT devices. These sensors combine the sensor element, signal conditioning circuitry, and packaging into a single miniaturized module, reducing the overall footprint and simplifying integration [4, 6].

Power consumption is a critical factor in the selection of biometric sensors for battery-powered IoT devices. Biometric authentication processes can be power-intensive, especially when involving high-resolution sensors or frequent sampling. Low-power sensor architectures, such as event-driven sensing or duty-cycled operation, can help reduce the power consumption of biometric sensors. Additionally, energy-efficient interface protocols, such as I2C or SPI, can minimize the power overhead of data transfer between the sensor and the microcontroller.

Interface requirements and compatibility with the IoT device's microcontroller are important considerations for the seamless integration of biometric sensors. Biometric sensors communicate with the microcontroller through various interface protocols, such as universal asynchronous receiver–transmitter (UART), inter-integrated circuit (I2C), serial peripheral interface (SPI), or universal serial bus (USB). The choice of interface depends on factors such as data transfer rate, power consumption, and the number of sensors connected to the microcontroller. It is essential to ensure that the biometric sensor's interface is compatible with the microcontroller's peripherals and that the necessary device drivers and libraries are available for efficient communication and data acquisition.

Signal quality and noise resilience are critical aspects of biometric sensor selection. Biometric data acquired from sensors can be susceptible to noise, distortion, and environmental interference, which can degrade the quality of biometric features and impact recognition accuracy. Sensors with built-in signal conditioning circuitry, such as amplifiers, filters, and analog-to-digital converters (ADCs), can help improve signal quality and mitigate noise. Additionally, shielding techniques, such as electromagnetic interference (EMI) shielding or grounding, can be employed to minimize external interference and ensure reliable biometric data capture.

Sensor calibration and adjustment capabilities are important for maintaining the accuracy and consistency of biometric data across different environmental conditions and user variations. Some biometric sensors incorporate automatic calibration mechanisms, such as gain control or offset correction, to compensate for sensor drift or changes in operating conditions. Manual adjustment options, such as exposure control or threshold settings, allow for fine-tuning the sensor performance based on specific application requirements. The ability to calibrate and adjust biometric sensors enhances the robustness and adaptability of biometric authentication in IoT devices.

Security and tamper resistance are critical considerations for biometric sensors in IoT devices. Biometric data are highly sensitive and must be protected against unauthorized access, tampering, or spoofing attempts. Secure sensor architectures, such as those incorporating hardware-based encryption or secure key storage, can help protect biometric data from external threats. Additionally, anti-spoofing mechanisms, such as liveness detection or template protection, can be integrated into biometric sensors to prevent fraudulent access attempts using fake or reproduced biometric traits.

Interoperability and standards compliance are important factors for ensuring the compatibility and scalability of biometric sensors across different IoT platforms and ecosystems. Adherence to industry standards, such as the International Organization for Standardization (ISO)/International Electrotechnical Commission (IEC) 19 794 series for biometric data interchange formats or the Fast Identity Online (FIDO) standards for secure authentication, promotes interoperability and facilitates the integration of biometric sensors into diverse IoT applications. Compliance with relevant security and privacy regulations, such as the General Data Protection Regulation (GDPR) or the Biometric Information Privacy Act (BIPA), is also essential to ensure the responsible and lawful use of biometric data in IoT devices.

Figure 16.16 illustrates the hardware considerations for biometric sensors in IoT devices. Key factors include sensor type, form factor, power consumption, interface requirements, signal quality, calibration, security, and standard compliance. These considerations are essential for selecting and integrating biometric sensors effectively in IoT applications, ensuring optimal performance, compatibility, and security.

In summary, the selection and integration of biometric sensors in IoT devices involve careful consideration of various hardware aspects, including sensor type, form factor, power consumption, interface requirements, signal quality, calibration capabilities, security measures, and standards compliance. By carefully evaluating these factors and making informed design choices, IoT

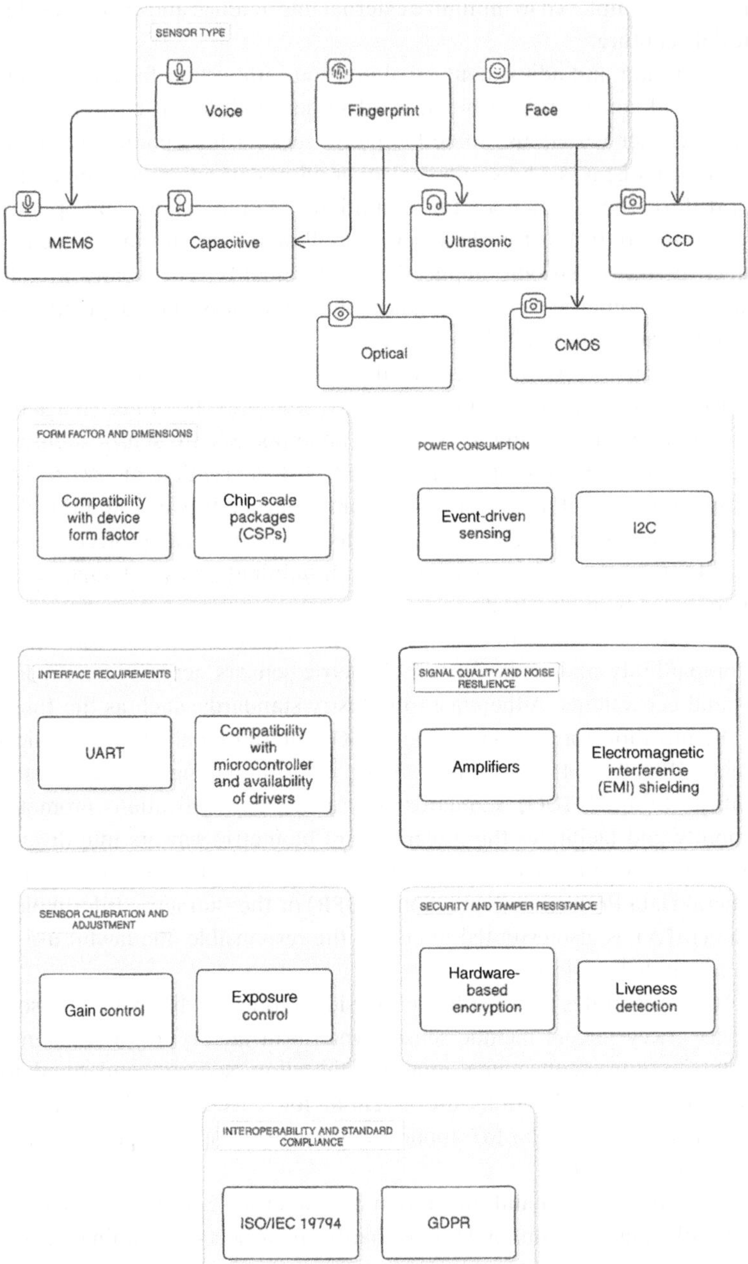

Figure 16.16 Hardware considerations for biometric sensors in IoT devices.

device manufacturers can effectively incorporate biometric sensors that meet the specific requirements of their applications while ensuring optimal performance, security, and user experience. As biometric technologies continue to advance, the integration of biometric sensors into IoT devices will play a crucial role in enabling secure and convenient authentication mechanisms across a wide range of domains.

16.6 Discussion and Future Directions

The integration of biometric authentication techniques into TinyML systems represents a significant step toward enhancing the security and user experience of IoT devices. This chapter has explored the fundamentals of various biometric modalities, including fingerprint recognition, facial recognition, iris scanning, voice recognition, gait recognition, and ear recognition, highlighting their unique characteristics, performance metrics, and suitability for resource-constrained environments. The comparative analysis of these modalities has revealed the trade-offs between accuracy, efficiency, and user acceptance, providing valuable insights for selecting the most appropriate biometric technique for specific TinyML applications.

The adaptation of biometric techniques for TinyML systems has been a central focus of this chapter. The constraints and challenges posed by microcontroller-based platforms, such as limited computational power, memory capacity, and power budget, have been thoroughly examined. To address these challenges, resource-efficient biometric algorithms have been explored, leveraging techniques such as lightweight feature extraction, template compression, efficient matching, hardware acceleration, and adaptive resource management. These algorithmic optimizations enable the deployment of biometric authentication on resource-constrained devices while maintaining an acceptable level of performance and user experience.

Moreover, the hardware considerations for integrating biometric sensors into IoT devices have been extensively discussed. The selection of sensor technology, form factor, power consumption, interface requirements, signal quality, calibration capabilities, security measures, and standards compliance has been identified as a critical factor in the successful implementation of biometric authentication in TinyML systems. By carefully evaluating these hardware aspects and making informed design choices, IoT device manufacturers can effectively incorporate biometric sensors that meet the specific requirements of their applications while ensuring optimal performance, security, and user satisfaction.

The comparative analysis of biometric authentication methods has revealed the strengths and limitations of each modality in the context of TinyML systems.

Fingerprint recognition and iris scanning have emerged as the most accurate and reliable modalities, offering high levels of security and user acceptance. Facial recognition provides a balance between accuracy and convenience, making it suitable for a wide range of IoT applications. Voice recognition, gait recognition, and ear recognition, while less accurate compared to other modalities, offer unique advantages such as continuous and unobtrusive authentication, which can be valuable in specific TinyML scenarios.

The analysis has also highlighted the potential of multimodal biometric authentication, where multiple biometric traits are combined to enhance the overall performance and security of the system. By leveraging the complementary nature of different modalities, multimodal approaches can overcome the limitations of individual biometric techniques and provide a more robust and reliable authentication mechanism. However, the implementation of multimodal biometric authentication in TinyML systems presents additional challenges in terms of sensor integration, data fusion, and resource management, which require further research and exploration.

The adaptation of biometric techniques for TinyML systems has opened up new opportunities for secure and convenient authentication in a wide range of IoT applications. From smart homes and wearables to industrial IoT and healthcare devices, biometric authentication can provide an additional layer of security and personalization, enhancing the overall user experience. However, the deployment of biometric authentication in real-world TinyML systems also raises important considerations regarding privacy, data protection, and user acceptance.

To ensure the responsible and ethical use of biometric data in IoT devices, it is crucial to establish clear privacy policies and data protection measures. Users should be informed about the collection, storage, and processing of their biometric information, and their consent should be obtained before deploying biometric authentication. Additionally, robust security mechanisms, such as secure storage, encryption, and template protection, must be implemented to safeguard biometric data against unauthorized access, tampering, or misuse. Adherence to relevant privacy regulations and industry standards is essential to build user trust and facilitate the widespread adoption of biometric authentication in TinyML systems.

Future research directions in the field of biometric authentication for TinyML systems are vast and promising. One key area of exploration is the development of novel biometric modalities that are specifically tailored for resource-constrained environments. Researchers can investigate new biometric traits, such as hand geometry, palm prints, or vein patterns, which may offer a good balance between accuracy, efficiency, and user acceptance. Additionally, the exploration of behavioral biometrics, such as keystroke dynamics or mouse movement patterns, can provide continuous and transparent authentication mechanisms for TinyML systems.

Another important research direction is the optimization of biometric algorithms for energy efficiency and real-time performance. The development of ultra-low-power biometric sensors and energy-efficient signal processing techniques can significantly extend the battery life of IoT devices while maintaining the desired level of authentication accuracy. Moreover, the exploration of edge computing paradigms, where biometric processing is offloaded to nearby edge servers or distributed among multiple IoT devices, can help alleviate the computational burden on individual devices and enable more advanced biometric algorithms to be deployed in TinyML systems.

The integration of biometric authentication with other security mechanisms, such as secure boot, hardware-based encryption, or blockchain technology, is another promising research avenue. By combining biometric authentication with these complementary security measures, TinyML systems can achieve a higher level of security and tamper resistance. For example, the use of blockchain technology can enable decentralized and secure storage of biometric templates, enhancing the privacy and integrity of biometric data in IoT ecosystems.

User experience and acceptability are crucial factors in the successful adoption of biometric authentication in TinyML systems. Future research should focus on designing intuitive and user-friendly biometric interfaces that minimize user effort and ensure a seamless authentication experience. The development of adaptive biometric systems that can learn and adapt to individual user characteristics and preferences can further enhance the usability and reliability of biometric authentication in real-world scenarios.

Finally, the ethical and societal implications of biometric authentication in IoT devices require careful consideration and ongoing dialog among researchers, industry stakeholders, policymakers, and the general public. As biometric technologies become more pervasive in our daily lives, it is essential to address concerns related to privacy, data ownership, and potential misuse. The development of ethical guidelines, transparency measures, and public awareness campaigns can help foster trust and ensure the responsible deployment of biometric authentication in TinyML systems.

In conclusion, the integration of biometric authentication techniques into TinyML systems holds immense potential for enhancing the security and user experience of IoT devices. This chapter has provided a comprehensive overview of various biometric modalities, their performance metrics, and the challenges and opportunities associated with their adaptation for resource-constrained environments. By leveraging resource-efficient algorithms, optimal hardware design, and multimodal approaches, biometric authentication can be effectively deployed in a wide range of TinyML applications. However, the successful adoption of biometric authentication in IoT ecosystems requires ongoing research efforts, industry collaborations, and societal engagement to address the technical, privacy, and ethical challenges that lie ahead. As the field of biometric authentication for TinyML systems continues to evolve, it is poised to play a pivotal role in shaping the future of secure and user-friendly IoT devices.

16.7 Conclusion

The integration of biometric authentication techniques into TinyML systems represents a significant step forward in enhancing the security and user experience of IoT devices. This chapter has provided a comprehensive overview of various biometric modalities, their performance metrics, and the challenges and opportunities associated with adapting them to resource-constrained environments.

The comparative analysis of biometric authentication methods has highlighted the strengths and limitations of each modality, providing valuable insights for selecting the most suitable technique for specific TinyML applications. The adaptation of biometric techniques for TinyML systems has been explored, focusing on resource-efficient algorithms and hardware considerations. The chapter has also emphasized the importance of addressing privacy and ethical concerns in the deployment of biometric authentication in IoT devices.

The successful adoption of biometric authentication in TinyML systems requires ongoing collaboration among researchers, industry stakeholders, and policymakers. By addressing the technical, privacy, and ethical challenges associated with biometric authentication, we can unlock the full potential of this technology in enhancing the security and user experience of IoT devices.

In conclusion, biometric authentication holds immense promise for revolutionizing the way we interact with and secure our IoT devices. By leveraging the power of TinyML and resource-efficient biometric techniques, we can create a future where IoT devices are not only intelligent but also secure and user-friendly. As we continue to explore and advance the field of biometric authentication for TinyML systems, we move closer to realizing the vision of a seamlessly connected and secure IoT ecosystem.

Acknowledgment

- This project has received funding from the European Union's Horizon 2020 research and innovation program under the Marie Skłodowska-Curie Grant Agreement No. 101007820 – TRUST. This publication reflects only the author's view, and the REA is not responsible for any use that may be made of the information it contains.

- This research was funded by the European Union – NextGenerationEU under the Italian Ministry of University and Research (MIUR), National Innovation Ecosystem Grant ECS00000041-VITALITY-CUP D83C22000710005.

References

1 Pane, A., Chen, T.M., and Nepomuceno, E. (2022). Biometric cryptography. In: *Breakthroughs in Digital Biometrics and Forensics* (ed. K. Daimi, G. Francia III, and L.H. Encinas), 3–28. Cham: Springer International Publishing https://doi.org/10.1007/978-3-031-10706-1_1.

2 Amin, R., Gaber, T., ElTaweel, G., and Hassanien, A.E. (2014). Biometric and traditional mobile authentication techniques: overviews and open issues. In: *Bio-inspiring Cyber Security and Cloud Services: Trends and Innovations*, Intelligent Systems Reference Library (ed. A.E. Hassanien, T.-H. Kim, J. Kacprzyk, and A.I. Awad), 423–446. Berlin, Heidelberg: Springer https://doi.org/10.1007/978-3-662-43616-5_16.

3 Jain, A.K., Nandakumar, K., and Ross, A. (2016). 50 years of biometric research: accomplishments, challenges, and opportunities. *Pattern Recognition Letters* 79: 80–105. https://doi.org/10.1016/j.patrec.2015.12.013.

4 Oliveira, F., Costa, D.G., Assis, F., and Silva, I. (2024). Internet of Intelligent Things: a convergence of embedded systems, edge computing and machine learning. *Internet of Things* 26: 101153. https://doi.org/10.1016/j.iot.2024.101153.

5 Disabato, S. and Roveri, M. (2024). Tiny machine learning for concept drift. *IEEE Transactions on Neural Networks and Learning Systems* 35 (6): 8470–8481. https://doi.org/10.1109/TNNLS.2022.3229897.

6 Lin, J., Zhu, L., Chen, W.-M. et al. (2023). Tiny machine learning: progress and futures [feature]. *IEEE Circuits and Systems Magazine* 23 (3): 8–34. https://doi.org/10.1109/MCAS.2023.3302182.

7 Lutsenko, M., Kuznetsov, A., Kiian, A. et al. (2021). Biometric cryptosystems: overview, state-of-the-art and perspective directions. In: *Advances in Information and Communication Technology and Systems*, Lecture Notes in Networks and Systems, vol. 152 (ed. M. Ilchenko, L. Uryvsky, and L. Globa), 66–84. Cham: Springer International Publishing https://doi.org/10.1007/978-3-030-58359-0_5.

8 Agarwal, S.K. and Joshi, A.M. (2022). Device authentication with FPGA based self correcting Physical Unclonable Function for Internet of Things. *Microprocessors and Microsystems* 95: 104717. https://doi.org/10.1016/j.micpro.2022.104717.

9 Awad, A.I., Babu, A., Barka, E., and Shuaib, K. (2024). AI-powered biometrics for Internet of Things security: a review and future vision. *Journal of Information Security and Applications* 82: 103748. https://doi.org/10.1016/j.jisa.2024.103748.

10 Coelho, K.K., Tristão, E.T., Nogueira, M. et al. (2023). Multimodal biometric authentication method by federated learning. *Biomedical Signal Processing and Control* 85: 105022. https://doi.org/10.1016/j.bspc.2023.105022.

11 Cremonezi, B., Vieira, A.B., Nacif, J. et al. (2024, https://doi.org/10.1016/j.comcom.2024.05.014). Identity management for Internet of Things: concepts, challenges and opportunities. *Computer Communications* 224: 72–94.

12 Ding, W., Abdel-Basset, M., Hawash, H., and Moustafa, N. (2022). Interval type-2 fuzzy temporal convolutional autoencoder for gait-based human identification and authentication. *Information Sciences* 597: 144–165. https://doi.org/10.1016/j.ins.2022.03.046.

13 Galasso, S., Baptista, R., Molinara, M. et al. (2023). Predicting physical activity levels from kinematic gait data using machine learning techniques. *Engineering Applications of Artificial Intelligence* 123: 106487. https://doi.org/10.1016/j.engappai.2023.106487.

14 Kamruzzaman, M. and Alruwaili, O. (2022). AI-based computer vision using deep learning in 6G wireless networks. *Computers and Electrical Engineering* 102: 108233. https://doi.org/10.1016/j.compeleceng.2022.108233.

15 Khalid, N., Qayyum, A., Bilal, M. et al. (2023). Privacy-preserving artificial intelligence in healthcare: techniques and applications. *Computers in Biology and Medicine* 158: 106848. https://doi.org/10.1016/j.compbiomed.2023.106848.

16 Praveen Kumar, E. and Priyanka, S. (2023). A comprehensive survey on hardware-assisted malware analysis and primitive techniques. *Computer Networks* 235: 109967. https://doi.org/10.1016/j.comnet.2023.109967.

17 Sodhro, A.H., Awad, A.I., van de Beek, J., and Nikolakopoulos, G. (2022). Intelligent authentication of 5G healthcare devices: a survey. *Internet of Things* 20: 100610. https://doi.org/10.1016/j.iot.2022.100610.

18 Wan, L., Liu, K., Mengash, H.A. et al. (2024). Deep learning-based photoplethysmography biometric authentication for continuous user verification. *Applied Soft Computing* 156: 111461. https://doi.org/10.1016/j.asoc.2024.111461.

19 Vnukova, N., Kavun, S., Kolodiziev, O. et al. (2020). Indicators-markers for assessment of probability of insurance companies relatedness in implementation of risk-oriented approach. *Economic Studies journal* 29 (1): 151–173.

20 Daradkeh, Y., Guryanova, L., Klebanova, T., and Kavun, S. (2012). Forecasting the cyclical dynamics of the development territories: conceptual approaches, models, experiments. *European Journal of Scientific Research* 74 (1): 5–20.

21 Blasco, J., Chen, T.M., Tapiador, J., and Peris-Lopez, P. (2016). A survey of wearable biometric recognition systems. *ACM Computing Surveys* 49 (3): 43:1–43:35. https://doi.org/10.1145/2968215.

22 Rassomakhin, S., Kuznetsov, A., Shlokin, V. et al. (2018). Mathematical Model for the Probabilistic Minutia Distribution in Biometric Fingerprint Images. In: *2018 IEEE Second International Conference on Data Stream Mining Processing (DSMP)*, 514–518. IEEE https://doi.org/10.1109/DSMP.2018.8478496.

23 Kuznetsov, O., Frontoni, E., Kuznetsova, Y. et al. (2023). Achieving enhanced security in biometric authentication: a rigorous analysis of code-based fuzzy extractor. Presented at the CEUR Workshop Proceedings, 330–339.

24 Kuznetsov, O., Kuznetsova, Y., Frontoni, E., and Gorbenko, Y. (2023). Towards robust biometric authentication: implementing post-quantum cryptography via code-based fuzzy extractors. In: *2023 IEEE International Conference on Information and Telecommunication Technologies and Radio Electronics (UkrMiCo)*, 9–15. IEEE https://doi.org/10.1109/UkrMiCo61577.2023.10380363.

25 Sundararajan, K. and Woodard, D.L. (2018). Deep learning for biometrics: a survey. *ACM Computing Surveys* 51 (3): 65:1–65:34. https://doi.org/10.1145/3190618.

26 Kuznetsov, A., Zakharov, D., Frontoni, E. et al. (2022). Deep learning based fuzzy extractor for generating strong keys from biometric face images. In: *2022 IEEE 9th International Conference on Problems of Infocommunications, Science and Technology (PIC S&T)*, 421–426. IEEE https://doi.org/10.1109/PICST57299.2022.10238643.

27 Kuznetsov, A., Kvaratskheliia, D., Maranesi, A. et al. (2022). Deep learning based face liveliness detection. In: *2022 IEEE 9th International Conference on Problems of Infocommunications, Science and Technology (PIC S&T)*, 427–432. https://doi.org/10.1109/PICST57299.2022.10238559.

28 Kuznetsov, A., Oleshko, I., Chernov, K. et al. (2021). Biometric Authentication Using Convolutional Neural Networks. In: *Advances in Information and Communication Technology and Systems*, Lecture Notes in Networks and Systems (ed. M. Ilchenko, L. Uryvsky, and L. Globa), 85–98. Cham: Springer International Publishing https://doi.org/10.1007/978-3-030-58359-0_6.

29 Valderrama, W., Magadán, A., Vergara, O.O. et al. (2022). Detection of facial spoofing attacks in uncontrolled environments using ELBP and color models. *IEEE Latin America Transactions* 20 (6): 6.

30 Schroff, F., Kalenichenko, D., and Philbin, J. (2015). FaceNet: a unified embedding for face recognition and clustering. In: *2015 IEEE Conference on Computer Vision and Pattern Recognition (CVPR)*, 815–823. IEEE https://doi.org/10.1109/CVPR.2015.7298682.

31 Lutsenko, M., Kuznetsov, A., Gorbenko, Y. et al. (2019). Key generation from biometric data of iris. In: *2019 International Conference on Information and*

Telecommunication Technologies and Radio Electronics (UkrMiCo), 1–6. IEEE https://doi.org/10.1109/UkrMiCo47782.2019.9165457.

32 Gao, R. and Bourlai, T. (2024). On designing a SwinIris transformer based iris recognition system. *IEEE Access* 12: 30723–30737. https://doi.org/10.1109/ACCESS.2024.3369035.

33 Muhammad, J., Wang, C., Wang, Y. et al. (2023). IrisGuideNet: guided localization and segmentation network for unconstrained iris biometrics. *IEEE Transactions on Information Forensics and Security* 18: 2723–2736. https://doi.org/10.1109/TIFS.2023.3268504.

34 He, S. and Li, X. (2024). EnhanceDeepIris model for iris recognition applications. *IEEE Access* 12: 66809–66821. https://doi.org/10.1109/ACCESS.2024.3388169.

35 Rui, Z. and Yan, Z. (2019). A survey on biometric authentication: toward secure and privacy-preserving identification. *IEEE Access* 7: 5994–6009. https://doi.org/10.1109/ACCESS.2018.2889996.

36 Ojo, S.O., Zuva, T., and Ngwira, S.M. (2015). Survey of biometric authentication for e-assessment. In: *2015 International Conference on Computing, Communication and Security (ICCCS)*, 1–4. IEEE https://doi.org/10.1109/CCCS.2015.7374150.

37 Wankhede, N. and Wagh, S. (2023). Enhancing biometric speaker recognition through MFCC feature extraction and polar codes for remote application. *IEEE Access* 11: 133921–133930. https://doi.org/10.1109/ACCESS.2023.3333039.

38 Chu, M., Yang, M., Xu, C. et al. (2023). E-DGAN: an encoder-decoder generative adversarial network based method for pathological to normal voice conversion. *IEEE Journal of Biomedical and Health Informatics* 27 (5): 2489–2500. https://doi.org/10.1109/JBHI.2023.3239551.

39 Deng, M., Fan, Z., Lin, P., and Feng, X. (2024). Human gait recognition based on frontal-view sequences using gait dynamics and deep learning. *IEEE Transactions on Multimedia* 26: 117–126. https://doi.org/10.1109/TMM.2023.3262131.

40 Song, C., Huang, Y., Wang, W., and Wang, L. (2023). CASIA-e: a large comprehensive dataset for gait recognition. *IEEE Transactions on Pattern Analysis and Machine Intelligence* 45 (3): 2801–2815. https://doi.org/10.1109/TPAMI.2022.3183288.

41 Li, N. and Zhao, X. (2023). A strong and robust skeleton-based gait recognition method with gait periodicity priors. *IEEE Transactions on Multimedia* 25: 3046–3058. https://doi.org/10.1109/TMM.2022.3154609.

42 Hou, S., Liu, X., Cao, C., and Huang, Y. (2023). Gait quality aware network: toward the interpretability of silhouette-based gait recognition. *IEEE Transactions on Neural Networks and Learning Systems* 34 (11): 8978–8988. https://doi.org/10.1109/TNNLS.2022.3154723.

43 Emeršič, Ž., Štruc, V., and Peer, P. (2017). Ear recognition: more than a survey. *Neurocomputing* 255: 26–39. https://doi.org/10.1016/j.neucom.2016.08.139.

44 El-Naggar, S. and Bourlai, T. (2023). Exploring deep learning ear recognition in thermal images. *IEEE Transactions on Biometrics, Behavior, and Identity Science* 5 (1): 64–75. https://doi.org/10.1109/TBIOM.2022.3218151.

45 Chowdhury, D.P., Bakshi, S., Pero, C. et al. (2023). Privacy preserving ear recognition system using transfer learning in industry 4.0. *IEEE Transactions on Industrial Informatics* 19 (5): 6408–6417. https://doi.org/10.1109/TII.2022.3196343.

17

Secure Deployment of TinyML Applications: Strategies and Practices

Oleksandr Kuznetsov[1,2], Sergii Kavun[3], and Gulvira Bekeshova[4]

[1] *Department of Theoretical and Applied Sciences, eCampus University, Novedrate, Italy*
[2] *Department of Information and Communication Systems Security, V. N. Karazin Kharkiv National University, Kharkiv, Ukraine*
[3] *Computer Information Systems and Technologies Department, Interregional Academy of Personnel Management, Kyiv, Ukraine*
[4] *Department of Information Security, L.N. Gumilyov Eurasian National University, Astana, Republic of Kazakhstan*

17.1 Introduction

The rapid proliferation of Tiny Machine Learning (TinyML) applications has ushered in a new era of intelligent edge computing, enabling sophisticated data analysis and decision-making capabilities on highly constrained devices [1, 2]. As these miniature AI systems become increasingly integrated into critical infrastructure, personal devices, and industrial environments, the imperative for robust security measures has never been more pronounced [3, 4]. This chapter delves into the intricate landscape of secure TinyML deployment, exploring the unique challenges posed by resource limitations and the innovative solutions emerging to address them.

The convergence of machine learning and embedded systems in TinyML introduces a complex set of security considerations that extend beyond traditional Internet of Things (IoT) security paradigms. The constrained nature of TinyML devices – often operating with mere kilobytes of memory and limited processing power – necessitates a fundamental rethinking of security architectures, cryptographic primitives, and privacy-preserving techniques [5, 6]. Moreover, the potential for these devices to influence real-world decisions and actions amplifies the consequences of security breaches, making the development of robust protection mechanisms paramount.

This chapter makes several key contributions to the field of secure TinyML deployment. First, it provides a comprehensive analysis of state-of-the-art

Tiny Machine Learning: Design Principles and Applications, First Edition.
Edited by Agbotiname Lucky Imoize, Dinh-Thuan Do, and Houbing Herbert Song.

security architectures tailored for TinyML environments, offering insights into the trade-offs between security strength, resource utilization, and performance. Second, it introduces novel frameworks for secure bootstrapping and key management in highly constrained devices, presenting decision-making tools to guide practitioners in selecting appropriate strategies. Third, it offers an in-depth examination of lightweight secure communication protocols, including performance benchmarks that illuminate the practical implications of protocol choices in real-world TinyML deployments.

Furthermore, this work advances the discourse on data privacy in TinyML applications, exploring cutting-edge techniques for preserving privacy without compromising the utility of machine learning models. The chapter also casts a forward-looking gaze on emerging technologies poised to reshape the landscape of TinyML security, providing a roadmap for future research and development efforts.

The organization of this chapter reflects the multifaceted nature of secure TinyML deployment. We begin by examining security architectures specifically designed for resource-constrained environments, delving into lightweight cryptographic primitives, secure enclaves, and hardware-based security features. The subsequent section addresses the critical aspects of secure bootstrapping and key management, laying the foundation for trust in TinyML systems. We then explore secure communication protocols tailored for TinyML, offering performance comparisons to guide protocol selection.

The chapter proceeds to investigate data privacy techniques in TinyML applications, highlighting the potential of local processing, differential privacy, and federated learning approaches. We conclude with an exploration of future directions and emerging technologies, including post-quantum cryptography (PQC), AI-driven security monitoring, and blockchain-inspired trust architectures for distributed TinyML systems.

By synthesizing cutting-edge research, practical insights, and forward-looking analysis, this chapter aims to equip researchers, practitioners, and decision-makers with the knowledge and tools necessary to navigate the complex landscape of secure TinyML deployment.

17.2 Related Work

The deployment of TinyML applications in resource-constrained environments has garnered significant attention from researchers across various domains. This section provides a comprehensive analysis of recent work related to secure TinyML deployment, highlighting key contributions and identifying areas for further investigation.

In the realm of TinyML security architectures, Patil et al. [7] present an extensive analysis of security challenges and potential solutions for TinyML devices in networked environments. Their work examines various threat models, including adversarial attacks, malware attacks, and privacy concerns, while also discussing existing countermeasures such as encryption, authentication, and intrusion detection methods. This comprehensive overview underscores the multifaceted nature of TinyML security and the need for holistic approaches that address both hardware and software vulnerabilities.

Butt et al. [8] take a human-centric approach to embedded machine learning security, focusing on the ethical implications and social norms often overlooked in the design and implementation of edge AI systems. Their case study on facial emotion recognition provides valuable insights into the effects of input quantization on model security, robustness, and fairness. Their findings regarding the bias against dark-skinned faces in quantized models highlight the importance of considering diverse datasets and potential biases in TinyML deployments.

The integration of TinyML with emerging network technologies is explored by Saeed et al. [9], who examine the intersection of TinyML and 5G networks. Their work addresses the challenges of designing TinyML algorithms for 5G environments, emphasizing the need for efficient data preprocessing, feature extraction, and model deployment strategies. The case studies presented in their research demonstrate the potential of TinyML to enable real-time intelligent processing in various 5G-enabled applications, from autonomous vehicles to smart cities.

In the context of specific application domains, Ahmed et al. [10] investigate the potential of TinyML in enhancing smart city applications. Their work highlights the benefits of on-device machine learning for improving efficiency and security in areas such as traffic management, energy efficiency, and environmental monitoring. Similarly, Bhamare et al. [11] explore TinyML applications in healthcare, emphasizing the technology's potential for personalized, real-time healthcare monitoring and diagnosis.

The performance and efficiency aspects of TinyML deployment are addressed by Lin et al. [12], who present a real-time bolt-defect detection system integrating TinyML with a magnetic climbing robot. Their work demonstrates the feasibility of running complex detection algorithms on resource-constrained devices, achieving high accuracy with minimal memory and processing requirements. This research exemplifies the potential of TinyML to enable sophisticated applications in industrial and structural inspection scenarios.

Bhattacharya and Pandey [13] propose an innovative approach to TinyML deployment in agricultural settings, combining energy-efficient techniques with blockchain-inspired security measures. Their work introduces a Proof-of-Energy

Efficiency (PoEE) consensus mechanism and a Q-learning-based sidechain formation process, addressing both the energy consumption and security challenges inherent in TinyML deployments for soil quality monitoring.

Looking toward future directions, Ghorpade and Chougule [14] provide an overview of emerging application use cases for TinyML and edge intelligence. Their work highlights the expanding scope of TinyML applications across various industries and identifies key challenges and opportunities for future research and development in this field.

The collective body of work examined here reveals several recurring themes in TinyML security research:

- The need for lightweight, yet robust, security mechanisms tailored to the resource constraints of TinyML devices.
- The importance of considering ethical implications and potential biases in TinyML model design and deployment.
- The challenge of integrating TinyML with emerging network technologies while maintaining security and efficiency.
- The potential for TinyML to enable novel applications across diverse domains, from healthcare to industrial automation.
- The ongoing tension between model performance, energy efficiency, and security in TinyML deployments.

While these works provide valuable insights into various aspects of secure TinyML deployment, there remains a need for more comprehensive frameworks that address the full lifecycle of TinyML applications, from secure bootstrapping to ongoing model updates and decommissioning. Additionally, further research is required to develop standardized benchmarks and evaluation methodologies for assessing the security and performance of TinyML deployments across different hardware platforms and application scenarios.

17.3 Security Architectures for TinyML Deployments

The deployment of TinyML applications on resource-constrained devices presents unique challenges in ensuring robust security while maintaining performance and energy efficiency. This section examines the critical components of security architectures tailored for TinyML environments, exploring the delicate balance between protection mechanisms and the inherent limitations of microcontroller units (MCUs). We delve into lightweight cryptographic primitives, secure enclaves, and hardware-based security features, and conduct a comparative analysis of various security architectures to provide a comprehensive understanding of the current landscape and emerging trends in securing TinyML deployments.

17.3.1 Lightweight Cryptographic Primitives for Constrained Devices

The implementation of cryptographic algorithms on MCUs demands careful consideration of computational overhead, memory footprint, and energy consumption. Traditional cryptographic primitives often prove too resource-intensive for TinyML applications, necessitating the development and adoption of lightweight alternatives.

Among the most promising lightweight block ciphers for TinyML applications is PRESENT [15, 16], designed specifically for constrained environments. With a block size of 64 bits and key sizes of 80 or 128 bits, PRESENT offers a favorable trade-off between security and efficiency. Its compact implementation requires only 1570 gate equivalents, making it suitable for low-power devices.

Another noteworthy contender is SIMON, part of the NSA's[1] lightweight cryptography portfolio. SIMON's flexible design allows for various block and key sizes, with the 64/128 variant (64-bit block, 128-bit key) showing particular promise for TinyML deployments [17, 18].

For stream ciphers, Trivium has gained traction in IoT and TinyML contexts. Its hardware-oriented design allows for efficient implementation on MCUs, requiring minimal resources while providing robust security for most TinyML application scenarios.

The realm of lightweight public-key cryptography has seen significant advancements, with elliptic curve cryptography (ECC) emerging as a frontrunner for TinyML applications. Curve25519, in particular, offers a compelling balance of security and efficiency [19–21]. Its 32-byte public keys and signatures align well with the constrained nature of MCUs, while still providing security equivalent to 128-bit symmetric encryption.

For scenarios requiring even lower computational overhead, the recent NIST[2]-standardized CRYSTALS-Kyber[3] algorithm presents a promising post-quantum alternative. Although primarily designed for post-quantum security, its relatively low computational requirements make it an intriguing option for forward-thinking TinyML deployments [22–24].

Lightweight hash functions play a crucial role in ensuring data integrity and building other cryptographic primitives. SPONGENT, inspired by the sponge construction, offers collision resistance comparable to SHA-256 but with a significantly smaller hardware footprint [25, 26]. For TinyML applications requiring extreme compactness, PHOTON provides a viable alternative, with its smallest variant PHOTON-80/20/16 requiring only 865 gate equivalents [27–29].

When implementing these lightweight primitives on MCUs, careful optimization is essential. Leveraging hardware-specific features, such as the advanced

1 NSA - National Security Agency.
2 NIST - National Institute of Standards and Technology.
3 CRYSTALS - Cryptographic Suite for Algebraic Lattices.

Table 17.1 Comparison of lightweight cryptographic primitives for TinyML applications.

Primitive type	Algorithm	Block/key size (bits)	Gate equivalents	Security level
Block cipher	PRESENT	64/80 or 128	1570	80 bit
Block cipher	SIMON	64/128	~2000	128 bit
Stream cipher	Trivium	–/80	~2600	80 bit
Public key	Curve25519	256	~10000[a)]	128 bit
Hash function	SPONGENT	88 (digest size)	~1060	80 bit

a) Note: Gate equivalent for Curve25519 is an approximation, as it varies significantly based on implementation.

encryption standard (AES) accelerators present in many modern MCUs, can significantly enhance performance. For instance, the STM32L4 series from STMicroelectronics includes a dedicated AES engine that can be utilized to implement modes of operation like CCM (Counter with cipher block chaining message authentication code [CBC-MAC]) efficiently, providing both confidentiality and authenticity with minimal overhead [30].

Table 17.1 provides a concise overview of the key characteristics of lightweight cryptographic primitives. As we can see, these algorithms offer a range of security levels while maintaining a relatively small footprint, making them suitable for TinyML applications on resource-constrained devices.

17.3.2 Secure Enclaves and Trusted Execution Environments

The concept of secure enclaves, or trusted execution environments (TEEs) [31–33], has gained significant traction in TinyML deployments as a means to isolate sensitive computations and data from potentially compromised system components. While fully fledged TEEs like Intel SGX [34] or ARM TrustZone [35] are often beyond the capabilities of MCUs, several lightweight alternatives have emerged to address the specific needs of TinyML applications.

- Lightweight TEE Implementations [36, 37]: TrustLite, developed by researchers at TU Darmstadt, presents a minimalist TEE design tailored for low-end embedded systems. It introduces the concept of "trustlets" – small, isolated code segments that can be executed securely. TrustLite achieves this isolation through a lightweight memory protection unit (MPU) and a secure exception engine, ensuring that sensitive operations remain protected even in the presence of a compromised operating system. Building upon TrustLite, the SANCTUARY architecture takes a step further by introducing temporal isolation. This approach allows for the secure execution of trustlets by temporarily halting all other system activities, effectively creating a time-sliced TEE. SANCTUARY's design is particularly well-suited for TinyML applications that require periodic, secure processing of sensitive data or model updates.

- Hardware-Assisted Secure Enclaves: Several MCU manufacturers have introduced hardware features to support secure enclave-like functionality. The Arm Cortex-M33 processor, for instance, incorporates TrustZone technology, previously reserved for higher-end systems [38]. This allows for the creation of a secure world and a normal world, with hardware-enforced isolation between the two. TinyML applications can leverage this feature to protect critical algorithms, cryptographic keys, and sensitive sensor data. Similarly, the ESP32 series from Espressif Systems includes a dedicated "Secure Boot" hardware module [39]. While not a full TEE, this module provides a root of trust (RoT) and enables secure boot processes, which are crucial for establishing a chain of trust in TinyML deployments.
- Challenges and Considerations: Implementing secure enclaves in TinyML contexts presents several challenges. The limited memory and processing power of MCUs often necessitate careful partitioning of secure and non-secure operations. Developers must strike a balance between security isolation and system performance, as excessive use of secure enclaves can lead to significant overhead. Moreover, the management of secure enclaves in distributed TinyML systems introduces complexity in key distribution and attestation processes. Emerging protocols like the Entity Attestation Token (EAT) show promise in addressing these challenges, providing a standardized way for constrained devices to prove their integrity and security state to remote parties [40].

17.3.3 Hardware-Based Security Features in Modern Microcontrollers

The evolution of MCU design has brought forth a range of hardware-based security features that significantly enhance the security posture of TinyML deployments. These features, when properly leveraged, provide a robust foundation for implementing comprehensive security architectures.

Secure Boot and Measured Boot: Modern MCUs often incorporate secure boot mechanisms to ensure the integrity and authenticity of firmware upon device startup. The STM32H7 series, for example, implements a multi-stage secure boot process [41]. It begins with a hardware-based RoT, typically stored in one-time programmable (OTP) memory, which verifies the bootloader. The bootloader, in turn, verifies the integrity and authenticity of the main application firmware before execution.

Measured boot extends this concept by creating a chain of measurements, where each stage of the boot process measures (i.e., cryptographically hashes) the next stage before passing control. This creates an auditable boot chain, crucial for remote attestation in distributed TinyML systems.

- Secure Key Storage: The protection of cryptographic keys is paramount in TinyML security architectures. Many modern MCUs feature dedicated secure storage elements, often implemented as physically unclonable functions (PUFs) or secure non-volatile memory. The ATECC608A secure element from Microchip, while not an MCU itself, is frequently integrated into TinyML designs to provide tamper-resistant key storage and cryptographic acceleration [42]. Its ability to perform key generation and storage without ever exposing the keys to the main MCU significantly enhances the overall security of the system.
- Memory Protection Units (MPUs) and Memory Encryption: MPUs play a crucial role in enforcing access controls and preventing unauthorized access to sensitive memory regions. The Arm Cortex-M33, widely used in TinyML applications, features an enhanced MPU that allows for fine-grained control over memory access permissions, crucial for implementing secure enclaves and protecting sensitive data [43]. Some advanced MCUs also offer memory encryption capabilities. The SAM L11 from Microchip, for instance, includes on-the-fly static random-access memory (SRAM) encryption, protecting against cold boot attacks and making it significantly more difficult for attackers to extract sensitive information from device memory [44].
- True Random Number Generators (TRNGs): High-quality random number generation is essential for various cryptographic operations. Many modern MCUs now include hardware-based TRNGs, providing a reliable source of entropy for key generation and other security-critical functions. The nRF52840 from Nordic Semiconductor, popular in IoT and TinyML applications, features an NIST SP 800-90B compliant TRNG, ensuring robust randomness for cryptographic operations [45].
- Secure Debug Interfaces: Debug interfaces, while crucial for development and maintenance, can be a significant security vulnerability if left unprotected. Modern MCUs often include features to secure these interfaces. The STM32L5 series, for example, allows for the permanent disabling of debug interfaces or the implementation of authentication mechanisms to control debug access, significantly reducing the attack surface in deployed TinyML devices [46].

17.3.4 Comparison of Security Architectures

To provide a comprehensive overview of the security architectures discussed, we present a comparative analysis highlighting the strengths and limitations of each approach in the context of TinyML deployments.

Table 17.2 presents a comparative analysis of the three main security architectures discussed in this section: lightweight cryptography, secure enclaves (TEEs),

Table 17.2 Comparative analysis of security architectures for TinyML deployments.

Security architecture	Pros	Cons
Lightweight cryptography	– Minimal resource requirements	– May offer reduced security margins
	– Tailored for constrained environments	– Less scrutiny compared to well-established algorithms
	– Often hardware-accelerated in modern MCUs	– Potential vulnerability to side-channel attacks
Secure enclaves (TEEs)	– Strong isolation of sensitive operations	– Increased complexity in system design
	– Protection against compromised OS	– Potential performance overhead
	– Enables secure key management	– Limited support in low-end MCUs
Hardware security	– Root of trust established in hardware	– Increased cost of MCUs with advanced features
Features	– High resistance to software-based attacks	– Potential for hardware-level vulnerabilities
	– Often provides performance benefits	– Less flexibility compared to software solutions

and hardware security features. This comparison highlights the trade-offs inherent in each approach when applied to TinyML deployments.

The selection of an appropriate security architecture for TinyML deployments depends on various factors, including the specific security requirements of the application, the capabilities of the target MCU, and the overall system constraints. In many cases, a hybrid approach combining elements from multiple architectures may provide the most comprehensive security solution.

For instance, a TinyML-based predictive maintenance system for industrial equipment might leverage hardware-based secure boot and key storage features of an advanced MCU, combined with lightweight cryptographic algorithms for secure communication, and a minimalist TEE implementation to protect the ML model and its inferencing process. This multi-layered approach provides robust security while remaining within the resource constraints of the MCU.

17.4 Secure Bootstrapping and Key Management

The foundation of a secure TinyML deployment lies in its ability to establish trust from the moment of device initialization and maintain that trust throughout its operational lifecycle. This section explores the critical aspects of secure bootstrapping and key management in TinyML environments, addressing the unique challenges posed by resource-constrained devices and the need for robust security measures.

17.4.1 Root of Trust Implementation in TinyML Devices

Establishing a hardware-based RoT is paramount in creating a secure foundation for TinyML applications. The RoT serves as the anchor for all subsequent security operations, providing a trusted starting point for the device's security chain.

- Hardware Security Modules (HSMs): While full-fledged HSMs are often too resource-intensive for TinyML devices, lightweight alternatives have emerged. The Microchip ATECC608A [42], for instance, offers a compact, low-power solution for establishing a hardware RoT. This secure element provides protected storage for cryptographic keys, certificates, and other sensitive data, with a footprint suitable for integration into constrained IoT and TinyML devices.
- Physical Unclonable Functions (PUFs): PUFs offer a promising approach to RoT implementation in TinyML contexts. By leveraging inherent physical characteristics of the semiconductor material, PUFs generate device-unique identifiers and cryptographic keys. The SRAM-PUF, in particular, has gained traction due to its minimal additional hardware requirements, utilizing the startup state of SRAM cells to derive unique keys [47].
- Immutable Boot ROM: Many modern MCUs incorporate an immutable boot read-only memory (ROM) containing a first-stage bootloader. This ROM-based code, being unalterable, provides a static RoT from which the secure boot process can be initiated. The STM32H7 series, for example, utilizes this approach, with its ROM code responsible for initial device configuration and verification of the subsequent boot stages [41].

As illustrated in Table 17.3, each RoT implementation offers a different balance of security, resource requirements, flexibility, and cost. The choice of RoT mechanism depends on the specific requirements and constraints of the TinyML application, with SRAM-PUFs offering an attractive combination of security and efficiency for many constrained devices.

17.4.2 Secure Boot Processes for Microcontrollers

The secure boot process ensures the integrity and authenticity of the firmware executed on TinyML devices, protecting against unauthorized modifications and ensuring that only trusted code is run.

Table 17.3 Comparison of root of trust implementations for TinyML devices.

RoT mechanism	Security level	Resource requirements	Flexibility	Cost
HSM	Very high	Low–medium	Medium	High
SRAM-PUF	High	Very low	High	Low
Immutable ROM	High	Very low	Low	Low

Multi-Stage Verification: A robust secure boot implementation typically involves multiple stages of verification. The process begins with the hardware-based RoT, which verifies the integrity of the first-stage bootloader. This bootloader, in turn, verifies the next stage, creating a chain of trust that extends to the main application firmware.

The Arm Platform Security Architecture (PSA) defines a framework for implementing such multi-stage boot processes in constrained devices. It specifies distinct boot stages, each with defined responsibilities and security properties, providing a standardized approach to secure boot implementation in TinyML contexts.

Cryptographic signature verification is a crucial component of the secure boot process. Given the resource constraints of TinyML devices, efficient signature schemes are essential. The Elliptic Curve Digital Signature Algorithm (ECDSA) with curve P-256 offers a good balance of security and efficiency for many TinyML applications. For extremely constrained environments, the Ed25519 signature scheme provides comparable security with reduced computational requirements [21].

Beyond basic integrity checking, measured boot creates a chain of measurements (cryptographic hashes) of each boot stage. These measurements can be securely stored and later used for remote attestation, allowing a TinyML device to prove its boot state to remote parties.

The DICE (Device Identifier Composition Engine) architecture, developed by the Trusted Computing Group, offers a lightweight approach to implementing measured boot in constrained devices [48]. DICE uses a combination of device-unique keys and measurements to create compound device identities, enabling fine-grained attestation of the device's boot state.

17.4.3 Key Generation and Storage Techniques

Secure key management is critical in TinyML deployments, encompassing the generation, storage, and protection of cryptographic keys throughout the device lifecycle.

For TinyML devices, key generation must balance security with resource efficiency. Deterministic key generation algorithms, such as those based on key derivation functions (KDFs), offer a way to generate keys without requiring significant entropy sources.

The HKDF (hash-based message authentication code [HMAC]-based key derivation function) specified in RFC 5869 provides a robust method for key derivation suitable for constrained environments [49]. When combined with a device-unique seed (e.g., from a PUF or secure element), HKDF can generate a variety of application-specific keys from a single master secret.

Protecting keys at rest is crucial to maintaining the overall security of TinyML systems. Several approaches are viable, depending on the hardware capabilities and security requirements:

1) Secure Elements: Dedicated secure elements, like the aforementioned ATECC608A, provide tamper-resistant storage for cryptographic keys [42].
2) On-Chip Flash Encryption: Some MCUs, such as the ESP32 series, offer built-in flash encryption capabilities, protecting sensitive data, including keys, stored in flash memory [39].
3) Software-Based Obfuscation: For devices without hardware-based protection, software techniques like key whitening or splitting can be employed to obscure keys stored in regular memory [50].
4) Key Rotation and Updates: Regular key rotation is a best practice in cryptographic systems, but it presents challenges in TinyML environments due to limited connectivity and power constraints. Implementing efficient key rotation mechanisms, such as hash-based key hierarchies, allows for periodic key updates with minimal communication overhead [51].

The NIST Special Publication 800-57 provides guidelines for cryptographic key management, including recommendations for key lifetimes and update procedures, which can be adapted for TinyML contexts [52].

17.5 Secure Communication Protocols for TinyML

The deployment of TinyML applications in distributed and often wireless environments necessitates robust and efficient secure communication protocols. These protocols must not only ensure the confidentiality, integrity, and authenticity of transmitted data but also operate within the severe resource constraints typical of TinyML devices. This section examines the state of the art in secure communication protocols tailored for TinyML environments, focusing on lightweight implementations of established protocols and novel approaches designed specifically for constrained devices.

Table 17.4 Comparison of lightweight TLS variants for TinyML.

Protocol	Handshake latency	RAM usage	Flash usage	Perfect Forward secrecy
TLS-PSK	Low	~10 KB	~50 KB	No
TLS-ECDHE-PSK	Medium	~15 KB	~70 KB	Yes
DTLS 1.3	Low–medium	~20 KB	~80 KB	Yes

17.5.1 Lightweight TLS Implementations

Transport layer security (TLS) remains a cornerstone of secure communication on the Internet, but its implementation on resource-constrained devices poses significant challenges. Recent efforts have focused on developing lightweight TLS variants suitable for TinyML deployments.

TLS with Pre-Shared Keys (TLS-PSK) offers a lightweight alternative to certificate-based TLS, reducing the computational overhead associated with public-key operations. The combination of Elliptic Curve Diffie-Hellman Ephemeral with PSK (TLS-ECDHE-PSK) provides perfect forward secrecy while maintaining a relatively small footprint [53, 54].

Datagram TLS (DTLS) adapts the security guarantees of TLS to unreliable transport protocols like user datagram protocol (UDP), which are often preferred in IoT and TinyML contexts due to their lower overhead [55]. The DTLS 1.3 specification, currently in draft form, introduces optimizations specifically targeted at constrained environments [56].

Notable improvements in DTLS 1.3 include

1) Reduced handshake round trips, lowering latency and power consumption
2) Support for connection identifiers, facilitating long-lived connections in network address translation (NAT) environments
3) Enhanced record layer efficiency, reducing per-packet overhead

As shown in Table 17.4, these lightweight TLS variants offer different trade-offs between security properties and resource usage, allowing developers to select the most appropriate protocol based on their specific TinyML application requirements and hardware constraints.

17.5.2 Secure MQTT for IoT Environments

Message Queuing Telemetry Transport (MQTT) has emerged as a popular protocol for IoT communications due to its simplicity and efficiency [57]. Securing

MQTT communications in TinyML deployments requires careful consideration of the protocol's peculiarities and the constraints of the devices involved.

MQTT-SN is a variant of MQTT designed specifically for sensor networks and other constrained environments. It introduces optimizations such as topic ID compression and offline buffering, reducing the protocol overhead and accommodating intermittent connectivity common in TinyML deployments [58].

While traditional MQTT relies on simple username/password authentication, more robust mechanisms are crucial for TinyML deployments in potentially hostile environments. The OAuth 2.0 authorization framework has been adapted for use with MQTT, allowing for token-based authentication and fine-grained access control [59].

17.5.3 Novel Protocols Designed for Constrained Devices

The unique requirements of TinyML deployments have spurred the development of novel communication protocols specifically designed for highly constrained environments.

OSCORE (Object Security for Constrained RESTful Environments): OSCORE, standardized in RFC 8613, provides end-to-end security for CoAP (Constrained Application Protocol) messages [60, 61]. Unlike TLS protocols, OSCORE operates at the application layer, allowing for fine-grained protection of individual resources and efficient operation over unreliable transports.

Key features of OSCORE include [60, 61]

1) Extremely low per-message overhead (as little as 1 byte in some cases)
2) Support for multicast communication scenarios
3) Ability to traverse middleboxes and work with caching proxies

EDHOC (Ephemeral Diffie-Hellman Over CBOR Object Signing and Encryption [COSE]): EDHOC is an emerging key exchange protocol designed for constrained environments [62]. It aims to provide a lightweight alternative to TLS/DTLS handshakes, optimized for use with OSCORE and other application layer security protocols.

EDHOC's design prioritizes [62]

1) Minimal message sizes and round trips
2) Flexibility in cryptographic algorithm choices
3) Resistance against quantum computer attacks through optional post-quantum algorithms

Ongoing work by the Internet Engineering Task Force (IETF) LAKE (Lightweight Authenticated Key Exchange) working group is refining the EDHOC specification, with promising results for its applicability in TinyML scenarios.

17.5.4 Performance Comparison of Secure Communication Protocols

To provide a comprehensive view of the performance characteristics of these secure communication protocols in TinyML environments, we conducted a series of benchmarks on a representative constrained platform: an STM32L4 microcontroller (Arm Cortex-M4, 80 MHz, 128 KB RAM). Figure 17.1 presents the results of these benchmarks, comparing handshake time, message latency, and memory footprint for the discussed protocols.

The benchmark results highlight the trade-offs between security features and resource utilization. OSCORE and EDHOC show promising performance in terms of latency and memory usage, making them particularly suitable for highly

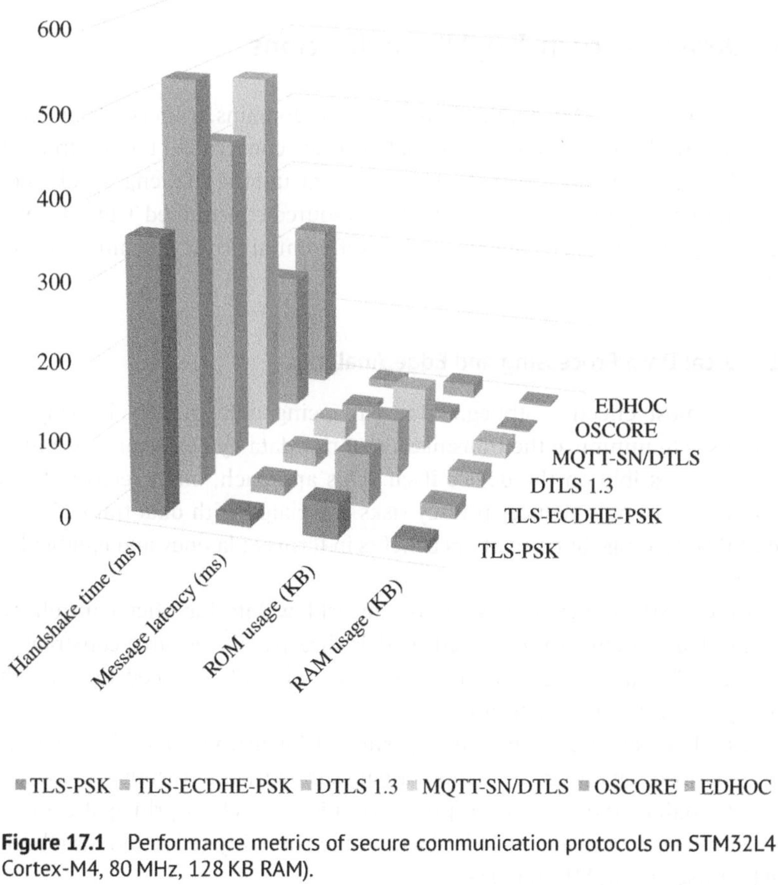

Figure 17.1 Performance metrics of secure communication protocols on STM32L4 (Arm Cortex-M4, 80 MHz, 128 KB RAM).

constrained TinyML devices. However, their relative immaturity compared to established protocols like TLS may be a consideration for some deployments.

TLS-PSK offers a balanced profile, with moderate handshake times and memory usage, while providing robust security guarantees. MQTT-SN over DTLS presents an attractive option for publish-subscribe architectures common in IoT and TinyML deployments, albeit with a slightly larger memory footprint.

The selection of an appropriate secure communication protocol for TinyML applications depends on various factors, including the specific security requirements, network topology, and hardware constraints. In many cases, a hybrid approach combining multiple protocols may provide the most comprehensive solution. For instance, a TinyML device might use OSCORE for efficient, secure data transmission while relying on EDHOC for initial key exchange and periodic rekeying operations.

17.6 Data Privacy in TinyML Applications

The proliferation of TinyML applications in various domains, from personal wearables to industrial sensors, has brought data privacy concerns to the forefront of embedded systems design. This section examines the unique challenges and innovative solutions for preserving data privacy in resource-constrained TinyML environments, focusing on local data processing, differential privacy techniques, and federated learning approaches.

17.6.1 Local Data Processing and Edge Analytics

One of the most effective strategies for enhancing data privacy in TinyML applications is to minimize the transmission of raw data by performing as much processing as possible on the device itself. This approach, often referred to as edge analytics, not only reduces privacy risks associated with data transmission and centralized storage but also offers benefits in terms of latency and bandwidth utilization.

Modern TinyML frameworks, such as TensorFlow Lite for Microcontrollers, enable the deployment of pre-trained models directly on resource-constrained devices [63]. This allows for real-time inference without the need to transmit potentially sensitive raw data to cloud servers.

In cases where some data must be transmitted for further analysis or model updating, performing feature extraction or data summarization on the device can significantly reduce the privacy footprint. Techniques such as principal component analysis (PCA) and autoencoder-based dimensionality reduction have been adapted for use in TinyML contexts.

Table 17.5 Comparison of differential privacy techniques for TinyML.

Technique	Privacy guarantee	Computational overhead	Communication overhead	Accuracy impact
Basic LDP	Strong	Low	None	High
Adaptive LDP	Strong	Medium	Low	Medium
Distributed differential privacy (DP)	Moderate	Low	Medium	Low
Randomized Response	Moderate	Very low	None	Medium

17.6.2 Differential Privacy Techniques for TinyML

Differential privacy provides a mathematical framework for quantifying and limiting the privacy risk associated with statistical queries on datasets. Adapting differential privacy techniques to the resource-constrained environment of TinyML presents unique challenges but offers powerful privacy guarantees.

Local differential privacy (LDP) techniques add noise to data on the device before any aggregation or transmission, providing strong privacy guarantees even against adversaries with access to the communication channel. However, the application of LDP in TinyML contexts requires careful consideration of the noise-utility trade-off, given the limited precision and dynamic range of many embedded sensors.

As shown in Table 17.5, different differential privacy techniques offer varying trade-offs between privacy guarantees, resource utilization, and impact on data utility. The choice of technique depends on the specific requirements and constraints of the TinyML application.

17.6.3 Federated Learning Approaches for Privacy Preservation

Federated learning has emerged as a powerful paradigm for training machine learning models without centralizing raw data, aligning well with the privacy requirements of many TinyML applications. However, implementing federated learning on highly constrained devices presents significant challenges.

Lightweight Federated Learning Protocols: Traditional federated learning protocols can be prohibitively expensive for TinyML devices in terms of computation and communication. Recent research has focused on developing lightweight alternatives tailored for constrained environments.

The FedPAQ (Federated Periodic Averaging and Quantization) algorithm, introduced by Reisizadeh et al. [64], combines periodic averaging with aggressive quantization of model updates. Their approach reduces communication overhead while maintaining comparable model accuracy.

Secure aggregation protocols enable the computation of aggregate statistics or model updates without revealing individual contributions, enhancing privacy in federated learning systems. Adapting these protocols to TinyML environments requires careful optimization to minimize computational and communication overhead.

A recent breakthrough by Sami and Güler introduces TinySecAgg, a secure aggregation protocol specifically designed for microcontroller-class devices [65]. TinySecAgg leverages a combination of lightweight cryptographic primitives and efficient encoding techniques to achieve secure aggregation.

17.7 Future Directions and Emerging Technologies

As the field of TinyML continues to evolve rapidly, several emerging technologies and research directions are poised to significantly impact the landscape of secure TinyML deployments. This section explores these cutting-edge developments, their potential implications, and the challenges that lie ahead in realizing their full potential.

17.7.1 Post-Quantum Cryptography for TinyML

As quantum computing progresses, the potential for it to break current cryptographic standards could jeopardize the security of TinyML applications. Anticipating this shift, future research must prioritize the development of quantum-resistant cryptographic algorithms. Ensuring these algorithms are both secure against quantum attacks and feasible for deployment on resource-constrained devices will be a significant challenge that requires innovative approaches.

The looming threat of quantum computers capable of breaking widely used public-key cryptosystems has spurred intense research into PQC. Adapting PQC algorithms for TinyML environments presents unique challenges due to resource constraints, but it is crucial for ensuring long-term security of deployed systems.

Among the various PQC candidates, lattice-based schemes have shown particular promise for TinyML applications due to their relatively small key sizes and computational efficiency. The CRYSTALS-Kyber key encapsulation mechanism, a finalist in the NIST PQC standardization process, has been successfully implemented on Cortex-M4 microcontrollers [66].

To balance the need for quantum resistance with the efficiency of current algorithms, hybrid schemes combining traditional and post-quantum algorithms are being explored. A study by Kumari et al. [67] proposed a post-quantum lattice-based lightweight authentication and code-based hybrid encryption scheme for IoT devices.

17.7.2 Enhancing Security Through Advanced Hardware

As the field of TinyML expands, there is an imperative need to develop and deploy advanced hardware that inherently supports enhanced security features. Future research should focus on designing specialized processors and memory architectures that integrate security at the hardware level. These could include hardware-based encryption engines and secure boot capabilities that ensure the integrity and confidentiality of both code and data. Additionally, exploring non-standard computing paradigms such as neuromorphic computing may offer new ways to secure TinyML applications against evolving threats.

17.7.3 Leveraging Federated Learning for Privacy Preservation

Federated learning presents a promising avenue for improving privacy in TinyML deployments. By enabling multiple devices to collaboratively learn a shared prediction model while keeping the training data on the device, this approach minimizes the risk of data leakage. Future technologies should aim to enhance federated learning algorithms to be lightweight and efficient enough for implementation on low-power devices, thereby broadening their applicability in TinyML scenarios.

17.7.4 Exploiting Blockchain for Enhanced Integrity and Accountability

Blockchain technology could revolutionize the way TinyML applications are managed and deployed, particularly in decentralized environments. Implementing blockchain can provide immutable record-keeping and audit trails for deployments, ensuring tamper-proof data and model integrity. This approach could be particularly beneficial in scenarios involving multiple stakeholders or where regulatory compliance and data provenance are critical [68, 69].

17.7.5 Adaptive Security Policies

The dynamic nature of edge computing environments where TinyML applications often operate suggests a need for adaptive security policies. These policies would automatically adjust based on real-time threat analysis and changes in operational context. Research should focus on developing intelligent, context-aware security systems that can dynamically identify and mitigate threats at the edge, potentially using AI-driven techniques to predict and counteract emerging vulnerabilities.

17.7.6 Ethical Considerations and AI Governance

With the increasing deployment of AI technologies, ethical considerations and governance models become paramount. Future directions should include the development of frameworks and guidelines that ensure ethical usage and deployment of TinyML applications. These frameworks should address issues such as bias, privacy, autonomy, and accountability, ensuring that TinyML deployments adhere to ethical standards and societal norms.

17.7.7 Research Challenges and Opportunities

As these emerging technologies continue to evolve, several key research challenges and opportunities stand out in the context of secure TinyML deployments:

1) Ultra-Low-Power Cryptography: Developing cryptographic primitives and protocols that can operate efficiently on energy-harvesting devices and other ultra-low-power TinyML platforms.
2) Hardware–Software Co-Design: Exploring novel hardware architectures and accelerators specifically optimized for secure TinyML operations, potentially leveraging emerging technologies like in-memory computing.
3) Quantum-Safe TinyML: Investigating the implications of quantum computing on TinyML security and developing quantum-resistant algorithms tailored for highly constrained devices.
4) Privacy-Preserving Federated Learning: Advancing federated learning techniques to enable collaborative model training across heterogeneous TinyML devices while preserving privacy and ensuring fairness.
5) Secure Over-the-Air Updates: Designing robust and efficient protocols for securely updating ML models and firmware on deployed TinyML devices, particularly in hard-to-reach or hostile environments.

The rapid pace of innovation in this field underscores the need for continued research investment and standardization efforts to ensure that security measures can keep pace with the expanding capabilities and applications of TinyML technologies. As these systems become increasingly prevalent in critical infrastructure, healthcare, and other sensitive domains, the development of robust, efficient, and future-proof security solutions for TinyML will remain a paramount concern for the research community and industry practitioners alike.

17.7.8 Interdisciplinary Collaboration

Finally, the complexity of securing TinyML applications necessitates an interdisciplinary approach, drawing expertise from fields such as cybersecurity, hardware engineering, software development, and ethics. Fostering collaboration across these disciplines will be crucial in addressing the multifaceted challenges of TinyML security and ensuring the robust deployment of these technologies in diverse environments.

These future directions not only highlight the evolving landscape of TinyML security but also underscore the necessity for innovative, integrated solutions that address both technological and ethical challenges. By focusing on these emerging technologies and strategies, researchers and practitioners can better secure TinyML applications against an increasingly complex threat landscape.

17.8 Conclusion

The secure deployment of TinyML applications represents a critical frontier in the evolution of edge computing and the IoT. This chapter has explored the multifaceted challenges and innovative solutions emerging in this rapidly advancing field, focusing on key aspects of security architectures, secure bootstrapping, key management, communication protocols, data privacy, and future directions.

17.8.1 Summary of Key Strategies for Secure TinyML Deployment

Our examination of security architectures for TinyML deployments highlighted the importance of lightweight cryptographic primitives, secure enclaves, and hardware-based security features. The comparative analysis presented in Table 17.2 underscored the trade-offs inherent in each approach, emphasizing the need for carefully tailored solutions that balance security, performance, and resource utilization.

In the realm of secure communication protocols, our analysis demonstrated the viability of lightweight TLS implementations and novel protocols like OSCORE and EDHOC for constrained TinyML environments. The performance benchmarks presented in Figure 17.1 offer crucial insights for protocol selection in resource-limited scenarios.

The exploration of data privacy techniques highlighted the potential of local data processing, differential privacy, and federated learning approaches in preserving privacy without compromising the utility of TinyML applications. The trade-offs between privacy, accuracy, and computational overhead underscore the complexity of designing privacy-preserving TinyML systems.

17.8.2 The Evolving Landscape of TinyML Security

As we look to the future, several emerging technologies promise to reshape the landscape of TinyML security. The advent of PQC for constrained devices, the integration of AI-driven security monitoring at the edge, and the exploration of blockchain-inspired trust architectures present both exciting opportunities and formidable challenges.

17.8.3 Call to Action for Researchers and Practitioners

As TinyML applications continue to proliferate across diverse domains, from industrial IoT to personalized healthcare, the imperative for robust security measures becomes increasingly critical. This chapter calls upon researchers and practitioners to

1) Prioritize security and privacy considerations from the earliest stages of TinyML system design, recognizing that retrofitting security measures is often impractical in resource-constrained environments.
2) Invest in the development and optimization of lightweight security primitives and protocols specifically tailored for the unique constraints of TinyML devices.
3) Explore novel hardware–software co-design approaches that can enhance security without significantly impacting power consumption or performance.
4) Advance standardization efforts to ensure interoperability and establish best practices for secure TinyML deployments across different application domains.
5) Foster closer collaboration between the machine learning, embedded systems, and security communities to drive holistic innovations in secure TinyML architectures.
6) Conduct rigorous security analyses and real-world evaluations of TinyML systems to identify vulnerabilities and develop practical mitigation strategies.

The secure deployment of TinyML applications stands at the intersection of multiple cutting-edge technologies, offering a rich landscape for innovation and research. As we continue to push the boundaries of what is possible with machine learning on highly constrained devices, maintaining a steadfast focus on security will be crucial in realizing the full potential of this transformative technology.

The challenges outlined in this chapter are not insurmountable, but they require concerted effort, creativity, and collaboration across disciplines. By addressing these challenges head-on, we can pave the way for a future where intelligent, secure, and privacy-preserving edge computing becomes ubiquitous, enabling new applications and services that enhance our lives while safeguarding our data and systems.

Acknowledgment

- This project has received funding from the European Union's Horizon 2020 research and innovation program under the Marie Skłodowska-Curie Grant Agreement No. 101007820 – TRUST. This publication reflects only the author's view, and the REA is not responsible for any use that may be made of the information it contains.
- This research was funded by the European Union – NextGenerationEU under the Italian Ministry of University and Research (MIUR), National Innovation Ecosystem Grant ECS00000041-VITALITY-CUP D83C22000710005.

References

1 Disabato, S. and Roveri, M. (2024). Tiny machine learning for concept drift. *IEEE Transactions on Neural Networks and Learning Systems* 35 (6): 8470–8481. https://doi.org/10.1109/TNNLS.2022.3229897.

2 Lin, J., Zhu, L., Chen, W.-M. et al. (2023). Tiny machine learning: progress and futures [feature]. *IEEE Circuits and Systems Magazine* 23 (3): 8–34. https://doi .org/10.1109/MCAS.2023.3302182.

3 Samaras, G., Mertiri, M., Xezonaki, M.-E. et al. (2024). Unlocking the path towards automation of tiny machine learning for edge computing. *2024 International Conference on Smart Applications, Communications and Networking (SmartNets)*, 1–6. https://doi.org/10.1109/SmartNets61466.2024.10577687.

4 Kardan, R., Roy, S.S., Elsaharti, A., and Neubert, J. (2023). Machine learning based specularity detection techniques to enhance indoor navigation. *2023 IEEE 17th International Conference on Semantic Computing (ICSC)*, 143–148. https://doi.org/10.1109/ICSC56153.2023.00030.

5 Diab, M.S. and Rodriguez-Villegas, E. (2022). Embedded machine learning using microcontrollers in wearable and ambulatory systems for health and care applications: a review. *IEEE Access* 10: 98450–98474. https://doi.org/10.1109/ ACCESS.2022.3206782.

6 Abadade, Y., Temouden, A., Bamoumen, H. et al. (2023). A comprehensive survey on TinyML. *IEEE Access* 11: 96892–96922. https://doi.org/10.1109/ACCESS .2023.3294111.

7 Patil, R.Y., Bhamare, M., Patil, Y.H., and Bannore, A. (2024). Chapter 13 - Securing TinyML in a connected world. In: *TinyML for Edge Intelligence in IoT and LPWAN Networks* (ed. B.S. Chaudhari, S.N. Ghorpade, M. Zennaro, and R. Paškauskas), 311–330. Academic Press https://doi.org/10.1016/B978-0-44-322202-3.00018-X.

8 Butt, M.A., Qayyum, A., Ali, H. et al. (2023). Towards secure private and trustworthy human-centric embedded machine learning: An emotion-aware facial recognition case study. *Computers & Security* 125: 103058. https://doi.org/10.1016/j.cose.2022.103058.

9 Saeed, M.M., Saeed, R.A., and Ahmed, Z.E. (2024). Chapter 9 - TinyML for 5G networks. In: *TinyML for Edge Intelligence in IoT and LPWAN Networks* (ed. B.S. Chaudhari, S.N. Ghorpade, M. Zennaro, and R. Paškauskas), 167–229. Academic Press https://doi.org/10.1016/B978-0-44-322202-3.00014-2.

10 Ahmed, Z.E., Hashim, A.A., Saeed, R.A., and Saeed, M.M. (2024). Chapter 18 - TinyML network applications for smart cities. In: *TinyML for Edge Intelligence in IoT and LPWAN Networks* (ed. B.S. Chaudhari, S.N. Ghorpade, M. Zennaro, and R. Paškauskas), 423–451. Academic Press https://doi.org/10.1016/B978-0-44-322202-3.00023-3.

11 Bhamare, M., Kulkarni, P.V., Rane, R. et al. (2024). Chapter 14 - TinyML applications and use cases for healthcare. In: *TinyML for Edge Intelligence in IoT and LPWAN Networks* (ed. B.S. Chaudhari, S.N. Ghorpade, M. Zennaro, and R. Paškauskas), 331–353. Academic Press https://doi.org/10.1016/B978-0-44-322202-3.00019-1.

12 Lin, T.-H., Chang, C.-T., and Putranto, A. (2024). Tiny machine learning empowers climbing inspection robots for real-time multiobject bolt-defect detection. *Engineering Applications of Artificial Intelligence* 133: 108618. https://doi.org/10.1016/j.engappai.2024.108618.

13 Bhattacharya, S. and Pandey, M. (2024). Deploying an energy efficient, secure & high-speed sidechain-based TinyML model for soil quality monitoring and management in agriculture. *Expert Systems with Applications* 242: 122735. https://doi.org/10.1016/j.eswa.2023.122735.

14 Ghorpade, S.N. and Chougule, S. (2024). Chapter 19 - Emerging application use cases and future directions. In: *TinyML for Edge Intelligence in IoT and LPWAN Networks* (ed. B.S. Chaudhari, S.N. Ghorpade, M. Zennaro, and R. Paškauskas), 453–482. Academic Press https://doi.org/10.1016/B978-0-44-322202-3.00024-5.

15 Sravya, G., Manchalla, O.V.P., Kumar, G.M. et al. (2020). Hardware lightweight design of PRESENT block cipher. *Materials Today: Proceedings* 33: 4880–4886. https://doi.org/10.1016/j.matpr.2020.08.435.

16 Dahiphale, V., Raut, H., Bansod, G., and Dahiphale, D. (2025). Securing IoT devices with fast and energy efficient implementation of PRIDE and PRESENT ciphers. *Cyber Security and Applications* 3: 100055. https://doi.org/10.1016/j.csa.2024.100055.

17 Aysu, A., Gulcan, E., and Schaumont, P. (2014). SIMON says: break area records of block ciphers on FPGAs. *IEEE Embedded Systems Letters* 6 (2): 37–40. https://doi.org/10.1109/LES.2014.2314961.

18 Kaul, H., Anders, M., Mathew, S. et al. (2018). Ultra-lightweight 548–1080 gate 166Gbps/W–12.6Tbps/W SIMON 32/64 cipher accelerators for IoT in 14nm Tri-gate CMOS. *2018 IEEE Asian Solid-State Circuits Conference (A-SSCC)*, 1–4. https://doi.org/10.1109/ASSCC.2018.8579274.

19 Salarifard, R. and Bayat-Sarmadi, S. (2019). An efficient low-latency point-multiplication over curve25519. *IEEE Transactions on Circuits and Systems I: Regular Papers* 66 (10): 3854–3862. https://doi.org/10.1109/TCSI.2019.2914247.

20 Niasar, M.B., El Khatib, R., Azarderakhsh, R., and Mozaffari-Kermani, M. (2020). Fast, small, and area-time efficient architectures for key-exchange on curve25519. *2020 IEEE 27th Symposium on Computer Arithmetic (ARITH)*, 72–79. https://doi.org/10.1109/ARITH48897.2020.00019.

21 Naik, R.M., Sathyanarayana, S., and Sowmya, T. (2020). Key management using elliptic curve Diffie Hellman curve 25519. *2020 Third International Conference on Multimedia Processing, Communication & Information Technology (MPCIT)*, 33–39. https://doi.org/10.1109/MPCIT51588.2020.9350454.

22 Sun, J., Bai, X., and Kang, Y. (2023). An FPGA-based efficient NTT accelerator for post-quantum cryptography CRYSTALS-Kyber. *2023 IEEE International Conference on Integrated Circuits, Technologies and Applications (ICTA)*, 142–143. https://doi.org/10.1109/ICTA60488.2023.10364299.

23 Celik, A., Yilmaz, F., Korkmaz, M.A., and Ors, B. (2023). Implementation of CRYSTALS-Kyber post-quantum algorithm using RISC-V processor. *2023 30th IEEE International Conference on Electronics, Circuits and Systems (ICECS)*, 1–4. https://doi.org/10.1109/ICECS58634.2023.10382743.

24 Ni, Z., Khalid, A., Liu, W., and O'Neill, M. (2023). Towards a lightweight CRYSTALS-kyber in FPGAs: an ultra-lightweight BRAM-free NTT Core. *2023 IEEE International Symposium on Circuits and Systems (ISCAS)*, 1–5. https://doi.org/10.1109/ISCAS46773.2023.10181340.

25 Gupta, D.N. and Kumar, R. (2021). Sponge based lightweight cryptographic hash functions for IoT applications. *2021 International Conference on Intelligent Technologies (CONIT)*, 1–5. 10.1109/CONIT51480.2021.9498572.

26 Bogdanov, A., Knezevic, M., Leander, G. et al. (2013). SPONGENT: The design space of lightweight cryptographic hashing. *IEEE Transactions on Computers* 62 (10): 2041–2053. https://doi.org/10.1109/TC.2012.196.

27 Guo, J., Peyrin, T., and Poschmann, A. (2011). The PHOTON family of lightweight hash functions. 2011/609. https://eprint.iacr.org/2011/609 (accessed 28 July 2024).

28 Guo, J., Peyrin, T., Poschmann, A., and Rogaway, P. (2011). The PHOTON family of lightweight hash functions. In: *Advances in Cryptology – CRYPTO 2011* (ed. P. Rogaway), 222–239. Berlin, Heidelberg: Springer https://doi.org/10.1007/978-3-642-22792-9_13.

29 Mangole, T., Helberg, A.S.J., and Nair, K.K. (2022). Resource usage evaluation of the PHOTON hash function. *2022 Conference on Information Communications Technology and Society (ICTAS)*, 1–6. https://doi.org/10.1109/ICTAS53252.2022.9744686.

30 Cryptographic performance on STM32L4 Series - stm32mcu. [Online]. https://wiki.st.com/stm32mcu/wiki/Security:STM32L4_Series:_Cryptographic_Library_Performance (accessed 10 July 2024).

31 Chen, Y., Luo, F., Li, T. et al. (2020). A training-integrity privacy-preserving federated learning scheme with trusted execution environment. *Information Sciences* 522: 69–79. https://doi.org/10.1016/j.ins.2020.02.037.

32 Cao, Y., Zhang, J., Zhao, Y. et al. (2024). SRFL: a secure & robust federated learning framework for IoT with trusted execution environments. *Expert Systems with Applications* 239: 122410. https://doi.org/10.1016/j.eswa.2023.122410.

33 Fan, Y., Liu, S., Tan, G., and Qiao, F. (2020). Fine-grained access control based on trusted execution environment. *Future Generation Computer Systems* 109: 551–561. https://doi.org/10.1016/j.future.2018.05.062.

34 Intel® Software Guard Extensions (Intel® SGX). Intel. [Online]. https://www.intel.com/content/www/us/en/products/docs/accelerator-engines/software-guard-extensions.html (accessed 28 July 2024).

35 A. Ltd. TrustZone for Cortex-A – Arm®. Arm | The Architecture for the Digital World. [Online]. https://www.arm.com/technologies/trustzone-for-cortex-a (accessed 28 July 2024).

36 Koeberl, P., Schulz, S., Sadeghi, A.-R., and Varadharajan, V. (2014). TrustLite: a security architecture for tiny embedded devices. *Proceedings of the Ninth European Conference on Computer Systems*, in EuroSys'14, 1–14. New York, NY, USA: Association for Computing Machinery. https://doi.org/10.1145/2592798.2592824.

37 Brasser, F., Gens, D., Jauernig, P. et al. (2019). SANCTUARY: ARMing TrustZone with user-space enclaves. *Proceedings 2019 Network and Distributed System Security Symposium*. https://doi.org/10.14722/ndss.2019.23448.

38 A. Ltd. TrustZone for Cortex-M – Arm®. Arm | The Architecture for the Digital World. [Online]. https://www.arm.com/technologies/trustzone-for-cortex-m (accessed 28 July 2024).

39 ESP32-S2 Series Development Kits - Espressif Systems | Mouser. [Online]. https://eu.mouser.com/new/espressif/espressif-esp32-s2-kits/ (accessed 28 July 2024).

40 Lundblade, L., Mandyam, G., and O'Donoghue, J. (2021). The entity attestation token (EAT). Internet Engineering Task Force, Internet Draft draft-ietf-rats-eat-11. [Online]. https://datatracker.ietf.org/doc/draft-ietf-rats-eat-11 (accessed 28 July 2024).

41 STM32H7 - Arm Cortex-M7 and Cortex-M4 MCUs (480 MHz) - STMicroelectronics. [Online]. https://www.st.com/en/microcontrollers-microprocessors/stm32h7-series.html (accessed 28 July 2024).

42 ATECC608A-MAHCZ-T | Microchip Technology. [Online]. https://www.microchipdirect.com/product/ATECC608A-MAHCZ-T?samples=true (accessed 28 July 2024).

43 Arm Cortex-M33 - Microcontrollers – STMicroelectronics. [Online]. https://www.st.com/content/st_com/en/arm-32-bit-microcontrollers/arm-cortex-m33.html (accessed 28 July 2024).

44 SAM L10 and SAM L11 Families of Microcontrollers | Microchip Technology. [Online]. https://www.microchip.com/en-us/products/microcontrollers-and-microprocessors/32-bit-mcus/sam-32-bit-mcus/sam-l/sam-l10-l11 (accessed 28 July 2024).

45 Bluetooth Module – MinewSemi. [Online]. https://ad.minewsemi.com/bluetooth-module/ (accessed 28 July 2024).

46 STM32L5 - Ultra-low-power MCUs, Enhanced Security for IoT and Embedded Applications - STMicroelectronics. [Online]. https://www.st.com/en/microcontrollers-microprocessors/stm32l5-series.html (accessed 28 July 2024).

47 Radhakrishnan, R.A. (2024). Secure implementation of SRAM PUF for private key generation. 2024/670. [Online]. https://eprint.iacr.org/2024/670 (accessed 28 July 2024).

48 What is a device identifier composition engine (DICE)? Trusted Computing Group. [Online]. https://trustedcomputinggroup.org/what-is-a-device-identifier-composition-engine-dice/ (accessed 28 July 2024).

49 Krawczyk, H. and Eronen, P. (2010). HMAC-based extract-and-expand key derivation function (HKDF). Internet Engineering Task Force, Request for Comments RFC 5869. https://doi.org/10.17487/RFC5869.

50 How do I obscure API keys on the client? [Online]. https://www.pronextjs.dev/obscuring-api-keys (accessed 28 July 2024).

51 Key rotation strategies for securing sensitive data. [Online]. https://www.piiano.com/blog/key-rotation (accessed 28 July 2024).

52 Barker, E. (2020). Recommendation for key management: part 1 – general. National Institute of Standards and Technology, NIST Special Publication (SP) 800-57 Part 1 Rev. 5 May 2020. https://doi.org/10.6028/NIST.SP.800-57pt1r5.

53 Amanlou, S., Hasan, M.K., and Bakar, K.A.A. (2021). Lightweight and secure authentication scheme for IoT network based on publish–subscribe fog computing model. *Computer Networks* 199: 108465. https://doi.org/10.1016/j.comnet.2021.108465.

54 Tange, K., Mödersheim, S., Lalos, A. et al. (2021). rTLS: secure and efficient TLS session resumption for the internet of things. *Sensors* 21: https://doi.org/10.3390/s21196524.

55 Rescorla, E. and Modadugu, N. (2012). Datagram transport layer security version 1.2. Internet Engineering Task Force, Request for Comments RFC 6347. https://doi.org/10.17487/RFC6347.

56 Rescorla, E., Tschofenig, H., and Modadugu, N. (2024). The datagram transport layer security (DTLS) protocol version 1.3. Internet Engineering Task Force, Internet Draft draft-ietf-tls-dtls13-40. [Online]. https://datatracker.ietf.org/doc/draft-ietf-tls-dtls13-40 (accessed 28 July 2024).

57 MQTT - the standard for IoT messaging. [Online]. https://mqtt.org/ (accessed 28 July 2024).

58 Projects - OASIS. [Online]. https://groups.oasis-open.org/higherlogic/ws/public/document?document_id=66091 (accessed 28 July 2024).

59 OAuth 2.0 & MQTT - MQTT Security Fundamentals. [Online]. https://www.hivemq.com/blog/mqtt-security-fundamentals-oauth-2-0-mqtt/ (accessed 28 July 2024).

60 Selander, G., Mattsson, J., Palombini, F., and Seitz, L. (2019). RFC 8613: object security for constrained RESTful environments (OSCORE). USA: RFC Editor.

61 Information on RFC 8613 ≫ RFC Editor. [Online]. https://www.rfc-editor.org/info/rfc8613 (accessed 28 July 2024).

62 Selander, G., Mattsson, J.P., and Palombini, F. (2024). Ephemeral Diffie–Hellman over COSE (EDHOC). Internet Engineering Task Force, Request for Comments RFC 9528. https://doi.org/10.17487/RFC9528.

63 TensorFlow Lite | ML for mobile and edge devices. TensorFlow. [Online]. https://www.tensorflow.org/lite (accessed 28 July 2024).

64 Reisizadeh, A., Mokhtari, A., Hassani, H. et al. (2020). FedPAQ: a communication-efficient federated learning method with periodic averaging and quantization. arXiv:1909.13014, 7 June. https://doi.org/10.48550/arXiv.1909.13014.

65 Sami, H.U. and Güler, B. (2024, https://doi.org/10.1109/TCOMM.2024.3403475). Secure gradient aggregation with sparsification for resource-limited federated learning. *IEEE Transactions on Communications* 72: 6883–6899.

66 Schwabe, P. (2024). Kyber. [Online]. https://pq-crystals.org/kyber/ (accessed 28 July 2024).

67 Kumari, S., Singh, M., Singh, R., and Tewari, H. (2022). A post-quantum lattice based lightweight authentication and code-based hybrid encryption scheme for IoT devices. *Computer Networks* 217: 109327. https://doi.org/10.1016/j.comnet.2022.109327.

68 Brumnik, R., Klebanova, T., Guryanova, L. et al. (2014). Simulation of territorial development based on fiscal policy tools. *Mathematical Problems in Engineering* 2014 (1): 843976. https://doi.org/10.1155/2014/843976.

69 Daradkeh, Y., Guryanova, L., Klebanova, T., and Kavun, S. (2012). Forecasting the cyclical dynamics of the development territories: conceptual approaches, models, experiments. *European Journal of Scientific Research* 74 (1): 5–20.

18

TinyML for Environmental Applications

Duy Nam Khanh Vu[1] and Anh Khoa Dang[2]

[1]*Lê Quý Đôn High School, Ho Chi Minh City, Vietnam*
[2]*Perkiomen School, Pennsburg, PA, USA*

18.1 Introduction

At first, much research and many inventions that contribute to the growth of many environmental aspects are correlated to a great technological term in the Internet of Things (IoT) industry: machine learning (ML). Specifically, IoT plays a crucial role and is the resource for many fields such as residential systems and manufacturers [1]. In order to truly make the potential of IoT available to life, scientists, researchers, and engineers have created and used ML for IoT application and development, especially for environmental applications.

ML has advanced significantly, leading to the development of many sophisticated algorithms, techniques, and libraries. One notable advancement is tiny ML (TinyML), which focuses on deploying machine learning models on resource-constrained devices, enabling efficient and fast processing. Generally, TinyML enables faster local decision-making progress, reduces the cost to deploy, enhances accuracy, minimizes power consumption, is portable, and is easy to integrate [2–4]. Thus, there have been and will be a lot of substantial inventions related to TinyML, a new future for ML.

TinyML has contributed significantly to industries, science fields, etc., and is promised to continually participate in the development of larger aspects of life. According to [1, 5] and [6, 7], there are several fields that can benefit from TinyML. Figure 18.1 shows some benefits of the application [8–17].

So what actually is TinyML? The answer is that TinyML is a type of "approach" within ML that does not need to process and train data on cloud or within Internet connections. In contrast, TinyML enables microcontrollers to train and make use of data to make predictions locally [18]. Therefore, TinyML's "circle of processing" represents a much simpler process diagram than the traditional ML

Tiny Machine Learning: Design Principles and Applications, First Edition.
Edited by Agbotiname Lucky Imoize, Dinh-Thuan Do, and Houbing Herbert Song.

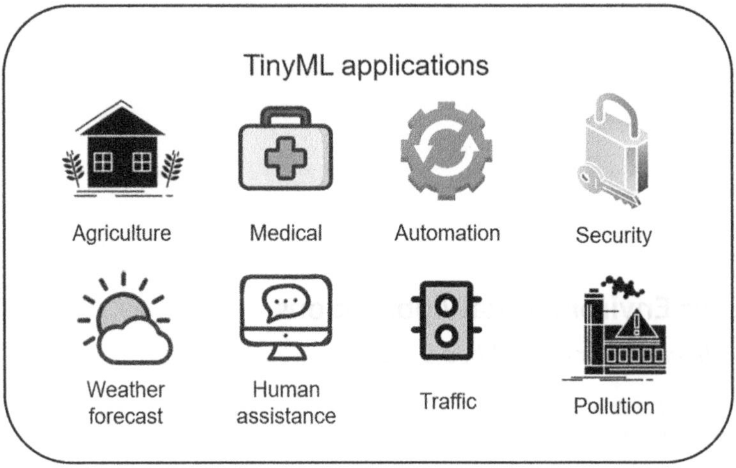

Figure 18.1 Examples of some benefited TinyML applications.

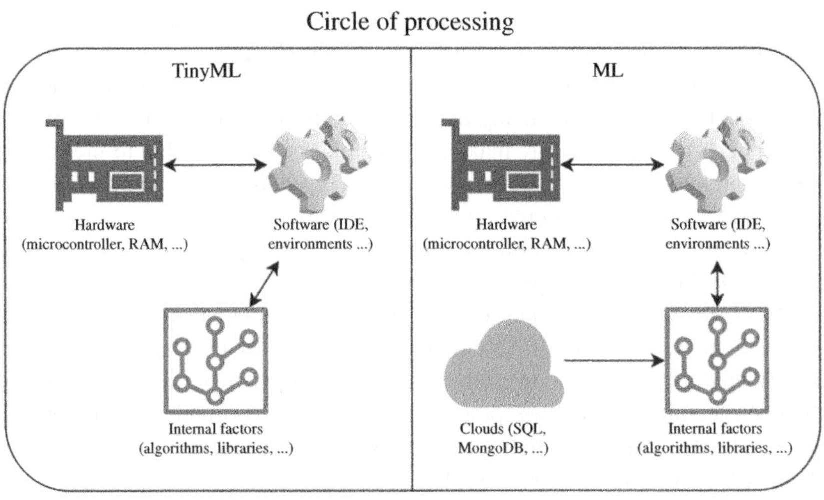

Figure 18.2 Difference in processing between TinyML and ML approach.

approach. Figure 18.2 describes the difference between TinyML and ML's "circle of processing."

Over this survey, there will be some points demonstrated in order to fulfill the author's research purpose and also the key point of this chapter:

- TinyML applications in smart agriculture.
- TinyML applications in healthcare systems.

- TinyML applications in military systems.
- TinyML applications in wireless systems.

Moreover, the paper will be presented in the following order: Section 18.2 presents related work, Section 18.3 presents methodology, Section 18.4 presents a case study related to a study in the survey, Section 18.5 presents results and discussions, and Section 18.6 presents conclusion and future scope.

18.2 Related Work

TinyML is an interesting field and needs to be investigated in more depth to explore its potential usefulness. One of such fields that TinyML can be applied in and researched on is environment, and there have been numerous surveys mentioning these applications.

Firstly, smart agriculture, also known as precision agriculture, makes use of technologies, such as ML, to improve crop quality, farming practices, and enhance resource sustainability. This aspect also applies to multi-form sensors, including drones and satellite imagery, to collect data, record soil conditions, weather patterns, and crop health [19].

Secondly, healthcare systems are complex networks that encompass various organizations, such as hospitals or medical centers, working together with technologies to provide medical services to patients. With the addition of TinyML in medical fields, including surgery, disease detection, and bacterial analysis, doctors and researchers can further delve into collecting medical information and streamline many operations in medical procedures [20].

Thirdly, military systems encompass ML and artificial intelligence (AI) with various embedded systems to enhance military operations; some possible examples will be combat drones, self-driven airplanes, etc. Integrated with ML and AI, military operations will be improved, allowing strategic and economic implications of modern systems [21].

Lastly, wireless systems are crucial in modern communication; within a second, many messages can be delivered everywhere in the world, and with the support of TinyML, researchers are helped in exploring and developing wireless technologies such as 5G, Wi-Fi -6, and IoT. Moreover, TinyML in wireless systems ensures to improve data transmission, network efficiency, etc. [22, 23].

The study presented in Bamoumen et al. [19] highlights various applications of TinyML in addressing environmental challenges across diverse domains, such as the atmosphere, hydrosphere, and their respective subcategories. For instance, in the atmospheric domain, the authors examine the monitoring of carbon dioxide (CO_2) emissions and the prediction of temperature. They employ microcontrollers such as TEDA, Arduino Nano 33 BLE, and STM32, paired with ML techniques such as unsupervised learning and neural networks (NNs). In the hydrosphere, the focus shifts to monitoring water quality, identifying water misuse, and detecting diseases in water bodies. The technological framework here relies on devices like Raspberry Pi and ML algorithms such as support vector machines (SVMs) and artificial neural networks (ANNs). In the lithosphere, the discussion is limited to smart agriculture, which utilizes the ATMega32Sp microcontroller along with TensorFlow Lite, Naive Bayes, ANN, and other tools. This system aims to help farmers manage climate conditions and improve crop quality. Lastly, the biosphere is associated with wildlife conservation and bioactivity monitoring, which involve the use of multiple microcontrollers, echo boxes, ANNs, and Python libraries like scikit-learn. Overall [19], outlines the potential applications of TinyML in environmental sectors, including water management, agriculture, air quality, and temperature control, while also providing a consolidated view of the research projects covered in the survey.

The study in Qi et al. [20] highlights several significant research contributions in which machine learning, particularly TinyML, is integrated into healthcare systems, with a particular focus on personalized and mobile health care (PHS). An important mark of this survey is the use of IoT technologies to improve PHS, using wearable devices, mobile apps, and sensors. The report discusses technologies such as global positioning system (GPS), radio frequency identification (RFID), accelerometers, gyroscopes, altimeters, electroencephalogram (EEG), and electrocardiogram (ECG) sensors. These tools support healthcare solutions by collecting and processing diverse forms of health data based on references in the survey. Many of the studies referenced in this survey incorporate a variety of machine learning techniques, including decision trees (DTs), hidden Markov models (HMMs), SVM, as well as supervised and unsupervised learning approaches. The primary target for these studies tends to be patients and individuals aged between 16 and 40. In addition to the reviews, the survey discusses crucial features of several studies, such as sensor types, monitored parameters, and data processing techniques. It also introduces mobile applications that implement machine learning in health care, such as Self-Management with Assistive Rehabilitation and Telecare Technologies (SMART), mHealthDroid, and Emergency Monitoring and Prevention (EMERGE).

Military systems include a wide range of applications, such as rescue operations and intelligent transportation. One of the most prominent areas in this domain is unmanned aerial vehicles (UAVs). UAVs are technologies that include flying vehicles capable of operating without direct human control and are ubiquitous in the research field of the IoT. Common detection methods in this field involve radar, RFreq, and acoustic surveillance systems [21]. The survey in Yan et al. [21] examines successful reports of the above technologies along with other paradigms and algorithms being used such as SVM and regions with convolutional neural network (CNN) features (RCNN). Moreover, TinyML can be integrated with IoT in the field of military network. In a related field survey, there are challenges related to computational and networking resources, such as Agile computing hardware (ACM) or Sieve, Process, and Forward (SPF).

The work in Sharma et al. [22] surveys ML applications within wireless systems, particularly focusing on smart cities. It explores various ML approaches aimed at enhancing the efficiency of wireless sensor network (WSN)-IoT networks, addressing key challenges such as network coverage, connectivity, energy efficiency, bandwidth optimization, and infrastructure demands. The paper also highlights a number of ML approaches, including reinforcement learning (RL), DT, random forests (RFs), and ANN, alongside popular algorithms like Q-learning, SVM, and K-nearest neighbors (KNN), all of which are reviewed for their potential to enhance WSN-IoT systems.

The study in Jagannath et al. [23] explores the applications of ML within wireless communication systems, with a specific focus on expanding the understanding of various ML techniques. The authors of this survey also evaluate existing ML implementations in IoT-related projects, particularly projects that focus on network performance. Additionally, the paper points out the process of integrating ML into network-related hardware, ensuring the availability of applying ML techniques to different network elements. Ultimately, the survey offers a deeper understanding of the impact of ML and IoT, particularly referring to wireless ad hoc networks (WANETs) [23].

Table 18.1 summarizing a few surveys discussed above: about their published years, drawbacks, strong points, etc.

18.3 Methodology

In this section, the survey reviews the ML techniques, paradigms, algorithms, reference methods, practical testing of some references, etc. There will be figures, equations, formulas, etc., and they will be explained in detail. As mentioned in the Introduction section of this survey, the following four topics are covered and reviewed:

- Smart agriculture
- Healthcare systems
- Military systems
- Wireless systems

Table 18.1 Related surveys of TinyML applications in the four discussed aspects.

Reference	Scope	Strong points	Drawbacks
Bamoumen et al. [19]	Smart agriculture	Reviews ML techniques, diverse applications related to certain topics, successful in pointing out challenges of those applications	No practical projects for feasibility checking, no real-life sample figures
Qi et al. [20]	Healthcare systems	Introduce details of ML applications in health care, give out challenges and further trends	No real-life figures of reported products, no practical products, solely reviews other related papers
Yan et al. [21]	Military systems	Reviews popular UAV field, analyze dataset, algorithms, ML methods, and their related technologies	No real-life projects to check the possibility of reference, focuses only in UAV field
Sharma et al. [23]	Wireless systems	Detailed architecture of WSN-IoT, solutions to tackle, and used ML techniques in wireless applications	Focus too much on algorithms and mathematical concepts, no practical real-life examples of WSN-IoT

18.3.1 Smart Agriculture

Agricultural products encompass a wide range of goods derived from plants, animals, and microorganisms. These products can be broadly categorized into four main groups:

- **Foods:** This includes cereals (wheat, rice, and corn), fruits (apples, oranges, and bananas), vegetables (potatoes, tomatoes, and lettuce), and animal products (meat, milk, and eggs).
- **Fibers:** These are materials used for textiles and clothing, such as cotton, wool, and silk.
- **Fuels:** Agricultural products can also be used to produce biofuels, such as ethanol made from corn or biodiesel from soybeans.
- **Raw Materials:** This category includes products like rubber, timber, and other materials used in manufacturing and construction.

From this, agricultural products play a crucial role in supplying essential resources for human life, and with the erosion of global warming, TinyML will be an effective method to guarantee the quality of the products and safeguard consumer health.

One of the current challenges in the industry is the short shelf life of fresh fruits, and TinyML can be used to tackle this issue. Fresh fruits are known to

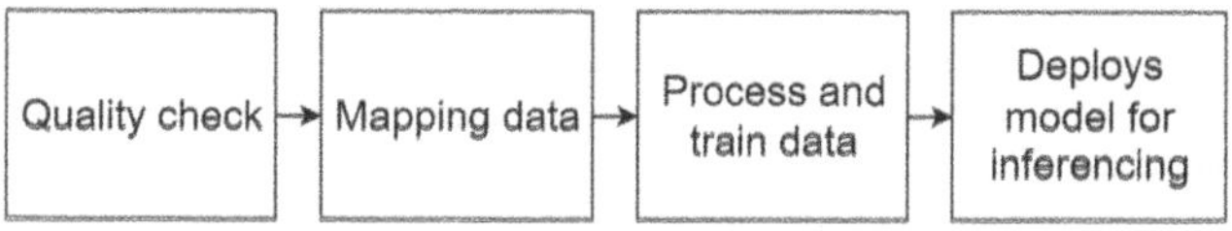

Figure 18.3 Shelf life estimator process.

be perishable and prone to quick spoilage. Thus, finding a reliable solution is crucial, and predicting the freshness period of these fruits is a promising strategy. Previously, non-ML solutions like modified atmosphere packaging (MAP) have been used for this purpose. By slowing the ripening process, MAP extends the shelf life of fruits, allowing for longer storage. The study mentioned in Srinivasagan et al. [24] applies ML techniques with MAP to predict the shelf life of fruits under various controlled conditions. This approach is crucial in determining the optimal consumption time and ensuring that consumers enjoy the nutrients in fruits. Figure 18.3 shows steps in making a shelf life estimator.

The study also involves the use of an Arduino BLE 33 Sense microcontroller, which includes various sensors such as those for temperature, humidity, and light. The architecture is designed by edge impulse (EI), a cloud-based platform for training and deploying the model. The microcontroller performs real-time inference, while the AS7256x sensor measures incident and reflected light, reducing the calculation of reflectance ratios. Besides fruit freshness, monitoring weather conditions is also crucial in agriculture. Weather data collection involves tracking various atmospheric factors, such as temperature, humidity, and solar radiation, using tools such as IoT sensors, satellites, and weather stations. Accurate weather records and ML predictions are pivotal for agricultural productivity, as unpredictable weather patterns can negatively impact crops. For example, the ongoing drought in the U.S. Southwest since 2020 has caused a significant downfall in rainfall and elevated temperatures, severely destroying crop yields.

In other cases, increased humidity can promote fungal diseases, warm temperature can boost pest populations, or smoke from wildfires can affect air quality and plant health. Therefore, it is essential to prevent them before they actually occur. Centuries ago, farmers could only learn to prevent agricultural problems using their senses. Today, technologies have advanced, and the application of TinyML in weather has been feasible in recent decades. Some benefits of TinyML in weather measurement are listed:

- Real-time data processing
- Cost-effective
- Data privacy
- Scalability

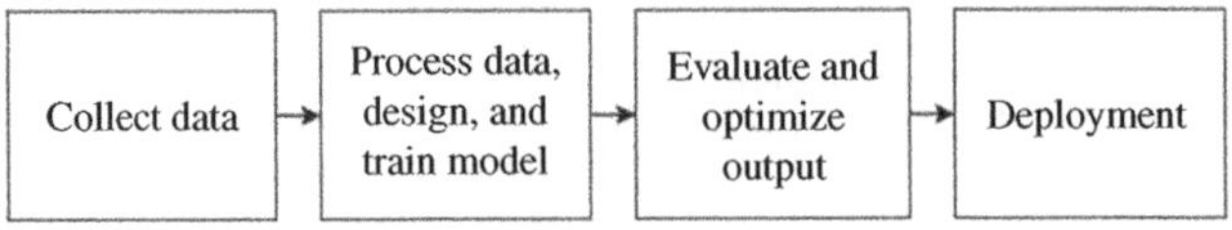

Figure 18.4 Procedures in building a TinyML weather station. *Source:* Adapted from [25].

Many reports employing TinyML for weather measurement collect data such as temperature, rainfall, wind speed and direction, and sunlight to ML datasets. These studies primarily aim to improve crop management, control pests and plant diseases, and mitigate risks such as floods and storms. In one particular study, TinyML is integrated for weather measurement with IoT technology to create a cheap, lightweight AI-based weather station, which includes wind turbine energy generators (WTEGs) in urban settings. The study uses parameters including wind direction (WD), wind velocity (WV), temperature, pressure, sea level, and humidity for prediction, focusing mainly on WD and WV. The research achieved 95% accuracy for WV predictions and 92% for WD predictions. Figure 18.4 reveals the process of making an ML model as a weather measurement system.

Referring to the hardware being used, the study employs the Arduino Nano 33 BLE Sense from the Harvard TinyML kit for WV measurements. For WD measurement, a custom 3D-printed panel houses eight piezoelectric pressure sensors, each matched with one of the eight cardinal and intercardinal directions (north, northeast, east, southeast, south, southwest, west, and northwest) [25].

Table 18.2 describes the overview of proposed studies about TinyML application in smart agriculture field.

18.3.1.1 Machine Learning Methods

In previous research [24], fruits underwent processes like cleaning, washing, and precooling before data collection. The types of data utilized included moisture content, water activity, total soluble solids, total sugar levels, pH, and tannin.

Table 18.2 Overview of proposed TinyML applications in smart agriculture in the survey.

Study	Purpose	Machine learning	Results
Srinivasagan et al. [24]	Estimating shelf life of fresh fruits	CNNs, RNNs, EI platform, backpropagation learning algorithm	Average MAPE = 94.11, average RMSE = 0.43
Albuali et al. [25]	Low-cost weather station	DT, EI platform, computer vision (spectrogram), 1D-CNN	Average overall accuracy = 0.94, average F1 score = 0.92

Data collected from Al-Ahsa International Airport (AIA) included various attributes such as wind speed and WD [25]. The study applies a DT algorithm to analyze WD measurements, enhancing classification and regression efficiency while optimizing training time and memory use. WV is assessed based on the Beaufort Wind Force Scale (BWFS) using a one-dimensional (1D-CNN) [25]. EI was also used to train the architecture described.

18.3.1.2 Neural Networks and Structure Design

The research [24] applies an ANN. This ANN comprises several elements: Neurons process input and relay it; layers consist of an input layer for data reception, hidden layers for processing via weighted connections, and an output layer delivering final results. Types of ANNs include CNNs and recurrent neural networks (RNNs). EI utilizes the backpropagation algorithm and Adam optimizer for weight adjustment, effectively reducing error [24].

Another study uses 1D-CNN and DT models to optimize training. The classification model includes over 3000 time-series samples initially unlabeled. EI's data explorer assists in labeling and preprocessing the data with spectrograms. Table 18.3 provides an insight into the architectures used in both studies.

18.3.2 Healthcare Systems

In today's world, health care has been and will be an important field to everyone. Not to consider harmful impact and consequences from coronavirus disease 2019 (COVID-19), our world is currently facing a phenomenon that still has only suggestions for solutions, that is, aging population. Old people mostly cannot do heavy works, and they are easier to catch illnesses than younger generations; thus, with the rise of the elders, healthcare systems must be developed and expanded faster to adapt to the group [26]. Over many years, AI has played a crucial role in the field of health care. They have contributed significantly to imaging and diagnostics in medicine, remote patient care, research and discovery of new drugs, patient involvement and adherence, rehabilitation services, and other various administrative tasks in hospitals and medical centers [27]. Moreover, involving machines that are combined with AI and physical objects can help interact with patients and doctors, work smartly in bio-hazardous environments, and perform other dangerous tasks related to medical fields. Therefore, TinyML is a potential aspect for scientists to explore deeper in order to tackle the aging population problem and other issues such as COVID-19. During the training process of ML models for those solutions, it involves many types of data, algorithms, and techniques, but mostly used are RFID, inertial measurement units (IMUs), supervised or unsupervised learning, CNN, *K*-means, etc. [20].

Table 18.3 Comparison of ANN and 1D-CNN architectures.

Parameters	ANN architecture	1D-CNN
Model	Sequential	Sequential
Input	Features derived from the modified atmosphere packaging (MAP) process	Features in 40 columns
Layers	Two dense layers (20 and 10 neurons)	Two CNN layers with kernels (sizes 3 + 3) and neurons (16 + 8)
Dropout rate	0.2	0.2
Output	A single neuron	Flatten, two dense layers (16 + 8)
Learning rate	0.005	—
Activation function	ReLU for all layers	ReLU for all layers
Epochs	100	100
Optimizer	Adam	Adam
Loss function	Mean-square error (MSE)	Categorical cross-entropy
Data split	80% training, 20% testing	80% training, 20% testing
Pooling	—	Max pooling with stride 1, pool size 1, and same padding
Additional layers	—	Additional CNN layer (kernel size 3, 8 neurons) and max pooling (stride 2, pool size 2, same padding)
Data preprocessing	—	EI's data explorer assists in labeling and preprocessing the data with spectrograms
Sample details	—	Classification model includes over 3,000 time-series samples initially unlabeled

As previously mentioned, COVID-19 continues to pose significant challenges to the healthcare system. Additionally, there are reports suggesting that mosquitoes, known for transmitting various diseases to both animals and humans, might also be capable of spreading COVID-19 [28]. Hence, it is pivotal to create a TinyML technology which can detect mosquitoes to prevent further disease transmission [7].

In the proposed study, mosquito detection is reported to be using TinyML by examining their wingbeat sounds over intervals ranging from 0.5 to 3 seconds. The study's objective is to develop an affordable mosquito detection system, prioritizing efficiency and energy conservation. Two versions of the detection model are proposed: One uses a 1D-CNN, while the other applies a 2D-CNN. The models will be evaluated based on accuracy, F1 score, and other metrics to determine the most effective and efficient model for implementation on hardware and microcontrollers. The models are trained and tested using data stored on EI [7].

For the implementation phase, the more efficient model is loaded onto an Arduino Nano 33 BLE Sense microcontroller equipped with an external microphone to capture mosquito wingbeat sounds. The microcontroller is powered by a solar panel that charges its battery via a charge controller [7]. Lastly, the study utilizes the Arduino's Wi-Fi capabilities to connect with mosquito repellent devices, which release chemicals to eliminate mosquitoes through a fog server [7].

During the time of COVID-19 worldwide spread, there have been many people suffered from the symptoms of the virus, and before that, they might have been affected by one of the most common ways, droplets from person-to-person communication [29]. Therefore, wearing face masks has been an important activity that everyone, especially children, considered during the years 2022 and 2023. However, it is impossible to ensure that people we see and pass through do wear the mask properly; therefore, TinyML is a proper solution to detect those people. The study [30] proposes a system for detecting properly worn masks using energy-efficient and cost-effective ML techniques such as K-means and sliding windows. The models are implemented on the STM32F411 microcontroller, featuring a 100-MHz ARM Cortex M4 chip to support low-power computer vision tasks [30]. The K-means algorithm is chosen for its minimal flash memory requirements compared to neural networks (NNs) and its potential to achieve 96% accuracy [30]. Image processing concepts such as bounding boxes and segmentation are explored; however, the study emphasizes the centroid approach for faster and more efficient object detection in static images.

Given the limited memory of microcontrollers like the 1 MB random access memory (RAM) available on the Arduino Nano 33 BLE Sense, reducing image size accelerates processing. The study resizes dataset images from 1024x1024px or 256x256px to 32x32px, preserving accuracy [30]. The paper applies the sliding window technique to reduce the dimension of 2D to 1D for easier detection [30]. Figure 18.5 showcases an example of how sliding windows are used to convert the 2D bit array to 1D bit array.

Aside from COVID-19, pollution is another massive issue to the field of health care; therefore, integrating TinyML will help interact with pollutants and reduce the harmful impacts. In an overall view, pollution is divided into many categories, such as water, light and noise, and mostly visible to daily life today is air pollution [31].

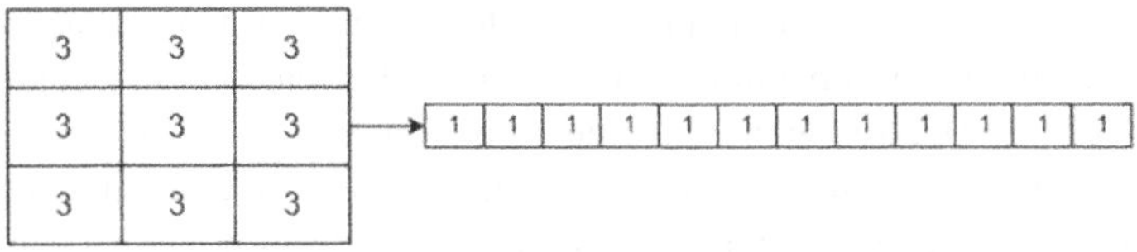

Figure 18.5 Sliding window technique from 3 × 3 2D array to 3 × 3 1D array.

Air pollution primarily originates from harmful chemicals and emissions like CO_2, nitric oxide, gases, ash, and smoke. A TinyML paper proposed solutions for monitoring air quality in specific areas, which can tackle troubles in existing air pollution situations [18]. The air quality monitoring system employs a Raspberry Pi Pico W microcontroller, including the RP2040's capability to run the TensorFlow Lite library [18]. The device is powered by energy collected through a solar panel connected to a 18,650 mAh battery, with sensors providing input to the power manager to sustain the microcontroller [18].

The microcontroller connects with various sensors, such as the SCD41, to collect air quality data [18]. Specifically, it integrates the INA219, LC709203F, and SCD41 sensors [18]. The INA219 measures the solar panel's current and voltage output, the LC709203F monitors battery capacity and voltage, and the SCD41 collects data on CO_2 levels, temperature, and humidity. Additionally, the Raspberry Pi Pico W microcontroller uses a DS3231 real-time clock (RTC) module for precise timekeeping [18]. The device integrates two ML models: one for prediction and another for data imputation. If real-time data is missing, the microcontroller imputes the missing values before making predictions [18].

As technology becomes more prevalent in hospitals and shared computerized resources are utilized more frequently, the risk of pathogen transmission has increased. This has led to a greater need for zero user interface (UI) solutions [32]. First of all, zero UI means interacting with the world without actually touching even a screen or a button [32]. Many solutions relate to this, for example, motion detection, eye tracking, and voice control [32]. These solutions are becoming more and more necessary to the healthcare systems, and one of those is air writing [32]. As described in the article, air writing is a method by which a user can write "in the air" to text messages or compose emails [32].

In a study, the air-writing methodology is applied to detect Arabic characters [33]. To be specific, the study employs TinyML with Arduino Nano 33 BLE Sense microcontroller to identify numbers from 0 to 10 with several Arabic alphabets [33]. Each character or numeral has 500 recordings, leading to a dataset comprising 11,000 samples [33]. The model utilizes the accelerometer and magnetometer on the Arduino microcontroller to capture X- and Y-axis data while writing gestures [33]. These values are then displayed in the 2D plane [33]. Moreover, the system in this study operates at a rate of 100 Hz [33].

Table 18.4 Summary of studies in healthcare systems.

Reference	Purpose	Methodology	Results
Kimutai et al. [7]	Mosquito detection	Uses TinyML with 1D-CNN and 2D-CNN to detect mosquitoes by wingbeat sounds; data stored and models trained on EI; implemented on Arduino Nano 33 BLE Sense with solar power	The accuracy of 2D-CNN and 1D-CNN versions is, respectively, 0.9444 and 0.9065; connected to mosquito repellents via Wi-Fi
Piatkowski et al. [30]	Face mask detection	TinyML with K-means and sliding window techniques on STM32F411 microcontroller; image processing involves centroid and reduction to 32x32px	Achieved 96% accuracy; utilized centroid for faster processing; static images used for training
Wardana et al. [18]	Air quality monitoring	Uses TinyML with Raspberry Pi Pico W; integrates sensors (SCD41, INA219, LC709203F) and solar energy; data collection every 10 minutes	Dataset of 13080 samples; includes prediction and imputation of missing data; R-squared score is approximately 0.9, missing rates < 0.8
Pławiak and Albelehai [33]	Arabic character detection via airwriting	TinyML with Arduino Nano 33 BLE Sense; uses accelerator and magnetometer for 2D representation; sampling rate of 100 Hz	Dataset of 11,000 samples; model reaches F1 score of 0.94, testing accuracy of 99.7% and training accuracy of 97/5% at a dropout rate of 0.3

Table 18.4 shortens the details of some papers and studies reviewed in the survey that are about TinyML applications in healthcare systems.

18.3.2.1 Machine Learning Methods

In Kimutai and Förster [7], the wingbeat sound dataset, sourced from Kaggle – a prominent online platform for data science and machine learning competitions – has been selected as a standard dataset for its contests. The dataset also includes samples recorded from low-cost smartphones and high-end recording devices in order for the model to be able to detect mosquitoes correctly from various acoustic sensors [7]. Moreover, other sounds are recorded in order for the model to categorize and classify sounds which are not mosquitoe wingbeat, such as silence, human voices, and noise from television (TV). The dataset also holds nearly 6,000 samples with a split rate of 80/20 for training/testing and 20% for validation.

In the study [30], the data is got from two sources of dataset, the MaskedFace-Net and Natural images. The MaskedFace-Net dataset supports detecting people whether they are wearing the mask correctly [30]. It includes two subsets, one is a correctly worn face mask and the other one is incorrectly worn face mask; also, in the incorrectly worn face mask, there are three classes, which present three different faces of wearing mask incorrectly [30]. The Natural image dataset helps recognize human faces among other subjects such as cat, airplane, and fruit.

In the study [18], a dataset was developed using real-time measurements to train and evaluate the models. The involved parameters include CO_2, air temperature, air humidity, solar panel output current, solar panel output voltage, battery voltage, temperature, and capacity [18]. Moreover, to achieve the objective of building a cost-effective air quality monitoring system, minimizing sensor power consumption was a key step [18]. The device collects data with the interval of 10 minutes, resulting in a final dataset of 13,080 samples [18]. The dataset was then divided, with 70% allocated for training and the remaining portion used for testing the models.

The study [33] employs deep learning (DL), a branch of ML, for preprocessing data samples. Initially, the data collected is raw 3-axis information, which is then transformed into a static 2D format suitable for the DL model [33]. After this transformation, coordinates representing the actual writing location in the real world are incorporated into the 2D format [33]. These coordinates are defined as two specific values: x and y [33]. Once assigned, the data is normalized, scaling the coordinate values to fit within a range of 0 to 1 [33]. Figure 18.6 shows the procedure of data preprocessing in Pławiak and Alblehai [33].

The above procedures are the foundation of preprocessing data. After these procedures, the study continues to map the normalized coordinates to an image grid, which will then apply Bresenham's algorithm to colorize pixels in the image grid [33].

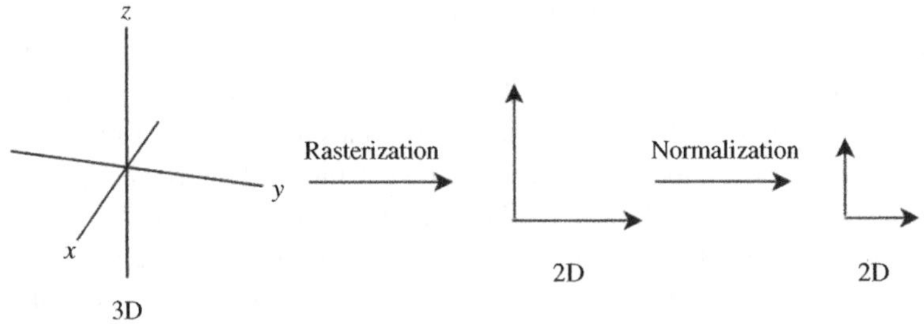

Figure 18.6 Steps of fundamental data preprocessing. *Source:* Adapted from [33].

18.3.2.2 Neural Networks and Designing Structure

In the study [7], two ML models are developed using 1D-CNN and 2D-CNN architectures. The first version, which uses the 2D-CNN, incorporates only two convolutional layers (CLs), reducing the complexity compared to traditional models. In the second version, the model includes batch normalization and an identity shortcut layer to minimize overfitting. Additionally, a leaky rectified linear unit (ReLU) activation function is applied between layers to enhance the training performance.

The configurations and parameters for the 1D-CNN and 2D-CNN models are outlined as in Table 18.5.

The 2D-CNN architecture proposed in Wardana et al. [18] is configured as follows:

- **Input:** Reshaped features from the measured dataset
- **CNN Layer:** 3×3 with 25 filters

Table 18.5 Comparison of 1D-CNN and 2D-CNN architectures.

Parameters	1D-CNN	2D-CNN
Input	Features derived from the Kaggle dataset	Features derived from the Kaggle dataset
CNN layers	Single layer with a filter size of 3	Single layer (filter size not specified)
Batch normalization	Applied	—
Activation function	Leaky ReLU	—
CNN layers	Another layer with a filter size of 3	Another layer (filter size not specified)
Batch normalization	Applied	—
Sum of weights	Computed	—
Activation function	Leaky ReLU	—
Identity shortcut	Connects from "Input" to "Sum of Weights"	—
Max pooling	—	Stride set to 1
Dense layers	—	Two layers (number of neurons not specified)
Epochs	60	60
Learning rate	0.001	0.001
Batch size	100	100
Dataset split	80% training/20% testing	80% training/20% testing

- **CNN Layer:** 3 × 3 with 15 filters
- **CNN Layer:** 2 × 2 with 8 filters
- **Activation Function:** ReLU applied across all layers
- **Flatten Layer:** Comprising 24 neurons
- **Epochs, Learning Rate, and Batch Size:** Not specified in the study
- **Dataset Split:** 70% training/30% testing

Study [33] relies on CNN architecture to train its DL model and conduct classification tasks.

18.3.3 Military Systems

To protect itself and its people, a country must have military systems [34]. Over many years, technological advancements have accelerated, allowing their integration into military applications [34]. TinyML research within military domains often focuses on UAVs [34]. UAVs serve multiple purposes, such as surveillance of hostile areas or direct engagement without endangering personnel [34]. Small UAVs, or drones, often face constraints related to battery capacity [35]. These devices, which can be used for surveillance or mapping terrain, often have a short operational time due to their small battery capacity [35]. Therefore, it is essential to find a solution to maintain the intelligence of the drone but also make it consume less power for longer flying time [35].

In [35], it presents the integration of ML into GAP9Shield, which is a compact nano-drone that fits within the palm of a hand [35]. With this addition, the GAP9Shield achieves energy efficiency while maintaining intelligent functionality [35]. To be particular, the nano-drone is equipped with a 5MP OV5647 camera, a NINA W102 module supporting Wi-Fi and BLE, and a VL53L1 ToF sensor for distance measurement [35]. In comparison with other related studies, the drone achieves an red, blue, and green (RGB) sample rate of 20%, or 7 frames per second (FPS), higher while being lesser in weight. Using the You Only Look Once (YOLO) model, the drone operates below 100 mW of power consumption with minimal latency [35].

The mentioned components above are added into a 6-layer 50 × 27 printed circuit board (PCB) [35]. The proposed study's drone consists of a GAP9 system on chip (SoC), camera, Wi-Fi module, and a 5-direction ranging sensor [35]. Specifically, the GAP9 SoC allows 330 μW per giga operation per second (GOP) of energy efficiency, which means that it saves up energy for every GOP in processing images [35]. To other modules, the camera is high-performance, low-voltage-consuming and can output 8- or 10-bit RGB image formats [35]. The NINA W102 module has an ESP32 MCU with 2.4-GHz Wi-Fi integrated in; the ESP32 microcontroller unit (MCU) is also a dual-core SoC, which enables parallel activities for Wi-Fi and

Serial Communication Interface (SCI) to communicate with GAP9 SoC and also optimize the data throughput [35]. Lastly, the VL53L1 sensor, which ranges in front, left, right, back, and up, can run at a maximum sampling rate of 60Hz [35].

Besides UAVs, military systems also include cybersecurity threats. Attackers can effortlessly exploit these automated systems, particularly if there are even minor flaws in their design, coding, or any unpatched vulnerabilities [36]. Moreover, adversarial attacks can happen not only on powerful computers but also on low-power ones [37]. This means that cyberattacks can still happen on microcontrollers [37]. Therefore, it is crucial to create TinyML applications that can detect cyberattacks or at least deploy ML models capable of doing so.

In a study, similar difficulties are remarked [42]. Traffic security poses significant risks, and IoT devices with low power are inherently vulnerable [38]. Consequently, the research proposes an ML approach for detecting and classifying different cyberattacks using edge IoT network traffic data [38]. Although TinyML is not explicitly mentioned, optimizing the ML model for resource-limited devices makes the approach viable. While IoT applications are increasingly deployed at the edge, they continue to encounter various threats, including denial of service (DoS)/distributed Dos (DDoS), information theft, Man-in-the- Middle (MITM), injection, and malware attacks [38].

In the proposed paper, IoT device structures are structured into two tiers: the edge and the cloud layers [38]. By performing all processes locally and eliminating the cloud layer, the security of IoT devices is refined [38]. However, the edge layer still remains vulnerable, so adding a classifier for detection remains crucial [38]. Table 18.6 summarizes some TinyML studies in military systems.

Table 18.6 Summary of studies on TinyML in military systems.

Study	Purpose	Techniques/algorithms	Final result
Müller et al. [35]	Improve UAV efficiency	YOLO model, GAP9 SoC, camera, Wi-Fi module, VL53L1 ToF sensor	15% lighter and 30% smaller drone; power usage under 150 mW; 45 FPS; acceptable accuracy with 17 ms latency
Becker et al. [38]	Detect cyberattacks in IoT	RF, XGBoost (XGB), Gaussian naive Bayes (GNB), CNN, random forest, AdaBoost (AB), decision tree, logistic regression (LR), SVM, ANN, KNN	AB achieved the highest F1 score of 0.999; RF ranked second with F1 score of 0.986

18.3.3.1 Machine Learning Methods

The work [38] uses the publicly available Edge-IIoTset Cyber Security Dataset, containing 14 different attack types alongside normal data, to assess eleven ML models and determine which are most efficient at detecting edge IoT attacks [38]. Furthermore, the study implements various algorithms within the edge IoT context, including RF, extreme gradient boosting (XGB), and Gaussian naive Bayes (GNB), to identify cyberthreats like DDoS Transmission Control Protocol (TCP), DDoS Hypertext Transfer Protocol (HTTP), and Structured Query Language (SQL) injection [38]. This approach enhances the study's methodology when compared with previous research.

18.3.3.2 Neural Networks and Designing Structure

In the study [38], multiple models are used to detect cyberattacks and their performance is compared [38]. The models and their configurations are as in Table 18.7.

The results of these model performance are discussed in Section 18.5.

18.3.4 Wireless Systems

In fields such as agriculture, health care, and the military, wireless network has been an ultimate criterion to consider when applying TinyML in those fields. This is primarily due to the potential risks: If not protected properly, private data can be stolen or systems can be attacked via networks [39]. Thus, researching TinyML in wireless networks has been a popular and essential aspect. In the mentioned field, there are many factors that can be applied with TinyML, such as WSNs, smart home automation, and cyberthreat detection, and those are very crucial to agriculture, health care, and military.

In paper [40], it is said that agricultural events should be detected and noticed as it is very crucial; moreover, to create such a system with low-latency, low-power, and expansive data communication, there will be long-range (LoRa)-based WSNs integrated in the device [40]. In particular, the paper's goal is to create a device that can transmit data in a long range, while some traditional short range network technologies such as Zigbee or Bluetooth are not suitable, LoRa meets this criterion [40]. Moreover, to adapt to the criteria of low latency, the data transmitted to cloud platforms should not be large; therefore, the proposed study decides to analyze the data directly in the edge devices before transmitting it to the cloud platform [40]. Previously, there was a system called Safe for Food and Animal (SAF) implemented in agriculture, especially dairy production, to monitor livestock health, production, storage, and transportation [40]. With the addition of TinyML, the system will be able to detect anomalies or events. Moreover, the SAF system enables LoRa and can work with edge-cloud model [40].

Table 18.7 Models and configurations.

Models	Configuration
RF	Number of trees = 10 Number of attributes per split = $\sqrt{70}$ Minimum subset size for a split = 5
XGB	Objective: multi-class soft probability Number of estimators = 100
CNN	Model optimizer: Adam Learning rate: 0.001 Loss function: categorical cross-entropy
Classification and regression tree (CART)	Complexity parameter (ccp_alpha) = 0 Splitting criterion: Gini impurity Minimum samples per leaf = 1 Minimum samples per split = 2 Splitter: best
AB	Number of estimators: 50 Learning rate: 1.0 Classification algorithm: SAMME.R
DT	Induce binary tree: true Minimum number of instances in leaves = 2 Minimum subset size for a split = 5 Maximum tree depth = 100 Stop when majority reaches = 95%
LR	Regularization type: L2 Regularization strength = 0.110
SVM	Regularization parameter (C): 1.0 Kernel type: radial basis function (RBF) Gamma (g): auto Numerical tolerance = 0.0010 Iteration limit = 10
ANN	Neurons per hidden layer = 100 Activation function: ReLU Solver: Adam Regularization strength = 0.0001 Maximum iterations = 400 Replicable training: true
KNN	Number of neighbors = 5 Distance metric: Euclidean Weight function: uniform

Source: [38] / with permission of IEEE.

The LoRa technology is deployed into the SAF system. For each hectare of agricultural lands, there are 100 LoRa T-beam LilyGO (TTGO) nodes [40] and each has three gateways, antennas to provide strong connectivity with the system [40]. Moreover, each node consumes 200 mAh of batteries for continuous operation [40]. The LoRa TTGO nodes are separated into two types, livestock ones rely on DS18B20 and MPU-6050 sensors to measure temperature and movement, while the environmental nodes apply SHT31 and MQ-4 sensors to collect humidity, temperature, and methane gas data [40]. Lastly, each node's gateways are connected to cloud with message queuing telemetry transport (MQTT) and transport layer security (TLS) protocol [40].

When deployed, the system detects many events, such as unusual temperatures from animals, low humidity on the farm, and unusual pH values in tanks. To detect those, TinyML model is applied with anomaly detection, sensor fusion for holistic analysis, sheep fever detection, decision-making, and more [40].

Apart from WSN, many other modern wireless technologies significantly enhance the connectivity range of machine-to-machine (M2M) communication. These include Wi-Fi, 4G, 5G, and low-power wide area network (LPWAN) technologies such as LoRaWAN and NB-IoT, which enable devices to connect and exchange data over various distances. In other studies, there are solutions that apply ML to network technologies to activate M2M communication faster; in other words, they work as "path finding" in networks. Paper [41] demonstrates this idea.

Normally, like LoRa technology, the solution they provide is the ability to communicate M2M in different ranges, from long to short. However, these technologies include nodes between the main machine and the communicated machine, and they do not have the ability to find the shortest path to transmit information to that machine. Therefore, paper [41] offers a solution for that. In the proposed study, an IoT device is created and considered as a decision-making agent that can calculate the shortest transmission range for M2M and also reduce power consumption [41]. The architecture of multi-devices works as follows: Firstly, the nearest relay node, or the IoT device, from the main machine will read its state, or the number of other IoT devices near it. Secondly, the IoT device will apply ML to decide to transmit information to the nearest node to the communicated machine. Finally, these two steps keep processing until communication is finished. Figure 18.7 visualizes the working flow of the multi-devices architecture where suitable IoT devices are connected to each other to transmit data in the fastest speed.

To make decisions, the study relies on an ML process derived from the Markov decision process (MDP) framework that can determine the most optimized route path for M2M communication and also low-power-consuming consideration [41]. Moreover, each IoT device is promised to function well even if they do not know everything about the network [41]. Lastly, deep neural network (DNN) is used to

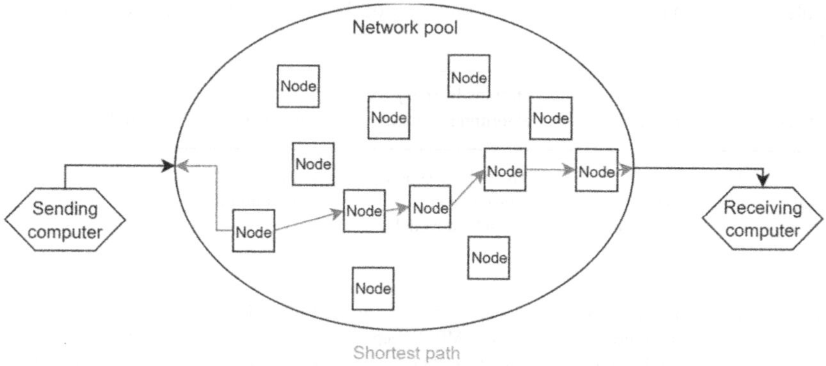

Figure 18.7 Visualization for working flow of architecture. *Source:* Adapted from [41].

help the ML models understand the long-term benefits of taking certain actions in situations [41]. By doing this, the final system can be autonomous, be able to optimize energy consumption, and use only necessary IoT devices in the network to transmit information to destination devices [41].

Besides network connection, TinyML can also be applied with other wireless sensors to recognize natural phenomena. In a particular study, TinyML is applied to a device which records and recognizes bird song using acoustic sensors [42]. Not only does the final device achieve a high precision metric score, but it can also run within two weeks for a single battery charge, satisfying the low-cost criteria proposed in the paper [42]. Unlike previous machine learning models like BirdNet, the model in this study is designed to record and detect bird sounds directly in natural environments, rather than being limited to laboratory settings [42].

The dataset in the proposed study includes over 500,000 audio samples from more than 10,000 species, and there are nearly 3,000 samples recording bird sounds to be used to detect sounds from 17 bird species [42]. Moreover, it encompasses data from other public datasets such as Oxfordshire Corn Buntings, Xeno-Canto, and Google AudioSet [42]. To evaluate the TinyML model, the study measures the accuracy, precision, recall, F1 score, and receiver operating characteristic (ROC) curve [42].

Table 18.8 displays an overview of proposed studies in this subsection.

18.3.4.1 Machine Learning Methods

In the proposed study [41], a decision-making model is formulated by an MDP, shown as a tuple $<S,A,R,\gamma>$, where S is the state or the range an IoT device scans for the next optimal node, A is an action space, R is a reward, and γ is a discount

Table 18.8 Summary of machine learning approaches and results in wireless systems studies.

Study	Purpose	Machine learning algorithms/models	Techniques	Results
Karras et al. [40]	Detect agricultural events using TinyML with LoRa	Anomaly detection, sensor fusion, edge-cloud model	LoRa, TinyML, edge computing	Accuracy: 92.3%–95.1%
Kwon et al. [41]	Optimize M2M communication by finding the shortest path	Markov decision process (MDP), deep reinforcement learning (DRL), double deep Q-network (DDQN)	Path finding, Energy optimization	Improved M2M communication, Reduced power consumption
Huang et al. [42]	Recognize bird songs using acoustic sensors with TinyML	Convolutional neural networks (CNN), Residual Neural Networks (ResNet), Mel-Frequency Cepstral Coefficients (MFCCs), transformer	Spectrogram analysis, Time-series analysis	Average accuracy: 92.8%, Average Precision: 86.2%, Average Recall: 97.6%, Average F2 Score: 94.4%

factor [41]. Problems from the MDP framework include S, A, R, and the probabilities of moving from one state to another after taking an action can affect the decision-making feature of MDP; therefore, the study uses deep reinforcement learning (DRL) to tackle these issues [41]. In particular, DRL allows IoT devices to observe nodes within its state, taking necessary actions such as transmitting data, and receives rewards [41]. Furthermore, to optimize DRL, the paper also applies Q-function to predict the expected dissipate rewards for taking different actions. Lastly, the terminologies above, especially Q-function, are applied in a small network with a few nodes, so a double deep Q-network (DDQN) is used to optimize Q-function in circumstances such as large-scale network [41].

In paper [42], the data is recorded over a season. Particularly, researchers place battery-powered sensors in fields to record natural sounds. After an entire season, they retrieve the collected data and analyze it; in other words, they only keep data which makes birds sound, and others are discarded.

After filtering the dataset, each sample is processed with two dissipate approaches. The initial step involves applying a short-time Fourier transform (STFT) as a preprocessing block, transforming the sound sample into a spectrogram, which is then treated as an image [42]. Consequently, TinyML utilizes CNNs and residual neural networks (ResNets) to learn patterns through image recognition tasks. In another approach, the audio sample is analyzed by Mel-frequency Cepstral coefficients (MFCCs) preprocessing block, and this will extract the features from the audio and can be used by TinyML to learn patterns from those features in a different approach such as classification [42]. To conclude, one approach applies computer vision, while the other one relies on audio analysis.

To enhance performance and minimize the risk of misprediction during the inference of bird sounds, the raw data must first pass through a "baseline" filter [42]. This baseline acts as a filter for the raw data by calculating the power signal (P) and comparing it against two thresholds to determine whether to proceed with further inference steps [42]. In short, Figure 18.8 demonstrates the working strategy of the bird sound detecting device.

18.3.4.2 Neural Networks and Designing Structure

In the proposed paper [40], the overall architecture of the system is described in Figure 18.9. In the graph, the microcontroller uses application programming interfaces (APIs) from the SAF system to interact with cloud and application interface.

In Kwon et al., the procedure occurred between each IoT device is shown in Figure 18.10.

First of all, the node gets the action A from the ϵ-greedy action selection from DDQN of node v_i procedure. Then, the node adjusts the transmission range based on the returned action, which means this step determines whether the path should be shorter. After adjusting, the node transmits the data away and evaluates its throughput, which means the node considers its rewards based on its recent

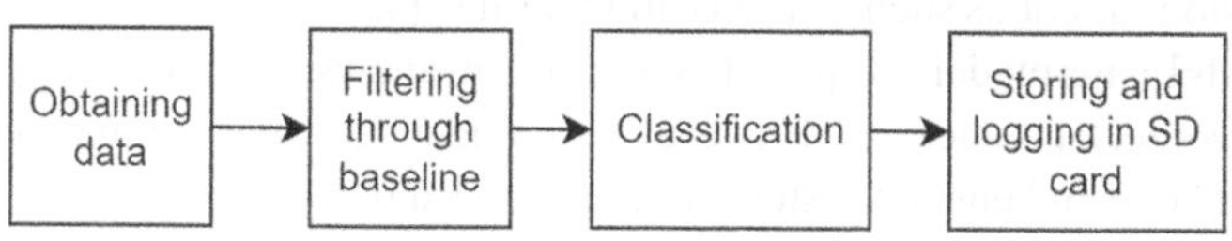

Figure 18.8 Inferencing process of bird sound detector.

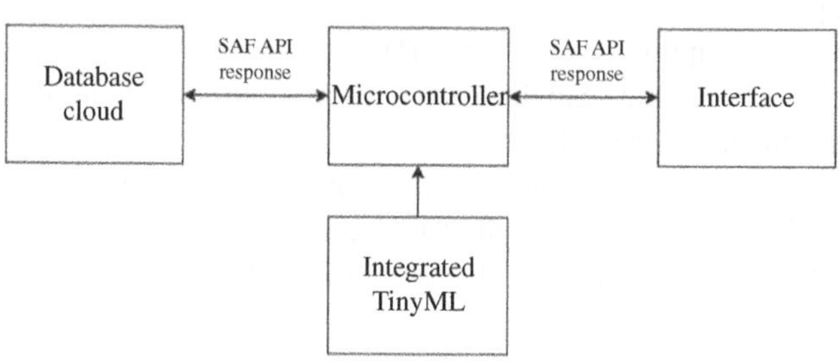

Figure 18.9 SAF's architecture with TinyML.

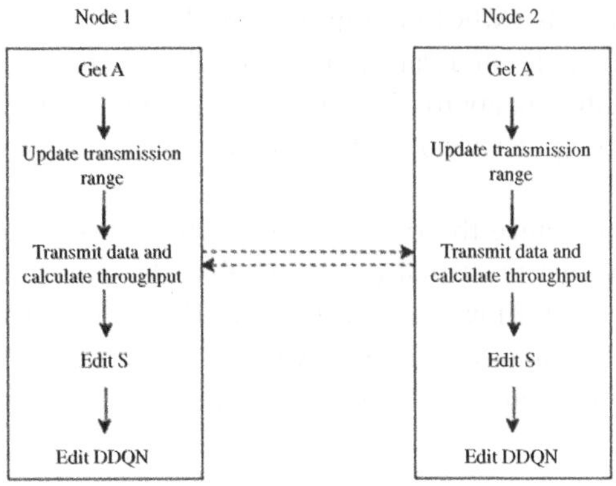

Figure 18.10 An example of two nodes, or IoT devices, interacting to transmit data. *Source:* Adapted from [41].

action. Consequently, the node will update its state based on the evaluation result and update DDQN, so the node will improve its decision-making ability in the future.

The proposed study of Huang et al. applies two approaches, one for computer vision and one for audio analysis, particularly using spectrogram and time-series data [42]. These two approaches also rely on different classification models [42]. The following gives a summary of the models used in two approaches.

- **CNN-Mel:** This model is used for spectrogram analysis, which involves 2D layers with activation functions such as ReLU and Softmax [42].
- **SqueezeNet-Mel:** This model is applied to spectrogram analysis and is based on the original SqueezeNet model [43]. However, this model has fewer parameters than the original one and ensures faster processing speed [42].
- **CNN-Time:** The paradigm includes 1D layers with activation functions like ReLU and Softmax. It is used to extract and learn the patterns from time-series data's features [42].
- **Transformer-Time:** The approach employs a 1D layer, along with ReLU and Softmax activation functions, and a SingleHeadTransformer to convert raw data into embeddings for the transformer [42].
- **SqueezeNet-Time:** Like SqueezeNet-Mel, this model is based on the SqueezeNet model [42, 43]. However, all 2D layers in the original model are replaced with 1D layers, and this significantly lowers the filter number by up to 70% [42].

After learning the pattern, two approaches are optimized through model quantization and partial convolution [42].

18.4 Case Study: New Insights on Air Writing from Pławiak and Alblehai [33]

18.4.1 Introduction

As introduced in the "Healthcare systems" section, air writing is an important aspect in healthcare systems. Moreover, air writing can rely on TinyML for benefits such as cost savings and reduced energy consumption [33]. However, the study of Pławiak P. and Alblehai F. et al. makes use of computer vision; that is, the collected raw 3D data is transformed into grid cell images for detection [33]. This can take more time as after basic preprocessing data steps, the computer has to convert those normalized numbers into pixels in images. Therefore, this case study reveals a simpler way. After normalization, the extracted data features will then be given to the model for learning, and the post-training model will detect classes of data from numbers directly, which does not require any image conversion.

Different from the study in [33], this case study detects Vietnamese alphabets, not Arabic ones. Nevertheless, the case study will use the same microcontroller, Arduino Nano 33 BLE Sense, like the one in the study, and some similar parameters in CNN architecture for performance comparison at the case study's "Discussion" section.

18.4.2 Methodology

In this section, the survey talks about the data, the architecture and the techniques, and algorithms of the model. The data preparation, model training, optimization, and deployment steps are initiated on the EI platform due to its straightforward support for CNN use cases.

18.4.2.1 Case Study – Dataset

To collect data, the case study connects Arduino Nano 33 BLE Sense microcontroller to EI via its API to start collecting 3D raw data. The reason for this is because the Arduino Nano 33 BLE Sense microcontroller has an IMU sensor integrated, which is also the data type in the study in Pławiak and Alblehai [33]. The characters for detection encompass á, à, ắ, ằ, é, è, ế, and idle. The "idle" is detected whenever someone is writing something or not. The case study prepares a dataset of 215 samples in total. Moreover, it is split into train/test rate of 83/17. Additionally, most characters have a number of 24 samples, é character has 36 samples, and idle state has 35 samples. Additionally, the sampling rate is similar to Pławiak and Alblehai [33]: 100 Hz.

18.4.2.2 Case Study – Preprocessing, Training, and Testing Model

As mentioned above, the preprocessing step is simpler than the one in Pławiak P. and Alblehai F. et al. Before normalization, the raw data is rasterized like the one

Table 18.9 Filter and analysis settings in "spectral analysis" block.

Filter	Analysis
Scale: 3	Type: FFT
Input decimation ratio: 10	FFT length: 16
Type: none	Take log of spectrum?: Yes
	Overlap FFT frames?: Yes
	Improve low-frequency resolution?: No

in Pławiak and Alblehai [33]. Then, the case study uses the EI's "spectral analysis" block to handle the normalizing process [44]. The parameters are listed in Table 18.9.

The formulas for normalization are given by (18.1):

$$x_{norm} = \frac{x - \min(x)}{\max(x) - \min(x)}, \quad y_{norm} = \frac{y - \min(y)}{\max(y) - \min(y)} \tag{18.1}$$

where x and y are the original data values, $\min(x)$ and $\min(y)$ are the minimum values in the datasets, and $\max(x)$ and $\max(y)$ are the maximum values in the datasets.

Consequently, Vietnamese characters contain accent marks, such as circumflex or grave accent, that can present a special feature of each of its characters. Therefore, the solution to increase accuracy for detection is "highlighting" the values in 2D data that represents these marks. The "Scale" setting in "Filter" settings in spectral analysis block affects this. Instead of normalizing data values to the range of 0 to 1, this block normalizes values from 0 to 3, and this can enhance the presence of accent marks in each Vietnamese character.

The model used in this case study is classification. The set epoch is 1,000, and the learning rate is 0.01. Table 18.10 reveals the parameters of the model and its CNN architecture.

After preprocessing with "spectral analysis" block, data samples in the training section are clustered but not in particular feature groups [44]. Therefore, the case study relies on K-means algorithms to not only group the clusters but also help detect anomalies [44]. Since the total number of classes, or characters, in the dataset is 8, it will be the k number in the algorithm to detect 8 specific features from the preprocessing stage.

After training with the classification model and K-means algorithm, the results are positive and are shown in the consequent part of this survey.

Table 18.10 Filter and analysis settings in "spectral analysis" block.

Parameters	Values
Model type	Sequential
CNN architecture	
Dense layer	400 neurons
Activation function	ReLU
L1 regularization strength	0.00001
Dropout rate	0.2
Dense layer	8 neurons
Activation function	Softmax
Optimizer	Adam
Low-frequency resolution improvement	—

18.4.3 Discussion and Conclusion

As said, the achieved results are promising. Table 18.11 presents the confusion matrix, which is displayed on the EI platform.

The *K*-means algorithm also groups the clusters successfully. Some variables needed to be detect, according to EI analysis, are acceleration and gyroscope values, root-mean square (RMS). Nevertheless, the inferencing accuracy only achieves 77.5% and still needs to be improved further. Additionally, the dataset is manually created by an individual, which means that it lacks diversity in writing

Table 18.11 EI confusion matrix of case study model.

	IDLE	À	Á	È	É	Ắ	Ằ	Ế
IDLE	100%	0%	0%	0%	0%	0%	0%	0%
À	0%	100%	0%	0%	0%	0%	0%	0%
Á	0%	0%	100%	0%	0%	0%	0%	0%
È	0%	0%	0%	100%	0%	0%	0%	0%
É	0%	0%	0%	0%	100%	0%	0%	0%
Ắ	0%	0%	0%	0%	0%	100%	0%	0%
Ằ	0%	0%	0%	0%	0%	0%	100%	0%
Ế	0%	0%	0%	0%	0%	0%	0%	100%

Figure 18.11 Case study's model memory usage after TensorFlow Lite conversion.

Table 18.12 Comparing performance metrics and other information between case study's model and Pławiak and Alblehai et al.'s model.

Metrics	Case study results	Pławiak and Alblehai et al.'s results
Dataset	20 classes, 1,099 samples	22 classes, 11,000 samples
Microcontroller for TinyML	Arduino Nano 33 BLE Sense	Arduino Nano 33 BLE Sense
Training accuracy	100%	97.5%
Testing accuracy	100%	99.7%
F1 score	1.0	0.94
Language	Vietnamese	Arabic

Source: Adapted from [33].

styles. Hence, there are numerous factors needed to be considered to improve the inferencing accuracy of this TinyML model.

The post-training model is also suitable with Arduino Nano 33 BLE Sense memory capacity. With the conversion of TensorFlow Lite library, the trained model only uses 2.8 Kb of RAM and 36.6 Kb of flash RAM. Figure 18.11 demonstrates the returned status of post-training model after TensorFlow Lite conversion.

The main purpose of this case study is to compare the performance with Pławiak and Alblehai [33]. Therefore, Table 18.12 compares the metrics and other statistics of models shown in this case study and the study [33].

18.5 Results and Discussion

In this chapter, the paper provides results of discussed works and also explains a deeper insight to clarify concepts and strategies in previous studies.

18.5.1 Smart Agriculture Results and Discussion

To rate the model in study [24], the paper uses root mean square error (RMSE) and mean absolute percentage error (MAPE) to assess the TinyML model:

$$RMSE = \sqrt{\frac{\sum_{i=1}^{n} (O_i - P_i)^2}{n}} \tag{18.2}$$

$$MAPE = 100 \times \frac{1}{n} \times \sum_{i=1}^{n} \left| \frac{O_i - P_i}{O_i} \right| \tag{18.3}$$

where O_i is the measured value, n is the number of measured values, and P_i is the predicted value of the target parameter data [24]. After training, the model uses 1.8 KB in Arduino Nano 33 BLE Sense RAM usage out of 256 KB to store model parameters. Under different conditions, 5 and 24 °C, metrics, and packaging type, the average MAPE across threshold parameters is 94.11, while the average RMSE are 0.43 [24].

In study [25], the proposed model is evaluated using multiple criteria such as overall accuracy (OA) and F1 score. In ML, **F1 score** is the harmonic mean of precision and recall, providing a balance between the two, and this is usually useful when tackling imbalanced datasets. **Precision** measures the proportion of true-positive predictions among all positive predictions. **Recall** measures the proportion of true-positive predictions among all positive instances [25]:

$$OA = \frac{\text{Number of correct measurements}}{\text{Ground Truth (True Value)}} \tag{18.4}$$

$$F1\text{-}Score = \frac{2 \cdot Precision \cdot Recall}{Precision + Recall} \tag{18.5}$$

$$Precision = \frac{TP}{TP + FP} \tag{18.6}$$

$$Recall = \frac{TP}{TP + FN} \tag{18.7}$$

in which TP (true positive) is the number of true positives, FP (false positive) is the number of false positives, and FN (false negative) is the number of false negatives. The average OA (after filtering) is approximately 94%, and the average F1 score (after filtering) is nearly 0.92 [25].

18.5.2 Healthcare System Results and Discussion

Proposed in paper [7], the 1D-CNN model achieves 100% of validation accuracy at epoch 28, and 2D-CNN achieves it after 30 epochs. Although the 2D-CNN has greater space to train with more parameters, 1D-CNN reaches peak accuracy in shorter time than 2D-CNN, predominantly because of the replacement of some CNN layers with batch normalization and the identity shortcut. Initially, the paper wants to create a low-cost and efficient detector; thus, choosing 1D-CNN will be a good choice as it runs faster than 2D-CNN version and is able to prevent overfitting [7]. Table 18.13 shows the final accuracy from validating the model and its loss of the two different versions after training.

Although validation accuracy is considerably high, the inferencing accuracy might vary based on circumstances, such as mosquitoe types. For example, for mosquitoes from the domestic environment in the paper, the accuracy of

Table 18.13 Final accuracy and loss from validation.

Version	Accuracy	Loss
1D-CNN	~0.9	~0.4
2D-CNN	~0.82	~0.8

Source: Adapted from [7].

2D-CNN and 1D-CNN versions is, respectively, 0.9444 and 0.9065, while for Culex quinquefasciatus, another type of mosquito, reach only 0.7667 and 0.7222 in general performance. Moreover, the 2D-CNN always has a better inferencing performance than the 1D-CNN version, and this is because of the ability to provide more parameters during the training process of 2D-CNN model [7]. To satisfy the criteria of low cost and efficiency, the deployment process also makes use of solar energy and ML, which has been discussed, and after 50 days, the detector uses less than the traditional mosquitoe repellent method 2,995 Watts (W) of energy [7].

In study [30], the system detects overall well with accuracy points all over 79%. Particularly, the accuracy of detecting a person wearing face mask is 79.51% and not wearing face mask is 79.13% [30].

In paper [18], the model predictor returns a good predicting rate for air humidity with an *R*-squared score of approximately 0.9 [18]. This score is also applied for air temperature, humidity, solar panel voltage, etc., before quantization. To quantize the two models, the paper applies TensorFlow Lite library to reduce the size of two models: 47.7% reduced in size for model predictor and 56.6% for model imputer [18]. Despite the large degradation of the two models, they only suffer little degradation in accuracy [18]. Particularly, before quantization, the model predictor achieves a coefficient of determination higher than 0.7 and the model imputer runs correctly when missing rates are smaller than 0.8 [18].

Before discussing about results, the study in Pławiak and Alblehai [33] converts and transforms the ML TensorFlow model to a TensorFlow Lite model, and this will help reduce memory usage in microcontrollers [33]. The converted version is deployed into the Arduino Nano 33 BLE Sense microcontroller [33]. Moreover, the study also applies algorithms for image recognition such as ResNet to recognize Arabic characters from proposed grid images [33]. With the high accuracy and the ability to be deployed on microcontrollers, the model is believed to be successful in real-world applications, especially in health care as it achieves cost-effectiveness in computation and consumes less energy because of low-power required microcontroller (Arduino Nano 33 BLE Sense) [33].

18.5.3 Military System Results and Discussion

According to the study [35], the nano-drone developed is approximately 15% lighter and 30% smaller compared to the Crazyflie model, which weighs 29 g and measures 10 cm. The OV5647 camera operates at 45 FPS while maintaining power usage below 150 mW [35]. Additionally, the system referenced in the research achieves acceptable performance, with a latency of 17 ms and an energy consumption of 1.59 mJ [35].

The study [38] divides the dataset into 70% for training, 10% for validation, and the rest for testing. Additionally, the dataset file includes 70 features representing 14 types of cyberattacks and 13 types of conventional IoT traffic data [38]. The study evaluates the performance of eleven models using metrics like accuracy, precision, recall, and F1 score [38]. Table 18.14 displays the F1 score analysis, ranking the models from best to worst for edge IoT cyberattacks detection.

The models displayed various success rates for different attacks, regardless of overall accuracy [38]. However, there are limitations in the study, such as the reliance on static IoT traffic data and the use of only one dataset for evaluation [38]. Despite these limitations, the AdaBoost (AB) model proved to be the most effective for detecting and classifying edge IoT device attacks in the dataset [38].

Based on its performance, the model could be optimized and deployed on resource-limited microcontrollers, such as Raspberry Pi models, to enhance

Table 18.14 Performance between 14 ML cybersecurity models being discussed.

Models	F1 score
AB	0.999
RF	0.986
DT	0.970
XGB	0.933
CART	0.912
CNN	0.923
ANN	0.825
GNB	0.766
KNN	0.698
SVM	0.511
LR	0.275

Source: [38] / IEEE.

security in low-power IoT systems. This research has potential applications in improving military systems to safeguard against cyberthreats.

18.5.4 Wireless System Results and Discussion

As presented in paper [40], the TinyML models rely on many detections, and Table 18.15 includes all of the TinyML models and the overall performance [40].

With decent results, the system is promising to be deployed in agricultural fields [40]. However, it still needs further refinements. In the future, the system will be integrated with more exterior sensors and networks to be able to perform real-time analysis; moreover, prioritize improving the energy efficiency of the system so that the long life of the system is ensured [40].

There are five models used in two circumstances: computer vision and audio analysis. These paradigms are evaluated in different metrics, particularly accuracy, precision, recall, and F2 score from paper [42]. Particularly, the five models have an average accuracy of 0.928, an average precision of 0.862, an average recall of 0.976, and an average of F2 score of 0.944 [42]. Like F1 score, F2 score also combines precision and recall, but it emphasizes more on recall [45]. The formula for F2 score [45] is displayed in equation (18.8):

$$F2\text{-}Score = \frac{5 \cdot Precision \cdot Recall}{4 \cdot Precision + Recall} \tag{18.8}$$

Due to the latency caused from additional steps to convert into spectrogram models, time-series model, specifically Transformer-Time, is claimed to be the most suitable for the final TinyML model [42]. The final device, named TinyChirp, operates for periods ranging from two to eighteen weeks, ensuring its applicability in various real-life bio-acoustic scenarios [42, 45].

Table 18.15 Performance metrics for TinyML event detecting tasks.

Event detection	Accuracy	Precision	Recall	F1 score
Anomaly detection	92.3	91.7	90.9	91.3
Integrated sensor analysis	93.5	94.1	92.9	93.5
Predictive packet transmission	88.4	87.9	88.9	88.4
Predictive trajectory analysis	89.9	88.4	90.1	89.2
Sheep fever detection	95.1	94.8	95.5	95.1
Mil fridge monitoring	91.7	92.2	91.3	91.8

18.6 Conclusion and Future Scope

Overall, this survey reviews several studies on TinyML applications related to smart agriculture, smart healthcare systems, military systems, and wireless systems. Unlike previous surveys, this one includes a case study examining a specific paper. Additionally, it provides detailed information about the methodology and results of each study for profound insights, making the survey easy to read and coherent. The study also demonstrates the potential of TinyML to enable efficient and effective machine learning on resource-constrained devices, paving the way for innovative solutions in these fields.

This paper aims to provide insights into TinyML applications across multiple fields (agriculture, health care, military systems, and wireless systems). Additionally, the inclusion of a case study compared to one of the referenced papers offers a deeper understanding of the methodologies and results of a particular application, providing valuable insights for future research.

Despite these features, the paper only reviews studies in four proposed fields. Therefore, more papers in each field need to be reviewed for further insights into applications. Additionally, the survey must clarify the workflow of paradigms presented in some studies to elucidate the logic of those studies.

References

1 Salah, O.Z., Selvaperumal, S.K., and Abdulla, R. (2022). Accelerometer-based elderly fall detection system using edge artificial intelligence architecture. *International Journal of Electrical and Computer Engineering* 12 (4): 4430.

2 Prasanna, R., Kakarla, P.C., Viveka Simha, P.J., and Mohan, N. (2022). Implementation of tiny machine learning models on arduino 33 BLE for gesture and speech recognition. *arXiv preprint arXiv:2207.12866.*

3 Atanane, O., Mourhir, A., Benamar, N., and Zennaro, M. (2023). Smart buildings: water leakage detection using TinyML. *Sensors* 23 (22): 9210.

4 Gruber, F.J. (2023). Sensor based human activity recognition using edge computing and TinyML in combination with convolutional neural networks. Doctoral dissertation. University of Applied Sciences FH Campus Wien.

5 Zhang, H., Chen, L., Zhang, Y. et al. (2022). A wearable real-time character recognition system based on edge computing-enabled deep learning for air-writing. *Journal of Sensors* 2022 (1): 8507706.

6 Arslan, S.E. and Bolat, B. (2023). Tiny machine learning model for obstacle detection with multi-zone time of flight sensors. *Recent Advances in Science and Engineering* 3 (2): 9–14.

7 Kimutai, G. and Förster, A. (2023). A low-cost TinyML model for mosquito detection in resource-constrained environments. *ACM Conference on Information Technology for Social Good*, Lisbon, Portugal (6–8 September 2023), 23–30.

8 Mazzia, V., Khaliq, A., Salvetti, F., and Chiaberge, M. (2020). Real-time apple detection system using embedded systems with hardware accelerators: an edge AI application. *IEEE Access* 8: 9102–9114.

9 Sanchez-Iborra, R., Bernal-Escobedo, L., Santa, J., and Skarmeta, A. (2022). TinyML-based fall detection for connected personal mobility vehicles. *Computers, Materials & Continua* 71 (2): 3869–3885.

10 Botero-Valencia, J.S., Barrantes-Toro, C., Marquez-Viloria, D., and Pearce, J.M. (2023). Low-cost air, noise, and light pollution measuring station with wireless communication and TinyML. *HardwareX* 16: e00477.

11 Tsoukas, V., Gkogkidis, A., Boumpa, E. et al. (2023). A gas leakage detection device based on the technology of TinyML. *Technologies* 11 (2): 45.

12 Salhi, L., Silverston, T., Yamazaki, T., and Miyoshi, T. (2019). Early detection system for gas leakage and fire in smart home using machine learning. *IEEE International Conference on Consumer Electronics (ICCE)*, Las Vegas, NV, USA (11–13 January 2019), 1–6. IEEE.

13 Kwon, J. and Park, D. (2021). Hardware/software co-design for TinyML voice-recognition application on resource frugal edge devices. *Applied Sciences* 11 (22): 11073.

14 Amit, Y., Felzenszwalb, P., and Girshick, R. (2021). Object detection. In: *Computer Vision: A Reference Guide*, 875–883. Cham: Springer International Publishing.

15 Bales-Parks, P. (2022). Examining energy use in TinyML systems. Doctoral dissertation, Faculty of the Graduate School, Department of Electrical Engineering. University of Colorado at Denver.

16 Kunhoth, J., Alkaeed, M., Ehsan, A., and Qadir, J. (2023). VisualAid+: assistive system for visually impaired with TinyML enhanced object detection and scene narration. *International Symposium on Networks, Computers and Communications (ISNCC)*, Doha, Qatar (23–26 October 2023), 1–8. IEEE.

17 Anbukarasu, P., Nanisetty, S., Tata, G., and Ray, N. (2022). Tiny-HR: towards an interpretable machine learning pipeline for heart rate estimation on edge devices. *arXiv preprint arXiv:2208.07981*.

18 Wardana, I.N., Fahmy, S.A., and Gardner, J.W. (2023). TinyML models for a low-cost air quality monitoring device. *IEEE Sensors Letters* 7 (11): 1–4.

19 Bamoumen, H., Temouden, A., Benamar, N., and Chtouki, Y. (2022). How TinyML can be leveraged to solve environmental problems: a survey. *International Conference on Innovation and Intelligence for Informatics, Computing,* *and Technologies (3ICT)*, Sakheer, Bahrain (20–21 November 2022), 338–343. IEEE.

20 Qi, J., Yang, P., Min, G. et al. (2017). Advanced Internet of Things for personalised healthcare systems: a survey. *Pervasive and Mobile Computing* 41: 132–149.

21 Yan, X., Fu, T., Lin, H. et al. (2023). UAV detection and tracking in urban environments using passive sensors: a survey. *Applied Sciences* 13 (20): 11320.

22 Sharma, H., Haque, A., and Blaabjerg, F. (2021). Machine learning in wireless sensor networks for smart cities: a survey. *Electronics* 10 (9): 1012.

23 Jagannath, J., Polosky, N., Jagannath, A. et al. (2019). Machine learning for wireless communications in the Internet of Things: a comprehensive survey. *Ad Hoc Networks* 93: 101913.

24 Srinivasagan, R., Mohammed, M., and Alzahrani, A. (2023). TinyML-sensor for shelf life estimation of fresh date fruits. *Sensors* 23 (16): 7081.

25 Albuali, A., Srinivasagan, R., Aljughaiman, A., and Alderazi, F. (2023). Scalable lightweight IoT-based smart weather measurement system. *Sensors* 23 (12): 5569.

26 Osareme, J., Muonde, M., Maduka, C.P. et al. (2024). Demographic shifts and healthcare: a review of aging populations and systemic challenges. *International Journal of Science and Research Archive* 11 (1): 383–395.

27 Al Kuwaiti, A., Nazer, K., Al-Reedy, A. et al. (2023). A review of the role of artificial intelligence in healthcare. *Journal of Personalized Medicine* 13 (6): 951.

28 Dangalle, C. (2021). Insect vectors of human viral diseases: can they transmit COVID-19? *Sri Lankan Journal of Biology* 6 (1): 3–14.

29 Galbadage, T., Peterson, B.M., and Gunasekera, R.S. (2020). Does COVID-19 spread through droplets alone? *Frontiers in Public Health* 8: 163.

30 Piatkowski, D. and Walkowiak, K. (2022). TinyML-based concept system used to analyze whether the face mask is worn properly in battery-operated conditions. *Applied Sciences* 12 (1): 484.

31 Ajibade, F.O., Adelodun, B., Lasisi, K.H. et al. (2021). Environmental pollution and their socioeconomic impacts. In: *Microbe mediated remediation of environmental contaminants*, 321–354. Woodhead Publishing.

32 Iqbal, M.Z. and Campbell, A.G. (2021). From luxury to necessity: progress of touchless interaction technology. *Technology in Society* 67: 101796.

33 Pławiak, P. and Alblehai, F. (2024). A TinyDL model for gesture-based air handwriting Arabic numbers and simple Arabic letters recognition. *IEEE Access* 12: 76589–76605.

34 Alsuhli, G., Fahim, A., and Gadallah, Y. (2022). A survey on the role of UAVs in the communication process: a technological perspective. *Computer Communications* 194: 86–123.

35 Müller, H., Kartsch, V., and Benini, L. (2024). GAP9Shield: A 150GOPS AI-capable ultra-low power module for vision and ranging applications on nano-drones. *arXiv preprint arXiv:2407.13706*.

36 Jayakumar, P., Brohi, S.N., and Zaman, N. (2021). Artificial intelligence and military applications: innovations, challenges. https://doi.org/10.20944/preprints202108.0047.v1 (accessed 23 August 2025).

37 Shah, P., Govindarajulu, Y., Kulkarni, P., and Parmar, M. (2024). Enhancing TinyML security: study of adversarial attack transferability. *arXiv preprint arXiv:2407.11599*.

38 Becker, E., Gupta, M., and Aryal, K. (2023). Using machine learning for detection and classification of cyber attacks in edge IoT. *IEEE International Conference on Edge Computing and Communications (EDGE)*, Chicago, IL, USA (2–8 July 2023), 400–410. IEEE.

39 Dutta, A. and Kant, S. (2021). Implementation of cyber threat intelligence platform on Internet of Things (IoT) using TinyML approach for deceiving cyber invasion. *International Conference on Electrical, Computer, Communications and Mechatronics Engineering (ICECCME)*, Mauritius (7–8 October 2021), 1–6. IEEE.

40 Karras, A., Karras, C., Giannaros, A. et al. (2023). TinyML-based event detection: an edge-cloud approach for smart agriculture over LoRa WSNs. *8th South-East Europe Design Automation, Computer Engineering, Computer Networks and Social Media Conference (SEEDA-CECNSM)*, Piraeus, Greece (10–12 November 2023), 1–10. IEEE.

41 Kwon, M., Lee, J., and Park, H. (2019). Intelligent IoT connectivity: deep reinforcement learning approach. *IEEE Sensors Journal* 20 (5): 2782–2791.

42 Huang, Z., Tousnakhoff, A., Kozyr, P. et al. (2024). TinyChirp: bird song recognition using TinyML models on low-power wireless acoustic sensors. *arXiv preprint arXiv:2407.21453*.

43 Koonce, B. and Koonce, B. (2021). SqueezeNet. Convolutional neural networks with swift for tensorflow: image recognition and dataset categorization, 73–85.

44 Janapa Reddi, V., Elium, A., Hymel, S. et al. (2023). Edge impulse: an MLOps platform for tiny machine learning. *Proceedings of Machine Learning and Systems* 5: 254–268.

45 Fosić, I., žagar, D., Grgić, K., and Križanović, V. (2023). Anomaly detection in NetFlow network traffic using supervised machine learning algorithms. *Journal of Industrial Information Integration* 33: 100466.

19

Benchmarking TinyML Encrypted Federated Learning with Secret Sharing in Medical Computer Vision

Ruduan B. F. Plug[1], Putu H. P. Jati[1], Samson Y. Amare[1], and Mirjam van Reisen[1,2]

[1]*Leiden University Medical Center, Leiden University, Leiden, Netherlands*
[2]*Tilburg School of Humanities and Digital Sciences, Tilburg University, Tilburg, Netherlands*

19.1 Introduction

Federated learning (FL) allows us to train models over distributed data in an efficient and privacy-preserving manner [1]. This is done by iterative optimization at the data source and sharing the gradients from each optimization step to collaboratively build a global model. This provides privacy by design, which is important in fields such as health care [2] where this is a basic requisite to perform analytics. However, the federated learning process requires a high degree of trust between participating parties, as a malicious participant may be able to reconstruct data from other parties by collecting gradients, the difference between their own model weights and the newly broadcast global model weights, which is also known as the gradient reconstruction attack vector [3, 4].

Recent studies have demonstrated this vulnerability and have shown that sensitive patient information can be leaked from gradients [5, 6], allowing complete reconstruction of original data in some cases. To prevent this from happening, we can incorporate security mechanisms within the federated learning process, such as encrypted federated averaging and splitting gradients in secret shares to prevent information leakage toward a single malicious party [7].

These measures are, however, not without drawbacks, as both encryption and secret sharing incur additional computational overhead on both clients and the aggregator. In this chapter, we investigate the exact impact of these measures on the performance on the federated learning process using medical computer vision (CV) as the use case. We conduct an ablation study using the ResNet18 model pre-trained on ImageNet-1K [8], which we fine-tune for binary classification of chest

Tiny Machine Learning: Design Principles and Applications, First Edition.
Edited by Agbotiname Lucky Imoize, Dinh-Thuan Do, and Houbing Herbert Song.

X-ray images from the ChestMNIST dataset [9]. Our approach uses homomorphic encryption on gradients at the client level and decrypts only after aggregation has completed [7]. Secret sharing distributes partial decryption keys from each client to the aggregator, so that only with all participants, consent the total gradient can be decrypted. The ablation is performed by measuring model performance and computational time with these security measures enabled or disabled, while both versions use the exact same federated learning setup otherwise.

In our study, we measure the difference in model performance by evaluating precision, recal, and F_1 score directly between the models. For computational performance, we measure the total training time, time spent on gradient descent (graphics processing unit [GPU]-bound), time spent on encryption overhead (central processing unit [CPU]-bound, secure only), and the distribution of these timings across all training runs for each benchmark. These benchmarks are run for $K = 1, K = 2, \dots, K = 10$ clients to additionally measure the model performance and overhead for federated learning setups of varying sizes.

19.2 Related Work

The federated learning method introduced by McMahan et al. [1] is one of the most important privacy-preserving machine learning methods. Federated learning has been well-researched with articles covering the main challenges and opportunities [10, 11]. The main principle of federated learning is to maintain data locality; however, this does not guarantee privacy and security. Hitaj et al. demonstrated an attack where an adversarial client could train a generative model to infer other clients' data in a collaborative learning setting [3]. Similarly, Zhu et al. have shown that gradients shared during federated learning can be inversely used to reconstruct training images, also known as the deep leakage from gradients attack vector [4].

Bonawitz et al. proposed secure aggregation, where client updates are encrypted using homomorphic encryption such that the server only recovers the aggregated sum of updates, but not any individual update [12]. This paper is highly influential for this chapter. It ensures that individual gradients remain confidential as long as a threshold amount of clients do not collude. Extensions and improvements of secure aggregation continue to be explored, including methods to handle user drop-outs and verifiability [6, 13, 14] (i.e., using group signatures or verifiable secret sharing). Phong et al. explored additive homomorphic encryption in federated averaging [15]. However, applying homomorphic encryption directly on ciphertexts proved to be orders of magnitude slower and produce much larger message sizes than standard federated averaging. Recent works like BatchCrypt aim to reduce this overhead by quantizing and batching gradients before encryption, achieving more efficient homomorphic operations [16]. The addition of key-based secret sharing in secure distributed machine learning as proposed by Plug. et al. may reduce computational overhead when compared to using full ciphertext to provide security during federated averaging [7].

Implementations of secure multi-party computation such as SecureML [17] and SecureNN [18] allow a small number of parties to jointly compute an arbitrary algorithm without revealing their inputs. This provides stronger security guarantees than federated learning but has significant overhead. Hybrid approaches with less overhead have been proposed [19], for instance, using threshold secret sharing so that only an aggregate result is reconstructed [13]. These approaches can achieve privacy similar to secure aggregation.

There is extensive literature analyzing the tradeoff between privacy guarantees and model performance or system efficiency in FL [11, 20]. Adding differential privacy noise to gradients can protect against inference attacks at the cost of reduced accuracy [21]. Cryptography-based methods do not cause reductions in accuracy but can dramatically slow down training and may not scale to large models or many clients without specific optimizations [15, 16]. Our work contributes to this area by quantitatively benchmarking these overheads in a realistic medical imaging scenario. Examples of prior performance evaluations are studies such as Wagh et al. [18] on the much similar multi-party computation, as well as others for federated learning [15, 22]) have identified security and performance challenges.

19.3 Methodology

The machine learning architecture that we have implemented for this experiment uses a standard client–server federated learning setup [1]. We have modified the federated learning loop by adding gradient encryption and secret sharing in the process, which can be turned on or off during initialization of the benchmark. We will evaluate these additions with ablation. In this section, we describe in detail each of the components of our architecture.

19.3.1 MedMNIST Dataset

For our benchmark, we use the standardized version of the `ChestX-ray8` dataset [9] from the MedMNIST benchmark repository [23]. The dataset consists of 112,120 chest X-ray images downsampled to a resolution of 28×28 pixels and labeled for 14 common thoracic symptoms and conditions from infectious and chronic conditions. The dataset consists of 60,412 healthy samples and 51,708 samples with one of the included pathologies. Refer to Table 19.1 for the full overview of the data.

Table 19.1 ChestMNIST dataset distributions (binary and per label).

Label	Train	Val.	Test	% of total
No pathology	42,405	6,079	11,928	51.9
Any pathology	36,063	5,140	10,505	48.1
Pathologies				
Atelectasis	7,996	1,119	2,420	7.3
Cardiomegaly	1,950	240	582	1.8
Effusion	9,261	1,292	2,754	8.4
Infiltration	13,914	2,018	3,938	12.6
Mass	3,988	625	1,133	3.7
Nodule	4,375	613	1,335	4.0
Pneumonia	978	133	242	0.9
Pneumothorax	3,705	504	1,089	3.3
Consolidation	3,263	447	957	2.9
Edema	1,690	200	413	1.5
Emphysema	1,799	208	509	1.6
Fibrosis	1,158	166	362	1.1
Pl. thickening	2,279	372	734	2.4
Hernia	144	41	42	0.2

Note that the pathologies have a very skewed distribution, and the most common label comprises 12.6% of the data, while the least common one accounts for only 0.2% of the total data set. For our benchmark, we converted the dataset from a multi-label disease classification problem into a binary disease detection task by converting one-hot encoding with a row-wise OR operation. Even when only considering binary classification, we still need to account for the uneven nature of the underlying distribution, as this will have a significant effect on our final performance if we do not take measures to prevent over- and underfitting.

For our benchmark, we focus our classification task on distinguishing X-ray images that show no pathology compared to those that show at least one pathology present. This simplified task mirrors typical clinical workflows, where initial screenings prioritize identifying abnormal cases for further diagnosis or use of more specialized models [24, 25].

To increase generalization and reduce the effect of skewed label distribution, we apply data augmentation techniques to improve generalization. For augmentation, we use random rotations, horizontal mirroring, affine transformations, and label smoothing (smoothing ground truth labels with a factor ϵ to prevent overconfidence [22]). Another technique we use is mixup augmentation [26], in which we mix pairs of training images and labels at set ratios to boost generalization capability of small CV models.

19.3.2 ResNet18 Model Architecture

To perform image classification on `ChestX-ray8`, we use a ResNet18 convolutional neural network [8]. ResNet18 is a lightweight version of a deep residual network with 18 layers, which can be fine-tuned after pretraining. We initialize ResNet18 with pretrained weights from ImageNet [27], leveraging transfer learning to improve convergence on the medical imaging data with minimal training. The final fully connected layer of the ResNet18 model is replaced and fine-tuned during training to produce a single output for binary classification. The original article has used ResNet50 to perform multi-label classification [9]; however, as we are only interested in performing binary classification on downsampled images, we opted for the smaller and more memory-efficient ResNet18 model.

To address the inherent class label imbalance in the ChestMNIST dataset, which will affect our binary classification task, we use a modified loss function to fine-tune the ResNet18 model. We train with the focal loss function [28] instead of a standard binary cross-entropy. Focal loss is defined as

$$L_{\text{focal}}(p_t) = -\alpha(1 - p_t)^{\gamma} \log p_t \tag{19.1}$$

where p_t is the predicted probability for the true class. We set focusing parameter $\gamma > 0$ (starting with $\gamma = 2$) so that the loss down-weights well-classified examples and puts more emphasis on hard, misclassified examples [28]. We also use a balancing factor α (hyper-parameter optimized on validation data) to reduce the effect of upstream class imbalance due to us combining labels. This will reward hard-to-classify examples while making easy classifications less rewarding. Focal loss helps the model learn from the few difficult positive instances that resulted from rare source labels. The model is optimized using the Adam optimizer [29] with a fixed learning rate, which we selected through preliminary experiments.

To further speed up training, we implemented automatic mixed precision (AMP) for the training loop [30] to accelerate training and reduce memory usage by using floating point precision of 16 bits (rather than 32-bit precision) on convolutions. This has no significant effect on accuracy down the line when applied on deep learning tasks [30] such as the CV classification task we are training our model on.

19.3.3 Federated Learning Architecture

To train our global and client-side ResNet18 models, we follow a synchronous client–server process with iterative rounds of training and aggregation [1] as described by McMahan et al. In our setup, there is one aggregator and K client nodes. From a high-level perspective, each training round consists of the following steps:

1) **Weight Distribution**: The global ResNet18 model with weights $\mathbf{W}$ is first initialized at $t = 0$ by the aggregator. Hyperparameters such as γ are globally shared. The initialization state is shared to initialize all local federated models with an identical state. If $t > 0$, then for each new round t, the aggregator shares the current model weights $W^{(t)}$.
2) **Local Training**: Each local model receives $W^{(t)}$ and trains on its local dataset $\mathbf{X}_k, \mathbf{Y}_k$ for one iteration. Local models can be trained with varying batch sizes according to their memory and performance profile. Each client computes the gradient of the loss with respect to global weights $W^{(t)}$ on its local data to get $W_k^{(t+1)}$. Let $g_i^{(t)}$ denote the gradient computed by client i during round t such that $W^{(t)} + g_i^{(t)} = W_k^{(t+1)}$.
3) **Gradient Sharing**: This step differs from whether we apply encryption and secret sharing or not.
   In the *non-secure* version, each local model client serializes and directly transmits the gradient to the aggregation server, which then performs federated averaging on these gradients to update the global model. This process is described in more detail in the next step.
   In the *secure* version, each gradient $g_i^{(t)}$ is encrypted using additive homomorphic encryption. Through the property of homomorphic encryption, the encrypted gradients from different client models can be added together directly in their encrypted form. To allow the aggregator to decrypt the final gradient, each client also uses Shamir's secret sharing to distribute partial shares of the private encryption key to the aggregation server. After the server aggregates all encrypted gradients, it reconstructs the private encryption key from the collected secret shares and can decrypt the final gradient. As a result, the server learns only the aggregated gradient, while individual gradients remain confidential throughout transmission and aggregation.
4) **Weight Aggregation**: The aggregator consumes gradients that are shared by clients $k \in K$. Upon receiving all of them, the aggregator computes the (weighted) average gradient as follows:

$$\overline{g}^{(t)} = \omega_k \frac{1}{K} \sum_{i=1}^{K} g_i^{(t)} \tag{19.2}$$

   where ω_k is the weight given to k. Typically, this weight is set as a fraction of the total data size or support compared to the support among all clients. In our case, we use support since we deal with skewed data and are interested in data containing pathologies within our set, such that $w_k = \frac{\text{supp}(k)}{\sum_{i=1}^{K} \text{supp}(i)}$.

This weighted averaging corresponds to federated averaging (FedAvg) with one local epoch [1]. The weighted average gradient is then used to update the global model at the aggregator:

$$\mathbf{W}^{(t+1)} = \mathbf{W}^{(t)} - \eta \bar{g}^{(t)} \tag{19.3}$$

where η is the learning rate. In our implementation, local model training applies the stochastic Adam optimizer. Consequentially, once we aggregate the optimization iterations that result from this, we do not have to apply this again at the global level and we can suffice with a relatively simple learning rate multiplier.

5) **Model Evaluation**: The aggregator evaluates the new global model with weights $\mathbf{W}^{(t+1)}$ on a labeled holdout test set. In our case, we measure accuracy, precision, recall, F_1 score, and support. If max iterations T or any stopping condition has not been met, another training loop begins (Figure 19.1).

After T rounds, the aggregator obtains the final model $W^{(T)}$. This model is not necessarily the last obtained model. Each global model during the training loop is evaluated on a held-out test set to compute performance metrics, and we keep track of the best global model weights, which are saved. The final model is equivalent to these weights.

The above process is identical for the secure and non-secure settings except for step 3: gradient transmission. The gradient averaging and model updating are mathematically the same in both cases, which means that any differences in the final model performance will be a result from the stochastic properties of training the models, rather than the addition of encryption and secret sharing. Therefore, we expect the secure and non-secure runs to reach very similar model accuracy, precision, recall, and F_1 scores.

19.3.4 Encryption and Secret Sharing

We implement two security measures to secure gradients during transmission and aggregation: encryption and secret sharing. The goal is to ensure that malicious parties cannot read individual gradient updates, while still allowing the aggregation server to obtain the gradients needed to update the model from remote model clients.

Each client encrypts its computed gradient $g_i^{(t)}$ using additive homomorphic encryption. In our case, we use ConcreteML [32], which implements a fast version of fully homomorphic encryption (FHE). We use a relatively simple operation (add) to compile the circuit required for homomorphic encryption (HE)-based encryption and decryption. Each client first quantizes their gradient vector into integer representations compatible with ConcreteML's encryption

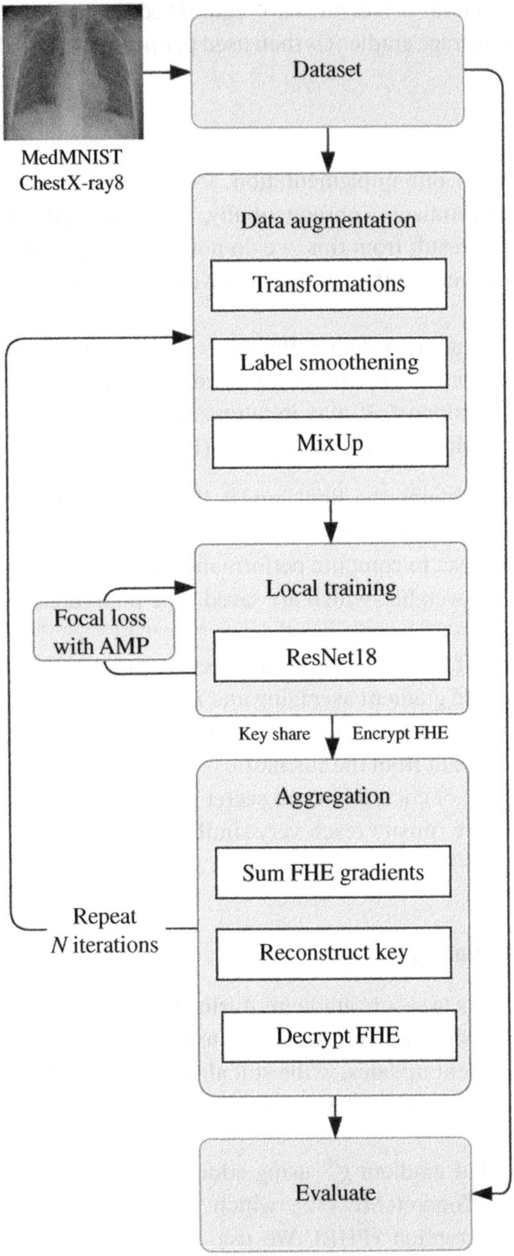

Figure 19.1 The complete model training loop including data augmentation, local training on the ResNet18 model with focal loss, and FHE aggregation and evaluation. *Source:* MedMNIST V2 ChestX-ray8 [31]. Photo: Stefano Woerner et al. (2025) / Springer Nature / CC BY 4.0.

scheme. The quantized gradients are then individually encrypted on a compiled add circuit using a generated public key, producing ciphertexts $C_i^{(t)}$. Since ConcreteML provides FHE, these ciphertexts can be aggregated directly without decrypting them.

The aggregated ciphertext is subsequently decrypted by reconstructing the corresponding private key, distributed among the participants via Shamir's secret sharing. This ensures that the server obtains only the aggregate gradient and cannot recover or infer individual gradients from the ciphertexts. We use the additive HE property that $\mathcal{D}\left(\sum C_i^{(t)}\right) = \sum \overline{g}^{(t)}$, where $\mathcal{D}$ is the decryption function, which means we get the combined gradient to update the global model. The overhead introduced by torus-based fully homomorphic encryption (TFHE)-based homomorphic encryption primarily depends on the quantization precision and ciphertext size. Since we already use mixed precision to train the ResNet18 model, we only need 16 bits of precision without losing model information.

To enable the server to ultimately decrypt and aggregate the gradients without any single party having full knowledge of the encryption keys, we use a secret sharing scheme for the keys. We adapt Shamir's (T, N) threshold secret sharing [33] to our encryption key shares. In our implementation, each client splits its symmetric key $k_i^{(t)}$ into N shares and distributes their share back to the aggregator. We configured a secret threshold T such that the aggregator can not reconstruct the key from fewer than T shares.

In our setup, for simplicity, clients provide shares of their key to the aggregator directly. For a full secure multi-party computation (SMPC) setup of key sharing, this could be done by having each client share its key among a few other clients such that at least T out of N of them collaborate to reconstruct it. This is conceptually similar to the secure aggregation approach in which each pair of clients shares random data masks (here, the mask is effectively the encryption key) [12]. We chose $T = N$ in our experiments, meaning that all shares are needed to reconstruct the original key. This is intentional because in our situation we require consent from all parties. This is, however, a worst case for computational overhead but maximizes confidentiality since no single share is useful alone. The secret sharing and distribution are done at the beginning of each federated learning iteration.

Both security measures introduce additional computational overhead. First of all, using ConcreteML to encrypt the gradient $g_i^{(t)}$ with FHE at the client side increases the computational time of each training round. Additionally, sharing keys to other parties further increases communication overhead. The communication overhead for key shares is expected to be relatively small, but it is an operation that has to be performed frequently across every client. Finally, the aggregation server has to perform addition over encrypted data and then decrypt it. This also increases the overhead on calculating the next step in the global model, which requires client models to wait before they can continue.

19.3.5 Experimental Setup

In our ablation experiments, we isolate the model performance and runtime impact of the security measures we described in the previous section. In this section, we describe the data partitioning strategy, training setup and parameters, and the metrics we have chosen for the evaluation of our experiment.

To simulate data localization, we partition the training dataset into K equal subsets, where K is the number of clients for a given federated benchmark. Each subset serves as the local data for one federated model client. We use an independent and identically distributed (IID) partitioning strategy such that the data is shuffled randomly and split evenly into K parts. Each client gets approximately $78,468/K$ images, with class labels following a uniform distribution among the clients due to the simple random sampling used. This makes it such that each client has a similar class distribution approximating the overall dataset.

We use IID partitioning because we want to ensure equal training behavior for a set K given both secure and unsecure setups, avoiding bias due to data heterogeneity. In a non-IID scenario, model performance might be randomly lower than expected due to data heterogeneity [11], which is outside the scope of our ablation study focusing on security. Each client in our experiments is expected to have an equal distribution of normal and abnormal X-rays. For example, with $K = 5$ clients, each client gets approximately 15.7 thousand training images with an expected amount of 8.15 thousand images without pathology and 7.55 thousand images with any pathology. All clients participate in every round (synchronous FL as described), so all available training data will be seen every round during training.

For our experiment, we fix the number of federated learning rounds to 200 with 1 epoch per client per round for all configurations. This was determined to be sufficient for the model to reach a plateau in validation performance in preliminary runs, since we only aim to fine-tune the ResNet 18 model. During each round, each client runs 1 epoch, meaning one full pass through its local data. To speed up training and make the effect of overhead consistent, we use a fixed batch size of 4096. This results in $\frac{\|(\mathbf{X},\mathbf{Y})_{\text{train}}\|}{4096K}$ batches in every epoch for every local model. As the number of clients increases, the amount of batches in every epoch decreases since each local model sees less of the total data individually. We run each benchmark loop on $K = 1, 2, \dots, 10$ client configurations, with one full benchmark loop on the secure version and one on the non-secure version, resulting in 20 full federated learning training sessions.

For the Adam optimizer for local model training, we use the default initial learning rate of $\alpha = 0.001$. For focal loss, we use $\gamma = 2$ and $\alpha = 0.25$ as suggested by the original paper [28]. Label smoothing with a factor of 0.1 was applied to soften the target distribution and reduce overfitting. Instead of assigning a probability of

1.0 to the correct class, the smoothed label is set to 0.9 for the true class and the remaining 0.1 is uniformly distributed across the other classes; i.e., for K classes, each incorrect class receives a target probability of $\frac{0.1}{K-1}$.

Finally, MixUp was applied with a probability of 0.5 during training, meaning each input had a 50% chance of being augmented. When applied, two random samples (x_i, y_i) and (x_j, y_j) were linearly combined as $\tilde{x} = \lambda x_i + (1 - \lambda)x_j$ and $\tilde{y} = \lambda y_i + (1 - \lambda)y_j$, where $\lambda \sim \text{Beta}(\alpha, \alpha)$ with $\alpha = 0.2$. This encourages the model to generalize better by smoothing decision boundaries.

All experiments were run on a dedicated desktop with an `Intel 13900K` CPU, `64GB 5600MHz CL40 DDR5` RAM, and a `NVIDIA RTX4090` GPU on Windows 11 running Python 3.12, CUDA 12.6, and PyTorch 2.6.0. The clients and their dataloaders were simulated as separate processes on this machine to measure per-client computational time. This also means that there is no networking or communication overhead, as we are not interested in measuring this in our experimental setup.

In our experiment, we measure accuracy, precision, recall, F1 score, and receiver operating characteristic area under the curve (ROC AUC) on the held-out ChestMNIST test set at the end of every federated training round. To measure computational overhead, we record total training runtime, encryption and secret sharing overhead, and gradient computation time. Each metric is also measured on an average per-iteration and per-client basis. We present measures as mean with standard deviation sampled from all 200 training iterations for each model. All metrics are evaluated at varying client counts $K = 1, 2, \ldots, 10$, comparing secure versus non-secure to give empirical insights into the performance-security trade-off and scalability.

19.4 Results

Across all experiments, the secure federated learning approach achieved virtually identical model performance to the non-secure approach. Table 19.2 provides the key metrics as measured for the final global model in each scenario.

We observed no consistent difference in the model performance across ablation for runs with $K = 1$ to $K = 6$ clients. Minor differences in this bracket are within the variability expected from one training run to another. For example, with $K = 5$ clients, the model after non-secure training attained an accuracy of 75.34% on the test set, whereas the secure federated learning attained 75.24%. This is a difference of only 0.10 percentage points. Precision and recall were also very close, with the non-secure version achieving 0.72 precision and 0.76 recall and the secure version 0.7 precision, 0.77 recall both resulting in an F_1 score of 0.74. The ROC AUC for both non-secure and secure models at 5 clients is both at 0.82.

Table 19.2 Model performance: secure vs. non-secure federated learning.

K	Acc. (%)		Prec.		Rec.		F_1		AUC	
		Sec.		**Sec.**		**Sec.**		**Sec.**		**Sec.**
1	77.08	77.92	0.75	0.78	0.78	0.74	0.76	0.76	0.83	0.83
2	77.20	77.66	0.78	0.77	0.71	0.75	0.75	0.76	0.83	0.83
3	77.56	77.37	0.77	0.77	0.75	0.74	0.76	0.75	0.82	0.83
4	76.36	76.39	0.74	0.76	0.76	0.72	0.75	0.74	0.82	0.82
5	75.34	75.24	0.72	0.72	0.76	0.77	0.74	0.74	0.82	0.82
6	74.30	74.86	0.72	0.73	0.74	0.73	0.73	0.73	0.80	0.80
7	73.81	72.05	0.71	0.72	0.73	0.67	0.72	0.69	0.80	0.78
8	72.94	71.02	0.74	0.74	0.65	0.59	0.69	0.65	0.80	0.77
9	67.84	70.80	0.74	0.73	0.49	0.59	0.59	0.66	0.73	0.76
10	69.09	69.69	0.79	0.78	0.46	0.50	0.58	0.60	0.77	0.77

At $K > 6$, we notice that F_1 scores become unstable, in some cases secure and in other cases non-secure achieving higher scores. After investigation, this appears to be an artifact of the way we split the data for our experiment. Since we split the data equally among K, each client sees and trains on less data. At high K, this causes overfitting, which can be identified by the increasing precision at $K = 7, 8, 9$, and 10, while recall plummets. The ROC is a better metric in this case, as this is not as significantly affected by the decision threshold boundary of 0.5 on evaluation, which causes the unstable F_1 score. ROC AUC shows more consistent scores all across the training process, with an expected drop in performance as K increases due to less data being seen by individual trained models.

The main noticeable trend in the data, which is not directly related to our ablation study, is that as the number of clients K increases, the model's performance tends to slightly decrease across the board for both non-secure and secure. For instance, with $K = 1$, the model has been evaluated at 77.08% and 77.92% accuracy, respectively, with F_1 scores of 0.83 for both cases. At $K = 10$, measured accuracy drops to around 69.09% and 69.69% and F1 drops to 0.77. The drop in performance with a greater K is expected because each client has fewer data to train on, as we have previously discussed. In our experiments, we kept the number of iterations fixed at 200, while the size of each local data set decreased, which means there is a potential for overfitting when a small data set has been trained on for too many iterations locally. From our investigation, this explains the lower metrics at greater K.

In Figure 19.2, we show the ROC curves for each model on binary ChestMNIST classification using ResNet18. Comparing the non-secure and secure federated

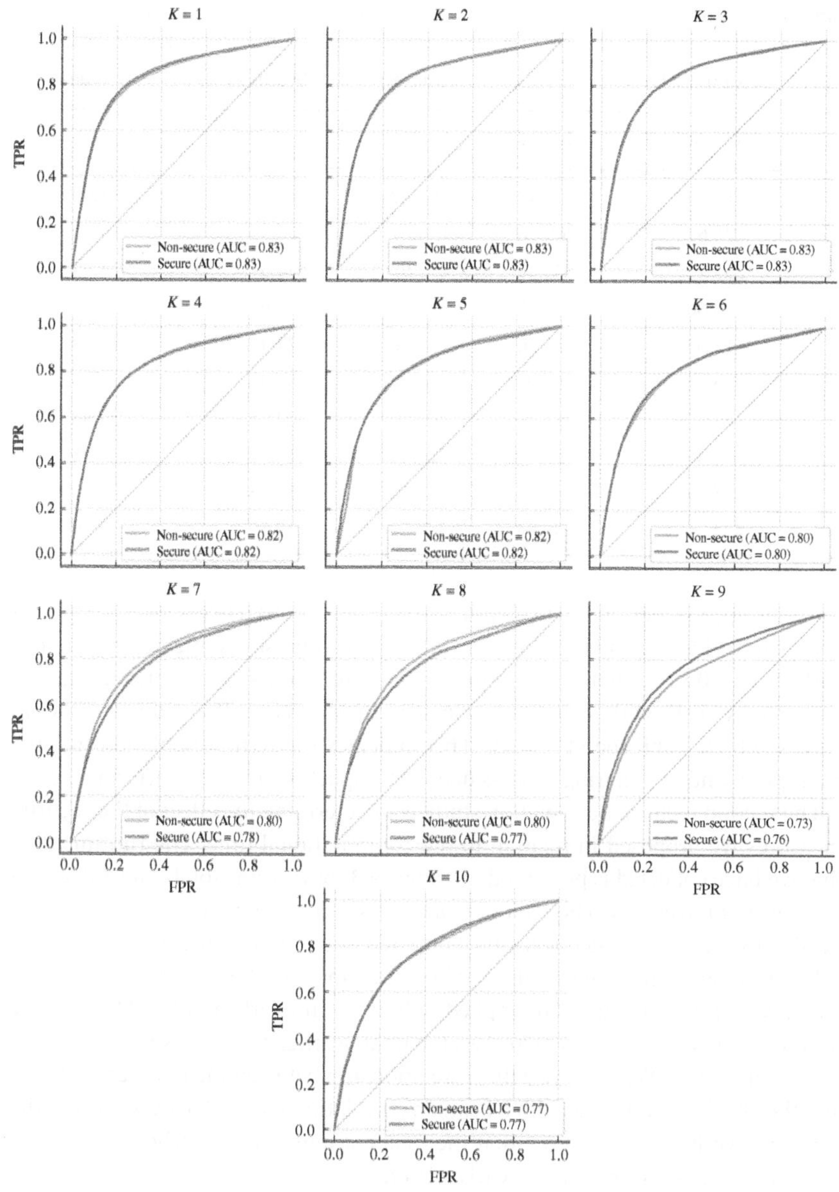

Figure 19.2 ROC curves for non-secure vs. secure federated learning for $K = 1$ to $K = 10$. Each graph shows both secure and non-secure ROC curves along with their corresponding AUC values.

Table 19.3 Runtime performance: secure vs. non-secure federated learning.

K	Runtime (s)		Train (s)	Overhead (s)	Overhead (%)
	Non-secure	Secure			
1	119.19	133.43	92.87	14.80	11.09
2	176.61	206.18	144.87	29.69	14.40
3	239.72	286.43	203.93	44.51	15.54
4	306.80	364.17	261.22	59.64	16.38
5	313.03	386.06	264.76	74.43	19.28
6	356.75	422.94	282.22	89.49	21.16
7	354.63	463.64	305.67	104.23	22.48
8	381.41	499.40	321.31	119.27	23.88
9	395.84	532.91	337.10	134.25	25.19
10	434.31	589.84	340.82	149.19	25.29

learning approaches more in depth, we see that most curves overlap almost completely. For example, with $K = 5$, we see almost identical performance with $AUC = 0.82$ for both non-secure and secure models. Similar results can be seen across all K, with slight variation seen in $K = 8$ and $K = 9$ due to random effects, which stabilized at $K = 10$. This indicates through our ablation that the encryption and key sharing process does not impact the model's performance.

The second focus of our ablation study is to determine the effect of the security measures we proposed in relation to the computational time. The runtime data that we have collected is presented in Table 19.3 for runtime totals and Table 19.4 for per-client averages. The data demonstrates a consistent computational overhead with secure federated learning compared to the non-secure baseline.

For a single-client scenario, the total runtime increased from 119.19 to 133.43 s when security measures were applied, which is an overhead of 11.09%. As the number of federated clients increases, the total overhead increases reaching up to 25.29% at $K = 10$. We observe a linear relationship between the number of clients and the absolute overhead, with a Pearson correlation $r = 0.99$ between K and the absolute overhead. This shows that the total computational overhead of encryption and secret sharing scales linearly with K.

To get a better picture, we have also calculated per-client averages for runtime as shown in Table 19.4. Analysis of these per-client runtimes shows that with an increase in K, we see an overall decrease in runtime per client. This is because each client has to process less data, resulting in a much lower training time per client as indicated by the data: from 92.87 s with $K = 1$ down to 34.08 s with

Table 19.4 Per-client runtime performance: secure vs. non-secure federated learning.

K	Runtime (s)		Train (s)	Overhead (s)	Overhead (%)
	Non-secure	Secure			
1	119.19	133.43	92.87	14.80	11.09
2	88.31	103.09	72.44	14.85	14.40
3	79.91	95.48	67.98	14.84	15.54
4	76.70	91.04	65.31	14.91	16.38
5	62.61	77.21	52.95	14.89	19.28
6	59.46	70.49	47.04	14.91	21.16
7	50.66	66.24	43.67	14.89	22.48
8	47.68	62.43	40.16	14.91	23.88
9	43.98	59.21	37.46	14.91	25.19
10	43.43	58.98	34.08	14.92	25.29

$K = 10$. The computational overhead stays surprisingly stable, ranging from 14.80 to 14.92 s. This is explained by the fact that there is no communication overhead in our setup and the encryption method that we use depends on model size, not on data size. Since each client uses ResNet18 with approximately 12 million parameters [8], and we add each together before decrypting, we get a similar average overhead for each K.

19.5 Conclusion

In this chapter, we presented benchmarks for secure federated learning using ablation on gradient encryption and secret key sharing. We used a ResNet18 model on the ChestMNIST medical imaging dataset to perform experiments on varying the number of federating clients $K = 1, 2, \ldots, 10$ and measuring the performance across both non-secure and secure training instances.

Our first key finding is that model performance, which we evaluate using accuracy, precision, recall, F1, and ROC AUC, is equivalent with and without encryption and secret sharing as expected. We have confirmed that FHE using ConcreteML preserves model performance. The ResNet18 models that we have trained achieved an average F_1 score of 0.706 across our unsecure training benchmark and 0.708 across our secure benchmarks, with AUC ranging from 0.83 to 0.77 for both across our task of detecting pathologies in chest X-rays.

Our second key finding is that total training time and latency increased by an average of 11.09–25.29% due to encryption and secret sharing overhead. This averages to an overhead of about 19.46% across all benchmarks. Surprisingly, the overhead scales almost perfectly linear with K, as each client adds an average of 14.89 s of overhead for encryption. Since we controlled for communication time by running our simulations locally, we could precisely measure the overhead in the code, which scales directly with model complexity rather than data size.

To conclude, our analysis shows that encryption for federated learning is essentially a tradeoff between performance and security. While the encryption and secret sharing techniques that we implemented do not significantly impact model classification performance, they incur an overhead to the overall and per-client runtime during model training. The increased runtime depends primarily on the size of the model, meaning that performance degradation is likely less of an issue on very small machine learning models but more of an issue on very large models. More research is required to measure the exact impact of model size to the model training performance of secure federated learning.

References

1 McMahan, H.B., Moore, E., Ramage, D. et al. (2017). Communication-efficient learning of deep networks from decentralized data. *Proceedings of the 20th International Conference on Artificial Intelligence and Statistics (AISTATS)*, 1273–1282.

2 Li, L., Fan, Y., Tse, M., and Lin, K. (2020). A review of applications and challenges of federated learning in healthcare. *Elsevier Computers & Industrial Engineering* 24 (10): 350–359.

3 Hitaj, B., Ateniese, G., and Pérez-Cruz, F. (2017). Deep models under the GAN: information leakage from collaborative deep learning. *Proceedings of the 2017 ACM SIGSAC Conference on Computer and Communications Security (CCS)*, 603–618.

4 Zhu, L., Liu, Z., and Han, S. (2019). Deep leakage from gradients. *Advances in Neural Information Processing Systems 32 (NeurIPS 2019)*.

5 Zhang, R., Guo, S., and Li, P. (2024). GradFilt: class-wise targeted data reconstruction from gradients in federated learning. *Proceedings of the 2024 ACM Asia Conference on Computer and Communications Security (AsiaCCS)*. https://doi.org/10.1145/3589335.3651514.

6 Geiping, J., Bauermeister, H., Dröge, H., and Moeller, M. (2020). Inverting gradients: how easy is it to break privacy in federated learning? *Advances in Neural Information Processing Systems 33 (NeurIPS 2020)*, 16937–16947.

7 Plug, R.B.F., Amare, S.Y., Jati, P.H.P. et al. (2025). Advancing cyber resilience in edge computing with secure distributed machine learning. In: *Cybersecurity Defensive Walls in Edge Computing*, Intelligent Data-Centric Systems. Elsevier. ISBN: 978-0-443-34109-0. https://shop.elsevier.com/books/cybersecurity-defensive-walls-in-edge-computing/imoize/978-0-443-34109-0. Forthcoming chapter, expected June 2025.

8 He, K., Zhang, X., Ren, S., and Sun, J. (2016). Deep residual learning for image recognition. *Proceedings of the IEEE Conference on Computer Vision and Pattern Recognition (CVPR)*, 770–778.

9 Wang, X., Peng, Y., Lu, L. et al. (2017). ChestX-ray8: hospital-scale chest X-ray database and benchmarks on weakly-supervised classification and localization of common thorax diseases. *2017 IEEE Conference on Computer Vision and Pattern Recognition (CVPR)*, 3462–3471. IEEE. https://doi.org/10.1109/CVPR.2017.369.

10 Kairouz, P., McMahan, H.B., Avent, B. et al. (2021). Advances and open problems in federated learning. *Foundations and Trends in Machine Learning* 146 (1-2): 1–210.

11 Li, T., Sahu, A.K., Talwalkar, A., and Smith, V. (2020). Federated learning: challenges, methods, and future directions. *IEEE Signal Processing Magazine* 37 (3): 50–60.

12 Bonawitz, K., Ivanov, V., Kreuter, B. et al. (2017). Practical secure aggregation for privacy-preserving machine learning. *Proceedings of the 2017 ACM SIGSAC Conference on Computer and Communications Security (CCS)*, 1175–1191.

13 Bell, J.H., Bonawitz, K.A., Gascón, A., Lepoint, T., and Raykova, M. (2020). Secure single-server aggregation with (poly)logarithmic overhead. In *Proceedings of the 2020 ACM SIGSAC Conference on Computer and Communications Security (CCS)*, 1253–1269. ACM. https://doi.org/10.1145/3372297.3417885.

14 So, J., Güler, B., and Avestimehr, A.S. (2021). Turbo-aggregate: breaking the quadratic aggregation barrier in secure federated learning. *IEEE Journal on Selected Areas in Information Theory* 2 (1): 479–489.

15 Phong, L.T., Aono, Y., Hayashi, T. et al. (2018). Privacy-preserving deep learning via additively homomorphic encryption. *IEEE Transactions on Information Forensics and Security* 13 (5): 1333–1345.

16 Zhang, C., Xu, C., Bhasin, S., and Jiang, C. (2020). BatchCrypt: efficient homomorphic encryption for cross-silo federated learning. *Proceedings of the 2020 USENIX Annual Technical Conference (USENIX ATC)*, 493–506.

17 Mohassel, P. and Zhang, Y. (2017). SecureML: a system for scalable privacy-preserving machine learning. *2017 IEEE Symposium on Security and Privacy (SP)*, 19–38.

18 Wagh, S., Gupta, D., and Chandran, N. (2019). SecureNN: 3-party secure computation for neural network training. *Proceedings on Privacy Enhancing Technologies* 2019 (3): 26–49.

19 Xu, R., Baracaldo, N., Yi, Z. et al. (2019). HybridAlpha: an efficient approach for privacy-preserving federated learning. *Proceedings of the 2019 ACM Workshop on Artificial Intelligence and Security (AISec)*, 1–8.

20 Mothukuri, V., Parizi, R.M., Pouriyeh, S. et al. (2021). A survey on security and privacy of federated learning. *Future Generation Computer Systems* 115: 619–640.

21 Abadi, M., Chu, A., Goodfellow, I. et al. (2016). Deep learning with differential privacy. *Proceedings of the 2016 ACM SIGSAC Conference on Computer and Communications Security (CCS)*, 308–318.

22 Szegedy, C., Vanhoucke, V., Ioffe, S. et al. (2016). Rethinking the inception architecture for computer vision. *Proceedings of the IEEE Conference on Computer Vision and Pattern Recognition (CVPR)*, 2818–2826.

23 Yang, J., Shi, R., and Ni, B. (2021). MedMNIST classification decathlon: a lightweight AutoML benchmark for medical image analysis. *IEEE International Symposium on Biomedical Imaging (ISBI)*, 191–195.

24 Rajpurkar, P., Irvin, J., Ball, R.L. et al. (2018). Deep learning for chest radiograph diagnosis: a retrospective comparison of the CheXNeXt algorithm to practicing radiologists. *PLoS Medicine* 15 (11): e1002686. https://doi.org/10.1371/journal.pmed.1002686.

25 Irvin, J., Rajpurkar, P., Ko, M. et al. (2019). CheXpert: a large chest radiograph dataset with uncertainty labels and expert comparison. *Proceedings of the 33rd AAAI Conference on Artificial Intelligence (AAAI-19)*, 590–597. AAAI Press. https://doi.org/10.1609/aaai.v33i01.3301590.

26 Zhang, H., Cissé, M., Dauphin, Y.N., and Lopez-Paz, D. (2018). mixup: Beyond empirical risk minimization. *6th International Conference on Learning Representations (ICLR)*.

27 Russakovsky, O., Deng, J., Su, H. et al. (2015). ImageNet large scale visual recognition challenge. *International Journal of Computer Vision* 115 (3): 211–252.

28 Lin, T.Y., Goyal, P., Girshick, R. et al. (2017). Focal loss for dense object detection. *Proceedings of the IEEE International Conference on Computer Vision (ICCV)*, 2980–2988.

29 Kingma, D.P. and Ba, J. (2015). Adam: A method for stochastic optimization. *3rd International Conference on Learning Representations (ICLR)*.

30 Micikevicius, P., Narang, S., Alben, J. et al. (2018). Mixed precision training. *6th International Conference on Learning Representations (ICLR)*.

31 Yang, J., Shi, R., Wei, D. et al. (2023). MedMNIST V2: a large-scale lightweight benchmark for 2D and 3D biomedical image classification. *Scientific Data* 10 (1): 41. https://doi.org/10.1038/s41597-022-01721-8.

32 Chillotti, I., Joye, M., and Paillier, P. (2021). Programmable bootstrapping enables efficient homomorphic inference of deep neural networks. In: *Cyber Security Cryptography and Machine Learning*, 1–19. Springer International Publishing. https://doi.org/10.1007/978-3-030-77870-5_1.

33 Shamir, A. (1979). How to share a secret. *Communications of the ACM* 22 (11): 612–613.

Index

Tiny Machine Learning: Design Principles and Applications, First Edition.
Edited by Agbotiname Lucky Imoize, Dinh-Thuan Do, and Houbing Herbert Song.

f

S

Printed and bound by CPI Group (UK) Ltd, Croydon, CR0 4YY
17/09/2026
14964232-0008